INTERNATIONAL SOCIETY FOR ROCK MECHANICS
SOCIÉTÉ INTERNATIONALE DE MÉCANIQUE DES ROCHES
INTERNATIONALE GESELLSCHAFT FÜR FELSMECHANIK

International Congress on Rock Mechanics
Congrès international de mécanique des roches
Internationaler Kongress über Felsmechanik

PROCEEDINGS / COMPTES-RENDUS / BERICHTE

VOLUME / TOME / BAND 3

MELBOURNE / 1983

Proceedings
Comptes-rendus
Berichte

Fifth Congress of the International Society for Rock Mechanics

Cinquième congrès de la Société Internationale de Mécanique des Roches

Fünfter Kongress der Internationalen Gesellschaft für Felsmechanik

MELBOURNE (AUSTRALIA) 1983

Rock Mechanics for Resource Development, Mining and Civil Engineering

La mécanique des roches en rapport avec l'exploitation de ressources naturelles, l'industrie minière et le génie civil

Felsmechanik für den Aufschluss von Bodenschätzen, Bergbau und Tiefbau

VOLUME 3 General Reports, Discussions and Lectures

TOME 3 Rapports généraux, discussions et conférences

BAND 3 Generalberichte, Diskussionen und Kurzreferate

A. A. BALKEMA / ROTTERDAM / 1987

and Australian Geomechanics Society

et Société Australienne de Géomécanique

und Australische Gesellschaft für Geomechanik

For the complete set of three volumes, ISBN 90 6191 236 9
For volume 1, ISBN 90 6191 237 7
For volume 2, ISBN 90 6191 238 5
For volume 3, ISBN 90 6191 239 3

Published by A.A.Balkema, P.O.Box 1675, Rotterdam, Netherlands
Printed in the Netherlands

VOLUME 3

General Reports, Discussions and Lectures

TOME 3

Rapports généraux, discussions et conférences

BAND 3

Generalberichte, Diskussionen und Kurzreferate

Themes and contents of the volumes
Thèmes et contenu des tomes
Themen und Inhalt der Bände

Volume 1

Theme A Site Exploration and Evaluation
Theme B Surface and Near-surface Excavations

Volume 2

Theme C Deep Underground Excavations
Theme D Rock Dynamics
Theme E Special Topics in Rock Mechanics

Volume 3

General Reports, Discussions and Lectures

Tome 1

Thème A Exploration et evaluation in situ
Thème B Excavation en surface et à faible profondeur

Tome 2

Thème C Excavation à grande profondeur
Thème D Dynamique des roches
Thème E Aspects particuliers de la mécanique des roches

Tome 3

Rapports généraux, discussions et conférences

Band 1

Thema A Untersuchung und Beurteilung des Betriebspunktes
Thema B Übertägige und oberflächennahe Felsbauwerke

Band 2

Thema C Tiefe unterirdische Hohlräume und Abbaue
Thema D Felsdynamik
Thema E Sonderthemen der Felsmechanik

Band 3

Generalberichte, Diskussionen und Kurzreferate

Foreword

The first two volumes of the Proceedings of the 5th International Congress on Rock Mechanics were published by the Australian Geomechanics Society before the Congress. The Congress Organizing Committee set a target for the total number of pages of preprinted papers, based upon consideration of the numbers of pages of pre-prints at each of the preceding four Congresses. The Advisory Committee determined the formula for the distribution of pages between nations. This was based partly upon the number of ISRM members in each national group, and partly upon the total fees paid by each national group to the ISRM Secretariat during the preceding 3 years. In addition the President of the ISRM and the Congress Organizing Committee were each allocated a number of pages, for their discretionary allocation. These latter allocations were used for authors from countries having no national groups, or to satisfy appeals from national groups with too many worthwhile papers to fit into their page allocation. All papers were refereed, not by the Organizing Committee, but by the National Groups of the country of residence of the authors. In the case of the few authors not resident in a country having a national group, the President of the ISRM was responsible for refereeing.

The General Reporters prepared their texts for distribution at the relevant Congress sessions. They are published in Volume 3, which also contains most of the verbal presentations by dignitaries and by people discussing the papers after they had been presented by the authors. These verbal presentations have been transcribed by the speakers' notes in recollection. They may not always have the polished style of the written contributions, but they provide a valuable record of many opinions and facts which emerged from the impromptu discussions. The Organizing Committee thanks the authors, the Congress participants, and the International Society for Rock Mechanics, for entrusting the Australian Geomechanics Society with the onerous but interesting and rewarding task of mounting the 5th Congress of the Society.

W. E. Bamford

Avant-propos

Les deux premiers tomes des Comptes-Rendus du 5e Congrès International de Mécanique des Roches ont été publiés par la Société Australienne de la Mécanique des Roches avant Congrès. Le Comité d'organisation du Congrès a fixé une limite au nombre de pages de communications imprimées avant Congrès en se référant aux nombres de pages imprimées avant Congrès lors de chacun des quatre Congrès précédents. Le Comité consultatif a fixé la formule de diffusion de pages entre les pays membres. Cette formule a été calculée en partie sur le nombre de membres de la SIMR appartenant à chaque groupe national et en partie sur le montant global des cotisations versées au Secrétariat de la SIMR par chaque groupe national au cours des trois années précédentes. Par ailleurs, un certain nombre de pages a été attribué au Président de la SIMR et au Président du Comité d'organisation pour diffusion comme bon leur semblera. Les pages ainsi attribuées l'ont été soit au bénéfice des auteurs des pays où il n'y a pas de groupe national, soit pour satisfaire aux demandes des groupes nationaux ayant contribué un trop grand nombre de communications de valeur pour qu'il soit possible de les limiter à leur quota de pages. Toutes les communications ont été sélectionnées non par le Comité d'organisation mais par les groupes nationaux des pays de résidence des auteurs. Dans le cas des quelques auteurs domiciliés dans un pays sans groupe national, le Président de la SIMR a effectué la sélection.

Les Rapporteurs généraux ont préparé leurs textes pour diffusion lors des séances s'y rapportant; Ces textes sont publiés dans le tome 3 qui contient également la plupart des communications orales faites par des dignitaires et par des personnes ayant commenté sur les communications après que celles-ci eurent été présentées par les auteurs. Ces communications orales ont été reproduites à partir d'enregistrements sur bande magnétique et, dans certains cas, d'après les notes des présentateurs. Ces textes n'auront peut-être pas toujours le style châtié des communications écrites, mais ils fournissent une documentation précieuse contenant maintes opinions et données qui ont jailli de ces discussions impromptues.

Le Comité d'organisation remercie vivement les auteurs et les personnes ayant assisté au Congrès, ainsi que la Société Internationale de la Mécanique des Roches d'avoir confié à la Société Australienne de la Géomécanique la tâche ardue mais intéressante et enrichissante de réaliser le 5e Congrès de la Société.

W. E. Bamford

Vorwort

Die beiden ersten Bände der Berichte des 5. Internationalen Kongresses über Felsmechanik wurden von der Australischen Gesellschaft für Geomechanik vor dem Kongreß veröffentlicht. Das Organisationskomitee des Kongresses bestimmte die Gesamtzahl der Vorveröffentlichungen aufgrund der Gesamtseitenzahl der Vorveröffent-lichungen von jedem der vier vorhergehenden Kongresse. Das Beratende Komitee bestimmte die Formel für die Aufteilung der Seiten unter den verschiedenen Ländern. Die Grundlage dafur war einerseits die Zahl der IGFM Mitglieder jeder Nationalgruppe und andererseits die Gesamthöhe der finanziellen Beiträge der betreffenden Nationalgruppe an das IGFM Sekretariat während der vorhergehenden 3 Jahre. Außerdem wurden dem Präsidenten der IGFM und dem Organisationskomitee eine Anzahl von Seiten zur Verfügung gestellt, um nach eigenem Ermessen verteilt zu werden. Diese letzteren Seiten waren für Autoren von Ländern ohne Nationalgruppe bestimmt, oder um Einspruche von Nationalgruppen zu befriedigen, die zu viele qualitativ gute Beiträge für ihre Seitenzuteilung hatten. Ein jeder wissenschaftlicher Beitrag wurde nicht von dem Organisationskomitee, sondern von der Nationalgruppe des Aufenthaltslandes des Autoren beurteilt. Bei den wenigen Autoren aus Ländern ohne Nationalgruppe mu te der Präsident der IGFM den Beitrag beurteilen.

Die Hauptreferenten bereiteten ihr Schriftmaterial für Verteilung bei den entsprechenden Kongreß-Sitzungen vor. Dieses Textmaterial ist im 3. Band veröffentlicht worden, welcher gleichzeitig mündliche Beiträge von Fachleuten enthält, sowie auch von Zuhörern, die Beiträge besprachen, nachdem diese von ihren Autoren vorgetragen worden waren. Diese mündlichen Beiträge sind von Tonbandaufnahmen niedergeschrieben worden, manchmal unter Zuhilfenahme von schriftlichen Anmerkungen des Redners. Nicht immer werden diese Beiträge den guten Stil der schriftlichen Beiträge aufweisen, aber sie sind trotzdem ein wertvolles Protokoll der mannigfaltigen Meinungen und Tatsachen, die aus solchen improvisierten Besprechungen hervorgingen.

Das Organisationskomitee dankt den Autoren, den Kongreßteilnehmern und der Internationalen Gesellschaft für Felsmechanik dafür, daß sie der Australischen Gesellschaft für Geomechanik das Vertrauen schenkte, den 5. Kongreß dieser Gesellschaft zu organisieren, was wohl eine mühsame, aber doch hochinteressante und dankbare Aufgabe war.

W. E. Bamford

Organization of the fifth ISRM Congress

CONGRESS ORGANISING COMMITTEE

W. E. Bamford (Chairman of Committee)
Professor L. A. Endersbee, AO (Congress Chairman)
Dr. J. R. Barrett
Dr. A. G. Bennet
W. J. Cuming
Professor I. B. Donald
J. R. Enever
Dr. R. S. Evans
Dr. I. W. Johnston
Dr. M. Kurzeme
Professor H. G. Poulos
W. M. G. Regan
E. D. J. Stewart
W. E. Vance (Executive Secretary)
Mrs Judy Webber (Secretary to the Committee)

ADVISORY COMMITTEE

President of the ISRM: W. Wittke
Secretary General of the ISRM: A. Silverio

Vice-President for

Africa:	A.Chaoui
Asia:	M. Yoshida
Australasia:	W. E. Bamford
Europe:	S. Uriel Romero
North America:	T. C. Atchison
South America:	O. Moretto

GENERAL REPORTERS

Theme A

Professor David H. Stapledon, South Australian Institute of Technology, Adelaide, Australia
Dr. Peter Rissler of Ruhrtalsperrenverein, Essen, West Germany

Theme B

Professor Richard E. Goodman, University of California, Berkeley, U.S.A.
Professor Klaus W. John, Ruhruniversität, Bochum, West Germany

Theme C

Professor Charles Fairhurst and Dr. Barry H. G. Brady, University of Minnesota, Minnesota, U.S.A.
Professor Yoshio Hiramatsu, Kyoto University, Kyoto, Japan

Theme D

Dr. Per Anders Persson, Director, Research and Development, Nitro Nobel AB, Sweden
Roger Holmberg, President, Swedish Detonic Research Foundation Stockholm, Sweden

Theme E

Professor François G. Cornet, University of Pierre and Marie Curie, Paris, France

Organisation du cinquième Congrès de la SIMR

COMITE D'ORGANISATION DU CONGRES

W. E. Bamford (Président du Comité)
Professeur L. A. Endersbee, AO (Président du Congrès)
Dr. J. R. Barrett
Dr. A. G. Bennet
W. J. Cuming
Professeur I. B. Donald
J. R. Enever
Dr. R. S. Evans
Dr. I. W. Johnston
Dr. M. Kurzeme
Professeur H. G. Poulos
W. M. G. Regan
E. D. J. Stewart
W. E. Vance (Secrétaire Exécutif)
Mme Judy Webber (Secrétaire du Comité)

COMITE CONSULTATIF

Président de la SIMR: W. Wittke
Secrétaire Général de la SIMR: A. Silverio

Vice-Président pour

l'Afrique:	A.Chaoui
l'Asie:	M. Yoshida
l'Australasie:	W. E. Bamford
l'Europe:	S. Uriel Romero
l'Amérique du Nord:	T. C. Atchison
l'Amérique du Sud:	O. Moretto

RAPPORTEURS GÉNÉRAUX

Thème A

Professeur David H. Stapledon, Institut de Technologie de l'Australie du Sud, Adelaïde, Australie
Dr. Peter Rissler, Ruhrtalsperrenverein, Essen, République Fédérale d'Allemagne

Thème B

Professeur Richard E. Goodman, Université de Californie, Berkeley, E.U.
Professeur Klaus W. John, Université de la Ruhr, Bochum, République Fédérale d'Allemagne

Thème C

Professeur Charles Fairhurst et Dr. Barry H. G. Brady, Université de Minnesota, E.U.
Professeur Yoshio Hiramatsu, Université de Kyoto, Japan

Thème D

Dr. Pers Anders Persson, Directeur, Recherche et Développement, Nitro Nobel AB, Suède
Roger Holmberg, Président, Fondation Suèdoise pour la Recherche Détonique, Stockholm, Suède

Thème E

Rapporteur Général:
Professeur François G. Cornet, Université de Pierre et Marie Curie, Paris, France

Organisation des fünften Kongresses der IGFM

ORGANISATIONSKOMITEE DES KONGRESSES

W. E. Bamford (Vorsitzender des Komitees)
Professor L. A. Endersbee, AO (Vorsitzender des Kongresses)
Dr. J. R. Barrett
Dr. A. G. Bennet
W. J. Cuming
Professor I. B. Donald
J. R. Enever
Dr. R. S. Evans
Dr. I. W. Johnston
Dr. M. Kurzeme
Professor H. G. Poulos
W. M. G. Regan
E. D. J. Stewart
W. E. Vance (Geschäftsführer)
Frau Judy Webber (Schriftführerin des Komitees)

BERATENDES KOMITEE

Der Präsident der IGFM: W. Wittke
Der Generalsekretär der IGFM: A. Silverio

Vizepräsident für

Afrika:	A.Chaoui
Asien:	M. Yoshida
Australien und Ozeanien:	W. E. Bamford
Europa:	S. Uriel Romero
Nordamerika:	T. C. Atchison
Südamerika:	O. Moretto

HAUPTREFERENTEN

Thema A

Professor David H. Stapledon, South Australian Institute of Technology, Adelaide, Australien
Dr. Peter Rissler des Ruhrtalsperrenvereins, Essen, Westdeutschland

Thema B

Professor Richard E. Goodman, University of California, Berkeley, U.S.A.
Professor Klaus W. John, Ruhruniversität, Bochum, Westdeutschland

Thema C

Professor Charles Fairhurst und Dr. Barry H. G. Brady, University of Minnesota, Minnesota, U.S.A.
Professor Yoshio Hiramatsu, Kyoto Universität, Kyoto, Japan

Thema D

Dr. Per Anders Persson, Direktor, Forschung und Entwicklung, Nitro Nobel AB, Schweden
Roger Holmberg, Präsident der Swedish Detonic Research Foundation, Stockholm, Schweden

Thema E

Professor François G. Cornet, Université de Pierre et Marie Curie, Paris, Frankreich

Contents / Contenu / Inhalt
Volume / Tome / Band 3

B Theme / Thème / Thema
Surface and Near-surface Excavations
Excavation en surface et à faible profondeur
Übertägige und oberflächennahe Felsbauwerke

Theme / Thème / Thema — Deep Underground Excavations / Excavation à grande profondeur / Tiefe unterirdische Hohlräume und Abbaue

GENERAL REPORTS / RAPPORTS GÉNÉRAUX / GENERALBERICHTE:

DISCUSSIONS / DISCUSSIONS / DISKUSSIONEN:

D Theme / Thème / Thema

Rock Dynamics
Dynamique des roches
Felsdynamik

GENERAL REPORTS / RAPPORTS GÉNÉRAUX / GENERALBERICHTE:

DISCUSSIONS / DISCUSSIONS / DISKUSSIONEN:

E Theme / Thème / Thema

Special Topics in Rock Mechanics
Aspects particuliers de la mécanique des roches
Sonderthemen der Felsmechanik

GENERAL REPORT / RAPPORT GÉNÉRAL / GENERALBERICHT:

DISCUSSIONS / DISCUSSIONS / DISKUSSIONEN:

LECTURES / CONFÉRENCES / KURZREFERATE:

Index to Authors
Index des Auteurs
Inhaltsverzeichnis nach Schriftstellern

List of Exhibitors
Liste des exposants
Ausstellerliste

Australian Ground Instruments
PO Box 195
GLEN IRIS VIC 3146

Barrett, Askew, Fuller & Partners
466 High Street
PRAHRAN VIC 3151

Chemie Linz AG
8th Floor, Watkins Centre
225 Wickham Terrace
BRISBANE QLD 4000

D A Book Depot Pty. Ltd.
PO Box 163
MITCHAM VIC 3132

Dames and Moore
Suite 5
65 Mary Street
BRISBANE QLD 4000

Geomech Pty. Ltd.
31 Cremorne Street
RICHMOND VIC 3121

Irad Gage
Etna Road
LEBANON, NH 03766
UNITED STATES OF AMERICA

Rock Engineering
77 Parkhill Road
KEW VIC 3103

AND

Rock Instruments Pty. Ltd.
PO Box 209
WARWICK QLD 4370

Snowy Mountains Engineering Corporation
PO Box 356
COOMA NORTH NSW 2630

List of Participants
Liste des participants
Teilnehmerliste

ARGENTINA/ARGENTINE/ARGENTINIEN

Moretto O.Prof.	Bolognesi-Moretto	Luis S. Pena 250 6, Buenos Aires (1110), ARGENTINA

AUSTRALIA/AUSTRALIE/AUSTRALIEN

Alexander L.G.	CSIRO Division of Applied Geomechanics	38 Headingly Road, Mt. Waverley, Vic. 3149, AUSTRALIA
Ameratunga J.J.P.	Monash University	Department of Civil Engineering, Monash University, Clayton, Vic. 3168, AUSTRALIA
Anderson D.R.	VSL Engineering	5 Hanwell Court, Glen Waverley, Vic 3150 AUSTRALIA
Ashley J.J.	Maunsell & Partners	6 Claremont Street, South Yarra, Vic. 3141, AUSTRALIA
Attebo K.A.G.	Atlas Copco MCT	P.O. Box 444, Blacktown, N.S.W. 2148, AUSTRALIA
Bailey J.D.	AMIRA	63 Exhibition Street, Melbourne, Vic. 3000, AUSTRALIA
Bamford W.E.	University of Melbourne	Department of Civil Engineering, Parkville, Vic. 3052, AUSTRALIA
Banda D.M.M.	Fluor Australia	Fluor Place, 616 St. Kilda Road, Melbourne, Vic. 3004, AUSTRALIA
Beer D.M.M.	University of Queensland	Dept. of Civil Engineering, St. Lucia, Qld. 4067, AUSTRALIA
Bell A.E.	Public Works Department, N.S.W.	State Office Block, Phillip Street, Sydney, N.S.W. 2000, AUSTRALIA

Bennet A.G. Dr.	John Connell Grp.	60 Albert Road, South Melbourne Vic. 3205 AUSTRALIA
Bhattacharyya A.K. Dr.	University of New South Wales	C/- School of Mining, Kensington, N.S.W. 2033, AUSTRALIA
Blackwood R.L. Dr.	University of New South Wales	P.O. Box 1, Kensington N.S.W. 2033, AUSTRALIA
Bowling A.J.	The Hydro-Electric Commission	Birdwood Avenue, Moobah, TAS. 7009, AUSTRALIA
Boyce B.T.	Queensland Institute of Technology	Civil Engineering Department, G.P.O. Box 2434, Brisbane, Qld. 4001, AUSTRALIA
Boyd G.L.	CSR Coal Division	P.O. Box 281, Brisbane, Qld. 4001, AUSTRALIA
Brady B.H.G. Dr.	CSIRO - Division of Applied Geomechanics	P.O. Box 54, Mt. Waverley, VIC. 3149, AUSTRALIA
Brady J.T.	Australian Mining & Smelting Limited	G.P.O. Box 384D, Melbourne, Vic. 3001, AUSTRALIA
Brennan S.	Electricity Trust of South Australia	220 Greenhill Road, Eastwood, S.A. 5063 AUSTRALIA
Brewin J.L.B.	Reinforced Earth	84 Cummins Street, Unaderra, N.S.W. 2526 AUSTRALIA
Bridges M.C.	MIM Holdings Limited	160 Ann Street, Brisbane, Qld. 4000, AUSTRALIA
Brophy C.J.	N.S.W. Dept. Ind. Rel.	P.O. Box 30, Wagga Wagga, N.S.W. 2650, AUSTRALIA
Brown J.R.	Snowy Mountains Engineering Corporation	S.M.E.C. P.O. Box 356, Cooma North, N.S.W. 2630, AUSTRALIA
Brumley J.C.	State Electricity Commission of Victoria	Fuel Department, 15 William Street, Melbourne, Vic. 3000, AUSTRALIA
Bywater S.	Mount Isa Mines Limited	Mount Isa, Qld. 4825, AUSTRALIA
Carter J.P. Dr.	University of Sydney	School of Civil Engineering, Sydney, N.S.W. 2006, AUSTRALIA
Chandler K.R.	Soilmech Pty. Ltd.	31 Cremorne Street, Richmond, Vic. 3121, AUSTRALIA
Chappell B.A. Dr.	Snowy Mountains Engineering Corporation	Geomechanics Division, P.O. Box 356, Cooma North, N.S.W. 2630, AUSTRALIA

Chase K.J.	Utah Development Company	Goonyella Mine, P.M.B. Moranbah, Qld. 4744, AUSTRALIA
Chilman K.J.	Mines Inspection	32 Sulphide Street, Broken Hill, N.S.W. 2880, AUSTRALIA
Choi S.K.X.	Monash University	Department of Civil Engineering, Clayton, Vic. 3168, AUSTRALIA
Clark D.A.	The Broken Hill Proprietary Company Ltd	P.O. Box 1239, Wollongong, N.S.W. 2500, AUSTRALIA
Collins J.L.	Shell Company of Aust. Ltd.	Metals Division, 155 William Street, Melbourne, Vic. 3000, AUSTRALIA
Conor C.H.H.	Roxby Management Services	P.O. Box 405, Unley, S.A. 5061, AUSTRALIA
Cooney A.M.	Department of Minerals and Energy,	G.P.O. Box 1683P, Melbourne, Vic. 3001, AUSTRALIA
Coulsell J.B.	Melbourne Metropolitan Board of Works	P.O. Box 64, Chadstone Centre, Vic. 3148, AUSTRALIA
Coulthard M.A. Dr.	CSIRO - Division of Applied Geomechanics	P.O. Box 54, Mount Waverley, Vic. 3149, AUSTRALIA
Cowling R.	Mount Isa Mines Limited	Mount Isa, QLD. 4825. AUSTRALIA
Cox F.	State Rivers & Waters Supply	590 Orrong Road, Armadale, Vic. 3145, AUSTRALIA
Cox R.H.T.	CSIRO	PO Box 54 Mt. Waverley, Vic. 3149 AUSTRALIA
Crotty J.M. Miss	CSIRO- Division of Applied Geomechanics	PO Box 54 Mt Waverley, VIC. 3149 AUSTRALIA
Crozier P.J.	Esso Australia Ltd.	G.P.O. Box 4047, Sydney, N.S.W. 2001, AUSTRALIA
Cuming W.J.		The Aus.I.M.M. 81 Mont Albert Road, Canterbury, Vic. 3126, AUSTRALIA
Curnow G.R.	John Holland Constructions	Box 199C G.P.O. Melbourne, Vic. 3001, AUSTRALIA
Davidson D.E.	Collinsville Coal,	P.O. Box 60, Collinsville, Qld. 4804, AUSTRALIA
Davis P.A.J.	Ullman & Nolan	P.O. Box 354, Spring Hill, Qld. 4000, AUSTRALIA

de Ambrosis L.P. Dr.	Longworth & McKenzie Pty Ltd	P.O. Box 785, North Sydney, N.S.W. 2060, AUSTRALIA
Dight P.M. Dr.	The Broken Hill Proprietary Company Ltd	140 William Street, Melbourne, Vic. 3000, AUSTRALIA
Donald I.B. Prof.	Monash University	Dept. of Civil Eng. Clayton, Vic. 3168, AUSTRALIA
Doyle N.F.	Esso Australia Ltd.	Unit 3, 10 -12 Ross Street, Sale, Vic. 3850, AUSTRALIA
Dunbavan M. Dr.	CSIRO - Division of Applied Geomechanics	P.O. Box 54, Mount Waverley, Vic. 3149, AUSTRALIA
Eades G.W.	Queensland Water Resources Commission	G.P.O. Box 2454, Brisbane, Qld. 4001, AUSTRALIA
Eckersley J.D.	James Cook University	Dept. of Civil & Systems Engineering, Townsville, Qld. 4811, AUSTRALIA
Endersbee L.A. Prof.	Monash University	Faculty of Engineering, Clayton, Vic. 3168, AUSTRALIA
Enever J.R.	CSIRO - Division of Applied Geomechanics	59 Campbell Street, Glen Waverley, Vic. 3150, AUSTRALIA
Ervin M.C.	Coffey & Partners	1503 Malvern Road, Glen Iris, Vic. 3146, AUSTRALIA
Evans R.S. Dr.	State Electricity Commission of Victoria	Fuel Department, 15 William Street, Melbourne, Vic. 3000, AUSTRALIA
Fabjanczyk M.W.	CSIRO - Division of Applied Geomechanics	P.O. Box 54, Mount Waverley, Vic. 3149, AUSTRALIA
Flintoff W.T.	Country Roads Board	1 Scott Street, Canterbury, Vic. 3126, AUSTRALIA
Forrester K.	Department of Main Roads	P.O. Box 198, Haymarket, N.S.W. 2000, AUSTRALIA
Fowler J.C.W.	University of New South Wales	3/191 Lawrence Hargrave Drive, Thirroul, N.S.W. 2515, AUSTRALIA
Fraser C.J.	SEC of Victoria	P.O. Box 195, Morwell, Vic. 3840, AUSTRALIA
Fuller P.G. Dr.	Barrett, Askew, Fuller & Partners	456 High Street, Prahan, Vic. 3181, AUSTRALIA
Furnell G.W.G.	Rocsol Pty. Ltd.	72 Second Avenue, Mt. Lawley, W.A. 6050, AUSTRALIA

Galvin J.M. Dr.	Newcom Collieries	Myuna Colliery, P.O. Box 1, Wangi Wangi, N.S.W. 2267, AUSTRALIA
Gerrard C.M. Dr.	CSIRO	P.O. Box 56, Highett, Vic. 3190, AUSTRALIA
Grant B.P.M.	Cobar Mines	P.O. Box 31, Cobar, N.S.W. 2835, AUSTRALIA
Gray P.A.	BHP Steel Division Collieries	PO BOX 1239 WOLLONGONG, N.S.W. 2500, AUSTRALIA
Green N.P.	Australian Anglo American Ltd.	581 Little Collins Street, Melbourne, Vic. 3000, AUSTRALIA
Guest R.D.	Seltrust Mining	P.O. Box R1274, GPO Perth, W.A. 6001, AUSTRALIA
Gurtunca G.	University of New South Wales	P.O. Box 1, Kensington, N.S.W. 2033, AUSTRALIA
Hagan T.N. Dr.	Golder Associates	3 Albany Court, Wantirna, Vic. 3152, AUSTRALIA
Hargraves A.J. Prof.	University of Wollongong	P.O. Box 1007, Wollongong, N.S.W. 2500, AUSTRALIA
Hattersley P.J.	Sydney Water Board	282 Sylvania Road, Miranda, N.S.W. 2228, AUSTRALIA
Henderson A.D.	Sydney Water Board	12 Redan Street, Mosman, N.S.W. 2088. AUSTRALIA
Heslop T.G.	Ullman & Nolan	P.O. Box 165, MacKay, Qld. 4740, AUSTRALIA
Hilleard P.R.	State Rail, N.S.W.	3/20 Bassett Street, Hurstville, N.S.W. 2220, AUSTRALIA
Hollingsworth P.C.	Hollingsworth Consultants	P.O. Box 251, Spring Hill, Qld. 4000, AUSTRALIA
Holt G.E.	ACIRL Limited	P.O. Box 83, North Ryde, N.S.W. 2113, AUSTRALIA
Hutchins W.B.	Rock Engineering	77 Parkhill Road, Kew, Vic. 3101, AUSTRALIA
Johnston I.W. Dr.	Monash University	Department of Civil Engineering, Clayton, Vic. 3168, AUSTRALIA
Jordan D.W.	Snowy Mountains Engineering Corporation	9 Norris Street, Cooma, N.S.W. 2630, AUSTRALIA

Kaesehagen F.E.	MIM Holdings Limited	7 Marney Street, Chapel Hill, Qld. 4069, AUSTRALIA
Kinstler F.L.	The Hydro-Electric Commission	20 Earl Street, Sandy Bay, Tas. 7005, AUSTRALIA
Knoop B.P.	Hydro Electric Commission	11 Ascot Avenue, Sandy Bay, Tas. 7005, AUSTRALIA
Kurzeme M. Dr.	Golder Associates Pty. Ltd.	466 Malvern Road, Prahran, Vic. 3181 AUSTRALIA
Lam S.K.T.	Monash University	Department of Civil Engineering Clayton, Vic. 3168, AUSTRALIA
Lama R.D. Dr.	Kembla Coal & Coke	P.O. Box 1770, Wollongong, N.S.W. 2500, AUSTRALIA
Lang A.M.	Renison Limited	P.O. Box 20, Zeehan, Tas. 7469, AUSTRALIA
Lee M.	Mount Isa Mines Limited	Mount Isa, Qld. 4825, AUSTRALIA
Legg D.J.	BHP Engineering	P.O. Box 1237, North Sydney, N.S.W. 2060, AUSTRALIA
Liebelt J.L.	The Aus.I.M.M.	C/- CRA Ltd., 55 Collins Street, Melbourne, Vic. 3000, AUSTRALIA
Lilly J.D.	Hamersley Iron Pty. Ltd.	P.O. Box 22, Tom Price, W.A. 6751, AUSTRALIA
Maconochie D.J. Dr.	Hollingsworth Consultants	P.O. Box 251, Spring Hill, Qld. 4000, AUSTRALIA
Madden B.	Chamber of Mines of South Africa	7 Stratford Avenue, East Bentleigh, Vic. 3165, AUSTRALIA
Mallett C.W. Dr.	CSIRO - Division of Applied Geomechanics	P.O. Box 54, Mount Waverley, Vic. 3149, AUSTRALIA
Maslen J.E.	Royal Melbourne Institute of Technology	Dept. of Applied Geology, 124 LaTrobe St., Melbourne, Vic. 3000, AUSTRALIA
Mather R.P.	Geological Survey of WA	66 Adelaide Terrace Perth, W.A. 6000, AUSTRALIA
Mathews R.E.	S.A. Dept. of Mines	Dept. Mines and Energy, P.O. Box 151, Eastwood, S.A. 5063, AUSTRALIA
May J.R.	AMIRA	63 Exhibition St., Melbourne, VIC. 3000, AUSTRALIA

McAnnally P.A.	Queensland Inst. of Technology	G.P.O. Box 2434, Brisbane, Qld., 4001 AUSTRALIA
McGill R.A.	Office of the Supervising Scientist	P.O. Box 461, Darwin, N.T. 5790, AUSTRALIA
McKenzie C.K.	J.K.M.R.C. University of Queensland	Isles Road, Indooroopilly, Qld. 4068, AUSTRALIA
McMahon M.D.	Dames and Moore	58 Bent Street, Lindfield, N.S.W. 2070, AUSTRALIA
McMahon B.K. Dr.	Dames and Moore	58 Bent Street, Lindfield, N.S.W. 2070, AUSTRALIA
Miller R.R. Dr.	Shell Co. of Australia Ltd.	Coal Division, P.O. Box 872K, Melbourne, Vic. 3001, AUSTRALIA
Mitchell P.W.	Pak-Poy & Kneebone	7 Verdale Avenue, Linden Park, S.A. 5065, AUSTRALIA
Mohtaji A.A.	University of Melbourne	Dept. of Mechanical Engineering Parkville, Vic. 3052, AUSTRALIA
Moorhouse P.R.	Coffey & Partners	P.O. Box 152, Fyshwick, A.C.T. 2609, AUSTRALIA
Morriss P.	Golder Associates	3 Ashburton Drive, Mitcham, Vic. 3132, AUSTRALIA
Mostyn G.R.	Dames and Moore	84 Alexander Street, Crows Nest, N.S.W. 2065, AUSTRALIA
Nag D.K.	King Island Scheelite	10 Blackwood Street, Grassy, King Island Tas. 7256, AUSTRALIA
Neville M.J.	Public Works Department NSW	State Office Block Phillip Street, Sydney, N.S.W. 2000, AUSTRALIA
Nizankowski A.M.	University of New South Wales	P.O. Box 245, Fairy Meadow, N.S.W. 2519, AUSTRALIA
Novello E.	Monash University	Dept. of Civil Engineering Clayton, Vic. 3168 AUSTRALIA
Paikopoulos J.	State Electricity Commission of Victoria	2 Albot Grove, Clifton Hill, Vic. 3068, AUSTRALIA
Paine G.G.	(Applied Blasting Research)	ICI Aust. Operations Pty. Ltd. 1 Nicholson Street, East Melbourne, Vic. 3002, AUSTRALIA
Parkin A.K. Dr.	Monash University	109 Locksley Road, Ivanhoe, VIC. 3079, AUSTRALIA

Pedler I.V.	State Electricity Commission of Victoria	15 William Street, Melbourne, Vic. 3000, AUSTRALIA
Pegrem B.S.	Pacific Coal Pty. Ltd.	Tarong Coal Project, P.O. Box 36, Nanango, Qld. 4315 AUSTRALIA
Pells P.J.N.	Coffey & Partners	12 Boronga Avenue, West Pymble, N.S.W. 2073, AUSTRALIA
Peou S. Dr.	Newlands Coal Pty. Ltd.	160 Ann Street, Brisbane, Qld. 4000, AUSTRALIA
Philp M.G.	Coffey & Partners	151 Wellington Road, East Brisbane, Qld. 4169, AUSTRALIA
Poulos H.G. Prof.	University of Sydney	School of Civil Engineering Sydney, N.S.W. 2006, AUSTRALIA
Rawlings C.D. Dr.	CSR Coal Division	G.P.O. Box 281, Brisbane, Qld. 4001, AUSTRALIA
Reed M.R.	Western Mining Corp.	PO Box 105 Fimiston, W.A. 6433, AUSTRALIA
Regan C.D. Dr.	State Electricity Commission of Vic.	65 Carnarvon Road, Strathmore, Vic. 3041, AUSTRALIA
Reid S.P.	Esso Australia Ltd.	G.P.O. Box 4047, Sydney, N.S.W. 2001, AUSTRALIA
Rosengren K.J. Dr.	Golder Associates	45 McLachlan Street Fortitude Valley, Qld. 4006, AUSTRALIA
Safk M.U.	Electrolytic Zinc Company of Australasia Limited	PO Box 433, Elura, N.S.W. 2835, AUSTRALIA
Seneviratne P.	University of New South Wales	School of Mining Engineering, P.O. Box 1, Kensington, N.S.W. 2033, AUSTRALIA
Shepherd J. Dr.	ACIRL	P.O. Box 83, North Ryde, N.S.W. 2113, AUSTRALIA
Simmons J.V. Dr.	James Cook University	Dept. Civil Engineering, Townsville, Qld. 4811, AUSTRALIA
Slepecki S.	Mt. Newman Mining	24 Culldorah Street, Newman, W.A. 6753, AUSTRALIA
Stafford G.	Blair Athol Coal	PO Box 391, Brisbane Qld. 4001, AUSTRALIA
Stapledon D.H. Prof.	SA Institute of Technology	North Terrace, Adelaide, S.A. 5000, AUSTRALIA

Thornton P.N.	Melbourne Metropolitan Board of Works	276 Burke Road, Gardiner, Vic. 3146, AUSTRALIA
Tillmann V.H.	The Zinc Corporation Ltd	P.O. Box 444, Broken Hill, N.S.W. 2880, AUSTRALIA
Trudinger J.P.	Dames & Moore	26 Lyall Street, South Perth, W.A. 6151, AUSTRALIA
Truscott E.G. Dr.	GHD Wood	87 Wickham Terrace, Brisbane, Qld. 4000, AUSTRALIA
Vance W.E.	The Aus.I.M.M.	P.O. Box 310, Carlton South, Vic. 3053, AUSTRALIA
Vutukuri V.S.	University of New South Wales	Broken Hill Division, P.O. Box 334, Broken Hill, N.S.W. 2880, AUSTRALIA
Warburton P.M. Dr.	CSIRO - Division of Applied Geomechanics	P.O. Box 54, Mount Waverley, Vic. 3149, AUSTRALIA
Ward B.	Shell Coal Company of Aust Ltd.	P.O. Box 872K, Melbourne, Vic. 3001, AUSTRALIA
Wardle L.J.	CSIRO Div App. Geomechanics	P.O. Box 54, Mount Waverley, Vic. 3149, AUSTRALIA
Warner K.R.	Geol. Survey Qld.	Mineral House, 41 George Street, Brisbane, Qld. 4000, AUSTRALIA
Washusen J.A.	State Electricity Commission of Victoria	P.O. Box 629, Moe, Vic. 3825, AUSTRALIA
Watson A.	W.A. School of Mines	12/244 Mill Point Road, South Perth, W.A. 6151, AUSTRALIA
Watts A.J.	Snowy Mountains Engineering Corporation	31 Jerrang Avenue, Cooma North, N.S.W. 2630, AUSTRALIA
Webster C.R.	MIM Holdings Ltd.	5 Trevose Street, The Gap, Brisbane, Qld. 4061 AUSTRALIA
Wesson V.C.	Phillip Institute of Technology	Seismology Research Centre, Plenty Road, Bundoora, Vic. 3083, AUSTRALIA
Whitfield L.M. Mrs.	Dams Safety Committee	Room 1122, Chief Secretaries Bldg., 121 Macquarie Street, Sydney, N.S.W. 2000, AUSTRALIA
Williams J.R.	Department of Main Roads, N.S.W.	309 Castlereagh Street, Sydney, N.S.W. 2000, AUSTRALIA

Wise N.A.		98 Iola Avenue, Farmborough Heights, Wollongong, N.S.W. 2526, AUSTRALIA
Wold M.B.	CSIRO - Division of Applied Geomechanics	P.O. Box 54, Mt Waverley, Vic. 3149, AUSTRALIA
Wood J.H.	B.H.P. Minerals Division	P.O. Box 481, Newcastle, N.S.W. 2300, AUSTRALIA
Worotnicki G.	CSIRO - Division of Applied Geomechanics	P.O. Box 54, Mount Waverley, Vic. 3149, AUSTRALIA
Yamazaki S.	Shimizu Construction	Suite 4303, Australia Square, Sydney, N.S.W. 2000, AUSTRALIA

AUSTRIA/AUTRICHE/OSTERREICH

Bonapace B.		Tiroler Wasserkraft Werke AG Landhausplatz 2, 6010 Innsbruck, AUSTRIA
Innerhofer G.H.	Illwerke	Batloggstrasse Illwerkehaus A-6780 Schruns AUSTRIA
Makovec F.F. Prof.	Techn. Univ. Vienna	Schinaweisa. 17 A-1190 Wien AUSTRIA
Mayr G.	Interfels	A-5020 Salzburg Schwarzstrasse 27, AUSTRIA
Mueller L. Prof.	Ingenieurburo	Parcelsustr. 2 A-5020 Salzburg, AUSTRIA
Wagner H. Dr.		Beton-U. Monierbau Zeughausgasse 3, Postfach 500, A-6021 Innsbruck AUSTRIA
Weiss E.H. Prof. Dr.	Universitat fur Bodenkultur	A1180 Wien, Gregor Mendel - Strasse 33, AUSTRIA
Wellacher J.O.H. Dr.	Kelag	Kirchengasse 37 A9020 Klagenfurt, AUSTRIA

BELGIUM/BELGIQUE/BELGIEN

Brych J. Prof. Dr.	University Mons.	Dept. Mining and Geology 53 Rue Joncquois, 7000 Mons, BELGIUM
Huergo P.J. Dr.	University Bruxelles	Av. Henri Bourghys 68, B-1710 Waterloo BELGIUM
Lousberg E. Prof.	University Louvain	Av Espinette 25, B 1348 Louvain - La - Neuve, BELGIUM

BRAZIL/BRESIL/BRASILEN

Napoles-Neto A.D.F.	Civil Engineer	Caixa Postal 11325, Pinheiros, 05499 Sao Paulo SP BRAZIL

CANADA/CANADA/KANADA

Benson R.P. Dr.	Klohn Leonoff Ltd.	10180 Shellbridge Way, Richmond, B.C., V6X 2W7 CANADA
Curran J.H. Prof.	University of Toronto	Dept. of Civil Engineering Toronto, Ontario M5S 1A4 CANADA
Franklin J.A. Dr.	Franklin Geotech.	The Stream, R.R. No. 1, Orangeville, Ontario L9W 2Y8 CANADA
Kaiser P.K. Prof.	University of Alberta	Dept. of Civil Engineering, Edmonton T6G 2G7 CANADA
Ladanyi B. Prof.	Ecole Polytechnique	Civil Engineering, CP 6079 Station A, Montreal, H3C 3A7 CANADA
Milne-Home W.A.	University of Alberta	Department of Geology, Edmonton, Alberta, T6G 2E3 CANADA
Mohanty Dr.	CIL Inc.	Explosive Res., McMasterville, Quebec, J3G 1T9 CANADA

CHINA/CHINE/CHINA

Bai S.W.	Institute of Rock and Soil Mechanics	Academia Sinica, Wuhan, CHINA
Baoshen L. Prof.	Chinese Society of Metals	Changsha Research Institute of Mining and Metallurgy, P.O. Box 67, Changsha, Hunan, CHINA
Chen J. Mrs.	Central Coal Mining Research Institute	Hepingli, Beijin, 100013, CHINA
Fu B.J.	Research Institute	P.O. Box 366, Beijing CHINA
Gau C.C.	University Sichuan	Civil Engineering Department, South West Jiaotong, University Sichuan CHINA
Jisongwu	Academia Sinica	Beijing CHINA
Li X. Prof.	China Mining Institute	Department of Mine Construction, Xuzhou Jiangsu, CHINA
Liao C.T.	Institute of Geomechanics	Chinese Academy of Geological Sciences Institute of Geomechanics, Chinese Academy of Geological Sciences, Fahuasi, Beijing CHINA
Lin Y. Prof.	North East Institute of Technology	Dept. of Mining Engineering, Shenyang, CHINA
Mei J. Prof.	Institute of Geophysics Academia Sinica	Beijing, CHINA
Pan Q. Mrs.	Central Coal Mining Research Institute	Beijing Research Institute of Coal Mining, Hepingli, Beijing, CHINA
Sun G.Z. Prof.	Institute of Geology Academia Sinica	Beijing, CHINA

Sun Y.Y. Dr.	Society for Rock Mechanics of China	Institute of Geology Academia Sinica, Beijing CHINA
Tan Tjong Kie Prof.	Institute of Geophysics Academia Sinica	Beijing, CHINA
Xu L.X.	Research Institute	Huang Pu Road, Wuhan CHINA
Zhu K.S. Prof.	CIAE	Chongqing Sichuan, CHINA
Zhu W.S. Prof.	Institute of Rock and Soil Mechanics	Academia Sinica, Wuhan, CHINA
Zhuo J.S.	College of Hydraulic Engineering	Xikang Road 1, Najing CHINA

FINLAND/FINLANDE/FINNLAND

Hakalehto K.O. Dr.	Tampella Ltd.	P.O. Box 256, 33101 Tampere 10, FINLAND
Saanio V.T.	Saanio & Riekkola	Mannerheimintie 31 A3, SF-00250 Helsinki 25, FINLAND
Saari K.H.O. Dr.		FINLAND
Sarkka P.S. Dr.	Academy of Finland	The Helsinki Uni. of Tech., Dept. of Min. & Met.,

FRANCE/FRANCE/FRANKREICH

SF-09150 ESPOO 15 FINLAND

Arcamone J.A. Dr.	Cherchar	BP No 2, 60550 Verneuil-En-Halatte, FRANCE
Cornet F.H. Dr.	University Paris	IPG Tour 14, Universite Paris 6, 4 Place Jussieu, 75230, Paris, Cedex 05. FRANCE
Duffaut P.M.J.	French Committee on Rock Mechanics	130 Rue de Rennes, 75006 Paris FRANCE
Fourmaintraux D.F. Dr. EDF.TEGG,	EDF-DGG	Direction de L'equipement- 3 Rue de Messine, F 75008 Paris FRANCE
Habib P.A. Prof.	Lab Meca Solides	Ecole Polytechnique, 91128 Palaiseau FRANCE
Lefin Y.	Cherchar	Boite Postale No 2, 60550 Verneuil-En-Halatte, FRANCE
Maury V.M.R. Dr.	Elf-Aquitaine	64018 Pau-Cedex FRANCE
Nguyen Minh N.M.D. Dr.	Engineer	Ecole Polytechnique, 91128 Palaiseau. FRANCE
Rochet L.	LPC	109 Av. S. Allende, 69672 Bron. FRANCE
Salembier M.S.	Coyne et Bellier	5 Rue D'Heliopolis,

FEDERAL REPUBLIC OF GERMANY/REPUBLIQUE FEDERALE D'ALLEMAGNE/BUNDESREPUBLIK DEUTSCHLAND

Borm G. Dr.	University of Karlsruhe	Lehrstuhl fuer Felsmechanik, Kaiserstr. 12, D-7500 Karlsruhe, GERMANY

Name	Institution	Address
Froehlich B.O. Dr.	University of Karlsruhe	Lehrstuhl fuer Felsmechanik, Kaiserstr. 12, D-7500 Karlsruhe GERMANY
Gross H.	University of Karlsruhe	Lehrstuhl fuer Felsmechanik, Kaiserstr. 12, D-7500 Karlsruhe, GERMANY
Haupt M.	University of Karlsruhe	Lehrstuhl fuer Felsmechanik, University of Karlsruhe, Kaiserstr. 12, D-7500 Karlsruhe GERMANY
John K.W. Prof. Dr.	Ruhr Universitaet	D-4630 Bohuh, GERMANY
Kiehl J.R.	RWTH Aachen	RWTH Aachen, Mies-van-der-Rohe-Str. 1, D-5001 Aachen GERMANY
Koerner U.E.V. Dr.	Geol. Landesamt	Herlinstr. 6, D 7830 Emmendingen, GERMANY
Lempp C.H. Dr.	University of Karlsruhe	Lehrstuhl fuer Felsmechanik, Kaiserstr. 12, D -7500 Karlsruhe, GERMANY
Muehlaus H.B. Dr.	University of Karlsruhe	Lehrstuhl fuer Felsmechanik, Kaiserstr. 12, D-7500 Karlsruhe GERMANY
Mutschler T.H.O.	University of Karlsruhe	Lehrstuhl fuer Felsmechanik, Kaiserstr. 12, D-7500 Karlsruhe GERMANY
Natau O.P. Prof.	University of Karlsruhe	Lehrstuhl fuer Felsmechanik, Kaiserstr. 12, D-7500 Karlsruhe, GERMANY
Plischke B.	RWTH Aachen	Institute f. Grundbau Bodenmechanik, Felsmechanick Mies-van-der-rohe-str. 1, D-5100 Aachen GERMANY
Rechtern J.	RWTH Aachen	Mies-van-der-Rohe -Str 1, D-5001 Aachen, GERMANY
Rissler P. Dr.	Ruhrtalsp. Verein	Uronprinzenstr 37, 4300 Essen GERMANY
Schuetz H. Prof.	Univ. Wuppertal	Hohlenscheidterstr 50, D 5600 Wuppertal 12, GERMANY
Sonntag G.H. Prof. Dr.	Technical University Munich	Arcis Str. 21, D-8000 Muenchen 2, GERMANY
Stock G.U.E.	Thyssen Bergbautechnik	46 Dortmund-Husen (13), Kohlkamp 21, GERMANY
Wenz E.	University of Karlsruhe	Lehrstuhl fuer Felsmechanik, Kaiserstr. 12, D-7500, Karlsruhe GERMANY
Werner H.U.	Dorsch Consult	Institut Fuer Erdu Grundbau, Elsenheimerstr 63, D 8000 Muenchen 21, GERMANY
Wittke W. Prof.	RWTH Aachen	Institut fur Grundbau, Bodenmechanik, Felshmechanik und Verkehrswasserbau, Mies-van-der-Rohe-Str. 1., D-5100 Aachen, GERMANY

Wullschlaeger D.	University of Karlsruhe	Lehrstuhl fuer Felsmechanik, Kaiserstr. 12, D-7500 Karlsruhe GERMANY

HONG KONG/HONG KONG/HONG KONG

Hencher S.R. Dr.	Hong Kong Govt.	GCO, 6th Floor, Empire Centre, 68 Mody Road, Tsim Sha Tsui, Kowloon HONG KONG
Malone A.W. Dr.	Hong Kong Government	The Geotechnical Control Branch Empire Centre 5th Floor 68 Mody Road, Kowloon HONG KONG
Roberts P.A.	Hong Kong Polytechnic	Dept. Civil Engineering, Hung Hom Kowloon HONG KONG

INDIA/INDE/INDIEN

Chitale V.M.	Madhyapradesh	E 4/17 Oarera Colony, BH Opal 462014, Madhya, Pradesh, INDIA

INDONESIA

Manaf M.H.	MTDC	Jl. Cisitu Lama Komp. PPTM Bandung, INDONESIA

ITALY/ITALIE/ITALIEN

Ribacchi R. Prof.	University of Rome	Instituto Arte Mineraria via Eudossiana 18 0018 ITALY

JAPAN/JAPON/JAPAN

Akimoto M.A. Assoc. Prof.	Kyushu Tokai University	Faculty of Engineering, Ohemachi 223, Kumamoto 862, JAPAN
Aoki K.	Kajima Institute of Construction Technology	2-19-1 Tobitakyu, Chofu-shi, Tokyo, JAPAN
Dohi M.	Kajima Corporation	1-2-7 Motoakasaka, Minato-ku, Tokyo JAPAN
Fukuda K.	Hokkaido University	Resources Dev. Eng. Faculty of Eng., North 13, West 8, Kita-ku, Sapporo #060 JAPAN
Haga T.	Ohbayashi-Gumi Co. Ltd.	640, 4-Chome, Shimo Kiyoto, Kiyose-shi Tokyo JAPAN
Hanamura T.	Taisei Corporation	Civil Engineering Dept., P.O. Box 4001, Shinjuku Cent. Bld., 160-91 Tokyo JAPAN
Hasegawa M.	Shimizu Construction Co Ltd	2-17-1, Kyobashi, Chuo-ku, Tokyo JAPAN

Hattori M.H.	Japan Pet. Expl. Co.	Japex Expl. Dept. 1-6-1, Ontemachi Chiyoda-ku Tokyo 100 JAPAN
Hibino S. Dr.	Criepi	1-42-17, Shin-toride, Toride-shi, Ibaraki-ken JAPAN
Hiramatsu Y. Prof.	Kyoto University	Yoshida-shimooji-cho 45, Sakyoku Kyoto 606 JAPAN
Hisatake M.H. Dr.	Osaka University	Yamada-Oka2-1, Suita, Osaka 565 JAPAN
Hoshiya M. Prof.	Musashi Institute of Technology	1-28, Tamazutsumi, Setagaya-ku, Tokyo JAPAN
Ishikawa K. Dr.	Chuo Kaihatsu Corp.	2-5-19 Nishihonmachi, Nishi-ku, Osaka-shi, Osaka JAPAN
ItoH. Prof.	University of Osaka	804, 4-Cho, Mozu-ume-machi, Sakai-shi Osaka-fu, #591 JAPAN
Kanda A.	Kokusai Kogyo Co. Ltd.	3-6-1 Asahigaoka Hino-shi Tokyo JAPAN
Kaneko K.	Electric Power Development Co. Ltd.	8-2 Marunouchi, 1-Chome, Chiyoda-ku, Tokyo #100 JAPAN
Kawamoto T.K. Prof.	Nagoya University	23 Takamine, Showa, Nagoya, 466, JAPAN
Kikuchi K.	Tokyo Electric Power Service Co Ltd	1-4-6 Nishi-Shinbashi, Minato-ku, Tokyo JAPAN
Kitagawa T.	Nishimatsu Construction Co. Ltd.	1-20-10 Toranomon, Minato-ku Tokyo JAPAN
Maeda H.	The Tokyo Electric Power Co., Inc.	1-3 Uchisaiwai-cho, Chiyoda-ku, Tokyo JAPAN
Matsue M.A.	Tobishima Co.	Sunakoseki, Nishikawa-cho, Murayamagun Yamagata-ken, JAPAN
Matsuki K. Dr.	Tohoku University	3-505, 18, 5-chome, Nakayama Sendai 980 JAPAN
Mitani S.	Kumagai-Gumi Co. Ltd.	17-1, Tsukudo-cho, Shinjuku-ku, Tokyo JAPAN
Nagata H.	Hokkaido Engineering Consultant Co. Ltd.	9, Tsukisamuhigashiyojo, Tokohira-ku, Sapporo-shi, Hokkaido JAPAN
Nakao K. Dr.	Taisei Corp.	344-1, Nase-cho, Totsuka-ku, Yokohama 245 JAPAN
Nishimatsu Y. Prof.	University of Tokyo	Dept. of Mineral Development Engineering Tokyo 113 JAPAN
Noto T.	Kiso-Jiban Consultants Co. Ltd.	11-5, 1-Chome, Kudan Kita, Chiyoda-ku, Tokyo JAPAN

Ochi H.	Honshi Bridge Authority	2631-283, Shintakayama, 2 Chyome, Onomichi-shi, Hiroshima-ken, JAPAN
Ohhashi S.	Suncoh Consultants Co Ltd	Mitsui-Daini 4-2 Nihombashihon, Gokucho, Chuo-ku Tokyo JAPAN
Ohnishi Y. Prof.	Kyoto University	School of Civil Engineering, Kyoto 606 JAPAN
Saito T. Dr.	Kyoto University	Dept. of Mineral Science & Tech., Kyoto 606 JAPAN
Sakurai S. Prof.	Kobe University	Dept. of Civil Engineering, Kobe 657 JAPAN
Shimizu T.	Kumagai-Gumi Co. Ltd.	17-1 Tsukudo-cho, Shinjuku-ku, Tokyo JAPAN
Shimizu K.	Nippon Koei Co. Ltd.	5-4, Koji-Machi, Chiyoda-ku, Tokyo JAPAN
Sugawara K.S.	Kumamoto University	Faculty of Engineering, Kurokami 2-39-1, Kumamoto 860, Tokyo, JAPAN
Tagawak H.	Takenaka Duboku Co. Ltd.	8-21-1, Ginza, Chuo-ku, Tokyo JAPAN
Takeuchi M.	(Nuclear Power Department)	Okumura Corporation Minato-ku Tokyo 107 JAPAN
Takewaki N.	Shimizu Construction Co Ltd	2-2-2, Uchisaiwai-cho, Chiyoda-ku, Tokyo JAPAN
Terada M.T. Prof.	Kyoto University	Dept. of Mineral Science & Technology, Sakyo-ku, Kyoto 606 JAPAN
Tsujita M.	Hazama-Gumi Ltd	17-23, 4-Chome, Honcho-nishi, Yono-city, Saitama-ken, JAPAN
Yamagata M.	Honshu-Shikoku Bridge Authority	Mori Building, 4-3-20 Toranomon, Minato-ku, Tokyo 105 JAPAN
Yamashita M.Y.	Kansai Electric	3-22 Nakanoshima, 3-Chome Kita-ku Osaka 530 JAPAN
Yoshida M.Y. Dr.	Kansai Electric	3-22 Nakanoshima, 3-Chome, Kita-ku Osaka 530 JAPAN
Yoshinaka R. Prof.	Department of Foundation Engineering, Saitama University	355 Shimo-okubo, Urawa-shi, Saitama-ken JAPAN
	KOREA	
Kim C.D.	Jungwoo Engineering Co. Ltd.	#212-2, Seachadong Kangnamku, Seoul 134-03 KOREA

KUWAIT

Al-Hussaini M.	Kuwait University	Kuwait Foundation for the Advancement of Sciences P.O. Box 25263, Safat - Kuwait, KUWAIT

MEXICO/MEXIQUE/MEXIKO

Bello-Maldonado A.A. Prof.	Geosistemas, S.A.	Aniceto Ortega No. 1306, Col del Valle, 03100 Mexico DF MEXICO
Ruiz-Vazquez M. Prof.	Fac. Ingen. Unam.	Av. San Francisco 551 Del. Magd. Contreras, 20 D.F. Cod. Postal 10500 MEXICO

NEW ZEALAND/NOUVELLE SELANDE/NEUSEELAND

Bracengirdle A.		NEW ZEALAND
Brown I.R. Dr.	DSIR	P.O. Box 30368, Lower Hutt, NEW ZEALAND
Hegan B.D.	NZ Geological Survey	4 Canterbury Street, Tamatea, Napier NEW ZEALAND
Pender M.J. Dr.	University of Auckland	Civil Engineering Dept., Private Bag, Auckland, NEW ZEALAND
Bratli R.K.	Cont. Shelf Institute	Haakon Magnussonsgt 18, P.O. Box 1883, N-7001 Trondheim NORWAY
By T.L. Dr.	Norwegian Institute of	Rock Blasting Techniques PB 341 Blindern, Oslo 3 NORWAY
Myrvang A.M. Dr.	Norwegian Institute of Technology	Mining Division, N 7034 Trondheim NORWAY

PAPUA NEW GUINEA

Lye G.N.	Bougainville Copper Limited	P.O. Box 942, Panguna, Bougainville Island, P.N.G.
Read J.R.L.		Bougainville Copper Limited P.O. Box 879, Arawa, Bougainville Island, P.N.G.

PORTUGAL/PORTUGAL/PORTUGAL

Barroso M.J.G.	LNEC	LNEC Av do Brasil, 1799 Lisboa Codex, PORTUGAL
Grossmann N.F.	LNEC	Laboratorio Nacional de Engenharia Civil, Av. do Brasil, P-1799 Lisboa Codex PORTUGAL
Oliveira R. Prof.	LNEC	Av do Brasil, 1799 Lisboa Codex PORTUGAL

Peres-Rodrigues F.	LNEC	Av do Brasil, 1799 Lisboa Codex PORTUGAL
MelloMendez F. Prof.	University Tec. Lisboa	R. Cidade Cabina 22-2, DTO 1800 Lisboa, PORTUGAL
Rodrigues F. Dr.	LNEC	Av do Brasil, 1799 Lisboa Codex PORTUGAL
Silverio A. Dr.	International Society for Rock Mechanics	Avenida do Brasil, P-1799, Lisboa Codex PORTUGAL

REPUBLIC OF SOUTH AFRICA/REPUBLIQUE OF D'AFRIQUE DUSUD/REPUBLIK SUD AFRIKA

De Jongh C.L.	Gold Fields of South Africa Ltd.	P.O. Box 1167, Johannesburg 2000 SOUTH AFRICA
Hillyard E.G.	Ninham Shand	P.O. Box 6323, Roggebaai 8012 SOUTH AFRICA
		Private Bag, Carletonville 2500, SOUTH AFRICA
Kotze T.J.	Gencor	P.O. Box 73, Evander 2280, SOUTH AFRICA
Ortlepp W.D.	Rand Mines Ltd.	P.O. Box 62370, Marshalltown, 2107 SOUTH AFRICA
Ras D.J.R.	Anglovaal Ltd.	P.O. Box 161, Allanridge 9490 Orange Free State, SOUTH AFRICA
Ryder J.A. Dr.	The Chamber of Mines	Mining Ops. Lab. Com. Research Organization, P.O. Box 91230, 2006 Auckland Park SOUTH AFRICA
Salamon M.D.G. Dr.	The Chamber of Mines	P.O. Box 809, Johannesburg 2000 SOUTH AFRICA
Stacey T.R. Dr.	SRK Inc.	P.O. Box 8856, Johannesburg, 2000 SOUTH AFRICA
Van Schalkwyk A. Prof.	University of Pretoria	Department of Geology, 0002 Pretoria, SOUTH AFRICA
Wagner H. Dr.	Chamber of Mines	P.O. Box 9130, Auckland Park, 2006 SOUTH AFRICA

SPAIN/ESPAGNE/SPANIEN

Romana M. Prof.	Pol. University Valencia	Cristobal Bordiu 23, Madrid-3 SPAIN
Uriel S. Prof.	Lab. Geotecnia	Alfonso XII 3, Madrid (7) SPAIN

SWEDEN/SUEDE/SCHWEDEN

Bergman S.G.A. Dr.		Foreningsvagen 19, S18274 Stocksund, SWEDEN

Bjurstrom S.A.G. Dr.	Swedish Rock Engineering Res. Foundation	Box 5501, S-114 85 Stockholm, SWEDEN
Fischer H.C. Prof.	Upsala University	Institute of Technology, Box 534 S-75121, UPSALA, SWEDEN
Holmberg R.	Svedefo	P.O. Box 32058, S-12611 Stockholm, SWEDEN
Holmgren B.J. Dr.	Swedish Rock Engineering Res. Foundation	Box 5501, S-114 85 Stockholm, SWEDEN
Krauland N.	Boliden Minerals A.B.	S-93600, BOLIDEN, SWEDEN
Lundstrom R.B.	Scandiaconsult	PO Box Y560, 10265 Stockholm, SWEDEN
Martna J.	Vattenfall	Swedish State Power Board, S-16287 Vallingby, SWEDEN
Paganus T.	LKAB	Iron-Ore Division, S-98186, Kiruna, SWEDEN
Roshoff N.K. Dr.	University of Lulea	Dept. of Rock Mechanics, S-95149 Lulea SWEDEN
Stephensson O.J. Prof.	University of Lulea	Division of Rock Mechanics, S-95167 SWEDEN
Thurner H.F. Dr.	Geodynamik AB	Regeringsgatan 111, 11139 Stockholm SWEDEN

SWITZERLAND/SUISEE/SCHWEIZ

Egger P. Dr.	Federal Institute Technology	EPF - L CH-1015 Lausanne, SWITZERLAND
Grob H. Prof.	Institute for Road Railroad and Rock Engineering	ETH - Hoenggerberg, 8093 ZURICH SWITZERLAND
Huder J. Prof.	Institute of Foundation Engineering and Soil Mechanics	ETH Hoenggerberg CH-8093 Zurich SWITZERLAND

TAIWAN

Chang S.C.	RSEA	207 Sung Chiang Road, Taipei Taiwan REPUBLIC OF CHINA
Ling T.		Taiwan Power Company Power Development Dept., 242 Roosevelt Road, Section 3 TAIPEI 107 Taiwan REPUBLIC OF CHINA
Wang W.T.	Taiwan Power Company	Design & Construction Dept., 242 Roosevelt Road, Section 3 TAIPEI 107 Taiwan REPUBLIC OF CHINA
Wong S.L.	RSEA	38 Chin-mun Street, Taipei-107 Taiwan, Taiwan REPUBLIC OF CHINA

THAILAND/THAILANDE/THAILAND

Bhucharoen V.	Electricity Generating Authority of Thailand	Hydro-Electric Construction Dept., Electricity Generating Authority of Thailand THAILAND
Yudhbir Prof.	Asian Institute of Technology	P.O. Box 2754, Bangkok 10501 THAILAND

UNITED STATES OF AMERICA/ETATS UNIS D'AMERIQUE/VEREINIGTE STAATEN VON AMERIKA

Amadei B.A. Prof.	University of Colorado	Dept. of Civil Engineering, Boulder, Co. 80309 U.S.A
Atchison T.C. Prof.	University of Minnesota	Dept. of Civil & Min. Eng., 500 Pillsbury Drive SE, Minneapolis Mn. 55455 U.S.A.
Einstein H.H. Prof.	MIT	Room 1-330, Cambridge, Massachusetts 02139 U.S.A.
Goodman R.E. Prof.	University of California	Dept. of Civil Engineering Berkeley 94720 U.S.A.
Gonano L.P. Dr.	Golder Associates	2950 Northup Way, Bellevue WA 98033 U.S.A.
Hadley K. Dr.	USNC/Rock Mech	Exxon Co. U.S.A. POB 2180, 800 Bell RM 3945, Houston, Texas 77001 U.S.A.
Hansen D.E.	Lachel Hansen	130th Pl Ne Bellevue WA 98005 U.S.A.
Hardy M.P. Dr.	J.F.T. Agapito & Associates	Suite 340, 715 Horizon Drive, Grand Junction Colorado U.S.A.
Heuze F.E. Dr.	Lawrence Livermore Nat. La.	Code L-200, P.O. Box 808, Livermore Ca. U.S.A.
Lang T.A.	Woodward Clyde	1 Doral Drive, Moraga, California 94556, U.S.A.
Lu P.H. Dr.	U.S. Bureau of Mines	Bldg 20, Denver Fed Ctr., Denver, CO. 80225, U.S.A.
Mohn J. Dr.	Ideal Basic Industries	Exploration Dept., P.O. Box 1949, Fort Collins, Co. 80522, U.S.A.
Rockaway J.D. Dr.	University of Missouri	Dept. Geol. Engineering Rolla MO 65401, U.S.A.
Roegiers J.C. Professor	Dowell	P.O. Box 2710, Tulsa, Oklahoma, U.S.A.
Scott G.A.	Bureau Reclamation	P.O. Box 25007, Code D-221 Denver, Colorado, 80225 U.S.A.
Tanimoto C. Dr.	Kyoto University	440 Davis Hall, Dept. Civil Engineering, University of California, Berkeley CA. 94720 U.S.A.
Thompson T.W. Prof.	University of Texas	Petroleum Engineering Dept., Austin Texas 78712 U.S.A.

Wagner R.A.W.	RE/Spec Inc.	P.O. Box 725, Rapid City South Dakota, 57709 U.S.A

UNITED KINGDOM/ROYAUME UNI/VEREINIGTES KONIGREICH

Brown E.T. Prof.	Imperial College of Science	Dept. of Mineral Resources Engineering, London SW7 2BP U.K.
Cooling C.M. Miss	Geotechnics Division Building Research Establishment	Garston NR Watford, Herts, WD2 7JR U.K.
Fowell R.J. Dr.	Newcastle University	Dept. of Geotechnical Engineering, Newcastle Upon Tyne NE1 7RU, ENGLAND, UNITED KINGDOM
Pande G.N. Dr.	University College Swansea	Dept. of Civil Engineering, Singleton Park, Swansea SA2 8PP U.K.
Parry R.H.G. Dr.	ISSMFE	University Engineering Dept., Trumpington Street, Cambridge, England CB2 1PZ UNITED KINGDOM
Tunnicliffe J.F. Prof.	Newcastle University	Dept. of Mining, Newcastle Upon Tyne NE1 7RU, ENGLAND, UNITED KINGDOM
Wareham B.F.	Wimpey Geoconsult	Wimpey Laboratories, Beaconsfield Road, Hayes, Middlesex UB4 OLS UNITED KINGDOM

YUGOSLAVIA/YOUGASLAVIE/JUGOSLAWIEN

Vujec S. Prof.	University of Zagreb	Mining Faculty, Pierottieva 6, 41000 Zagreb YUGOSLAVIA

ACCOMPANYING PERSONS

ARGENTINA/ARGENTINE/ARGENTINIEN

Dr. E. Moretto

AUSTRALIA/AUSTRALIE/AUSTRALIEN

Mrs N.J. Alexander
Mrs D. Bhattacharyya
Mrs J.T. Brophy
Mrs G.L. Cuming
Mrs M.A. Donald
Mrs M.J. Enever
Mr C. Galloway
Mrs M.A. Hollingsworth
Mrs C.R. Kinstler
Mrs T.A. Kurzeme
Mrs S.E Liebelt
Mrs D.A. Vance

Mrs A. Bailey
Mrs J.T. Brady
Mrs M.L. Collins
Mrs M.M. Curnow
Mrs M.R. Endersbee
Mrs M. Fuller
Mrs M.E. Hilleard
Mrs M.E. Hutchins
Mrs M.E.Knoop
Mrs G. Maconochie
Mrs R.J. McAnally
Mrs B. Watson

AUSTRIA/AUTRICHE/OSTERREICH

Mrs D.H. Innerhofer
Mrs I. Mayr

CANADA

Mrs N. Ladanyi
Ms A. Reid

FEDERAL REPUBLIC OF GERMANY/REPUBLIQUE FEDERALE D'ALLEMAGNE/BUNDESREPUBLIK DEUTSCHLAND

Mrs S. John
Miss S. Schuetz
Mrs L. Wittke
Mrs E. Koerner
Mrs G. Stock

FINLAND/FINLANDE/FINNLAND

Mrs A.M. Saanio

FRANCE/FRANCE/FRANKREICH

Mrs N. Arcamone
Mrs L. Rochet
Mrs A. Duffaut

JAPAN/JAPON/JAPAN

Mrs M.K. Kawamota

NORWAY/NORVEGE/NORWEGEN

Mrs E. Myrvang

PORTUGAL/PORTUGAL/PORTUGAL

Mrs R. Peres-Rodrigues

SOUTH AFRICA/REPUBLIQUE D'AFRIQUE DU SUD/REPUBLIK SUD AFRIKA

Mrs R. Ortlepp
Mrs J.H. Ryder

SPAIN/ESPAGNE/SPANIEN

Mrs M. Romana

SWEDEN/SUEDE/SCHWEDEN

Mrs E Bergman
Mrs A.C. Holmgren

SWITZERLAND/SUISSE/SCHWEIZ

Mrs C.H. Egger

UNITED STATES OF AMERICA/ETATS UNIS D/AMERIQUE/VEREINIGTE STRATEN VON AMERIKA

Mrs E.P.L. Hardy
Mrs A. Lang
Mrs R. Scott
Ms M. Kern
Dr S.C. Lu

Course of the Congress
Déroulement du Congrès
Kongressverlauf

EVENTS BEFORE AND AFTER THE CONGRESS

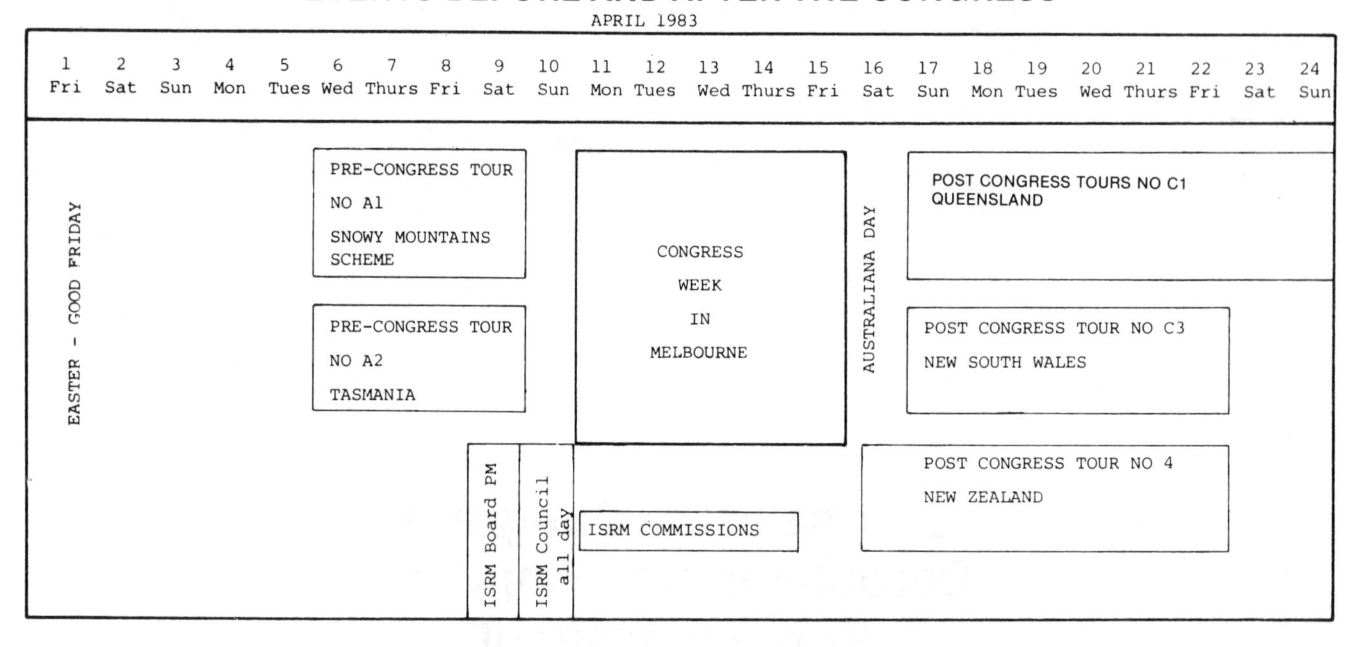

CONGRESS PROGRAMME

VENUE – DALLAS BROOKS HALL

Time	April 9 Saturday	April 10 Sunday	April 11 Monday	April 12 Tuesday	April 13 Wednesday	April 14 Thursday	April 15 Friday
08^{00}	ISRM BOARD MEETING	Registration open at 9.00	Registration open at 8.00 am		TECHNICAL EXCURSIONS	Registration open at 8.00 am	
09^{00}–10^{00}		ISRM COMMISSIONS	OFFICIAL OPENING General Report THEME C	TOPIC A3, Classification, Prediction Observation and Monitoring		General Report THEME D TOPICS D1 and D2	General Report THEME E TOPIC E2
11^{00}–12^{00}		ISRM COUNCIL MEETING	TOPIC C1 Mining Excavations	General Report THEME B, TOPIC B3 Near Surface Construction		TOPICS C3 and C4 Coal Mining and Subsidence	TOPIC E1 Fracture and flow of earth's crust
13^{00}			LUNCH	LUNCH		LUNCH	LUNCH
14^{00}–15^{00}		ISRM COUNCIL MEETING	General Report THEME A, TOPIC A1 Geophysical exploration	TOPIC B1, Stability of Rock Slopes		TOPIC C2, Permanent underground excavations (1)	Deep underground extraction, storage and disposal TOPICS A4, C5, D3
16^{00}–17^{00}			TOPIC A2, In Situ and Laboratory Testing	TOPIC B2, Foundations on Rock incl dam foundations		TOPIC C2, Permanent Underground Excavations (2)	TOPIC E3, Future Developments CLOSURE
18^{00}–19^{00}							
20^{00}–23^{00}		CONGRESS CHAIRMAN'S RECEPTION	PRESIDENT'S RECEPTION AND WELCOMING CONCERT	OPERA "DIE BERGKNAPPEN"	EVENING FREE	CONGRESS DINNER	WOOLSHED DANCE

EVENEMENTS AVANT ET APRES LE CONGRES

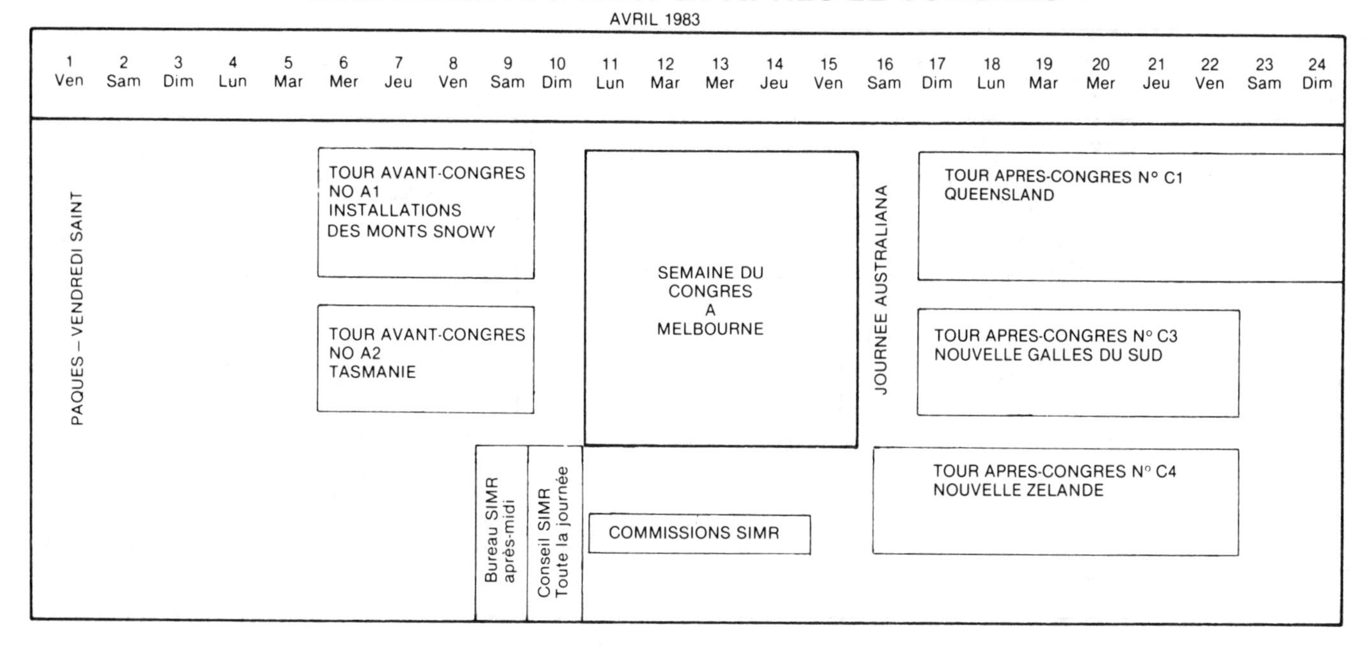

PROGRAMME DU CONGRES

LIEU – DALLAS BROOKS HALL

Avril 9 Samedi	Avril 10 Dimanche	Avril 11 Lundi	Avril 12 Mardi	Avril 13 Mercredi	Avril 14 Jeudi	Avril 15 Vendredi
	Inscription ouvert dès 9h	Inscription ouvert dès 8h			Inscription ouvert dès 8h	
	COMMISSIONS SIMR	OUVERTURE OFFICIELLE Rapport Général THEME C	TOPIQUE A3, Classification, Prédiction, Observation et Contrôle		Rapport Général THEME D TOPIQUES D1 et D2	Rapport Général THEME E TOPIQUE E2
	REUNION CONSEIL SIMR	TOPIQUE C1 Excavations Minières	Rapport Général THEME B, TOPIQUE B3 Construction à Faible Profondeur		TOPIQUES C3 et C4 Les Houillères et l'Affaissement	TOPIQUE E1 Fracture et écolement de l' écorce terrestre
		DEJEUNER	DEJEUNER	VISITES TECHNIQUES	DEJEUNER	DEJEUNER
REUNION DU BUREAU SIMR	REUNION CONSEIL SIMR	Rapport Général THEME A, TOPIQUE A1 Exploration geophysique	TOPIQUE B1, Stabilité des Pentes Rocheuses		TOPIQUE C2, L'Excavation Souterraine Permanente (1)	Extraction, stockage et enlèvement à grande profondeur TOPIQUES A4, C5, D3
		TOPIQUE A2, Essais In Situ et en Laboratoire	TOPIQUE B2, Fondations sur Rochers y comp. fondations de barrage		TOPIQUE C2, L'Excavation Souterraine Permanente (2)	TOPIQUE E3, Voies Futures CLOTURE
	RECEPTION DU PRESIDENT DU CONGRES	RECEPTION PRESIDENTIELLE ET CONCERT D'ACCUEIL	OPERA "DIE BERGKNAPPEN"	SOIRE LIBRE	DINER DU CONGRES	BAL CAMPAGNARD

08°° 09°° 10°° 11°° 12°° 13°° 14°° 15°° 16°° 17°° 18°° 19°° 20°° 21°° 22°° 23°°

VERANSTALTUNGEN VOR UND NACH DEM KONGRESS

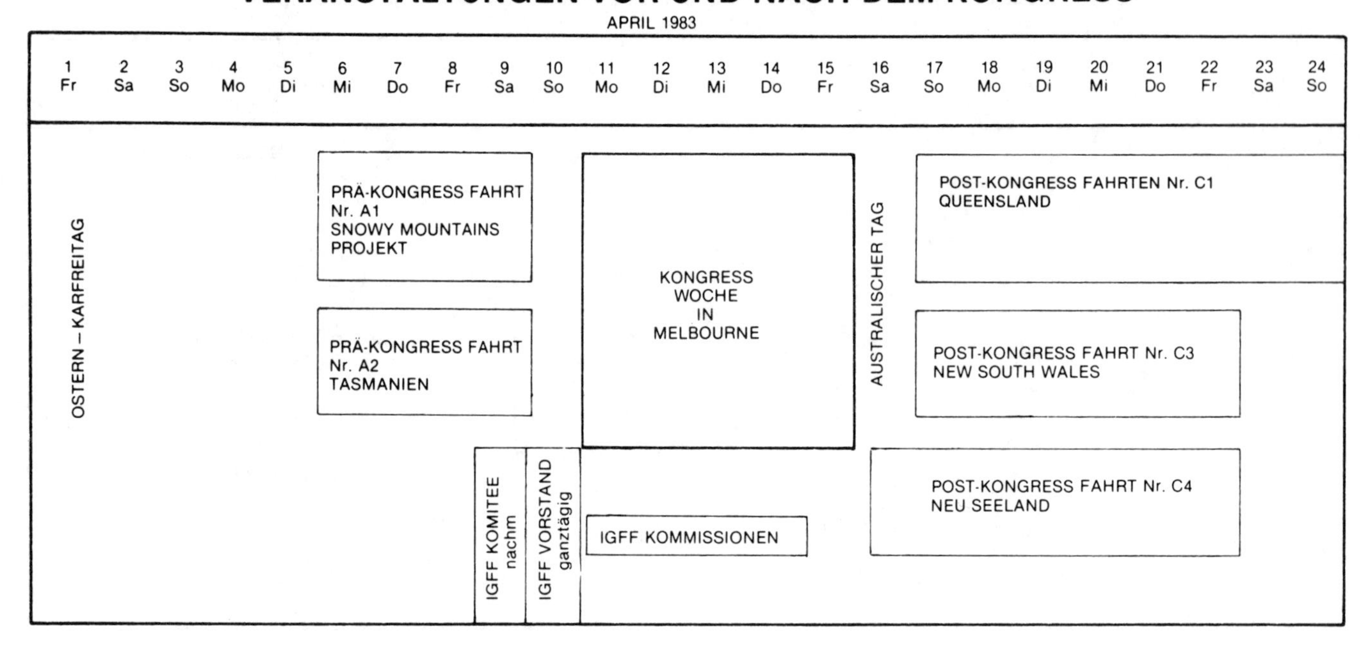

KONGRESS PROGRAMM

TREFFPUNKT – DALLAS BROOKS HALLE

	April 9 Samstag	April 10 Sonntag	April 11 Montag	April 12 Dienstag	April 13 Mittwoch	April 14 Donnerstag	April 15 Freitag
08°°	IGFF KOMITEE VERSAMMLUNG	Einschreibung offen um 9.00 Uhr	Einschreibung offen um 8.00 Uhr		TECHNISCHE BESICHTIGUNGEN	Einschreibung offen um 8.00 Uhr	
09°°		IGFF KOMMISSIONEN	OFFIZIELLE ERÖFFNUNG Generalbericht THEMA C	UNTERTHEMA A3, Klassifizierung, Voraussage, Beobachtung und Überwachung		Generalbericht THEMA D UNTERTHEMEN D1 und D2	Generalbericht THEMA E UNTERTHEMA E2
11°°		IGFF VORSTANDSVERSAMMLUNG	UNTERTHEMA C1 Tiefe unterirdische Hohlräume	Generalbericht THEMA B, UNTERTHEMA B3 Oberflächennahe Konstruktionen		UNTERTHEMEN C3 und C4 Kohlenbergbau und Absenkungen	UNTERTHEMA E1 Bruch und Kriechen der Erdkruste
13°°			MITTAGESSEN	MITTAGESSEN		MITTAGESSEN	MITTAGESSEN
14°°		IGFF VORSTANDSVERSAMMLUNG	Generalbericht THEMA A, UNTERTHEMA A1 Geophysische Untersuchung	UNTERTHEMA B1 Standsicherheit von Felsabhängen		UNTERTHEMA C2 Longlebige Grubenbaue (1)	UNTERTHEMEN A4, C5, D3 Tiefe Untergrund-Gewinnung, Lagerung und Beseitigung
16°°			UNTERTHEMA A2 Prüfung an Ort und Stelle und im Labor	UNTERTHEMA B2 Fundamente auf Fels, einschl. Stauwerkfundamente		UNTERTHEMA C2 Longlebige Grubenbaue (2)	UNTERTHEMA E3 Zukünftige Entwicklungen
17°°							SCHLUSS-SITZUNG
18°°							
19°°–23°°		EMPFANG BEIM KONGRESS VORSITZENDEN	EMPFANG BEIM PRASIDENTEN UND KONZERT	OPER "DIE BERGKNAPPEN"	FREIER ABEND	KONGRESS ESSEN	WOLLSCHUPPENTANZ

Pre and post Congress Tours
Excursions avant et après le Congrès
Exkursionen vor und nach dem Kongress

Many of the features of Australasia, of both tourist and technical interest, were quite different from that usually experienced by the overseas delegates to the Congress. For that reason, the pre and post congress tours were arranged to cover a large proportion of both Australia and New Zealand.

Only short tours could be organised before the Congress due to the Easter break so that incoming delegates were offered tours to either Tasmania or the Snowy Mountains area.

Tasmania is a separate island in the southern part of Australia and presents an area of rain forest, scenic mountains and both hydro-electric and mineral potential. Delegates were met in Melbourne and were flown to the northern part of Tasmania. On the first day they visited a new hydro-electric development at Tulla and saw two dams under construction by the Hydro-Electric Commission. On the second day they saw underground metalliferous mines at Rosebnery and Rension Bell. Both open stoping and cut and fill operations could be seen. On the third day they visited some of the scenic areas in the south-west of the Island before their return to Melbourne.

The tour to the Snowy Mountains commenced in Australia's capital city of Canberra. Delegates met at a dinner and then drove into the Snowy Mountains where they examined the facilities of the Snowy Mountains Engineering Corporàtion and several dams and tunnels. Construction work in this area ceased several years earlier but many of the projects achieved prominance due to their uniqueness and in their rate of construction. Delegates were then driven through the mountains, past many ski resorts, and on to Melbourne.

Three longer tours were held after the Congress. One tour visisted a variety of mining projects in New South Wales. Undergound metalliferous mines were visited at Broken Hill including stops at 1600 m depth. The tour then moved across pastoral country to Cobar where two further metalliferous mines were examined before entering the coal mining country closer to the coast. A new strip mining operation and an underground mine were visited before the visit terminated in Sydney.

The second post-congress tour was to Queensland. Delegates flew from Melbourne to the strip mining operations in the Bowen Basin. Highwall and spoil pile stability problems could be seen. The party then flew to Mt. Isa - the location of one of the world's largest underground metal mines. Delegates met the rock mechanics personnel and visited both open stoping and cut and fill operations. Some of the party moved on to the Great Barrier Reef before proceeding home.

The final post-congress tour was to New Zealand where the New Zealand Geomechanics Society organised an extensive tour. Delegates first visited the South Island where alluvial gold mining took place hydro-elelectric schemes are being developed and the mountains are extremely beautiful. The group then moved to the geothermal and coal mining areas of the North Island before departing to their various destinations from Auckland.

EXCURSION B1

LATROBE VALLEY BROWN COAL OPEN CUTS

Excursion B1 went to the Latrobe Valley 150km east of Melbourne, where there is a total brown coal reserve of 108,000 x 10^6 tonnes, of which approximately 30% is at present an economic resource. The coal is won from three major open cuts at Yallourn, Morwell and Loy Yang to fire power stations with a current generating capacity of 4,200MW. The largest power station under construction at Loy Yang will produce 4000MW by the mid 1990's.

The tour began at a vantage point in the hills west of the Latrobe Valley, with a view of the 100,000 hectare valley which is underlain by coal, the open cuts, power stations and the townships of Morwell and Traralgon. After a presentation by officers of the State Electricity Commission of Victoria at the Morwell visitor's centre, the Loy Yang open cut, power station and overburden removal operations were inspected. In the afternoon the tour went into the Morwell open cut. Commercial operation of the open cut began in 1955 and current production is 15 x 10^6 tonnes per year. Open cut batter stability is controlled by the strength of the clay seam which forms the base of the open cut, pore pressure in this clay/coal interface, and hydrostatic driving force of water contained in the vertical coal joints which divide the almost impervious coal seam into large blocks. Stabilising measures include dewatering of the aquifer beneath the open cut (750 litre/sec pumped out), horizontal drains in the batters and surface drainage to prevent water entering the vertical joints.

View of Morwell Open cut showing dredgers on the operating faces in the foreground, and the Hazelwood Power Station behind the open cut.

EXCURSION B2

THOMSON AND BLUE ROCK DAMS

The Blue Rock Dam is a 75 m high zone earth and rockfill dam having a central clay core with filters on the upstream and downstream sides and rockfill shoulders. The dam is on the Tanjil River which is one of the main northern tributaries of the La Trobe River. It is being constructed by the State Rivers and Water Supply Commission of Victoria.

Lower Devonian mudstones from the main foundation rock on both abutments. Older Volcanic basalts overlie the mudstone at higher elevations on both abutments and the basalts are overlain by the Haunted Hill Gravels. Rockfill for the dam was mudstone obtained from a quarry developed about 0.5 km upstream from the wall. Filter materials were produced by crushing basalt and the earth fill was obtained from sand and clay layers in the Haunted Hill Gravels.

The features of the dam were explained and then the embankment, diversion tunnel portals, spillway, quarries and borrow pit were inspected.

In the afternoon Thomson Dam was inspected. It is being constructed by the Melbourne and Metropolitan Board of Works to augment Melbourne's water supply. The main dam embankment is 166 high, has a crest length of 590 m and is of zoned earth and rockfill construction. The dam is founded on folded interbedded siltstone and sandstone adjacent to a granodiorite batholith.

The features of the project were explained and then the main dam, saddle-dam, spillway, diversion tunnel and outlet works were inspected.

EXCURSION B3

HISTORICAL GOLD MINING IN CENTRAL VICTORIA

The excursion to Ballarat was enjoyed by about thirty people. After a short bus tour of the City, taking in the Eureka Stockade display depicting the brief but bloody battle by gold miners in 1854 against the government, the mock gold mining village of Sovereign Hill was visited. All were impressed by the buildings and other displays faithfully reproducing the early gold mining era in Australia. The excursion included a conducted tour of underground mine workings by the Mine Manager.

Lunch at the United States Hotel in the village was followed by a visit to the Gold Museum to view an extensive display of gold, gold production and early Australiana. The excursion provided an informative and welcome diversion from the austere atmosphere of the congress' technical sessions.

EXCURSION B4

ISRM CONGRESS

King Island Scheelite Ltd. owns and operates two underground mines on the South East coast of the island. Tour participants visited the Dolphin Mine, to view the mining method and the rock mechanics program, and saw the milling and metallurgical plant associated with the production of high quality concentrate.

The mine areas visited were chosen to demonstrate ore extraction using the post-pillar method. The mining sequence is especially suited to wide, flat-lying orebodies as structural support is given to the newly exposed backs by leaving slender (post) pillars on a regular grid pattern. An explanation was given of the rock mechanics problems associated with using this method in a well jointed, low stress environment. The research programme, conducted jointly by King Island Scheelite and CSIRO Division of Geomechanics, involves monitoring the behaviour of, and the interaction between, the back-fill material, the pillars and the slope abutments.

The Dolphin mine portal, near the base of the original open cut. The orebody extends below the sea floor.

ISRM - TOUR OF CSIRO DIVISION OF GEOMECHANICS - APRIL 1983

On 13th April 1983 sixty one delegates at the 5th ISRM Congress meeting held in Melbourne undertook a tour of the laboratories of the CSIRO Division of Geomechanics (then Applied Geomechanics) located at Syndal a suburb of Melbourne 12 km east of the city. The tour commenced at 0900 hrs with a Welcome by Dr. B.H.G. Brady, Chief of the Division, after which the delegates were split into four groups which were guided through the Division's research facilities. Sixteen specific displays were arranged in eleven separate laboratories and at each the work was explained by a member of CSIRO staff involved. Research aspects presented included in-situ stress measurement and hydraulic fracture, geophysical characterization of rocks, mechanical rock testing, large scale physical modelling of coal mines and numerical modelling of stress analysis. Prominent among the displays were results from the many field based projects conducted by the Division at various mines around Australia.

Following the tour of the scheduled displays delegates were invited to re-visit any of the laboratories for more detailed discussions with CSIRO staff on matters of specific interest. They were also provided with a booklet describing most of the current, or recent past, research activities of the Division. The tour concluded with a luncheon after which Delegates returned to the city.

OPENING SESSION

Séance d'ouverture

Eröffnungssitzung

L. A. Endersbee, AO
Congress President

It is my pleasure to welcome you to the opening of the 5th International Congress on Rock Mechanics for the International Society for Rock Mechanics. It is a special pleasure for us in Australia to welcome you all to our country and we are very pleased to welcome our international colleagues. In Australia the challenge is of developing a vast continent with relatively few people: for example, Australia is the size of the continental United States.

Many of us in Australia have attended the earlier congresses and symposia of the International Society for Rock Mechanics and have felt privileged to meet distinguished colleagues from other countries and to share experiences with them. At the same time we have felt enriched by the opportunity to visit your countries and to come to understand your intellectual and cultural traditions. I should point out that in a new and developing country like Australia that your intellectual and cultural traditions are also ours and we feel that they are part of our own intellectual and cultural traditions.

Because of the many times we have attended past Congresses and visited your countries and enjoyed and respected your culture and way of life, the Australian participants in the Congress are especially pleased to welcome you all to our country. It is also our appreciation of the benefits of such international scientific co-operation that has led us in Australia to be strong supporters of the International Society for Rock Mechanics. On the other hand I should point out that our remoteness has sometimes encouraged an independence of spirit and we hope that your visit to Australia will give us the opportunity to let you see some of our own developments in our discipline. I extend a special welcome to President Professor Dr. Wittke and the members of the Board and Council of the International Society for Rock Mechanics and our overseas friends. In Australia the representative body for Rock Mechanics is the Australian Geomechanics Society which is the adhering body for the International Society for Rock Mechanics. It is also the adhering body for the International Society for Soil Mechanics and Foundation Engineering and the International Association for Engineering Geology. The Australian Geomechanics Society is a joint technical unit of two learned societies in Australia, The Australasian Institute of Mining and Metallurgy and The Institution of Engineers, Australia, and I shall shortly be introducing the Presidents of these major Australian learned bodies. I now would like to introduce to you Professor Harry Poulos who is Chairman of the Australian Geomechanics Society.

As I have indicated the Australian Geomechanics Society is the National learned body in our field. It offers a strong program of learned meetings and seminars for both mining engineers and civil engineers and geologists throughout Australia. I am pleased now to invite the President of the Australian Geomechanics Society, Professor Harry Poulos, to address you. Professor Poulos is a distinguished expert in geomechanics from the University of Sydney.

PROFESSOR H.G. POULOS
Chairman of the Australian Geomechanics Society:

My Lord Mayor, distinguished guests, ladies and gentlemen. It is my pleasant task to welcome you here to the 5th International Congress on Rock Mechanics on behalf of the Australian Geomechanics Society. In Montreux in 1979 our Society issued an invitation to host this Congress and were honoured to be chosen by the International Society for Rock Mechanics. Since then our Organizing Committee has worked very hard to ensure the success of this Congress and it is our hope that you will all find it to be valuable technically and enjoyable socially. It may perhaps be of interest to recall briefly the history of the Australian Geomechanics Society as I think it was one of the first national geotechnical societies to combine the three sister disciplines of soil mechanics, rock mechanics and engineering geology. The first step in the formation of the Society was taken in 1947 when an Australian National Committee for Soil Mechanics and Foundation Engineering was formed under the auspices of The Institution of Engineers, Australia. In 1957 the first Australasian Vice-President of the International Society for Soil Mechanics was appointed. Interest and activity in rock mechanics was also developing during this period, and in 1966 the first Australasian Vice-President of the International Society for Rock Mechanics was appointed, this being our present Congress Chairman, Professor Lance Endersbee. Following a number of symposia and conferences on various aspects of geotechnical engineering in the late 1960's it became clear that there was much to be gained by amalgamating the various geotechnical activities into a single society, therefore in 1970 the Australian Geomechanics Society became a reality. It was formed as Professor Endersbee has mentioned to you under the joint sponsorship of The Institution of Engineers, Australia and The Australasian Institute of Mining and Metallurgy and was affiliated with the International Societies for Soil Mechanics, Rock Mechanics and Engineering Geology. In subsequent years this society has continued to develop and has proved to be a focus for fruitful collaboration between civil engineers, mining engineers, and engineering geologists. It has also acted as a catalyst for the formation of The Australian Tunnelling Association as a separate organization in the mid 1970's and has organized a number of regional conferences and specialist symposia. Its activities have culminated in the organization of this present International Congress. It is our hope that delegates attending this Congress will derive a number of benefits including firstly the opportunity to hear general reporters of international stature summarize the latest developments in their area of specialization, to give them the opportunity to discuss technical details with authors of papers both during plenary sessions and during poster sessions, to renew old friendships and form new ones; this is surely one of the major benefits of any conference or congress such as this. Then to inspect some of the geotechnical activity within Victoria during the Congress and in other parts of the country after the Congress. And finally to sample and savour some of the Australian way of life - to this end the Organizing Committee has included a number of peculiarly Australian social functions in the program including a wool shed dance and a visit to an Australian rules football match. In return it is our hope that delegates will offer us something, primarily of course, the benefit of their experiences, but also, of great importance to us, the opportunity for geomechanics and rock mechanics in particular to gain some national exposure. Unhappily too few of the public in this country are aware of the work we do. In the majority of cases our work goes on underground or is hidden beneath a structure or a dam and therefore many of our spectacular technical achievements go unnoticed or are ignored by the public at large. Therefore I sincerely hope that this Congress will serve to make the Australian public aware of our positive contribution to the development of natural resources, energy requirements and the creation of the nation's infrastructure. International meetings such as this Congress provide an unequalled opportunity for people from all nations to meet on common ground and to develop not only increased knowledge in the field of rock mechanics but also friendship and understanding. The Australian Geomechanics Society will feel well pleased if at the end of this week delegates feel that Melbourne 1983 has fulfilled these dual functions of advancing scientific knowledge and developing international understanding and friendship.

PROFESSOR L.A. ENDERSBEE, AO:

President Wittke, I am pleased now to call on Mr. Jack Liebelt, President of The Australasian Institute of Mining and Metallurgy. Mr. Liebelt is a mining engineer and was for some time general manager of group operations for the Conzinc Riotinto Mine at Broken Hill. He is now a group executive of C.R.A. Ltd., one of the largest mining companies in Australia and in fact by its size it is one of the largest mining companies in the world. I am pleased now to call on Mr. Jack Liebelt, President of The Australasian Institute of Mining and Metallurgy.

MR. J.L. LIEBELT:
1983 President, The Aus.I.M.M.:

Chairman, My Lord Mayor, ladies and gentlemen; The Australasian Institute of Mining and Metallurgy which was established in 1893 is pleased to be associated with this very important Congress. Members of The Institute along with members of The Institution of Engineers, Australia, have been involved over many years with rock mechanics problems. From about 1960 significant work in rock mechanics has been carried out by specialist departments in the mines at Mount Isa and Broken Hill and some of you will be visiting these operations in the post-Congress tours. As the Chairman has said I was responsible some years ago for the mining operations at two of the Broken Hill mines, when we recruited the first rock mechanics engineer into the company from overseas. In those days the rock mechanics department in the mine was seen by the operating engineers as something of an unrealistic exotic backwater. A transfer to the rock mechanics department for the young mining engineer was usually seen as a vote of no confidence, but today that scene has changed dramatically. The rock mechanics department at those two mines now has about twelve people. It is totally accepted as an essential part of the mine planning and operating functions. It is now a well accepted possible career path for the young mining engineer. The future, I believe, will see more changes, and the next generation of underground mining systems will probably be the result of a technical joint venture between rock mechanics, engineering, including mining, and electronics. In recognition of the growing involvement of members of both The Institute and The Institution of Engineers, Australia, in rock mechanics the two organizations combined to sponsor the Australian Geomechanics Society in 1970 as Professor Poulos has already told us. We are pleased to be one of the sponsors of this society which has now more than 400 members and is the host for this Congress. I support the comments made by the Chairman of the Society in his welcome to you this morning. My Institute has been pleased to provide the Secretariat and organizing facilities for this Congress. The collaboration achieved through the Australian Geomechanics Society between civil engineers, mining engineers and engineering geologists has proved beneficial and fruitful for all concerned. The Australian mining industry in its rapid expansion has made great demands on rock mechanics in both underground and open pit operations. The developments taking place in Australia in Rock Mechanics, both in metalliferous and coal mining are reflected in some of the papers being presented at this Congress. Some of you will have had the opportunity of seeing some of these developments on the pre-congress tours and others will have the opportunity after the congress. The Australian mining industry collectively has sponsored through AMIRA (which is a mining industry research organization) major geomechanical research projects associated with both open pit and underground operations including coal. One of these projects is a continuing study of geomechanics of underground metalliferous mines which addresses such topics as the assessment of changing rock conditions induced by mining, stope design, the prediction of rock mass conditions, and ground support. The major coal project is a study of fragmentation in coal overburden. The main aim of this study, which is also continuing, is the improvement of blasting techniques for coal overburden removal. The major studies will continue and rock mechanics will remain an essential part of our operations. On behalf of The Australasian Institute of Mining and Metallurgy I welcome you all to this Congress. I particularly welcome you, Professor Wittke in your capacity as President of the International Society. We are pleased you have had the opportunity of seeing some of Australia over the last three weeks in your visits to some of our mining centres and in your talks to groups of the Australian Geomechanics Society. I wish you well in your deliberations and to our overseas visitors I wish you an enjoyable stay in Australia and that you will find your visit worthwhile and rewarding.

PROFESSOR L.A. ENDERSBEE, AO:

It is now my pleasure to call on Mr. John McIntyre, who is President of The Institution of Engineers, Australia. The Institution of Engineers, Australia is the learned body covering all engineering disciplines in Australia: civil engineering, mechanical engineering, electrical engineering, chemical engineering etc. and so on. There are about 36,000 members around Australia. Mr. McIntyre, their President, is a consulting civil engineer with major activities in Queensland servicing clients in mining, government and industry. I can also mention that our President Professor Walter Wittke, has recently been in Townsville and knows how congenial it would be to have an engineering practice in North Queensland. Mr. McIntyre's practice is the largest in North Queensland. I am pleased now to call on the President of The Institution of Engineers, Australia, Mr. John McIntyre.

MR. J. McINTYRE
President, The I.E. Aust.:

Congress Chairman, my Lord Mayor, ladies and gentlemen. Thank you for inviting me on behalf of The Institution of Engineers, Australia to participate in the Official Opening of this very important Congress in the world mining and civil engineering scene. This Congress, as you have already been told, is being jointly sponsored by The Institution of Engineers, Australia and

The Australasian Institute of Mining and Metallurgy, but I note that The Institute has done most of the hard work in providing administrative and secretarial services, and from my telephone conversations with Bill Vance in recent months I suspect that he has not had much time to spare. On the other hand Lance Endersbee, a recent President of our Institution is the Congress Chairman and a member of the substantial Organizing Committee. One must therefore encompass the whole of the Organizing Committee in offering congratulations for what appears to me to be a first class show. The holding of the 5th International Congress on Rock Mechanics in Melbourne represents an important milestone in the development of geomechanics in Australia and is testimony to the growing influence of this subject on the Australian scene, especially in relation to the mining industry. From The Institution's point of view this connection dates back to 1948 when a group of interested engineers participated in the Rotterdam Conference of the International Society of Soil Mechanics and Foundation Engineering. This led to the establishment of what became known as the technical committee number 3 of The Institution which in turn gave way to the Australian National Committee on Soil Mechanics and Foundation Engineering. A major impetus was gained when Australasia was recognized as one of the six continental regions of the International Society of Soil Mechanics and Foundation Engineering in 1957. When, during the 1960's, the pressures for the establishment of a separate group to cater primarily for hard rock engineering emerged, the International Society for Rock Mechanics was formed with a similar regional structure and Australasia has played its part in the affairs of the Society from the beginning. The 1960's also saw an upsurge in interest in rock mechanics in the mining industry, and it became apparent that there was much to be gained from the fusion of soil and rock mechanics interests. After much discussion and deliberation the Australian Geomechanics Society was formed in 1970 under the joint auspices of The Institution and The Australasian Institute of Mining and Metallurgy. The Australian Geomechanics Society had initially, the responsibility of looking after the local affairs of the International Society for Soil Mechanics and Foundation Engineering, the International Society for Rock Mechanics and later performed a similar function for the International Association of Engineering Geology , and this position remains today. Recently, the Australian Geomechanics Society has reviewed its membership structure and introduced provision for a student membership category. It may well be said therefore that Australian geomechanics and in particular rock mechanics has now come of age. One of its major strengths lies in having one organization to consider and represent the diverse interests from civil to mining engineering. This is in keeping with the wider philosophy of The Institution of Engineers, Australia which represents all the branches of engineering in this country. The theme of the Congress: "Rock mechanics for resource development, mining and civil engineering", further emphasizes this wide diversity of topics which come under the geomechanics umbrella. Last night I was pleased to hear that Professor Ted Brown, now of the Imperial College, London, but previously of the James Cook University of North Queensland is to follow Professor Wittke as your International President. This news will be received with great pleasure by Ted's many friends and contemporaries in North Queensland and particularly by his old friend and mentor Professor Hugh Trollope. Finally, ladies and gentlemen, I hope that you all find this Congress a very worthwhile experience and I have much pleasure in joining my co-openers in welcoming you and particularly our visitors from around the world.

PROFESSOR L.A. ENDERSBEE, AO:

It is now my pleasure to introduce to you the Lord Mayor of Melbourne, Councillor Bill Gardner, who I am going to ask to formally open the Congress for us. Our Lord Mayor is a very popular choice of his colleagues on the Council of the City of Melbourne, the Councillors who have elected him to be Lord Mayor of this city. I should mention that Councillor Gardner had worked for a fair part of his life on the waterside and he became a Councillor of the City of Melbourne and because of the very great respect in which he was held by his colleagues on the Council, he is now the Lord Mayor of Melbourne. We are very pleased to have the Lord Mayor with us and I am pleased now to call upon him to open our Congress.

COUNCILLOR W. GARDNER
Lord Mayor of Melbourne:

Professor Endersbee, Professor Wittke, Professor Poulos, Mr. Liebelt, Mr. McIntrye, Mr. Bamford, Mr. Vance, official guests, delegates, ladies and gentlemen. In 1979 the then Lord Mayor of Melbourne supported the Australian Geomechanics Society in its invitation to Professor Habib, the then President of the International Society for Rock Mechanics, for the Society to hold your Congress in Australia and in this City of Melbourne. I understand that Professor Habib is in the audience today and I must say how pleased I am that on behalf of the International Society he accepted the invitation for the Congress to be held in Australia and in the City of Melbourne. I see from the literature I have read that previous Congresses have been held in Denver, Lisbon, Belgrade and Montreux. I understand Melbourne rates several firsts in the geomechanics field in Australia, it being the host city for the first Australia New Zealand Geomechanics Conference, the host city for the first Australian Tunnelling Conference and now the Australian city to host the first International Society for Rock Mechanics Congress held in Australia and the first in the Southern Hemisphere. I hope the hospitality and the facilities we can provde you in Melbourne are more than equal to that of your previous venues. I understand we have 31 countries represented by the 400 delegates present. I particularly extend the warmest of welcomes to the many visitors from overseas. Melbourne itself has very warm sister-city relationships with cities in both China and Japan and we in Melbourne hope to foster and create further sister relationships throughout the rest of the world. It is hoped that you will have the opportunity of visiting some of the attractions of Melbourne during your stay. We do to some extent follow the pattern of cities today where you have instant soup and plenty of stomach troubles, you have a type of instant city with all the resultant social problems and of course now we have engineers going into the business of genetic engineering, so I can say that Melbourne being in the forefront of those three things we can rest assured that at least we are going to have a very interesting city for many years to come. I understand that after the Congress some of you will be joining thousands of local people in watching one of our famous football games. To those visitors from overseas I would say that it is such a spectacular game that some of us are inclined to think it is perhaps the last of the gladiator sports, but for those who go to see it I hope you have a very enjoyable day, I know you will have an exciting day. Turning to the technical aspects of the Congress I see you have more than 200 technical papers to be presented. Unfortunately at the present time there is not much in the geomechanics area for you to inspect in Melbourne itself. We have recently completed our new underground railway loop, which I understand posed many problems to the geomechanics engineers at different times in its construction. Of course I would imagine as a layman that that would not be unusual and I would take this opportunity to say on behalf of the great majority of people in Melbourne that the area of geomechanics or rock mechanics is quite a mystery to all of us, but we do understand that without this great endeavour in this field that cities of today just would not exist. Melbourne is the headquarters of Federal Government Rock Mechanics research organization and some of you will be visiting their laboratories later this week. I know from your program that many of you will be travelling on technical excursions on Wednesday to such places as the Latrobe Valley to see the large scale brown coal operation, others will be visiting the Thomson and Blue Rock Dams, and travelling to these places will give you an opportunity to see some of the suburbs of Melbourne. Melbourne itself is what we would refer to as a sprawling type of city. The residential areas of it have not as most of the cities in the older parts of the world more or less gone upwards; Melbourne has spread outwards, and it was not until recent years that not only in Melbourne but throughout Australia in comparatively recent years that it was recognized that we have a distinctive landscape in Australia. Speaking from the Melbourne point of view I would as a point of interest point out to you that in your travels around Melbourne and its environment you will see that there is evidence of very deliberate attempts to change the Australian landscape to a European one. Our early painters for instance painted our landscape after the style of the European landscape. The gum treas were contorted on canvas to look like European trees, the native people of Australia were painted to look like people from the West Indies and so it went on. Melbourne is surrounded by beautiful gardens which are essentially in the European style. The gardens which surround the private homes in Melbourne always contained the plants from other parts of the world, and it is only in recent years that the people of Melbourne have recognized the worth and distinction of their own native flora here in Australia and we are planting natives as we say in our home gardens. You will notice this is a very distinctive part about the suburbs of Melbourne, this Europeanisation of the landscape, and it was not until comparatively recent years when a group of artists in Melbourne set themselves up in then what was an outer suburb, virtually in the countryside and the name of that outer suburb was Heidelberg. We referred to those painters as members of the Heidelberg School and it is generally accepted that those group of painters were the first ones to recognize that the Australian landscape was something entirely different from that of Europe and indeed other parts of the world and should be recognized as such, and from then on our wonderful landscape artists have been painting the Australian landscape as this strange harsh interesting landscape. You will see it as you travel around the country and although we are in the southern part of Australia we do not have these tremendous extremes, such as of snow and heat around Melbourne, but we do have nature at times I think taking a hand and lashing back at us. You will see evidence of that.

Chairman, Professor Wittke, members of the International Society for Rock Mechanics, I am very pleased you accepted the invitation to have the Congress in Melbourne and may I say to the companions who are visiting also that if you are in Melbourne and the heart of Melbourne is only about a kilometre away from this building that you remember that the Melbourne Town Hall is the very centre of Melbourne and we would be very pleased if you have a few minutes to drop in and see us and have a look at our lovely building. You will find the visit well worthwhile. Mr. Chairman it gives me great pleasure to formally declare your Congress open and to extend to you all the warmest of welcomes to Melbourne and may your Congress be successful and your stay in our fine city be pleasant and enjoyable. Thank you very much.

GENERAL REPORT / RAPPORT GÉNÉRAL / GENERALBERICHT

SITE EXPLORATION AND EVALUATION

Exploration et évaluation in situ

Untersuchung und Beurteilung des Betriebspunktes

David H. Stapledon
South Australian Institute of Technology, Adelaide, Australia

Peter Rissler
Ruhrtalsperrenverein, Essen, FR Germany

SUMMARY:

Site investigations cost money and time but are essential for the successful planning and construction of projects. Six papers on geophysical methods suggest that the seismic method is the most widely accepted. In thirty-five papers on laboratory and in-situ testing, emphasis is placed on anisotropic properties, behaviour under long-term or cyclic load, shear strength of discontinuities and scale effects. Fifteen papers on classification illustrate the value of classification systems, and the difficulties associated with designing any single system which would relate to all project types, in all geological environments. Five papers on hydrogeology deal mainly with permeability assessment and groundwater observations.

ZUSAMMENFASSUNG:

Geländeuntersuchungen kosten Geld und Zeit, sind jedoch wesentlich für erfolgreiches Planen und Bauen. Sechs Beiträge über geophysikalische Methoden deuten an, daß die seismischen Methoden überaus weit verbreitet sind. Im Bereich der Labor- und Feld-versuche liegt der Schwerpunkt von 35 Beiträgen auf der Auswertung anisotroper Eigenschaften, dem Verhalten unter Dauer- und zyklischer Last, auf der Bestimmung der Scherfestigkeit von Trennflächen und auf Maßstabeffekten. Fünfzehn Beiträge zum Thema Klassifikation beleuchten den Wert von Klassifikationssystemen und die Schwierigkeiten beim Entwurf eines umfassenden, für alle Projekttypen und alle geologischen Situationen geeigneten Systems. Fünf Aufsätze handeln von hydrogeologischen Problemen wie Bestimmung der Wasserdurchlässigkeit und Grundwasserbeobachtungen.

RESUME:

Les analyses de terrain coutent du temps et de l'argent mais sont d'une importance capitale pour des traces de plan et des constructions couronnes de succes. Six articles sur les methodes geophysiques donnet a entendre que les methodes sismiques sont tres repondues. Dans le domaine des recherches en laboratoire et sur le terrain, l'accent de 35 rapports est mis sur l'evaluation des proprietes anisotropes, sur le comportement sous une charge permanente et cyclique, sur la definition de la resistance au cisaillement des surface de separation et sur les effects de l'echelle. Quinze rapports sur le theme de la classification demontrent la valeur des systemes de classification et les difficultes rencontrees en essayant d'etablir un systeme unique approprie a tous les types de projets et a tous les terrains geologiques. Cinq articles traitent des problemes hydrogeologiques tels que la determination de la permeabilite et l'observation des eaux souterraines.

GENERAL REMARKS

Site exploration and evaluation cover a vast field in rock mechanics. These activities are presupposition for all papers delivered for the General Themes B, C, D and partly E.

Site exploration and evaluation often seem to be costly, but in comparison to the construction budget these costs are small. Indeed, they virtually disappear when compared with those of a failure caused by inadequate or omitted investigations.

Also, site exploration and evaluation usually need a lot of time and they cannot be time-scheduled and costed with the same degree of precision as construction work. This is because our sites have been built by natural processes, during vast periods of time, and Nature has left us an incomplete and often well-concealed record of her activities, and no "as constructed" drawings. Therefore we often have to start exploring a site with no clear picture of what we will find. The exploratory programme must therefore proceed in stages with the activities in each stage designed to answer questions which arise from evaluation of the results of the preceding stage.

Both the cost and time requirements should be understood by the owners of projects and should be accepted.

In 1975 this Society published a Commission report "Recommendation on Site Investigation Techniques" which included a suggested systematic approach to exploratory programmes.

In his general report to Theme 2 at the ISRM-Congress in Montreux Franklin (1979) described in detail the use of tests and monitoring in the design and construction of rock structures and showed very clearly the inter-relationships between geotechnical and other engineering activities throughout the planning and construction periods.

Due to its central importance in all geotechnical work, site exploration and evaluation has been a topic in several former ISRM-Congresses, for example:

In 1974 (Denver) Theme 1

"Physical Properties of Intact Rock and Rock Masses",

in 1974 (Montreux) Theme 1 (partly)

"Rheological Behaviour of Rock Masses"

and Theme 2

"Use of Tests and Monitoring in the Design and Construction of Rock Structures"

were dedicated to exploration and evaluation work.

As decided by the Organizing Committee the general topic of this Congress dealing with site investigations has been subdivided into four sub-themes,

A.1 Geophysical Testing and Exploration
A.2 In-Situ and Laboratory Testing
A.3 Classification, Prediction, Observation and Monitoring
A.4 Hydro-Geology

Our comments are presented under those headings.

Sub-Theme A.1

GEOPHYSICAL TESTING AND EXPLORATION

It was a little surprising for us, that only 6 papers have been submitted for sub-theme A.1 compared to 34 for sub-theme A.2. One could interpret this as an expression of the value of geophysical testing in comparison to direct methods for in-situ and laboratory testing. We are convinced that this interpretation would not be justified. Geophysical methods are advantageous and have - on the other hand - some handicaps. Their advantage is their ability to provide an overview of large rock mass volumes. With regard to this they are superior to all direct methods. Their disadvantage: in most cases they give only index values for the characterization of the rock mass. On the other hand it seems to be a very special advantage that their application is rather cheap. This makes it possible to provide much information, during all stages of development of projects, and during their operation.

The degree of success achieved by any exploratory method must depend not only on the intrinsic value of the method, but also on the skill and knowledge of the persons applying it, in particular, on their understanding of the limitations of the method.

Experience has shown that the limitations of geophysical methods are frequently not well understood and that their application in unsuitable situations, often without the support of direct methods, has led to a lower success rate than perhaps they deserve.

The seismic method, being the most commonly used, is also the most commonly misused. The following are what we believe to be two important limitations of that method:

(a) Important engineering properties of rock masses (particularly shear strength) are commonly governed by quite minor geological defects (minor faults or weathered seams) which occur in critical positions and orientations in relation to the stress to be applied. The seismic method generally is unable to locate such defects.

(b) The seismic method applies minute loads to the rock mass for minute periods. The dynamic modulus of elasticity (E_{dyn}) as calculated from the shear wave velocity (V_s) or assessed from the longitudinal wave velocity (V_p) will therefore reflect the response of the rock mass to such minute transient loads. Thus E_{dyn} should not be expected to provide a reliable indication of the deformation of a jointed and seamy rock mass under long-term heavy loads.

Also, we have heard it argued that the longitudinal wave velocity does not necessarily reflect the properties of the "average" rock, as commonly believed. This is because the longitudinal waves, being the first to arrive at the geophones, must take the shortest path through the best rock, that is, there is a "short-circuit" effect.

If we accept the above argument, then the belief that electrical resistivity methods always provide numbers representing the "average" rock, appears ill-founded. Current flow will tend to concentrate in the rock with the highest conductivity, tending to give a short circuit effect through weathered or faulted zones.

It would be useful to have comments on the above and any other possible limitations of geophysical methods, from the Authors and the Congress.

Laboratory and field research relating to seismic methods

In laboratory tests described by Tanimoto and Ikeda the longitudinal wave velocity V_p was found to be insensitive to the spacing of joints with apertures of less than 0.01mm and to show relatively small decreases (5 to 15 percent) with joint apertures of 2 to 4mm. Rock with joint apertures of less than 0.04mm was indicated to be similar to intact rock, in seismic behaviour.

These results appear to be in conflict with the views of other workers. For example Anstey (1977) states

> "The effects of fractures on seismic velocity is major ......; as soon as there is a void, lateral deformation of the particles is possible, and the appropriate modulus of elasticity changes. We recall also how small is the displacement of a particle as a seismic wave passes across it - typically at least two orders of magnitude smaller than the wavelength of light; consequently cracks far too thin to be seen under an optical microscope can produce major changes in the elastic properties".

Also, Savich et al, in Figure 6 of their paper, present the results of both geophysical (ultrasonic and seismic) and geological (visual) determinations of "joint void ratios". The geophysical methods yield joint void ratios 1.7 to 2 times larger than those produced by the geological method. The Authors' explanation of this discrepancy includes the statement

> "the ultrasonic measurements take account of not only visible joints but also of microjoints which cannot be identified in the boreholes visually."

We suggest that the differences in the effects observed by different workers may result from scale effects related to differences in the amplitudes, and lengths of the waves. Comments from the Authors are invited.

We note that in their synopsis, Tanimoto and Ikedo mention field studies at surface sites and at 104 underground sites. These studies are not described in the paper. It would be useful if the Authors could tell us whether their laboratory results agreed with the results of these field studies.

In their travel path experiments (Figures 8 and 9) Tanimoto and Ikedo simulated rock with discontinuous joints, or joints with local contact points, by cutting slits into 50mm plaster cubes. From the travel-time plots the Authors concluded that although there was an apparent overall reduction in V_p, this was caused only by lengthening of the wave propagation path as it avoided passing through the slits. It seems to us that Figure 8 of this paper is a good illustration of the "short-circuit" effect mentioned above in our discussion of limitations. It seems from this figure that unrealistically high values of V_p might be obtained in intensely fractured and/or weathered rock masses if "short-circuit" paths through unfractured or fresh rock exist between shot-points and receivers.

Although the practical significance of their results may not yet be fully understood, we consider that Tanimoto and Ikeda have produced valuable findings with respect to amplitude attenuations, the effects of joint contact areas and contact pressures, and wave travel paths. Of particular interest is the sensitivity of amplitude attenuations to joint apertures which was mentioned already by Kaneko et al (1979) at the 1979 Congress in Montreux.

The paper of By deals specifically with attenuation of longitudinal waves in jointed rock masses. A theoretical model is proposed and verified by an elaborate field blasting experiment. The results are presented on the one hand as a relationship between maximum stress amplitude and distance from the shock source (Fig. 8) and on the other as a nomograph (Fig. 9) from which the "damping factor" can be obtained if the fracture width (joint aperture) and "layer thickness" are known. We assume that layer thickness refers to the spacing of the joints. It should be noted that Plischke (1980) has published work concerning the evaluation of shock wave propagation by numerical methods. Co-operation between these two authors should assist further development of this subject. The results of By referring to joint influence are in general agreement with those of Tanimoto and Ikeda.

Estimation of the effectiveness of grouting

Rodrigues et al present convincing evidence for the success of the seismic method in evaluating the effectiveness of grouting at Cabril Dam. The geophysical work was adequately supported by down-hole permeability tests and was carried out with a high degree of precision. The use of both down-hole and cross-hole techniques should have drawn attention to any anomalous V_p values due to short-circuit effects. In this regard we ask, were there any significant differences between the V_p values obtained by the different techniques in the same portion of rock mass? If so, what were these differences interpreted to mean (in terms of the structure of the rock) and what V_p value was adopted as representative of such portions of rock mass?

From our experience, the recorded V_p values and grouting results are consistent with a rock mass of relatively high mechanical quality before grouting. The data in Tables I and II would be more useful if the Authors could provide further details, e.g. a geological profile, an indication of the defect pattern in the granite, details of the patterns of both consolidation and curtain grout holes, and the lengths of the Lugeon test stages.

Savich et al, referring to grouting quality control studies at Inguri arch dam, show as their Figure 11, a nomograph produced at that site, for the estimation of grouting effectiveness. Termination of the scale for V_o (initial longitudinal wave velocity) at about 5km/sec suggests that rock with V_p of more than this value was found to be ungroutable? We plotted the results of the Cabril Dam tests onto this nomograph, in our Figure 1 below. It can be seen that, as would be expected, the mean results for Blocks V and VI lie in this "virtually ungroutable" area. The results for the remaining blocks lie in the "satisfactory" Zone III and at its boundary with the "good" Zone II. As stated in the preceding paragraph, we consider this Cabril data to be consistent with a rock mass of high mechanical quality. It appears from Figure 11 of the Savich et al paper, that in some areas of the Inguri site, rock with V_o values comparable to those at the Cabril site showed much larger increases in V_p values after grouting (final V_p values of more than 6km/sec in the "excellent" grouting Zone I). It appears likely that the reason for this large difference is related to differences in the character of the rock masses, prior to grouting. It would be useful therefore if the Authors could provide details of those parts of the rock mass at the Inguri site, where "good" and "excellent" grouting results were indicated by V_p values of around 6km/sec after grouting. In particular, could the Authors explain the increases of more than 20 percent in V_p after grouting of rock with V_p values of around 5km/sec, i.e. rock which should have been of high mechanical quality prior to grouting?

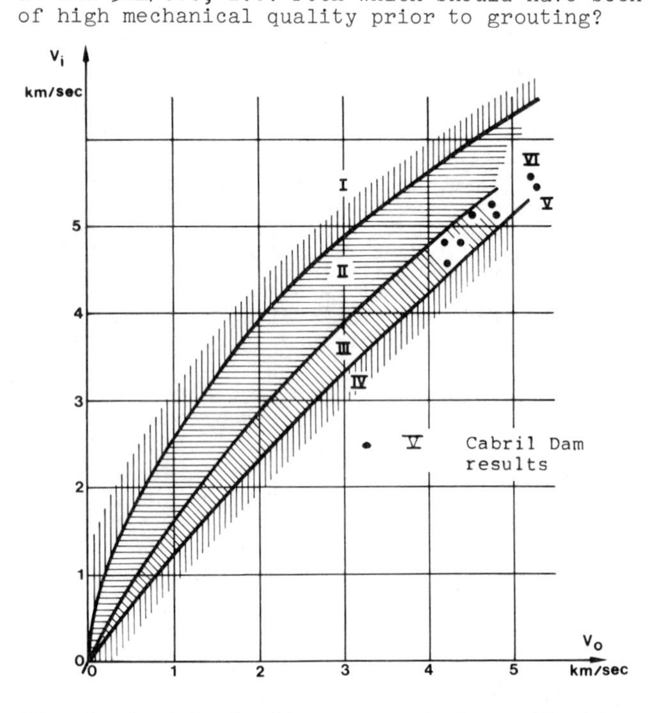

Fig. 1 Savich et al's nomograph for estimating grouting quality, with the Cabril Dam results plotted

Assessment of elastic and deformation properties

Dhawan et al obtained V_p and V_s values from seismic spreads in tightly folded and jointed quartzitic phyllites at a dam site in the Himalayas. Some of the spreads were in an exploratory drift in which plate bearing and flat jack tests were carried out to obtain values of deformation modulus (D) and elastic modulus (E). The figures they present show reasonable correlation between V_s and V_p, but rather poor correlations between V_p and the D and E values. The poor correlations appear to be at least partly caused by the wide variation in the results of the in-situ plate bearing and flat jack tests, as seen in their Table 3. Also because of limitation (b) stated at the beginning of this General Report, we would not expect the values of either D or E from insitu loading tests to be related in any simple fashion to the E_{dyn} values.

Savich et al state that seismoacoustic methods have become "the leading methods of quantitative studies of elastic and deformation properties of rock masses". They describe how systematic studies using seismoacoustic methods plus engineering geology, static modulus determinations and statistical processing of the data, have enabled the development of generalized equations for the computation of the static moduli D and E, from the dynamic modulus E_{dyn}. Four sets of values for the coefficients used in these formulae are given in their Table II, there being one set for each of four very broad classes of rocks. It is claimed by the Authors that the use of these generalized formulae in any rock mass allows computation of values of D and E with a 90 percent probability of less than 30 percent error.

The General Reporters applaud the thorough and systematic approach to the studies outlined in the paper, but we have reservations about the general applicability of the equations, to all rock masses, throughout the world. We say this because of the large number of geological variables likely to influence the compressibility of rock masses. For example, a rock mass comprising rock substance from any of the four broad groups in Table II may contain, as well as joints, localized wide or narrow zones of crushed rock, and/or zones of weathered or altered rock of regular or irregular shape. Part or all of the crushed, weathered or altered materials may have "soil" properties and be highly compressible. The positions and orientations of such compressible zones in relation to the joint pattern and to the loads applied must have a significant effect on the deformability of the rock mass.

We would appreciate answers to the following:

(a) Can the Authors justify the number of significant figures quoted for the parameters in Table II, bearing in mind the large range in physical properties likely to be represented by each of the broad groups of rock types.

(b) Have equations (2) and (3) been applied in rock masses which have been seriously weakened by the effects of chemical weathering or chemical alteration, as well as by fractures?

Rock mass monitoring during dam construction and reservoir filling

Savich et al describe the results of some geophysical studies conducted beneath the 271.5m high Inguri arch dam and its reservoir, to monitor changes in the rock mass during construction, and during and after filling of the reservoir. Both detailed and regional studies were made, in conjunction with other instrumental deformation measurements and geodetic surveys. The studies showed increases in the longitudinal wave velocity V_p and electrical resistivity ρ, down to about 80m below the base of the dam, during its construction. Marked reductions in V_p and ρ occurred during filling of the reservoir, as the joints became subjected to hydrostatic pressure from the reservoir. Beneath the dam these effects were detected down to a depth of 250m, and near the head of the storage, to about 500m.

Could the Authors tell us whether the closure and later opening of joints beneath the dam were monitored also by borehole extensometers, and if so the amounts of strain indicated?

We consider that monitoring of the kind described for Inguri dam represents a very useful application of geophysics.

Estimation of deformation behaviour of weak rocks during earthquakes

For estimating the influence of earthquakes on structures it is important to know the deformation behaviour of the foundation rock. Ishikawa et al describe the results of low-frequency cyclic vertical loading tests which show various relationships between the dynamic modulus E_{dyn}, loading amplitudes, loading velocities, vertical strains and damping ratios. Because the authors demonstrated results from only two sites it appears that the main purpose of their work was to establish a new test procedure. The results are mainly influenced by the rock conditions within 1.5m below the loading device, where the rock is supposed to be always more or less loosened and weathered, i.e. not a usual situation. Therefore we do not know how to apply the test results to design. The authors are invited to explain this.

On the other hand these tests are related to seismic and ultrasonic tests. Therefore it is felt that the work of Ishikawa et al should be considered with the paper by Tanimoto and Ikeda and could give us a better understanding of decrease in amplitudes. For the evaluation of such and similar tests a lot of work has already been carried out. Langer summarized it in his General Report for Theme 1 of the 1979 ISRM-Congress in Montreux.

Other applications of geophysical methods in site investigations

The seismic refraction method has been used traditionally for assessing the depths of alluvial or weathered materials overlying bedrock at the sites for dams and other surface structures. Dhawan et al present samples of the results of work of this kind at their project in the Himalayas and report that the boundaries given by the seismic work agreed with the results of boreholes, within 10 percent.

In their section 2.1 "Rock masses mapping" Savich et al suggest that the reliability of the results of both geological and geophysical work can be increased by the application of several geophysical methods at the same time. Their Figure 1 is a cross section through a dam site which has been explored by seismic, electrical and magnetometer surveys. Geophysically defined geological boundaries are shown, together with faults ("zones of tectonic dislocation") and boreholes which were drilled after the completion of the geophysical work. The Authors claim that the boundaries as defined by the geophysical work agree with the results of drilling to within 5 percent.

Savich et al give brief statements about a number of other applications of geophysical methods, including estimation of the state of stress in rocks, studies of de-stressed zones in slopes and around man-made excavations, and studies to classify rocks in terms of their "degree of preservation" (see also Savich et al (1979)).

Sub-Theme A.2

IN-SITU AND LABORATORY TESTING

In contrast to previous ISRM-Congresses where the properties of intact rock and rock masses were discussed, we will now examine the techniques of testing and methods of evaluation of test results. Therefore our interest is directed to the following main questions:

- How should in-situ and laboratory tests be performed?
- How should they be evaluated?
- Which main and which second order parameters should be measured?
- How are these parameters to be introduced in the evaluation?

In-situ and laboratory tests can be aimed either at providing data for the design of a structure, or at improving our understanding of some aspect of rock behaviour. In the first case one will try to apply well known and proven test procedures whereas in the other case exotic and very sophisticated tests may be designed. But we learn from the papers of this sub-theme that even so-called standard tests and their evaluation can be improved.

In addition the 35 papers submitted for this topic have focussed our attention on the boundary conditions of each test. Only if we take care to define them well, and afterward to introduce them into the evaluation, can we be confident that the test results are realistic.

In this context much emphasis must be laid upon good and undisturbed sampling and an adequate preservation of the samples. It is common to record the sampling conditions and in some cases sketches or photographs are made of the test specimens. However we are sure that this is insufficient; also important are at least the following: the weathering condition of the sample, its water content and density, the length of core in relation to the average, details of joints or other defects, orientation of bedding, and all evidence of disturbance.

The papers submitted under in-situ and laboratory testing may be subdivided into several main areas:

- consideration of anisotropic and non-elastic properties during the evaluation of tests
- investigation of strength and deformability under short-term loads
- determination of strength and deformability under long-term loads and under cyclic loads; studies of creep and of fracture initiation
- investigation of the shear strength of discontinuities
- scale effects
- rock anchor testing

and some additional topics, like

- development of special test devices
- sample preparation standards
- design procedures
- special applications of known test procedures
- theoretical model of joint deformation behaviour
- erosion test for grouting material
- statistical description of discontinuities
- exploration of rock mass quality ahead of a tunnel face
- real time evaluation of tests by computer

The General Reporters will concentrate their comments on the main areas

Consideration of anisotropic and nonelastic properties during the evaluation of tests

During the last ten years numerical methods have proved to be a powerful tool in rock mechancics. In principle there are very few of the important properties of rock masses which at present cannot be simulated numerically, but in practice the value of the methods is usually limited by the poor quality or availability of input parameters.

We consider it unfortunate that even the most important anisotropic characteristics of the intact rock and of the rock mass have not been taken sufficiently into account. Even when a lot of money was spent on the tests, allowing anisotropic deformations to be measured, in most cases the evaluation has been based on the assumption of isotropy.

In a physical sense for the description of fully anisotropic conditions 21 parameters are needed. It is considered unlikely that this requirement will ever be entirely met. But - speaking as engineers - on the other hand such an exact and extended description will not be justified because of the heterogeneous nature of rock mass. But nevertheless the most important feature of anisotropy, called transversal anisotropy or cross-anisotropy, must be considered in the evaluation of tests because it may be the most distinguishing feature of the rock mass as well as of the intact rock.

The transversal anisotropic deformation behaviour may be described by only five parameters, E_1, E_2, ν_1, ν_2, G_2. As the deformation properties of sedimentary as well as of metamorphic rock masses usually are transversally anisotropic to a high degree it is of considerable importance for the practice in rock mechanics to develop methods for the evaluation of these parameters. We are very glad that there are some papers showing efforts to do this.

Since the early days of the science of rock mechanics, the laboratory unconfined compression test has been used to find out the uniaxial compressive strength and the moduli of deformation parallel and normal to the bedding. But core samples in many cases cannot be provided in such special orientations to the bedding. Therefore arbitrarily oriented specimens must be tested. The results have to be considered in comparison to the angle α between load direction and bedding. Work on this field has been carried out in the past e.g. by John (1969) for the uniaxial compressive strength, by Barla and Goffi (1974) for the tensile strength and elastic modulus. Now this development has been continued by Kiehl and Wittke who show that the shear modulus G_2 may be evaluated from the results of uniaxial compression tests (Fig. 2).

Another interesting test is introduced by Amadei et al who use thin intact rock discs to measure and to evaluate the anisotropic deformation parameters and the tensile strength (Fig. 2).

It is even more important to consider the anisotropic properties of rock masses in field tests, than in laboratory tests. Field tests are rather expensive and time costly. Therefore less field tests will be performed than laboratory tests. So they should be evaluated with much more effort. At the ISRM-Congress in 1979 Oberti et al were the first who took anisotropy into consideration. They derived a theory for evaluation of the plate bearing test oriented normal and parallel to the stratification.

Now a lot of work in this field is presented. Oberti et al show how transversally anisotropic conditions of rock mass may be assumed when evaluating hydraulic chamber tests. Kiehl and Wittke describe procedures which improve the evaluation of borehole dilatometer tests and large flat-jack tests. Additional comments are made about anisotropic conditions when applying the in-situ triaxial test (Fig. 3).

Gonano and Sharp report on extended stress measurements at a project in South Africa. Of special interest is that these tests have been performed in weak rock using overcoring techniques. Their developments include the critical appraisal of overcoring strain data and the determination of stress-strain transformation

Test		Author (s)
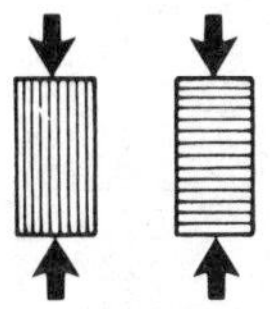	E_1, E_2, ν_1, ν_2	
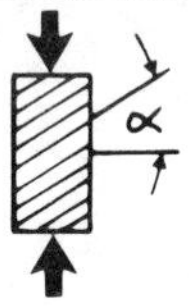	uniaxial compression strength related to α	John (1969)
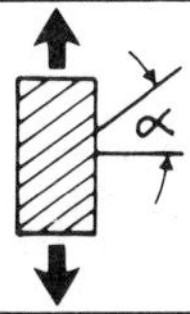	tensile strength	Barla/Goffi (1974)
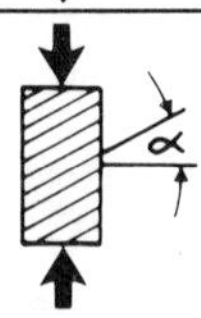	deformability related to α	Barla/Goffi (1974)
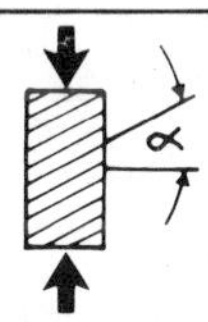	shear modulus G_2	Kiehl / Wittke(1983)
	$E_1, E_2, G_2, \nu_1, \nu_2$ and tensile strength from diametral compression test	Amadei et al (1983)

Fig. 2 Determination of anisotropic properties in the laboratory

Test		Author (s)
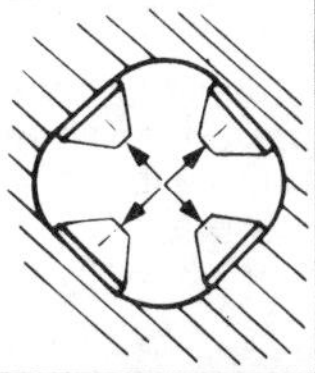	$E_1, E_2, G_2, \nu_1, \nu_2$ from plate bearing test	Oberti et al (1979)
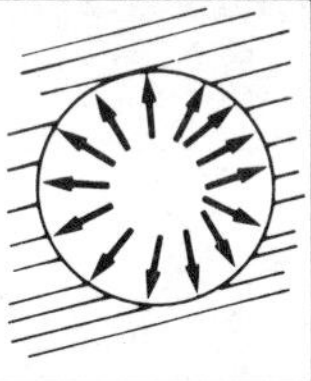	$E_1, E_2, G_2, \gamma_1, \gamma_2$ from hydraulic chamber test	Oberti et al (1983)
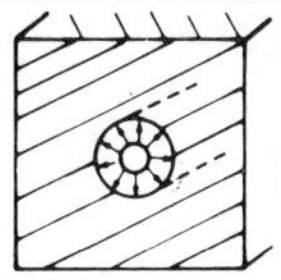	E_1, E_2, G_2 from borehole test	Kiehl / Wittke(1983)
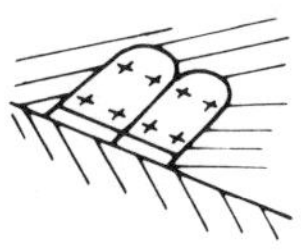	E_1, E_2, G_2 from LFJ - test	Kiehl/Wittke(1983)
	$E_1, E_2, G_2, \nu_1, \nu_2$ from in - situ triaxial test	Lögters/ Voort (1974) evaluation im - proved by Kiehl/Wittke (1983)
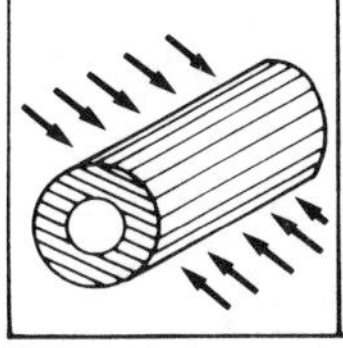	in - situ stresses and non-linear deformability	Gonano/Sharp (1983)

Fig. 3 Determination of anisotropic properties in field tests

parameters for transversely anisotropic nonlinear rock, and the formulation of a numerical solution. Considerable improvement in the scatter of results (magnitudes and directions of principal stresses) shows very clearly the value of these more sophisticated methods.

When using the large flat-jack tests in a test adit or near the wall of any underground opening we normally assume elastic conditions over the whole area subjected to the test. We ignore the fact that the portions near the wall, even when attempts are made to avoid unnecessary disturbances, are loosened or beyond the elastic range in most cases. Borsetto et al present an interpretation procedure assuming elastoplastic behaviour of the rock mass. Their aim is to prevent systematic errors in the evaluation of the residual cohesion, and also of original stresses and deformability. The work so far has been limited to a tunnel of circular cross section. It is suggested that the authors extend their work using numerical evaluation techniques.

Investigation of strength and deformability under short-term loads

Strength and deformability characteristics of intact rock normally are investigated in the laboratory. Loyal to our intention to emphasise exploration and evaluation methods rather than the properties of rock we will discuss only the tests and the influences of test conditions on the results.

The compressive strength of intact rock may be investigated in uniaxial (σ_c), biaxial or triaxial compression tests (Fig. 4). The triaxial tests are performed with two principal stresses being equal ($\sigma_2 = \sigma_3$) or unequal. We are accustomed to compare uniaxial and triaxial ($\sigma_2 = \sigma_3$) test results by plotting both together as Mohr circles in a shear stress-normal stress-diagram and to fit a straight or curved envelope onto these circles (Fig. 5). But most of us do not know how to evaluate the third (mean) principal stress if σ_2 is different from σ_3.

But we remember from the ISRM-Congress 1974 that Brown (1974) showed us the influence of a second principal stress σ_2 on the strength characteristics of Wombeyan Marble (Fig. 6). We remember too that the test results were more influenced by end-effects than by σ_2 (Fig. 6). Best results there were obtained by using brush platens, as introduced by Hilsdorf for testing concrete. It is rather interesting to see what different results have been obtained by different researchers in the past (Fig. 7). It must be assumed that many but not all of these differences may be caused by friction problems.

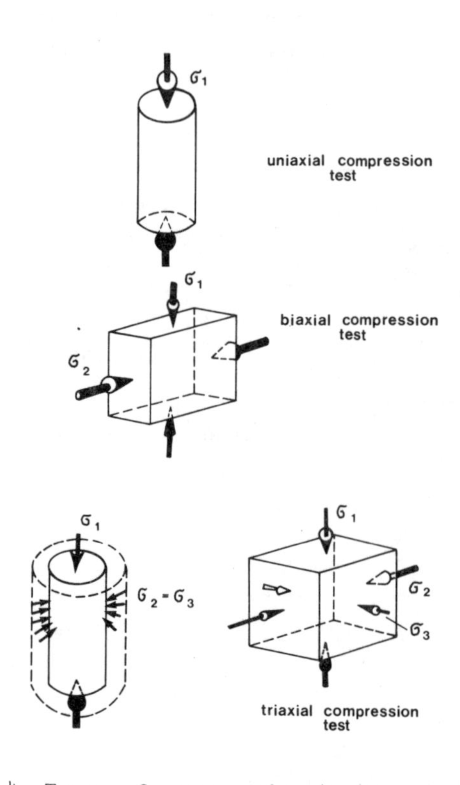

Fig. 4 Types of compression tests

Now we learn from the triaxial compression tests of Gau and Cheng performed with $\sigma_1 \neq \sigma_2 \neq \sigma_3$ that for a given relation σ_1/σ_3 the strength of the intact rock does not depend very much on the intermediate principal stress σ_2 (Fig. 1 and 2 of their paper). Here too, an influence of the boundary conditions can be seen. But it is much smaller than shown in Fig. 7. The work done by Brown and by Gau and Cheng suggests that designs should not assume an enhanced strength for intact rock when the intermediate principal stress σ_2 is greater than σ_3. This corresponds to the theory (Fig. 8).

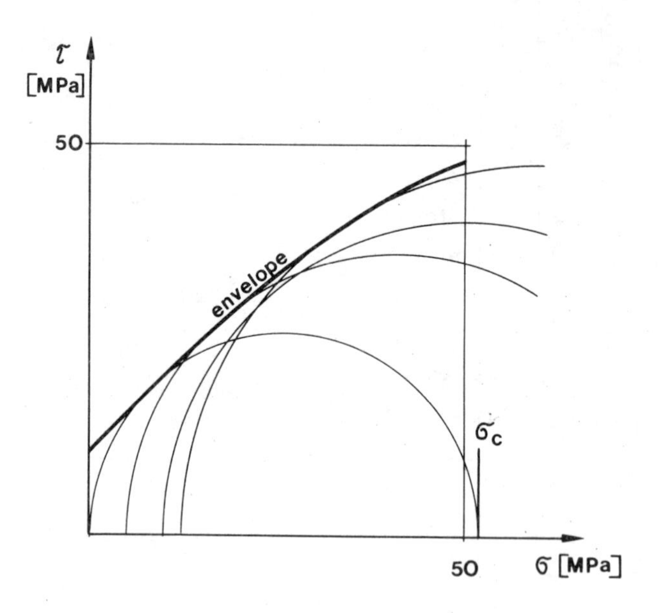

Fig. 5 Failure criterion (example)

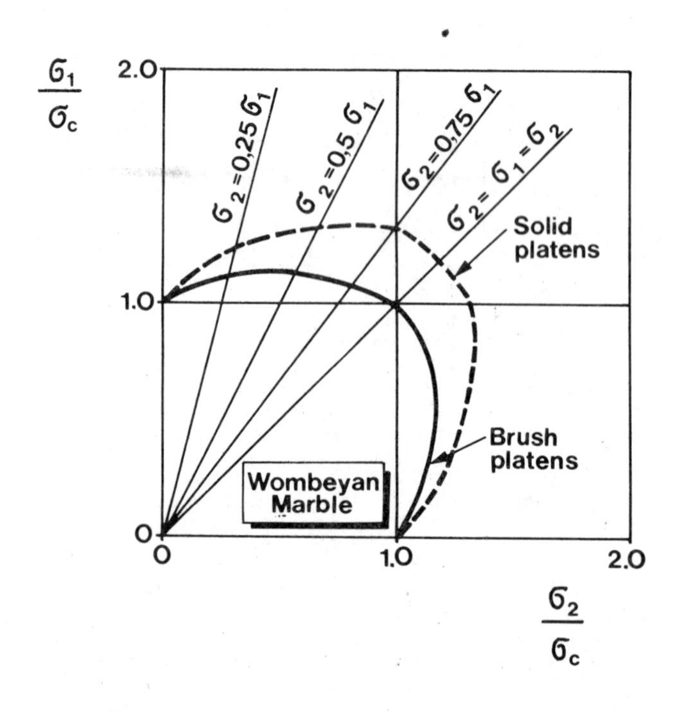

Fig. 6 Influence of the mean principal stress (after Brown, 1974)

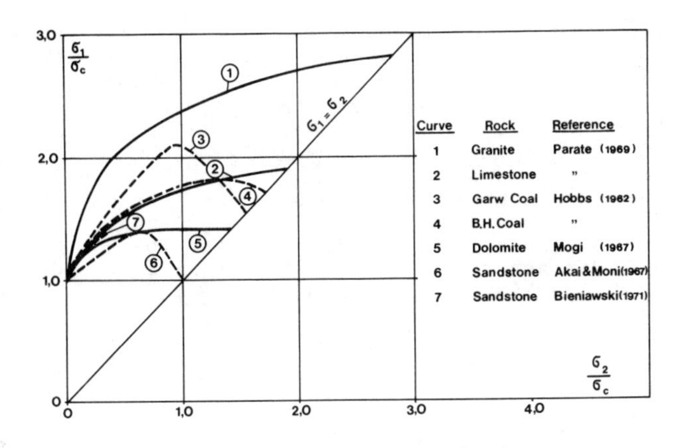

Curve	Rock	Reference
1	Granite	Parate (1969)
2	Limestone	"
3	Garw Coal	Hobbs (1962)
4	B.H.Coal	"
5	Dolomite	Mogi (1967)
6	Sandstone	Akai&Moni(1967)
7	Sandstone	Bieniawski(1971)

Fig. 7 Biaxial compression test results as reported in the literature (after Brown, 1974)

Other boundary conditions which influence the results of compression tests, especially for soft rock, are the water content and the loading rate. As pore pressure decreases when the loading rate is small enough, the ISRM Suggested Methods recommend a loading rate, so that failure occurs after 5 to 10 minutes. These recommendations have been made for hard rock. They may not be applied without critical examination to soft rock which is marked by much lower strength and higher water content. Here the loading rate has to be accommodated to the strain rate.

Also with soft rock there is often the phenomenon, that the water content of apparently similar samples may be very different, (e.g. caused by increased porosity due to weathering). In such cases, even when the loading rate is diminished, variations in the strength results will be found. Chiu and Johnston show this using Melbourne mudstone specimens subjected to uniaxial compression tests (Fig. 9).

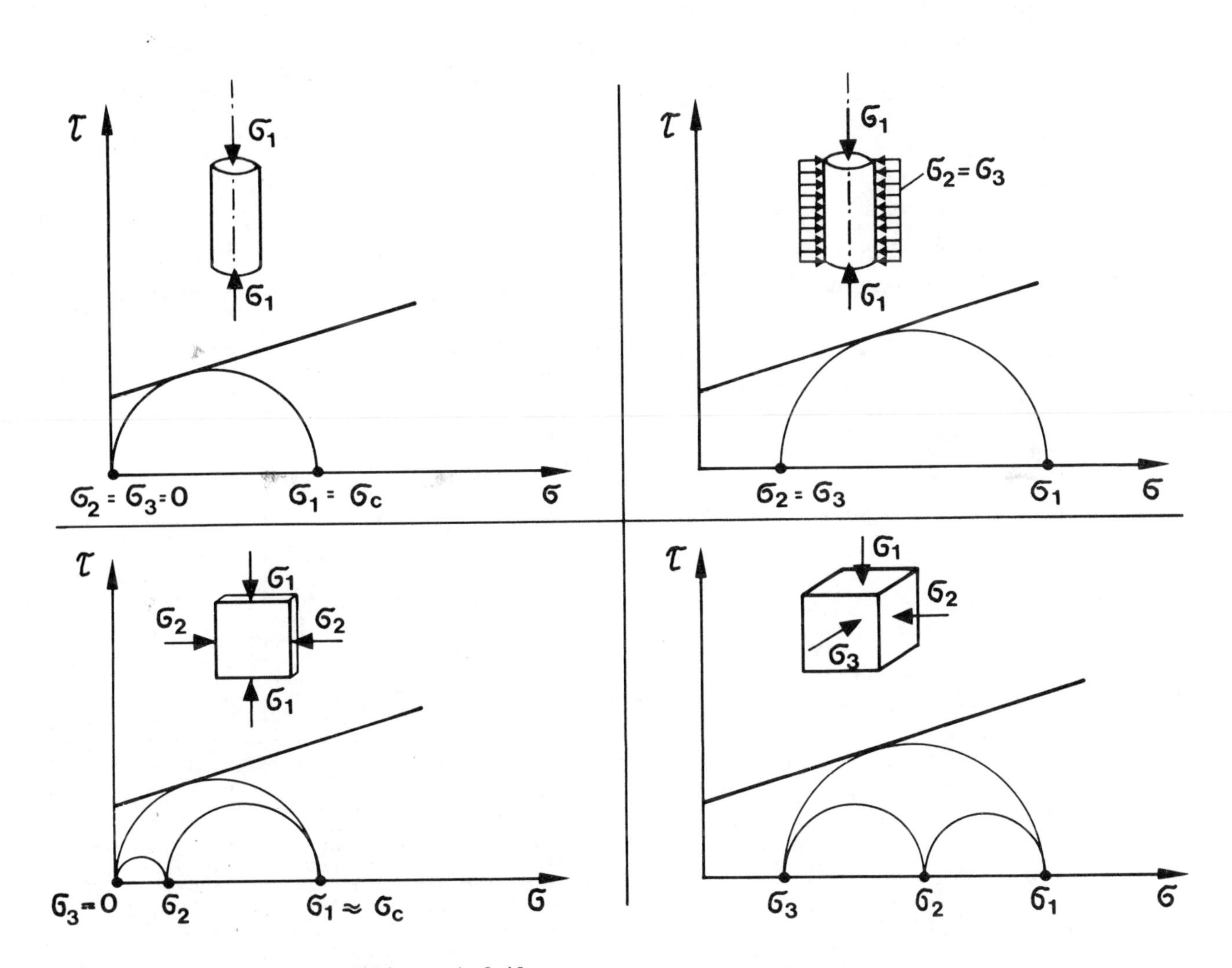

Fig. 8 Different stress conditions at failure

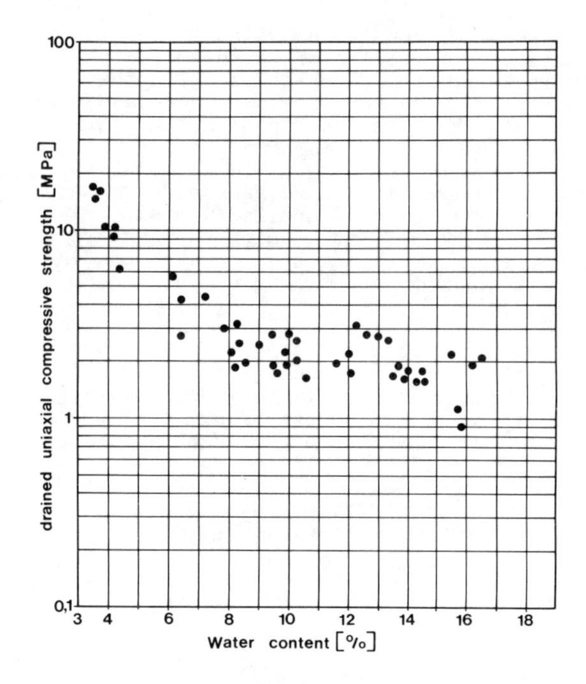

Fig. 9 Influence of water content of Melbourne mudstone on the compressive strength (after Chiu and Johnston)

During compression strength testing one usually also investigates the deformability of the intact rock. In a previous section of this paper the influence of anisotropy has already been discussed. When considering now only isotropic material, it is clear that the axial deformability measured during the test depends upon the confining pressure. One has to distinguish between an increase which is caused by the confining pressure (similar to the relation between Young's modulus and stiffness modulus in soil mechanics) and an increase which has to be considered as a specific property of the particular rock. For the latter Chiu and Johnston give us an example (Figure 10), where, in addition, the influence of water content is shown. Similar results are presented by Kwasniewski in his Figure 7 gained from sandstone specimens.

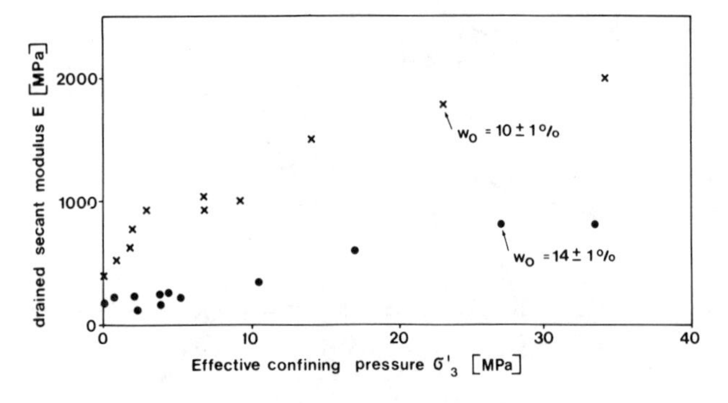

Fig. 10 Influence of confining pressure on the deformability of Melbourne mudstone (after Chiu and Johnston)

It is the aim of many studies to get reliable equations for the strength of the intact rock. Here it is necessary to distinguish between those being developed from theoretical considerations (i.e. Mohr-Coulomb, Griffith, and others) and those which are derived from test results. Examples of the latter can be seen in the papers submitted to this congress by Gau and Cheng, Chiu and Johnston, and Kwasniewski. These formulae can be valid only for the particular intact rock tested. They are influenced by the particular test conditions and therefore can be applied with other materials and in other places only to a very limited extent. Therefore it is rather difficult to make critical comparisons between strength equations developed by different authors.

Until now our comments only concern laboratory tests performed with intact rock specimens. However it seems to be important to point to a special development which allows jointed rock mass samples to be tested triaxially in the laboratory. A large amount of construction work in rock takes place in tightly jointed soft rock, in which it is often difficult to determine values for strength and deformability. For this reason, large scale triaxial laboratory tests and a special technique to take undisturbed jointed samples from the rock mass, have been developed in Karlsruhe. This technique has repeatedly been reported on in the past and the German Society for Soil Mechanics and Foundation Engineering has published recommendations for it. In their paper to this Congress, Natau et al present recent experiences.

Determination of strength and deformability under long-term load and under cyclic load; creep, development of failure

During short-term load intact rock behaves more or less linear-elastic, but when it is stressed for longer periods very different mechanisms can be observed. These may be described with different rheological models.

It is not generally known that even hard rock can show viscous behaviour at low stresses. Ito reports on long-term tests with granite beams. These tests are remarkable because of their duration of about 24 years. They seem to verify the Nabarro and Herring prediction that granite has no yield stress. The tested specimens show strains which are independent from the stresses, indicating viscous flow (curve 1 in Figure 11). Ito reports also on supplementary tests with gabbro specimens, from which it may be seen that both increased confining pressure and raised temperature reduce the viscosity. From this work it may be inferred that for strain rates $> 10^{-13}$/s, intact rock behaves like a Hooke-body, whereas at lower strain rates it behaves like a Newtonian fluid. It seems to us to be important for the understanding of geological processes that the values of viscosity determined in the laboratory be made available to geologists and geophysicists involved in studies of the earth's crustral deformations.

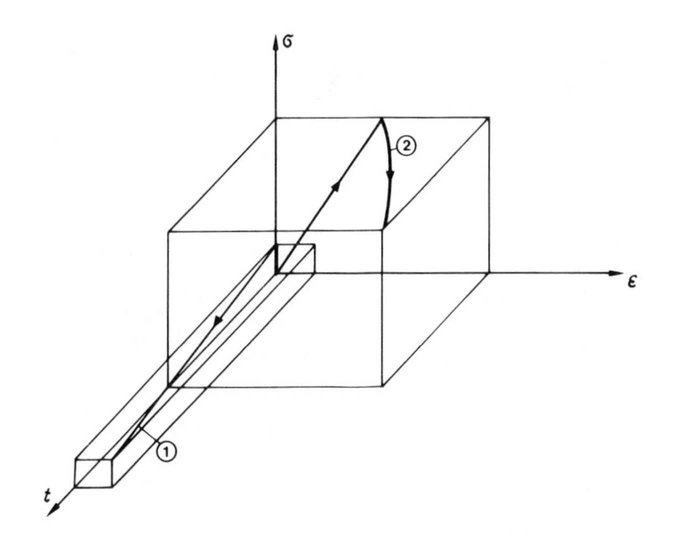

Fig. 11 σ - ε - t - diagram. ① viscous flow, ② creep

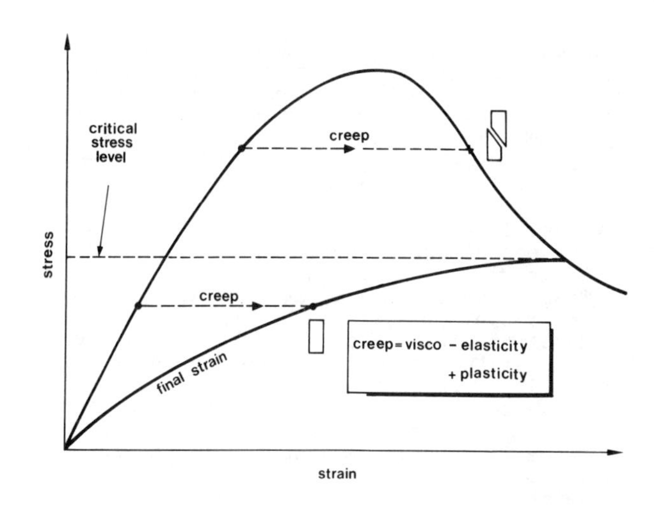

Fig. 12 Stress-strain behaviour at long-term

Viscous events in hard rock of course take place very slowly and therefore are scarcely of importance for construction work in rock. However the loading rate may not be neglected. It is well known that in the uniaxial compression test the loading rate influences the strength. Also, the strain at failure depends upon the loading rate, because the strain is composed of elastic, visco-elastic (reversible creep) and plastic (irreversible creep) portions. The slower the loading rate the more the creep influences the total deformation. Such behaviour is described by curve 2 in Figure 11. It is well known that beyond a so-called criticall stress level the time-dependent strain leads to failure. Therefore one has to distinguish between primary (decreasing velocity), secondary (constant velocity) and tertiary (increasing velocity) creep. At lower stresses the creep finally disappears; the strain reaches a final value (Fig. 12). Intact rock reacts in a similar manner during cyclic loading. We are pleased to have on this topic two very interesting papers, one by Homand-Etienne et al (granite and marble) and another by Akai and Ohnishi (Funyu Tuff). As in the case of creep, one may define three separate phases of behaviour under cyclic loading (Fig. 13a). Therefore we can infer that these behaviour patterns are related.

Both phenomena may be explained by the progressive development of microcracks, which increase with the duration (or the number of cycles) and magnitude of loading. Below the critical stress level the cracks stop, beyond it they accelerate. Homand-Etienne et al found in their tests that the long-term strengths of granite and marble were 85% and 82% respectively, of the short-term strength. Similar results are known for concrete.

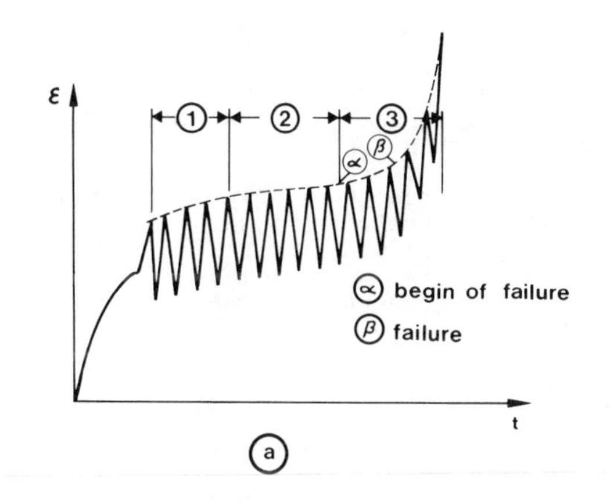

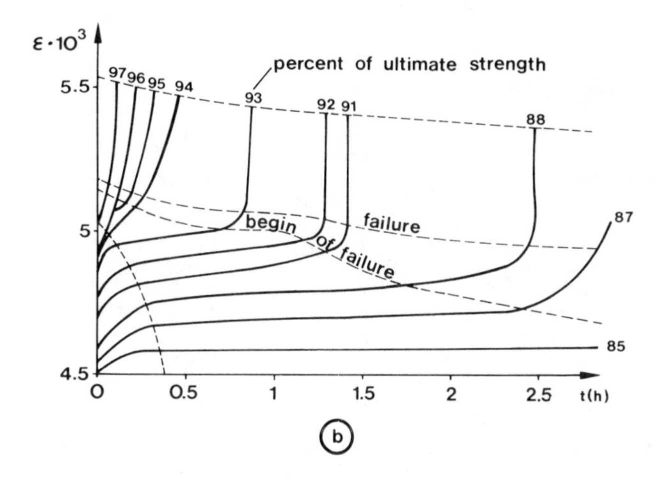

Fig. 13 Behaviour under cyclic load (after Homand-Etienne et al)
① decreasing ② constant
③ increasing rate)

The theory of the development of microcracks may be used within a statistical, mathematical model. Weber and Saint-Lot introduce such a model in their paper. It should be mentioned that, despite of some simplifications, this model is able to simulate stress-strain-curves measured in tests.

The developments leading up to failure, which have been discussed previously may be observed also in rock mass specimens. Here too, different phases may be defined during long-term or cyclic loading. Müller and Ge demonstrate this clearly with the behaviour of some in-situ specimens tested for the Kurobe dam. The first phase is connected with the initial compaction of the jointed rock mass. Müller and Ge show the second phase to be one of relative stabilization, and in the third phase the failure is initiated. It is considered important that these stages be recognized, especially for large size tests, because the judgement of safety based on such tests should be based mainly on the deformation behaviour of the rock mass.

Investigation of the shear strength of joints

At the ISRM-Congress 1974 a whole sub-theme was dedicated to the shear test. In Montreux, 1979, too, fundamental contributions were submitted on this topic.

The shear strength of joints is a rather complicated topic which is certainly not yet entirely clarified. To this, first some general remarks will be made: if a continuum is subjected to shear stresses, the behaviour shown schematically in Figure 14a is observed. From the τ - Δh-curve a distinct peak shear strength τ_t can be seen. In the τ-σ-diagram τ_t represents one point of the failure criterion. If shear stresses occur along a smooth joint (Fig. 14b) a very different behaviour may be observed. Here the shear displacements increase continuously but progressively with increasing shear stresses τ. At the shear stress τ_r the displacement Δh increases without further increase in shear stress. In both idealized models, as mentioned before, there is no vertical displacement Δv.

On the contrary, vertical displacement does occur in the models in Fig. 14c and d which look more like natural joints. Here the behaviour is determined by - rather idealized - asperities or teeth. As long as the teeth are not damaged, displacements are possible only when the upper body is gliding along the slopes of the teeth of the lower body. Hence besides Δh a displacement Δv normal to the main shear plane may be observed. Corresponding to the slope angle i (Fig. 14c) the failure criterion in the σ-τ-diagram is inclined to the σ-axis by the angle $\phi + i$. Beyond certain stresses τ and σ, the teeth will fail. This state and the resulting consequences are shown in Fig. 14d. It can be seen that the displacements at that time are growing more and more parallel to the main shear plane. The failure criterion (in the τ-σ-diagram) becomes flatter and approaches the value ϕ.

Of course, in practical cases the shear surfaces are very irregular. Therefore one cannot expect such idealized curves. On the contrary rather different results are gained, depending upon the roughness, the strength of the intact rock, the weathering conditions, coatings, mineralogy, water content and water saturation of the rock mass. Figure 15 shows an example which has been published by Barton (1971). Of course the behaviour of joints is dependent, also, on whether they have previously been subjected to relative displacements, in other words, whether their teeth or asperities have already been sheared. Therefore with our present state of knowledge we must expect the shear strength conditions in each case to be different. Examples of this are presented in the papers of Tinoco and Salcedo, Everitt and Goldfinch, and Bollo et al.

Scale effects

In jointed hard rock often only the shear strength and tensile strength are influenced significantly by the joints. The deformability is, in comparison, only slightly increased. But in soft rock, especially when it is closely jointed, the influence of the joints is dominating. Spacings of joints of only some centimeters cause the rock mass to behave like masonry without water, sometimes even with soft bricks and lubricated fissures. Each "brick"-body has a noticeable space for displacements and rotations. Therefore the strength is diminished and deformability is increased, compared with the intact rock.

But it is rather difficult to determine these changes reliably, because such a jointed body may be, in a kinematic sense, a very movable system. The strength and deformability of such a system then depends very much upon the boundary conditions of the test. Therefore it is felt that uniaxial compression tests (for example) are likely to provide results deviating from those of triaxial tests and the results of both would differ more or less from the natural rock behaviour. It would be appreciated if the authors (Natau et al, Sun and Zhou) contributing to this theme, would make comments on this question.

While noticeable scale effects are observed in compression (strength and deformability) tests these seem to be sometimes much smaller during the investigation of the shear strength of joints. Pasamehmetoglu et al. and Bollo et al. demonstrate this effect very clearly. From their papers and furthermore from previous work done by Leichnitz and Natau (1979) it must be concluded that scale effects are mostly observed with regard to the peak shear strengths of rough joints. Here the roughnesses are, no doubt, arbitrarily distributed over the joint plane, but in a small shear specimen there may be by chance a sharp tooth which does not correspond to the average roughness.

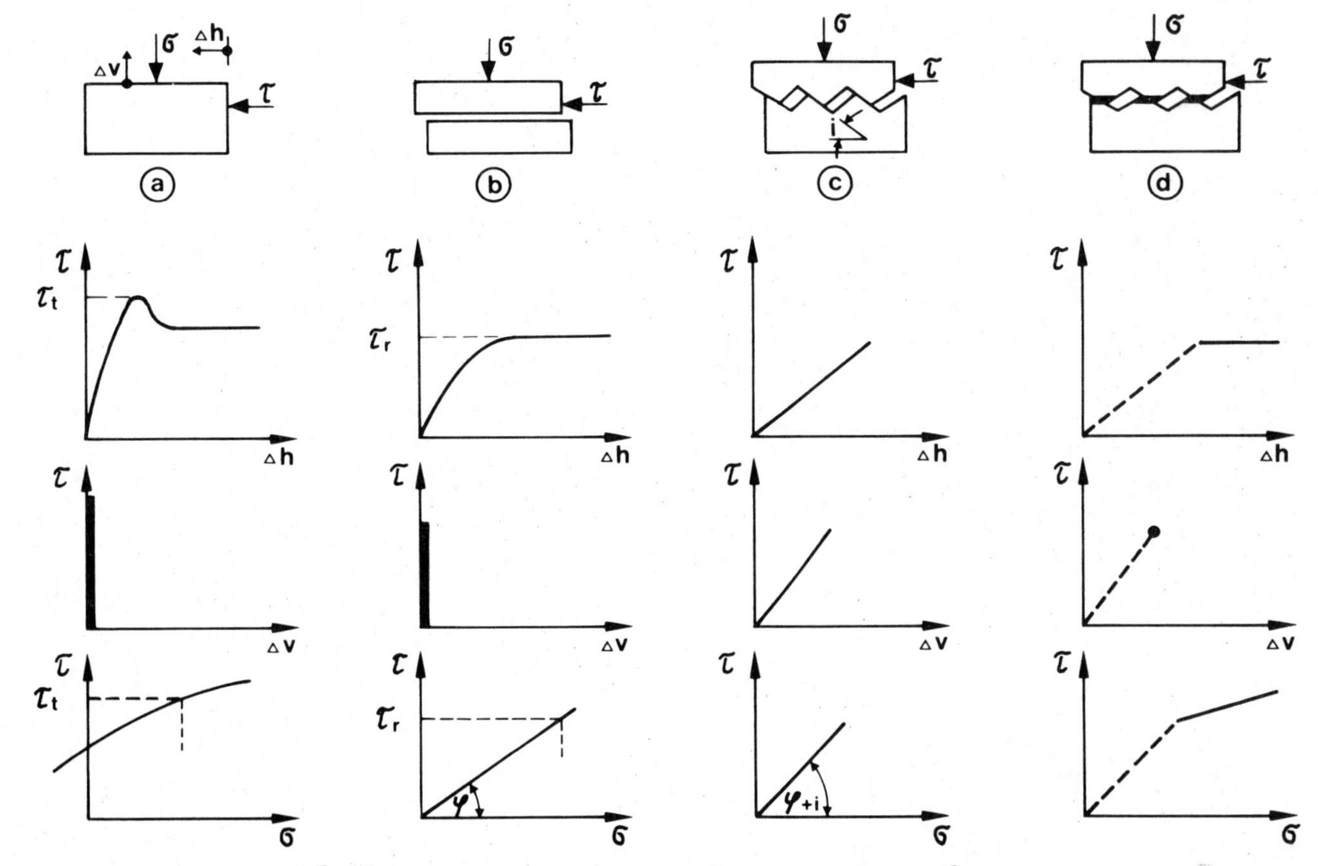

Fig. 14 Shear behaviour under different conditions (ⓐ intact rock, ⓑ smooth joint, ⓒ rough joint before, ⓓ rough joint after failure of teeth)

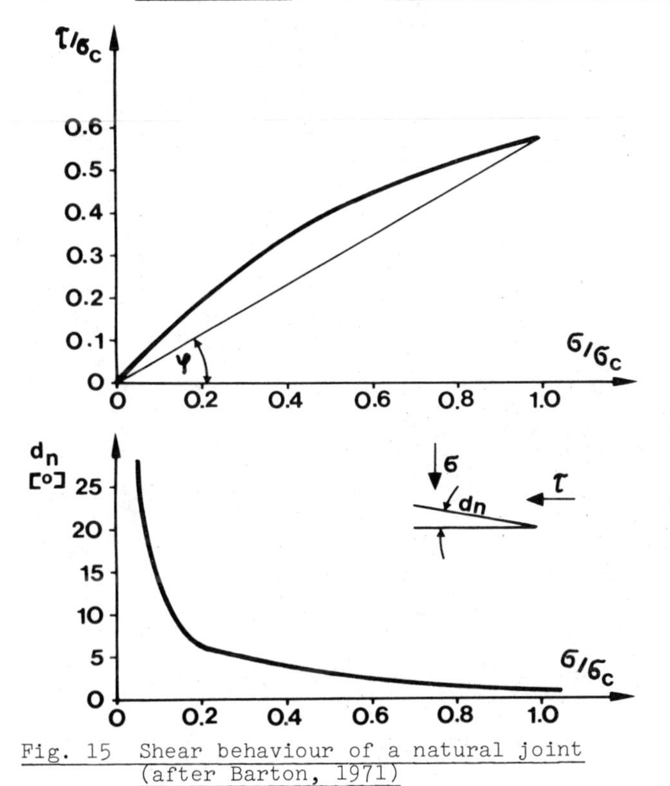

Fig. 15 Shear behaviour of a natural joint (after Barton, 1971)

Testing of rock anchors

Rock anchor testing may have different characters. It may be destructive or non-destructive. It may be performed as a qualification test or for inspection purposes, or beyond this, for fundamental investigations. The papers submitted to this topic deal with fundamental investigations for large rock anchors (Natau and Wullschläger) and small scale laboratory models (Ballivy et al.), developed for quality inspections of short grouted rock anchors.

It is essential for the long term performance of a rock anchor that the bond length (steel bars and grouting) remains undamaged. The state of stress within the bond length is especially important as it must remain below a certain level if the grout is to remain uncracked. The stresses are influenced by the deformability of the surrounding rock mass. Natu and Wullschläger simulate the rock mass in their tests by a concrete-filled steel tube. Ballivy et al. used intact rock cores of small diameter.

An interesting development is introduced by Bergman et al. The boltometer, a device for the test of short grouted rock anchors, is based on the measurement of velocity and damping of waves in the steel bar. The amplitude is more or less diminished, depending on the quality of the grout. We think that the boltometer will become a good tool but it seems to us, from the demonstrated examples, that it is not yet fully developed.

Sub-Theme A.3

CLASSIFICATION, PREDICTION, OBSERVATION AND MONITORING

Since the system published by Terzaghi (1946) there has been a proliferation of classification schemes for rock masses. These have been designed to assist communication between geotechnical workers and to assist in the prediction of rock mass behaviour. However because of the large number of classification schemes now in the literature there is a danger that engineers wishing to apply such schemes may have difficulty in deciding which to adopt for their specific circumstances. Indeed it is evident from papers within this theme that such difficulties do exist. It seemed therefore to your General Reporters that we should try at this Congress to achieve a better understanding of the strengths and weaknesses of the existing systems, and if possible to achieve some degree of concensus on their applicability.

Having discussed the existing "proliferation", it is with some trepidation that we point out the necessity for yet another classification - this time to consider the classifications themselves in different groups, that is to classify them!

Categories of classification

In the following we will refer to the two main components of all rock masses as substance (material) and defects (discontinuities) while recognizing that fluids in the defects and/or substance pores are also of importance and taken into account in some of the systems.

A. Component property classifications

In these classifications, ranges of values of a single property of a component are characterized by names, letters or numbers. The ranges of values adopted are usually those which practical experience has shown to be useful. The user of the classification chooses the relevant number from the results of simple tests or measurements. Examples of this type of classification are the classification of rock substance by Point Load Strength Index (Broch and Franklin, 1972) and the four component classifications in this Society's "Basic Geotechnical Description of Rock Masses" (ISRM, 1980, Tables 1 to 4).

B. Multi-property classifications

These are systems whereby an "effectively uniform" zone of a rock mass may be characterized with regard to its general engineering properties, by means of the listing of a number of component descriptive terms. The BGD "Basic Geotechnical Description of Rock Masses" is a system in which a rock mass is characterized by a genetic rock name plus descriptive letters and numbers describing four characteristics of the components. The main purpose of this and other such systems is to provide a standard language for use by geotechnicians.

C. Multi-property (weighted) classifications

In this type, the rock mass properties are described by several component properties, as in B above, and these component properties are then given "importance ratings" according to their judged contribution to the strength of the rock mass. The rating judgements are made on the basis of the analysis of experience, and tests and calculations. The sum of the ratings is the total rating for the rock mass.

The systems of Barton et al (1974) and Whickam et al (1972) have been designed specifically for the estimation of the support needs for underground openings. Bieniawski's RMR (Rock Mass Rating) system (Bieniawski, 1974a and 1979) is intended for more general use. All three systems have been used with success for the prediction of the support requirements of tunnels. Both Barton's Q and Bieniawski's RMR have been shown to correlate quite well with insitu deformation moduli (Bieniawski, 1978).

D. Weathered rock mass classifications

There are many classifications which have developed in particular geological environments, to describe weathered rock masses for various purposes. One example of these is the system in use here in Melbourne in which the weathered profile in the Silurian mudstone is classified into zones (Neilson, 1970). Saunders and Fookes (1970) describe various systems developed for rocks weathered in different climates, and Sancio and Brown (1980) describe a system of classifying weathered rock masses for predicting their potential for slope instability.

The proliferation of classifications gives us cause for some concern. The situation is similar to that at the beginning of this century when petrologists produced the proliferation of igneous rock classifications and rock type names which still today cause confusion amongst geologists, let alone engineers! At a time when it is fashionable to "rationalize", can we rationalize our use of the existing geotechnical classifications, and our approach to the development of new classifications?

Looking first at Category A, the General Reporters suggest that it should be possible to adopt world-wide standard groupings for the following component properties:

- rock substance strength
- layer thickness
- joint spacing
- joint strength or condition
- groundwater conditions

If the Category A properties were standardized, Category B systems would be readily standardized. We would then have a language for rock masses perhaps equivalent in usefulness to the Unified Soils Classification. Most persons involved seriously in design and construction in soils will agree with us that the Unified Soils Classification is an extremely valuable international language.

Undoubtedly, Category D is where most of the proliferation has occurred. This is probably due partly to the extreme variability of weathered rock masses, both within a particular geological environment, and from one geological environment to another. Another contributing factor could be the fierce desire in many geologists, to classify. We accept that many of these classifications are useful in the restricted geological environments in which they evolved but consider that none should be adopted for universal use.

The potential value of Category C classifications is beyond doubt, as the following statement by Hoek and Brown (1980) will illustrate:

> *"In view of the scarcity of reliable information on the strength of rock masses and the very high cost of obtaining such information, the authors consider it unlikely that a comprehensive analysis of rock mass strength will ever be possible. Since this is one of the key questions in rock engineering, it is clear that some attempt should be made to provide some form of general guidance on reasonable trends in rock mass strength.*
>
> *"Having considered several alternatives, the authors have turned to the rock mass classification schemes ...."*

In the pages following that quotation, Hoek and Brown provide approximate strength equations for intact and jointed masses of the most common rock types, relating these to the full range of RMR (Bieniawski) and Q (Barton et al) indices, in each case. Hoek and Brown point out that their equations should be used only for preliminary design and sensitivity studies. However, for the reasons given in the above quotation it seems likely that such empirical relationships and rock mass classifications of the RMR and Q type will find wide usage in rock engineering in the future.

Papers relating directly to rock mass classification and prediction

Five papers relate to these topics. Kendorski et al describe in some detail, the basis and method of application of a Category C system for predicting the temporary and production stage support requirements for production drifts (tunnels) in block caving mines. This excellent paper illustrates the benefits to be gained from research carried out by a well-balanced team; it reflects the Authors' good understanding of mine geology and theoretical rock mechanics, and their appreciation of real rock behaviour and mining practice which can only have been obtained by first-hand experience underground.

It is particularly pleasing to see that the characterization of the strength of the rock mass is carried out in a separate, initial operation, and that the effects of loadings induced by engineering activities are considered later, by means of a series of adjustments. It can be seen readily on their Figure 1, that this concept has allowed the Authors to apply their system logically to predict both the initial (development stage) support requirements, and the later (production stage) requirements. It appears to us that Kendorski et al may have provided the kind of "new approach" which Lokin et al are looking for, in their paper. We will discuss this matter further when considering the Lokin et al paper.

Referring again to Figure 1 of the Kendorski paper, we note that logically "fracture orientation" is now allowed for in the second stage adjustments, rather than being considered as part of the rock mass characterization. What the authors are saying here is that the fracture pattern is an unchangeable characteristic of the rock mass, but the fracture orientations in relation to the axis of the drifts depend upon the directions in which the drifts are driven, i.e. they are a part of the input which is imposed by the engineering activities.

We were unable to fully understand Table IV, "Fracture Orientation Rating A_O Based on Direct Observation in Drift". We assume that the "Number of Fractures Defining Block" refers to a two-dimensional exposure of a joint block, in a drift. If so, the orientation of the exposure surface would presumably be required, and this is not stated. We suggest that the Authors provide a diagram to explain the derivation of A_O from Table IV.

Referring again to the A_O rating, we note that the Authors give precedence to the use of Table IV, if direct observations of exposed rock are available. Table V, which contains much more comprehensive data, is only to be used "If no drift exposures exist but fracturing trends are known....". We cannot see the logic in the Authors' choice here. We consider that any experienced engineer or geologist could readily determine A_O from drift exposures and Table V. This would be done at the same time as the observations to assess blasting damage (Table III), and experienced persons would no doubt take defect orientation effects into account when deciding upon which values for A_B to adopt from the types of observation listed in this table.

The General Reporters consider that the MBR system has considerable potential and we look forward to future publications providing feedback on its performance.

Lokin et al have produced as their Table I, a useful listing of the elements of nine existing rock mass classifications of our Category C. They then discuss seven "shortcomings" which they see as limiting the applicability of the existing systems. While we agree that some of the simpler classifications are insufficiently comprehensive, we suggest that most of the "shortcomings" listed do not necessarily apply directly to the more comprehensive systems (e.g. RMR and Q) but rather to the organization of their application. As stated previously during our discussion of the MBR system, it seems likely to us that the first example of the "new approach" which Lokin et al have foreshadowed, has been provided to us at this Congress, by Kendorski et al.

After commenting on the shortcomings, Lokin et al proceed to discuss the principles of a proposed new approach to classification. Most engineers with experience in project planning should agree with most of the stated principles relating to geotechnical studies and their integration with other studies during investigation, design, construction and monitoring.

In this "Basic Principles" section the Authors explain that the classification of rock for a project must include differentiation of the relevant area into quasi-homogeneous zones, preferably to the "rock class" level. The Authors then state, when referring to the zoning procedure ... "To be meaningful it must cover the entire investigated volume to approximately the same level of reliability". The General Reporters suggest that both the detail and level of reliability required should often vary, according to requirements dictated by the engineering project and the geological environment. For example the investigation of the route for a long pressure tunnel, particularly in a region with steep slopes and a long history of past weathering, would normally require the portal areas to be covered in considerably more detail than the remainder of the route. This is because of the need to locate the portals so as to avoid areas of disturbed or deeply weathered rock, and to assess the stability and strength of the shallow cover zones, in relation to the requirements for watertightness of the tunnel lining.

In Table III the Authors list all of the data which they propose to include in their classification, at various stages of development of projects. This is a rather complete list of geotechnical and related aspects requiring investigation for major projects, covering such items as seismic (neo-tectonic) activity, rock temperature, groundwater hydrology including density and temperature of waters, etc. We consider that the Authors, when stating that their classification will be based on all of these types of data, may not be referring to a single Category C type classification, but an approach to design, using a series of classifications?

Nakao et al discuss possible shortcomings in the established system of Barton et al, Whickham et al and Bieniawski when used to predict tunnel support requirements in Japan. They refer to earlier work which had suggested that these established systems, being developed in "geomechanically stable hard rocks" were not readily applicable in Japan where there is an abundance of low strength rock substances in a highly fractured state. They describe the method and results of statistical studies on data collected from completed tunnels, aimed at preparing a rock mass classification scheme suited to Japanese rock conditions. The rock behaviour in the tunnels was assessed by means of convergence measurements after excavation.

The results showed that all of the parameters used in the three established rock mass classification systems are reasonably applicable to the conditions met in the tunnels studied. However, the Authors conclude that the "thickness" of overburden or the tunnel width, is of more importance in its effect on tunnel behaviour than the other factors. In particular, it replaces the degree of fracturing (by RQD and joint condition) which heads the list of ratings in the established systems.

The General Reporters can appreciate that variability in the spacing and condition of joints will be of less significance in Japan where as the Authors say, most of the rock is virtually in a "fault zone". Also we can see that the width of the tunnel will have an important effect on its deformation, where both rock substance and mass strengths are generally low. However, we suggest that the Authors might follow the approach of Kendorski et al, and treat the tunnel width or overburden thickness separately, as part of the engineering activities, which imposes loads on the rock mass. We endorse the Authors' intention to include performance parameters other than tunnel deformation, in future studies.

Mizuno et al outline a Category B rock mass classification used to characterize the foundations for dams in Japan. They then attempt to determine relationships between various rock classes, and the results of direct shear tests carried out at the sites of a number of concrete dams. The fairly wide scatter in results of direct shear tests within each class examined is reduced slightly when the data is replotted after further subdivision into geological origin classes. However the scatter of shear test results is still fairly large particularly in sedimentary and metamorphic rocks, and the Authors relate this variation to the influence of the configurations of rock layering and fractures at the individual direct shear test sites. We agree with this conclusion.

Yudbir et al have carried out laboratory tests on model specimens aimed at simulating both intact and fractured rock. They have used their own test results and those of others to examine the ability of the shear strength equations of Bieniawski (1974b) and Hoek and Brown (1980) to predict failure envelopes, in both the brittle and ductile failure ranges. They conclude that the Bieniawski equation

$$\frac{\sigma_1}{\sigma_c} = A + B\left({}^{\sigma_3}/_{\sigma_c}\right)^{\alpha}$$

might be considered for such prediction, with Parameter A ranging from 0 for completely disintegrated rocks, to 1.0 for intact rock, and Parameter B being a property of rock material (substance), presumably related to mineral composition and fabric (microstructure). They suggest some typical values for Parameter B, and tentative relationships between Parameter A and Barton et al's Q and Bieniawski's RMR.

Papers which relate rock micro-structure to rock properties

It is clear that the physical properties of any rock substance will have been developed during the history of its formation, and that the properties of the substance as determined by tests will depend upon its present mineral composition and microstructure. It is clear also that in most practical engineering situations, variations in the physical properties of the rock substance are of small significance compared to variations in the pattern of defects or discontinuities in the mass. However in particular situations, for example where the rock stresses or applied loads approach the strength of the substance, it can be important to determine the properties of the substance fairly precisely and to understand the relationships between, these properties and the microstructure. The four papers in this group discuss such relationships.

The paper by Pan et al first describes relationships between the mineral composition and microstructure of two varieties of arkose (felspar-rich sandstone), and index properties including specific gravity, porosity, water absorption, unconfined compressive strength and strength reduction after soaking in water. Carbonate-cemented arkose was found to be stronger, denser, and less subject to strength reduction, than the clay cemented arkose. The influences of dissolved oxygen in the water and of different durations of soaking were also tested in the laboratory. Dissolved oxygen was found to cause the oxidation of pyrites in the sandstone, to form sulphuric acid which attacked the carbonate cement, thus increasing the strength reduction on soaking.

In the light of these laboratory results, underground tests were conducted in collieries, and showed that water injection into the roof prior to mining had a number of beneficial effects, including improved caving, less dust at the face, less damage to supports and reduction in the incidence of bumping.

Rippa and Vinale describe the origin, mineral composition and microstructure of samples of tuff from two geological units and then relate them to the strength and deformation properties of the rock substances, in uniaxial compression. Both yellow and grey tuff (units) contain tuff substances which can be classified into four groups on the basis of the abundance and sizes of inclusions of pumice.

The testing programme showed that the unconfined compressive strength of the tuff from both units

- decreases with increasing proportions of pumice inclusions

- decreases (up to 30 percent) on saturation of samples with water

Test results suggest also that the variability of the yellow tuff is both within the horizontal layers, and vertically from layer to layer.

The microstructure of the tuff varieties, as determined using a scanning electron microscope, is shown to have a marked influence on their deformation properties.

Baklashov et al report on studies of mechanisms of deformation and fracture in rock substances and masses. They first list what they consider to be ten generally accepted facts, resulting from observation of rock failure mechanisms. They then discuss existing theories of rock failure, and find that they do not explain many of the observations. The Authors then discuss in outline, various laboratory tests and analytical work aimed at providing a better understanding of the observed mechanisms.

Test methods outlined include the following

- conventional unconfined compression, with close to zero friction at the platens, showing the mechanisms of axial splitting or columnar failure to be due to propagation of microcracks parallel to the maximum principal stress as observed by previous workers, e.g. Gramberg (1965).

- triaxial with pre-set deformation levels rather than the conventional system of pre-set load levels.

- special triaxial tests in which both axial and lateral loads were applied by independent rigid systems.

- special triaxial tests in which non-uniform three axis compressions ($\sigma_1 > \sigma_2 > \sigma_3$) were applied.

The samples tested triaxially included salt, oil shale, marble, limestone, coal, sandstone, and various rocks containing coal seams. The test arrangements enabled thorough studies to be made of volume changes, longitudinal and lateral strains, strength anisotropy and plastic and creep deformations. Rock failure concepts developed as a result of these tests have been utilized in underground mines and permanent underground works.

In his paper "Petrophysics: the petrographic interpretation of the physical properties of rocks", Montoto states that petrographic studies of rock substances are of growing importance in engineering. In support of this he cites nuclear waste disposal and geothermal projects in which prediction is required of the behaviour of microcracks under stress and temperature changes.

Montoto first discusses relationships which have been established by others, between rock substance properties including strength, permeability, durability and abrasiveness, and the following "rock forming components":

- minerals

- void spaces (pores and cracks)

- texture

The particular influence of the nature and number of voids on rock properties, is emphasized.

The paper lists various petrographic aspects of these components, which require study and quantification. Methods for sample preparation are discussed briefly followed by an outline of computer analogue systems which are being developed to obtain quantitative data on the various petrographic aspects. The Author claims that these systems will provide better data than the conventional, time-consuming "point-counter" methods.

Papers describing statistical or theoretical studies

In their paper Nishimatsu and Okubo first point out some of the difficulties in evaluating the stability of jointed rock slopes. They then assume a very simple rock mass, containing only one set of discontinuities, the spacing of which follows an exponential distribution. Assuming further that there is no interaction between the discontinuities, i.e. they do not curve and intersect, the Authors derive statistically a formula for the probability that failure will not occur along a discontinuity.

The Authors then derive a statistical distribution of shear strength for rock joints, from a series of direct shear tests carried out on joints produced in the laboratory, in 50mm diameter core samples of shale.

Using the results of all of the above, the Authors present an example of a design calculation of the probability that failure will not occur, in a slope with one set of joints striking parallel to it, with negligible water pressure.

The General Reporters consider that there are other factors, not covered by the analyses of Nishimatsu and Okubo, which have greater influence on the stability of real rock slopes than the factors studied in this paper. These more significant factors have been considered by others, for example, Patton and Deere (1971) and McMahon (1971 and 1974). These factors include the following:

- strain-softening of the joints or discontinuities, so that their strengths are reduced to residual values. The tests described in the paper appear to have produced peak values only, for which wide variations are usual, in rock joints.
- the continuity of joints.
- the shape, or waviness of joints.
- the number of sets of joints.

The paper by Grossman describes the development of a numerical method for the definition of sets of discontinuities in rock masses. The method is stated to be of most value when only a small number of defect measurements is available, and can be applied to data obtained by either uniform or non-uniform sampling of the discontinuity (joint) orientations.

The General Reporters agree that this method may assist in the definition of joint sets, but as alluded to by the Author in the introduction, complex joint patterns developed during folding of rocks will always require detailed field study of the spatial distribution and character of joints as well as their orientations.

The paper by Pineau and Thomas describes statistical studies of the degree of heterogeneity of portions of intact rock from an iron ore deposit, at various scales. The study involved 482 unconfined compression tests on cylindrical saturated samples of various sizes cut at various measured intervals, from large blocks of rock. From statistical analysis of the results, the Authors conclude that the degree of variability in the test results is related to variations in the fabric of the rocks which are in turn related to the regional geological (tectonically produced) structure. They point out that the use of simple, statistical methods e.g. arithmetical averaging, would have given incorrect conclusions about the strength of the tested rocks.

While not disputing these conclusions the General Reporters point out that the amount of variation in strength shown by the samples, e.g. Example 1 (range 15 to 29MPa), Example 3 (range 19 to 30MPa), and in all except three samples in Example 2 (range 12 to 25MPa) would be insignificant in most practical engineering and mining situations where structural discontinuities (joints etc.) have an overwhelming influence on the rock behaviour. We also ask the Authors, did they determine the amount of variation in strength values which could be attributed to minor variations in preparation of the test samples, and in the testing procedure?

Optimization of insitu investigations of rock masses

The General Reporters found the paper of Fishman et al to be refreshingly practical. The Authors point out that the improved analytical capability provided by numerical methods has created a demand for measurement of a greater variety of geotechnical parameters, and for improved quality of measurements. Accepting that insitu investigations must give the most valid results, the Authors state and illustrate five principles for their conduct, aimed at ensuring that the maximum value is obtained from this necessarily expensive work. We have summarized their five principles, as follows:

1. The type of test chosen should be that which most closely simulates the loading conditions to be imposed by the proposed structure.
2. The tests should allow for any effects on the rock mass, caused by construction activities.
3. The tests should be large enough to realistically test a portion of the rock mass (including substance and discontinuities) and in the interpretation of results, the effects of scale must be considered.
4. The tests should be interpreted in the light of the results of observational studies of the fracture pattern, and of other tests, so that the applicability of the test results to the whole construction site and to future projects, can be assessed.

5. The scope and precision of the tests should be tailored closely to the requirements of the project.

The Authors then illustrate this last principle by what we would call "sensitivity studies" in the design of the lining for a pressure tunnel. Their Figure 1 shows the sensitivity of significant crack formation in both mass and reinforced concrete, to the modulus of deformation of the rock mass surrounding the lining. In the sound rock area (Zone I on Figure 1), expensive in situ tests are clearly not warranted, but in the lower modulus areas (within and near Zone II) relatively precise measurements may be justified, if such low modulus rock is of widespread occurrence. For localized fault zones such precise tests may not be justified economically.

Various methods of insitu tests are then described and discussed. The tests include plate loading, radial loading (in tunnels), unconfined compression, tensile, and shear.

Predictions from mechanical and mathematical models

The paper by Langof describes the construction of a mechanical block model in which a jointed rock mass was simulated by rectangular plastic blocks. The blocks were arranged so as to simulate a layered rock mass, with one set of continuous joints separating the layers, and another set of discontinuous joints within each layer, normal to the first set. After determination of the deformation properties of this model rock mass by biaxial testing of a 30cm square sample, the model was used to predict the rock mass deformations around a circular tunnel, for four cases in which the rock layers were respectively horizontal, vertical, and dipping 30 degrees and 60 degrees. Also predicted for each case and for different primary stress fields, were the deformations and radial stresses in the tunnel lining and the secondary stresses developed in the rock adjacent to the tunnel crown and spring line.

A reasonable correlation between the predictions of the mechanical model, and a finite element analysis was obtained for the horizontally layered case only. The finite element analysis used was designed for a continuum and could not therefore model the pattern of joints.

The General Reporters point out that there are finite element programmes now available which could model these joint patterns, for example Burman, Trollope and Philp (1975) and Burman and Hammett (1975).

Monitoring

Habenicht and Behensky have provided the only paper on this topic. They describe a hydraulic pressure measuring cell in which the pressure on the body of the cell is transmitted from the contained hydraulic fluid, directly to a spring-loaded indicator rod. Displacements of the rod are proportional to the hydraulic pressure, and can be measured by a dial gauge or recorded by a transducer system. The Authors claim that their direct mechanical readout system has the following advantages over conventional hydraulic pressure gauges:

- higher sensitivity and accuracy
- the pressure-deflection relationship is linear
- there is almost zero hysteresis effect

Sub-Theme A.4

HYDROGEOLOGY

Hydrogeology is a vast field encompassing many topics other than those normally treated within the scope of rock mechanics or rock hydraulics. The topics discussed here are limited to those of direct interest to the rock mechanics engineer, namely:

- permeability of rock masses
- groutability of rock masses
- pressure distribution around underground cavities
- influence of ground water on the stability of slopes
- water flows into underground cavities

Only five papers have been submitted to this theme.

Engineering judgements on these topics require values for the permeability of the rock mass, as well as knowledge of the geological conditions.

The permeability of the rock mass is governed by the permeability of both the intact rock and of the fissures. Often the intact rock is so tight that it can be assumed that only the fissures are permeable. But it must be remembered that rock mass is not homogeneous. In particular, faults may intersect a rock mass sporadically and require special treatment.

The permeability of homogeneous regions may be determined by various methods. Without doubt the most widely accepted is the water pressure test because it may be applied above and below the ground water table. Each water pressure test investigates only a small section of the rock mass and so a large number of tests is necessary to estimate the rock mass permeability.

Water pressure tests may be evaluated in different ways. The oldest and still used method is to consider the flow rate related to the pressure and the length of the test section as a measure of the permeability. This evaluation is connected with the name Lugeon (Lugeon 1933). Other methods have since been developed which attempt to determine permeability factors or permeability tensors, i.e. directed permeability properties (e.g. Baker (1955), Maini (1971), Snow (1966), Snow (1969), Rissler (1977)). These usually assume steady flow and the tests must be performed in an appropriate manner. The paper by Schneider presents an evaluation method which is based upon unsteady flow conditions and which obviously leads to results which differ from those of the previously developed methods. It is felt that these different results might be caused by changes in the test procedure which are introduced by the author (control of flow rate instead of pressure). In any case the General Reporters are familiar with many results where steady flow has been arrived at much more quickly than in the tests described by the author. Comments of the author and of the auditorium are requested.

It is well known that joints with wide apertures allow both laminar and turbulent flow conditions. In this case one observes that the flow rate increases at a lesser rate than the rising pressure. Rissler (1977) has investigated this case theoretically for radially symmetrical conditions. In their paper Cruz and Quadros present in-situ measurements made in a basaltic rock mass, which confirm these characteristic curves.

As well as permeability, interest is directed often to groutability of rock masses. High permeability must not be assumed to indicate good groutability. The permeability may be caused by many joints with small apertures, which refuse to take grout. Blinde et al describe investigations of this topic performed in disintegrated granites and consisting of fabric studies, optical examination of bore holes, joint tracing techniques, water pressure tests, geoelectrical self-potential measurements and seismic experiments. The authors found that not all apparently existing joints contributed to the permeability.

Huder and Amberg report on the influence of ground water on the stability of slopes and underground cavities and on the effectiveness of relief measures. In addition various piezometer devices are discussed.

Bello-Maldonado presents an analytical procedure for estimation of water inflows into tunnels during and after construction. Such calculations are certainly interesting but the General Reporters fear that in practice rock conditions will seldom be sufficiently homogeneous to justify the considerable mathematical effort involved.

CONCLUSIONS

To the theme "Site exploration and evaluation" a total of 61 contributions have been submitted. They deal with geophysical methods, in-situ and laboratory investigations, classification, prediction, observation and monitoring and hydrogeological questions.

Since the last ISRM Congress the most important advances appear to have been made in the following areas:

- the application of seismic methods to the assessment of grouting results
- the development of procedures to evaluate anisotropic deformation properties from tests
- the further development of classification systems.

REFERENCES

Anstey, N.A. (1977). Seismic interpretation, the physical aspects. Int. Human Resources Development Corporation Boston, U.S.A., p. 81.

Baker, W.J. (1955). Flow in fissured formations. Proc. 4. World Petroleum Congr., Rome, Section II/E, Paper 7, pp. 379-393.

Barla, G. et al. (1974). Direct tensile testing of anisotropic rocks. Proc. Third Int. Congr. Rock Mech. ISRM, Vol. 2A, pp. 93-98.

Barton (1971). A relationship between joint roughness and joint shear strength. Report I 8, Int. Symp. on Rock Mech., Nancy.

Barton, N., Lien, R., and Lunde, J. (1974). Engineering classification of rock masses for the design of tunnel support. Rock Mechanics. Vol. 6, No. 4, pp. 189-236.

Bieniawski, Z.T. (1974a). Geomechanics classification of rock masses and its application in tunnelling. Proc. Third Int. Conf. Rock Mech., ISRM. Vol. IIA, pp. 27-32.

Bieniawski, Z.T. (1974b). Estimating the strength of rock materials. J. S. African Inst. Min. and Met., Vol. 74, No. 8.

Bieniawski, Z.T. (1978). Determining rock mass deformability, experience from case histories. Int. J. Rock Mech. and Min. Sci., Vol. 15, pp. 237-247.

Bieniawski, Z.T. (1979). The geomechanics classification in rock engineering applications. Proc. Fourth Int. Conf. Rock Mech., ISRM, Vol. 2, pp. 41-48.

Broch, E. and Franklin, J.A. (1972). The Point Load strength test. Int. J. Rock Mech. and Min. Sci., Vol. 9, pp. 669-697.

Brown, E.T. (1974). Fracture of rock under uniform biaxial compression. Proc. Third Int. Congr. Rock Mech., ISRM, Vol. 2A, pp. 111-117.

Burman, B.C., Trollope, D.H., and Philp, M.G. (1975). The behaviour of excavated slopes in jointed rock. Aust. Geomech. J., Vol. G5, No. 1, pp. 26-31.

Burman, B.C. and Hammett, R.D. (1975). Design of foundations in jointed rock masses. Proc. Second Aust.-N.Z. Conf. on Geomechanics, pp. 83-88.

Franklin, J.A. (1979). Use of tests and monitoring in the design and construction of rock structures. General Report, Theme 2, ISRM Congress, 1979, Montreux, Switzerland, Vol. 3, p. 163.

Gramberg, J. (1965). Axial cleavage fracturing a significant process in mining and geology. Eng. Geol. Vol. 1, No. 1, pp. 31-72.

Hoek, E. and Brown, E.T. (1980). Underground excavations in rock. The Institution of Mining and Metallurgy, London.

International Society for Rock Mechanics (1980). Basic geotechnical description of rock masses. Int. J. Rock Mech. Min. Sci. and Geomech. Abstr. Vol. 18, pp. 85-110.

John, K.W. (1969). Festigkeit und Verformbarkeit von druckfesten, regelmäßig gefügten Diskontinuen. Ph.D. Thesis, Karlsruhe.

Kaneko, K., Inoue, I., Sassa, K. and I. Ito (1979). Monitoring the stability of rock structures by means of acoustic wave attenuation. ISRM Congress, 1979, Montreux, Switzerland, Vol. 2, p. 287.

Langer, M. (1979). Rheological Behaviour of Rock Masses. General Report, Theme 1, ISRM Congress, 1979, Montreux, Switzerland, Vol. 3, p. 29.

Leichnitz, W. and Natau, O. (1979). The influence of peak shear strength determination on the analytical rock slope stability. Proc. Fourth Int. Congr. Rock Mech. ISRM, Vol. 2, pp. 335-341.

Lögters, G. and Voort, H. (1974). In-situ determination of the deformational behaviour of a cubical rock mass sample under triaxial load. Rock Mechanics (6), pp. 65-79.

Lugeon, M. (1933). Barrages et Geologie. Dunod, Paris.

Maini, Y.N. (1971). In situ hydraulic parameters in jointed rock - their measurement and interpretation. Ph.D. Thesis Imperial College, London.

McMahon, B.K. (1971). A statistical method for the design of rock slopes. Proc. First Aust.-N.Z. Conf. on Geomechanics. Vol. 1, pp. 314-321.

McMahon, B.K. (1974). Design of rock slopes against sliding on pre-existing fractures. Proc. Third Conf. of Int. Soc. Rock Mech., Vol. IIB, pp. 803-808.

Neilson, J.L. (1970). Notes on weathering of the Silurian rocks of the Melbourne district. J. I. E. Aust. Vol. 42, 1-2, pp. 9-12.

Oberti, G., Carabelli, E., Goffi, L. and Rossi, P.P. (1979). Study of an orthotropic rock mass: Experimental techniques, compara tive analysis of results. Proc. Fourth Int. Congr. Rock Mech. ISRM, Vol. 2, pp. 485-491.

Patton, F.D. and Deere, D.U. (1971). Significant geologic factors in rock slope stability. In: Planning Open Pit Mines. A. A. Balkema, Amsterdam.

Pinto (1979). Determination of the elastic constants of anisotropic bodies by diametral compression test. Proc. Fourth Int. Congr. Rock Mech. ISRM, Vol. 2, pp. 359.

Plischke, b. (1980). Ein Berechnungsverfahren zur Untersuchung der dynamischen Beanspruchung von Felsbauten. 4. Nationale Tagung über Felsmechanik, 5./6. Mai 1980, Aachen, Deutsche Gesellschaft für Erd-und Grundbau, p. 259-277.

Rissler, P. (1977). Determination of the water permeability of jointed rock. Aachen, Institute for Foundation Engineering, Soil Mechanics, Rock Mechanics and Water Ways Construction.

Rodrigues (1979). The anisotropy of the moduli of elasticity and of the ultimate stress in rocks and rock masses. Proc. Fourth Int. Congr. Rock Mech. ISRM, Vol. 2, pp. 517.

Sancio, R.T. and Brown, I. (1980). A classification of weathered foliated rocks for use in slope stability problems. Proc. Third Aust.-N.Z. Conf. on Geomechanics, Vol. 2, pp. 81-86.

Saunders, M.K., Fookes, P.G. (1970). A review of the relationship of rock weathering and climate and its significance to foundation engineering. Eng. Geol. Vol. 4, pp. 289-325.

Savich, A.I., Koptev, V.I., Iljin, M.M. and Zamakhajev, A.M. (1979). Distinguishing features of deformation development in the foundation of the Inguri Arch Dam. ISRM Congress, 1979, Montreux, Switzerland, Vol. 2, p. 585.

Snow, D.T. (1966). Three-hole pressure test for anisotropic foundation permeability. Felsmechanik und Ingenieurgeologie 4, Nr. 6, pp. 298-316.

Snow, D.T. (1969). Steady flow from cylindrical cavities in saturated, infinite anisotropic media. Association of Engineering Geologists, Sacramento, California.

Terzaghi, K. (1946). Rock defects and loads on tunnel supports. In: Rock Tunnelling with Steel Supports. Commercial Shearing and Stamping Co., Youngstown.

Whickham, G.E., Tiedemann, H.R. and Skinner, E.H. (1972). Support determinations based on geological predictions. Proc. First North American Tunnelling Conference, A.I.M.E., New York, pp. 43-64.

Wichter, L. (1979). On the geotechnical properties of a Jurassic clay shale. Proc. Fourth Int. Congr. Rock Mech. ISRM, Vol. 1, pp. 319-326.

EXPLORATION ET ÉVALUATION IN SITU

Site exploration and evaluation

Untersuchung und Beurteilung des Betriebspunktes

David H. Stapledon
Institut de Technologie de l'Australie du Sud, Adelaïde, Australie

Peter Rissler
Ruhrtalsperrenverein, Essen, RF Allemagne

RESUME :

Les reconnaissances de sites demandent du temps et de l'argent, mais sont d'une importance capitale pour le dimensionnement des projets et le succès des travaux. Six communications sur les méthodes géophysiques donnent à entendre que les méthodes sismiques sont très répandues. En ce qui concerne les essais en laboratoire et in situ, 35 communications mettent l'accent sur la détermination de l'anisotropie des propriétés, du comportement sous charge constante et cyclique, de la résistance au cisaillement des discontinuités et sur les effets d'échelle. Quinze communications sur le thème "classification" illustrent les divers systèmes et la difficulté d'établir un système unique adapté à tous les types de projets et tous les cadres géologiques. Cinq communications traitent des problèmes hydrogéologiques tels la détermination de la perméabilité et l'étude des eaux souterraines.

REMARQUES GENERALES

La reconnaissance des sites et leur évaluation occupe un vaste domaine de la mécanique des roches. Ces activités sont sous entendues pour toutes les communications concernant les thèmes généraux B C D et partiellement E.

Reconnaitre et évaluer les conditions d'un site semble souvent coûteux, mais en comparaison du montant des travaux de constructions, ce coût est faible. De plus, il disparait complètement en regard avec celui d'un accident causé par une reconnaissance inadéquate ou omise.

D'autre part, la reconnaissance des sites nécessite habituellement beaucoup de temps et elle ne peut être programmée et budgétisée avec autant de précisions que les travaux de construction. Ceci est dû au fait que nos sites ont été modelés par des processus naturels durant des durées très longues et la Nature nous laisse en héritage un enregistrement de ses activités incomplet et bien souvent fortement oblitéré, et non pas un plan "B.P.E. (Bon pour exécution)". Ainsi l'exploration d'un site doit être lancée sans avoir aucune idée claire de ce que l'on trouvera. Le programme des reconnaissances doit donc procéder par étapes, les actions réalisées à chaque étape étant destinées à répondre à des questions formulées à partir des résultats de l'étape précédente.

Les délais comme les financements nécessaires doivent être compris et acceptés par le maître d'oeuvre du projet.

En 1975 notre Société publia le rapport de la commission chargée d'établir "les Recommandations pour la reconnaissance des sites" qui suggérait une approche systématique pour les programmes de reconnaissance.

Dans son rapport général du thème 2 au Congrès de Montreux de la S.I.M.R., John Franklin (1979) décrivait en détails l'utilisation des essais et des mesures dans le dimensionnement et la réalisation des constructions au rocher, et démontrait vraiment très clairement les interrelations existantes entre la géotechnique et les autres activités d'inginierie tout au long des études et des travaux.

A cause de cette importance primordiale pour toute activité géotechnique, la reconnaissance et l'évaluation des sites ont été le thème de maints Congrès passés de la S.I.M.R., par exemple :

En 1974, à Denver, le thème 1 qui était

> "Propriétés physiques de la roche intacte et du massif rocheux", et

En 1979, à Montreux, une partie du thème 1 qui concernait

> "Le comportement rhéologique des massifs rocheux"

et le thème 2 à propos de

> "L'utilisation des essais et les mesures

dans le dimensionnement et la réalisation des constructions au rocher"

furent dévolus aux travaux de reconnaissance et d'évaluation des sites.

Comme il a été décidé par le comité d'organisation de ce Congrès, le thème général de ce Congrès consacré à la reconnaissance des sites a été subdivisé en 4 sous-thèmes :

A.1 Reconnaissance et essais géophysiques
A.2 Essais en laboratoire et in situ
A.3 Classification, prévisions, observations et mesures
A.4 Hydro-géologie

Nos commentaires seront présentés dans cet ordre.

Sous-thème A.1

RECONNAISSANCE ET ESSAIS GEOPHYSIQUES

Ce fut pour nous une surprise de ne trouver que 6 communications pour ce sous-thème comparativement aux 34 soumises pour le thème A.2 . On peut interpréter ceci comme une expression de la validité relative des essais géophysiques par rapport aux essais in-situ ou en laboratoire. Nous sommes convaincus que cette interprétation n'est pas correcte. Les méthodes géophysiques présentent beaucoup d'avantages et d'un autre côté quelques inconvénients. Leurs avantages résident dans leur possibilité de fournir une information concernant un large volume du massif rocheux. Sous cet angle, elles sont supérieures à toutes les méthodes directes. Leur inconvénient : dans de nombreux cas elles ne fournissent qu'une valeur indicative pour la caractérisation du massif rocheux. D'une autre façon, il semble qu'un avantage particulier soit que leur utilisation est largement meilleur marché. Ceci permet d'obtenir plus d'informations durant tous les stades de développement des projets et leurs mises en route.

Le niveau de réussite atteint par toute méthode d'exploration ne dépend pas seulement de la valeur intrinsèque de la méthode, mais aussi de la compétence et de la connaissance de la personne qui l'applique, en particulier de sa compréhension des limites de cette méthode. L'expérience a déjà montré que les limites des méthodes géophysiques sont fréquemment mal comprises et que leur application dans des situations inadaptées, souvent sans le support de méthodes directes, a donné une si faible réussite qu'elle les a desservies.

La méthode sismique, la plus communément utilisée est également la plus communément utilisée de façon erronée. Ce qui suit représente ce que nous estimons être les deux limitations importantes de cette méthode :

(a) D'importantes propriétés mécaniques des massifs rocheux (en particulier la résistance au cisaillement) sont communément sous la dépendance de quelques accidents géologiques mineurs (tels failles de faible importance ou banc altéré) qui occupent des positions et des orientations critiques en relation avec les contraintes appliquées. La méthode sismique est en général incapable de localiser de tels défauts.

(b) La méthode sismique applique des charges faibles et pendant de faibles durées. Les modules d'élasticité dynamique (E_{dyn}) calculés à partir des célérités des ondes transversales (V_s) ou estimés à partir des célérités des ondes longitudinales (V_p) reflètent ainsi la réponse du massif rocheux à des chargements transitoires. De tels modules ne peuvent être considérés pour fournir une indication utilisable pour la déformation d'un massif rocheux stratifié et diaclasé sous fortes charges à long terme.

Nous avons relevé un argument selon lequel la célérité des ondes longitudinales ne reflètent pas nécessairement les propriétés d'une roche "moyenne", comme souvent supposé. Ceci parce que les ondes longitudinales, qui arrivent les premières aux géophones, peuvent emprunter la trajectoire la plus courte à travers la roche la meilleure, par une sorte d'effet de "court-circuit".

Si nous acceptons l'argument ci-dessus, l'hypothèse que les méthodes électriques fondées sur les mesures de résistivités fournissent elles aussi une indication quantitative "moyenne" à propos de la roche, apparaitrait fausse également. Les lignes de courant tendent à se focaliser dans la roche de conductivité la plus forte, tendant ainsi à fournir un effet de court-circuit par les zones altérées ou faillées.

Il serait très utile d'obtenir quelques commentaires à propos de ce qui précède et d'autres éventuelles limitations des méthodes géophysiques, de la part des auteurs et des participants au Congrès.

Recherches en laboratoire et sur le terrain relatives aux méthodes sismiques

Dans des essais en laboratoire décrits par Tanimoto et Ikeda, la célérité des ondes longitudinales est trouvé être insensible à l'espacement de discontinuités d'ouverture inférieure à 0,01 mm et montrer une relativement faible décroissance (5 à 15 %) avec des discontinuités d'ouverture inférieure à 2 à 4 mm. Est indiquée comme étant similaire à une roche intacte, du point de vue de son comportement sismique, une roche à discontinuités d'ouverture $< 0,04$ mm.
Ces résultats apparaissent en contradiction avec ceux d'autres auteurs. Par exemple Anstey (1977) énonce

> "Les effets des fractures sur la célérité sismique est majeur ... ; sitôt qu'il y a un vide, la déformation transversale des particules est possible et le module d'élasticité associé change. Nous rappelons aussi combien est faible le déplacement d'une particule lorsqu'une onde sismique passe par elle - classiquement au moins 2 ordres de grandeurs plus petits que la longueur d'onde de la lumière ; par conséquent des fissures bien trop fines pour être visible au microscope optique peuvent produire une modification majeure des propriétés élastiques".

De même, Savich et al. à la figure 6 de leur communication présentent les résultats de déterminations de ratios discontinuités-vides, par méthodes géophysiques (ultra-sons, sismique) et géologiques (visualisation). Les méthodes géophysiques fournissent des ratios discontinuités-vides 1,7 à 2 fois plus élevés que la méthode géologique. Les auteurs expliquent cette divergence à partir de l'approximation suivante :

> "Les mesures ultra-soniques prennent en compte non seulement les discontinuités visibles mais aussi des micro discontinuités qui ne peuvent être identifiées visuellement dans les sondages".

Nous suggérons que les différences d'effets observées par les différents chercheurs peuvent résulter d'effets d'échelle en relation avec les différences dans les amplitudes et les longueurs d'ondes. Des commentaires de la part des auteurs sont réclamés.

Nous notons que dans leur résumé, Tanimoto et Ikeda font mention d'études in situ sur des sites de surface et sur 104 sites souterrains. Ces études ne sont pas décrites dans la communication. Il serait très instructif que les auteurs puissent nous préciser si leurs résultats en laboratoire sont en accord avec ces résultats d'étude in situ.

Dans leurs expérimentations sur les trajectoires (Figures 8 et 9) Tanimoto et Ikeda simulent la roche avec discontinuités non continues, ou des discontinuités avec ponts rocheux, en réalisant des fentes dans des cubes de plâtre de 50 mm. A partir des mesures de temps de transit, les auteurs concluent que la réduction apparente de la célérité V_p n'est seulement dûe qu'à l'augmentation de la longueur de la trajectoire des ondes qui ne peuvent passer à travers les fentes. Il nous semble que la figure 8 de cette communication est une bonne illustration de l'effet de "court circuit" mentionné ci avant dans notre discussion sur les limites. Il semble à partir de cette figure que des valeurs élevées non réalistes de V_p peuvent être obtenues dans des massifs rocheux fortement fissurés et/ou altérés si des "courts circuits" peuvent exister à travers des zones saines ou non fissurées du rocher entre le point d'émission et les récepteurs.

Bien que la signification pratique de ces résultats ne soit pas encore bien totalement compris, nous considérons que Tanimoto et Ikeda exposent des résultats intéressants à propos de l'atténuation des amplitudes, de l'effet de l'aire de contact des joints, des pressions de contacts, et des trajectoires. La sensibilité de l'atténuation d'amplitudes à l'ouverture des discontinuités est particulièrement intéressante, comme l'avait mentionné déjà Kaneko et al. (1979) à Montreux.

La communication de By se rapporte spécifiquement à l'atténuation des ondes longitudinales dans les roches discontinues. Un modèle théorique est proposé et vérifié par une expérimentation élaborée autour de sautage à l'explosif in situ. Les résultats sont présentés d'une part sous forme d'une relation entre l'amplitude de la contrainte maximale et la distance au point d'émission (figure 8) et d'autre part sous forme d'un abaque à partir duquel un "facteur d'amortissement" peut être déterminé si on connait l'ouverture des discontinuités et "l'épaisseur des bancs". Nous supposons que "l'épaisseur des bancs" doit être entendue comme "l'intervalle entre discontinuités". On peut noter que Plischke (1980) a publié des travaux concernant l'estimation de la propagation des ondes de choc par des méthodes numériques. Une coopération entre ces deux auteurs pourrait aider à des développements nouveaux sur ce sujet. Les résultats de By à propos de l'influence des discontinuités sont en général en bon agrément avec ceux de Tanimoto et Ikéda.

Estimation de l'efficacité des injections

Rodrigues et al. présentent des résultats convaincants sur l'utilisation réussie de la méthode sismique pour l'évaluation de l'efficacité des injections du barrage de Cabril. Les mesures géophysiques sont bien recoupées par les essais de perméabilité en forages et sont menées avec un haut degré de précision. L'attention est attirée sur des anomalies dans les résultats des mesures de V_p utilisant les techniques "fond de trou" (down-hole) ou "trou à trou" (cross-hole), dues à des effets de "court-circuit". Dans cet ordre d'idées, nous posons les questions suivantes : des valeurs significativement différentes de V_p ont-elles été obtenues par différentes techniques dans une même zone d'un massif rocheux ? Si oui, à quelle interprétation de ces différences peut-on penser (en termes de structure du massif rocheux) et quelle valeur de V_p doit être adoptée comme représentative de cette portion du massif rocheux ?

Leurs expériences donnent à penser que les valeurs de V_p et les résultats des injections sont associés dans le cas d'injections dans un massif rocheux de qualité mécanique avant injections relativement élevée. Les données des tableaux I et II seraient plus utilisables si les auteurs fournissaient plus de détails, p.e. les profils géologiques, une indication sur la structure du réseau de discontinuités du granite, des détails sur la disposition des forages des injections de consolidation et ceux du voile d'injection, et les hauteurs sur lesquelles furent réalisées les essais Lugeon.

Savich et al. faisant référence aux études pour le contrôle de la qualité des injections au barrage-voûte de l'Inguri, montrent à leur figure 11 un abaque établi sur ce site pour l'estimation de l'efficacité de l'injection. L'arrêt de l'échelle des célérités initiales des ondes, V_o, à une valeur d'environ 5 km/s suggère-t-elle que des roches dotées d'une célérité V_p supérieure ou égale à cette valeur ne sont pas injectables ? Nous avons reporté sur cet abaque les résultats des essais du Barrage de Cabril. On peut alors voir que comme nous le supposions, les valeurs moyennes pour les blocs V et VI se situent dans la zone "pratiquement non injectable". Les valeurs des autres blocs se situent dans la zone III "injection satisfaisante" et à la limite avec la zone II "bonne injection". Ainsi que nous l'avons établi au paragraphe précédent, nous considérons que les données du barrage de Cabril correspondent à un massif rocheux de bonne qualité mécanique. Il apparait à partir de la fi-

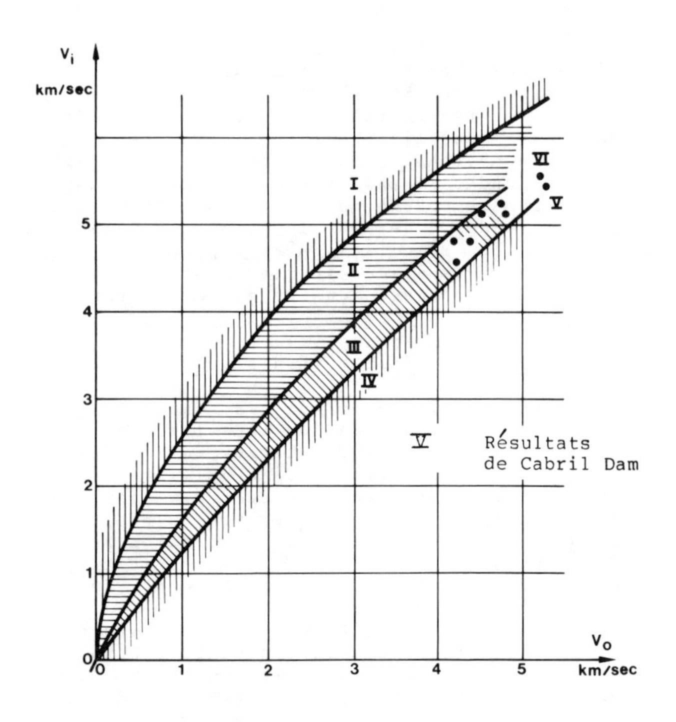

Fig. 1 Abaque de Savich et al. pour l'évaluation de la qualité des injections sur lequel sont reportés les résultats du site du barrage de Cabril

gure 11 de Savich et al., que en queques secteurs du site de l'Inguri, des roches présentent des valeurs initiales V_o de la célérité, comparables à celles du site du Cabril, montrent une plus forte augmentation de la célérité après injection (des valeurs finales de V_p atteignent plus de 6 km/s : zone I "excellente injection". Il apparait également que la cause de cette large différence est en relation avec les caractéristiques différentes des massifs rocheux avant l'injection. Il serait utile que les auteurs puissent fournir des détails sur ces secteurs du massif rocheux du site de l'Inguri où des célérités V_p après injection approchent 6 km/s qualifient les injections de "bonne" et "excellente". En particulier, les auteurs peuvent-ils expliciter l'augmentation de plus de 20 % de la célérité V_p par injection dans un rocher doté d'une célérité initiale V_p d'environ 5 km/s, rocher qui aurait été d'une qualité mécanique élevée avant injection?

Evaluation des propriétés élastiques et de déformation

Dhawan et al. obtiennent des valeurs de V_p et V_s à partir de "trainées" sismiques dans une quartzite phylliteuse plissée et discontinue sur un site de barrage dans l'Himalaya. Certains de ces trainées sont réalisés dans un puits de reconnaissance où furent réalisés également des essais de chargement à la plaque rigide et des essais au vérin plat pour obtenir des valeurs du module de déformation (D) et du module élastique (E). Les figures présentées par les auteurs, montrent une corrélation raisonnable entre les valeurs de V_p, D et E. Les faibles corrélations pour les résultats des essais au vérin plat et les essais à la plaque apparaissent pour partie au moins dûs à la forte dispersion des résultats (voir leur tableau 3). En raison même de la limitation (b) exposée au début de ce Rapport Général, nous considérerons que les valeurs de D et de E obtenus par ces essais de chargement in situ ne sont pas reliées de façon simple aux valeurs de E_{dyn}.

Savich et al. exposent que les méthodes séismoacoustiques sont devenues "les méthodes de pointe pour les études quantitatives des propriétés élastiques et de déformabilité des massifs rocheux". Ils décrivent comment des études systématiques utilisant les méthodes séismoacoustiques associées à la géologie de l'ingénieur, à des mesures de module statique et à un traitement statistique des données ont rendu possible le développement d'équations généralisées pour le calcul des modules statiques D et E à partir du module dynamique E_{dyn}. Quatre séries de valeurs des coefficients utilisés dans ces formules sont donnés dans leur tableau II; il y a une série de valeurs pour chacune des 4 classes très larges de roches. Les auteurs indiquent que l'utilisation de ces formules générales dans un massif rocheux quelconque permet de calculer les valeurs de D et E avec une probabilité de 90 pour 100 de faire une erreur inférieure ou égale à 30 %.

Les Rapporteurs Généraux apprécient les études minutieuses et systématiques présentées dans cette communication, mais ils émettent cependant des réserves sur la possibilité d'application générale de ces équations à tous les massifs rocheux à travers le monde ; ceci en raison du grand nombre de variables géologiques susceptibles d'influencer la compressibilité d'un massif rocheux. Par exemple, un massif rocheux contenant des types de roches de quelques uns des 4 larges groupes du tableau II peut contenir en plus des discontinuités, des zones localisées plus ou moins larges de roches broyées, et/ou des zones de roches altérées de forme plus ou moins régulière. Tout ou partie des matériaux broyés ou altérés peut avoir un comportement du type "sol" et être fortement compressible. La position et l'orientation de telles zones compressibles par rapport à celles du réseau de discontinuités et des charges appliquées peuvent avoir un effet significatif sur la déformabilité du massif rocheux.

Nous serions heureux d'obtenir des réponses aux questions suivantes :

(a) Les auteurs peuvent-ils justifier d'un nombre de cas de figures significatifs pour la fixation des paramètres du tableau II, en gardant à l'esprit le large domaine des propriétés physiques susceptibles d'être représentées par chacun des 4 larges groupes de types de rocher.

(b) Les équatios (2) et (3) ont-elles été appliquées à un massif rocheux fortement affaibli tant par une altération météorique ou chimique que par une fracturation ?

Auscultation des massifs rocheux durant la construction des barrages et le remplissage de la retenue

Savich et al. décrivent les résultats d'études géophysiques menées autour du barrage voûte de 271,5 m de hauteur d'Inguri et de sa retenue, pour suivre les modifications dans le massif rocheux durant la construction et pendant et après le remplissage de la retenue. Des études régionales et des études de détails furent réalisées en jonction avec des mesures de déformation et une surveillance géodésique. Les études mettent en évidence une augmentation de la célérité des ondes longitudinales V_p et de la résistivité électrique jusqu'à environ 80 m sous la base du barrage, durant sa construction. Des réductions remarquables de la célérité V_p et de la résistivité apparurent durant le remplissage de la retenue, lorsque les discontinuités furent soumises à la charge hydrostatique de la retenue. A proximité du barrage ces effets furent décelés jusqu'à environ 250 m de profondeur et près du maximum de la retenue jusqu'à environ 500 m.

Les auteurs peuvent-ils indiquer si la fermeture et l'ouverture des discontinuités à proximité du barrage furent suivies également à l'aide d'extensomètres en forages et si oui, l'amplitude des déformations observées ?

Nous considérons qu'une auscultation du type décrite pour le barrage de l'Inguri représente une application extrêmement riche de la géophysique.

Estimation du comportement en déformation des roches de faibles caractéristiques ("weak rocks") durant les séismes

Pour estimer l'influence des séismes sur les structures il est important de connaître le comportement de la roche de fondation. Ishikawa et al. décrivent les résultats d'essais cycliques de chargement vertical à basse fréquence qui montrent diverses relations entre le module dynamique E_{dyn}, l'amplitude du chargement, la vitesse de chargement, les déformations verticales et les coefficients d'amortissement. Comme les auteurs exposent des résultats ne concernant que deux sites, le principal intérêt de leurs travaux est d'établir une nouvelle procédure d'essai. Ces résultats sont principalement influencés par les conditions dans lesquelles se présente le rocher dans les 1,50 m situés sous l'appareillage d'essai, où il est supposé être toujours plus ou moins perturbé ou altéré, c'est à dire dans des conditions anormales. C'est pourquoi nous ne voyons pas comment utiliser les résultats de cet essai dans un dimensionnement. Les auteurs sont invités à expliciter ce point.

D'un autre point de vue, ces tests sont associés à des tests sismiques et ultra-soniques. C'est pourquoi le papier d'Ishikawa et al. devrait être lu en association avec celui de Tanimoto et Ikéda et pourrait nous donner une meilleure compréhension de la décroissance des amplitudes. Pour l'analyse de ces tests et d'autres similaires, de nombreux travaux ont déjà été menés à bien. Langer les a résumé dans son Rapport Général sur le thème I du 4ème Congrès International de la S.I.M.R. à Montreux en 1979.

Autres applications des méthodes géophysiques à la reconnaissance des sites

La sismique réfraction est utilisée traditionnellement pour déterminer l'épaisseur de la couverture d'alluvions ou de matériaux altérés recouvrant le bed-rock sur les sites de barrage et autres structures superficielles. Dhawan et al. présentent des exemples de l'utilisation de cette méthode pour des projets dans l'Himalaya et indiquent une concordance des résultats à 10 % près avec ceux obtenus par sondages.

Dans leur paragraphe "2.1 Cartographie des massifs rocheux", Savich et al. suggère que la précision des résultats des investigations géologiques et géophysiques peut être améliorée en mettant en oeuvre plusieurs méthodes géophysiques en même temps. Leur figure 1 est un profil en travers du site du barrage qui a été reconnu par des méthodes sismiques, électriques et magnétométriques. Les limites géologiques définies par les méthodes géophysiques sont indiquées ainsi que les failles ("zones de dislocation tectonique") et les sondages qui ont été forés après l'instrumentation géophysique. Les auteurs annoncent que les limites déterminées par la géophysique sont en accord avec celles déduites des sondages à moins de 5 %.

Savich et al. exposent brièvement un certain nombre d'applications des méthodes géophysiques, telles l'estimation de l'état de contrainte dans les roches, l'analyse des zones décomprimées dans les versants et autour des excavations artificielles et des études pour classer les roches en termes de "degré de pérennité" ("degree of preservation") déjà évoqué dans Savich et al. (1979).

Sous-Thème A.2

ESSAIS IN SITU ET EN LABORATOIRE

Par contraste avec le précédent Congrès de la S.I.M.R. où les propriétés de la roche intacte et des massifs rocheux furent l'objet des discussions, nous allons maintenant examiner les techniques d'essais et les méthodes d'évaluation des résultats d'essais. Notre démarche sera donc guidée par les questions suivantes :

- Comment les essais doivent-ils être exécutés in situ ou en laboratoire ?
- Comment doivent ils être évalués ?
- Quels paramètres principaux et de second ordre doivent être mesurés ?
- Comment introduire ces paramètres dans l'évaluation ?

Les essais in situ et en laboratoire sont réalisés tant pour fournir des données pour le dimensionnement d'une structure que pour améliorer notre connaissance de quelque aspect du comportement de la roche. Dans le premier cas on s'efforce d'appliquer des modes opératoires bien connus tandis que dans le second cas des méthodes d'essais originales et très sophistiquées peuvent être mises au point. Mais nous apprenons à partir des communications retenues pour ce sous-thème que ces essais réputés classiques peuvent être améliorés ainsi que leur évaluation.

En plus, ces 35 communications ont attiré notre attention sur les conditions limites de chaque essai. Ce n'est que si nous prenons soin de bien les définir et de les prendre en considération pour l'évaluation, que nous pouvons avoir confiance dans le réalisme des résultats d'essais.

Dans ce contexte plus d'importance doit être donnée à un bon échantillonnage ne remaniant pas les échantillons et à une conservation adéquate de ceux-ci. Il est courant de noter soigneusement les conditions de prélèvement et de faire des photographies des éprouvettes d'essais. Cependant nous estimons cela insuffisant ; il est au moins aussi important de noter ce qui suit : l'état d'altération de l'échantillon, sa teneur en eau et sa masse volumique, la longueur de la carotte comparée à la longueur moyenne des carottes, des détails concernant les discontinuités et autres défauts, l'orientation du litage et toute anomalie visible.

Les communications traitant des essais in situ et en laboratoire peuvent être réparties en plusieurs domaines :

- Prise en compte de l'anisotropie et de propriétés non élastiques dans l'évaluation de l'essai.
- Détermination de la résistance et du comportement à court terme.
- Détermination de la résistance et du comportement à long terme et sous charges cycliques, étude du fluage et de l'initiation de la rupture.
- Recherches sur la résistance au cisaillement des discontinuités.
- Effets d'échelle.
- Essais de boulon d'ancrage

et des sujets additionnels tels :

- Le développement d'appareillages pour essais spéciaux.
- Les normes de préparation des échantillons.
- Procédures de dimensionnement.
- Applications spéciales de processus d'essais classiques.
- Modèles théoriques de comportement des discontinuités.
- Essais d'érosion de produit d'injection.
- Description statistique des discontinuités.
- Reconnaissance de la qualité du massif rocheux en avant du front en tunnel.
- Evaluation des essais en temps réel assisté par ordinateur.

Les Rapporteurs Généraux ont principalement centrés leurs commentaires sur les domaines principaux.

Prise en compte de l'anisotropie et des propriétés non élastiques dans l'évaluation des essais

Durant les dix dernières années les méthodes numériques ont prouvé être un outil puissant en mécanique des roches. En principe, il y a peu d'importantes propriétés des massifs rocheux qui à présent ne puissent être modelisées numériquement, mais en pratique la valeur des méthodes est habituellement limitée par la qualité médiocre et la faible disponibilité des valeurs des paramètres à introduire. Nous considérons très regrettable le fait que l'anisotropie, caractéristique la plus importante des roches et des massifs rocheux, ne soit pas suffisamment prise en compte. Même lorsque beaucoup d'argent est dépensé pour des essais autorisant ainsi la détermination de l'anisotropie des déformations, l'interprétation, dans bien des cas, reste basée sur une hypothèse d'isotropie.

En toute rigueur, la description complète de l'anisotropie la plus importante nécessite la détermination de 21 paramètres indépendants. Il est fort peu probable que cette exigence soit jamais complètement rencontrée. Mais, du point de vue de l'ingénieur, une telle description si précise et complète ne sera justifiée en raison de l'hétérogénéité même des massifs rocheux. Cependant, l'orthotropie, cas le plus important d'anisotropie, (dite aussi "anisotropie transversale") doit être prise en considération dans l'interprétation dans la mesure où elle constitue la caractéristique distinctive la plus importante tant du massif rocheux que de la roche intacte.

Le comportement en déformation d'un milieu orthotropique peut être décrit par 5 paramètres indépendants à savoir E_1, E_2, ν_1 , ν_2 et G_2. Les formations sédimentaires comme certaines formations métamorphiques présentent habituellement un fort degré d'orthotropie et il est considérablement important pour la mécanique des roches appliquée de développer des méthodes de détermination de ces paramètres. Nous sommes heureux d'avoir trouver quelques communications consacrées à ce problème.

Depuis les premiers ages de la mécanique des roches, l'essai de compression non confinée est utilisé pour déterminer la résistance en compression uniaxiale et le module de déformation parallèlement et perpendiculairement au litage. Mais dans bien des cas, les éprouvettes ne peuvent être confectionnées selon ces orientations. C'est ainsi que les éprouvettes arbitrairement orientées doivent être testées. Les résultats sont à considérer en fonction de l'angle α entre la direction d'application de la charge et le litage. Des travaux dans ce domaine ont été menés dans le passé par John (1969) sur la résistance en compression uniaxiale et par Barla et Goffi (1974) sur la résistance en traction et le module d'élasticité. Actuellement ces travaux sont poursuivis par Kiehl et Wittke qui montrent que le module de cisaillement G_2 peut être calculé à partir des résultats d'essais en compression uniaxiale (figure 2). Un autre essai intéressant est introduit par Amadei et al. qui utilisent des disques fins de roche intacte pour déterminer les paramètres d'anisotropie et la résistance en traction (figure 2).

Il est encore plus important de considérer les propriétés anisotropes du massif rocheux par des essais in situ qu'en laboratoire. Ces essais in situ sont aussi longs que coûteux et ainsi il s'en réalise moins que d'essais en laboratoire. Aussi ils sont plus difficiles à interpréter. Au Congrès SIMR de Montreux (1979), Oberti et al. furent les premiers à prendre en compte l'anisotropie. Ils en tirèrent une théorie pour l'interprétation des essais à la plaque de charge réalisés parallèlement et perpendiculairement à la stratification.

Aujourd'hui de nombreux travaux consacrés à ce domaine sont présentés. Oberti et al. montrent comment l'orthotropie d'un massif rocheux peut être décelée lors de l'interprétation d'essais hydrauliques en caverne. Kiehl et Wittke décrivent des méthodologies qui améliorent l'interprétation des essais au dilatomètre en forage et au vérin plat de grandes dimensions. Des compléments se rapportent aux conditions anisotropes dans l'essai triaxial in situ (figure 3).

La communication de Gonano et Sharp rend compte de nombreuses mesures de contraintes pour un projet en Afrique du Sud. L'intérêt particulier de ces mesures est d'avoir été réalisées dans une roche de faibles caractéristiques ("weak rock") selon la méthode du sur-carottage. Leur analyse comprend une appréciation critique des données de déformation en surcarottage et la détermination des paramètres de la relation contrainte-déformation pour une roche orthotrope à comportement non linéaire, ainsi que la formulation d'une solution numérique.

Des améliorations considérables de la dispersion des résultats (amplitudes et directions des contraintes principales) montrent bien clairement la validité de ces méthodes plus sophistiquées.

Lors de la réalisation d'essais au vérin plat de grandes dimensions dans un tunnel d'essai ou près de la paroi d'une excavation souterraine, on suppose normalement que toute zone autour du point de mesure se trouve dans le domaine élastique. Nous négligeons le fait que la zone proche de la paroi, même si on a tout fait pour éviter toute perturbation inutile est dans bien des cas affaiblie voire sortie du domaine élastique. Borsetto et al. proposent une interprétation basée sur un comportement élastoplastique du massif rocheux. Ils visent à prévenir les erreurs systématiques dans l'interprétation de la cohésion résiduelle et aussi celle des contraintes initiales et de la déformabilité.

Ce travail a été limité à un tunnel circulaire. Il est suggéré aux auteurs de l'étendre en utilisant des méthodes d'interprétation numériques.

Détermination de la résistance et de la déformabilité sous charge à court-terme

Les caractéristiques de résistance et de comportement de la roche intacte sont normalement étudiées en laboratoire. Respectant notre intention initiale de développer plutôt les méthodes de reconnaissance et d'interprétation plutôt que les propriétés des roches, nous discuterons essentiellement des essais et de l'influence des conditions d'essais sur les résultats même des essais.

La résistance en compression de la roche intacte peut être étudiée en conditions uniaxiales (σ_c), biaxiales ou triaxiales (figure 4).

Les essais triaxiaux sont effectués avec deux contraintes principales égales ($\sigma_2 = \sigma_3$) ou différentes. Nous sommes accoutumés à comparer les essais monoaxiaux et les essais triaxiaux (où $\sigma_2 = \sigma_3$) en les reportant sur un même diagramme de Mohr sous forme de cercles de Mohr et à tracer une enveloppe tangente à ces cercles droite ou courbe. Mais plus d'un d'entre nous ne sait pas comment évaluer la 3ème contrainte principale (contrainte moyenne) si σ_2 est différent de σ_3. Rappelons-nous le Congrès SIMR de 1974, où Brown (1974) montra l'influence de la seconde contrainte principale σ_2 sur la résistance du marble de Wombeyant (figure 6), et nous rappelons aussi que les résultats des essais étaient plus influencés par les effets de tête (frettage des extrémités de l'éprouvette) que par σ_2. Les meilleurs résultats furent obtenus en utilisant des plateaux d'appuis en forme de peignes comme l'introduit Hilsdorf pour des essais sur béton. On peut admettre que beaucoup de ces différences mais pas toutes ces différences, peuvent provenir de problèmes de frottement.

Maintenant nous apprenons à partir des essais triaxiaux en compression de Gau et Cheng réalisés avec $\sigma_1 \neq \sigma_2 \neq \sigma_3$ que pour un rapport donné σ_1 / σ_3, la résistance de la roche intacte ne dépend pas énormément de la contrainte principale intermédiaire σ_2 (figures 1 et 2 de leur communication). Ici aussi, l'influence des conditions aux limites peut être sentie, mais elle est bien moindre que celle montrée figure 7.Les travaux de Brown et de Gau et Cheng suggèrent que les dimensionnements ne doivent pas prendre en compte une augmentation de la résistance de la roche intacte liée à l'existence d'une contrainte intermédiaire principale σ_2 supérieure à σ_3. Ceci correspond à la théorie (figure 8).

D'autres conditions limites qui influencent le résultat des essais de compression, surtout dans les roches tendres, sont la teneur en eau et la vitesse de mise en charge. Comme la pression interstitielle diminue quand la vitesse de mise en charge est suffisamment faible, les recommandations de la SIMR préconisent que la rupture apparaisse après 5 à 10 minutes. Ces recommandations ont été élaborées pour des roches résistantes. Elles ne peuvent être appliquées sans réflexions aux roches tendres, dotées d'une plus forte teneur en eau et d'une plus faible résistance : la vitesse de mise en charge doit céder le pas à la vitesse de déformation.

Ainsi dans les roches tendres on observe souvent que deux échantillons apparemment similaires ont des teneurs en eau très différentes (dûe souvent à une porosité plus forte provoquée par l'altération). Dans de tels cas, même si la vitesse de chargement est réduite, des variations

Test		Author (s)
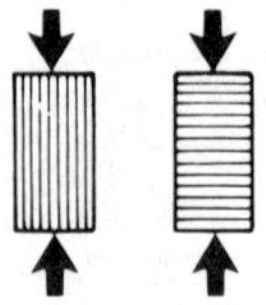	E_1, E_2, ν_1, ν_2	
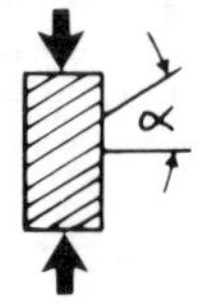	uniaxial compression strength related to α	John (1969)
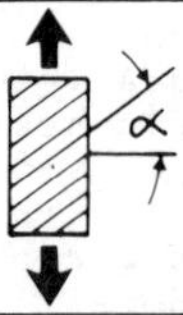	tensile strength	Barla/Goffi (1974)
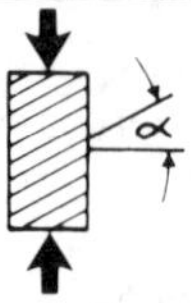	deformability related to α	Barla/Goffi (1974)
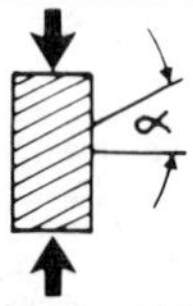	shear modulus G_2	Kiehl/Wittke (1983)
	$E_1, E_2, G_2, \nu_1, \nu_2$ and tensile strength from diametral compression test	Amadei et al (1983)

Fig. 2 Détermination de l'anisotropie en laboratoire

Test		Author (s)
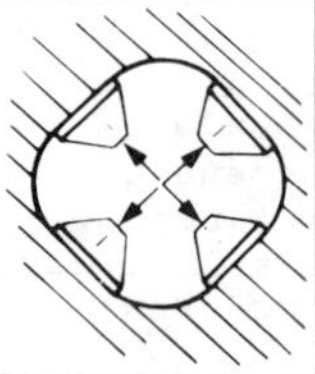	$E_1, E_2, G_2, \nu_1, \nu_2$ from plate bearing test	Oberti et al (1979)
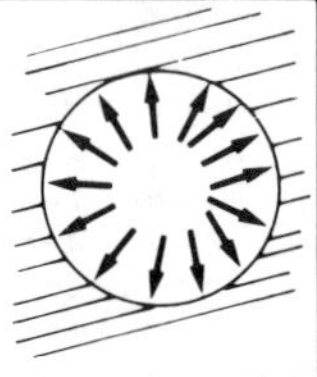	$E_1, E_2, G_2, \nu_1, \nu_2$ from hydraulic chamber test	Oberti et al (1983)
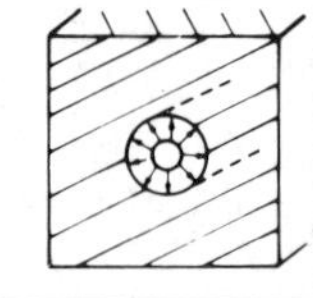	E_1, E_2, G_2 from borehole test	Kiehl/Wittke (1983)
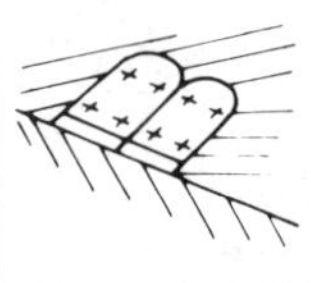	E_1, E_2, G_2 from LFJ-test	Kiehl/Wittke (1983)
	$E_1, E_2, G_2, \nu_1, \nu_2$ from in-situ triaxial test	Lögters/Voort (1974) evaluation improved by Kiehl/Wittke (1983)
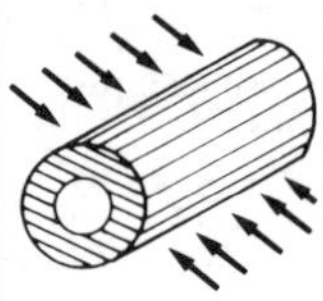	in-situ stresses and non-linear deformability	Gonano/Sharp (1983)

Fig. 3 Détermination de l'anisotropie in situ

dans les résultats d'essais seront observées. Chiu et Johnston montrent ces phénomènes dans l'argilite de Melbourne en compression simple (figure 9).

Durant les essais de compression, on étudie habituellement la déformabilité de la roche intacte. Dans une partie ci-dessus de ce rapport, l'influence de l'anisotropie a déjà été discutée. Si on ne considère maintenant que les matériaux isotropes, il est clair que la déformabilité axiale déterminée pendant l'essai de compression dépend de la pression de confinement. Il faut cependant faire la distinction entre une augmentation due à la contrainte de confinement (similaire à la relation entre module de Young et module de rigidité en mécanique des sols) et une augmentation qui doit être regardée comme une propriété spécifique à une roche particulière. Chiu et Johnston donnent un exemple de ce dernier cas, avec en plus l'influence de la teneur en eau. Des résultats semblables sont fournis par Kwasniewski dans sa figure 7 obtenue pour des échantillons de grès.

Le but de beaucoup d'études est de fournir des équations valables pour exprimer la résistance de la roche intacte. Il faut distinguer entre celles développées à partir de considérations théoriques (tels les critères de Mohr, Coulomb, Griffith et autres) et celles dérivées de résultats d'essais. Des exemples de ce dernier type sont fournis à ce Congrès par Gau et Cheng, Chiu et Johnston, Kwasneiwski. Ces formules

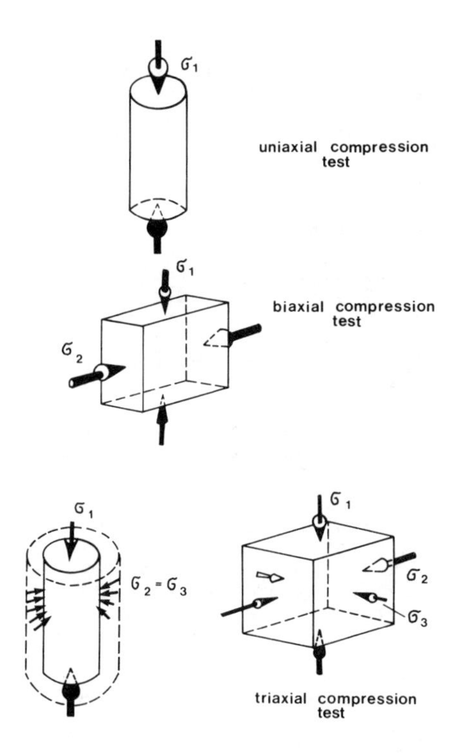

Fig. 4 Types d'essais de compression

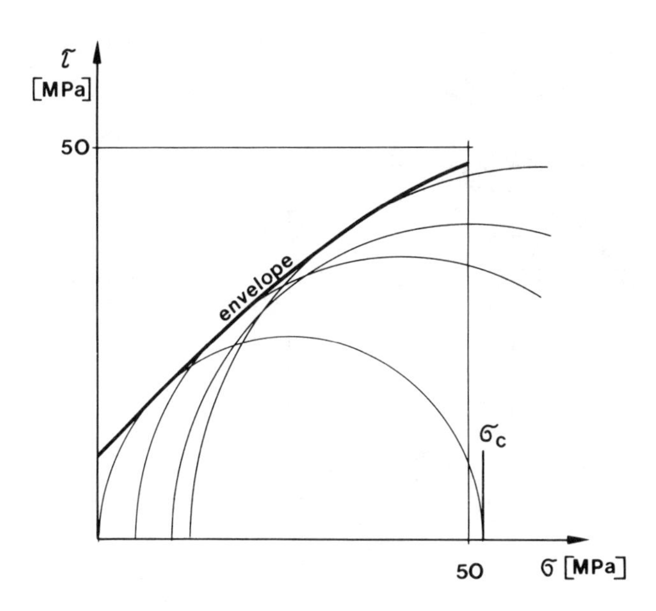

Fig. 5 Critère de rupture (exemple)

peuvent n'être valables que pour les roches particulières testées. Elles sont influencées par les conditions d'essai particulières et c'est pourquoi elles ne peuvent être utilisées pour d'autres matériaux sur d'autres sites que dans certaines limites. Les comparaisons critiques entre les équations de la résistance formulées par divers auteurs se révèlent ainsi extrêmement difficiles. Jusqu'ici, nous n'avons commenté que des essais en laboratoire sur échantillons de roche intacte.

Cependant, il semble important de souligner les développements particuliers concernant les essais triaxiaux en laboratoire d'échantillons de massifs rocheux stratifiés. Un grand nombre de projets sont situés dans des roches tendres finement discontinues dans lesquelles il est souvent difficile de déterminer des valeurs de résistance et de déformation. Pour cette raison, des essais triaxiaux à grande échelle en laboratoire et une technique de prélèvement d'échantillons non perturbés du massif rocheux ont été développés à Karlsruhe. Ces techniques ont été maintes fois présentées dans le passé et le Comité Allemand de Mécanique des Sols et de Fondations en a publié les recommandations d'utilisation. Dans leur papier, Natau et al. présentent des expérimentations récentes.

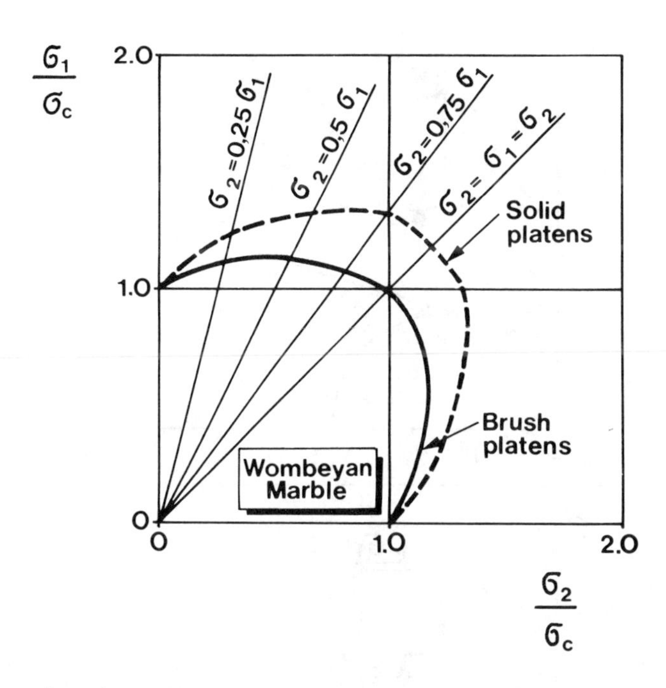

Fig. 6 Influence de la contrainte principale moyenne (d'après Brown, 1974)

<u>Détermination de la résistance et de la déformabilité sous l'action de charges de longue durée et de charges cycliques ; fluage, développement de la rupture</u>

Le rocher intact, sous l'action d'une charge de courte durée, se comporte plus ou moins comme un matériau linéaire élastique, mais lorsqu'il est soumis à des contraintes pendant un temps plus long, on observe des mécanismes très différents. Ceux-ci peuvent être décrits au moyen de différents modèles rhéologiques.

On ignore en général que même le rocher dur peut avoir un comportement visqueux sous l'action de faibles contraintes. Ito a écrit un rapport sur des essais à long terme réalisés sur des poutres en granite. Leur durée d'environ 24 ans en fait des essais remarquables. Il semble que ceux-ci vérifient la prédiction de Nabarro et Herring selon laquelle le granite n'a pas

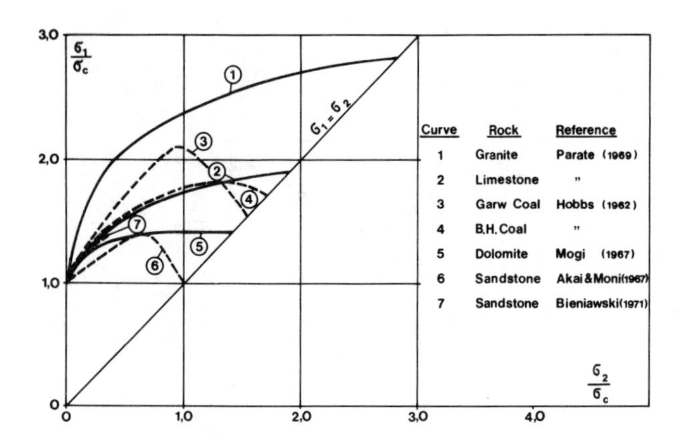

Fig. 7 Résultats d'essais de compression biaxiale compilés dans la littérature (d'après Brown, 1974)

de contrainte limite. Les échantillons testés ont des déformations indépendantes des contraintes, ce qui indique un écoulement visqueux (courbe 1 de la figure 11). Ito a aussi écrit un rapport concernant des essais supplémentaires sur des échantillons de gabbro ; ceux-ci montrent qu'un accroissement de la pression de confinement comme une augmentation de la température réduisent la viscosité. On peut déduire de ces résultats que pour des taux de déformation de 10^{-13}/s, le rocher intact se comporte comme un solide de Hooke alors que pour des taux de déformation moindres, il se comporte comme un fluide newtonien. Il nous semble important, pour la compréhension des processus géologiques, que les valeurs de la viscosité déterminées en laboratoire soient mises à la disposition des géologues et géophysiciens chargés de l'étude des déformations de la croûte terrestre.

Naturellement, les phénomènes visqueux dans le rocher dur se développent très lentement et n'ont donc que peu d'importance pour les travaux de construction dans le rocher. Cependant, on ne peut négliger le taux (vitesse) de chargement.

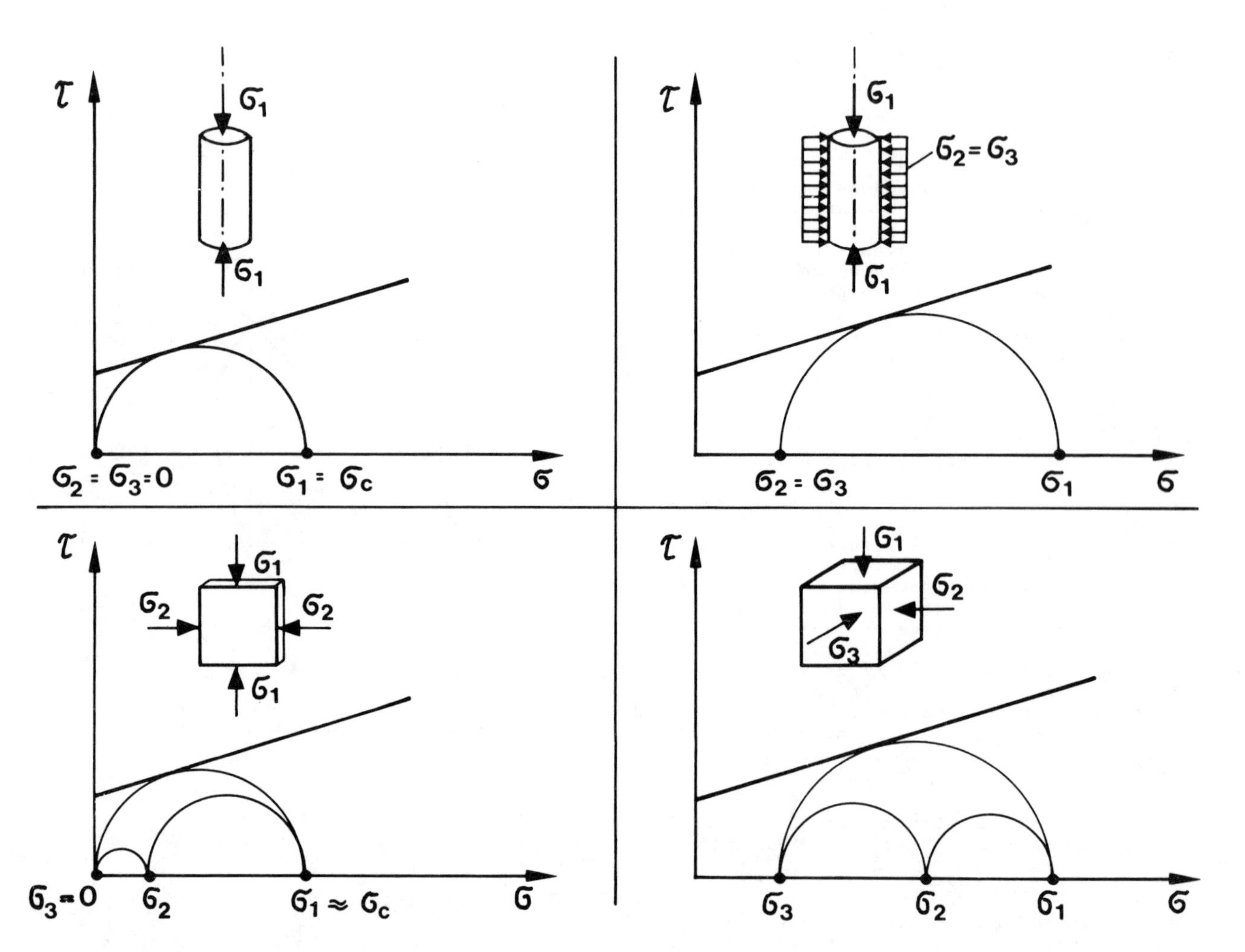

Fig. 8 Diverses états de contraintes à la rupture

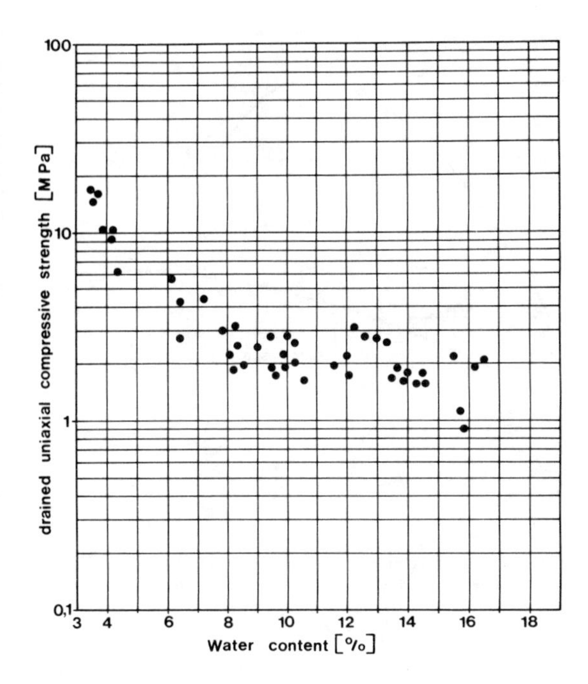

Fig. 9 Influence de la teneur en eau de l'argilite de Melbourne sur la résistance en compression uniaxiale (d'après Chiu et Johnston)

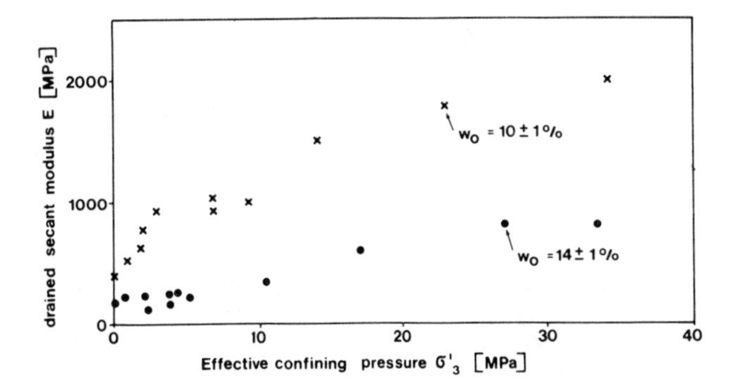

Fig. 10 Influence de la pression de confinement sur la déformabilité de l'argilite de Melbourne (d'après Chiu et Johnston)

Il est bien connu que le taux de chargement a une influence sur la résistance dans un essai de compression uniaxiale. De plus, la déformation à la rupture dépend du taux de chargement puisque la déformation a une composante élastique, une composante visco-élastique (fluage réversible) et une composante plastique (fluage irréversible). Plus le taux de chargement est lent, plus le fluage influence la déformation totale. La courbe 2 de la figure 11 illustre un tel comportement. Il est bien connu qu'au delà d'un niveau de contrainte dit critique, la déformation qui dépend du temps, conduit à la rupture. On doit par conséquent distinguer le fluage primaire (vitesse décroissante) du fluage secondaire (vitesse constante) du fluage tertiaire (vitesse croissante). Le fluage cesse finalement à des contraintes plus faibles et la déformation atteint une valeur finale (fig. 12). Le rocher intact réagit d'une manière similaire lors d'un chargement cyclique. Nous sommes heureux de posséder sur ce sujet deux articles très intéressants, l'un de Homand-Etienne et al. (granite et marbre) et l'autre de Akai et Ohnishi (Tuf de Funyu). Ainsi que pour le fluage, on peut définir trois phases distinctes de comportement sous chargement cyclique (fig. 13 a). On peut, par conséquent, supposer que ces deux modes de comportement sont liés.

On peut expliquer ces deux phénomènes par le développement progressif de microfissures qui s'amplifient avec la durée (ou le nombre de cycles) et l'amplitude du chargement. La fissuration stoppe en dessous d'un niveau de contrainte critique, elle accélère au-delà de ce niveau. Homand-Etienne et al. ont découvert, à travers leurs essais, que les résistances à long terme du granite et du marbre étaient respectivement égales à 85 % et 82 % des résistances à court terme. On sait que l'on obtient des résultats analogues avec le béton.

La théorie du développement de microfissures peut être utilisée dans une modélisation mathématique, statistique. Weber et Saint-Lot introduisent un tel modèle dans leur article. Il faut mentionner, que, malgré certaines simplifications, ce modèle est capable de simuler les courbes contrainte-déformation mesurées au cours d'essais.

Les phénomènes conduisant à la rupture, discutés précédemment peuvent aussi être observés avec des échantillons de masse rocheuse. Là aussi, différentes phases peuvent être définies durant un chargement de longue durée ou cyclique. Müller et Ge le démontrent clairement par le comportement de quelques échantillons in situ testés pour le barrage de Kurobe. La première phase est liée au compactage initial de la masse rocheuse jointée. Müller et Ge montrent que la seconde phase est une phase de stabilisation relative, la rupture s'amorçant au cours de la troisième phase. On estime très important que ces trois phases soient reconnues, en particulier pour des essais sur des échantillons de grandes dimensions car une estimation de la sécurité basée sur de tels essais devrait découler principalement du comportement relatif aux déformations de la masse rocheuse.

<u>Analyse de la résistance au cisaillement des joints</u> (= des discontinuités)

Au Congrès SIMR de 1974, un sous thème entier a été consacré aux essais de cisaillement. A Montreux aussi en 1979, des contributions fondamentales furent soumises sur ce sujet.

La résistance au cisaillement des joints est un sujet assez compliqué qui n'est certes pas encore entièrement clarifié. Quelques remarques générales doivent tout d'abord être faites à ce sujet : si un milieu continu est soumis à des contraintes de cisaillement, on observe le comportement schématisé dans la figure 14 a. La courbe τ-Δ h montre une résistance au cisaillement maximale τ_t très distincte. Dans le diagramme τ-σ , τ_t représente un point du critère de rupture. Si les contraintes de cisaillement appa-

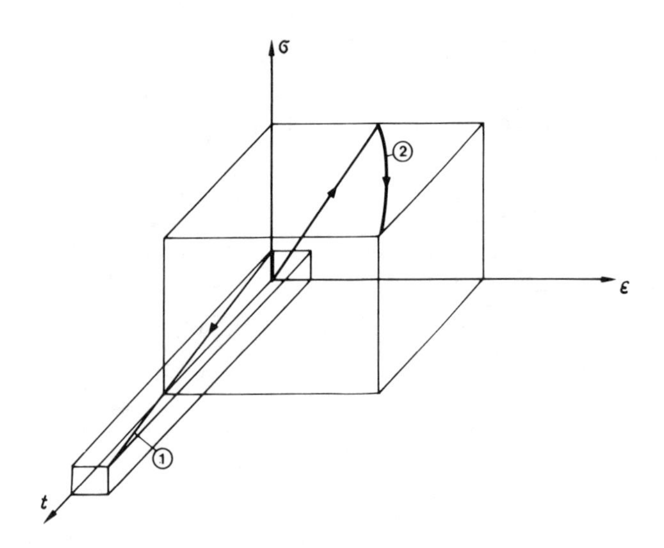

Fig. 11 Diagramme σ - ε - t ; ① écoulement visqueux ; ② fluage

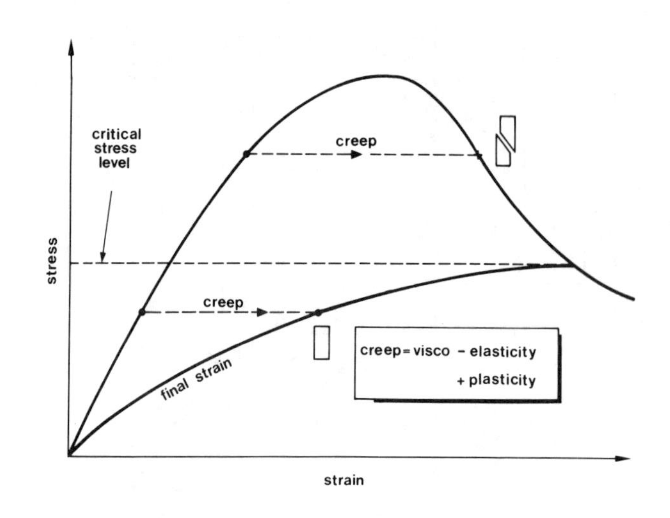

Fig. 12 Comportement contrainte-déformation sous une charge de longue durée

raissent le long d'un joint régulier (fig. 14 b), on peut observer un comportement très différent. Là, les déplacements dûs au cisaillement augmentent de façon continue mais progressive lorsque les contraintes de cisaillement τ augmentent. A la contrainte τ_r , le déplacement Δ h augmente sans accroissement additionnel de la contrainte de cisaillement. Ainsi qu'il a été mentionné auparavant, il n'y a pas de déplacement vertical Δv dans chacun des deux modèles idéalisés.

Au contraire, un déplacement vertical s'effectue dans les modèles des figures 14 c et d qui ressemblent plus à des joints naturels. Ici le comportement est déterminé ou plutôt idéalisé. - par des aspérités ou des dents. Aussi longtemps que les dents ne sont pas abimées les déplacements sont possibles que lorsque le solide supérieur glisse le long des pentes des dents du solide inférieur. On observe donc, en plus de Δ h, un déplacement Δ v normal du plan principal de cisaillement. Le critère de rupture dans le diagramme σ-τ , correspondant à une pente d'angle i (fig. 14 c) est incliné à un angle ϕ + i par rapport à l'axe Au delà de certaines contraintes τ et σ , la dent rompra. La figure 14 d montre cet état et ses conséquences. On peut voir que les déplacements à cet instant tendent à s'effectuer dans des plans de plus en plus parallèles au plan principal de cisaillement. Le critère de rupture (dans le diagramme τ-σ) devient de plus en plus plat et se rapproche de la valeur ϕ .

Naturellement, les surfaces de cisaillement sont très irrégulières dans la réalité. On ne peut donc s'attendre, à obtenir ces courbes idéales. On obtient au contraire des résultats assez différents selon la rugosité, la résistance de la roche intacte, les conditions d'altération, les revêtements, la minéralogie, la teneur en eau et le degré de saturation en eau de la masse rocheuse. La fig. 15 montre un exemple publié par Barton (1971). Naturellement, les comporte-

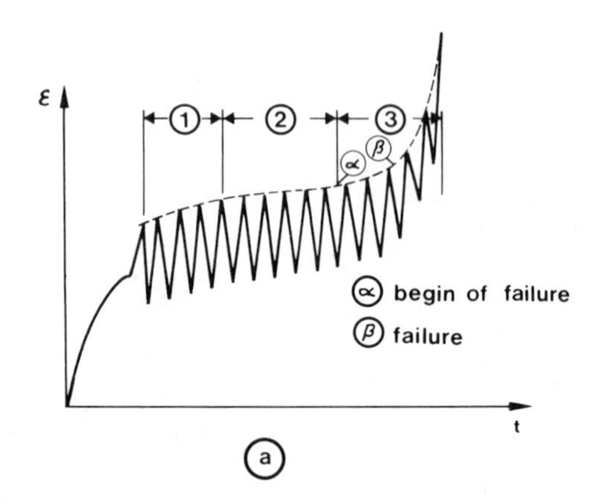

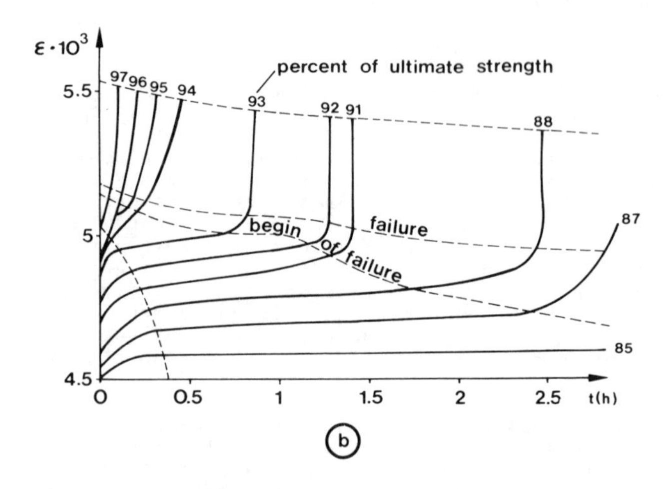

Fig. 13 Comportement sous charge cyclique (d'après Homand-Etienne et al.)
① taux décroissant
② constant ③ croissant

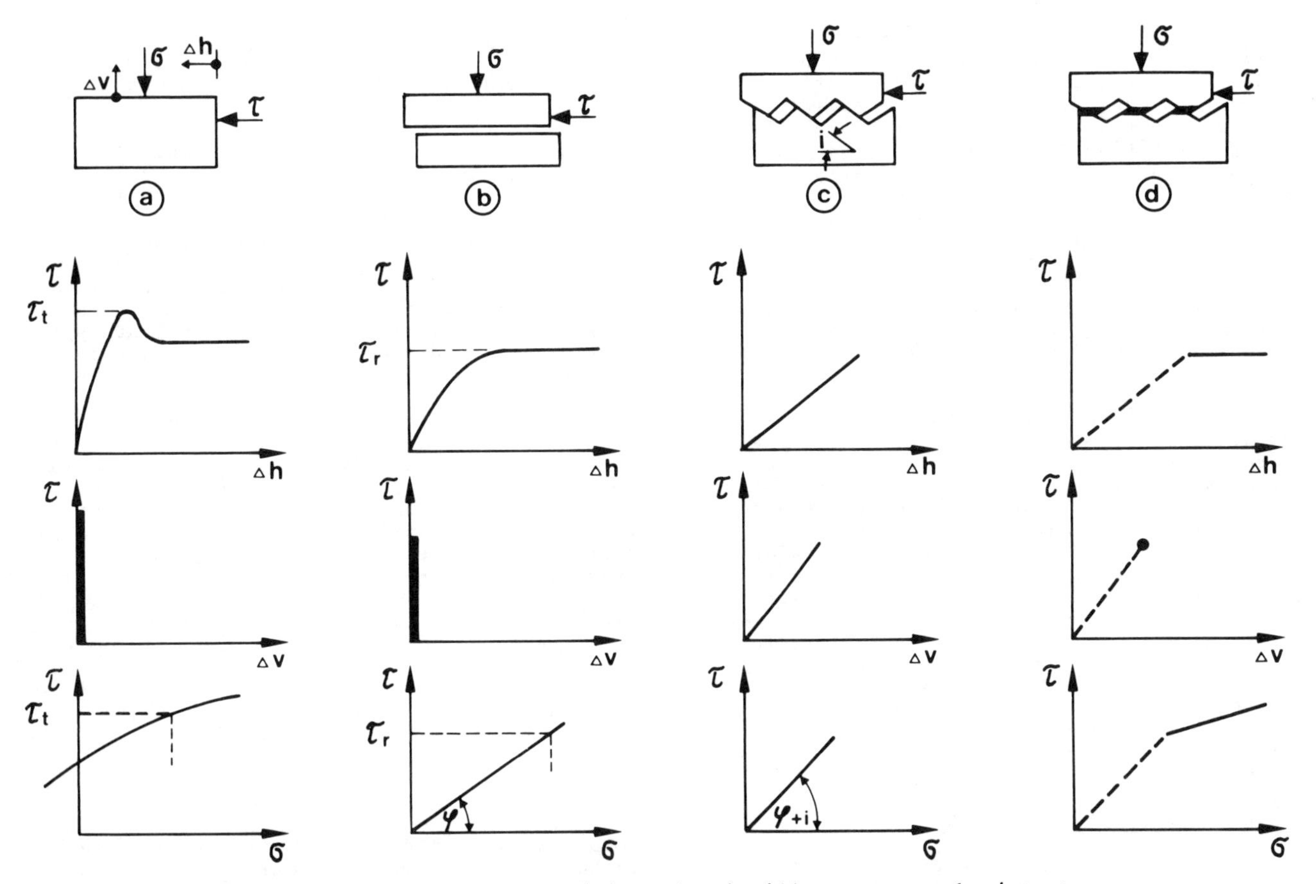

Fig. 14 Comportement dans différentes conditions de cisaillement a roche intacte
b joint lisse c joint rugueux avant d joint rugueux après rupture des dents)

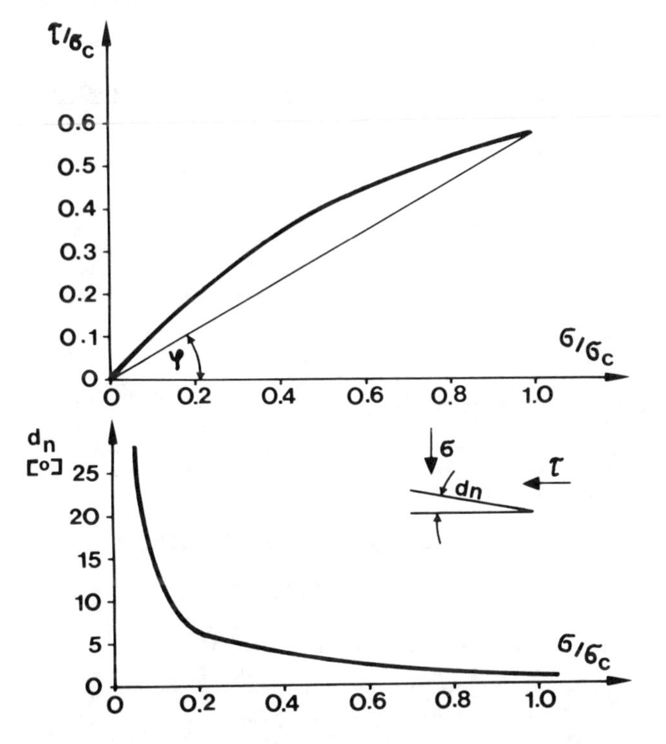

Fig. 15 Comportement au cisaillement d'un joint naturel (d'après Barton, 1971)

ments des joints ne seront pas non plus les mêmes selon que les joints auront ou non été soumis auparavant à des déplacements relatifs, c'est-à-dire selon que leurs dents ou aspérités auront ou non déjà été cisaillées. Par conséquant, dans le présent état de nos connaissances, nous devons nous attendre à des conditions différentes de résistance au cisaillement dans chaque cas.Des exemples de ce qui précède sont présentés dans les articles de Tinoco et Salcedo, Everitt et Goldfind et Bollo et al.

Les effets d'échelle

Dans une roche dure jointée, il n'y a souvent que les résistances au cisaillement et à la traction qui sont influencées de façon significative par les joints. La déformabilité n'est, en comparaison, que faiblement accrue. Mais dans une roche tendre, surtout si elle a des joints très rapprochés, l'influence des joints est dominante. Des espacements entre joints de quelques centimètres seulement font que la masse rocheuse se comporte plutôt comme de la maçonnerie sans eau, et même quelquefois comme de la maçonnerie faite avec des briques tendres et des fissures lubrifiées. Chaque "brique" dispose d'un espace notable pour se déplacer et tourner. La résistance est donc diminuée et la déformabilité accrue, comparées à celles de la roche intacte.

Mais il est assez difficile de déterminer ces variations de façon sûre car un tel corps jointé peut être, au sens de la cinématique, un système très "mobile". La résistance et la déformabilité d'un tel système dépendent alors beaucoup des conditions aux limites de l'essai. On pense donc que des essais de compression uniaxiale (par exemple) donneront probablement des résultats déviant énormément de ceux d'essais triaxiaux et que les résultats de ces deux types d'essais différeront plus ou moins selon le comportement de la roche naturelle. On apprécierait que les auteurs (Natau et al., Sun et Zhou) qui ont fait des contributions à ce sujet, fassent des commentaires sur cette question.

Alors qu'on observe des effets notables d'échelle dans des essais de compression (résistance et déformabilité), ces effets semblent être parfois bien moindres dans les analyses de la résistance au cisaillement des joints. Pasamehmetoglu et al. et Bollo et al. le montrent très clairement. On doit conclure des articles et encore plus du travail antérieur de Leichnitz et Natau (1979) que les effets d'échelle sont observés essentiellement en liaison avec les résistances de cisaillement maximales des joints rugueux. Dans le cas présent, les aspérités sont, sans aucun doute, distribuées arbitrairement sur le plan du joint, mais dans un petit échantillon à cisailler, il peut y avoir éventuellement une dent pointue qui ne correspond pas à la rugosité moyenne.

Essais d'ancrage dans le rocher

Les essais d'ancrage dans le rocher peuvent être de natures très différentes. Ils peuvent être destructifs ou non. On peut les faire en tant qu'essais d'aptitude ou dans un but d'inspection ou au-delà dans un but de reconnaissances fondamentales. Les articles qui ont été soumis sur ce sujet traitent des reconnaissances fondamentales pour d'importants ancrages dans le rocher (Natau et Wullschläger) et de modèles réduits de laboratoire (Ballivy et al.) mis au point pour des inspections de la qualité d'ancrages courts injectés dans le rocher.

Il est essentiel pour la performance à long terme d'un ancrage dans le rocher que la partie scellée (barres d'acier et injection) ne subisse aucun dommage. L'état de contraintes sur la longueur de scellement est particulièrement important puisqu'il doit rester inférieur à un certain niveau afin qu'aucune fissure ne se développe dans le coulis. Les contraintes sont influencées par la déformabilité de la masse rocheuse environnante. Natau et Wullschläger simule la masse rocheuse dans leurs essais par un tube en acier rempli de béton. Ballivy et al. utilisèrent des carottes de roche intacte de petit diamètre.

Bergman et al. introduisent un développement intéressant. Le "boltometer", un appareil utilisé dans l'essai d'ancrages courts injectés dans le rocher, est basé sur la mesure de la vitesse de propagation et de l'amortissement des ondes dans la barre d'acier. L'amplitude décroit plus ou moins selon la qualité du coulis. Nous pensons que le "boltometer" deviendra un bon outil mais comme le démontrent les exemples, il ne nous semble pas encore s'être totalement développé.

Sous-thème A.3

CLASSIFICATION, PREDICTION, OBSERVATION ET AUSCULTATION

Depuis le système publié par Terzaghi (1946), il y a une prolifération de méthodes de classification des masses rocheuses. Celles-ci ont été mises au point pour aider à la communication entre géotechniciens et faciliter les prédictions du comportement des masses rocheuses. En raison toutefois du grand nombre de méthodes de classification existant aujourd'hui dans la littérature, les ingénieurs qui désirent appliquer de telles méthodes risquent d'avoir des difficultés pour choisir celle qui convient à leur cas particulier. Il est très clair d'après les articles sur ce sujet que ces difficultés existent réellement. Vos Rapporteurs Généraux ont toutefois l'impression que nous devrions essayer durant ce Congrès, de parvenir à une meilleure compréhension des forces et faiblesses des systèmes existants et si possible arriver à un certain degré de consensus sur leur applicabilité.

Ayant discuté de la prolifération actuelle, c'est avec quelque impatience que nous soulignons la nécessité d'une autre classification traitant de l'appartenance des classifications elles-mêmes à divers groupes, c'est à dire en classifiant les classifications !

Catégories de classifications

Nous nous référerons dans ce qui suit aux deux composantes principales de toutes les masses rocheuses en tant que substances (matériaux) et défauts (discontinuités) tout en reconnaissant que les fluides dans les défauts et/ou les pores de la substance ont aussi leur importance et sont pris en compte dans certains des systèmes.

A. Classification des propriétés des composantes

Dans ces classifications, des noms, lettres et nombres caractérisent des intervalles de valeurs d'une seule propriété d'un composant.

Les intervalles adoptés sont généralement ceux qui se sont avérés utiles dans l'expérience pratique. L'utilisation de la classification choisit le nombre approprié à partir des résultats d'essais simples ou de mesures. Comme exemples de ce type de classification, il y a la classification des substances rocheuses par l'index de résistance à une charge ponctuelle (Broch et Franklin, 1972) et les quatre classifications de composantes dans la "Description géotechnique fondamentale des massifs rocheux" de cette Société (SIMR, 1980, Tables 1 à 4).

B. Classification de propriétés multiples

Ce sont des systèmes dans lesquels une zone réellement uniforme de la masse rocheuse peut être caractérisée, du point de vue de ses propriétés générales, par la liste d'un nombre de termes décrivant les propriétés des composants. Le "BGD" description géotechnique fondamentale des masses rocheuses" est un système dans lequel la masse rocheuse est caractérisée par un nom de roche génétique et des lettres et nombres descriptifs décrivent quatre caractéristiques des composantes. Le but essentiel de ce système et d'autres systèmes analogues est de fournir aux géotechniciens un langage standardisé.

C. Classifications pondérées de propriétés multiples

Dans cette catégorie, les propriétés de la masse rocheuse sont décrites par plusieurs propriétés des composantes comme en B ci-dessus ; puis en associe à ces propriétés des composantes une "note d'évaluation" fonction du jugement que l'on porte sur leur degré de contribution à la résistance de la masse rocheuse. Ces jugements sont basés sur l'analyse de l'expérience, des essais et des calculs. La somme de ces notes constitue la note totale attribuée à la masse rocheuse.

Les systèmes de Barton et al. (1974) et Whickam et al. (1972)ont été spécialement conçus pour évaluer les nécessités de soutenir des ouvertures souterraines. Le système RMR (Jugement de valeur des masses rocheuses) de Bieniawski (Bieniawski, 1974 et 1979) recherche une utilisation plus générale. Ces trois systèmes ont été utilisés avec succès pour prédire la nécessité de soutenir des tunnels. On a montré que le Q de Barton et le RMR de Bieniawski sont tous deux bien corrélés aux modules de déformation in situ (Bieniaswski, 1978).

D. Classification des masses rocheuses altérées

De nombreuses classifications ont été mises au point dans des environnements géologiques particuliers, pour décrire, dans des buts divers, les masses rocheuses altérées. Un exemple en est le système, en usage ici à Melbourne, dans lequel le profil d'altération de l'argilite du Silurien est classé par zones (Neilson, 1970). Saunders et Fookes (1970) décrivent divers systèmes développés pour les roches qui se sont altérées sous différents climats, et Sancio et Brown (1980) décrivent un système de classification des masses rocheuses altérées servant à prédire leur potentiel à l'instabilité dans des pentes.

La prolifération de ces classifications nous cause certains soucis. La situation est analogue à celle du début de ce siècle, lorsque les pétrologues produisirent une prolifération de classifications des roches cristallines et des noms de types de roches qui, aujourd'hui encore, créent des confusions parmi les géologues, sans parler des ingénieurs ! A une époque où il est à la mode de rationaliser, pouvons-nous rationaliser l'usage que nous faisons des classifications géotechniques existantes ainsi que notre approche du développement de classifications nouvelles ?

Se penchant tout d'abord sur la catégorie A, les Rapporteurs Généraux suggèrent qu'il devrait être possible d'adopter des regroupements standardisés, pour le monde entier, des propriétés suivantes des composants :

. résistance de la matrice rocheuse

. épaisseur des bancs

. intervalle entre les discontinuités

. résistance des discontinuités

. eaux souterraines.

Si les propriétés de la catégorie A étaient standardisées, les systèmes de la catégorie B le seraient alors automatiquement. Nous aurions ainsi un langage relatif aux masses rocheuses d'une utilité peut être équivalente à la "classification des sols unifiée". La plupart des gens sérieusement impliqués dans les études de sols et travaux d'exécution sont d'accord avec nous sur le fait que la classification des sols unifiée est un langage international de très grande valeur.

La catégorie D est sans aucun doute celle qui a le plus donné lieu à la prolifération. Cela est probablement dû en partie à l'extrême variabilité des masses rocheuses altérées, à la fois au sein d'un environnement géologique particulier et d'un environnement géologique à un autre. Un autre facteur ayant contribué à cette prolifération pourrait être le désir forcené de nombreux géologues de classifier. Nous reconnaissons que beaucoup de ces classifications sont utiles dans les environnements géologiques restreints dans lesquels elles se sont développées mais nous considérons qu'aucune ne devrait être adoptée pour une utilisation universelle.

La valeur potentielle des classifications de la catégorie C ne fait aucun doute ainsi que l'illustre la déclaration suivante de Hoek et Brown (1980) :

"Etant donné la rareté des renseignements dignes de confiance dont on dispose sur la résistance des masses rocheuses et le coût très élevé que l'on paie pour obtenir de tels renseignements, les auteurs sont d'avis qu'une analyse complète de la résistance des masses rocheuses ne sera probablement jamais possible. Puisque ceci est une des questions clés de l'étude des roches appliquées au Génie Civil, il faudra s'efforcer de définir les grandes tendances permettant d'estimer la résistance du massif rocheux. "Considérant plusieurs alternatives, les auteurs se sont tournés vers les méthodes de classifications des masses rocheuses...".

Dans les pages qui suivent cette citation, Hoek et Brown donnent des équations approchées de la résistance des massifs intacts et jointés pour les types les plus usuels de roches, reliant celles-ci à l'ensemble des indices du RMR (Bieniawski) et Q (Barton et al.) dans chaque cas. Hoek et Brown soulignent le fait qu'on ne devrait appliquer leurs équations que pour les études de conception préliminaires et de sensitivité. Toutefois, pour les raisons exprimées dans la citation ci-dessus, de telles relations empiriques et classifications des masses rocheuses du type RMR et Q trouveront, probablement dans l'avenir un large usage pour l'étude des roches et constructions dans le rocher.

Articles liés directement à la classification et aux prédictions

Cinq articles se rattachent à ces sujets. Kendorski et al. décrivent, de façon relativement détaillée, les bases et la méthode d'application d'un système de la catégorie C aux prédictions des conditions dans lesquelles des galeries de production (tunnels) réalisées dans des mines sujettes à éboulement requièrent un blindage pendant les phases provisoires et en service. Cet excellent article illustre les avantages qu' apporte la recherche lorsqu'elle est menée par une équipe bien conçue, il reflète la bonne compréhension des auteurs de la géologie minière et de la mécanique des roches (Théorie) ainsi que leur appréciation du comportement réel des roches et de la pratique minière, qui ne peuvent s'acquérir que par une expérience propre des travaux souterrains.

Il est particulièrement satisfaisant de voir que la caractérisation de la résistance de la masse rocheuse se fait par une opération initiale indépendante et que seulement après on considère les effets des charges induites par les activités de génie civil au moyen d'une suite de réajustements. On voit immédiatement sur leur figure 1, que ce concept a permis aux auteurs d'appliquer leur système de manière logique à la prédiction des nécessités d'un blindage aussi bien dans la phase initiale (phase de construction) que dans la phase suivante (phase de service). Il nous semble que Kendorski et al. peuvent avoir fourni le type de "nouvelle approche" que Lokin et al. recherchent, dans leur article. Nous en discuterons plus loin lorsque nous parlerons de l'article de Cokin et al.

Nous reférant à nouveau à la figure 1 de Kendorski et al., nous notons que de façon logique l'orientation des fractures est maintenant prise en compe dans les réajustements plutôt que regardée comme un élément caractérisque de la masse rocheuse. Ce que les auteurs expriment ici est que le réseau de fractures est une caractéristique intrinsèque de la masse rocheuse, mais que les orientations des fractures liées à l'axe des galeries dépendent des directions suivant lesquelles on a foré ces galeries, c'est-à-dire qu'elles sont une donnée imposée par les travaux de génie civil.

Nous sommes incapables de comprendre totalement la table IV, "Evaluation A_o de l'orientation des fractures d'après l'observation directe dans la galerie". Nous supposons que le "nombre de fractures définissant le bloc" se rapporte à un affleurement bi-dimensionnel dans une galerie. S'il en est ainsi il serait probablement nécessaire de connaître l'orientation de l'affleurement, mais les auteurs n'affirment rien de tel. Nous suggérons que les auteurs fournissent un diagramme pour expliquer comment on obtient A_o à partir de la table IV.

Se référant à nouveau à l'évaluation A_o, nous remarquons que les auteurs recommandent d'utiliser la table IV en priorité, si on ne dispose pas d'observations directes d'affleurement. La table V qui contient beaucoup plus de données complètes ne doit être utilisée que s'il n'existe aucune exposition mais que l'on connait les zones tendant à se fracturer". Nous ne voyons pas où se trouve la logique dans ce choix des auteurs. Nous considérons que tout ingénieur ou géologue expérimenté pourrait déterminer immédiatement A_o à partir des expositions et de la table V. Ceci se ferait simultanément avec les observations qui servent à évaluer les dégâts provoqués par le dynamitage (table III) et les personnes expérimentées, sans aucun doute, prendraient en compte les effets néfastes de l'orientation lorsqu'ils décideraient quelles valeurs de B choisir d'après les types d'observations listés dans cette table.

Les Rapporteurs Généraux considèrent que le système MBR a un potentiel considérable et nous attendons avec impatience les publications futures qui fourniront les réactions sur sa performance.

Loking et al. ont mis dans leur table 1 une liste utile des éléments de neuf classifications existantes des masses rocheuses, de notre catégorie C. Ils discutent ensuite de sept limitations qui d'après eux restreignent l'applicabilité des systèmes existants. Alors que nous reconnaissons que certaines des classifications simples ne sont pas suffisamment complètes, nous suggérons que la plupart des limitations listées ne concernent pas nécessairement directement les systèmes plus complets (par exemple RMR et Q), mais plutôt l'organisation de leur application. Comme nous l'avons affirmé auparavant dans notre discussion sur le système MBR, il nous semble probable que nous devons à Kendorski et al. le 1er exemple présenté dans ce Congré et pressenti par Lokin et al, de cette nouvelle approche.

Après un commentaire sur les limitations, Lokin et al. continuent en discutant les principes d'une nouvelle approche du problème de classification. La plupart des Ingénieurs ayant de l'expérience dans la planification de projets devraient être d'accord avec la plupart des principes énoncés liés aux études géotechniques et à leur intégration à d'autres études faites pendant la reconnaissance, l'étude, la construction et la surveillance.

Dans cette section "Principes de base", les auteurs expliquent que la classification des roches pour un projet précis doit inclure une séparation de l'aire concernée en zones quasiment homogènes, de préférence au niveau de la classe rocheuse. Ils affirment ensuite qu'en ce qui concerne la procédure de séparation en zones : "elle doit pour avoir tout son sens, couvrir tout le volume soumis à la reconnaissance avec à peu près le même niveau de crédibilité". Les Rapporteurs Généraux suggèrent que la précision comme le niveau requis de crédibilité devraient souvent varier selon les exigences du projet de génie civil et de l'environnement géologique. Par exemple, la reconnaissance du tracé d'un long tunnel sous-pression, tout spécialement dans une zone de pentes abruptes et un long passé d'altération devrait normalement exiger que les extrémités du tunnel soient étudiées plus amplement que les autres sections. Ceci provient de la nécessité de choisir l'implantation des zones d'entrée et la sortie du tunnel afin d'éviter les zones où la roche est perturbée ou fortement altérée et d'évaluer la stabilité et la résistance des zones superficielles de couverture en relations avec les conditions d'étanchéité du revêtement.

Les auteurs listent dans la table III toutes les données qu'ils proposent d'inclure dans leur classification, à différents stades de réalisation des projets. C'est une liste plutôt complète des aspects géotechniques et autres aspects en relation avec la géotechnique qui nécessitent une reconnaissance pour les projets importants, couvrant des sujets tels que l'activité sismique (néotectonique) température de la roche, hydrologie souterraine comprenant la densité et la température des eaux, etc... A notre avis lorsque les auteurs disent que cette classification sera basée sur tous ces types de données, ils ne se réfèrent peut être pas à la seule classification de type catégorie C, mais pensent à une approche de l'étude au moyen d'une suite de classification.

Nakao et al. discutent des limitations possibles du système reconnu de Barton et al., Wickham et al. et Bieniawski lorsqu'on l'utilise pour prédire les nécessités de blindage de tunnel au Japon. Ils se réfèrent à des travaux antérieurs suggérant que ces systèmes confirmés développés pour des roches dures, stables du point de vue de la géomécanique, n'étaient pas directement applicables au Japon où il existe de très nombreux matériaux rocheux de faible résistance, dans un état très fracturé. Ils décrivent la méthode et les résultats d'études statistiques de données obtenues pour des tunnels terminés ; par cette approche, ils visent à préparer un schéma de classification des masses rocheuses adapté à l'état des roches au Japon.

Le comportement des roches dans les tunnels fut évalué au moyen de mesures de convergence après excavation. Les résultats montrèrent que tous les paramètres utilisés dans les trois systèmes de classification des masses rocheuses sont raisonnablement applicables aux conditions rencontrées dans les tunnels étudiés. Toutefois les auteurs concluent que l'épaisseur du terrain de couverture ou la largeur du tunnel a plus d'effet sur le comportement du tunnel que les autres facteurs. Cette épaisseur ou largeur remplace, en particulier, le degré de fracture (avec le RQD et la condition de joint) qui est en tête de liste dans les systèmes confirmés.

Les Rapporteurs Généraux sont à même de juger que la variabilité dans l'espacement et la condition des joints aura une plus faible signification au Japon où, comme le disent les auteurs, la plupart du rocher est en fait situé dans une zone de failles. Nous pouvons voir aussi que la largeur du tunnel aura un effet important sur les déformations du tunnel là où les résistances de la substance comme de la masse rocheuse sont généralement faibles. Nous suggérons cependant que les auteurs pourraient suivre l'approche de Kendorki et al., et traiter séparément la largeur du tunnel ou l'épaisseur du terrain de couverture comme une partie des activités de génie civil, qui imposent des charges sur la masse rocheuse. Nous approuvons l'intention des auteurs d'inclure, dans les études à venir, des paramètres de performance autres que les déformations du tunnel.

Mizuno et al. esquissent une classification des masses rocheuses de la catégorie B, utilisée pour caractériser les fondations de barrages au Japon. Ils tentent ensuite de déterminer les relations entre les diverses classes de roches et les résultats d'essais de cisaillement direct réalisés sur les sites de nombreux barrages en béton. La dispersion relativement élevée des résultats d'essais de cisaillement direct dans chaque classe examinée est légèrement réduite lorsqu'on retrace les données après avoir subdivisé les classes selon l'origine géologique. Cette dispersion reste cependant encore assez élevée, surtout pour les roches sédimentaires et métamorphiques et les auteurs relient cette variation à l'influence des configurations des couches rocheuses et des fractures sur les sites individuels. Nous sommes d'accord avec cette conclusion.

Yudbir et al. ont réalisé des essais en laboratoire sur des échantillons qui visaient à simuler à la fois le rocher intact et le rocher fracturé. Ils se sont servis de leurs propres résultats d'essais et d'autres résultats pour examiner si les équations de résistance au cisaillement de Bieniawski (1974 b) et Hoek et Brown (1980) pouvaient prédire les surfaces de rupture à la fois dans le domaine des ruptures fragiles et dans le domaine des ruptures ductiles. Ils concluent que l'équation de Bieniawski :

$$\frac{\sigma_1}{\sigma_c} = A + B\left(\sigma_3/\sigma_c\right)^{\alpha}$$

pourrait être envisagée pour faire de telles prédictions, le paramètre A variant de 0 pour les roches complètement désintégrées à 1.0 pour les roches intactes et le paramètre B étant une propriété du matériau rocheux (substance), supposée reliée à la composition du minéral et à la structure de l'édifice (microstructure). Ils proposent quelques valeurs types du paramètre B et des relations possibles entre le paramètre A, le Q de Barton et al. et le RMR de Bieniawski.

Articles reliant la microstructure aux propriétés des roches

L'article de Pan et al. décrit tout d'abord les relations entre la composition minéralogique et la microstructure de 2 variétés d'arkose (grès riche en feldspath) et les propriétés caractéristiques y compris masse volumique, porosité, absorption d'eau, résistance à la compression simple et la diminution de la résistance après immersion dans l'eau. L'arkose avec liant de carbonate s'est révélée plus résistante, plus dense et moins sujette à la diminution de résistance que l'arkose avec liant d'argile. L'influence d'oxygéne dilué dans l'eau et de différentes durées de trempage a aussi été testée en laboratoire. On a découvert que l'oxygéne dilué entraine l'oxydation des pyrites, du grès et la formation d'acide sulfurique qui attaque le liant de carbonate, d'où une plus forte diminution de la résistance après immersion dans l'eau.

A la lumière de ces résultats de laboratoire, des essais souterrains ont été réalisés dans des houillères, ils montèrent qu'une injection d'eau par le toit faite avant l'exploitation des mines avait un certain nombre d'effets bénéfiques, y compris une diminution des éboulements moins de poussière au front d'attaque, moins de dégâts dans les blindages, et moins d'écaillages.

Rippa et Vinale décrivent l'origine, la composition minéralogique et la microstructure des échantillons de tuf à partir de deux unités géologiques qu'ils relient alors aux propriétés de résistance et de déformation des substances rocheuses en compression uni directionnelle. Les tufs jaunes et gris (unités) contiennent tous deux des substances de tuf qui peuvent être rangées dans 4 groupes selon l'abondance et les dimensions des inclusions de pierre ponce.

Le programme d'essai montra que la résistance à la compression simple du tuf de chacune des 2 unités :

. décroit lorsque la proportion d'inclusions de pierre ponce augmente ,

. décroit (jusqu'à 30 %) lorsque les échantillons sont saturés en eau.

Les résultats d'essais suggèrent aussi que le tuf jaune a une variabilité horizontale et une variabilité verticale de couche à couche.

On montre que la microstructure des variétés de tuf, déterminée avec un microscope à balayage a une influence marquée sur les propriétés de déformation.

Baklashov et al ont écrit des rapports sur leurs études des mécanismes de déformation et ruptures dans la matrice et le massif rocheux. Ils font une liste, tout d'abord, de ce qu'ils regardent comme dix faits généralement acquis et résultant d'observations de mécanismes de rupture des roches. Ils discutent ensuite des théories existantes des ruptures des roches et découvrent qu'elles n'expliquent pas un grand nombre de leurs observations. Les auteurs discutent alors, dans les grandes lignes, divers essais de laboratoire et travaux analytiques visant à fournir une meilleure compréhension des mécanismes observés.

Les méthodes d'essais concernent les essais suivants :

. L'essai de compression simple conventionnelle avec un frottement quasiment nul sur les plateaux qui montre les mécanismes de scission axiale ou rupture en colonne dus à la propagation de microfissures parallèles à la contrainte principale maximale. Des travaux antérieurs, par exemple ceux de Gramberg (1965) avaient déjà mis ce phénomène en évidence.

. L'essai triaxial avec des niveaux de déformation fixés à l'avance plutôt que les niveaux de chargements comme pour l'essai conventionnel.

. Les essais triaxiaux spéciaux dans lesquels les charges à la fois axiales et latérales sont appliquées par des systèmes rigides indépendants.

. Les essais triaxiaux spéciaux dans lesquels on applique des efforts de compression non uniformes sur les trois axes ($\sigma_1 > \sigma_2 > \sigma_3$).

Les échantillons testés en triaxial comprenaient du sel, du schiste bitumeux, du marbre, du calcaire, du charbon, du grès et diverses roches contenant des veines de charbon. Les dispositifs d'essai permettaient une étude complète des variations de volume, déformations longitudinales et axiales, anisotropie de la résistance et déformations plastiques et de fluage. Les concepts de rupture des roches développés à partir de ces essais ont été appliqués aux activités minières souterraines et aménagements souterrains définitifs.

Articles décrivant les études statistiques et théoriques

Dans leur article, Nishimatsu et Okubo signalent tout d'abord quelques unes des difficultés rencontrées pour évaluer la stabilité des pentes rocheuses avec joints. Ils considèrent ensuite une hypothétique masse rocheuse très simple, ne présentant qu'une famille de discontinuités dont l'espacement suit une loi de distribution exponentielle. Supposant de plus qu'il n'y a aucune intéraction entre les discontinuités, c'est à dire que ces discontinuités ne sont pas courbes et ne se rencontrent pas, les auteurs décrivent statistiquement une formule pour calculer la probabilité que la rupture ne se développe pas le long d'une discontinuité.

Puis les auteurs dérivent une loi de distribution statistique de la résistance au cisaillement des joints rocheux à partir d'une suite d'essais de cisaillement direct réalisés sur des joints créés en laboratoire dans des carottes de schiste de 50 mm de diamètre.

Utilisant tous les résultats précédents, les auteurs présentent un exemple de calcul de la probabilité que la rupture ne se développe pas dans le cas d'une pente avec une seule famille de joints parallèles à la pente et soumise à des pressions interstitielles négligeables.

Les Rapporteurs Généraux pensent qu'il y a d'autres facteurs que ne couvrent pas l'analyse de Nishimatsu et Okubo et qui ont plus d'influence sur la stabilité des pentes rocheuses existantes que les facteurs qu'ils ont étudiés dans leur article. D'autres auteurs, par exemple, Patton et Deere (1971) et Mac Mahon (1971 et 1974) ont considérés ces facteurs plus significatifs. Ces derniers comprennent :

. Radoucissement lié aux déformations des joints (ou discontinuités) de sorte que leur résistance diminue aux valeurs résiduelles. Les essais décrits dans l'article semblent avoir mis en évidence que des valeurs maximales, qu'il est courant de voir largement varier, dans des joints rocheux.

. Continuité des joints.

. Forme ou ondulation des joints.

. Nombre de familles de joints.

L'article de Grossman décrit le développement d'une méthode numérique pour définir les familles de discontinuités dans les masses rocheuses. Il affirme que cette méthode présente un grand intérêt lorsque on ne dispose que d'un petit nombre de mesures de la roche
qu'elle est applicable aux données obtenues par

échantillonnage uniforme ou non des orientations des discontinuités (joints).

Les Rapporteurs Généraux sont d'accord avec le fait que cette méthode peut aider à définir les familles de joints. Mais ainsi que l'auteur y fait allusion dans l'introduction, des réseaux complexes de joints développés lors du plissement des roches nécessiteront toujours une étude in situ détaillée de la distribution spatiale et de la nature des joints ainsi que de leurs orientations.

L'article de Pineau et Thomas décrit les études statistiques du degré d'hétérogénéité des portions de rocher intact d'un gite de minerai de fer, à différentes échelles. L'étude a concerné 482 essais de compression simple, sur des échantillons cylindriques saturés de tailles diverses, pris à des intervalles choisis et mesurés, dans de large blocs rocheux. Les auteurs concluent, de leur analyse statistique des résultats, que le degré de variabilité des résultats d'essais est lié aux variations de l'édifice rocheux qui sont à leur tour liées à la structure géologique régionale, (issue de processus tectoniques). Ils signalent que l'utilisation des méthodes statistiques simples, par exemple la moyenne arithmétique, aurait conduit à des conclusions incorrectes pour la résistance des roches testées.

Bien que ne contestant pas ces conclusions, les Rapporteurs Généraux signalent que l'intervalle de variations de résistance mis en évidence par les exemples, soit exemple 1 (intervalle 15 à 28 MPA), exemple 3 (intervalle 19 à 30 MPA) et tous les échantillons sauf trois de l'exemple 2 (intervalle 12 à 25 PA) serait insignifiant dans la plupart des cas pratiques de génie civil où les discontinuités de structures (joints, etc...) ont une influence écrasante sur le comportement des roches. Nous posons aussi aux auteurs la question suivante : ont-ils déterminé l'intervalle de variations de résistance qui pourraient être attribuées à des modifications mineures dans la préparation des échantillons et dans la conduite de l'essai ?

Optimisation des reconnaissances in situ des masses rocheuses

Les Rapporteurs généraux trouvèrent que le côté pratique de l'article de Fishman et al. était rafraichissant. Les auteurs signalent que la possibilité d'améliorations analytiques offerte par les méthodes numériques a rendu la mesure d'un grand nombre de paramètres géotechniques nécessaire et implique donc une meilleure qualité des systèmes de mesures. Reconnaissant que les reconnaissances in situ doivent fournir les résultats les plus valables, les auteurs posent et illustrent cinq principe de leur conduite qui cherche à garantir le résultat optimum de cette tache nécessairement onéreuse. Nous avons résumé leurs cinq principes comme suit :

1. Le type d'essai choisi doit être celui qui simule le mieux les conditions de chargement imposées par la structure proposée.

2. Les essais doivent tenir compte de tous les effets, sur la masse rocheuse, des activités de construction.

3. Les essais doivent être réalisés à une échelle suffisamment grande pour tester, de façon réaliste, une portion de la masse rocheuse (y compris la matrice et les discontinuités) et l'influence de l'échelle doit être prise en compte dans l'interprétation des résultats.

4. Les essais doivent être interprétés à la lumière des résultats d'observations du réseau de fracture et des résultats d'autres essais, de sorte que l'on puisse estimer si les résultats de ces essais sont ou non applicables à tout un chantier de construction et à des projets futurs.

5. La portée et la précision de ces essais doivent être le plus adaptées aux exigences du projet.

Les auteurs illustrent ensuite ce dernier principe par ce que nous appelons "études de susceptibilité" dans la conception du revêtement d'un tunnel sous-pression. Leur figure 1 montre combien le développement de fissures non négligeables dépend aussi bien dans la masse rocheuse que dans le béton armé, du module de déformation de la masse rocheuse entourant le revêtement. Il est clair que des essais in situ, onéreux ne sont pas justifiés dans la zone de rocher sain (zone I de la figure 1); par contre dans les zones de faible module (dans la zone II et à proximité), des mesures relativement précises peuvent être justifiées si les zones de faible module sont étendues. Il est possible que de tels essais ne soient pas justifiés, au plan économique, pour des zones de failles localisées.

Ils décrivent et discutent ensuite de diverses méthodes d'essais in situ. Les essais comprennent les essais de chargement de plaque, de chargement radial (dans les tunnels), de compression simple, de traction et de cisaillement.

Prédictions à partir de modèles mécaniques et mathématiques

L'article de Langof décrit la construction d'un modèle mécanique dans lequel des blocs rectangulaires en plastique simulent la masse rocheuse jointée. Les blocs étaient disposés de façon à simuler une masse rocheuse stratifiée avec une famille de joints continus à l'interface des couches et une autre famille de joints discontinus dans chaque couche, perpendiculaires aux précédents. Une fois que les propriétés de déformation de ce modèle de masse rocheuse furent déterminées par des essais biaxiaux sur un échantillon carré de 30 cm de côté, le modèle fut utilisé pour prédire les déformations de la masse rocheuse autour d'un tunnel circulaire, dans quatre cas : couches rocheuses horizontales, verticales, inclinées à 30 degrés et à 60 degrés, respectivement. On prédit aussi dans chacun des cas et pour différents champs de contraintes initiales les déformations et contraintes radiales dans le revêtement du tunnel et les contraintes secondaires développées dans le rocher adjacent à la clé de voûte et aux naissances du tunnel.

Ils obtinrent une corrélation raisonnable entre les édictions du modèle mécanique et une analyse aux éléments finis dans le seul cas des couches horizontales. L'analyse aux éléments finis fut conçue pour un milieu continu et ne pouvait donc modéliser le réseau des joints.

Les Rapporteurs Généraux signalent qu'il y a maintenant des programmes aux éléments finis disponibles qui pourraient modéliser les réseaux des joints, par exemple ceux de Burman, Trollope et Philp (1975) et de Burman et Hammett (1975).

Auscultation

Habenicht et Behensky sont les seuls à avoir présenté un article sur ce sujet. Ils décrivent une cellule mesurant la pression hydraulique : le fluide hydraulique qu'elle contient transmet directement la pression exercée sur le corps de la cellule à une tige indicative munie d'un ressort. Les déplacements de la tige sont proportionnels à la pression hydraulique et peuvent être mesurés manuellement ou enregistrés. Les auteurs soutiennent que leur système mécanique de mesure directe présente les avantages suivants par rapport aux jauges conventionnelles de pression hydraulique :

. sensitivité et précision hydraulique,

. loi linéaire pression-déflection,

. pratiquement aucun effet d'hystérésis.

Sous-thème 4

HYDROGEOLOGIE

L'hydrogéologie est un vaste domaine couvrant de nombreux sujets autres que ceux traités normalement dans les domaines de mécanique et hydraulique des roches.

Les sujets discutés ici sont limités à ceux d'un intérêt immédiat pour le mécanicien des roches soit :

. perméabilité des masses rocheuses,

. injectabilité des masses rocheuses,

. distribution de la pression autour de cavités souterraines,

. influence de la nappe phréatique sur la stabilité des pentes,

. écoulements d'eau dans des cavités souterraines.

Cinq articles seulement traitant de ce sujet ont été soumis.

L'évaluation de ces sujets par l'ingénieur nécessite la connaissance de valeurs de la perméabilité de la masse rocheuse ainsi que la connaissance des conditions géologiques.

La perméabilité de la masse rocheuse est gouvernée par la perméabilité à la fois de la roche intacte et des fissures. La roche intacte est souvent si compacte qu'on peut supposer que seules les fissures sont perméables. Mais il faut se souvenir que la masse rocheuse n'est pas homogène. En particulier, les failles peuvent intersecter une masse rocheuse de façon sporadique et nécessiter un traitement spécial.

Diverses méthodes permettent de déterminer la perméabilité d'une zone homogène. La plus reconnue est sans aucun doute l'essai de pression d'eau car il peut être réalisé au dessus comme au dessous de la nappe phréatique. Comme un essai ne permet l'étude que d'une petite partie de la masse rocheuse, il est nécessaire de réaliser un grand nombre d'essais pour évaluer la perméabilité de la masse rocheuse.

Les essais de pression d'eau peuvent être évalués de diverses manières. La plus vieille méthode, encore utilisée aujourd'hui, consiste à prendre le débit d'écoulement en rapport avec la pression et la longueur de la section d'essai comme les éléments de mesure de la perméabilité. Cette évaluation est liée au nom de Lugeon (Lugeon, 1933). On a développé, depuis, d'autres méthodes qui tentent de déterminer les facteurs ou tenseurs de perméabilité, c'est à dire les amplitudes, directions et sens des perméabilités (par exemple Baker (1955), Maini (1971), Snow (1966), Snow (1969) Rissler (1977). Ces méthodes supposent en général un écoulement stationnaire et les essais doivent être réalisés de manière appropriée. L'article de Schneider présente une méthode d'évaluation dans le cas d'écoulements variables ; cette méthode conduit donc naturellement à d'autres résultats que les méthodes développées antérieurement. On sent que ces différences dans les résultats pourraient venir de modifications introduites par l'auteur dans la conduite de l'essai (contrôle du débit d'écoulement et non de la pression). Quoi qu'il en soit, les Rapporteurs Généraux connaissent de nombreux cas où l'on a atteint un écoulement stationnaire beaucoup plus rapidement que dans les essais décrits par l'auteur. Ils exigent des commentaires de l'auteur et de l'auditoire.

Il est bien connu que des joints largement ouverts permettent des écoulements laminaires aussi bien que turbulents. On constate dans ce cas que le débit d'écoulement croît plus lentement que l'augmentation de pression. Rissler (1977) a examiné ce cas au plan théorique pour des conditions de symétrie radiale. Dans leur article, Cruz et Quadros présentent des mesures in situ réalisées dans une masse rocheuse basaltique, qui confirment ces courbes caractéristiques.

On s'intéresse aussi souvent à l'injectabilité des masses rocheuses. Il ne faut pas supposer qu'une perméabilité élevée est synonyme d'une bonne injectabilité. La perméabilité peut provenir d'une grande quantité de joints faiblement ouverts qui refusent de laisser passer du coulis. Blinde et al. décrivent des études sur ce sujet réalisées sur des granites désintégrés et consistant en études de l'édifice granitique, examen optique des forages, techniques de tracés des joints, essais de pression d'eau, mesures géoélectriques de potentiel et expériences sismiques. Les auteurs trouvent que les joints apparamment existants ne contribuent pas tous à la perméabilité.

Huder et Amberg ont écrit un rapport qui traite de l'influence de la nappe phréatique sur la stabilité des pentes et cavités souterraines et de l'efficacité des "relief measures" (?) Ils discutent en outre divers procédés pour piézomètres.

Bello-Maldonado présente une procédure analytique pour évaluer le débit d'eau entrant dans un tunnel, avant et après se construction. De tels calculs sont certainement intéressants, mais les Rapporteurs Généraux craignent qu'en pratique les roches aient trop rarement une homogénéité suffisante pour justifier le considérable effort mathématique qu'ils nécessitent.

CONCLUSIONS

Soixante et un article au total ont été soumis sur le thème "Exploration et évaluation de sites". Ils traitent de méthodes géophysiques, de reconnaissances in situ et d'essais en laboratoire, de classification, prédiction, observation, auscultation et questions hydrogéologiques.

Les progrès les plus importants depuis le dernier Congrès SIMR semblent avoir été faits dans les domaines suivants :

- l'application de méthodes sismiques à l'évaluation de résultats d'injection ,
- le développement de procédures pour évaluer les propriétés de déformations anisotropiques à partir d'essais,
- la continuation du développement de systèmes de classification.

BIBLIOGRAPHIE

Anstey, N.A. (1977). Seismic interpretation, the physical aspects, Int. Human Resources Development Corporation Boston, U.S.A., p. 81.

Baker, W.J. (1955). Flow in fissured formations, Proc. 4. World Petroleum Congr., Rome, Section II/E, Paper 7, pp. 379-393.

Barla, G. et al. (1974). Direct tensile testing of anisotropic rocks, Proc. Third int. Congr. Rock Mech. ISRM, Vol. 2A, pp. 93-98.

Barton (1971). A relationship between joint roughness and joint shear strength. Report I 8, Int. Symp. on Rock Mech., Nancy.

Barton, N., Lien, R., and Lunde, J; (1974). Engineering classification of rock masses for the design of tunnel support, Rock Mechanics. Vol. 6, N) 4, pp. 189-236.

Bieniaswski, Z.T. (1974a). Geomechanics classification of rock masses and its application in tunnelling. Proc. Third Int. Conf. Rock Mech., ISRM. Vol. IIA, pp. 27-32.

Bieniawski, Z.T. (1974b). Estimating the strength of rock materials, J.S. African Inst. Min. and Met., Vol. 74, No. 8.

Bieniawski, Z.T. (1978). Determining rock mass deformability, experience from case histories, Int. J. Rock Mech. and Min. Sci., Vol. 15, pp. 237-247.

Bieniawski, Z.T. (1979). The geomechanics classification in rock engineering applications, Proc. Fourth Int. Conf. Rock Mech.n ISRM, Vol. 2, pp. 41-48.

Broch, E. and Franklin, J.A. (1972), The point Load strength test. Int. J. Rock Mech. and Min. Sci., Vol. 9, pp.669-697.

Brown, E.T. (1974). Fracture of rock under uniform biaxial compression, Proc. Third Int. Congr. Rock Mech.n ISRM, Vol. 2A, pp. 111-117.

Burman, B.C., Trollope, D.H., and Philp, M.G. (1975). The behaviour of excavated slopes in jointed rock, Aust. Geomech. J., Vol. G5, NO. 1, pp. 26-31.

Burman, B.C., and Hammet, R.D. (1975). Design of foundations in jointed rock masses, Proc. Second Aust.-N.Z. Conf. on Geomechanics, pp. 83-88.

Franklin, J.A. (1979). Use of tests and monitoring in the design and construction of rock structures, General Report, Theme 2, ISRM Congress, 1979, Montreux, Switzerland, Vol. 3, p. 163.

Gramberg, J. (1965). Axial cleavage fracturing a significant process in mining and geology, Eng. Geol. Vol. 1n No1, pp. 31-72.

Hoek, E. and Brown, E.T. (1980). Underground excavations in rock, The Institution of Mining and Metallurgy, London.

International Society for Rock Mechanics (1980). Basic geotechnical description of rock masses, Int. J. Rock Mech. Min. Sci. and Geomech. Abstr. Vol. 18, pp. 85-110.

John, K.W. (1969). Festigkeit und Verformbarkeit von druckfesten, regelmäßig gefügten Diskontinuen, Ph. D. Thesis, Karlsruhe.

Kaneko, K., Inoue,I., Sassa, K. and I. Ito (1979) Monitoring the stability of rock structures by means of acoustic wave attenuation, ISRM Congress, 1979, Montreux, Switzerland, Vol. 2, p. 287.

Langer, M. (1979). Rheological Behaviour of Rock Masses. General Report, Theme 1, ISRM Congress, 1979, Montreux, Switzerland, Vol. 3, P. 29.

Leichnitz, W. and O. Natau (1979). The influence of peak shear strength determination on the analytical rock slope stability, Proc. Fourth Int. Congr. Rock Mech. ISRM, Vol. 2, pp. 335-341.

Lögters, G. and H. Voort (1974). In-situ determination of the deformational behaviour of a cubical rock mass sample under triaxial load, Rock Mechanics (6), pp. 65-79.

Lugeon, M. (1933). Barrages et Géologie. Dunod, Paris

Maini, Y.N. (1971). In situ hydraulic parameters in jointed-rock - their measurement and interpretation, Ph. D. Thesis Imperial College, London.

Mc Mahon, B.K. (1971). A statistical method for the design of rock slopes, Proc. First Aust.-N.Z. Conf. on Geomechanics. Vol. 1, pp. 314-321

McMahon, B.K. (1974). Design of rock slopes against sliding on pre-existing fractures, Proc. Third Conf. of Int. Soc. Rock Mech.n Vol. IIB, pp. 803-808.

Neilson, J.L. (1970). Notes on weathering of the Silurian rocks of the Melbourne district, J. I.E. Aust. Vol. 42, 1-2, pp. 9-12.

Oberti, G., Carabelli, E., Goffi, L. and Rossi, P.P. (1979). Study of an orthotropic rock mass : Experimental techniques, comparative analysis of results, Proc. Fourth Int. Congr. Rock Mech. ISRM, Vol. 2, pp. 485-491.

Patton, F.D. and Deere, D.U. (1971). Significant geologic factors in rock slope stability. In : Planning Open Pit Mines. A.A. Balkema, Amsterdam.

Pinto (1979). Determination of the elastic constants of anisotropic bodies by diametral compression test, Proc. Fourth Int. Congr. Rock Mech. ISRM, Vol. 2, pp. 359.

Plischke, B. (1980). Ein Berechnungsverfahren zur Untersuchung der dynamischen Beanspruchung von Felsbauten. 4. Nationale Tagung über Felsmechanik, 5./6. Mai 1980, Aachen, Deutsche Gesellscharft für Erd-und Grundbau, p. 259-277.

Rissler, P. (1977). Determination of the water permeability of jointed rock. Aachen, Institute for Foundation Engineering, Soil Mechanics, Rock Mechanics and Water Ways Construction.

Rodrigues (1979). The anisotropy of the moduli of elasticity and of the ultimate stress in rocks and rock masses, Proc. Fourth Int. Congr. Rock Mech. ISRM, Vol. 2, pp. 517.

Sancio, R.T. and Brown, I. (1980). A classification of weathered foliated rocks for use in slope stability problems, Proc. Third Aust._N.Z. Conf. on Geomechanics, Vol. 2, pp. 81-86.

Saunders, M.K., Fookes, P.G. (1970). A review of the relationship of rock weathering and climate and its significance to foundation engineering, Eng. Geol. Vol. 4, pp. 289-325.

Savich, A.I., Koptev, V.I., Iljin, M.M. and Zamakhajev, A.M. (1979). Distinguishing features of deformation development in the foundation of the Inguri Arch Dam, ISRM Congress, 1979, Montreux, Switzerland, Vol. 2, p. 585.

Snow, D.T., (1966). Three-hole pressure test for anisotropic foundation permeability. Felsmechanik und Ingenieurgeologie 4, Nr. 6, pp. 298-316.

Snow, D.T. (1969). Steady flow from cylindrical cavities in saturated, infinite anisotropic media, Association of Engineering Geologists, Sacramento, California.

Terzaghi, R. (1946). Rock defects and loads on tunnel supports. In: Rock Tunnelling with Steel Supports. Commercial Shearing and Stamping Co., Youngstown.

Whickham, G.E., Tiedemann, H.R. an Skinner, E.H. (1972). Support determinations based on geological predictions, Proc. First North American Tunnelling Conference, A.I.M.E., New York, pp. 43-64.

Wichter, L. (1979). On the geotechnical properties of a Jurassic clay shale, Proc. Fourth Int. Congr. Rock Mech. ISRM, Vol. 1, pp. 319-326.

UNTERSUCHUNG UND BEURTEILUNG DES BETRIEBSPUNKTES

Site exploration and evaluation

Exploration et évaluation in situ

David H. Stapledon
Südaustralisches Technologisches Institut, Adelaide, Australien
Peter Rissler
Der Ruhrtelsperrenverein, Essen, BR Deutschland

ZUSAMMENFASSUNG:

Geländeuntersuchungen kosten Geld und Zeit, sind jedoch wesentlich für erfolgreiches Planen und Bauen. Sechs Beiträge über geophysikalische Methoden deuten an, daß die seismischen Methoden überaus weit verbreitet sind. Im Bereich der Labor- und Feldversuche liegt der Schwerpunkt von 35 Beiträgen auf der Auswertung anisotroper Eigenschaften, dem Verhalten unter Dauer- und zyklischer Last, auf der Bestimmung der Scherfestigkeit von Trennflächen und auf Maßstabeffekten. Fünfzehn Beiträge zum Thema Klassifikation beleuchten den Wert von Klassifikationssystemen und die Schwierigkeiten beim Entwurf eines umfassenden, für alle Projekttypen und alle geologischen Situationen geeigneten Systems. Fünf Aufsätze handeln von hydrogeologischen Problemen wie Bestimmung der Wasserdurchlässigkeit und Grundwasserbeobachtungen.

SUMMARY:

Site investigations cost money and time but are essential for the successful planning and construction of projects. Six papers on geophysical methods suggest that the seismic method is the most widely accepted. In thirty-five papers on laboratory and in-situ testing, emphasis is placed on anisotropic properties, behaviour under long-term or cyclic Load, shear strength of discontinuities and scale effects. Fifteen papers on classification illustrate the value of classification systems, and the difficulties associated with designing any single system which would relate to all project types, in all geological environments. Five papers on hydrogeology deal mainly with permeability assessment and groundwater observations.

RÉSUMÉ:

Les analyses de terrain coûtent du temps et de l'argent mais sont d'une importance capitale pour des tracés de plan et des constructions couronnés de succès. Six articles sur les méthodes géophysiques donnent à entendre que les méthodes sismiques sont très répondues. Dans le domaine des recherches en laboratoire et sur le terrain, l'accent de 35 rapports est mis sur l'évaluation des propriétés anisotropes, sur le comportement sous une charge permanente et cyclique, sur la définition de la résistance au cisaillement des surfaces de séparation et sur les effects de l'échelle. Quinze rapports sur le thème de la classification démontrent la valeur des systèmes de classification et les difficultés rencontrées en essayant d'établir un système unique approprié à tous les types de projets et à tous les terrains géologiques. Cinq articles traitent des problèmes hydrogéologiques tels que la détermination de la perméabilité et l'observation des eaux souterraines.

ALLGEMEINES

Baugrunduntersuchungen im Fels und deren Auswertungen nehmen in der Felsmechanik einen breiten Raum ein. Sie sind Voraussetzungen für alle Beiträge, die zu den Hauptthemen B,C,D und teilweise auch zu E eingereicht wurden.

Baugrunduntersuchungen und deren Auswertungen erscheinen oft kostspielig; aber im Vergleich zu den Baukosten sind diese Aufwendungen gering. Insbesondere treten sie gänzlich in den Hintergrund, wenn man sie mit den Kosten eines Versagensfalles vergleicht, der durch fehlende bzw. unzureichende Untersuchungen verursacht worden ist.

Die Untersuchungen und ihre Auswertungen erfordern viel Zeit und können zeitlich und kostenmäßig nicht mit der gleichen Genauigkeit vorhergesagt werden, wie die eigentlichen Bauarbeiten. Dies kommt daher, daß der anstehende Fels durch natürliche Vorgänge und während langer Zeiträume entstanden ist, und die Natur uns nur unvollkommene und oft nur versteckte Aufzeichnungen über ihr Wirken und keine „Bestandspläne" hinterlassen hat. Deshalb müssen wir oftmals diese Untersuchungen beginnen ohne klare Vorstellung, was wir finden werden. Das Untersuchungsprogramm muß deshalb stufenweise ablaufen, wobei die Aktivitäten jeder Stufe danach entworfen werden, daß Fragen beantwortet werden, die bei der Auswertung der vorhergehenden Stufe aufgeworfen wurden.

Beides, Kosten- und Zeitbedarf, sollte vom Bauherrn verstanden und akzeptiert werden.

1975 veröffentlichte diese Gesellschaft den Bericht einer Kommission "Empfehlungen über Techniken zur Baugrunduntersuchung im Fels", die einen Vorschlag für eine systematische Vorgehensweise beim Untersuchungsprogramm enthält .

In seinem Generalbericht zu Thema 2 des IGFM-Kongresses in Montreux beschrieb Franklin (1979) detailliert den Einsatz von Versuchen und Überwachungsmaßnahmen beim Entwurf und beim Bau von Felsbauwerken und zeigte klar die Wechselbeziehungen zwischen geotechnischen und anderen ingenieurmäßigen Tätigkeiten während der Planungs- und Bauphase auf.

Wegen der zentralen Bedeutung für alle geotechnischen Bauvorhaben waren Baugrunduntersuchungen im Fels und deren Auswertungen Gegenstand mehrerer früherer IGFM-Kongresse. So waren z.B.

1974 (Denver) Thema 1

„Physikalische Eigenschaften von Gestein und Gebirge",

1979 (Montreux) Thema 1 (teilweise)

„Rheologisches Verhalten von Gesteinen und Fels"

und Thema 2

„Verwendung von Versuchen und Kontrollmessungen bei Entwurf und Bau von Felsbauten"

diesem Problemkreis gewidmet.

Wie vom Organisationskomitee entschieden, soll das Baugrunduntersuchungen und deren Auswertung behandelnde Generalthema dieses Kongresses in vier Unterthemen unterteilt werden, nämlich in

A.1 Geophysikalische Versuche und Erkundungen
A.2 Feld- und Laborversuche
A.3 Klassifikation, Vorhersage, Beobachtung und Überwachung
A.4 Hydrogeologie

Die Generalberichter folgen mit ihren Ausführungen dieser Gliederung.

Unterthema A.1

GEOPHYSIKALISCHE VERSUCHE UND ERKUNDUNGEN

Es war für uns etwas überraschend, daß nur 6 Aufsätze zum Unterthema A.1 eingereicht worden sind, gegenüber 35 zum Unterthema A.2. Man könnte dies als Ausdruck der Wertschätzung geophysikalischer Versuche im Vergleich zu Feld- und Laborversuchen deuten. Wir sind überzeugt, daß diese Deutung nicht gerechtfertigt wäre. Geophysikalische Methoden sind einerseits vorteilhaft, haben andererseits aber einige Nachteile. Ihr Vorteil ist das Vermögen, einen Überblick über große Gebirgsbereiche zu verschaffen. Hier sind sie allen direkten Methoden überlegen. Ihr Nachteil: in den meisten Fällen verschaffen sie nur Indexwerte zur Einteilung des Gebirges. Andererseits scheint es ein besonderer Vorteil zu sein, daß sie in der Anwendung ziemlich billig sind. Dies ermöglicht es, viele Informationen und zwar während aller Phasen der Entwicklung eines Projektes und während des Betriebes zu beschaffen.

Der Erfolg bei der Anwendung jeder Erkundungsmethode hängt nicht nur von der Leistungsfähigkeit der Methode selbst ab, sondern auch vom Geschick und Kenntnisstand der Anwender, insbesondere von deren Wissen um die Grenzen der betreffenden Methode.

Die Erfahrung hat gezeigt, daß die Grenzen der geophysikalischen Verfahren häufig nicht recht verstanden werden, und daß ihre Anwendung in ungeeigneter Situation, oft noch ohne Unterstützung durch direkte Methoden, unnötigerweise zu Mißerfolgen geführt hat.

Das seismische Verfahren, welches als häufigstes angewandt wird, ist auch das am häufigsten mißbrauchte. Nach unserer Ansicht sind die folgenden zwei Begrenzungen wesentlich:

(a) Aus ingenieurmäßiger Sicht wesentliche Felseigenschaften (insbesondere die Scherfestigkeit) werden gewöhnlich durch ganz schmale geologische Defekte (schmale Störungen oder Verwitterungsränder) an kritischen Punkten und - bezogen auf den Spannungszustand - in kritischen Raumstellungen bestimmt. Die Seismik ist gewöhnlich nicht geeignet, solche Defekte ausfindig zu machen.

(b) Bei den seismischen Verfahren wird der Fels nur ganz kurzzeitig und nur ganz gering belastet. Der dynamische Elastizitätsmodul (E_{dyn}), wie er aus der Geschwindigkeit der Scherwellen (V_S) oder der Längswellen (V_p) errechnet wird, kann daher nur die Reaktion des Gebirges auf diese vorübergehende Belastung darstellen. Daher sollte nicht erwartet werden, daß E_{dyn} ein zuverlässiges Maß für die Verformungen eines geklüfteten und zerfurchten Gebirges unter hoher Dauerlast vermittelt.

Wir hörten auch den Einwand, die Geschwindigkeit der Longitudinalwellen würde nicht notwendigerweise, wie gewöhnlich geglaubt wird, die Eigenschaften des „durchschnittlichen" Felsens widerspiegeln. Dies deshalb, weil die Longitudinalwellen, welche als erste an den Geophonen ankommen, den kürzesten Weg durch den besten Fels nehmen, hier also ein Kurzschlußeffekt besteht.

Wenn wir diesen Einwand als wahr unterstellen, dann erscheint auch der Glaube, daß elektrische Widerstandsmethoden Werte für den „mittleren" Fels ergeben, schlecht begründet. Der Stromfluß wird dazu neigen, sich in den Felsbereichen mit der höchsten Leitfähigkeit zu konzentrieren, und dazu, in Verwitterungszonen und in Störungsbereichen einen Kurzschlußeffekt auszulösen.

Es wäre nützlich, zu diesen und anderen möglichen Begrenzungen der geophysikalischen Methoden von den Autoren und vom Kongreß Kommentare zu erhalten.

LABOR- UND FELDFORSCHUNG ZU SEISMISCHEN METHODEN

In Laborversuchen, die von Tanimoto und Ikeda beschrieben werden, wurde festgestellt, daß die Geschwindigkeit der Längswellen V_p gegenüber dem Abstand von Klüften mit Spaltweiten kleiner 0,01 mm unempfindlich ist und daß sie bei Spaltweiten von 2 bis 4 mm nur wenig abnimmt (5-15%). Es zeigte sich, daß Fels mit Kluftspaltweiten kleiner als 0,04 mm sich im seismischen Verhalten Gestein sehr ähnlich verhält.

Diese Ergebnisse scheinen im Widerspruch zu stehen mit der Ansicht anderer Forscher. Anstey (1977) stellt z.B. fest:

> „Der Effekt von Rissen auf die seismische Geschwindigkeit ist bedeutender....;sobald ein Hohlraum vorhanden ist, seitliche Verformung der Partikel möglich ist, und der entsprechende Elastizitätsmodul sich ändert. Wir erinnern auch daran, wie klein die Verschiebung eines Teilchens ist, wenn die seismische Welle hindurch geht - i.a. zwei Größenordnungen kleiner als die Wellenlänge des Lichts; dementsprechend können Risse, die viel dünner sind, als daß sie unter einem optischen Mikroskop gesehen werden können, wesentliche Änderungen der elastischen Eigenschaften bewirken."

Auch Savich et al bringen in Abb.6 ihres Beitrags Ergebnisse sowohl geophysikalischer (Ultraschall und Seismik) als auch geologischer (nach dem Augenschein) Bestimmungen des Hohlraumanteils. Die geophysikalischen Methoden ergeben 1,7 bis 2 mal größere Hohlraumanteile als die geologische Methode. Die Erklärung der Autoren für diese Unstimmigkeit enthält die Feststellung

> „Die Ultraschallmessungen berücksichtigen nicht nur sichtbare Klüfte, sondern auch Mikrorisse, die in Bohrlöchern mit dem Auge nicht festgestellt werden können."

Wir vermuten, daß die von verschiedenen Forschern beobachteten Unterschiede aus Maßstabeffekten hinsichtlich Amplituden und Wellenlängen herrühren. Die Autoren sind hierzu zu Anmerkungen eingeladen.

Wir halten fest, daß Tanimoto und Ikeda in ihrer Zusammenfassung Feldstudien an der Geländeoberfläche und an 104 untertägigen Bauwerken erwähnen. Diese Studien sind in dem Beitrag nicht beschrieben. Es wäre nützlich, wenn die Autoren uns mitteilen könnten, ob ihre Laborergebnisse mit den Ergebnissen ihrer Feldstudien übereinstimmten.

In ihren Laufwegexperimenten (Abb.8 und 9) simulierten Tanimoto und Ikeda Fels mit diskontinuierlichen Klüften oder Fels mit örtlichen Kontaktpunkten, indem sie Schlitze in 50 mm-Würfel aus Mörtel schnitten. Aus den Laufzeitaufzeichnungen folgerten die Autoren, daß dies zwar eine augenscheinliche, allgemeine Verminderung von V_p bewirkte, dies aber nur auf die Verlängerung des Laufweges zurückzuführen war, da die Welle nicht durch die Schlitze lief. Es scheint uns, daß Abb.8 dieses Beitrags eine gute Veranschaulichung des „Kurzschluß"-Effektes darstellt, wie er vorstehend bei unserer Diskussion der Verfahrensgrenzen erwähnt wurde. Nach dieser Abbildung scheint es, daß man unrealistisch hohe Werte für V_p erhalten könnte, wenn in stark geklüftetem und/oder verwittertem Fels „Kurzschluß"-Pfade durch ungeklüfteten oder frischen Fels zwischen Schußpunkt und Empfänger existieren.

Obwohl die praktische Bedeutung ihrer Ergebnisse noch nicht ganz zu verstehen ist, glauben wir, daß Tanimoto und Ikeda wertvolle Entdeckungen zur Dämpfung der Amplitude, zum Effekt von Kluftkontaktflächen und -drücken und zu den Laufwegen der Wellen gebracht haben. Von besonderem Interesse ist die Empfindlichkeit der Dämpfung auf die Kluftspaltweite. Dieser Effekt wurde bereits 1979 anläßlich des IGFM-Kongresses in Montreux von Kaneko et al (1979) erwähnt.

Der Beitrag von By handelt insbesondere von der Dämpfung der Längswellen in geklüftetem Gebirge. Ein theoretisches Modell wird vorgeschlagen und in einem Sprengversuch in situ bestätigt. Die Ergebnisse werden einerseits als Beziehung zwischen größter Spannungsamplitude und Abstand von der Schockquelle (Abb.8), andererseits als Diagramm (Abb.9) dargestellt, aus dem der „Dämpfungsfaktor" entnommen werden kann, wenn die Spaltweite und die „Schichtdicke" bekannt sind. Wir vermuten, daß „Schichtdicke" sich auf den Abstand der Klüfte untereinander bezieht. In diesem Zusammenhang sollte erwähnt werden, daß Plischke (1980) eine Arbeit veröffentlicht hat, die sich mit der Bestimmung der Schockwellenausbreitung durch numerische Methoden befaßt. Es erscheint wertvoll für die weitere Entwicklung, wenn beide Autoren ihre Arbeit gegenseitig befruchten würden. Die Ergebnisse hinsichtlich des Klufteinflusses stimmen mit denen von Tanimoto und Ikeda allgemein überein.

ABSCHÄTZUNG DER WIRKSAMKEIT VON INJEKTIONEN

Rodrigues et al präsentieren überzeugende Beweise für den Erfolg der Seismik bei der Ermittlung der Wirksamkeit von Injektionen am Cabril Dam. Die geophysikalische Arbeit wurde entsprechend unterstützt durch Durchlässigkeitsuntersuchungen und wurde mit großer Genauigkeit durchgeführt. Die Verwendung sowohl von Untersuchungen in Bohrlochlängs- als auch in Querrichtung sollte auf eventuelle, auf Kurzschlußeffekte zurückzuführende, anomale V_p-Werte aufmerksam machen. In diesem Zusammenhang stellen wir die Frage, ob es irgendwelche wesentliche Unterschiede zwischen den V_p-Werten aus unterschiedlichen Techniken aber im gleichen Felsbereich gab? Wenn ja, wie wurden diese Unterschiede interpretiert (im Hinblick auf die

Felsstruktur) und welcher V_p-Wert wurde als repräsentativ für solche Felsbereiche angenommen?

Nach unserer Erfahrung sind die berichteten V_p-Werte und die Verpreßergebnisse einem Fels mit hoher mechanischer Qualität bereits vor dem Vergüten zuzuordnen. Der Wert ihrer Daten in den Tabellen I und II würde gesteigert, wenn die Autoren weitere Einzelheiten über den Fels angeben könnten, wie z.B. ein geologisches Profil und einen Überblick über das Trennflächengefüge des Granits. Einzelheiten sowohl über die Anordnung der Vergütungs- und der Dichtungsschleierbohrungen und über die Längen der WD-Testbereiche wären ebenfalls hilfreich.

Savich et al berichten über Verpreßkontrollen an der Inguribogenmauer und zeigen als ihre Abb.11 ein auf der Baustelle entworfenes Diagramm zur Abschätzung der Wirksamkeit von Injektionen. Da die Skala für V_p (Wellenlängsgeschwindigkeit vor Injektion) bei etwa 5 km/s endet, läßt dies vermuten, daß Fels mit einem größeren V_p für unverpreßbar gehalten wird. Wir zeichneten die Ergebnisse der Cabril Dam Versuche in dieses Diagramm (unsere Abb.1) ein. Wie erwartet, kann hieraus entnommen werden, daß die mittleren Ergebnisse für die Blöcke V und VI in diesem „eigentlich unvergütbaren" Bereich liegen. Die Ergebnisse der übrigen Blöcke liegen in der „befriedigenden" Zone III und an ihrer Grenze zur „guten" Zone II. Wie bereits vorstehend ausgeführt, betrachteten wir die Cabril Daten als zu einem Fels ziemlich hoher mechanischer Qualität gehörend. Es scheint nach der Abb.11 von Savich et al, daß in manchen Bereichen des Inguribauvorhabens Fels mit V_o vergleichbar jener am Cabrilbauwerk nach dem Verpressen einen wesentlich größeren Anstieg in V_p zeigte (Endwerte V_p mit über 6 km/s in der „ausgezeichneten" Verpreßzone I). Es erscheint wahrscheinlich, daß der Grund für diese große Differenz mit den Felseigenschaften vor der Vergütung zusammen hängt. Es wäre deshalb zweckmäßig, wenn die Autoren von jenen Bereichen des Inguriuntergrunds, in denen „gute" und „ausgezeichnete" Injektionsergebnisse mit V_p-Werten von etwa 6 km/s angezeigt werden, Einzelheiten vortragen würden. Könnten die Autoren insbesondere die gezeigten Erhöhungen von V_p um mehr als 20% erklären bei Fels mit V_p um 5 km/s, d.h. bei Fels, der bereits vor dem Verpressen hohe mechanische Qualität gehabt haben müßte.

SCHÄTZUNG VON ELASTISCHEN EIGENSCHAFTEN UND VERFORMUNGSEIGENSCHAFTEN

Dhawan et al erhielten V_p und V_s-Werte aus der Seismik in eng gefalteten und geklüfteten quarzitischen Phylliten an einer Staumauerbaustelle im Himalaya. Einige der Versuche fanden in einem Erkundungsstollen statt, in dem Lastplatten- und Druckkissenversuche durchgeführt wurden, um Werte für Deformations- (D) und Elastizitätsmoduln (E) zu erhalten. Die von den Autoren vorgestellten Abbildungen zeigen beachtliche Übereinstimmung zwischen V_s und V_p, aber ziemlich schwache Beziehungen zwischen V_p und den D- und E-Werten. Die schwachen Übereinstimmungen scheinen zumindest teilweise durch die große Streuung in den Ergebnissen der Lastplatten- und Druckkissenversuche verursacht zu sein, wie aus ihrer Tabelle 3 zu

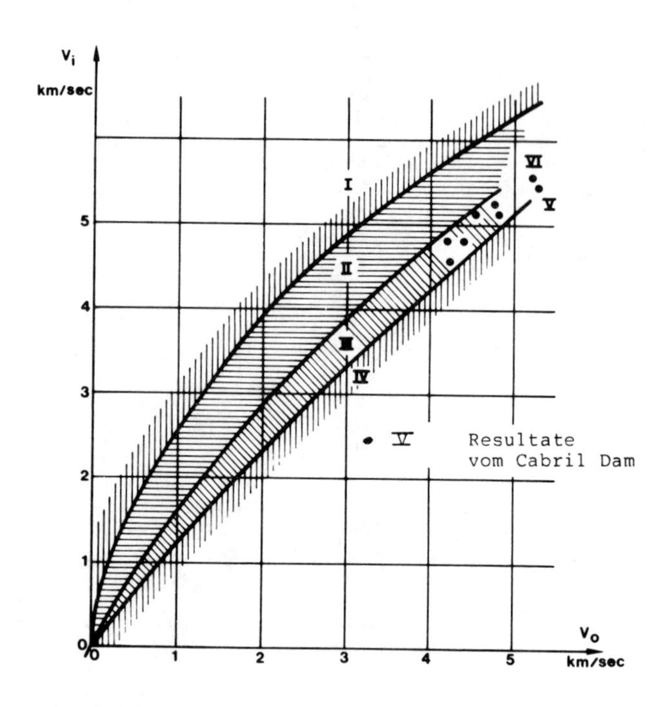

Abb. 1 Diagramm nach Savich et al zur Abschätzung des Injektionserfolges mit eingetragenen Ergebnissen vom Cabril Dam

entnehmen ist. Wegen der am Anfang dieses Generalberichts genannten Begrenzung (b) würden wir darüberhinaus auch nicht erwarten, daß D oder E aus Belastungsversuchen in irgendeiner einfachen Weise mit E_{dyn} aus der seismischen Wellengeschwindigkeit im Zusammenhang stehen.

Savich et al stellen fest, daß seismoakustische Methoden „die führenden Methoden für quantitative Studien der elastischen Eigenschaften und Verformungseigenschaften des Gebirges" geworden sind. Sie beschreiben, wie systematische Studien unter Verwendung seismoakustischer Methoden unter Einbeziehung der Ingenieurgeologie, der Bestimmung des statischen Moduls und einer statistischen Auswertung der Daten die Entwicklung verallgemeinerter Gleichungen für die Berechnung der statischen Moduln D und E aus dem dynamischen Modul E_{dyn} ermöglicht haben. Vier Wertegruppen für die verschiedenen in diesen Formeln genutzten Koeffizienten werden in Tabelle II angegeben, wobei jede Gruppe für eine sehr weite Klasse unterschiedlicher Gebirgstypen steht. Die Autoren erheben den Anspruch, daß die Verwendung dieser verallgemeinerten Formeln erlaubt, in jedem Gebirge die Werte für D und E mit einer Toleranz ± 30% bei 90% Wahrscheinlichkeit zu errechnen.

Die Generalberichter spenden der genauen und systematischen Arbeit dieser Studien Beifall, haben jedoch Vorbehalte bezüglich der allgemeinen Anwendbarkeit dieser Gleichungen für alle Gebirgsarten dieser Welt. Wir sagen dies wegen der großen Zahl geologischer Parameter, welche die Zusammendrückbarkeit beeinflussen können. Beispielsweise kann Fels aus jeder der vier weitgefaßten Gruppen der Tabelle II

Klüfte, lokale, breite oder schmale Zerrüttungszonen und/oder Zonen verwitterten oder veränderten Gesteins von regelmäßiger oder unregelmäßiger Form aufweisen. Das zerrüttete, verwitterte oder veränderte Material kann Bodeneigenschaften haben und sehr zusammendrückbar sein. Die Lage und Richtung solch zusammendrückbarer Zonen im Verhältnis zum Kluftgefüge und zu den einwirkenden Lasten muß bedeutenden Einfluß auf die Verformbarkeit des Gebirges haben.

Wir hätten gerne von den Autoren Antworten auf folgende Fragen:

(a) Können Sie die Zahlenwerte für die Parameter in Tafel II rechtfertigen, insbesondere bei Berücksichtigung der großen Streubreite physikalischer Eigenschaften, die von jeder der weitgefaßten Felsgruppen umfaßt wird?

(b) Sind die Gleichungen (2) und (3) auf Fels angewandt worden, der durch chemische Verwitterung bzw. Veränderung oder durch Brüche ernstlich geschwächt war?

FELSÜBERWACHUNG WÄHREND DER BAUZEIT VON STAUBAUWERKEN UND WÄHREND DER FÜLLUNG DES SPEICHERS

Savich et al beschreiben Ergebnisse in der Umgebung der 271,5m hohen Inguribogenmauer und des Speichers durchgeführter Arbeiten zur Überwachung etwaiger Veränderungen im Gebirge während des Baus und während und nach der Füllung des Speichers. In Verbindung mit anderen Verformungsmessungen und geodätischen Überwachungen wurden sowohl klein- als auch großräumige Messungen durchgeführt. Diese zeigten während der Bauzeit bis 80m unter die Gründungssohle eine Zunahme der Geschwindigkeit der Längswellen V_p und des spezifischen elektrischen Widerstands ρ .

Während der Füllung des Speichers, als die Klüfte vom Speicher her hydrostatischen Drücken unterworfen wurden, verringerten sich V_p und ρ beachtlich. Neben der Mauer wurden diese Einflüsse bis zu einer Tiefe von 250m festgestellt und am oberen Ende des Speichers bis zu 500m.

Könnten die Autoren uns mitteilen, ob die Zusammendrückung und spätere Öffnung von Klüften in der Nähe der Mauer auch durch Extensometer überwacht wurden und falls ja, welche Dehnungsbeträge angezeigt wurden?

Wir glauben, daß Überwachungsmaßnahmen in der Art, wie sie für die Ingurimauer beschrieben wurden, eine sehr nützliche Anwendung der Geophysik darstellen.

ABSCHÄTZUNG DES VERFOMUNGSVERHALTENS VON GESCHWÄCHTEM FELS BEI ERDBEBEN

Um den Erdbebeneinfluß auf Bauwerke einschätzen zu können, ist es wichtig, das Verformungsverhalten des Gründungsfels zu kennen. Ishikawa et al beschreiben die Ergebnisse niederfrequenter, vertikaler, zyklischer Belastungsversuche, die unterschiedliche Beziehungen zwischen dem dynamischen E-Modul E_{dyn}, den Amplituden, den Belastungsgeschwindigkeiten, den vertikalen Dehnungen und den Dämpfungsraten erkennen lassen. Da die Autoren Ergebnisse nur von zwei Versuchsfeldern vorführen, scheint es, daß der Hauptzweck ihrer Arbeit darin bestand, eine neue Versuchstechnik zu entwickeln. Da die Ergebnisse hauptsächlich von den Bedingungen des Gebirges knapp unterhalb (bis zu 1,5m) der Belastungsvorrichtung beeinflußt werden, und dieser Bereich immer als mehr oder weniger aufgelockert - also nicht im natürlichen Zustand - angesehen werden muß, wissen wir nicht, wie man die Versuchsergebnisse auf den Entwurf übertragen soll. Die Autoren sind eingeladen, dies zu erklären.

Andererseits sind diese Versuche mit den seismischen und Ultraschallversuchen verwandt. Deshalb sollte die Arbeit von Ishikawa et al im Zusammenhang mit jener von Tanimoto und Ikeda gesehen werden. Sie könnte unser Verständnis für die Verminderung der Amplituden verbessern. Zur Auswertung dieser und ähnlicher Versuche gibt es bereits eine Menge Vorarbeiten. Langer hat diese in seinem Generalbericht zum Thema 1 des IGFM-Kongresses 1979 in Montreux zusammengefaßt.

ANDERE ANWENDUNGEN GEOPHYSIKALISCHER METHODEN BEI GELÄNDEUNTERSUCHUNGEN

Die Refraktionsseismik ist seit jeher benutzt worden, um die Mächtigkeit der Überlagerung oder der Verwitterungszone bei Dammbaustellen oder bei anderen Bauwerken an der Geländeoberfläche festzustellen. Dhawan et al präsentieren Auszüge aus Ergebnissen dieser Art von ihrem Projekt im Himalaya und berichten, daß die aus der Seismik ermittelten Schichtgrenzen mit Bohrkernbefunden innerhalb von 10% übereinstimmten.

In ihrem Abschnitt 2.1 "Gebirgskartierung" schlagen Savich et al vor, daß die Verläßlichkeit geologischer und geophysikalischer Arbeiten verbessert werden kann, wenn verschiedene geophysikalische Methoden gleichzeitig angewendet werden. Ihre Abb.1 ist ein Querschnitt durch den Maueruntergrund, der durch seismische, elektrische und Magnetometeruntersuchungen erforscht wurde. Dort sind geophysikalisch definierte geologische Grenzen gezeigt, zusammen mit Störungen ("Zonen tektonischer Verschiebungen") und Bohrlöchern, die nach Beendigung der geophysikalischen Arbeiten niedergebracht wurden. Die Autoren erheben den Anspruch, daß die ingenieurgeologischen Grenzen, wie sie durch geophysikalische Verfahren ermittelt wurden, mit den Bohrergebnissen innerhalb von 5% Abweichung übereinstimmen.

Savich et al streifen kurz eine Anzahl anderer Anwendungen geophysikalischer Methoden, wie die Abschätzung des Spannungszustandes im Gebirge, das Studium entspannter Bereiche in Böschungen und an künstlichen Hohlräumen und Studien, um den Fels hinsichtlich seines gegenwärtigen in bezug auf den ursprünglichen Zustand zu klassifizieren (vgl. Savich et al (1979)).

Unterthema A.2

FELD- UND LABORVERSUCHE

Im Gegensatz zu ähnlichen Themen bei vorhergehenden IGFM-Kongressen stehen jetzt nicht die Eigenschaften von Gestein und Gebirge, sondern die Versuchstechniken und die Fortschritte bei der Auswertung im Vordergrund. Deshalb ist unser Interesse nun hauptsächlich auf die Fragen gerichtet, wie Feld- und Laborversuche durchgeführt werden sollten, wie sie ausgewertet werden sollten, welche erst- und welche zweitrangigen Parameter gemessen werden sollten, wie diese in die Auswertung eingehen sollten usw.

Feld- und Laborversuche zielen entweder dahin, notwendige Informationen für einen speziellen Entwurf zu liefern, oder dahin, unser Verständnis für die Natur, beispielsweise für das Verformungs- oder Bruchverhalten zu verbessern. Im ersten Fall wird man versuchen, bekannte und bewährte Versuchstechniken anzuwenden, während im anderen Fall oft exotische und sehr ausgetüftelte Versuche entworfen werden. Aber wir erfahren von den zu diesem Unterthema eingereichten Beiträgen, daß auch sogenannte Standardversuche und deren Auswertung verbessert werden können.

Daneben lenkten die 35 Beiträge, welche zu diesem Unterthema eingereicht wurden, unsere Aufmerksamkeit auf die Randbedingungen jedes Versuchs. Nur wenn wir dafür Sorge tragen, diese klar zu definieren und diese hinterher in die Auswertung einzubeziehen, können wir genügend sicher sein, daß die Versuchsergebnisse realistisch sind.

In diesem Zusammenhang muß Nachdruck auf die Gewinnung guter und ungestörter Probekörper und auf die entsprechende Konservierung der Proben gelegt werden. Es ist üblich, die Probenahmebedingungen aufzuzeichnen und zusätzlich in manchen Fällen Skizzen und Fotographien der Prüfkörper herzustellen. Aber wir sind sicher, daß dies nicht ausreichend ist. Zumindest der Verwitterungsgrad der Probe, ihr Wassergehalt und ihre Dichte, die Länge des Kernstücks im Verhältnis zum Durchschnitt, Einzelheiten von Klüften oder anderen Defekten, die Raumstellung der Schichtung und alle Anzeichen einer eventuellen Störung müssen ebenfalls festgehalten werden.

Die zu Feld- und Laborversuchen eingereichten Arbeiten können in mehrere Hauptgebiete eingeteilt werden

- Berücksichtigung anisotroper und nichtelastischer Eigenschaften bei der Auswertung von Versuchen
- Untersuchung der Festigkeit und der Verformbarkeit bei Kurzzeitbelastung
- Bestimmung von Festigkeit und Verformbarkeit bei Dauerlast und zyklischer Last, Studien zum Kriechen und zur Brucheinleitung
- Untersuchung der Scherfestigkeit von Trennflächen
- Maßstabseffekte
- Felsankerversuche

und in einige zusätzliche Themen, wie

- Entwicklung besonderer Versuchseinrichtungen
- Überlegungen zu den Vorschriften bezüglich Vorbereitung der Probekörper
- Vorgehen beim Entwurf
- besondere Anwendungen bekannter Versuchseinrichtungen
- theoretisches Modell des Verformungsverhaltens von Klüften
- Erosionsversuche an Injektionsmaterial
- statistische Beschreibung von Trennflächen
- vorauseilende Erkundung bei Tunneln
- Echtzeitauswertung von Versuchen mittels Rechnern.

Die Generalberichter konzentrieren ihre Ausführungen auf die Hauptgebiete.

BERÜCKSICHTIGUNG ANISOTROPER UND NICHTELASTISCHER EIGENSCHAFTEN BEI DER AUSWERTUNG VON VERSUCHEN

In den letzten 10 Jahren haben sich numerische Methoden als leistungsfähige Hilfsmittel in der Felsmechanik erwiesen. Im Prinzip gibt es nur wenige wesentliche Gebirgseigenschaften, die gegenwärtig nicht numerisch simuliert werden können. Aber gewöhnlich ist der Wert der Methoden in der Praxis begrenzt wegen der mäßigen Qualität oder Verfügbarkeit der Eingabedaten.

Wir empfinden es als unglücklich, daß sogar die wichtigsten Eigenschaften der Anisotropie von Gestein und Gebirge nicht genügend in Rechnung gestellt wurden. Auch wenn man für Versuche viel Geld ausgab, und wenn man anisotrope Verformungen messen konnte, erfolgte die Auswertung zumeist auf der Grundlage isotroper Bedingungen.

Aus physikalischer Sicht werden für die Beschreibung voll anisotroper Verhältnisse 21 Parameter benötigt. Vermutlich wird man diese Forderung nie vollständig erfüllen. Aber - aus der Sicht des Ingenieurs - dürfte eine so exakte und ausgedehnte Beschreibung wegen der Heterogenität des Gebirges andererseits auch nicht gerechtfertigt sein. Aber nichtsdestoweniger muß die bedeutenste Erscheinung der Anisotropie, die transversale Anisotropie, in die Versuchsauswertung einbezogen werden, da dies sowohl beim Gestein als auch beim Gebirge das ausgeprägteste Unterscheidungsmerkmal darstellt.

Transversal anisotropes Verhalten kann durch fünf Parameter, E_1, E_2, ν_1, ν_2, G_2 beschrieben werden. Da das Verformungsverhalten von sedimentärem wie von metamorphem Gebirge gewöhnlich stark transversal anisotrop ist, ist es für die felsmechanische Praxis von beträchtlicher Bedeutung, Methoden zur Bestimmung dieser Parameter zu entwickeln. Wir freuen uns sehr, daß hier einige Beiträge vorgestellt werden, die Anstrengungen erkennen lassen, dies zu tun.

Seit den frühen Tagen der Felsmechanik wird der einaxiale Druckversuch im Labor durchgeführt, um die einaxiale Festigkeit und den Verformungsmodul parallel und normal zur Schichtung herauszufinden. Aber in vielen Fällen können Bohrkerne nicht in diesen speziellen Orientierungen zur Schichtung zur Verfügung gestellt werden. Dann müssen Probekörper mit anderen Orientierungen untersucht werden. Die Ergebnisse müssen dann im Verhältnis zu dem Winkel $90^o - \alpha$ zwischen Belastungsrichtung und Schichtung betrachtet werden. Arbeiten auf diesem Gebiet sind in der Vergangenheit u.a. von John (1969) für den Einaxialversuch, von Barla und Goffi (1974) für die Zugfestigkeit und den Elastizitätsmodul ausgeführt worden. Nun ist diese Entwicklung von Kiehl und Wittke fortgesetzt worden, die zeigen, daß der Schermodul G_2 aus den Ergebnissen des einaxialen Druckversuchs ermittelt werden kann (Abb.2).

Ein weiteres interessantes Versuchsverfahren wird von Amadei et al vorgestellt, die dünne Gesteinsscheiben dazu benutzen, die anisotropen Verformungseigenschaften und die Zugfestigkeit zu ermitteln (Abb.2).

Sogar wichtiger als bei Laborversuchen ist es, die anisotropen Eigenschaften des Gebirges bei Feldversuchen zu betrachten. Feldversuche sind ziemlich teuer und zeitraubend. Man wird deshalb insgesamt weniger Feld- als Laborversuche durchführen. Deshalb sollte man in der Auswertung mehr investieren. Allem Anschein nach waren Oberti et al am IGFM-Kongreß 1979 die ersten, welche die Anisotropie in die Betrachtung einbezogen. Sie leiteten dort eine Theorie zur Auswertung des Lastplattenversuchs her, wobei die Belastung normal und parallel zur Schichtung ausgerichtet war.

Nun werden auf diesem Gebiet eine Anzahl Arbeiten vorgestellt. Oberti et al zeigen, wie man anisotrope Verhältnisse bei der Auswertung des Druckkammerversuchs annehmen kann.

Kiehl und Wittke beschreiben Verfahren zur Verbesserung der Auswertung des Dilatometerversuchs und des LFJ-Versuchs. Zusätzlich werden Ausführungen gemacht zu anisotropen Verhältnissen bei der Auswertung des in-situ Dreiaxialversuchs (Abb.3).

Genano und Sharp berichten über ausgedehnte Spannungsmessungen bei einem Projekt in Südafrika. Von besonderem Interesse ist, daß diese Versuche in mürbem Fels mit Überbohrtechnik durchgeführt wurden. Ihre Entwicklungen beinhalten die kritische Abschätzung der gemessenen Dehnungen und die Bestimmung von Transformationskoeffizienten für Spannungs-Dehnungsbeziehungen bei transversal anisotropem, nichtlinearem Fels und die Formulierung einer numerischen Lösung. Beachtliche Reduzierungen in der Streuung der Ergebnisse (Größe und Richtung der Hauptspannungen) zeigen sehr deutlich den Wert dieser komplizierten Auswertemethode.

Versuch		Autor(en)
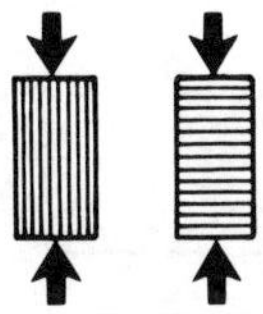	E_1, E_2, ν_1, ν_2	
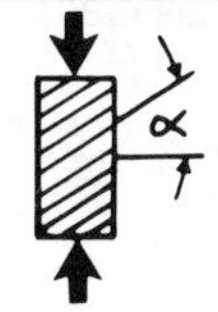	einaxiale Druckfestigkeit in Abhängigkeit von α	John (1969)
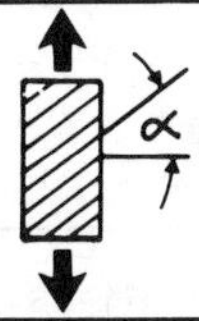	Zugfestigkeit	Barla/Goffi (1974)
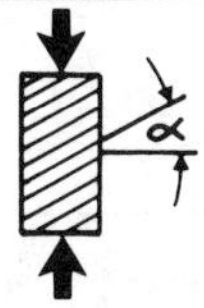	Verformbarkeit in Abhängigkeit von α	Barla/Goffi (1974)
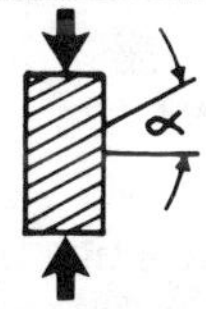	Schubmodul G_2	Kiehl/Wittke (1983)
	$E_1, E_2, G_2, \nu_1, \nu_2$ u. Zugfestigkeit aus Druckversuchen bei diametraler Belastung	Amadei et al (1983)

Abb. 2 Bestimmung anisotroper Eigenschaften im Labor

Bei der Anwendung der großen LFJ-Versuche in einem Versuchsstollen oder an der Wand irgendeines unterirdischen Hohlraums gehen wir normalerweise von elastischen Bedingungen im gesamten Versuchsbereich aus.

Dabei wird vernachlässigt, daß die Bereiche in Wandnähe in den meisten Fällen - auch wenn versucht wird, unnötige Störungen zu vermeiden - aufgelockert oder jenseits des elastischen Bereichs sind. Borsetto et al präsentieren eine Auswertemethode, die elastoplastisches Verhalten des Gebirges unterstellt. Ihr Ziel ist, systematische Fehler bei der Auswertung der Restscherfestigkeit, des ursprünglichen Spannungszustandes und der Verformbarkeit zu ver-

Versuch		Autor (en)
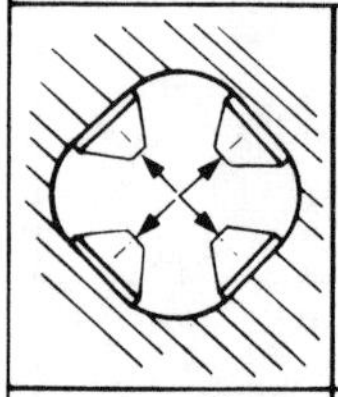	$E_1, E_2, G_2, \nu_1, \nu_2$ aus dem Lastplatten-versuch	Oberti et al (1979)
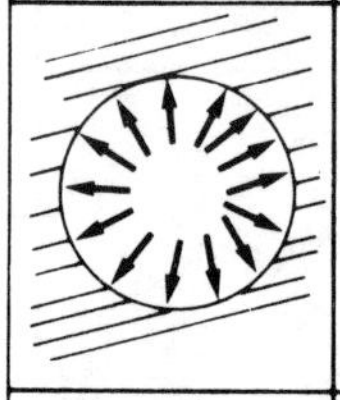	$E_1, E_2, G_2, \gamma_1, \gamma_2$ aus dem Druckkammer-versuch	Oberti et al (1983)
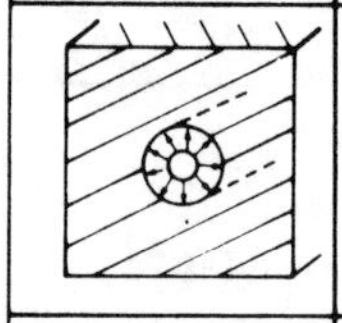	E_1, E_2, G_2 aus dem Dilatometer-versuch	Kiehl / Wittke (1983)
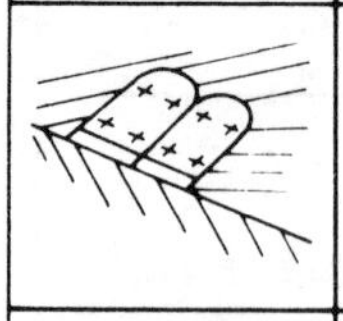	E_1, E_2, G_2 aus dem LFJ-Versuch	Kiehl/Wittke (1983)
	$E_1, E_2, G_2, \nu_1, \nu_2$ aus dem in-situ Dreiaxial-versuch	Lögters / Voort (1974) Auswertung verbessert durch Kiehl / Wittke (1983)
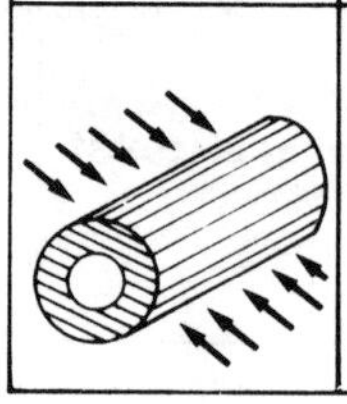	in-situ Spannungen und nicht lineare Verformbarkeit	Gonano/Sharp (1983)

Abb. 3 Bestimmung anisotroper Eigenschaften im Feldversuch

meiden. Da diese Arbeit auf einen Tunnel mit Kreisquerschnitt beschränkt ist, wird vorgeschlagen, daß die Autoren ihre Arbeiten weiterführen und dabei numerische Methoden verwenden.

UNTERSUCHUNG DER FESTIGKEIT UND DER VERFORMBARKEIT BEI KURZZEITBELASTUNG

Festigkeit und Verformbarkeit des Gesteins werden üblicherweise im Labor untersucht. Getreu unserer Absicht, das Schwergewicht mehr auf Versuchs- und Auswertemethoden als auf die Gebirgs- und Gesteinseigenschaften zu legen, werden

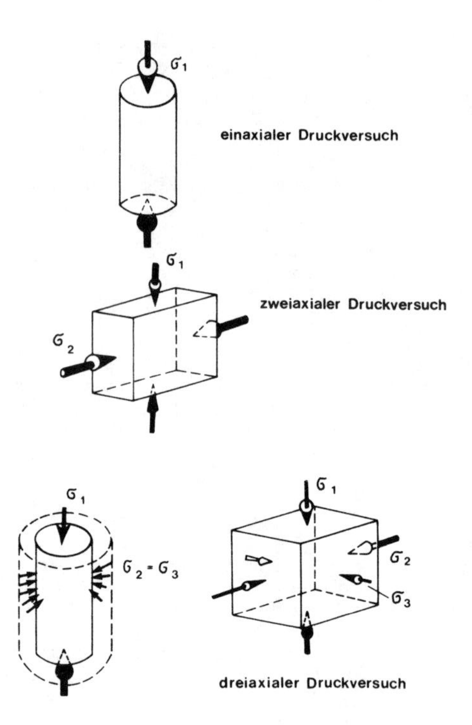

Abb. 4 Arten des Druckversuchs

wir hier mehr die Versuche und die Einflüsse der Versuchsbedingungen als die Ergebnisse diskutieren.

Die Druckfestigkeit des Gesteins kann im Einaxial-(σ_C), Zweiaxial- oder Dreiaxialversuch untersucht werden (Abb.4). Letzterer wird durchgeführt mit zwei gleichen ($\sigma_2 = \sigma_3$) oder ungleichen Hauptspannungen.

Wir sind daran gewöhnt, Ergebnisse des Einaxial- und des Dreiaxialversuchs ($\sigma_2 = \sigma_3$) miteinander zu vergleichen, indem wir beide zusammen als Mohrsche Spannungskreise in ein Schubspannungs-Normalspannungsdiagramm einzeichnen und eine gerade oder gekrümmte Umhüllende an diese Kreise anpassen (Abb.5).

Aber die meisten von uns wissen nicht, wie wir die dirtte, die mittlere Hauptspannung auswerten sollen, wenn σ_2 von σ_3 abweicht. Aber wir erinnern uns, daß Brown (1974) am IGFM-Kongreß 1974 uns den Einfluß der zweiten Hauptspannung σ_2 auf die Festigkeitseigenschaften von Wombeyan Marmor zeigte (Abb.6).

Wir erinnern uns auch, daß die Versuchsergebnisse mehr vom Lasteintragungsproblem beeinflußt waren als von σ_2. Die besten Resultate wurden mit Bürstenplätten erzielt, wie sie von Hilsdorf für Betonuntersuchungen eingeführt worden sind. Es ist interessant zu sehen, welch unterschiedliche Ergebnisse in der Vergangenheit von verschiedenen Forschern erhalten worden sind (Abb.7).

Es ist zu vermuten, daß ein großer Teil, aber nicht all diese Unterschiede von der Reibung verursacht worden ist.

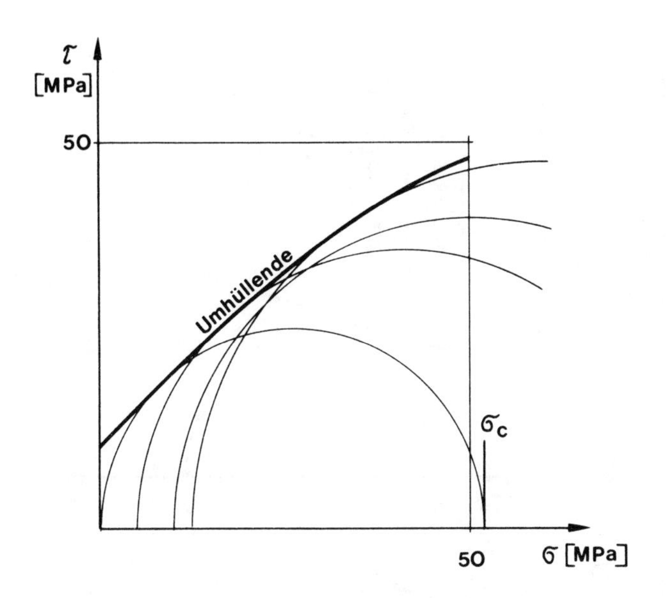

Abb. 5 Bruchkriterium (Beispiel)

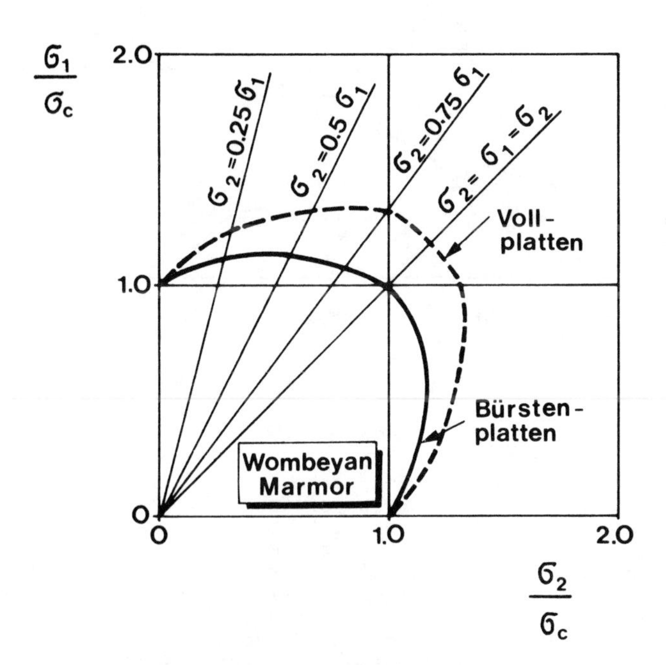

Abb. 6 Einfluß der mittleren Hauptspannung (nach Brown, 1974)

Nun lernen wir aus den Dreiaxialversuchen von Gau und Cheng, die mit $\sigma_1 \neq \sigma_2 \neq \sigma_3$ durchgeführt worden sind, daß die Gesteinsfestigkeit bei vorgegebenem Verhältnis σ_1/σ_3 nicht viel von der mittleren Hauptspannung σ_2 abhängt (Abb.1 und 2 ihres Beitrags). Auch hier kann ein Einfluß der Randbedingung festgestellt werden. Aber er ist viel kleiner als in Abb.7 dargestellt. Beide Arbeiten, die von Brown und die von Gau und Cheng lassen es für den Entwurf nicht angebracht erscheinen, von einer erhöhten Festigkeit auszugehen, wenn die mittlere (σ_2)

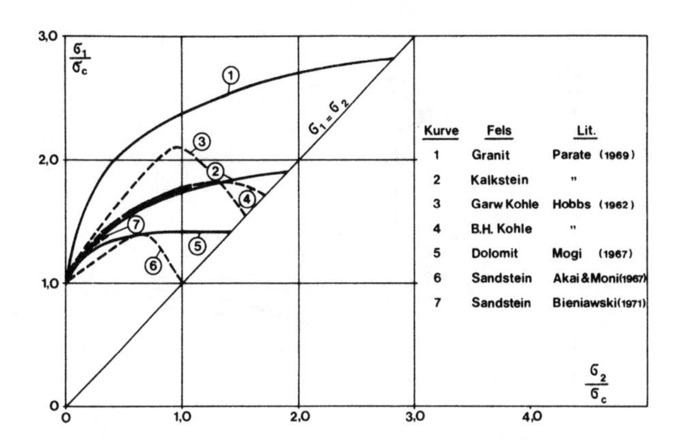

Abb. 7 Ergebnisse von Bialxialversuchen aus dem Schrifttum (nach Brown, 1974)

die kleinste Hauptspannung σ_3 überschreitet. Dies entspricht der Theorie (Abb.8).

Die Versuchsergebnisse von Druckversuchen werden außer von den Belastungsbedingungen auch von der Belastungsgeschwindigkeit beeinflußt. Die ISRM-Suggested Methods empfohlen, die Belastungsgeschwindigkeit so einzurichten, daß der Bruch nach 5 bis 10 Minuten Versuchsdauer eintritt. Diese Empfehlung ist entstanden im wesentlichen unter dem Blickwinkel harter Gesteine. Für Weichgesteine, die durch geringere Festigkeit und höheren Wassergehalt gekennzeichnet sind, darf diese Vorschrift nicht unbesehen übernommen werden. Hier muß man sich mit der Belastungsgeschwindigkeit an die Dehnungsgeschwindigkeit anpassen.

Bei Weichgesteinen gibt es oftmals zusätzlich das Phänomen, daß der Wassergehalt innerhalb sonst gleicher Proben (z.B. bedingt durch erhöhte Porosität infolge Verwitterung) sehr unterschiedlich sein kann. Man wird dann, auch bei entsprechender Verminderung der Belastungsgeschwindigkeit unterschiedliche Festigkeiten ermitteln. Chiu und Johnston zeigen dies am Beispiel einaxialer Druckversuche an Schluffsteinproben aus der Melbourner Gegend (Abb.9).

Parallel zur Druckfestigkeit wird meistens die Verformbarkeit des Gesteins festgestellt. Auf den Einfluß der Anisotropie zu diesem Punkt wurde bereits in einem früheren Abschnitt hingewiesen. Betrachtet man nur isotropes Material, so ist klar, daß auch die im Versuch gemessene axiale Verformbarkeit vom Seitendruck abhängt. Man muß hier unterscheiden zwischen einer Erhöhung, die sich nur scheinbar einstellt, und die durch teilweise behinderte Seitendehnung erzeugt wird (ähnlich zum Verhältnis zwischen E-Modul und Steifemodul bei Böden) und einer Erhöhung, die als markante Eigenschaft eines bestimmten Gesteins anzusehen ist. Für letztere geben Chiu und Johnston hier ein Beispiel (vgl. Abb.10), wobei zusätzlich der Einfluß des Wassergehalts gezeigt wird. Ähnliche Ergebnisse zeigt Kwasniewski in seiner Abb.7 aus Sandsteinproben.

Das Ziel vieler Überlegungen ist es, zuverlässige funktionale Beziehungen für das Bruchver-

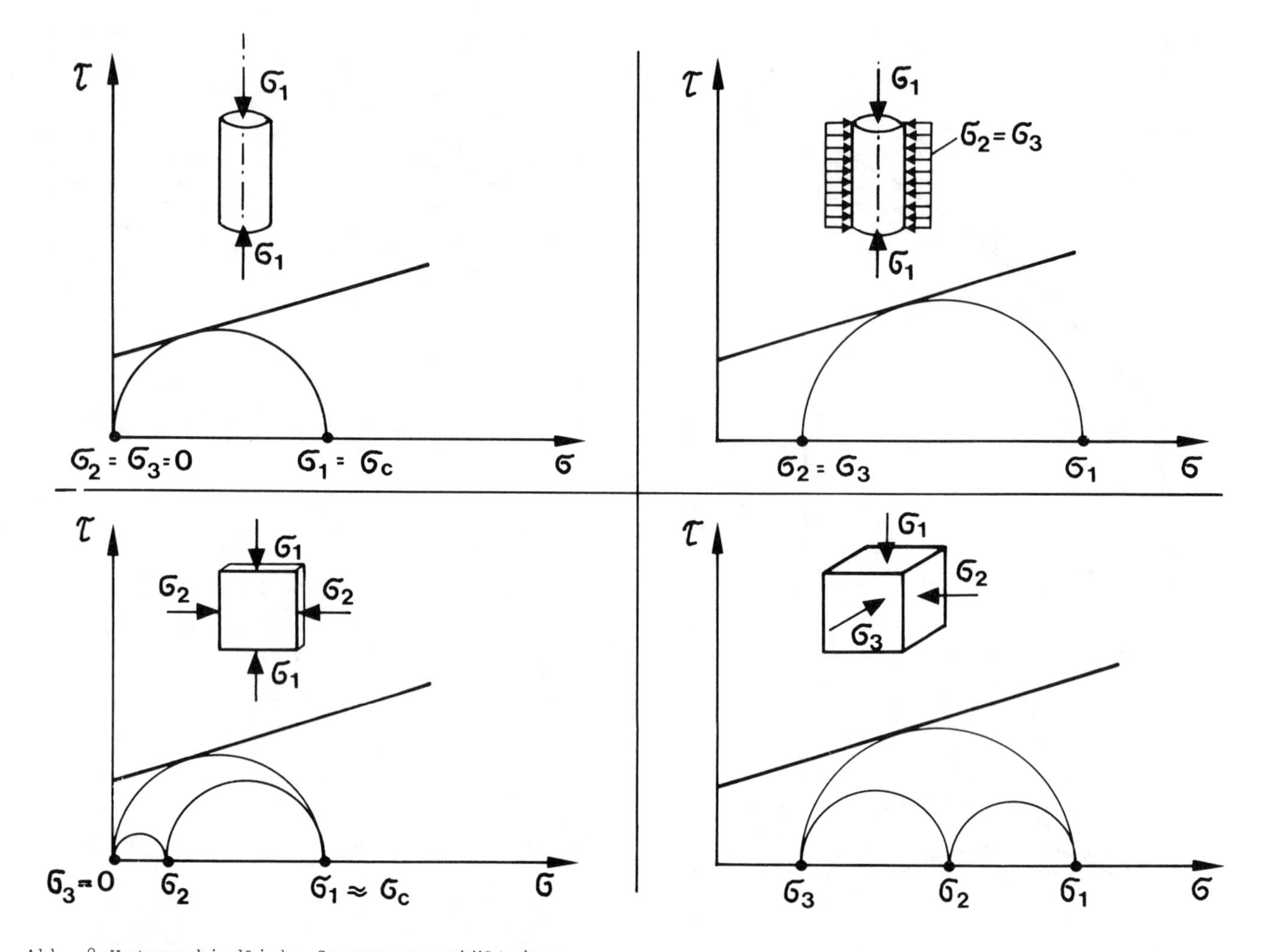

Abb. 8 Unterschiedliche Spannungsverhältnisse beim Bruch

halten des Gesteins zu finden. Man muß hier streng unterscheiden zwischen solchen, die aus theoretischen Überlegungen entwickelt wurden (z.B. Mohr-Coulomb, Griffith o.ä.) und solchen, die aus Versuchsergebnissen abgeleitet werden. Beispiele für letztere sind z.B. auch in den zu diesem Kongreß eingereichten Arbeiten von Gau und Cheng, Chiu und Johnston und Kwasniewski zu finden. Abgesehen davon, daß diese nur jeweils für ein bestimmtes Gestein, nämlich das untersuchte, gelten können, sind sie natürlich von den jeweils verwendeten Versuchsbedingungen abhängig, mithin an anderer Stelle nur begrenzt nachvollziehbar. Dies macht es auch schwierig, unterschiedliche Gesetze unterschiedlicher Autoren zu vergleichen und zu werten.

Die bisherigen Kommentare betrafen durchwegs nur Laborversuche an Gesteinsproben. Demgegenüber erscheint es wichtig, auf eine besondere Entwicklung hinzuweisen, die es erlaubt, im Labor auch geklüftete Felsproben dem Dreiaxialversuch zu unterwerfen. Es ist nicht zu leugnen, daß ein großer Teil der Felsbaumaßnahmen in dicht geklüftetem, weichem Fels stattfindet. Hier hat man zumeist Schwierigkeiten, die Festigkeit und Verformbarkeit zuverlässig zu bestimmen. Dies war in Karlsruhe der Anlaß, Großdreiaxialversuche im Labor zu entwerfen und eine eigene Technik zur Entnahme ungestörter Bodenproben zu entwickeln. Über die Technik ist bereits verschiedentlich berichtet worden (vgl. z.B. Wichter (1979)). Die Deutsche Gesellschaft für Erd- und Grundbau hat inzwischen eine Empfehlung für diesen Versuch veröffentlicht. Hier werden von Natau et al neuere Erfahrungen vorgestellt.

BESTIMMUNG VON FESTIGKEIT UND VERFORMBARKEIT BEI DAUERBELASTUNG UND ZYKLISCHER BELASTUNG, KRIECHEN, BRUCHENTWICKLUNG

Gestein verhält sich bei kurzzeitiger Belastung mehr oder weniger linear elastisch. Wird es aber über lange oder längere Zeit belastet, so treten sehr unterschiedliche Mechanismen in Erscheinung. Diese können mit verschiedenen rheologischen Modellen näherungsweise beschrieben werden.

Im allgemeinen wenig bekannt ist, daß sich sogar harte Gesteine und zwar selbst bei kleinen

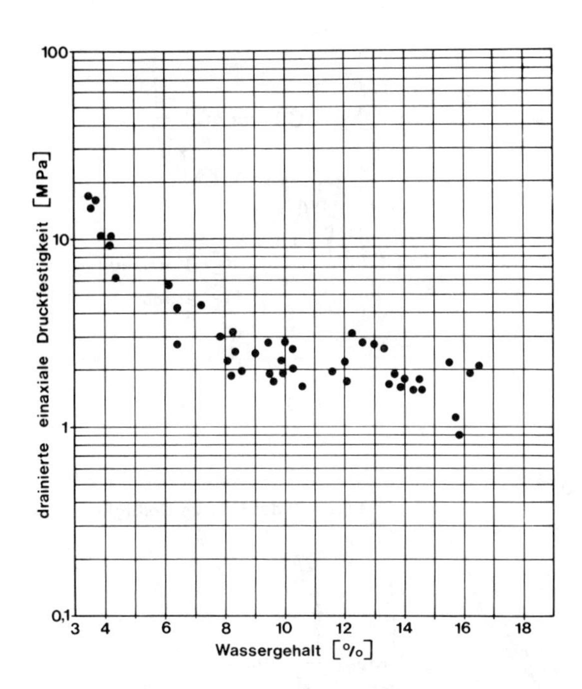

Abb. 9 Einfluß des Wassergehalts auf die Druckfestigkeit von Melbourner Schluffstein (nach Chiu und Johnston)

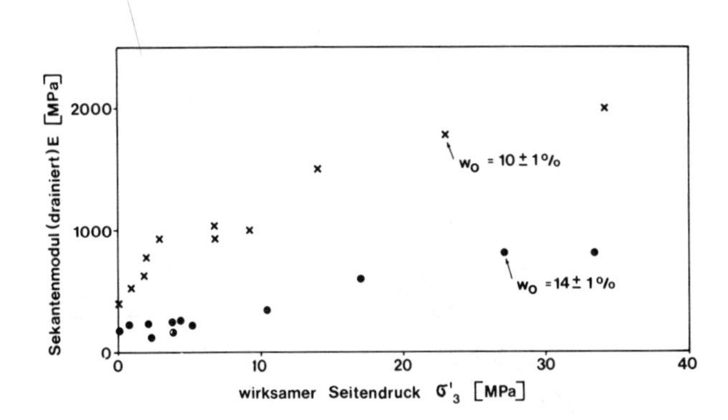

Abb. 10 Einfluß des Seitendrucks auf die Verformbarkeit von Melbourner Schluffstein (nach Chiu und Johnston)

Spannungen viskos verhalten. Ito berichtet von Langzeitversuchen an Granitbalken, die bereits wegen der Versuchsdauer von ca. 24 Jahren bemerkenswert sind. Diese Versuche scheinen die bereits von Nabarro und Herring geäußerte Ansicht zu bestätigen, daß Granit keine Fließgrenze hat. Die untersuchten Versuchskörper fließen nämlich nahezu unabhängig von den aufgebrachten Spannungen (Kurve 1 in Abb.11). Ito berichtet auch von ergänzenden Versuchen an Gabbroversuchskörpern, aus denen zusätzlich hervorgeht, daß sowohl der Seitendruck als auch erhöhte Temperatur die Viskosität merklich verringert. Diese Arbeit läßt vermuten, daß Fels sich bei Verformungsgeschwindigkeiten von $>10^{-13}$/s als Hooke-Körper, bei kleineren Verformungsgeschwindigkeiten als Newtonflüssigkeit verhält. Von grundlegender Bedeutung für das Verständnis mancher geologischer Vorgänge scheint uns zu sein, daß die im Labor ermittelten Viskositäten Geologen und Geophysikern, die mit Studien über die Verformungen der Erdkruste befaßt sind, bekannt werden.

Viskose Vorgänge bei Hartgesteinen laufen natürlich sehr langsam ab und sind daher für den Felsbau kaum von Bedeutung. Dafür ist die Belastungsgeschwindigkeit in der Regel nicht zu vernachlässigen. Es ist allgemein bekannt, daß die Belastungsgeschwindigkeit im einaxialen Druckversuch die Höhe der Bruchfestigkeit beeinflußt. Ebenso hängt die Bruchdehnung davon ab und zwar deshalb, weil sich die Dehnungen gewöhnlich aus einem elastischen, einem viskoelastischen sowie aus einem plastischen (irreversibles Kriechen) Anteil zusammensetzen. Je langsamer belastet wird, desto mehr fallen die Kriechanteile ins Gewicht. Ein solches Verhalten ist in Kurve 2 in Abb.11 beschrieben. Es ist bekannt, daß ab einer bestimmten kritischen Spannung die zeitabhängigen Verformungen zum Bruch führen. Dementsprechend unterscheidet man auch primäres (verlangsamtes), sekundäres (gleichmäßiges) und tertiäres (beschleunigtes) Kriechen. Darunter liegende Beanspruchungen führen letztlich zum Stillstand der Verformungszunahmen (Abb.12). Analog dazu reagiert das Gestein auf zyklische Belastung. Wir haben dazu zwei sehr interessante Beiträge, einen von Homand-Etienne et al (Granit und Marmor), einen von Akai und Ohnishi (poröser Tuff). Ähnlich wie beim Kriechen lassen sich auch bei zyklischer Belastung drei unterschiedliche Phasen definieren (Abb.13a), so daß sich der Eindruck aufdrängt, hier handle es sich im Prinzip um sehr verwandte Vorgänge im Gestein.

Erklären lassen sich beide Phänomene mit der progressiven Entwicklung von Mikrorissen, die mit der Belastungsdauer (bzw. der Anzahl der Lastzyklen) und der Belastungshöhe zunehmen. Unterhalb der kritischen Spannung kommt die Rißbildung zur Ruhe, oberhalb beschleunigt sie progressiv. Homand-Etienne et al fanden in den Versuchen, daß die Dauerfestigkeit des Granits bei 85%, die des Marmors bei 82% der Kurzzeitfestigkeit liegt. Ähnliche Ergebnisse sind vom Beton bekannt.

Die Theorie der Mikrorißbildung läßt sich auch als statistisch orientiertes mathematisches Modell verwirklichen. Weber und Saint-Lot stellen hier eine derartige Arbeit vor. Es ist bemerkenswert, daß es trotz mancher noch vereinfachender Annahmen möglich ist, damit gemessene Spannungs-Verformungslinien nachzuvollziehen.

Der zum Bruch führende Vorgang, welchen wir bisher an Gesteinsproben diskutiert haben, läßt sich auch an Gebirgskörpern in ähnlicher Form beobachten. Auch hier lassen sich unter Dauer- bzw. Wechsellast drei unterschiedliche Phasen irreversibler Verformungen definieren. Müller und Ge demonstrieren dies hier sehr deutlich am Verhalten einiger Versuchskörper bei den Feldversuchen zur Kurobestaumauer. Die erste Phase ist zu sehen in Verbindung mit einer Verdichtung des geklüfteten Materials. Die zweite Phase kennzeichnet nach Müller und Ge einen Zustand relativer Stabilisierung, während in der dritten Phase der Bruch eingeleitet wird. Eine Unterscheidung in diese drei

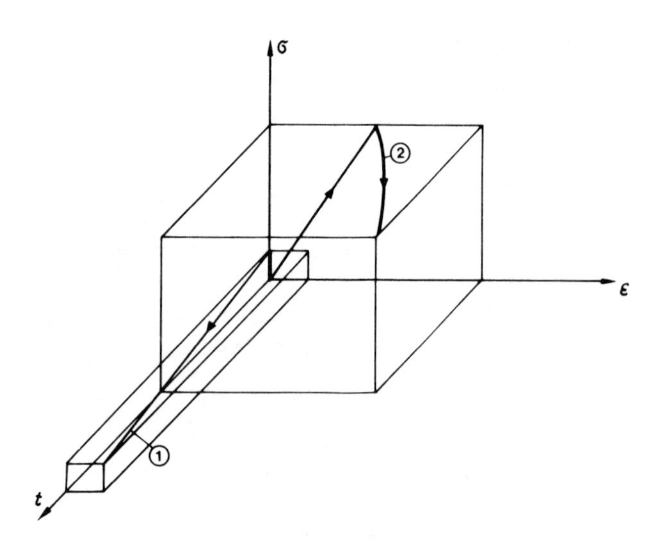

Abb. 11 σ – ε – t – Diagramm. ① viskoses Fließen, ② Kriechen

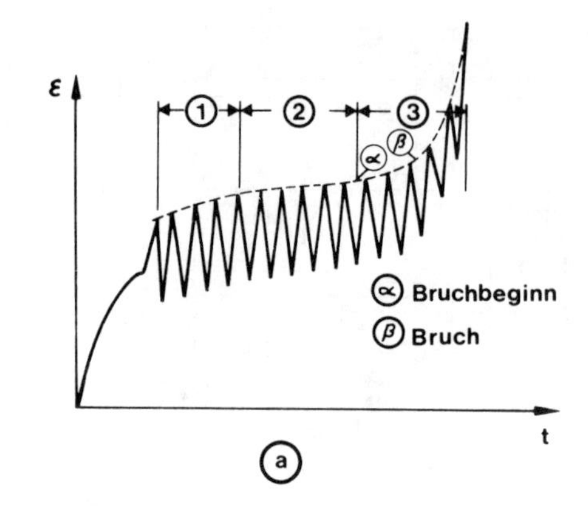

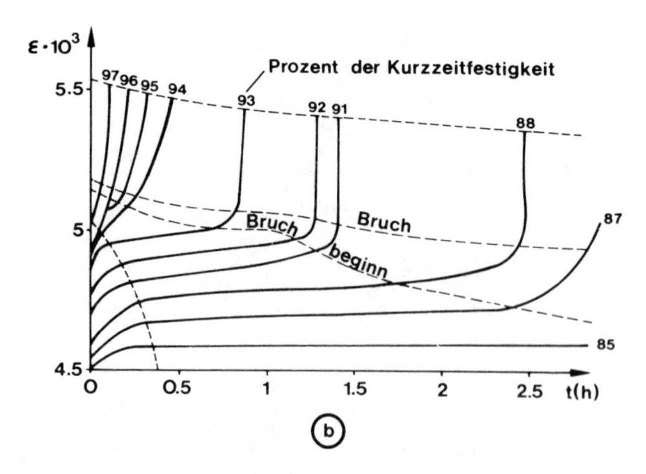

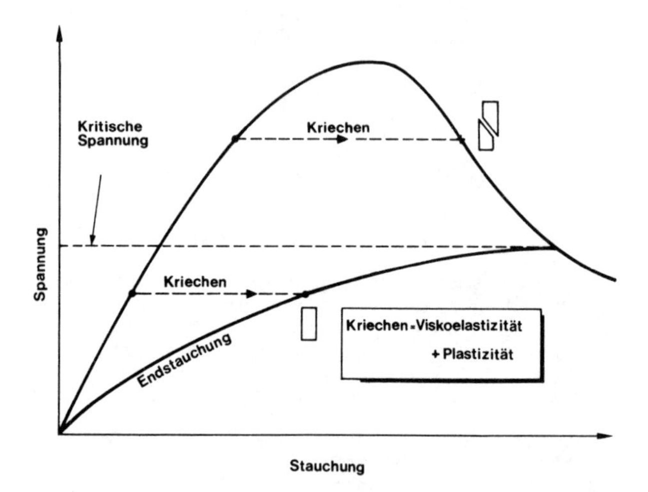

Abb. 12 Spannungs-Dehnungsverhalten bei Dauerlast

Abb. 13 Verhalten unter zyklischer Last (nach Homand-Etienne et al) (① abnehmende, ② konstante, ③ zunehmende Verformungsgeschwindigkeit)

Bereiche scheint gerade für Großversuche wesentlich zu sein. Denn gerade die Sicherheitsbeurteilung nach Großversuchen sollte sich vorwiegend auf das Verformungsverhalten des Gebirges abstützen.

UNTERSUCHUNG DER SCHERFESTIGKEIT VON KLÜFTEN

Beim IGFM-Kongreß 1974 war ein ganzes Unterthema dem Scherversuch gewidmet worden. Auch 1979 in Montreux wurden hierzu grundlegende Beiträge vorgelegt.

Die Scherfestigkeit von Trennflächen ist ein ziemlich komplexes und sicherlich noch nicht restlos geklärtes Thema. Hierzu sind zunächst einige allgemeine Bemerkungen angezeigt:

Wird ein Kontinuum einer Scherbeanspruchung unterworfen, so ergibt sich ein Verhalten, so wie es schematisch in Abb. 14a gezeigt ist. Die τ – Δh -Kurve läßt eine ausgeprägte Spitzenscherfestigkeit τ_t erkennen. Diese markiert im τ – σ -Diagramm einen Punkt der Bruchbedingung. Ganz anders läuft der Schervorgang entlang einer vorgegebenen glatten Trennfuge ab (Abb. 14b). Hier wachsen die Scherverformungen kontinuierlich aber progressiv mit zunehmender Schubspannung τ an. Bei der Schubspannung τ_r nimmt die Verformung Δh zu, ohne daß die Schubspannung weiter steigt. In beiden idealisierten Modellen ist das Abscheren nicht mit vertikalen Verschiebungen Δv verbunden.

Ganz anders bei dem in Abb.14 c und d gezeigten Modell, welches einer natürlichen Trennfläche bereits recht nahe kommt. Hier ist das Verhalten durch die gezackte, allerdings noch idealisierte Scherfläche bestimmt. Solange die Zähne nicht beschädigt sind, kann eine Deformation nur eintreten, wenn die beiden Teilflächen an den Zahnflanken gegeneinander gleiten.

Dies setzt in jedem Fall auch die gegenseitige Verschiebung senkrecht zur Scherebene voraus. Entsprechend dem Neigungswinkel i (vgl.Abb.14c) ist die Festigkeitsbedingung im τ – σ – Diagramm unter dem Winkel $\phi + i$ geneigt. Ab einer ge-

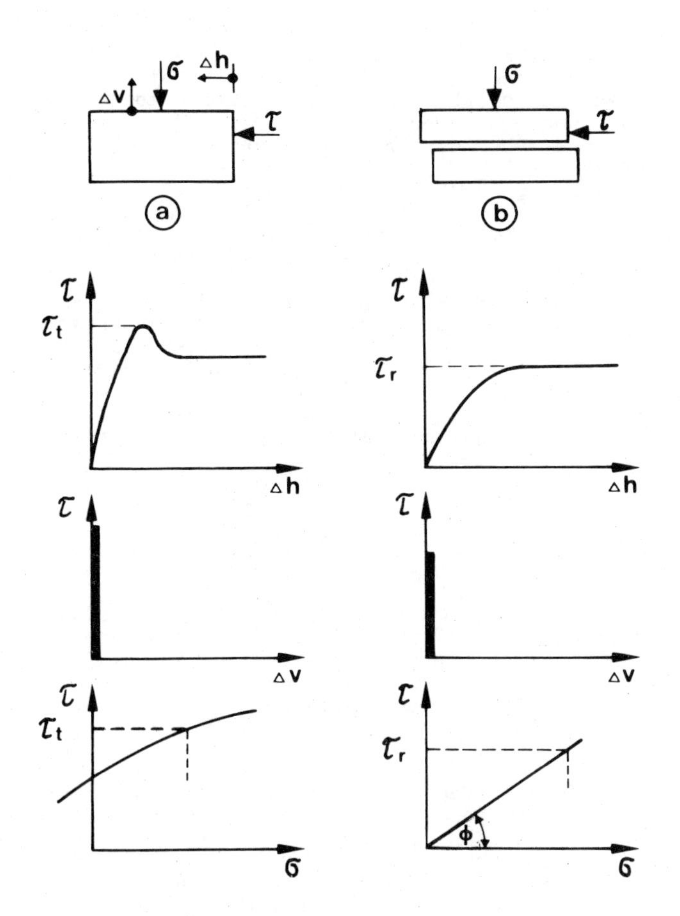

Abb. 14 Scherverhalten unter unterschiedlichen Bedingungen ((a) Gestein, (b) glatte Kluft, (c) rauhe Kluft vor, (d) rauhe Kluft nach dem Bruch der Zähne)

wissen Größe der Beanspruchung τ und σ gehen aber in Wirklichkeit die Zähne zu Bruch. Dieser Zustand und die daraus resultierenden Folgen sind - allerdings schematisch - in Abb.14d dargestellt. Man erkennt, daß die Scherbewegung dann in zunehmendem Maße in Scherrichtung erfolgt. Die Bruchbedingung (im τ - σ - Diagramm) wird flacher und nähert sich dem Wert ∅

In der Praxis sind die Scherflächen allerdings sehr unregelmäßig. Damit ergeben sich keineswegs die idealisierten Kurven. Je nach Rauhigkeit, Gesteinsbruchfestigkeit, Verwitterungsgrad, Füllungen, Mineralogie, Wassergehalt und Wassersättigung des Gebirges erhält man sehr verschiedene Ergebnisse. Abb.15 zeigt ein Beispiel, das Barton (1971) veröffentlicht hat. Das Verhalten von Trennflächen auf Scherbeanspruchung hängt natürlich auch davon ab, ob die Trennfläche zuvor bereits Bewegungen erlitten hat, ob also die Rauhigkeiten bereits abgeschert sind. Deshalb ist prinzipiell beim derzeitigen Kenntnisstand zu erwarten, daß die Festigkeitsbedingungen in jedem Fall unterschiedlich sind. Beispiele hierfür werden auch hier vorgestellt (z.B. von Tinoco und Salcedo, Everitt und Goldfinch, Bollo et al).

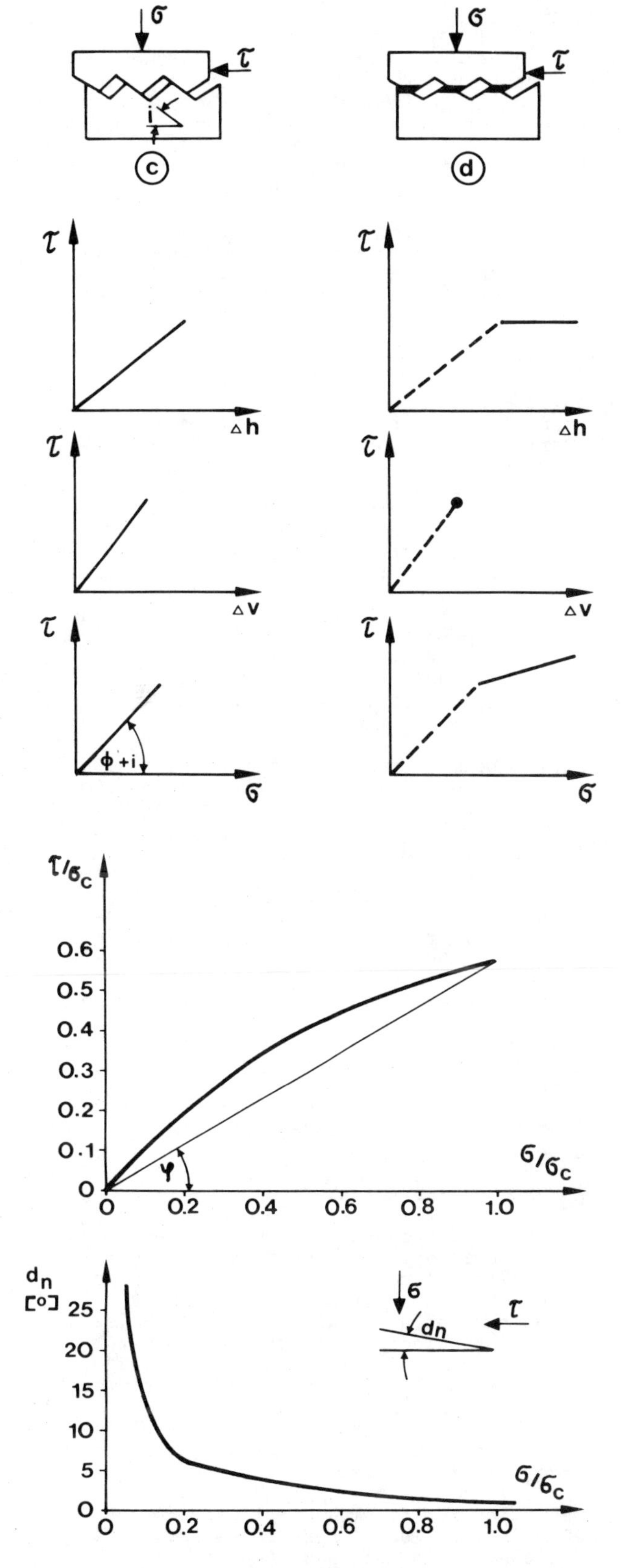

Abb. 15 Scherverhalten einer natürlichen Kluft (nach Barton, 1971)

MAßSTABSEFFEKTE

Bei geklüftetem Hartgestein wird oftmals nur die Scher- und Zugfestigkeit wesentlich von der Klüftigkeit beeinflußt. Die Verformbarkeit wird dagegen nur wenig vergrößert. Anders dagegen bei Weichgesteinen, insbesondere dann, wenn sie eng geklüftet sind. Hier ist der Einfluß der Klüfte oft von dominierender Größe. Kluftabstände von wenigen Zentimetern bedingen, daß sich der Fels mehr wie ein Trockenmauerwerk, oftmals sogar noch ein solches mit weichen Ziegeln und geschmierten Fugen, verhält. Jeder Kluftkörper hat einen beachtlichen Freiraum für Translation und Rotation. Demzufolge ist die Festigkeit gegenüber der des Gesteins wesentlich abgemindert, die Verformbarkeit erhöht.

Schwierigkeiten bereitet es allerdings, diese Abminderung zuverlässig zu bestimmen. Denn ein derartiger Kluftkörperverband kann unter Beanspruchung ein kinematisch hochgradig bewegliches System sein. Die Festigkeit und die Verformbarkeit eines solchen Systems hängt dann aber stark von den Randbedingungen des Versuchs ab. So dürfte beispielsweise ein Einaxialversuch andere Ergebnisse bringen als ein Dreiaxialversuch und beide dürften vom natürlichen Felsverhalten je nach dem dort herrschenden Spannungszustand noch um einiges entfernt sein. Es wäre zu begrüßen, wenn sich die Autoren (Natau et al, Sun und Zhou), welche hier zu diesem Thema beitragen, sich zu dieser Frage äußern könnten.

Während beim Versuch z.T. beachtliche Maßstabseffekte bei Druckfestigkeit und Verformbarkeit beobachtet werden, ist dieser Einfluß bei der Scherfestigkeit entlang von Trennflächen allen Anschein nach nicht in allen Fällen so groß. Dies wird auch aus den hier vorgestellten Arbeiten von Pasamehmetoglu et al und von Bollo et al deutlich. Hieraus und z.B. aus einer früheren Arbeit von Leichnitz und Natau (1979) ist zu schließen, daß Maßstabseffekte im wesentlichen bei der Spitzenscherfestigkeit und da wiederum bei rauhen Kluftflächen beobachtet werden. Deren Zähne sind zwar nach Größe und Steilheit mehr oder weniger zufällig über die Kluftfläche angeordnet. Jedoch gerade dadurch kann eine kleine Scherprobe zufällig eine überproportionale Verzahnung aufweisen, die der Gesamtheit nicht entspricht.

PRÜFUNG VON FELSANKERN

Felsankerprüfungen können verschiedenen Charakter besitzen. Sie können zerstörungsfrei und zerstörend sein. Sie können zur Eignungsprüfung durchgeführt werden oder zur Routineüberwachung an der Baustelle. Schließlich gibt es noch Grundlagenuntersuchungen. Die zu diesem Themenkreis eingereichten Beiträge beschäftigen sich z.T. mit Grundlagenuntersuchungen an Großankern (Natau und Wullschläger) und an verkleinerten Labormodellen (Ballivy et al), z.T. mit der Qualitätsüberwachung von kurzen einzementierten Felsankern.

Wesentlich für die Tragkraft eines Ankers auf Dauer ist, daß die Ankerstrecke (sowohl Injektionsmaterial wie auch Ankerstahl) unbeschädigt erhalten bleibt. Dabei ist insbesondere der Spannungszustand in der Verankerungsstrecke dafür maßgebend, ob die Umhüllung der Stahlstäbe rissefrei bleibt. Wesentlich hierbei ist u.a. die Verformbarkeit des umgebenen Gebirges. Natau und Wullschläger simulieren dies durch ein ausbetoniertes Stahlrohr. Ballivy et al verwenden Gesteinskörper, allerdings kleinen Durchmessers.

Eine interessante Entwicklung wird von Bergman et al vorgestellt. Das Boltometer, ein Gerät zur zerstörungsfreien Prüfung von kurzen einzementierten Felsankern beruht auf der Laufzeit- und Amplitudenmessung von Wellen im Ankerstahl. Je nach Grad der Einhüllung des Ankerstahls wird die Amplitude mehr oder weniger geschwächt. Wir glauben, daß dies eine interessante Entwicklung ist, doch scheint sie uns - auch nach den vorgeführten Versuchbeispielen - noch nicht voll ausgereift.

Unterthema A.3

KLASSIFIZIERUNG, VORHERSAGE, BEOBACHTUNG UND ÜBERWACHUNG

Seit dem von Terzaghi 1946 (Terzaghi,1946) veröffentlichten System hat es eine Wucherung von Gebirgsklassifizierungssystemen gegeben. Sie sind entworfen worden, um den Meinungsaustausch zwischen Geotechnikern und die Vorhersage des Gebirgsverhaltens zu unterstützen. Wegen der in der Literatur nun vorhandenen großen Zahl unterschiedlicher Klassifizierungssysteme besteht die Gefahr, daß die Ingenieure, welche solche Systeme anwenden wollen, Schwierigkeiten haben zu entscheiden, welches System bei ihren speziellen Verhältnissen angewandt werden soll. Auch aus den Beiträgen zu diesem Thema wird deutlich, daß derartige Schwierigkeiten bestehen. Es schien den Generalberichtern deshalb zweckmäßig, bei diesem Kongreß zu versuchen, ein besseres Verständnis für die Stärken und Schwächen der bestehenden Systeme herzustellen und, falls möglich, eine gewisse Übereinstimmung über ihre Anwendbarkeit.

Da wir die bestehende „Wucherung" bereits angesprochen haben, bestürzt es sicherlich, daß wir die Notwendigkeit einer weiteren Klassifizierung herausstellen - diesmal, um die Klassifizierungen selbst in verschiedenen Gruppen zu betrachten, sie also zu klassifizieren.

Gruppen von Klassifizierungssystemen

In der Folge werden wir uns auf die zwei wesentlichen Bestandteile von Fels beziehen, nämlich auf die Substanz (Material) und auf die Fehlstellen (Trennflächen), und dabei feststellen, daß auch Flüssigkeiten in den Fehlstellen und/oder in den Gesteinsporen wichtig sind und in einigen Systemen in Rechnung gestellt werden.

A. Klassifizierungen nach den Eigenschaften von Bestandteilen

In diesen Klassifizierungen werden Wertebereiche einer einzelnen Eigenschaft eines Bestandteils durch Namen, Buchstaben oder Zahlen bezeichnet. Die Wertebereiche sind gewöhnlich nach der Erfahrung festgelegt. Der Benutzer der Klassifizierung wählt die zutreffende Zahl nach den Ergebnissen einfacher Versuche oder Messungen. Beispiele dieses Klassifizierungstyps sind die Gesteinsklassifizierung nach dem Punktlastindex (Broch und Franklin, 1972) und die Klassifizierung unter Verwendung von vier Komponenten, wie sie von dieser Gesellschaft 1980 („Basic Geotechnical Description of Rock Masses", Tabellen 1 bis 4, International Society for Rock Mechanics, 1980) veröffentlicht wurde.

B. Klassifizierungen unter Einbeziehung mehrerer Eigenschaften

Dies sind Systeme, mit denen im Gebirge „Homogenbereiche" bezüglich allgemein technisch wichtiger Eigenschaften unter Zuhilfenahme mehrerer Beschreibungskriterien festgelegt werden können. In vorstehend, genannter Veröffentlichung (ISRM, 1980) wird das BGD-System ("Basic Geotechnical Description of Rock Mass") beschrieben, bei dem das Gebirge durch einen die Genese kennzeichnenden Namen und zusätzliche Buchstaben und Zahlen zur Beschreibung von vier Bestandteilen bezeichnet wird. Der Hauptzweck dieses und anderer, ähnlicher Systeme ist, eine standardisierte Sprache für den Geotechniker zu schaffen.

C. Klassifizierungen unter (gewichteter) Einbeziehung mehrerer Eigenschaften

Bei diesem Typ werden die Gebirgseigenschaften, wie vorstehend bei B, durch die Eigenschaften mehrerer Bestandteile beschrieben, und diesen werden Gewichte entsprechend ihrem vermutlichen Beitrag zur Gebirgsfestigkeit zugeordnet. Die Gewichte werden aus der Erfahrung, aus Versuchen und aus Berechnungen bestimmt. Aus der Summe der Gewichte errechnet sich die Gesamtwichtung für das Gebirge.

Die Systeme nach Barton et al (1974) und Whickam et al (1972) sind speziell für die Abschätzung von Auskleidungserfordernissen bei untertägigen Hohlräumen entworden worden. Bieniawskis RMR-System (Rock Mass Rating) (1974a,1979) ist mehr für allgemeinen Gebrauch gedacht. Alle drei Systeme sind mit Erfolg für die Vorhersage von Auskleidungsmaßnahmen bei Tunnels verwendet worden. Sowohl Bartons Q als auch Bieniawskis RMR haben gezeigt, daß sie ganz gut mit dem Verformungsmodul des Gebirges in Beziehung zu setzen sind (Bieniawski,1979).

D. Klassifizierungen der Gebirgsverwitterung

Es gibt viele Klassifizierungen, die in speziellen geologischen Umgebungen und für verschiedene Zwecke zur Beschreibung verwitterten Gebirges entwickelt wurden. Ein Beispiel dafür ist das hier in Melbourne benutzte, bei welchem das Verwitterungsprofil des aus dem Silur stammenden Schluffsteins in Bereiche eingeteilt wird (Neilson,1970). Saunders und Fookes (1970) beschreiben verschiedene, für in unterschiedlichem Klima verwitterten Fels entwickelte Systeme, und Sancio und Brown (1980) beschreiben ein Klassifizierungssystem für verwitterten Fels zur Vorhersage der Gefahr von Böschungsinstabilitäten.

Die Wucherung der Klassifizierungssysteme gibt uns einigen Anlaß zur Sorge. Die Situation ist ähnlich derjenigen am Beginn dieses Jahrhunderts, als Petrologen eine Wucherung von Klassifizierungen für Eruptivgesteine und von Gesteinsnamen zustande brachten, die noch heute unter Geologen, geschweige denn unter Ingenieuren, Verwirrung anrichtet. Können wir in einer Zeit, in der es Mode ist zu rationalisieren, unseren Gebrauch geotechnischer Klassifizierungen und unsere Einstellung zur Entwicklung neuer Klassifizierungen wirtschaftlicher gestalten?

Mit Blick zunächst auf die Gruppe A meinen die Generalberichter, daß es möglich sein müßte, weltweit standardisierte Gruppeneinteilungen für folgende Eigenschaften einzuführen:

- Gesteinsfestigkeit
- Schichtdicke
- Kluftabstand
- Kluftfestigkeit oder Kluftbeschaffenheit
- Grundwasserverhältnisse

Wenn die Eigenschaften nach Kategorie A standardisiert würden, so wären auch die Systeme nach B ohne Schwierigkeit zu vereinheitlichen. Wir hätten dann eine Sprache für Fels, wahrscheinlich gleichwertig dem "Unified Soils Classification"-System. Die meisten der mit Entwurf und Konstruktion von Bauwerken im Boden Beschäftigten werden uns zustimmen, daß das "Unified Soils Classification"-System eine äußerst wertvolle internationale Sprache ist.

Zweifellos sind die größten „Wucherungen" bei der Kategorie D zu verzeichnen. Dies hängt wahrscheinlich zusammen mit der äußerst großen Verschiedenartigkeit verwitterten Gebirges, einerseits innerhalb einer bestimmten, andererseits zwischen unterschiedlichen geologischen Umgebungen. Weiterhin könnte hierzu der ausgeprägte Wunsch vieler Geologen zu klassifizieren beitragen. Wir geben zu, daß viele dieser Klassifizierungen in begrenzten geologischen Bereichen nützlich sind, sind aber der Meinung, daß keine davon für den allgemeinen Gebrauch zu übernehmen ist.

Der mögliche Wert von Klassifizierungen der Gruppe C steht außer Frage, wie dies auch in der folgenden Feststellung von Hoek und Brown (1980) zum Ausdruck kommt:

> „In Anbetracht der Spärlichkeit zuverlässiger Informationen über die Gebirgsfestigkeit und die sehr hohen Kosten, die es verursachen würde, diese Informationen zu erhalten, halten es die Verfasser für unwahrscheinlich, daß je eine umfassende Analyse der Gebirgsfestigkeit möglich sein wird. Da dies eine der Schlüsselfragen der Felsmechanik ist, ist klar, daß Versuche unternommen werden sollten, um irgendwelche allgemeinen Angaben zum ungefähren Verlauf der Gebirgsfestigkeit machen zu können.
>
> Nachdem die Verfasser mehrere Alternativen betrachtet haben, haben sie sich den Gebirgsklassifizierungen zugewandt...."

In den diesem Zitat folgenden Seiten bringen Hoek und Brown für die am häufigsten vorkommenden Gebirgstypen Gleichungen zur näherungsweisen Bestimmung der Gesteins- und Gebirgsfestigkeit und verknüpfen diese mit dem gesamten Bereich der RMR- (Bieniawski) und der Q-Indizes (Barton et al). Hoek und Brown führen aus, daß ihre Gleichungen nur für den Vorentwurf und für Sensibilitätsstudien benutzt werden sollten. Wegen der im vorstehenden Zitat genannten Gründe ist es jedoch wahrscheinlich, daß derartige empirische Beziehungen und Felsklassifizierungen des Typs RMR und Q in Zukunft im Felsbau viel benutzt werden.

BEITRÄGE IN DIREKTEM ZUSAMMENHANG MIT FELSKLASSIFIZIERUNG UND VORHERSAGE

Fünf Aufsätze befassen sich mit diesen Themen. Kendorski et al beschreiben detailliert Grundlagen und Methode eines Systems nach Kategorie C zur Vorhersage der vorläufigen und während des Betriebs erforderlichen Ausbaumaßnahmen bei Strecken im Pfeilerbruchbau. Dieser ausgezeichnete Aufsatz zeigt deutlich den Nutzen, welcher aus Forschungsarbeiten eines personell ausgewogenen Teams gezogen werden kann. Er spiegelt das gute Verständnis der Verfasser für Bergbaugeologie und theoretische Felsmechanik wider. Ihre Einschätzung des tatsächlichen Gebirgsverhaltens und der Bergbaupraxis kann nur durch eigene untertägige Erfahrung erlangt worden sein.

Es ist insbesondere angenehm zu sehen, daß die Charakterisierung der Gebirgsfestigkeit in einem getrennten, ersten Schritt erfolgt und daß der aus dem Ingenieurbauwerk sich ergebende Belastungseinfluß später durch eine Reihe von Anpassungen berücksichtigt wird. Aus ihrer Abbildung 1 ist zu ersehen, daß dieses Konzept den Verfassern ermöglicht, ihr System logisch zur Vorhersage sowohl des anfänglichen (Aufschlußstadium) wie des späteren (Produktionsstadium) Ausbaubedarfs einzusetzen. Es scheint uns, daß Kendorski et al genau die „neue Vorgehensweise" vorstellen, die von Lokin et al in ihrem Beitrag gesucht wird. Wir werden dies später bei der Behandlung des Aufsatzes von Lokin et al noch besprechen.

Wir beziehen uns nochmals auf Abbildung 1 des Aufsatzes von Kendorski et al und stellen fest, daß die Raumstellung der Trennflächen nun in den Anpassungen des zweiten Schritts besser berücksichtigt wird, als wenn sie als Teil der Gebirgscharakterisierung angesehen werden würde. Die Verfasser sagen damit, daß das Trennflächengefüge zwar eine unveränderliche Eigenschaft des Gebirges ist, daß die Raumstellung der Trennflächen in Bezug auf die Stollenachsen jedoch von der Richtung der Strecken abhängt, also Teil der durch die ingenieurmäßigen Aktivitäten hervorgerufenen Bedingungen ist.

Wir sind nicht in der Lage, Tabelle IV „Fracture Orientation Rating A_O Based on Direct Observation in Drift" voll zu verstehen. Wir nehmen an, daß die „Number of Fractures Defining Block" sich auf die im Stollen sichtbare Fläche eines Kluftkörpers bezieht. Wenn das so ist, wäre vermutlich die Raumstellung dieser Fläche erforderlich, und diese ist nicht angegeben. Wir schlagen vor, daß die Verfasser ein Diagramm vorstellen, um die Ableitung von A_O aus Tabelle IV zu erläutern.

Wenn wir nochmals auf die Bewertung A_O zurückkommen, so stellen wir fest, daß die Verfasser einer Benutzung von Tabelle IV Vorzug geben, wenn unmittelbare Beobachtungen von aufgeschlossenem Fels vorliegen. Tabelle V, die umfassendere Daten enthält, wird nur dann verwendet, „wenn keine Stollenaufschlüsse vorhanden sind, aber die Trennflächenausbildung allgemein bekannt ist...." Wir können die Logik dieser Wahl nicht einsehen. Wir meinen, daß jeder erfahrene Ingenieur oder Geologe A_O aus Stollenaufschlüssen und aus der Tabelle V bestimmen könnte. Dies würde zur gleichen Zeit geschehen, wie die Beobachtungen, um Sprengbeschädigungen festzustellen (Tabelle III), und erfahrenes Personal würde ohne Zweifel Einflüsse der Trennflächenorientierung in Betracht ziehen, wenn es zu entscheiden gilt, welche Werte für A_B nach den aufgelisteten Beobachtungsklassen anzunehmen sind.

Die Generalberichter sehen das MBR-System als beträchlich entwicklungsfähig an und wir freuen uns auf weitere Veröffentlichungen über Erfahrungen bei der Anwendung.

Lokin et al haben in ihrer Tabelle I eine nützliche Liste der Elemente von neun existierenden Felsklassifizierungen des Typs C aufgestellt. Sie diskutieren danach sieben „Unzulänglichkeiten", die sie als Begrenzung der Anwendbarkeit dieser Systeme ansehen. Während wir zustimmen, daß einige der einfacheren Klassifizierungen zu wenig umfassend sind, vermuten wir, daß die meisten der aufgelisteten „Unzulänglichkeiten" den umfassenderen Systemen (z.B. RMR oder Q) nicht zuzuordnen sind, sondern eher sich auf die Organisation ihrer Anwendung beziehen. Wie vorstehend bei der Erörterung des MBR-Systems festgestellt, scheint das erste Beispiel einer „neuen Vorgehensweise", wie Lokin et al sie ahnen, bei diesem Kongreß von Kendorski et al bereits vorgestellt worden zu sein.

Nach der Erörterung der Unzulänglichkeiten diskutieren Lokin et al die Prinzipien einer neuen Vorgehensweise zur Klassifizierung. Die meisten in der Projektplanung erfahrenen Ingenieure werden den meisten der vorgetragenen Prinzipien über geotechnische Studien und ihrer Zuordnung zu anderen Studien während Untersuchung, Entwurf, Bau und Überwachung zustimmen.

In diesem „Grundlagen"-Abschnitt erläutern die Autoren, daß die Klassifizierung von Fels für ein Projekt die Einteilung des Gebietes in quasihomogene Bereiche, vorzugsweise nach „Felsklassen" beinhalten muß. Die Autoren stellen dann, im Hinblick auf diese Vorgehensweise, fest, „sie müsse, um sinnvoll zu sein, den gesamten Untersuchungsbereich mit etwa dem gleichen Anspruch auf Verläßlichkeit überdecken". Die Generalberichter meinen, daß sowohl die erforderlichen Einzelheiten wie auch der Grad der Zuverlässigkeit sich je nach den Anforderungen des Ingenieurprojekts und der geologischen Umgebung ändern sollten. So müßten beispielsweise die Untersuchungen für die Trasse eines langen Druckstollens, insbesondere in einer Region mit steilen Hängen und starker Verwitterung, die Portalbereiche wesentlich detaillierter abdecken als den Rest der Trasse. Dies deshalb, weil es notwendig ist, die Portale so anzuordnen, daß Bereiche gestörten oder tief verwitterten Gebirges vermieden werden und um die Standsicherheit und die Festigkeit in Bereichen flacher Überdeckung hinsichtlich der zu fordernden Wasserdichtigkeit der Tunnelauskleidung zu bewerten.

In Tabelle III listen die Verfasser alle Daten auf, die nach ihrem Vorschlag während der unterschiedlichen Entwicklungsstufen eines Projekts in die Klassifizierung eingehen sollten. Dies ist eine ziemlich vollständige Liste aller geotechnischer und verwandter Gesichtspunkte, die Untersuchungen wie für Großprojekte erfordern würde, mit solchen Dingen wie seismischen (neotektonischen) Aktivitäten, Felstemperatur, Grundwasserhydrologie einschließliche Dichte und Temperatur des Wassers etc. Wir meinen, daß die Verfasser, wenn sie feststellen. daß ihr Klassifizierungssystem auf all diesen Daten beruht, nicht ein einzelnes System vom Typ C meinen, sondern eine Vorgehensweise für den Entwurf, bei der ganze Serien von Klassifizierungen verwendet werden.

Nakao et al diskutieren mögliche Unzulänglichkeiten der eingeführten Systeme nach Barton et al, Whickham et al und Bieniawski, angewendet zur Vorhersage des Sicherungsbedarfs bei Tunneln in Japan. Sie beziehen sich auf frühere Arbeiten, die andeuteten, daß diese eingeführten Systeme, die in „geomechanisch stabilem, hartem Fels" entwickelt wurden, in Japan, wo wenig fester und stark geklüfteter Fels in Fülle vorkommt, nicht recht anwendbar sind. Sie beschreiben Methode und Ergebnisse statistischer Auswertungen von Daten fertiggestellter Tunnels, die zum Ziel hatten, ein für japanische Felsverhältnisse geeignetes Gebirgsklassifizierungssystem vorzubereiten. Das Gebirgsverhalten in den Tunneln wurde durch Konvergenzmessungen nach dem Auffahren abgeschätzt. Die Ergebnisse zeigten, daß alle Parameter, die in den drei eingeführten Klassifizierungssystemen benutzt werden, bei den in den Tunneln angetroffenen Bedingungen anwendbar waren. Die Autoren kommen jedoch zu dem Schluß, daß die Höhe der Tunnelüberdeckung oder die Spannweite des Tunnels für das Verhalten des Tunnels wichtiger sind als andere Faktoren. Insbesondere ersetzt dies den Zerklüftungsgrad, welcher bei den eingeführten Systemen die Liste der Bewertungen anführt.

Die Generalberichter erkennen an, daß die Unterschiede in Abstand und Ausbildung der Klüfte in Japan von geringerer Bedeutung sind, da dort, wie die Autoren ausführen, der größte Teil des Gebirges sich im Grunde genommen in einer Störung befindet. Wir sehen auch, daß die Spannweite des Tunnels dort, wo Gesteins- und Gebirgsfestigkeit gering sind, seine Verformungen stark beeinflußt. Wir möchten jedoch vorschlagen, daß die Verfasser der Vorgehensweise von Kendorski et al folgen und Spannweite und Überlagerungshöhe getrennt betrachten als Teil des Bauwerks, wodurch der Fels beansprucht wird. Wir pflichten den Autoren bei, wenn sie beabsichtigen, über die Verformungen hinaus auch Einzelheiten der Baudurchführung in zukünftige Studien einzubringen.

Mizuno et al umreißen eine Klassifizierung vom Typ B, die in Japan zur Kennzeichnung von Dammgründungen benutzt wurde. Sie versuchen dann, Beziehungen zwischen unterschiedlichen Felsklassen und den Ergebnissen von Scherversuchen im Untergrund von mehreren Staumauern herzustellen. Die ziemlich weite Streuung der Ergebnisse der Scherversuche innerhalb jeder der untersuchten Klassen wird ein wenig verringert, wenn die Daten getrennt nach dem geologischen Ursprung aufgetragen werden. Die Streuung der Scherversuchsergebnisse ist jedoch insbesondere in sedimentärem und metamorphem Fels noch ziemlich groß, und die Autoren bringen diese Abweichungen mit dem zufälligen Einfluß der Schichtung und der Klüfte bei den einzelnen Versuchsbereichen in Verbindung. Wir stimmen dieser Schlußfolgerung zu.

Yudbir et al haben Laborversuche an Modellprüfkörpern mit dem Ziel durchgeführt, Gestein und klüftigen Fels zu simulieren. Sie haben ihre eigenen Versuchsergebnisse und solche Anderer verwendet, um die Brauchbarkeit der Scherfestigkeitsgleichungen nach Bieniawski (1974b) sowie nach Hoek und Brown (1980) zur Vorhersage der Bruchbedingung sowohl bei sprödem wie bei duktilem Bruch zu prüfen. Sie kommen zu dem Schluß, daß die Gleichung nach Bieniawski

$$\frac{\sigma_1}{\sigma_c} = A + B\left(\frac{\sigma_3}{\sigma_c}\right)^{a}$$

für solche Vorhersagen in Betracht kommt. Die Parameter A reichen von 0 für vollständig entfestigten Fels bis 1,0 für das Gestein. Der Parameter B ist eine Eigenschaft des Gesteins, wahrscheinlich abhängig von der Mineralzusammensetzung und vom Gefüge (Mikrostruktur). Sie schlagen einige typische Werte für B und vorläufige Beziehungen zwischen A und Q nach Barton et al bzw. RMR nach Bieniawski vor.

BEITRÄGE, DIE MIKROSTRUKTUREN DES GEBIRGES MIT SEINEN EIGENSCHAFTEN IN VERBINDUNG BRINGEN

Es ist klar, daß die physikalischen Eigenschaften jedes Gesteins sich während seiner Entstehungsgeschichte entwickelt haben und daß seine Eigenschaften, so wie sie im Versuch festgestellt werden, von seiner gegenwärtigen Mineralzusammensetzung und den Mikrostrukturen abhängen. Es ist klar, daß für die meisten praktischen Fälle Unterschiede der physikalischen Gesteinseigenschaften vergleichsweise unwichtig sind gegenüber Unterschieden im Trennflächengefüge. In besonderen Fällen jedoch, z.B. wenn die Spannungen oder Beanspruchungen die Gesteinsfestigkeit erreichen, kann es wichtig sein, die Gesteinseigenschaften ganz genau zu bestimmen und die Beziehungen zwischen diesen Eigenschaften und der Mikrostruktur genau zu verstehen. Die vier Beiträge dieser Gruppe diskutieren solche Beziehungen.

Der Aufsatz von Pan et al bschreibt zunächst Beziehungen zwischen dem Mineralaufbau bzw. der Mikrostruktur zweier Arkosen (feldspatreiche Sandsteine) und Indexeigenschaften wie Wichte, Porosität, Wasseraufnahme, einaxiale Druckfestigkeit und Festigkeitseinbußen nach Wasserlagerung. Karbonatgebundene Arkose erwies sich als fester, dichter und weniger der Festigkeitsminderung unterworfen als tongebundene Arkose. Im Labor wurden ebenfalls die Einflüsse des im Wasser gelösten Sauerstoffes und der Dauer der Wasserlagerung untersucht. Es erwies sich, daß gelöster Sauerstoff den Pyrit im Sandstein oxydiert und Schwefelsäure bildet, welche die Karbonatbindung angreift und so die Festigkeitsminderung vorantreibt.

Ausgehend von diesen Laborergebnissen wurden in Kohlengruben Untertageversuche durchgeführt.

Diese zeigten, daß dem Abbau vorauseilende Wassereinpressungen im Hangenden eine Reihe von Vorteilen brachten u.a. eine Verbesserung des Ausbruchs, weniger Staub am Stoß, weniger Beschädigungen am Ausbau und eine Verringerung des Auftretens von Gebirgsschlägen.

Rippa und Vinale beschreiben Herkunft, Mineralzusammensetzung und Mikrostruktur von Tuffprobekörpern aus zwei geologischen Bereichen und stellen sie den Festigkeits- und Verformungseigenschaften des Gesteins bei einaxialem Druck gegenüber. Sowohl gelber wie grauer Tuff enthalten Bestandteile, die nach Menge und Größe der Bimseinschlüsse in vier Gruppen eingeteilt werden können.

Das Versuchsprogramm zeigte, daß die einaxiale Druckfestigkeit des Tuffs aus beiden Bereichen

* mit zunehmendem Anteil der Bimseinschlüsse abnimmt
* bei Wassersättigung der Proben (um bis zu 30%) abnimmt.

Die Versuchsergebnisse zeigen, daß Unterschiede im gelben Tuff sowohl horizontal innerhalb der Schichten, als auch vertikal von Schicht zu Schicht bestehen.

Die Mikrostrukturen des Tuffs, wie sie durch ein Elektronenmikroskop bestimmt wurden, haben wesentlichen Einfluß auf sein Verformungsverhalten.

Baklashov et al berichten über Studien zum Mechanismus von Verformung und Bruch bei Gestein und Gebirge. Zunächst listen sie das auf, was sie nach Beobachtungen der Bruchmechanismen als zehn gemeinhin anerkannte Tatsachen ansehen. Dann erörtern sie bestehende Bruchtheorien und finden, daß diese viele der Beobachtungen nicht erklären. Dann diskutieren die Autoren im Abriß verschiedene Laborversuche und Analysen, die auf ein besseres Verständnis der beobachteten Mechanismen abzielen.

Die vorgestellten Versuchsmethoden umfassen:

- konventionelle einaxiale Druckversuche mit nahezu aufgehobener Reibung an den Lastplatten, welche den Mechanismus axialer bzw. säulenförmiger Bruchentwicklung entsprechend der Ausbreitung der Mikrorisse parallel zur größten Hauptspannung zeigen, wie das bereits von anderen Forschern (z.B. Gramberg, 1965) beobachtet worden ist.
- Dreiaxialversuche vorzugsweise verformungsgesteuert
- besondere Dreiaxialversuche bei denen sowohl axiale wie seitliche Lasten durch unabhängige starre Systeme aufgebracht wurden
- besondere Dreiaxialversuche mit ungleichen Spannungen ($\sigma_1 > \sigma_2 > \sigma_3$)

Die Prüfkörper für die Dreiaxialversuche bestanden aus Salz, Ölschiefer, Marmor, Kalkstein, Kohle, Sandstein und verschiedenen Gesteinen mit Kohleeinschlüssen. Die Versuchseinrichtungen ermöglichten das Studium der Volumenänderung, der Längs- und Querdehnungen, der Festigkeitsanisotropie, des plastischen und des Kriechverhaltens. Aus diesen Versuchen entwickelte Bruchmodelle sind für Bergbauzwecke und für Felshohlraumbauten benutzt worden.

In seinem Beitrag "Petrophysik: die petrographische Interpretation der physikalischen Felseigenschaften" stellt Montoto fest, daß petrografische Studien im Ingenieurbau zunehmende Bedeutung gewinnen. Um dies zu stützen, führt er Deponien radioaktiver Abflälle und geothermische Projekte an, bei denen eine Vorhersage des Verhaltens der Mikrobrüche unter Spannung und unter Temperaturänderungen unumgänglich ist.

Montoto diskutiert zunächst von Anderen aufgestellte Beziehungen zwischen den Gesteinseigenschaften (einschließlich Festigkeit, Durchlässigkeit, Härte und Abriebfestigkeit) und folgenden „felsbildenden Bestandteilen":

- Mineralien
- Hohlräumen (Poren und Risse)
- Textur

Die besondere Bedeutung von Art und Zahl der Hohlräume auf die Felseigenschaften wird hervorgehoben.

Der Beitrag listet verschiedene petrografische Gesichtspunkte auf, die studiert und bestimmt werden müssen. Methoden zur Probengewinnung werden kurz erörtert und danach wird ein Überblick gegeben über analoge Computersysteme, die zur Ermittlung entsprechender Daten für verschiedene petrografische Fragestellungen entwickelt wurden. Der Verfasser behauptet, daß diese Systeme bessere Daten liefern als die üblichen, zeitraubenden Auszählmethoden.

BEITRÄGE ÜBER STATISTISCHE ODER THEORETISCHE STUDIEN

Nishimatsu und Okubo führen zunächst einige Schwierigkeiten bei der Bestimmung der Standsicherheit einer Felsböschung an. Dann gehen sie von sehr einfachen Gebirgsverhältnissen mit nur einer Kluftschar und exponential verteilten Kluftabständen aus. Unter der weiteren Annahme, daß es zwischen den Klüften keine gegenseitige Beeinflussung gibt, d.h. daß die Klüfte nicht verschwenkt sind und sich schneiden, leiten die Verfasser eine Formel für die Wahrscheinlichkeit her, daß ein Bruch nicht entlang einer Kluft eintritt.

Dann leiten die Verfasser aus einer Serie direkter Scherversuche an im Labor an Bohrkernen künstlich erzeugten Klüften eine statistische Verteilung für die Scherfestigkeit von Felsklüften her.

Unter Verwendung dieser Ergebnisse bringen die Verfasser ein Beispiel einer Entwurfsberechnung der Wahrscheinlichkeit, daß kein Bruch eintritt. Sie legen dabei eine Felsböschung mit einer Kluftschar zugrunde, die parallel zur Böschung streicht, und vernachlässigen den Wasserdruck.

Die Generalberichter meinen, daß es andere, in den Berechnungen von Niskimatsu und Okubo nicht enthaltene Faktoren gibt, die wesentlich größeren Einfluß auf die Standsicherheit echter Felsböschungen haben, als die in dem Beitrag beschriebenen. Diese bedeutenderen Faktoren sind von anderen Autoren, z.B. Patton und Deere (1971) und McMahon (1971 und 1974) untersucht worden.

Diese Faktoren beinhalten:

- Die Abnahme der Festigkeit bei Klüften und Trennflächen mit zunehmender Verschiebung, wodurch ihre Festigkeit bis auf die Restfestigkeit abnimmt. Die in dem Beitrag beschriebenen Versuche scheinen nur Spitzenwerte (peak shear strength) geliefert zu haben, für welche bei Klüften große Streuungen üblich sind

- die Stetigkeit der Klüfte

- Form oder Welligkeit der Klüfte

- die Anzahl der Kluftscharen.

Der Aufsatz von Grossmann beschreibt die Entwicklung einer numerischen Methode zur Definition von Kluftscharen im Fels. Es wird versichert, daß die Methode von größtem Wert ist, wenn nur wenige Trennflächen eingemessen werden können. Die Methode kann sowohl auf Kluftdaten aus einheitlichen oder aus uneinheitlichen Erhebungen angewendet werden.

Die Generalberichter sind auch der Meinung, daß diese Methode bei der Ermittlung der Kluftscharen helfen mag, aber wie vom Autor in der Einleitung erwähnt, bedürfen kompliziertere Trennflächenanordnungen immer eingehender Studien über räumliche Verteilung, Ausbildung und Raumstellung.

Der Aufsatz von Pineau und Thomas beschreibt statistische Studien zur Heterogenität des Gesteins eines Eisenerzlagers bei verschiedenen betrachteten Größenordnungen. Die Studie umfaßt 482 einaxiale Druckversuche an gesättigten zylindrischen Proben unterschiedlicher Größe, die aus unterschiedlichen Bereichen großer Felsblöcke geschnitten wurden. Aus einer statistischen Auswertung der Ergebnisse schließen die Autoren, daß das Ausmaß der Streuung bei den Versuchsergebnissen zusammenhängt mit Unterschieden in der Struktur des Gebirges, welches wiederum mit den regionalen geologischen (tektonisch bedingten) Strukturen in Beziehung steht. Sie führen aus, daß der Gebrauch einfacher statistischer Methoden, wie z.B. einer Mittelwertbildung, zu falschen Schlußfolgerungen über die Festigkeit des untersuchten Gebirges geführt hätte.

Während diese Schlußfolgerungen nicht angegriffen werden, weisen die Generalberichter darauf hin, daß die Streubreiten der Festigkeiten der Probekörper, z.B. Beispiel 1 (von 15 bis 28 MPa), Beispiel 3 (von 19 bis 30 MPa) und bei Beispiel 2 mit Ausnahme von drei Probekörpern (von 12 bis 25 MPa), für die meisten praktischen Bau- und Bergbausituationen unbedeutend wären, da dort Trennflächen einen überwältigenden Einfluß auf das Felsverhalten hätten. Wir fragen die Autoren, ob sie auch die Streuungen bestimmt haben, welche auf die Vorbereitung der Probekörper und auf das Versuchsverfahren zurückzuführen sind .

OPTIMIERUNG VON IN-SITU-UNTERSUCHUNGEN DES GEBIRGES

Die Generalberichter fanden den Aufsatz von Fishman et al erfrischend praktisch. Die Autoren führen aus, daß die verbesserten analytischen Möglichkeiten infolge der numerischen Methoden nach Messung von mehr geotechnischen Parametern verlangen, und daß diese Messungen von besserer Qualität sein sollen. Die Autoren gehen davon aus, daß in-situ-Untersuchungen die wertvollsten Ergebnisse bringen und stellen fünf Grundsätze für ihre Durchführung auf, die darauf abzielen, den größtmöglichen Nutzen aus den notwendigerweise teuren Arbeiten zu ziehen. Wir haben diese fünf Grundsätze wie folgt zusammengefaßt:

1. Es sollte derjenige Versuch gewählt werden, welcher so gut wie möglich die Belastungsbedingungen des Bauwerks simuliert.

2. Der Versuch sollte alle beim Bau auf das Gebirge wirkenden Einflüsse berücksichtigen.

3. Der Versuch sollte genügend groß sein, um auch wirklich einen Teil des Gebirges zu umfassen (Gestein und Trennflächen), und bei der Auswertung müssen Maßstabseffekte berücksichtigt werden.

4. Die Versuche sollten unter Berücksichtigung, des erkundeten Trennflächengefüges und anderer Versuche interpretiert werden, damit die Anwendbarkeit der Versuchsergebnisse auf den gesamten Gebirgsbereich am Bauwerk und auf spätere Projekte abgeschätzt werden kann.

5. Thematik und Genauigkeit der Versuche sollten auf die Bedürfnisse des Projekts zugeschnitten sein.

Die Verfasser erläuterten diesen letzten Grundsatz durch, wie wir sagen würden, „Sensibilitätsstudien" beim Entwurf eines Druckstollens. Ihre Abbildung 1 zeigt die Empfindlichkeit der wesentlichen Rißbildung im Fels und bewehrtem Beton gegenüber dem Verformungsmodul des umgebenden Gebirges. Im gesunden Gebirge (Zone I in Abbildung 1) sind teure Feldversuche natürlich nicht gerechtfertigt, aber in Bereichen mit kleinerem Verformungsmodul (innerhalb und in der Nähe von Zone II) können genauere Messungen gerechtfertigt sein, wenn solche Bereiche weitverbreitet vorkommen. Für örtliche Störungszonen können derart genaue Versuche jedoch wirtschaftlich nicht gerechtfertigt werden.

Anschließend werden verschiedene Feldversuche beschrieben und erörtert. Diese beinhalten Lastplattenversuche, Radialpressenversuche (in Tunnels), einaxiale Druck-, Zug- und Scherversuche.

VORHERSAGEN AUFGRUND MECHANISCHER UND MATHEMATISCHER MODELLE

Der Aufsatz von Langof beschreibt den Bau eines mechanischen Modells, in welchem geklüfteter Fels durch rechteckige Plastikblöcke simuliert wurde. Die Blöcke waren so angeordnet, daß ein geschichteter Fels simuliert wurde mit einer Kluftschar zwischen den Schichten und einer zweiten Schar innerhalb jeder Schicht, die senkrecht zur ersten angeordnet war. Nachdem die Verformungseigenschaften dieses Modellgebirges durch Zweiaxialversuche bestimmt worden waren, wurde das Modell verwendet, um in vier Fällen die Gebirgsverformungen um einen kreisförmigen Tunnel vorherzusagen, wobei die Schichtung horizontal, vertikal und unter 30° und 60° einfallend angenommen wurde. Für diese Fälle und unterschiedliche Primärspannungen wurden auch die Verformungen und die Radialspannungen in der Tunnelauskleidung und die Sekundärspannungen um den Tunnel vorhergesagt.

Eine vernünftige Übereinstimmung zwischen den Vorhersagen des mechanischen Modells und Finite Element Berechnungen wurde um für den Fall mit horizontaler Schichtung erzielt. Das benutzte Finite Element Verfahren war für ein Kontinuum entworfen und konnte daher das Trennflächengefüge nicht modellieren.

Die Generalberichter weisen darauf hin, daß es Finite Element Programme gibt, die das Trennflächengefüge modellieren können (Burmann et al, 1975 und Burmann und Hammett, 1975).

ÜBERWACHUNG

Habenicht und Behensky haben den einzigen Beitrag zu diesem Thema geliefert. Sie beschreiben eine hydraulische Druckmeßzelle, in welcher der auf die Zelle wirkende Druck direkt von der Druckflüssigkeit auf einen federnden Anzeigestab übertragen wird. Die Verschiebungen des Stabs sind proportional zum Druck und können über Meßuhr oder über Fernübertragung abgelesen werden. Die Autoren erheben den Anspruch, daß ihre direkte Ablesemethode folgende Vorteile gegenüber konventionellen Druckmeßzellen hat:

* höhere Empfindlichkeit und Genauigkeit
* die Druck-Anzeige-Beziehung ist linear
* es gibt fast keine Hystereseeffekte

Unterthema A.4

HYDROGEOLOGIE

Die Hydrogeologie ist ein weites Feld und geht weit über die Themenkreise hinaus, die im Rahmen der Felsmechanik bzw. der Felshydraulik gewöhnlich behandelt werden. Auch hier beschränken sich die behandelten Themen auf den kleinen Abschnitt, der den Felsmechaniker bei seiner Arbeit gewöhnlich interessiert.

Die hier zu behandelnden Grundaufgaben gliedern sich gewöhnlich in folgende Punkte

- Beurteilung der Wasserwegigkeit des Gebirges
- Beurteilung der Injizierbarkeit des Gebirges
- Beurteilung der Wirkung des Bergwassers auf die Standsicherheit von Böschungen
- Beurteilung der Wasserzuflüsse zu Felshohlraumbauten

Zu diesen Themenkreisen sind insgesamt nur fünf Arbeiten eingereicht worden.

Grundlage für alle ingenieurmäßigen Beurteilungen zu den vorstehend genannten Themen ist es, neben den geologischen Verhältnissen auch die Durchlässigkeit des betreffenden Gebirges zu kennen. Doch bereits hier gilt es zu differenzieren.

Die Durchlässigkeit des Gebirges wird generell bestimmt durch diejenige des Gesteins und diejenige der Trennflächen. Oftmals ist das Gestein selbst so dicht, daß man in guter Näherung nur die Trennflächen selbst als durchlässig annehmen kann. Daneben ist noch zu beachten, daß das Gebirge durchörternde Störungen immer eine besondere Behandlung erfordern.

Die Durchlässigkeit der Homogenbereiche kann auf verschiedene Art ermittelt werden. Am verbreitetsten ist zweifellos der Wasserdruckversuch (WD-Versuch) und zwar u.a. deshalb, weil er sowohl über als auch unter dem Grundwasserspiegel eingesetzt werden kann. Jeder WD-Versuch untersucht das Gebirge nur in einem geringen Ausschnitt. Erst die Vielzahl der Versuche erlaubt, die Durchlässigkeit abzuschätzen.

Der WD-Versuch kann auf unterschiedliche Art ausgewertet werden. Die wohl älteste und noch immer gebräuchliche Weise ist, die je Zeit- und Druckeinheit ins Gebirge abfließende Wassermenge, bezogen auf 1 m Abpreßstrecke, als Maß für die Durchlässigkeit anzusehen. Diese Art der Auswertung ist untrennbar mit dem Namen Lugeon (Lugeon 1933) verbunden. Später wurden dann Methoden entwickelt, mit denen versucht wird, einen Durchlässigkeitsbeiwert oder einen Durchlässigkeitstensor, also gerichtete Durchlässigkeitseigenschaften, zu bestimmen (z.B.: Baker (1955), Maini (1971), Snow (1966), Snow (1969), Rißler (1977)). Gewöhnlich werden hierbei stationäre Verhältnisse zugrunde gelegt. Dies bedingt natürlich entsprechende Durchführung der Versuche. Schneider stellt hier nun ein auf instationärem Fließen beruhendes Auswerteverfahren vor, welches offenbar zu anderen Ergebnissen führt als die früher entwickelten Verfahren. Diese unterschiedlichen Ergebnisse könnten aber auch teilweise dadurch erzeugt werden, daß der Author das Versuchsverfahren gegenüber herkömmlicher Praxis (Mengenregelung statt Druckregelung) geändert hat. Zumindest sind den Generalberichtern genügend Fälle bekannt, in denen der stationäre Zustand wesentlich schneller erreicht wurde, als dies vom Autor vorgetragen wird. Meinungsäußerungen des Authors und des Auditoriums hierzu würden begrüßt.

Es ist bekannt, daß bei weiten Klüften nicht nur laminare sondern auch turbulente Fließbedingungen auftreten. Man beobachtet dann mit zunehmendem Abpreßdruck einen unterproportionalen Anstieg der Wasseraufnahme. Rißler (1977) hat diesen Fall für radialsymmetrische Verhältnisse theoretisch untersucht. Nun präsentieren Cruz und Quadros hier Meßergebnisse aus dem Basalt, welche diese charakteristischen Kurven in der Natur bestätigen.

Parallel zur Wasserdurchlässigkeit ist oftmals die Injizierbarkeit des Gebirges von Interesse. Hohe Durchlässigkeit muß nicht zugleich auch gute Injizierbarkeit bedeuten, nämlich z.B. dann nicht, wenn viele enge Spalten vorhanden sind, die zwar das Wasser hindurchtreten lassen, einer Injektion aber großen Widerstand entgegensetzen. Blinde et al präsentieren hierzu Untersuchungen, durchgeführt an aufgelockertem Granit und bestehend aus einer Gefügestatistik, optischen Bohrlochsondierungen, Kluftmarkierungen, WD-Versuchen, geoelektrischen Eigenpotentialverfahren und seismischen Messungen. Sie fanden dabei auch heraus, daß nicht alle nach Augenschein vorhandenen Klüfte für die Wasserwegigkeit von Bedeutung sind.

Huder und Amberg berichten hier von Erfahrungen über den Einfluß des Bergwassers beim Bau von Böschungen und von Felshohlräumen und über entsprechende Entspannungsmaßnahmen. Weiterhin werden verschiedene Systeme von Piezometern diskutiert.

Bello-Maldonado stellt ein analytisches Rechenverfahren vor, welches es erlaubt, den Wasserzufluß zu Tunnels während des Baus und im Betrieb abzuschätzen. Derartige Berechnungen sind sicherlich interessant, aber die Generalberichter fürchten, daß in der Praxis nur selten so homogene Verhältnisse anzutreffen sein dürften, daß sich der beträchtliche mathematische Aufwand auch lohnt.

SCHLUSSFOLGERUNG

Zum Thema „Baugrunduntersuchungen und Auswertung" wurden insgesamt 61 Beiträge eingereicht. Sie befassen sich mit geophysikalischen Methoden, mit in-situ- und Laboruntersuchungen, mit Fragen der Klassifizierung, der Vorhersagen der Gebirgsqualität, mit Beobachtung und Überwachung, sowie mit hydrogeologischen Fragen.

Es war festzustellen, daß gegenüber dem letzten IGFM-Kongreß auf diesem Gebiet insbesondere folgende Entwicklungen bedeutungsvoll erscheinen:

- die Anwendung seismischer Methoden zur Überprüfung des Erfolges von Vergütungsmaßnahmen,
- die Entwicklung von Verfahren zur Berücksichtigung anisotroper Verformungseigenschaften bei der Auswertung von Versuchen,
- die Weiterentwicklung bereits vorhandener Klassifikationssysteme.

SCHRIFTTUM

Anstey, N.A. (1977). Seismic interpretation, the physical aspects. Int. Human Resources Development Corporation. Boston, U.S.A., p. 81.

Baker, W.J. (1955). Flow in fissured formations. Proc. 4. World Petroleum Congr., Rom, Section II/E, Paper 7, pp. 379 - 393

Barla, G. und L. Goffi (1974). Direct tensile testing of anisotropic rocks. Proc. Third Int. Congr.Rock Mech. ISRM, Vol. 2 A, pp. 93 - 98

Barton (1971). A relationship between joint roughness and joint shear strength. Report I 8, Int. Symp. on Rock Mech., Nancy

Barton, N., Lien, R. und Lunde, J. (1974). Engineering classification of rock masses for the design of tunnel support. Rock Mechanics. Vol. 6, No. 4, pp.189-236

Bieniawski, Z.T. (1974a). Geomechanics classification of rock masses and its application in tunnelling. Proc. Third. Int. Conf. Rock Mech., ISRM. Vol IIA, pp.27-32.

Bieniawski, Z.T. (1974b). Estimating the strength of rock materials. I.S.African Inst. Min. and Met., Vol. 74, No.8.

Bieniawski, Z.T. (1978). Determining rock mass deformability, experience from case histories. Int. J. Rock Mech. and Min. Sci. Vol 15, pp.237-247.

Bieniawski, Z.T. (1979). The geomechanics classification in rock engineering applications. Proc. Fourth Int. Conf. Rock Mech., I.S.R.M., Vol. 2, pp.41-48

Broch, E. und Franklin, J.A. (1972). The Point load strength test. Int. J. Rock Mech. and Min. Sci., Vol. 9, pp. 669-697.

Brown, E.T. (1974). Fracture of rock under uniform biaxial compression. Proc. Third Int. Congr. Rock Mech. ISRM, Vol. 2 A, pp. 111 - 117

Burman, B.C., Trollope, D.H. und Philp, M.G. (1975). The behaviour of excavated slopes in jointed rock. Aust. Geomech. J., Vol. G5, No. 1, pp. 26-31.

Burman, B.C. und Hammett, R.D. (1975). Design of foundations in jointed rock masses. Proc. Second Aust.-N.Z. Conf. on Geomechanics, pp.83-88.

Franklin, J.A. (1979). Use of tests and monitoring in the design and construction of rock structures. General Report, Theme 2, ISRM Congress, 1979, Montreux, Switzerland, Vol. 3, p. 163

Gramberg., J. (1965). Axial cleavage fracturing a significant process in mining and geology. Eng. Geol. Vol. 1, No.1, pp. 31-72.

Hoek, E. und Brown, E.T. (1980). Underground excavations in rock. The Institution of Mining and Metallurgy, London.

International Society for Rock Mechanics (1980). Basic geotechnical description of rock masses. Int. J. Rock Mech. Min. Sci. and Geomech. Abstr., Vol. 18, pp. 85 - 110.

John, K.W. (1969). Festigkeit und Verformbarkeit von druckfesten, regelmäßig gefügten Diskontinuen. Dissertation · Karlsruhe

Kaneko, K., Inoue, I., Sassa, K. und I. Ito (1979). Monitoring the stability of rock structures by means of acoustic wave attenuation. ISRM Congress, 1979, Montreux, Switzerland, Vol. 2, p. 287

Langer, M. (1979). Rheological Behaviour of Rock Masses. General Report, Theme 1, ISRM Congress, 1979, Montreux, Switzerland, Vol. 3, p. 29

Leichnitz, W. und O. Natau (1979). The influence of peak shear strength determination on the analytical rock slope stability. Proc. Fourth Int. Congr. Rock Mech. ISRM, Vol 2, pp. 335 - 341

Lögters, G. und H. Voort (1974). In-situ determination of the deformational behaviour of a cubical rock mass sample under triaxial load. Rock Mechanics (6), pp. 65 - 79

Lugeon, M. (1933). Barrages et Géologie. Dunod - Paris

Maini, Y.N. (1971). In situ Hydraulic Parameters in Jointed Rock - Their Measurement and Interpretation. Ph. D. Thesis Imperial College, London

McMahon, B.K. (1971). A statistical method for the design of rock slopes, Proc. First Aust.-N.Z. Conf. on Geomechanics. Vol. 1 pp. 314 - 321.

McMahon, B.K. (1974). Design of rock slopes against sliding on pre-existing fractures. Proc. Third Conf. of Int. Soc. Rock Mech., Vol. IIB, pp. 803 - 808.

Neilson, J.L. (1970). Notes on weathering of the Silurian rocks of the Melbourne district. J.I. E.Aust. Vol. 42, 1 - 2, pp. 9 - 12.

Oberti, G., Carabelli, E., Goffi, L. und P.P. Rossi (1979).
Study of an orthotropic rock mass: Experimental techniques, comparative analysis of results. Proc. Fourth Int. Congr.Rock Mech. ISRM,Vol.2,pp.485-491

Patton, F.D. und Deere, D.U. (1971). Significant geologic factors in rock slope stability. In: Planning Open pit mines. A.A. Balkema, Amsterdam.

Pinto (1979). Determination of the elastic constants of anisotropic bodies by diametral compression test. Proc. Fourth Int. Congr.Rock Mech. ISRM, Vol. 2, pp. 359

Plischke, B. (1980). Ein Berechnungsverfahren zur Untersuchung der dynamischen Beanspruchung von Felsbauten. 4. Nationale Tagung über Felsmechanik, 5./6. Mai 1980 , DGEG, p. 259 - 277

Rißler, P. (1977). Determination of the water permeability of jointed rock. Aachen, Institute for Foundation Engineering, Soil Mechanics, Rock Mechanics and Water Ways Construction

Rodrigues (1979). The anisotropy of the moduli of elasticity and of the ultimate stress in rocks and rock masses. Proc. Fourth Int. Congr. Rock Mech. ISRM, Vol. 2, pp. 517

Sancia, R.T. und Brown, I. (1980). A classification of weathered foliated rocks for use in slope stability problems. Proc. Third Aust-N.Z. Conf. on Geomechanics, Vol. 2, pp. 81-86.

Saunders, M.K., Fookes, P.G. (1970). A review of the relationship of rock weathering and climate and its significance to foundation engineering. Eng. Geol. Vol. 4, pp. 289-325.

Savich, A.I., Koptev, V.I., Iljin, M. M. und A. M. Zamakhajev (1979). Distinguishing features of deformation development in the foundation of the Inguri Arch Dam. ISRM Congress, 1979, Montreux, Switzerland, Vol. 2, p. 585.

Snow, D.T. (1966). Three - hole pressure test for anisotropic foundation permeability. Felsmechanik und Ingenieurgeologie 4, Nr. 6, pp. 298 - 316.

Snow, D.T. (1969). Steady flow from cylindrical cavities in saturated, infinite anisotropic media. Association of Engineering Geologists, Sacramento , California.

Terzaghi, K. (1946). Rock defects and loads on tunnel supports. In: Rock Tunnelling with Steel Supports. Commercial Shearing and Stamping Co., Youngstown.

Wichter, L. (1979). On the geotechnical properties of a Jurassic clay shale. Proc. Fourth Int. Congr. Rock Mech. ISRM, Vol. 1, pp. 319 - 326.

Whickham, G.E., Tiedemann, H.R. und Skinner, E.H. (1972). Support determinations based on geological predictions. Proc. First North American Tunnelling Conference, A.I.M.E., New York, pp.43-64.

DISCUSSIONS / DISCUSSIONS / DISKUSSIONEN

GEOPHYSICAL TESTING AND EXPLORATION

Exploration et essais géophysiques
Geophysikalische Prüfung und Untersuchung

W. Wittke
Chairman

GENERAL REPORT: Professor D. Stapledon and Dr. P. Rissler

PRESENTED BY: Dr. Rissler

PAPER: General Report

AUTHORS:

Question: To Dr. Rissler from Dr. John Simmons, Australia.

You mention the fact that numerical analyses can now be performed for most known aspects of rock behaviour. From the illustrations you presented there did not seem to be much consideration of post failure strength and deformation properties. For strain weakening materials the history to large deformations is required for sensible numerical modelling. Would you care to comment on the state of the art in evaluation of strength dilatancy properties beyond peak conditions towards the state of large deformation usually experienced somewhere in a highly stressed rock mass.

Answer: Dr. Rissler

It has not been the task of the general reporters to summarize numerical methods, but there are some papers dealing with the post failure behaviour of shear zones. These are the papers of Tinoco and Salcedo, Everitt and Goldfinch, and Bollo et.al. You should read them and you will find a lot of experience with such things; but as to the problem itself if you consider it as an engineer and not a scientist I would state the problem as follows. We have here a shear stress/normal stress diagram that means sigma/tau and on the other hand shear stress/deformation displacement diagram, now we have the points that we have here, the peak strength and afterwards we have some point where deformation goes up and we are beyond the peak strength of the shear plane. On the other hand this point here, maybe one of our failure criterion?, that means we have a cohesion anomaly or roughness of joint or something like that, and here we have an angle of friction, and when we go beyond this point we find here some more deformations and the strength in that part will be some of this here; we have more or less no friction because its displacement has grown up to a certain extent where all roughness will be sheared more or less and we get such a curve.

This curve now may have an angle of friction which is parallel to that or not, maybe a very flat angle of friction, that depends on the size of the shear stresses you can take there, and it is my opinion that such a model can be made without any complications numerically, and I think it is a question of the need you have for your special problems whether you succeed with such a rather simple model or you have to take some more investigations on more sophisticated models, but you should take one in mind.The large deformation has been here this part, the smaller the roughnesses are before they are sheared more or less and the fewer are the dilatancies, that means the normal displacement normal to the bedding.

Prof. Wittke has Chairman's priviledge to comment.

I just wanted to give my comment on your question if you don't mind. I do believe that for the post failure behaviour the testing techniques to determine the parameters are rather limited and not too informative at the moment, so I do believe that if you are interested in determining post failure and interpret the result. That is my position on that.

Thank you.

GENERAL REPORTERS: Prof. D. Stapledon and Dr. P. Rissler

PAPER: Cabril Dam - Control of the grouting effectiveness by Geophysical seismic tests

AUTHORS: L. Rodrigues, R. Oliveira and A. Correia de Sousa

PRESENTED BY: Oliveira

Question: From Mr Knoop, HEC Tasmania.

I would like to direct a question to Prof. Oliveira. I was very interested in his paper which described a very extensive foundation treatment and grouting treatment. What was the the change in leakage through the dam before and after the treatment and secondly was there any change in the deformation behaviour, or did you measure any of these parameters?

Answer: Dr. Oliveira

As you probably understood from the first reading of the paper, this is a three-author paper and two first authors are present here and are from the Civil Engineering Laboratory and the third author is from the owner of the works, the Electricity of Portugal, and he is not present here. He'd be the person who could supply you correctly with the following information so what I can do is to try to supply you with what I think is correct, but I can later give you more information in detail

With your first question you ask what was the dam and foundation before and after the work performed. Well I may tell you that the leakage was almost nil before the treatment and this we can see from the table we show from the Lugeon test made before the treatment. The Lugeon tests with one exception averaged 5-6 Lugeon and the leakage in volume is almost nil and it was no on account of any leakage that the problem was raised. The problem was raised based on both the joints which appear in the concrete and very minor leakage through it (very minor, almost nothing), and mainly based on geodetic observations of the dam, means the dam was moving down stream and this was clearly shown by the geodetic measurements, so the difference in leakage before and after treatment is very small, and it was not an important item.

For your second question you ask what was the change in foundation deformation.

Well I don't have the values with me but I think this is an interesting question because it touches two points and one of them was raised by both our Soviet colleagues of the paper which the general reporters commented upon before and from a general topic they also mention in their paper and that is the correlation between the static and dynamic moduli determined in situ. Well considering the bottom of the river I would say that there was absolutely no change in deformation.Concerning the upper levels of both banks we measured, as you saw from our paper, increases in velocities which may correspond to similar increases in dynamic moduli of the order of magnitude of up to 15-20%. So a similar trend has been observed in the study of moduli (I don't have the figures with me now), but this topic of correlation of dynamic moduli and static moduli is very interesting. We have done quite a lot of work on this topic and we find like the Soviet colleagues did that there is no meaning to try to correlate in a general way for any kind of rock type static and dynamic moduli. What we do normally is to make dynamic measurements in the same places where static tests (e.g. large flat jack tests) are made and then we tried for each site for each rock type a correlation between the static and dynamic moduli. To finalize I would say that there was an increase in the deformation in the upper part of the flank from the deformability moduli and this increase should be say of the same order of magnitude of the dynamic moduli measured by us.

Prof. Stapledon to make a comment on first question. Regarding General Report.

I believe that Dr. Simmons' question may be illustrated by my little diagram here on the bottom, which I think is what he was referring to. Certainly from the laboratory point of view we are limited in testing joints, in that the laboratory test gives displacements in millimetres over a period of hours whereas nature's shear test in faults is conducted in metres of displacement over perhaps millions of years and we get mineral changes and materials which are ground much finer than would happen in a short term laboratory test. Those of us who are going to Thomson Dam tomorrow should be able to engage in some interesting discussion on the tests of the faults at Thomson dam where in testing those faults we have picked up samples from here and tested their strength in a laboratory. Nature has already displaced them for many metres.

Dr. Tanimoto:

For the pointing out of general reporters I would like to have a short comment on two items. One is that conflict with the other authors. I myself was very suspicious on determination of dynamic elastic constants based on seismic survey because the original equation is derived from the very homogeneous and continuous body, so I don't think I can answer exactly to this conflict.

Also our general reporter pointed out the effect of short circuit, and I think this is another big problem because it can cover by selecting the measuring line and interval of the detectors.

Question: Dr. E. G. Truscott, Australia, to Rodrigues, Oliveira & De Sousa

The word control in the title implies a quantitative use of seismic velocity to ensure that adequate water tightness is achieved in the grouting of a dam foundation. A rough plot of change in permeability (i.e. difference in Lugeon value) against percent increase in seismic velocity from Table 11 of the paper gives a random distribution of points with, as far as I can see, no quantitative relationship between these two parameters. Could the authors please report any work they have done in attempting to derive quantitative relationships that could be used for control purposes and generally indicate how they envisage seismic velocity measurements would be used to ensure enough (but not too much) grouting has been performed to obtain a watertight dam foundation.

Answer: Not supplied to Organising Committee.

GENERAL REPORT THEME A - SITE EXPLORATION AND EVALUATION

With reference to paper "Geophysical Studies of Rock Masses"

ON ADVANTAGES AND LIMITATIONS OF SEISMIC METHODS

In the General Report made by Professor D.H. Stapledon and Dr. P. Rissler much attention is paid to consideration of the advantages and limitations of geophysical methods in studying rock masses and a number of points is raised to our paper "Geophysical studies of rock masses".

In this connection we would like to state the following.

1. It is well known that the geophysical methods including the seismic ones have certain advantages and limitations, a part of which was mentioned in the General Report. But the idea of the General Reporter that the radical limitation of the seismic methods is connected with the existance of the so-called "short-circuit effect" seems to be erroneous. In the General Report it is stated that the longitudinal waves do not necessarily reflect the properties of the "average" rock because they take the shortest part through the best rock, that is, there is a "short-circuit effect".

But it is well known from the theory of elastic waves propagation and from experiments (Ewing W.M., Jardetsky W.S., Press F. "Elastic waves in layered media", McGrow-Hill Inc., N.Y., 1957; White I.E. "Seismic waves radiation, propagation, absorption", McGrow-Hill Inc., N.Y., 1965; Savarensky E.F. "Seismic waves", Nauka Publishers, Moscow, 1972) that both the longitudinal and shear waves propagate by the shortest path, actually through strong blocks in case the length of the wave is considerably less than the sizes of these blocks. Therewith the so-called "ray" idea of propagation of elastic waves is justified. If the length of the wave is considerably more (ten times and up) than the sizes of non-uniformity, then the seismic parameters (velocity, amplitude) characterize the average, exactly, effective properties of the whole media. For example, the elastic waves of the acoustic range of frequencies (acoustic logging) are extensively used on an industrial scale to determine the porosity and jointing of oil collectors, sandstone, limestone etc. Here the length of waves is greater than the sizes of pores and joints, therefore it is possible to determine the porosity and jointing of the whole rock mass quite precisely and reliably. Moreover, a new tendency has been developed in the USSR over the recent years - multi-frequency seismic survey, in which the peculiarities of propagation of elastic waves of various frequencies (seismic, acoustic, ultrasound) in non-uniform media are purposely used. The comprehensive application of the multi-frequency survey has made it possible to solve very important problems, such as evaluation of the scale factor for elastic and strain parameters of rocks etc. This issue was considered in our paper in particular.

Thus, the "short-circuit effect" on the basis of which a number of papers on Theme A are evaluated becomes apparent only with a definite correlation between the wave length and the sizes of non-uniformity. To determine what kinds of properties of the rock mass are characterized by the elastic waves one should take into consideration this correlation. Otherwise the conclusion may be erroneous.

2. The results of evaluation of the effectiveness of grouting by value Vp at the Inguri Power Plant and at the Cabril Dam are compared in the seciton "Estimation of the effectiveness of grouting" of the General Report. The difference in the results is justly explained by the difference in the nature of the rocks explored. At the same time our nomograph for estimating grouting quality is criticised. In this connection it is necessary to point out that our paper presented a portion of the theoretical nomograph for estimating grouting quality, and this portion reflects the possible improvement of the properties of the limestones at the Inguri Power Plant at various degrees of filling the rock voids with cement. At higher values of V_0 (V_0 is greater than 5500 m/sec) the possible effect goes down sharply. The increase of velocities approximately by twenty per cent in the range of values V_0 = 5000 m/sec (at Vmax = 6800 m/sec) shown in our diagram corresponds to the case when all the existing voids in the rock are actually filled. The actual filling is usually less ($\sim$10%) and that is the evidence that the quality of the grouting due to impermeability of the voids in the rock corresponding to this range of velocities is not up to the mark.

All in all, our nomograph is certainly true only for rocks similar to the limestone of the Inguri Power Plant zone. However, the applied approach is of a general nature.

3. In the section "Assessment of elastic and deformation properties of rock" of the General Report is given an example of poor correlation between modulus (D) and modulus (E) for quartzitic phylites (Dhawan's paper), which is explained by the limitations of the seismic survey. That has given rise to doubts as to the possibility of wide application of geophysical methods for this purpose. In this connection it is necessary to point out that in our paper we present the results of the generalization of vast materials on application of seismic and static methods under various geological conditions. It is found that the reliable correlation between D and E values is possible only under certain conditions for measuring the said parameters. If these conditions are met, the relations between values D and E are well described by the correlation equations presented in our paper.

Unfortunately in the text of our paper is a misprint in the correlation equations. They should look as follows:

$$\lg E_e = (1{,}141 + A_E e^{-\alpha_E \sigma_{max}}) \lg E_{dyn} - (0.875 + B_E e^{-\beta_E \sigma_{max}}) \quad (2)$$

$$\lg D = (1{,}141 + A_D e^{-\alpha_D \sigma_{max}}) \lg D_{dyn} - (0.875 + B_D e^{-\beta_D \sigma_{max}}) \quad (3)$$

$$\lg D_\Sigma = (1{,}141 + 0{,}970 e^{-0.119 \cdot 80}) \lg E_{dyn} (0{,}375 + 6{,}153 e^{-0.110 \cdot 80}) = 1{,}515 \lg E_{dyn} - 3{,}427 \quad (4)$$

At the same time it should be taken into consideration that values α and β are shown in 10^5 N/m^2 in Table II. These equations are true for the rock mass with natural moisture content, the parameters of which are variable mainly due to jointing. The jointing is estimated automatically by introducing the respective value of the velocity Vp (or E_{dyn}). The rocks of various jointing degrees are featured by various standard coefficients in the correlation equations.

As a whole, the experience gained by us during a number of years in application of geophysical methods to geological engineering investigations shows that these methods, in particular seismic ones, even at present allow one to succeed in solving a great number of important problems of rock mechanics. Their importance, in our opinion, will be still greater in the future.

In conclusion, we would like to express deep appreciation to the General Reporters for comprehensive consideration of our paper on Theme A.

A.I. Savich, Dr.Sc.(Phys.-Math.),
Chief of Geophysical Department;

A.D. Mihkailov, Dipl.Eng.,
Chief Expert;
V.I. Koptev, Cand.Sc.(Tech.),
Chief Expert;

M.M. Iljin, Cand.Sc.(Phys.-Math.),
Senior Geophysicist;

Geophysical Department,
"Hydroproject" Institute, Moscow, USSR.

IN SITU AND LABORATORY TESTING

Essais in situ et en laboratoire

Prüfung an Ort und Stelle und im Labor

J. Franklin
Chairman

PAPER: Creep Of Rock Based On Long-Term Experiments

AUTHOR: H. Ito

PRESENTED BY: H. Ito

DISCUSSION

Question: Dr. G. Borm, Germany

The paper given is another demonstration of the well known fact that rock is visco-elastic if subjected to long term loads. As there is no explicit creep equation given I should like to ask Dr. Ito, do you agree that the viscosity in that Maxwell body is not a single valued constant but rather a highly non linear function of the strain rates applied in rock?

Answer: Dr. H. Ito

At first I proposed that a yield stress obtained in every day time is called the technical yield stress and a yield stress for the geological time is called the geological yield stress. I have insisted that the geological yield stress is zero. It is important to determine the lowest value of the technical yield stress which corresponds to theta prime (Θ') in figure 4. of my paper. I think the value may be estimated from stress released in an earthquake and the value ranges from 10 to 100 bars. This may be very small in the engineering sense. The relation between differential stress and creep strain rate is generally non linear. However, the relation has been obtained for very large stress, larger than the technical yield stress. I consider the relation between stress and strain rate is linear for a very small stress because the diffusion creep theory shows that relation is linear, and my model explains that relation is non linear for a larger stress.

Question: Dr. Roegiers, U.S.A.

Your gathering of data required a tremendous amount of time. The diffusion equation that you used suggested that we might run the test at higher temperatures to shorten the time and acquire more data. Do you have any comments on that?

Answer: Dr. H. Ito

When temperature is high the creep rate becomes faster. I agree with your opinion. But when temperature is high enough, the activation energy will change. Therefore the experiment should be done in the temperature range where the activation energy does not change. Therefore the experiment cannot be done faster.

Question: Dr. K. Saari, Finland

All your tests you have done are beam-bending tests where tensile stresses occur. Are your findings valid for cases where no tensile stresses occur, e.g. pure shear?

Answer: Dr. H.Ito

The deformation of rock under large stress may produce voids such as cracks along the boundaries or micro cracks in grains. In the case of the deformation of rock under low stress I think such voids are not produced and I think our results are valid for pure shear. In bending tests compression and tension occur. Compression and tension are also deviatoric. Therefore these include shear. The flow must be discussed under shear. I have discussed the flow of a beam under such consideration.

Question: Dr. Nguyen-Minh, France

Considering the long term behaviour of competent rock as that of a visco plastic body or fluid with a high viscosity constant is a philosophic point of view (as is for the case of rock salt). I am disturbed to imagine that granite can be a fluid at room temperature. I refer to the fact that we know that in a quick load test the rock strength in tension is much lower than in compression and that I look at beam tests in which the stress state is not homogeneous: the stress state in the beam under flexure is both compressive and tensile. I remark that the maximum compressive stress is small compared to the instantaneous compressive strength of the rock but the maximum tensile stress is not small compared to its tensile strengths. So I would rather think that the creep observed in the beams may be inferred to the creep in the tension zone, such a result could not be extended for rock in compression. Has the author considered this possibility?

Answer: Dr. H. Ito

The flow of solid depends directly on a yield stress and does not depend on a breaking strength. Your question may become reasonable, if "strength" is replaced by "yield stress". Let us assume that the yield stress in tension is much smaller than in compression. As an extreme case, let us suppose that the yield stress for tension is zero and the yield stress for compression is very large. In order that a beam is in mechanical equilibrium, the neutral layer of the beam tends to migrate toward the compression side, although at the beginning the neutral layer is at a middle. Therefore the compressional stress in the beam becomes larger and larger, but the tensile stress becomes rather smaller and smaller. When very long time elapses, the compression zone will become narrow and the compressional stress will become very large. Therefore yielding and fracturing will occur in the compression side. As discussed above, it seems to be difficult to think that the flow or creep of beam is due to the creep in the tension zone. I think the creep of beam takes place in the compression zone as well as in the tension zone.

Question: David Stapledon, Peter Rissler, General Reporters
To Dr. Ito

All rocks, even crystalline rocks like granite, contain micro-pores and micro-cracks which can be observed at different magnifications, by various methods.

For example, petrologists examine such features in very thin slices of rocks, using standard petrological microscopes. Much greater magnifications can be obtained, using a scanning electron microscope. Professor Montoto discusses some of these techniques in his paper to sub-theme A3, classification. We consider that in studies such as that undertaken by Professor Ito, indeed in all laboratory research studies involving rock substance deformation, it would prove of great value to examine and describe the changes in such rock microstructures by the methods of the pétrologist.

Answer: Dr. H. Ito

As a granite specimen contains micro-pores and micro-cracks initially, the proposal by the general reporters is important. A petrologist cuts a rock sample to examine its microstructures. An investigator of rock mechanics must not cut a test-piece to examine the change of micro-structures. I have started a new experiment in April 1980, using the same test-piece as Kumagai and Ito. The experiment involves an acoustic technique in order to examine a change of elasticity in creep process. But I can not find out that the elastic wave velocity changes with time. I have used a wave length longer than crystal grain size. If the wave length is planned to be very smaller than the grain size, the change in rock microstructures including pores and cracks may be detected. This is a future problem.

PAPER: Evaluation Of The Deformability Of Anisotropic Rock Masses From The Results Of Field Tests

AUTHORS: J.R. Kiehl And W. Wittke

PRESENTED BY: J.R. Kiehl

DISCUSSION

Question: Dr. P. J. Huergo, Belgium

Vous avez fait un developpement experimental et theorique dans le but de determiner la reponse deformationnelle d'un massif schisteux. Comment est-ce que vous tenez compte de l'etat des contraintes naturelles et de leur influence sur cette reponse deformationelle.

Answer: Mr. J.R. Kiehl

My work was upon anistropic types of rock masses not only shale. I dealt with schist and bedded rocks and the purpose of this paper is mainly to describe the overall deformational behaviour in the elastic domain or in the domain where failure doesn't occur. If you have stress deformation lines from the tests you can use this method on each line you want. You can use the first loading line and the unloading line or the repeated line in this elastic model. Long term deformational behaviour is not included.

PAPER: The Structural Effect In The Mechanical Behaviour Of Clay Shale

AUTHORS: G. Sun and R. Zhou

PRESENTED BY: G. Sun

DISCUSSION

Question: Dr. J. C. Roegiers, U.S.A.

In some work we have done recently we showed that if you plot elastic parameters as a function of water content, the results are extremely dependent on the loading rate on the specimen. You don't mention any loading rate in your paper. I would like to hear your comments, did you keep the loading rate (i.e. the speed at which you load the specimen) constant or not?

Answer: Mr. G. Sun

The relationship between the elastic parameters and the water content described in this paper was controlled manually. In the experiment the loading rate was controlled at 1×10^{-6} per second approximately.

PAPER: A Statistical Model Of Macroscopic Failure Under Compression

AUTHORS: Ph. Weber and P. Saint-Lot

PRESENTED BY: M. Duffaut

DISCUSSION

Comment: By Chairman, Dr. Franklin

One thing that impressed me about this and in fact the previous paper is that it seems to present a possibility of combining several parameters into a single parameter. We are always looking for ways to simplify our approaches in rock mechanics and if we can find that pre-peak strength Young's modulus and post-peak strength Young's modulus and the strength itself are interrelated by some sort of relationship that can be established then it very much simplifies our approach to modelling rock materials.

Question: Mr. P. Pells, Australia

It would appear to me that the slope of the stress-strain curve in the post-peak region is not a material property but is a function of the geometry of the specimen you are testing. You can almost get any shape post-peak curve you like, depending upon the particular geometric configuration you test. I don't think there are such materials as class 1 and class 2 materials. Obviously the post-peak behaviour is important but it appears to me that in order to get really valid results you have to test the geometric configuration applicable in the field. That may be possible with coal pillars as has been done by W. Van Heerden but it is very difficult to do in the context of tunnels or stopes or haulage ways. I wonder if you have some comment on that sir?

Answer: M.P. Duffaut

Il m'est difficile de repondre precisement a l'absence des auteurs. Je pense que je suis au moins partiellement en accord avec M. Pells sur sa facon de poser la question, mais probablement l'auteur aurait une meilleure reponse que je ne suis capable de faire ici.

Answer: Written answer from P. Saint-Lot

We agree with Mr. Pells when he says that the global behaviour of a rock structure is a function of its geometric configuration. This is (partly) because of the distribution of the stresses in the structure in question.

We think that the only way to avoid gigantic in-situ tests is to find stress-strain laws as a result of tests made in uniform stress states. These laws could be used in any computation scheme. In this method the heterogeneity of stress would of course be taken into account, so that the global behaviour of a coal pillar, for example, would be different from that of a single sample.

Of course the eternal problem of scale effect remains, which is more difficult when macrofractures appear. Further research will undoubtedly inform us about this topic.

Question: Dr. Rissler, Germany

Soweit ich verstanden habe, haben die Autoren den einaxialen Druchversuch nach vollzogen.

Besteht bei dieser Art des statistischen Modells auch die Moglichkeit, die Entwicklung des Bruches innerhalb der Probe zu lokalisieren? 2. Frage: Konnte man auf diese Weise auch die Bruchentwicklung bei zyklischer Belastung simulieren, wie das z.b. von Homand-Etienne experimentell utersucht wurde?

Answer: M.P. Duffaut

La je croiseue je peux repondre clairement qu on n'a pas essaiye de rentrer dans detail des phenomenes reels. On n'a pas du tout essaiye d'observer ou etaient et quels etaient, les microfissures qui se developpent. C'est un modele purement statistique et purement theorique.

Question: Prof. H. Grob, Switzerland

Est-ce que vous etes sur que la friction entre les plaques de la presse et la carotte qui est responsable pour le systeme de ruptur et le division des materiaux en deux classes.

Answer: M.P. Duffaut

Non, bien entendu, cette question est importante aussi, mais je pense que cela etait convenablement elucide dans les essais dont il s'agit. Car je pense que ce n'etait pas un probleme de frottement. Soli, ca, c'est un probleme qui est maintenant un petit peu depasse, j'espere, chez les bons experimentateurs.

Comment: Dr. J.A. Franklin

Like the preceeding paper by Mr Ito this research appears to make an important contribution by pointing to the interrelationship of several parameters hitherto considered independent. In this case the excellent fit of the selected statistical model appears to demonstrate an interrelationship between the pre- and post-peak deformability moduli of rock materials and their compressive strength, so that it might be possible to represent the uniaxial compressive behaviour of rock materials by a single parameter rather than three independent ones as before. Correlation between deformability modulus and compressive strength has been noted in earlier research work but has not previously been represented by a model for the complete stress-strain curve. Simplifications of this sort, if they can be confirmed as being valid and sufficiently accurate for engineering purposes, should be of great assistance in the characterization of rock behaviour.

PAPER: Needless Stringency In Sample Preparation Standards For Laboratory Testing Of Weak Rocks

AUTHORS: P.J.N. Pells and M.J. Ferry

PRESENTED BY: P.J.N. Pells

DISCUSSION

Question: M.P. Duffaut,

Il se trouve que je travaillair dans la commission de standardisation des essais de l'ISRM, il y'a un quinzaine d'annees, a l'epoque ou on travaillait sur la compression simple et sure ce genre de sujets, et je crois que la presentation de M. Pells e'claire tout a fait les problemes des ingenieurs d'une cote' et des universitaires de l'autre. Les ingenieurs ont besoin d'essais qui n'en coutent pas chers, et il est tout a fait normal, je suis tout a fait d'accord avec lui pour dire que lorsque qu'on a une certaine roche, on peut lui appliquer des criteres beaucoup plus simplifies a condition que cette roche le permette. C'esst la nature du gre's quia ete etudie par M. Pells qui permet cette simplification. Dans d'autrés types de roches il n'a pas ete possible de les accepter. Et je pense que M. Franklin doit avoir une opinion aussi.

Comment: Dr. J. Franklin

Yes, I certainly do. I'd like to very much congratulate Mr. Pells and Mr. Ferry on their work, I think it is very useful work and also very well undertaken. As previous President of the Commission on Testing methods of the Society (and I believe continuing President for the next 4 year period) I would like to point out that our object has been to get the methods into writing with the very point of encouraging evaluation, re-evaluation, criticism and modification with later drafts of the methods. The uniaxial compressive strength draft is one that I think should be rewritten quite soon along with several other of the earlier drafts that were published and we intend to try to do this. Another point worth noting is that approximately one person in 10 of the Membership of the ISRM has or is contributing actively to the Commission on Testing methods through approximately 30 different working groups of this commission, so its really not an elite who are doing this kind of work, its the Society itself that is busy preparing the standards which it will use for testing in routine applications. Again we very much encourage this kind of development and valuation work.

Question: Prof. Einstein, U.S.A.

You mentioned that clamped on strain measurement devices gave "disastrous" results. Could you expand a bit on these remarks.

Answer: Mr. P.J.N. Pells

Yes, I should have possibly been a bit more specific, but time was short. We made up a device, in fact we designed a device which we thought again would save us having to use strain gauges. It clamped directly to the specimen at the top and bottom 3rd points, and had a dial gauge indicator on it. Because of tilting problems within the device itself it gave us very erratic readings, so what I really said can't be taken as a generalization to all such devices and I apologise if I was a little bit naughty in that regard.

Question: Mr. Th.O. Mutschler, Germany

In soil mechanics greasing of the end faces is very common and also using length: diameter ratios of 1 or less. I would like to know your opinion about introducing this in rock mechanics for testing weak rocks too?

Answer:

My opinion is no. Because if you do introduce any material which can extrude laterally the specimen will fail in tension, so one has to be very careful with any capping or end material.

Question: Mr. A. J. Bowling, Australia

As a member of the sub-committee drafting the Australian Standard on Methods of Testing Rock for Engineering Purposes, I would like to thank Messrs. Pells and Ferry for the comments contained in their paper. However, whilst some standards for the preparation and testing of uniaxial specimens of weak rock such as Hawkesbury sandstone have been shown in this paper to be perhaps too stringent this may not be the case for all weak rocks. It is also noted that Pells & Ferry are concerned with the quick field testing of uniaxial specimens. I feel that there is a basic need for a rigid laboratory procedure for uniaxial testing which is what the ISRM Suggested Methods and the draft Australian Standard aim to provide. Should further investigations, such as these by Pells and Ferry, indicate that less rigid test procedures can be used for weak rocks then a note to this effect could be included in the test procedure or perhaps even an alternative uniaxial test procedure for weak rocks could be incorporated into the Australian Standard for example.

Comment: By Dr. John A. Franklin

As Co-President of the ISRM Commission on Testing methods, I would like to underline that the Commission greatly welcomes critical review of the ISRM Suggested Methods such as that presented by the authors. Indeed publication of these Suggested Methods of Testing is primarily intended to encourage such review, and periodic revisions to the Suggested Methods are planned that will take into account the results of critical evaluations such as these. From my personal experience I am in agreement with the authors that the method as presently published is in several respects too stringent when applied to the weaker rock materials. It is most important that we aim at simplification whenever possible in order to reduce the cost of routine rock mechanics testing.

PAPER: Critical Evaluation Of Rock Behaviour For In-Situ Stress Determination Using Overcoring Methods

AUTHORS: L.P. Gonano and J.C. Sharp

PRESENTED BY: L.P. Gonano

Question:

Have they had the opportunity to compare this South African overcoring cell with the CSIRO cell developed in Australia. From comments and discussions during this Congress it appears that the CSIRO cell, while similar in many respects to that of the CSIR, may have important practical advantages.

Answer: Dr. Gonano

I've used both types of cell and I think there are two types of errors which creep in. The first one is in the solution techniques and they are relevant in both cells. The second type of error refers to the practical exercise of obtaining the overcoring strain data in the field and I've found that the CSIRO cell is simply a technological advance and as a result the success rate is a lot higher.

Question: Prof. Dr. Natau, Germany

When you use triaxial-cells in small bore holes for stress measurements, the grain size of the host material is of importance. Please tell us how large the grain size was in your investigated material?

Answer: Dr. Gonano:

The idea with the laboratory testing is to effectively eliminate all experimental errors by doing a reverse test. You eliminate the effect of the grain size, in theory at least, irrespective of what the size of the grain is, because any particular minor errors or idiosyncrasies which exist because of strain gauge bridging two grains are effectively cancelled out when you back calculate the derformability factors. These factors are not just estimates of the modulus for example the Youngs Modulus and the Poissons ratio or any of the other factors that we use in cross anisotropic rock but also they represent the conversion factors from strain to stress which are applicable in this particular test with that piece of core and with that triaxial strain cell still in tact in it. Does that answer your question?

Question: Mr. P. Pells, Australia

When overcoring tests were done on the Orange Fish Tunnel in similar rock sequences they had a lot of trouble with the swelling of the rock during the actual test procedure. Did you have that problem at Drakensburg or did you overcome the problem in some way?

Answer: Dr. Gonano

We attempted to locate the tests in competent sandstone and we really didn't have any swelling problems at all.

Dr. Rissler Summarizing Comments To General Theme A.

Given in German.

Wir sind nun am Ende der Vortrage zum Thema A angekommen. Dies gibt mir Gelegenheit fur einige allgemeine Bemerkungen. Wir haben alle eine grosse Auzahl wertvoller Beitrage gehort. Weitere ausgezeichnete Arbeiten werden. Sie demnachst in den "Preprints" lesen. Einige wenige davon sind allerdings so, dass hier eine Kritische Anmerkung am Platz ist: Die Felsmechanik scheint sich einem Punkt zu nahern, wo side die Wege trennen. Der eine fuhrt zu einer verbesserten Sicherheit des Bauwerks und zu mehr Sicherheit, der andere zu mehr Theorie und zu Problemlosungen, wo niemand, mit Ausnahme des betreffenden Forschers, eine Frage gestellt hat. Diese Gefahr besteht nach meiner Ausicht besonders bei Beitragen zu diesem Haupthema. Wir sollten alle unsere vergangene Arbeit, aber auch unsere zukunftigen Vorhaben Kritisch daraufhin untersuchen, in welche Richtung sie fuhren und unnutze Dinge Lassen. Wir verschwenden sonst nur wertvolle Arbeitskraft und Geld.

Zum Schluss mochte ich mich an dieser Stelle bei Prof. David Stapledon bedanken, der in all den Monaten der Vorbereitung des General-berichts ein ausgezeichneter und fairer Kollege gewesen ist. Ich habe unsere Zusammenarbeit sehr genossen.
Nochmals herzlichen Dank!

CLASSIFICATION, PREDICTION, OBSERVATION AND MONITORING

Classification, prédiction, observation et contrôle

Klassifizierung, Voraussage, Beobachtung und Überwachung

A. Silverio
Chairman

DISCUSSION

Question: Mr. Pierre Duffaut makes short comment

Le Rapport General comporte un chapitre tres important.

Je vous racommande donc tres vivement de lire cette Partie du Rapport General.

Je ne peux pas donner ici l'opinion que j'ai moi-meme sur ce sujet; ce serait trop long. Mais, si vous permettez une comparaison tres simple, je dirais que toutes les universites du monde delivrent des grades, et que les reponsables qui doivent embaucher des ingenieurs ou des chercheurs ne se contentent pas des grades qui sont delivres et qu'ils utilisent des systemes d'evaluation beaucoup plus compliques et chacun le sien. Je crois que c'est un example qui n'est pas sans interet

General Comment by Prof. Stapledon.

Dr Rissler and I were concerned when reading these papers about classifications that workers around the world are going in many different directions in trying to develop rock classifications.

We think that it is not good to have too many classifications. We agree with the participant yesterday, who said that he wished we could talk in a common language when discussing rock masses.

We have therefore arranged a meeting of all persons interested in this question to be held in the discussion room on the second floor at 5.30 this evening.

PAPER: An empirical failure criterion for rock masses

AUTHORS: Prof. Yudhbir, W. Lemanza, F. Prinzl

PRESENTED BY: Prof. Yudhbir

DISCUSSION

Question: Dr Ian Johnston, Australia

My first question. The authors have developed a failure criterion which has as one of its three parameters, a constant α which has been assigned a value 0.65 "For all rock types". However by placing $\alpha = 1$, A = 1 and $B = \frac{1 + \sin \emptyset}{1 - \sin \emptyset}$ we have since the linear Mohr-Coulomb criterion. Since the latter is a reasonable representation of soil failure, would the authors care to comment on when α changes from 0.65 to 1 or most importantly when does a rock become soil.?

Secondly, have the authors used their failure criterion to predict the tensile strength of intact rock and if so what variations have they encountered in their ratio of uniaxial compressive strength to uniaxial tensile strength?

Answer: Dr. Yudhbir

I think regarding the first question: I have seen the question before and I thought I tried to answer it during the presentation. If you take A equal to 1 that represents the intact rock and what you say might well be true that one can represent that, but that's for one condition only as I pointed out during the presentation. In a way we try to put in the dilatancy effects which are controlled by jointing and alteration of the rock and I'm sure if you try to plot the failure, even the Mohr-Coulomb representation, the position of the Mohr-Coulomb line intercept on the normal stress equal to zero state will again shift so we have to alter these values in any case if you are going to encompass from intact rock to soft rock and at least with the limited information that we have reviewed as I said there are 6 rocks that we looked at and Bieniawski looked at 4 or 5 of them in intact state. He gave the alpha value .75 as the average for siltstone, sandstone, quartzite, norite and mudstone. So within these 7 to 8 rocks I admit this is quite limited but as I said earlier there are very few controlled laboratory tests available. I'm sure one has to check this consistancy of the slope within this narrow range of .65 to .75.

Regarding the second question, I think in our figure 9 is the answer. We did try, we haven't written about it in the paper as you will see here that the prediction of the tensile stress are measured values depending on which recommendation you use whether the Fairhurst or the ISRM. We get the ratio of the order of between 5 to 8 and of course its quite clear that there are numbers for rocks which are in that range up to 10 but then there are experimental values sometimes giving very high ranges but I think for soft rocks in our formulation we will predict a ratio between 5 to 8. I hope this answers both your questions.

PAPER: A Numerical method for the definition of discontinuity sets

AUTHOR: N.F. Grossmann

PRESENTED BY: N.F. Grossmann

DISCUSSION

Question: Dr Chappell, Australia

Dr Grossmann the thing I would like to do is pose a little comment and then ask a direct question. Breaking in discontinuity and the aspect of what you are discussing is to me very important, but in my mind I try to build up a geological model, quantify that model and then analyse it. In that model we have to quantify joint sets and the two things that come up are orientation and continuity. At the Snowy Mountains Engineering Corporation we have been using stereonets using normal distributions Schmidt and Wullff net and such, and from the frequency you can get a histogram, get the attitude and the frequencies and we get a block and represent that. The question is, in your approach there is a distortion that you bring in, and I can appreciate you can't get everything from one borehole you've got to go to three faces and all that sort of thing. How does your representation tie in with the standard system of statistically defining attitude in the normal Schmidt method and when you have to get a representation of histogram how would you try and bring in continuity and frequency. In other words, you define your attitude but you've still got the problem of continuity in it. Have you looked at the problem in relation to what you are doing there now?

Answer: Mr. Grossman

First of all Dr Chappell I would like to draw your attention to the three drawings which I hung up on a Poster in the upper floor, which correspond all to the same example. I obtained the data from a publication of the US Bureau of Mines (Mahtab M. et.al. - Analysis of fracture orientations for input to structural models of discontinuous rock - US Department of the Interior, Bureau of Mines, R. I. 7669, Washington DC USA, 1972), and the three drawings, which are all stereographic projections of the upper hemisphere, show (1) the pole diagram with the set domains (as defined in my paper), (2) the frequency diagram using the 1/N area counting circle, N being the number of considered discontinuities, and (3) the "classic" frequency diagram with the 1% area counting circle.

Secondly, I would like to point out that on the little paper, which hangs beside them, we have got the mathematical comparison of the results obtained with these data by different authors, using different clustering systems. This means, we compare the mean attitudes of the obtained sets, the standard deviations, and so on.

Lastly, I would like to point out that what I explained was only the first (but most difficult) step in a discontinuity study. What I have explained to you was the computer technique to get the sets. If the computer knows which discontinuity belongs to which set, it then can very easily calculate the different set characteristics, attitude, dispersion, spacing, opening, continuity, and all this.

The very important problem in a discontinuity study is to avoid the human manipulation of the data, which is always very time-consuming, and sometimes very subjective. My paper only describes how to find out from a huge collection of data which are the joints of set 1, which are the joints of set 2, which are the joints of set 3.

Just one other little remark.

You may stress very much the importance of the frequency, but I do not give it all this importance, and, as an explanation, I will pose you just one question - Which is more dangerous to the engineer: the isolated fault, which will let slide down into your cavern a big rock wedge (it is just one fault, no parallel discontinuities occur, but it does condition the stability of the rock wedge); or 20 small joints, which you find spread all along your cavern wall, but which have no consequences at all? - This is why I think that frequency is not such an important parameter.

At the LNEC we do this job like this:

We obtain the sets, we then calculate the set characteristics, attitude, dispersion, spacing, opening, continuity, and all this, and, finally, having got all these set data, we decide if yes or no it is an important set. Of course the number of the discontinuities will normally have a certain weight, but it is not the only important factor.

Question: Dr Chappell

In determining joint sets different representations from statistics is available. Assuming one has overcome the sampling problems and has obtained a good and reliable sample, how is bias put on choosing the joint sets. Is it from the technique of analysis; that is putting on the constraints of a particular method. For example a Poisson or Normal distribution. How does the various choices of distribution effect the sets of joints determined?

Answer Mr. Grossman

Your second question, Dr Chappell, may be partially answered through a glimpse at the little paper hanging at the Poster in the upper floor, which I mentioned in my previous answer.

The question you put, involves two main problems.

One, is the problem of what is the mathematical model which describes the distribution of the discontinuity attitudes.

Although several theoretical distributions have been proposed by different authors, none of them has, as far as I know and up to now, a physical justification. The only ones which take into consideration the nearly always existing anisotropy of the real attitude distributions, are the Bingham distribution, referred to by Shanley and Mahtab (Shanley R.J. and Mahtab M.A. - Fractan: a computer code for analysis of clusters defined on the unit hemisphere - US Department of the Interior, Bureau of Mines, I. C. 8671, Washington DC USA, 1975), and the bivariate normal distribution on the tangent plane at the mean attitude, referred to in my paper. The choice between these and any other distributions which may become available, has to be made, for the moment, by means of a goodness of fit test, like, for instance, the Kolmogorov test or the χ^2 test.

Due to the low precision of the χ^2 test when only a few data are available, I have been using the Kolmogorov test. As this test applies only to distributions of one variable, and the distribution under consideration depends on two variables, the test is first applied successively to the distributions of the two cartesian coordinates of the attitudes, and afterwards also to the distributions of the two polar coordinates. Up to now, no case of rejection (at the 95% level) has been detected for the bivariate normal distribution on the tangent plane at the mean attitude, for all cases in which the discontinuity diagram shows a unimodal set.

The second problem concerns the delimitation of the discontinuity sets, and is, in principle, independent of the first problem. It consists, basically, in finding the good engineering balance between having too many small, but not very disperse sets, or some few, but very variable sets.

This problem is satisfactorily answered by the method described in my paper, which I did expose some minutes ago.

PAPER: Development of a coupled hydromechanical model of rock joint behaviour.

AUTHOR: Dr. S. Bandis

PRESENTED BY: Dr. Bandis

DISCUSSION

Question: Dr J. Franklin, Canada

I would like to congratulate the author on a major step forward proceeding from the earlier major steps forward of Dr. Nick Barton. The question that I have really, relates to where do we go from here. It appears to me that extrapolating from the small to the large scale with regards to roughness will at some time in the future have to be linked, not only to roughness but also to other major aspects of joint characteristics, namely orientation and continuity or persistence of joint sets, because its not very often possible in a practical example to consider single joints. The question arises whether a shear failure is going to be jumping from one joint to another because of two closely similar joint orientations or because of a discontinuous joint which may step from one surface to another. I would like to ask the author whether in his research he has considered the possibility of an integrated approach taking into account joint roughness, continuity, persistence, spacing and closely similar orientations of separate joints and how this might be achieved?

Answer:

Continuity has been assumed. We have considered continuity to be persistent throughout the failure surface and what has been shown that in cases you have a continuous failure surface we can actually expect that the shear characteristics of an individual block which is defined by the average cross joint spacing on site can give us a quite good indication of the shear behaviour at full scale. The joint block size in situ is the minimum size of sample that has to be investigated in this case. It hasn't been considered in the case of as you mention different joints jumping into another and so on.

GENERAL REPORTS / RAPPORTS GÉNÉRAUX / GENERALBERICHTE

SURFACE AND NEAR-SURFACE EXCAVATIONS

Excavation en surface et à faible profondeur

Übertägige und oberflächennahe Felsbauwerke

R. E. Goodman
University of California, Berkeley, USA

SYNOPSIS

The papers offered in Theme B demonstrate that the profession of rock mechanics is making use of modern techniques for describing, measuring, and analyzing rock behavior, some of which were presented as new developments in previous Congresses. There is a wide use of an integrated approach to solutions of rock engineering problems, embracing geology, rock testing, structural mechanics, and instrumentation. The importance of statistics to rock mechanics is not apparent from the selection of papers. In the second half of this report, a new development, "Block Theory", is described. It permits 3-dimensional analysis for the orientation and support needs of surface and underground excavations.

RESUME

Les thèses qu'on présente dans la thème B montrent que le métier de mécanique des roches emploie des tecniques modernes pour décrire, mesurer, et analyser le comportement des roches. Certaines d'entre ces tecniques ont été presentées comme de nouveaux développements devant les Congrès précédents. Pour résoudre des problèmes d'application on utilise en général une approche intégrée qui combine: la géologie, les essais des roches, la mécanique structurelle, et l'auscultation. Il n'est pas évident, en considérant la ceuillaison des thèses, l'importance que porte la statistique pour la mécanique des roches. La deuxième moitié de ce rapport décrit un nouveau développement qui s'appele "Block Theory". Ce theorie permit une analyse en trois dimensions pour l'orientation, et pour les besoins de soutiennement des excavations.

ZUSAMMENFASSUNG

Die Arbeiten die zum Thema B gehören beweisen dass Felsmechanik als Beruf moderne Techniken zur Beschreibung, Messung, und des Analyse des Verhaltens von Felsen verwendet; etliche dieser Techniken wurden als Neuentwicklungen in vorhergehenden Tagungen vorgetragen. Viele der Arbeiten behandeln in integrierter Weise die Lösungen von Felsbauproblemen, so dass Geologie, Felsexperimentierung, Strukturmechanik und Instrumente umfasst werden. Die Auswahl der Arbeiten widerspiegelt ungenügend die Wichtigkeit einer statistischen Behandlung der Felsmechanik. In der zweiten Hälfte dieses Referates wird ein neuer Begrift, nämlich "Block Theory" eingeführt. Diese Theorie ermöglicht eine dreidimensionale Analyse der Orientierung und des Ausbau Bedarfs von unterdischen und Oberflächen Felsbauwerken.

1. INTRODUCTION

This report will discuss progress in rock mechanics for surface and near surface excavations and foundations, using the stimulus of papers submitted to the Congress. We know where the profession or rock mechanics has been, and this meeting demonstrates where it has arrived today, (although the adequacy of that statement could be questioned). I will go a step further and refer also to some new developments whose full applications are yet to come.

The contributions to theme B, contained in Preprint Volume C, were divided into three sub-themes -- Stability of rock slopes, Foundations on and in rock, and Near surface construction. Dam foundations are included in the second sub-topic.

The contents of the 46 papers printed under Theme B reflect a variety of viewpoints and concerns. They are rich in ideas, examples, and results ranging over a considerable portion of rock engineering. Consistent with the spirit of the Salzburg Colloquia most of these contributions avoid narrow specialization in sub-sub-elements but rather attempt to fuse geology, mechanics, and engineering practice in addressing real issues in what we know to be a complexly organized inter-disciplinary subject.

Tables 1, 2, and 3 list keywords for all the papers, to index their contents. These may serve to guide one's immediate browsing path but they were included mainly to chart the breadth of our material.

2. DISCUSSION OF SELECTED CONTRIBUTIONS

Among the papers of Theme B are many sorts of contributions; every paper has some merit and several will be enduring additions to the reference literature.

2.1 Revelations about practice in exotic places are contained in a number of papers. In an international Congress, what is "exotic" to one delegate will be common-place to the next. Nevertheless, I dare to mention the papers by Gaziev et al, Brand et al, and the several papers from China as particularly edifying to me. The second relates the difficulties of urbanization in Hong Kong, with steep slopes in decomposed rock, and shows how the profession has responded. Because problems of decomposed rock are wide-spread, around the whole world, and we have largely neglected their study in our laboratories, this paper is valuable.

2.2 Rock mechanics applied to construction in extreme environmental conditions occupies our fancy for remote projects connected with underground storage of toxic substances, energy storage and conversion schemes and so forth. Perhaps less hostile but surely troublesome is the environment at the site of a telecommunications station atop an Alpine peak, discussed by Hellerer and Ostermayer. Among the factors contributing to construction difficulty were highly fractured rock, permafrost, wind, and (last but not least) tourists. Novel methods of excavation and support are described.

2.3 Use of geological and rock-mechanics classification systems to subdivide complex rock masses. Zonation of rocks by means of classifications embracing rock properties continues to be a controversial subject. Some classification systems in vogue ask for data that can usually only be obtained after construction. Also geological factors, like rock type and age, are omitted from the input data. Nevertheless, the trend for the use of special rock mass classifications is increasing, as measured by the numbers and variety of applications discussed here. Gonzales de Vallejo discussed the zonation of a several km long open pit coal mine . In addition to fracturing and rock strength, the zonation included maximum possible amounts of geologic and hydrologic data and back-analysis. McCosh and Vladut applied a modified Q system of classification paying attention to particularly important variations in properties observed in the rocks of an open pit coal mine. Portillo applied classification by the RMR and Q systems on a massive scale for an iron mine developed in karstic rocks. He zoned the pit based upon rock class and mode of failure, as related to the analysis of structure. Perri performed a similar study, mainly on the basis of structural features, for natural slopes. Perhaps the most enlightening of the papers dealing with classification and zonation of rock masses was that by Einstein et al which compared predictions with actual developments in constructing a subway station in argillite. The predictions offered using different classifications were not dissimilar except that the Q system was better able to distinguish extremely bad conditions. Their very interesting paper explores the subjectivity and stage of project-development on the adequacy of predictions made using rock classification systems. But the questions asked have not been answered definitively. I predict the analysis of rock classification will continue to be a prominent issue in at least the next Congress.

2.4 Advances in handling difficult or weak rock materials are discussed by several authors. The paper by Brand, et al, on decomposed rock was previously mentioned. Grüter and Wittke presented a stimulating analysis of construction - deformations in a highly stressed mudstone. Cestelli, et al described the support of fractured cliffs of tuff. Soft, weak rocks are finding a place at last in rock mechanics congresses.

The Japanese, who have been very active in this field , are represented here by Maeda who presents original, and imaginative work on foundations in weak, tuff-breccia on a side hill. His is a complete case history with calculations for design based not only on the rock properties backed out of unusual field tests but on particular limit equilibrium analyses corresponding to the mode of failure observed and documented in the test.

2.5 Novel analysis of important problems is exemplified also by Egger's analysis of the stress state above shallow tunnels in weak rock. This clever mathematical development assumes limiting conditions corresponding to a chimney collapse tendency in the roof of the tunnel. The results are extended to the case of horizontally layered rock. For isotropic rock, the paper by Egger reaches the conclusion that stability without supports can be achieved only if $c/(\gamma a) > 1$ where a is the radius of a circular tunnel, γ is the unit weight of the rock, and c is the cohesion of the rock, assuming Mohr-Coulomb behavior. A number of other authors as listed in Table 3, have also discussed the analysis of shallow tunnels.

Another novel analysis was made by Pender, Graham, and Gray, who developed a dilatant constitutive model for finite element models of rock deformations. They have proposed a new deformability constant, D, to link volumetric strain and octahedral shear stress. The theory of elasticity would then be treated with a set of three constants, as opposed to the usual two. But such a material is really non-linear because the action of D should be to promote increase in volume regardless of the sign of the octahedral shear stress.

The dilatancy model was applied to the study of stress distributions around rock-socketed piers.

2.6 The use of numerical analyses to support engineering design in rock was the subject of a number of additional contributions. Jasarevic et al used finite element methods to study the support requirement for bridge foundation blocks. Columbet and Glories analyzed support requirements for slopes of a submerged portion of a quarry. And Scott and Dreher propose dymamic analysis of the equilibrium of foundation blocks of dams using finite element methods. Finite element analysis of piers and piles in weak rock was studied by Pender, as noted, and also by Chiu and Donald, who considered shear resistance of piles according to several non-linear deformability models for mudstone.

In the past, the use of finite element analysis for rock problems has not always yielded tangible results. A number of the papers in this Congress demonstrate that this method is now contributing significantly to the evaluation of the bearing capacity of rocks. Finite element analysis is also being used intelligently to forecast strains in man-made structures associated with movements within the rock. This is demonstrated effectively in the paper by Bailly et al, in which a quantitative assessment has been made of strains in the lining of an existing tunnel caused by driving a new tunnel nearby.

2.7 A number of meaningful case histories enrich these proceedings. Dam foundation papers by Abrahao et al, and Fu et al were particularly illuminating. The former discusses the evolution of values assigned to rock properties as information was acquired during construction and operation. The latter paper is a frank discussion of decision-making as affected by geological structure and rock properties. Grouting a karst foundation was discussed by Costa-Nuñes et al, and grouting a radio telescope foundation for dynamic stability is the subject of Werner's paper. An interesting case-history for open-pit mining is reported by Cox, who performed a detailed study of failure modes. The paper by Muzas and Elorza discusses the use of anchors to reduce the cost of a penstock and shows the basis for design calculations.

2.8 Improvements in techniques for in-situ testing of rock are the subject of several papers. This includes Costa-Nuñes et al's discussion of multi-stage in-situ direct shear tests in schist, Maeda's paper previously noted, and contributions by Hegan for weak sandstone and Milavanovich on test interpretation.

2.9 Discussion of non-deterministic bases for analysis and design or rock slopes are offered by Gaziev et al, by Read and Lye, and by Morriss and Stoter.

2.10 The costs of exploring rock are the subject of a paper by Van Schalkwyk. This is a novel contribution, although I did not fully appreciate the context in which it might be applied.

3. DISCUSSION

The impression derived from these 46 papers, as ambassadors for rock mechanics, is that we have arrived at a point where significant ideas of the past are now in regular use. There are many reports of integrated studies embracing exploration, lab and field testing, rock classification, analysis, monitoring and back analysis. It is now generally accepted that hard rock requires special study mainly by virtue of its defects. There is increasing focus, in addition, on inherently weak rocks.

Forty six papers is not a sufficiently large sample to measure much by its omissions, but it does seem that statistical approaches for measuring and describing defects are not sufficiently well represented here.

The application of structural mechanics and solid mechancis to problems of rock mechanics allows us to address problems of considerable complexity. That this approach can yield useful results is evidenced by the series of excellent papers in Theme B combining numerical analysis with geological and mechanical properties to answer practical questions. Numerical models are especially valuable for unprecedented problems where material properties are connected between conditions of stress, temperature and fluid pressure. These powerful numerical tools are also sometimes oversold for ill-defined problems related to design of excavations in rock. The ultimate numerical weapon is not always justifiable, or even suitable, and other, simpler approaches are in order. I will devote the balance of this paper to an introduction of one such approach, under development at Berkeley.

4. BLOCK THEORY

4.1 Block Theory is a new kinematic analysis technique to identify and describe the most critical blocks of rock around an excavation. The underlying assumption is that failure of a surface or underground excavation in hard rock will begin with the movement of a critically-located, unfavorably oriented block that has the excavation surface as one or more of its faces. For analyses using Block Theory, the primary input is the set of orientations of the various discontinuity planes of the rock mass. The intersections of these discontinuities define joint blocks, but these are of little interest per se as they do not intersect the opening. The keyblocks are formed by the intersection of discontinuities and free surfaces. The selection of the keyblock is the object of a rigorous search, facilitated by a body of theory about the finiteness, and removability of various intersection possibilities. This theory, and a series of interactive microcomputer programs enable an optimum choice of excavation shape, and direction and provides a good basis for designing supports, if needed. The subject is set forth in a book in press[1] and, is introduced in a series of short Proceedings Papers as listed in the References following this Report. Here I would like to present a short introduction to the methods and provide several examples.

The objective of Block Theory is to separate all possible rock blocks into five kinds (Figure 1): I - keyblocks, that are unstable without support; II - potential keyblocks, that are unstable only if the friction angles on the faces are below a given value; III - Safe, removable blocks that can not move even when the friction angle is zero; IV - Non-removable finite blocks; and V - infinite blocks. The procedure is to make a first separation into removable types (I, II, and III) versus non-removable types (IV and V) on the basis of three dimensional geometry, with the only required input being that needed to describe the orientations of joints and excavation surfaces. Subsequently, a division is made between Type III and Types I and II on the basis of a "mode analysis" that takes into account the direction of the resultant applied force in relation to the excavation. Finally, types I and II are separated on the basis of an equilibrium analysis, extending methods pioneered by Wittke, Londe, John, and others. The distinct advantage of this approach, over more formal solid mechanics analysis, is that it is fully three dimensional and does not require knowledge of the absolute locations of discontinuities as input to any computational mesh.

4.2 Removability of Blocks - Types I, II versus III, IV and V. A theorem developed by Gen-hua Shi provides the basis for the separation of blocks into removable and non-removable types. A block is determined by the intersection of $\underline{n}$ half-spaces, i.e. a block is the set of points that satisfies all n inequalities

$$\underset{nx3}{(A)}\ \underset{3x1}{(X)} > \underset{nx1}{(D)} \qquad (1)$$

where n is the number of faces of the block.

(A) is the $\underline{n}$ x $\underline{3}$ matrix of coefficients, representing the direction cosines to the normal to each face plane, and (D) is a column matrix establishing the location of each face plane. If all the faces are moved parallel to themselves to pass through a common origin, as in the convention when plotting various joint normals on a stereographic projection, the column matrix (D) becomes null and inequalities (1) are replaced by

$$\underset{(nx3)}{(A)}\ \underset{(3x1)}{(X)} \geq (0) \qquad (2)$$

Of the $\underline{n}$ planes, the first m are designated as those of discontinuities and the rest, $\underline{n-m}$, are those of excavation surfaces, -- "free planes". Then the system of inequalities (2) is separated into

[1] Goodman, R.E., and Shi, G.H., in press, Block Theory and its Application to Rock Engineering (Prentice-Hall, N.J., publication date in 1984).

$$\underset{(m\times3)}{(A)}\ \underset{(3\times1)}{(X)} \geq (0) \quad (3)$$

and

$$\underset{((n-m)\times3)}{(A)}\ \underset{(3\times1)}{(X)} \geq (0) \quad (4)$$

The set of points that satisfies (2) is termed a "block pyramid" (BP). It is the intersection of the set of points that satisfies (3) -- the joint pyramid (JP) -- and the set of points that satisfies (4) -- the excavation pyramid (EP).

$$BP = EP \cap JP$$

Shi's theorem states that a block is finite if and only if the intersection of JP and EP is empty, that is if its block pyramid has no edges. Additionally, the block is removable, i.e. not of class IV or V if and only if

$$JP \cap EP = \emptyset \quad (5)$$

and

$$JP \neq \emptyset \quad (6)$$

where $\emptyset$ represent the empty set.

The theorem can be applied by means of vector operations and we have developed a series of micro-computer programs to do this interactively. Alternatively, representation of equations 5 and 6 on the stereographic projection provides immediate graphical solution. In the stereographic projection a plane passing through the center of the reference sphere projects as a true circle, which crosses the reference circle (the projection of the equator) at the ends of a diameter. In the lower focal point projection, the area inside the equatorial circle represents the upper hemisphere. Similarly, the area inside the great circle for any plane represents the directions pointing from the center of the reference circle above that plane. The polygonal regions of intersection that are formed by the intersection of n non-parallel joint-plane great circles represent the joint pyramids.

The Joint Pyramids, JP's, are identified and described by a series of n digits, e.g. 0110. The digit 0 in the ith column represents the halfspace above joint plane i while the digit 1 in the ith column represents the half space below joint plane i. The digit 3 means that the block has both half-spaces, on either side of the joint plane. A joint pyramid with a single 3 in its code is a segment of a great circle.

Figure 2 shows an example of the coded joint pyramids for a joint system with the four joint sets listed in Table 4. The dotted circle is the reference circle (the projection of a horizontal plane). To analyze the removable blocks of the roof of an underground chamber, EP would be the region above the plane of the roof; therefore it is projected by the area inside the reference circle. A JP corresponds to a removable block in the roof if and only if it has no points in common with the EP. The removable blocks are 1011 and 1101. Equation (6) is satisfied if JP has a region in the projection. (It can be shown that there are half-space intersections that do not).

As an alternative to equation 5, we can write as the condition for finiteness of a block that its JP be contained inside SP, where SP stands for the Space Pyramid; SP is the complement to EP, i.e. $SP = \sim EP$.

$$JP \subset SP \quad (7)$$

Table 4

Plane	Dip(°)	Dip Direction (°)
Joint 1	71	163
Joint 2	68	243
Joint 3	45	280
Joint 4	13	343

4.4 The Space Pyramid

Problems of surface and underground excavations can be solved by finding an appropriate construction for the SP. Some examples are presented in Figures 3 to 7. The unshaded region in figure 3 is the SP for the interior edge of an underground gallery. JP 1101 is the only block that is removable in this edge. A block of this type can be very large before it bridges across the excavation and therefore such a block type is potentially dangerous. However, it is usually possible to avoid all such keyblocks by reorienting the gallery, searching for an SP that intersects all the important JP's. An even worse situation is created by a block that intersects a corner of the gallery. The SP's for the four roof corners of an underground chamber are plotted in Figure 4, and Figure 5 shows a block of this type. The excavation walls are planes 4 and 5, while plane 6 is the roof. A block that is removable in a corner of the chamber can be very large, with an edge of the same length as a diagonal of the chamber. Fortunately, it is easy to reorient an excavation to miss all important removable JP's of a corner because the SP is small.

Figure 6 shows the SP diagram for edges of the portal of a tunnel, where the ground surface intersects the interior surfaces of the tunnel. Portals are a problem in jointed rock because the SP is large and therefore the list of removable blocks is long. Also, the near surface weathering increases the number and severity of discontinuities.

Figure 7a shows a removable block - 1101, in relation to a cylindrical tunnel. The direction of the axis of the tunnel is A (and its opposite, A'). The block is removable in the tunnel between the extreme planes -- limit planes b and c -- that envelope JP 1101. These limit planes intersect the tunnel section at specific orientations with the horizontal, as read from the projection. The largest possible block intersects the tunnel cylinder between these limit planes as drawn in Figure 7b, a tunnel section, perpendicular to the tunnel axis; its appearance in the tunnel is shown in Figure 7c, a 3d view The block does not fill all the volume between the limit planes.

5. CONCLUSION

I believe that computer-oriented techniques like that offered by Block Theory, will make possible more economical design of tunnels. The input data required pertain to items that we know how to obtain reasonably well -- the orientations of joints, and, secondarily, the friction properties of the different joint sets. With additional information about the statistics of jointing and making use of simulation techniques it becomes possible to apply the theory with minimal exposure of rock, i.e. before actual construction. In contrast to structural mechanics computer codes, the analyses demand only stand-alone micro computers and can be performed at the site of the work in a construction trailer. If information on the state of stress can be obtained, it can be used to analyze the stability of important keyblocks. However, stress and deformability data are not essential to the performance of a keyblock analysis, whereas a structural mechanics analysis can not be performed without such information.

In conclusion, I believe that the current state of the art is making good use of the developments of the recent past. And there is every expectation that developments of today will become state of the art in the near future.

6. REFERENCES

Goodman, R.E.; and Shi, Gen-hua (1982) "Geology and rock slope stability -- application of the keyblock concept for rock slopes" Proc. 3rd International Conference on Stability in Surface Mining, (Society of Mining Engineers) pp. 347-374.

Shi, Gen-hua; and Goodman, R.E. (1981) "A new concept for support of underground and surface excavations in discontinuous rocks based on a keystone principle" Proc. 22nd Symposium on Rock Mechanics (M.I.T.) pp. 290-296.

Goodman, R.E.; Shi, Gen-hua; and Boyle, W. (1982) "Calculation of support for hard, jointed rock using the keyblock principle" Proc. 23rd Symposium on Rock Mechanics (Society of Mining Engineers) pp. 883-898.

Chan, Lap-Yan; and Goodman, R.E. (1983) Prediction of support requirements for hard rock excavations using keyblock theory and joint statistics; Proc. 24th Symposium on Rock Mechanics, (Association of Engineering Geologists and Texas A & M University) -- in press.

Shi, Gen-hua (1982)"A geometric method for stability analysis of discontinuous rocks" Scientia Sinica Vol. XXV, No. 1 Jan.

ACKNOWLEDGEMENT

I am deeply indebted to Gen-hua Shi, my collaborator on the development of Block Theory, for permission to include ideas and Figures from our book in preparation. Mr. Shi's brilliant computer programs were used on a micro-computer to construct the drawings. The Space Pyramid diagrams for edges of a gallery and of a portal to a tunnel have not previously been published. Above all, it is his general theorem about finiteness and removability of complex blocks that has allowed the development of Block Theory. The reference to the theorem presented in this report is only a special case of Shi's Theorem. In its broadest statement, his theory governs non-convex blocks with non-convex joint pyramids and non-convex excavation pyramids.

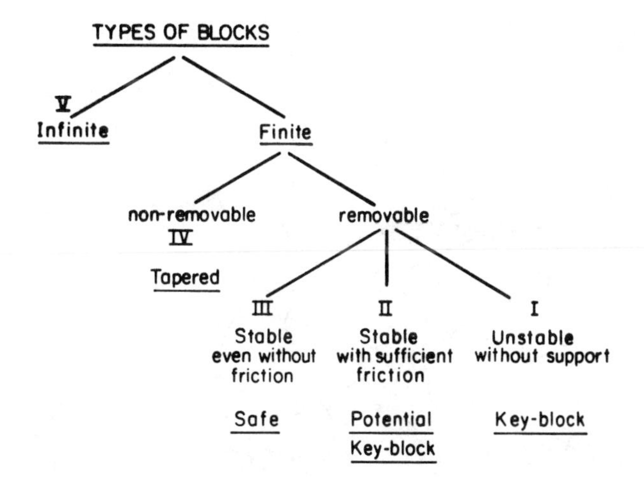

Fig. 1

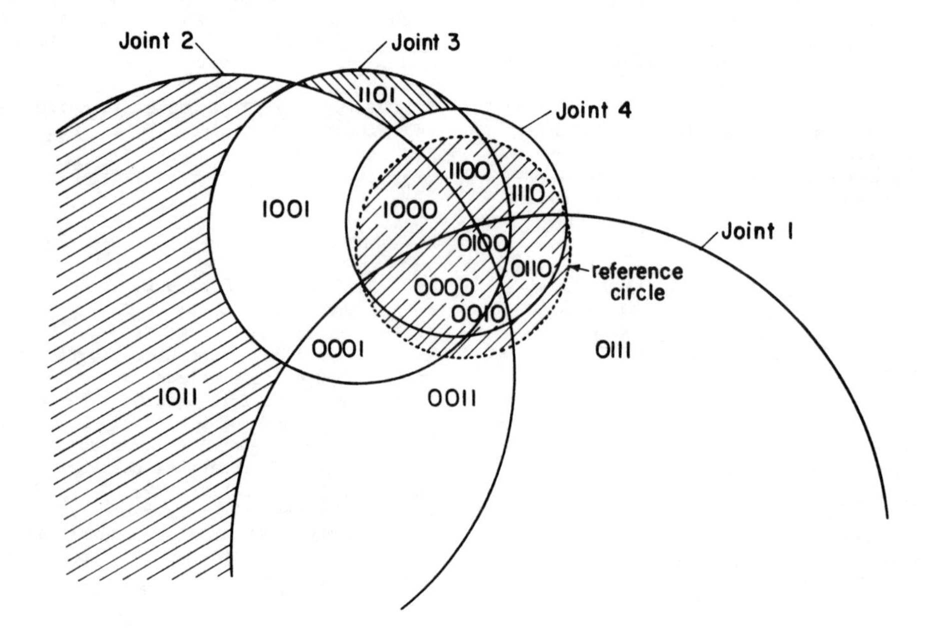
IIII
Joint 2
Joint 3
1101
Joint 4
1100
1110
1001
1000
0100
Joint 1
0110
reference circle
0000
0010
0001
0111
1011
0011

Fig. 2

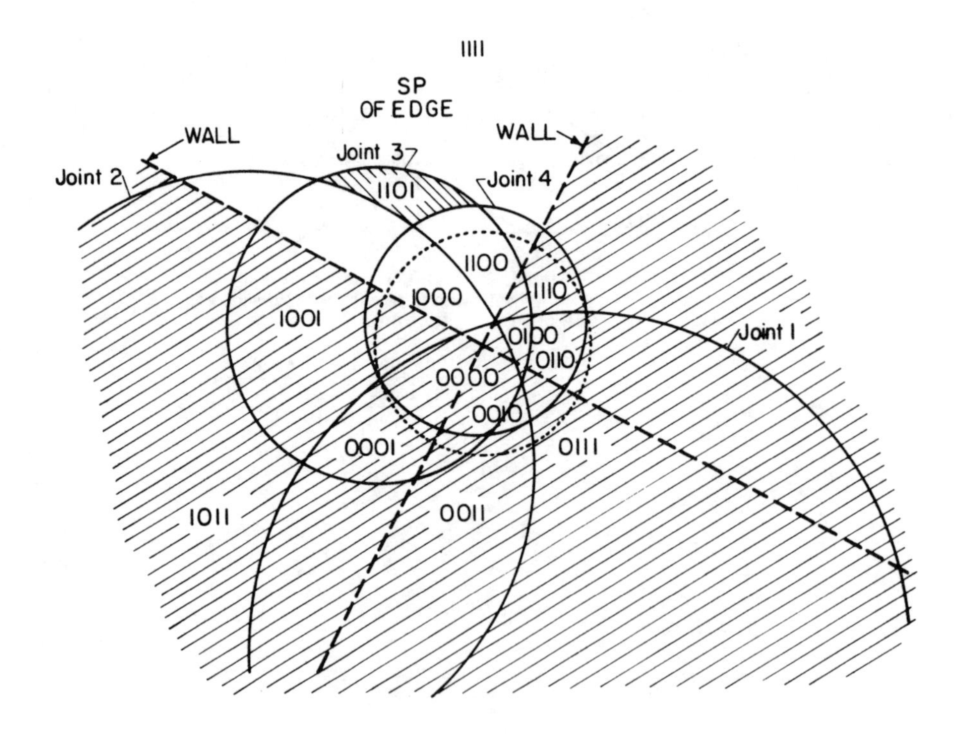
IIII
SP OF EDGE
WALL
WALL
Joint 3
Joint 2
1101
Joint 4
1100
1110
1000
1001
0100
Joint 1
0110
0000
0010
0001
0111
1011
0011

Fig. 3

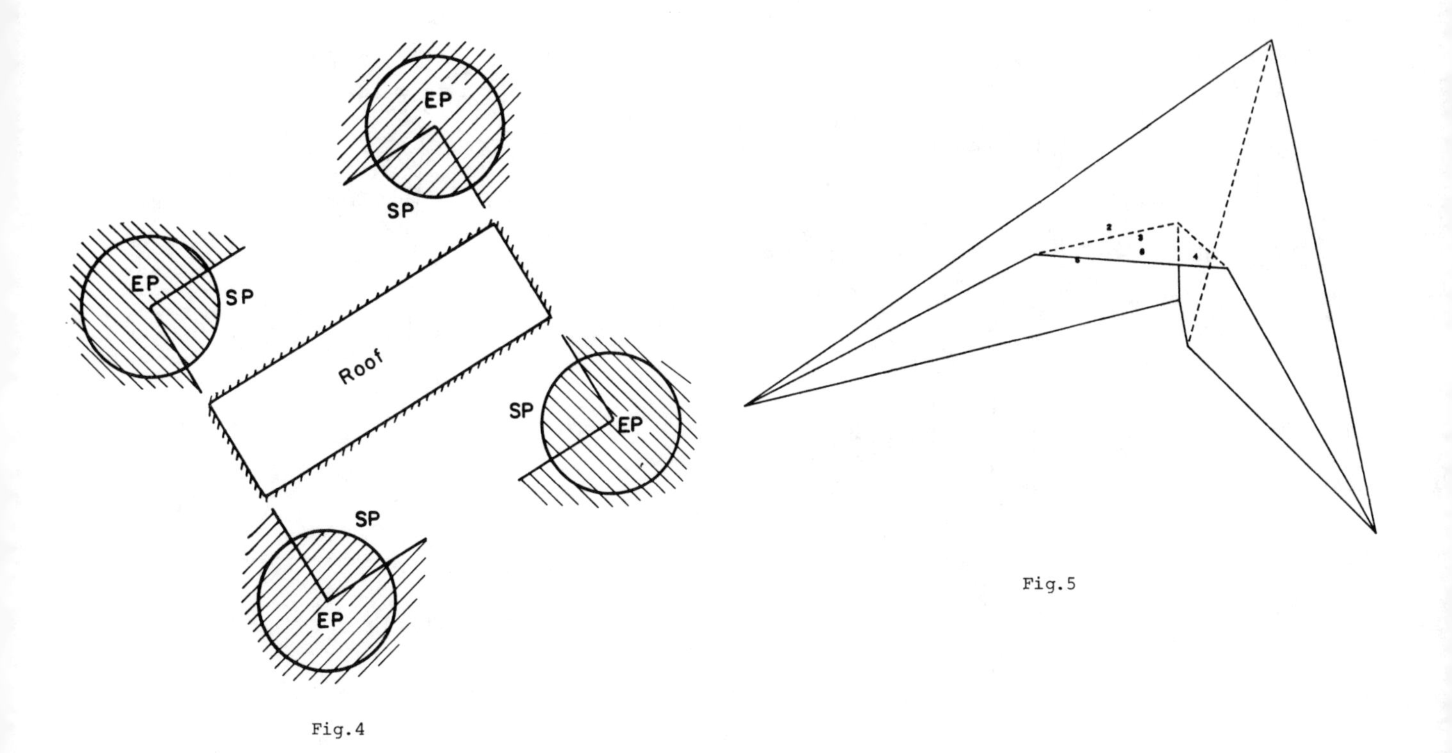

Fig.4

Fig.5

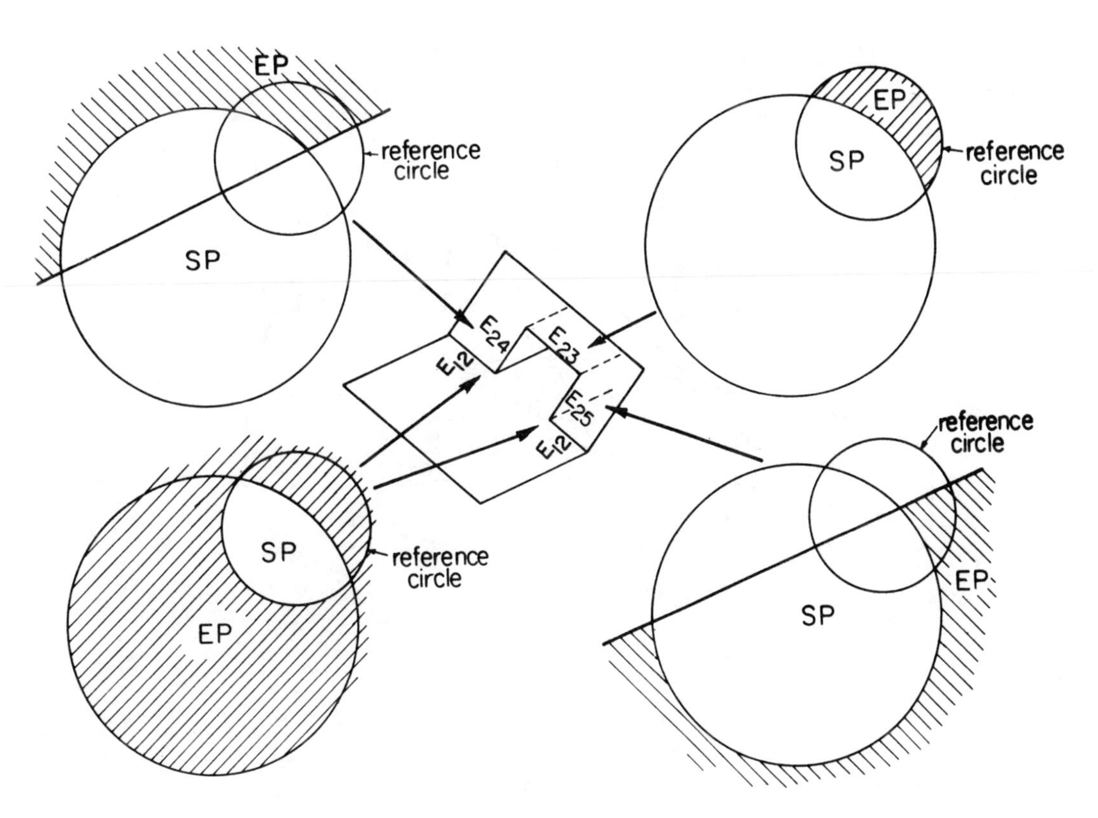

Fig. 6

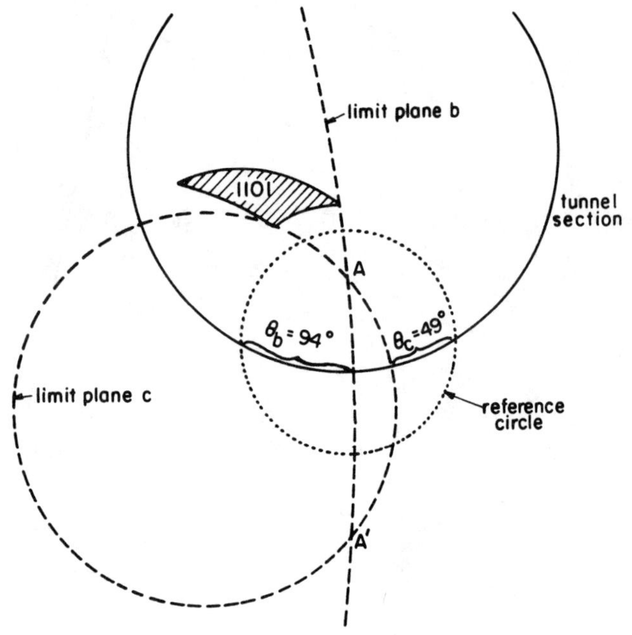

Fig. 7a

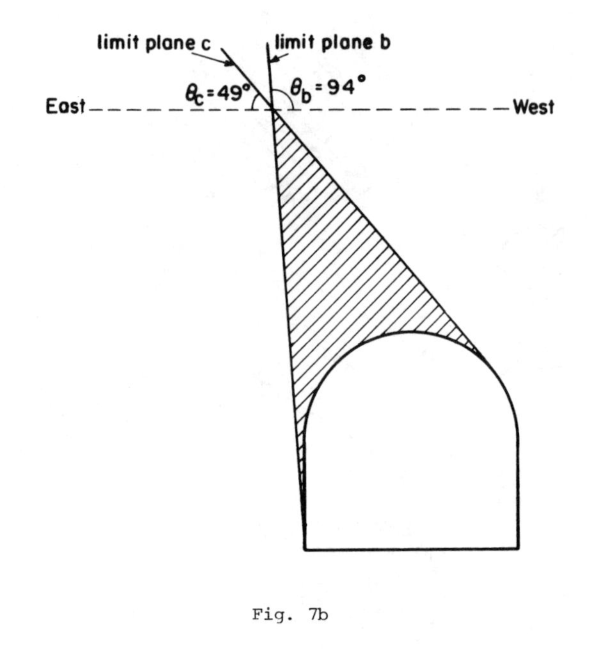

Fig. 7b

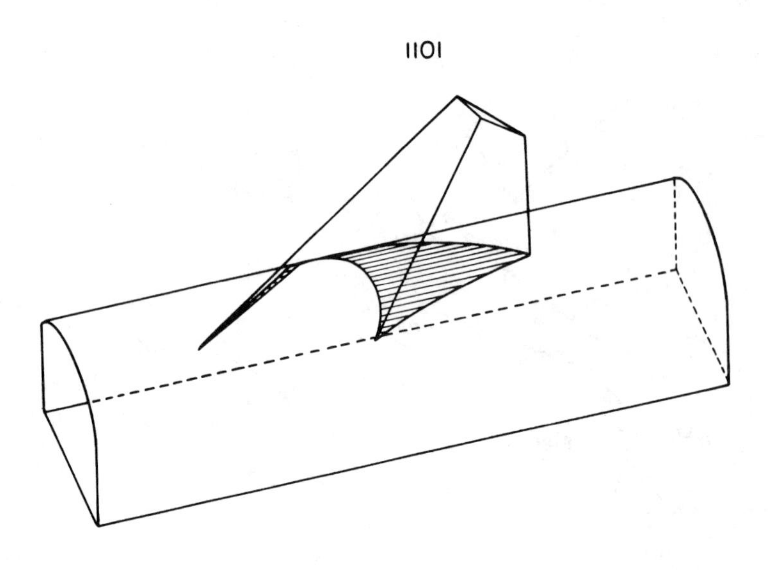

Fig. 7c

TABLE 1A THEME B1 STABILITY OF ROCK SLOPES

First Author:	Sugawara	Portillo	Gonzales	Brand	Columbet	Cestelli	Yuke	Rao	Pacher	daGama
granites				X	X				X	
slate, schist		X		X			X			
shale, mudstone			X				X	X	X	
limestones		X					X			
coal measures			X				X			
sandstones							X	X	X	X
volcanic flows				X						
tephra						X				
decomposed rock				X					X	X
exploration			X	X		X		X		X
rock classif.		X		X				X		
numerical models	X				X					X
physical models	X									
limit equil.	X									
back analysis		X								X
rock testing										
statistics										
factor of safety										
support					X	X			X	
monitoring					X	X				
blasting										
drainage										
failure modes	X									
toppling		X					X			
minepits		X	X				X	X		X
highways										
spillways					X					
space: bldgs				X						
natural slopes				X		X	X			
bridges									X	
regional planning				X						

TABLE 1B THEME B1 STABILITY OF ROCK SLOPES

First Author:	Costa-Nunes	Perri	Salcedo	Gaziev	Read	Cox	Morriss	McCosh	Solymar	delCorral
granites					X				X	
slate, schist	X	X	X						X	
shale, mustone							X			
limestones										
coal measures						X		X	X	
sandstones	X					X				
volcanic flows					X					
tephra					X					
decomposed rock	X	X	X		X					
exploration		X					X			
rock classif.								X		
numerical models				X						
physical models										
limit equil.		X		X		X				X
back analysis			X			X				
rock testing	X		X		X		X			
statistics				X			X			
factor of safety				X		X				
support				X						
monitoring				X		X				
blasting						X			X	
drainage				X				X		
failure modes				X						
toppling										X
minepits	X				X	X	X	X		
highways										
spillways										
space: bldgs										
natural slopes		X	X							
bridges										
regional planning		X								

TABLE 2A THEME B2 FOUNDATIONS ON OR IN ROCK

First Author:	Tan	Van Schalkwyk	Muzas	Bonaldi	Chitale	Pender	Hegan	Maeda	Costa-Nuñes	Abrahaõ
granitics	X									
shale, schist				X	X					
shale, mudstone					X					
limestone									X	
coal measures										
sandstones			X		X		X			
volcanic flows					X					
tephra							X			
decomposed rock	X				X					
exploration				X			X		X	
rock classification							X			
numerical models				X		X				
physical models										
limit equil.								X		
back analysis				X						
rock testing				X			X	X		
statistics										
support		X								
monitoring				X						X
dynamic loads										
drainage	X								X	
plasticity										
failure modes								X		
faults					X					X
highways										
earth/rock dams		X								
gravity dams		X		X	X					X
arch dams		X								
penstocks			X							
bridges/towers								X		
piers/piles						X				
materials										

TABLE 2B THEME B2 FOUNDATIONS ON OR IN ROCK

First Author:	Fu	Xu	Werner	Nieble	Gavarel	Scott	Chiu	Jasarevic	Milavanovic
granitics									
slate, schist	X		X	X	X				
shale, mudstone							X		
limestone								X	
coal measures		X							
sandstones									
volcanic flows				X				X	
tephra									
decomposed rock									
exploration			X						
rock classification									
numerical models		X				X	X	X	
physical models									
limit equil.		X			X	X			
back analysis									
rock testing			X		X	X			X
statistics									
support					X				
monitoring	X							X	
dynamic loads			X			X			
drainage	X		X						
plasticity							X		
failure modes									
faults	X	X		X				X	
highways									
earth/rock dams									
gravity dams	X	X				X		X	
arch dams					X				
penstocks									
bridges/towers			X					X	
piers/piles							X		
materials									

TABLE 3 B3 NEAR SURFACE CONSTRUCTION

First Author	Hellerer	Mühlhaus	Grüter	Bailly	Valore	Egger	Einstein
slate, schist							X
marl, mudstone	X		X	X	X		
limestone	X						
open-jointed	X						
sandstones					X		
exploration	X				X		X
rock classification							X
numerical models		X	X	X			
limit equil.						X	
support	X						X
monitoring	X		X	X			X
dynamic loads							
drainage					X		
shallow tunnels	X	X		X	X	X	
open cut			X	X			
construction	X			X	X		
initial stress			X				
shallow chamber							X
pilot tunnel							X

ENGINEERING PERSPECTIVES OF SURFACE AND NEAR-SURFACE EXCAVATIONS IN ROCK

Le point de vue de l'ingénieur sur les ouvrages au rocher, en surface ou en profondeur

Ingenieurtechnische Überlegungen zu übertägigen und oberflächennahen Felsbauwerken

Klaus W. John
Ruhr-University, Bochum, FR Germany

SYNOPSIS

This co-general report first exposes, from an engineering perspective, present trends in surface and near-surface construction in rock: system approach increasingly applied, with analysis being an integral part. Many case histories also cover weak rock indicating its engineering significance. Truly novel approaches appear to be rare. Present state of art is reviewed following system sequence planning/exploration/testing/idealization = modeling/analysis/design/construction. Considerable progress can be observed but report raises many still critical issues, such as supervision of exploration, over-reliance on routines, representative test results, realistic modeling of natural conditions, treatment of "ancient" landslides, input for analysis, limits of back-analysis, active vs. passive rock reinforcement, and flexibility in construction with specifications to provide for it.

ZUSAMMENFASSUNG

Dieser Hauptbericht stellt derzeitige Tendenzen im obertaegigen und oberflaechennahen Felsbau aus bautechnischer Sicht dar: zunehmend systematische Bearbeitung, wobei Berechnungen weitgehend integriert sind. Viele Beispiele behandeln auch wenig festes Gebirge, dessen bautechnische Bedeutung zunimmt. Wirklich neuartige Beitraege sind selten. Derzeitiger Kenntnisstand wird in folgender systematischer Abfolge behandelt: Planung, Erkundung, Versuche, Idealisierung = Modellvorstellung, Berechnung, Entwurf und Bauausfuehrung. Betraechtliche Fortschritte werden beobachtet, der Bericht weist auf zahlreiche, noch kritische Fragen hin, wie Ueberwachung von Erkundungen, uebertriebener Bezug auf Routinen, repraesentative Versuchsergebnisse, realistische Modellvorstellungen ueber natuerliche Verhaeltnisse, Behandlung alter Rutschungen, Eingangsdaten fuer Berechnungen, Grenzen der Rueckrechnung, aktive oder passive Felsbewehrung, Anpassungsfaehigkeit der Bauausfuehrung mit entsprechenden Ausschreibungsunterlagen.

RESUME

Ce co-rapport général présente le point de vue de l'ingénieur sur les récentes perspectives concernant les ouvrages en milieu rocheux, en surface ou en profondeur: on utilise de plus en plus fréquemment une approche systématique, dont les analyses deviennent part intégrale. De nombreux cas concrêts concernent les roches tendres, d'où leur importance pour l'ingénieur. Les approches totalement nouvelles semblent être rares. On examine ici l'état actuel des connaissances, puis on présente les phases successives de l'approche systématique: plannification/exploration/essais/idéalisation-simulation/analyses/conception/construction. De considérable progrès sont discernables, mais on attire l'attention sur plusieurs points critiques, notamment la supervision des explorations, l'excès de confiance en les méthodes routinières, la simulation réaliste des conditions en place, le traitement des "anciens" glissements, les limitations des rêtro-analyses, les conditions actives ou passives des soutènements au rocher, la flexibilité des méthodes de construction ainsi que les moyens de l'obtenir.

1. INTRODUCTION

1.1 Theme and sub-themes of this session

A total of 46 papers comprising more than 300 pages was accepted for this Session B which had been structured in:

- o Stability of rock slopes (20 papers),
- o Foundations on and in rock, including dam foundations (19 papers), and
- o Near-surface construction, especially in cities (6 papers).

Several papers received and reviewed could have been part of either Session A "Site Exploration and Evaluation" or Session C "Deep Underground Excavation." Such overlaps, on the other hand, are well exceptable as they counteract the potential compartmentalization of the different activities in the area of rock mechanics and rock engineering.

1.2 Objectives of this general report

The general reports of the sessions of this Congress were to:

- o Expose adequately the general trends as reflected by the papers submitted;
- o Present personal views of the state-of-the-art in the respective theme areas; and
- o Suggest particular areas for formal and also informal discussions at the Congress.

The Organizing Committee assigned two Co-General Reporters to this Session who decided that both would review all papers received but split assessment and discussions, with:

- o R.E. Goodman to cover more theoretical, analytical viewpoints, while
- o K.W. John would concentrate on more practice orientated engineering perspectives.

1.3 Contents of papers received

The co-general report by R.E. Goodman presents tables in which the contents of all papers of this session are analyzed and classed into the topics most relevant for theme and subthemes.

2. TRENDS REFLECTED BY PAPERS

2.1 General

When initially reviewing the papers submitted, this reporter was tempted to discuss all or selected papers at some length and then go on to add own contributions and/or case histories to round out his report. After getting more involved, however, he decided to endeavor in the presentation of a in-a-nutshell overall review of the broad theme defined by this Session, thus attempting to identify trends and issues believed to be most important, covered by the papers or still to be dealt with. In-depth coverage of selected topics is intended to be left for the discussions, hopefully, as this reviewer himself does not have convincing answers to many of the questions raised.

2.2 Trends observed

The review of the papers revealed certain trends ranging from quite definite to rather faint. In this order they are summarized as follows.

2.2.1 Stronger trends. In this reporter's opinion, maybe with a bit of his own wishful thinking, the papers submitted convey three trends quite strongly, quite convincingly.

- o Systems approach to solve and/or contribute to rock slope/foundation stability and design problems, consisting of the complete sequence of planning/exploration/testing/idealization (=modeling)/analysis/design (or design input). This trend is reflected in more than half of the papers received and in practically all case histories and more general discussions. Whereas there appears to be a widespread agreement as to the scope of the individual elements of the rock engineering (or geotechnical) contribution, the cost factor is discussed in only one paper, with the issue of cost effectiveness still more or less excluded.
- o Analysis in general, and geological and engineering analysis in particular, is clearly established as a mainstay of the rock engineering contribution. Just drilling and/or testing, without serious analytical follow-up, appears to be gone forever. The results of analysis performed and the contribution to design will, of course, always be tempered by our difficulties in deriving realistic but not more than reasonably conservative models (or idealizations of the natural conditions) and parameters. Back-calculations are often quite useful but sometimes not as productive as routinely believed, e.g., in cases in which the environment of the back-analyzed event is just guessed at.
- o Hard vs. weak rock: More than 50% of the case histories reported relate to slopes and foundations in weak, generally weathered rock, in quite many cases in combination with more competent rock masses. As to be expected the design of mine slopes has to deal more often with weak rock than that of dam foundations. This reviewer is quite pleased to detect this increasing interest in weak rock engineering which indicates the necessity for the rock (mechanics) engineering community to deal more systematically with weak rock masses than was perhaps done in the decades past. With continuing exposure to hard rock problems (and also non-problems) this reporter has more recently concluded that hard rock even if highly jointed but treated fairly during excavation in tunnels and otherwise, can perform surprisingly well, behaving much more "patiently" than could be expected solely relying on constitutive equations

and potential failure modes. Increased coverage of weak rock problems will enhance the overall contribution of rock mechanics to engineering in general.

2.2.2 Weaker trends. Two issues which have been discussed in the past are also reflected in some of the papers received.

- Rock classification systems and their systematic use not only in tunneling (one paper) but also in the design of rock slopes appears to be on the increase. Whereas this trend should be encouraged as it greatly improves communication between geological and engineering profession, the over-use of rock classification as a primary design tool, way beyond the original intentions of the respective systems, is observed with some skepticism.

- Failure modes other than sliding, several papers refer to toppling and toppling related failure modes in both analysis and design but the fine analytical concepts developed during the past decade are, obviously, not yet really widely applied in rock engineering practice.

2.2.3 Existing trends not or barely reflected. The present set of papers does not or barely reflects issues and/or trends believed to be important in rock slope (and foundation) design and construction.

- Probabilistic approach in assessing the practical stability of slopes (one paper only), promising but with acceptance in engineering practice lagging, maybe because many engineers prefer to deny even a slim possibility of failure of their designs.

- Novel approaches for the stabilization of rock slopes by (steel) reinforcement in its broadest sense, e.g., in-depth discussion of pros and contras of pre-tensioned vs. post-tensioned rock reinforcement. The trend favoring passive support because of simplicity in construction combined with obvious effectiveness as observed in tunneling should and is actually spilling over onto slope stabilization. This topic and its implications to actual rock slope construction should be discussed during this Congress, also possibly the reason why it was not covered by the papers received.

- Construction related problems, with only one paper giving more general assessment of blasting in rock slope stability and design, are barely touched on. Blasting and/or ripping deserves in this reviewer's opinion more coverage by the rock mechanics and geoscience professions, as perennial potential cause of real rock engineering problems in the field.

2.3 Conclusions

2.3.1 Earlier contributions. During the past two decades, at earlier congresses of this or related professional societies novel ideas "en mass" were presented, many relating to the subject of this Session, with few of them mentioned in the following.

- Geologic structure such as jointing often governing the engineering properties and actual behavior of rock masses, which led to increased interest of geologists of many shades in the field of rock engineering.

- Use of pre-stressed rock anchors for slope stabilization, first with very simple and then increasingly sophisticated designs, with continuing discussion of the permanency of these designs.

- Sophisticated testing of small samples of intact rock and rock masses, respectively, with critical discussion of questions as to representative sample sizes and locations, and extrapolation of test results.

- Application of the finite-element method of analysis for a very wide scope of rock mechanics and also rock engineering problems, with a very rapid succession of improved concepts which, however, did not always improve their engineering application.

- New concepts in tunnel design and tunnel construction which are only mentioned here as they convincingly have proven to a wide segment of the professions involved in rock engineering that treating rock well and motivating contractors doing just this, pays for all parties concerned.

2.3.2 Present contributions. The papers reviewed, together with other indicators observed by this reporter, show that the rock mechanics and rock engineering community is presently not in a period of dramatic new contributions. The time of exiting papers and lively discussions in field, lab, and office appears to be at least temporarily over, with the profession largely consolidating earlier achievements and finally putting to work concepts and ideas presented before, with actual productivity and cost effectiveness slowly but surely entering the considerations. It definitely appears that the engineering part of rock mechanics is finally gaining momentum -- for a somewhat old-fashioned engineer this observation is very gratifying to report. Still, novel concepts are already again available and waiting for discovery by the profession at large and integration into the engineering system: e.g., the keystone concept for general stability assessment, to name just one.

3 REVIEW OF STATE-OF-THE-ART

3.1 General

This overall review of the state-of-the-art in engineering of rock slopes and foundations in and on rock will range beyond the scope reflected by the papers submitted and reviewed. As indicated before this reporter will emphasize the engineering perspectives, with contributions toward the solution of problems and/or design tasks on hand. In the course of this review this reporter will venture to hopefully clearly express his concerns about the present state-of-the-art, concerns which have developed and only partially disappeared during his past thirty years in this field. Although certainly not in the foreground, the contributions and limits of theoretical approaches will be kept in mind throughout this review.

It is hoped that this review provides another basis for selection of truly relevant topics of discussion during this Congress. While an international congress will be a perfect occasion to show off past accomplishments and future possibilities, it is also the best opportunity to concentrate on the still weaker elements of the different systems in rock engineering in order to further improve the contribution of this specialty field to civil and mining engineering at large.

3.2 "Rock"

3.2.1 Rock mass. The differentiation between intact rock (or rock material) and rock mass has been clearly established for a long time, with its engineering significance having been discussed at great length. The pseudo-equation of "intact rock + geologic structure (ranging from microfractures to jointing and faults) = rock mass" is the basis of most of the present contributions in rock engineering.

3.3.2 Geologic structure. The importance of major geologic structures such as major joints (Grosskluefte) and faults in engineering decisions and designs is clearly established. However, the assessment of the engineering significance of minor structural elements, such as minor jointing (Kleinkluefte) and microcracks is less clear cut. Although there is available a wide variety of techniques to map, describe, and analyze all classes of structural features, the actual input for engineering analysis on scattered joint orientations and continuity, or discontinuity, of such joints is often less than convincing. The effect of construction on discontinuous rock appears to be another point of contention. Poor construction techniques, such as overblasting and excessive ripping tend to overemphasize the effect of e.g., jointing, while well-controlled construction techniques such as cushion-blasting and modern tunneling procedures greatly reduce the engineering significance of minor discontinuities in rock. Objective assessment is quite difficult, resulting in either over or under estimation. It is recommended that while analyzing the engineering significance of jointing in rock not just to look at the data basis on hand, e.g., stereo or equal-area plots of field data but also remain in touch with the actual field conditions as they actually are and as they are dealt with during construction. This applied particularly to design engineers. All too easily plots of any kind lead to abstract, unrealistic idealizations if severed from field conditions as-are.

3.2.3 Hard vs. weak rock. The early development of rock mechanics and rock engineering in Alpine countries, and also in Scandinavia, quite often in mining environments certainly favored hard rock mechanics and engineering early on. Based on repeated field experience this reviewer has concluded that hard rock treated "reasonably well" by controlled excavation techniques tends to be quite forgiving, at least for small to medium potential stability problems. Thus, he has seen interesting rock mechanics problems just fading away because of good construction procedures. Quite to the contrary, weak, i.e., younger and/or weathered rock appears to be less forgiving, even becoming quite treacherous if mistreated. Additionally, at least in routine civil engineering weak rock is much more often encountered than really hard rock. It is ventured to state that soft rock mechanics and engineering represent an engineering challenge definitely requiring the joining of forces of soil and rock engineering supported by geologists and engineering geologists.

3.3 Overall investigation of rock site

3.3.1 Systems approach. The systematic approach pursued in the overall investigation of rock sites for a variety of engineering designs is very well reflected in many papers of this session. The following sequence is now (almost) routine:

- Definition of rock engineering task within the overall planning process, i.e., the design of mine pit or dam foundation (or tunnel);
- Exploration of natural conditions, with geologic analysis;
- Testing, in field and/or in laboratory;
- Idealization of natural conditions to result in "model";
- Engineering analysis of this model;
- Design and/or contribution to design and specifications; and
- Construction, with "geotechnical" inspection of construction procedures applied.

3.3.2 Objectives and results to be obtained. The overall quality of the final contribution of above system depends on the consistence throughout the system. Excessive treatment of specific elements would amount to pursuing "hobbies," too scant a coverage would represent an engineering omission. The continuing review will follow largely the sequence within the investigative system outlined above.

More specifically the results to be fairly expected greatly depend on the engineering task posed. Two, and possibly three respective tasks can be defined in principle:

- Mine pit design, with slopes being near or at limit-equilibrium, the miners often coexisting with pending failure, with failure mechanisms actually being observed, with back-analyses becoming quite meaningful, resulting in this reporter's opinion in most convincing contributions.

- Dam (or other) foundations in and on rock, quite often combined with slope stability problems, with the overruling obligation to develop "absolutely safe" designs, with failure not acceptable at all, often even not being discussed. The often considerable margin of safety thus incorporated in the design very often obscures the actual contribution, although the question about the actual safety of such designs in rock can rarely be answered convincingly.

The third potential problem class, to be situated between the open pit mining and the foundation engineering problem, is:

- Slope stability problem in civil and also mining engineering involving natural slopes, with or without observed "landslides," with differentiation in:
  - "Potential" stability problem, with design to provide an increase in margin of safety as compared with the natural "safe" condition;
  - "Ancient landslide" as mapped by geologist, with potential re-activation by natural and/or man-made causes;
  - "Active landslide" requiring stabilization. The record of the rock mechanics community being able to positively deal with these problems is not perfect. Thus, this question will be returned to in a latter part of this discussion.

3.3.3 Cost of investigations and cost effectiveness. One paper only touches on the cost of investigations for dam foundations, while not relating to the problem solving power and the resulting cost effectiveness. The cost of exploration, testing, and analysis has greatly increased during the past decade, and not only because of more complex sites but also because of ever increasingly complicated technical, legislative, and administrative boundary conditions, some of which the professions themselves are to be blamed for. Simultaneously, within the framework of project financing the cost of time required for conducting investigations has also severely increased.

As it now stands this reviewer believes that this entire issue involving costs and effectiveness of the rock mechanics contribution should be of much greater concern to the profession than as reflected by the papers received. As a matter of fact, the profession, on one side, is being confronted with underbidding thus under-investigation due to competitive pressures, and on the other side, with the temptation of over-investigation when such such rare occasions arise, with well known diminishing returns. Both extremes tend to tarnish the contribution of the rock mechanics community.

In the following just two aspects related to the question of the cost of an investigation are mentioned in order to possibly initiate discussion.

- Budget dilemma: Investigations in the pre-construction phase, very often prior to the final commitment to a project, have to work within necessarily limited budgets resulting in reduced scopes of work while generally still expected to yield positive (and optimistic) conclusions. The resulting designs are not always realistic. The costs of delays during construction are often staggering. Unexpected conditions do easily require costly design modifications, less expensive than delays but generally much more costly than better investigations. The construction budget will have to absorb all that, very often resulting in dramatic cost overruns and related litigation.

- Flexible approach: Reasonable cost effectiveness can only be attained by compromising between "reasonable" pre-construction investigation, with investigative efforts continuing during construction, and "reasonable" flexibility during construction, reflected in design and specifications, to accommodate unavoidable "changes of conditions," often unduly blamed on geology when actually construction procedures selected prove to be ineffective. Of course, reasonable compromises for given overall conditions (engineering problems, geological conditions, local environment and/or usage, etc.) are only possible if the risks involved in design and construction and the resulting costs are also reasonably shared.

3.4 Field exploration, geologic analysis, and testing

3.4.1 General. The range of activities lumped together here leads to the qualitative and partially already quantitative model idealizing the natural conditions encountered combined with the engineering problem posed. This model, be it labeled physical or geotechnical model, is rarely drastically changed, only modified, during subsequent engineering analysis and design. The possibility of really drastic changes mainly exists during construction by exposure of unexpected conditions, involving geology or not. It has become quite obvious that the quality of the model as defined here and developed early on really governs the quality of engineering analysis, design, and design documents such as plans and specifications.

3.4.2 Exploration. At this time a few aspects which are of concern to this reporter are touched on.

- Quality of core drilling, still the most important technique for the investigation of major rock mass volumes, and generally consuming a major portion of the investigation budget, with possible problems seen in:

- Level of drilling supervision in the field, of course, is a result of contractural and other cost considerations. It is sad to state that the old argument between meters (or feet) drilled vs. information actually obtained is still too often decided in favor of the first criteria, particularly as competition in the field increases.

- Data acquisition in the field, with logging often conducted by barely qualified personnel, and widespread acceptance of low core recovery so well known to depend on a combination of geologic/geotechnical conditions and quality of drilling equipment, crew, and supervision.

- Data processing and assessing of results of exploratory drilling performed routinely relying on standard procedures, with too much costly information never used or picked up otherwise by geologic and engineering analysis and design.

- Over-reliance on and excessive extrapolation of rock classification, ranging from the very simple, straight forward RQD defining jointing in a most general way to more involved but still not really very sophisticated rock classification with different systems now available. Rock classification is believed to be very useful and is fully supported by this reviewer, provided it is used as a means to improve communication between the parties involved in a project. The use of a simple, or even not so simple, rock classification as the key criterion for rather complex design decisions is not supported by this author. This critique applies to the use of RQD as the sole geotechnical input in tunnel design, i.e., for selection of tunneling procedure and support required. It also applies to the use of classes determined by application of an eminently useful classification system as the basis for rock slope design, as suggested in one paper. It has been proven that the RQD is negatively affected by mediocre drilling procedures particularly in weaker sedimentary and metamorphic rocks, with inherent anisotrophy. Excessive reliance on global rock classification, in this reviewer's experience, has short-changed the actual potential of good ground control and other more refined construction procedures in medium to weaker, and closely jointed rock.

3.4.3 Geologic analysis. Any field exploration generally performed and/or supervised by geologists and engineering geologists will involve geologic analysis, often not clearly identified and even poorly documented but still essential in developing the "model" to form the basis for subsequent work. It is quite acceptable for this reporter that the geologic analyses will tend to be more qualitative than quantitative, with the exception of the analysis of geologic-geometric relationships. This reporter has learned the hard way that a perfectly documented, extremely quantitative analysis is utterly useless, even dangerous, if based on a simplistically assumed model or mechanism (sometimes just pulled from the drawer) and not the one rightly envisioned by the experienced geologist having lived at and with the site for a long time.

Geologic analysis utilizing the stereo (or equal/area) plots indicating and correlating joint orientations is now an established tool in rock slope and also other stability analyses. Fifteen years ago this reporter had considerable fun to discover this geologic tool, work with it in connection with stability of rock slopes and dam foundations, and also write about it. In the meantime he has more clearly recognized some shortcomings, even dangers of this general approach, such as:

- More often than not by the sole use of average joint orientations assumed to be "representative," the geometric scattering observed in practically all field conditions is suppressed, possibly resulting in non-conservative conclusions.

- If scattering joint data are considered the conclusions as to possible kinematic conditions and resulting (in)stabilities are likely to become a bit vague, resulting in rather involved graphical or numerical statements which translate to not much more than "no maybe yes" or vice versa.

- With real data sets, often not at all representing straight forward geologic structures, most techniques presently available become practically unworkable for other than real graphical or digital wizards.

At this time this reporter advises not to separate the as-actually-are conditions observed in the field from graphical or other models developed most likely in the office. Whenever possible the conclusions from analysis should critically and objectively be checked against actual field conditions, and vice versa. Such continuing critical comparison greatly helps to early on eliminate possible "paper wedges," as compared to the real "rock wedges" a realistic design should should take care of.

3.4.4 Testing. A discussion of the overall investigative process is certainly not complete without at least brief reference to testing. In rock slope design the shear strength of geological features representing weaknesses of the rock mass, such as major joints and faults, is of decisive importance. For rock foundations the deformability of rock masses enters the considerations. Laboratory and in situ testing aimed at the determination of these two key properties have been discussed extensively for the past two decades. However, 23 years after the pioneering large-scale in-situ testing at the site of the Kurobe IV arch dam in Japan, in which this reporter was actively involved, he is still hard-pressed to determine representative shear parameters for extended potential shear surfaces in jointed rock, with parameters to be realistic but not overly conservative. Of course, producing overly conservative values is quite easy. It appears that the following problems in rock testing are still being encountered as much as ever.

- Selection of representative sample locations, for cores and/or very limited number of in-situ tests, generally based on limited geological information and always facing access problems, one way or another.

- Scattering of results generally being obtained, particularly in more complex rock masses and for more irregular, often rather poorly defined weaknesses to be tested, respectively.

- Generally, the conversion of raw test data to design parameters requires considerable interpretation, not always very objective, often quite intuitive. It is always tempting to provide upper and lower bound input for analysis. But particularly for slope stability problems this does not always meet the job requirements, as the spread of input may very well cover the range from stable to unstable as far as the results are concerned.

- Areal extrapolation of test results: even most sophisticated and costly in-situ testing (except by geophysical methods) covers only some square meters or cubic meters (square feet or cubic yards) at the very most. Extrapolations to shear surfaces of several 10,000 m^2 for a typical pit slope or some millions of cubic meters for a large rock foundation are extremely difficult to make, to state it mildly. This reporter has become convinced that most such extrapolations end up near the unrealistically conservative end of the possible spectrum.

3.4.5 Idealization of natural conditions. The following listing summarizes this reviewer's concern about this all important station within the investigative process which is virtually the interface between geologists and engineers.

- The idealization, an essentially qualitative physical model of the natural conditions, is the basis for most or even all subsequent analysis and design. Thus, it should be a convincingly strong link within the overall system.

- The model is to be realistic, even if realistically conservative but not be dominated by either exploratory approach (e.g., multi-layer model inspired by geophysical methods) or methods of analysis at hand (sliding rock mass as compared to more complex failure mechanisms).

- During the past decade a multitude of potential model failure mechanisms have been defined:

  - Block/wedge/decks of blocks to
  - Slide/topple/buckle, with respective critical parameters and design solution reasonably well defined.

  Even with this quite convincing catalog of possibilities this reporter has more recently come to fully realize the still considerable difficulties to be overcome in the realistically conservative interpretation of natural conditions both geologists and engineers are expected to routinely perform. Either picking a too conservative model or overlooking hidden features and/or mechanisms appears to him easier than pinpointing the model truly and honestly corroborated by subsequent performance of the prototype slope.

- Dimensions: For small to medium stability problems the selection of the realistic model might be the task we can solve with fair confidence. For larger problems, a natural slope to support an arch dam or major portions of an open pit mine, the simple mechanisms listed above and in much detail in many papers generally do not occur individually but as parts of more complex potential or actual failure modes, often too complex for the most sophisticated methods of analysis.

- Parametric analysis: For some time, the panacea for the engineering treatment of more complex models has been the parametric analysis, or trend analysis, aimed at the determination of the input parameters most critical for the performance of a given model. Whereas it has always been emphasized to vary the input parameters, this reporter has concluded that, in many cases, the model forming the basis of the trend analysis should also be varied. This, of course, does complicate parametric studies as the numbers and complexities of the variables increase. But the different performance of different models, with comparable parameters, could easily be as shocking as the performance of an actual cut face in rock, in which the assumed rock wedge did not really materialize because of irregular and discontinuous jointing, and which deforms only due to lateral stress relief. Also, comparing analytically determined deformations based on an assumed model with the field measurements of the prototype reacting in a different mode could lead to disasterous conclusions which also would unfairly discriminate field monitoring as such.

3.4.6 "Ancient landslides". Planning, design, and construction engineers are but stopped dead in their tracks when confronted with geologic maps and plans of construction sites showing:

- "Ancient" landslides; and/or

- "Active" and "potentially active" faults.

The basis for assessment and resulting labeling

may range from very cursory to extremely comprehensive, depending on the project and local regulations and usage. What options are there left for the engineering design but to abandon the specific location and look for another one, a move increasingly difficult in more developed and geologically troubled regions, and of course, impossible when dealing with mining properties.

Of course, it is very well realized how difficult it is to define for earlier and "ancient" activities both:

- o Time of failure or movement, with the number of how many years ago it happened not being too useful if not usable for relating with other relevant events; and
- o Environment at that time, which was likely to be different and possibly more unfavorable than present field conditions.

Only reliable information about both aspects would provide a reasonable basis for engineering assessment of present stability by means of back-analysis of original event, analysis of present condition and design, and monitoring of present and future performance.

As it stands now any construction on or even near geological features labeled "ancient" or "potentially active" would be an unacceptable design liability, although present stability could well be higher than that of adjacent areas mapped to be "stable" which, very well, may presently be at limit equilibrium. It is strongly felt that both geologic and engineering analysts have to cooperate to establish a convincing line of thought and procedures on how to approach such a situation more realistically than done now. Of course, it is not advocated to construct a dam across a potentially active fault or knowingly found it on an ancient landslide. However, it will be necessary to assure that simple labeling possibly based on an unspecified basis will not close all and any possibilities for developing a site in spite of its geotechnical problems.

3.5 Engineering analysis

3.5.1 Approaches available. For rock slope stability and rock foundations two principal analytical approaches are available now:

- o Limit equilibrium concept; and
- o Finite element method of analysis.

The combination of both approaches is used more and more, for both static and dyanamic loading, with the stress field as determined by means of FEM forming the essential basis for determining either factors of safety or probabilities of failure based on limit equilibrium methods.

Both concepts and particularly the combination of the two are quite sensitive to variation in input as far as model and parameters are concerned. This very well known weakness is at least partially circumvented by parametric analysis. The fact that both model and parameters can gravely be affected by construction methods is rarely represented in the design analysis. The deformability of a rock mass as explored and tested may be considerably lower than that of the same mass after completion of an adjacent excavation in which blasting was not or only poorly controlled. On the other hand strata of sedimentary rock paralleling an inclined cut face may be considered to be a reasonable design basis until rock slabs undercut by overblasting and/or excessive ripping start sliding into the excavation, introducing a failure mode excluded in the design.

3.5.2 Keyblock analysis. An additional method of rock mechanics analysis developed by the Co-General Reporter and his Collaborator is believed to be just around the corner, the keyblock analysis. This is a most exciting and intellectually intriguing method of analysis used to identify the critical keyblock of a rock mass, based on geologic structure and the geometry of an excavation, to be supported before it fails, to stabilize the entire rock mass adjacent to the excavation. It should here be fairly noted that a similar although less formal approach has been applied in kind for a long time by e.g., limited rock bolting of rock cuts aimed at preserving the integrity of the surface rock, without directly stabilizing the innermost rock. The new approach is believed to be quite promising for clear-cut geological structures consisting of fully effective joints, maybe equivalent to blast-damaged rock masses.

The identification of the critical keyblock during actual excavation before it slips or falls will be on the more difficult side. Intuitively it is believed that the keyblock concept will be most useful as a basis for remedial design to be routinely applied in zones in which the prevalent geological structure makes likely the existence of keyblocks, without depending on their actual identification during excavation. It could well be that designs thus produced would be more economical than conventional designs, as local support would be provided where it would tend to maximize overall stability.

3.5.3 Probability of failure. Utilizing the conventional limit-equilibrium concept the engineering analyst is faced with the well-known dilemma to select reasonble factors of safety, unless this decision is taken out of his hand by regulations specifying factors of safety to be used for given conditions. And there is always the question if the value prescribed is a reasonable one after all. If the designer has his choice he has to ask, what is sufficient, what is safe, SF = 1.5 or 1.2 or a value in between. This decision will directly relate to the cost required to improve natural conditions, possibly assessed to have a SF = 1.0, up to a certain "safe" factor of safety by means of rock bolting or installation of rock tendons. The fact that an assumed value most directly determines the cost of remedial works has never satisfied this reviewer.

Most practitioners would agree that the most promising way out of this dilemma would be to finally adopt the probability of failure as the design criteria for slope stability. This possibility has been discussed for some time, with many quite promising approaches presentd, but it has not been accepted extensively, at least not by the civil engineering profession. The idea that failure be possible but unlikely is obviously too appalling to most civil engineers. As mining engineers are most often used to living with pending failure the entire concept appears to be more acceptable for this profession.

This reporter's experience from respective research relating to the rock wedge problem and application in engineering practice, considering statistical distribution of both geometric and strength parameters, indicated that the whole approach is clear-cut and convincing in principle, however, tends to produce rather vague results in real application. The first set of results relate to the probability that kinematic conditions congruous to shear failure exist. If they do, it is assessed if failure is actually probable in view of the shear strength most likely to be available at the location of concern. The results obtained were not yet acceptable for routine engineering design and related permitting.

3.5.4 Back calculations. Back-analysis is a most useful approach to derive realistic shear (or other) parameters from an observed failure known to have taken place in well recognized and quantitatively documented conditions. However, in most real cases lending themselves to back-analysis conditions at time of failure are not very well defined.

For "ancient" landslides as identified by geologists, with effect of the environmental conditions during the actual failure rarely being identifiable, the following possibilities could occur:

- Natural slopes performing well might have actual factors of safety in excess of 1.0, with the parameters to be produced by back-analysis based on present limit-equilibrium to be quite conservative, possibly unreasonably so.

- Man-made slopes also performing well, with comparative analysis possibly leading to too optimistic conclusions as the original construction techniques are not comparable with modern excavation methods. It is rather easy to see that slopes in comparable geological setting to be excavated now by uncontrolled blasting and/or excessive ripping will have to have a much flatter slope than the slopes excavated almost one hundred years ago by mild blasting using black powder, and performing well up to date. It is concluded that in such diverse conditions back-analysis would not be very productive.

3.6 Design of slopes and foundations, with respective stabilization

3.6.1 General. In the papers submitted and also in this discussion a wide variety of "structures in rock" is covered, with rock slopes and rock foundations representing the two extremes, and the rock foundation of an arch dam to be located in a steeply sloping valley often combining slope stability and foundation support problems.

The design is to utilize the results of and conclusions from the exploration-through-analysis sequence, with other aspects added such as cost effectiveness, ease of construction plus general experience closely linked with the intuition of the senior designer. The actual problem solving power of the investigative system contributing to the design question on hand is difficult to define. In some of the papers reviewed the conclusions for design appear to be somewhat detached from the studies actually performed.

This reviewer has experienced rather disappointing cases in which slope and foundation designs were developed in spite of geotechnical work performed to support them. In other cases the support of designs of structures in rock by the preceding exploration/testing/analysis was quite convincing. As indicated before it appears that the combination of problem solving power and cost effectiveness of rock mechanics investigations deserve increased attention by the rock engineering community in order to assure the long-term viability of this professional endeavor.

3.6.2 Drainage. To most of the profession the following remark will appear to be completely superfluous, as everybody knows it, or should know it but actual experience in the design practice sometimes tends to indicate otherwise: systematic and positive dewatering of rock masses affords the statically most effective, very reliable, and also most cost-effective rock stabilization. Remedial works by other means, such as rock bolting or anchoring, without fully integrated dewatering is believed to be rather poor rock engineering practice.

3.6.3 Slope stabilization. The following topics are of considerable concern for the design of remedial works on rock slopes:

- Pre-tensioned or post-tensioned rock support, the question if rock reinforcement should, at the time of installation, be preferably pre-tensioned (active) or post-tensioned (passive) has been increasingly discussed not only in tunneling but also for rock slope stabilization. This discussion is not yet reflected in the papers received.

  Twenty years ago the active support by pre-tensioned elements was strongly emphasized leading to quite sophisticated rock bolt and rock tendon designs. More recently rock reinforcement unstressed at the time of installation, to be activated by rock mass deformation, is seriously considered for a wide spectrum of applications, because of surprisingly positive performance in tunneling. Additionally, simplicity of installation, with resulting savings in cost and time, are very attractive for practical designs. Paralleling those aspects, the possibility of deteriation of rock support with time due to corrosion, with resulting need for long-term monitoring, maintenance and repair and/or replacement are being discussed.

- The analytical treatment of rock support by active (pre-tensioned) or passive (post-tensioned) elements is exclusively based on the limit-equilibrium concept, with or without input from FEM. Generally, there is no really clear-cut differentiation of analytical treatment of active and passive support. Kovari's persistent discussion of the definition of factors of safety for rock slope stability, however, greatly contributed toward clarification of this issue. Gradually, general agreement is developing to assume that pre-tensioned support would reduce the active forces in the denominator of the equation defining the factor of safety. In cases in which the first passive rock support is gradually activated by deformation within the rock mass, it would tend to add to the passive forces in the numerator of the SF quotient.

- The actual stressing of rock reinforcement fully grouted into place up and above pure tension is being discussed, with some additional shear stressing leading to doweling of jointed rock, which appears to be highly effective in tunneling, being gradually accepted. However, doweling across potential failure planes, by rebars installed normal to such surfaces, is well known to be quite ineffective.

3.6.4 Monitoring. The role of monitoring, both of rock mass deformation and stressing of rock support, as an integral part of rock slope design has clearly been established in mining and civil engineering applications during the past two decades. This acknowledges the fact that the results of even the most sophisticated stability analyses very often do not yet provide the assurance required for actual design.

Although extensively discussed at many times certain dilemmas are still being faced when monitoring the performance of a rock slope:

- What are the deformations actually permissible for given conditions and a given design, with rate of deformations and changes of it not always providing sufficient criteria for monitoring.

- Duration of monitoring, during excavation only and, if longer, how much longer, with most complex rock slope stabilization designs most likely to require performance monitoring over the full life of the structure.

- Reaction on the results of the monitoring, with respective decision making to be required often quite some time after construction is completed, with or without sharing of liabilities for performance of overall design, or only of rock supports.

3.7 Construction

3.7.1 Specifications. The design of any structure in rock and the specification for its construction should realistically reflect the geotechnical input, the extent of which will mirror strength or weakness of this contribution. The following specification related issues are believed to be most important for really successful construction in rock.

- Flexible specifications: Even most comprehensive investigations cannot guarantee to provide entirely realistic and thus final models and parameters, thus, "change" of geological conditions can never be avoided altogether. Such changes are to be anticipated and provided for by contract and reasonable flexibility in construction procedures, without having to reach beyond the original contractual agreement, and hopefully without having to resort to litigation. Such reasonable flexibility of construction is much easier to accomplish in mining but most difficult in civil engineering works. Unreasonable flexibility should not be expected, custom-tailoring of given designs to local geological conditions to be disclosed only during construction is believed to be a greatly unrealistic illusion in most construction environments.

- Simple vs. complex specifications: Too many projects involving rock excavation are based on the premise that keeping specifications as simple as possible will obtain the lowest bids, even if they would represent unreasonable figures. This is not realistic. Excessive claims based on "change conditions" not covered by such simplistic specifications, with resulting delays and/or litigation, easily result in total construction costs becoming much higher than what would have been obtained while starting out with more realistic specifications allowing for changed conditions without immediately necessitating claims. Of course, the question of most reasonable and realistic specifications cannot be separated from the question of one-sided or shared liability. This reporter strongly believes that the geological risk involved with major excavations in rock will have to remain with the owner. The transfer of this risk to the contractor will directly lead to claims aimed at covering the cost of such risk, particularly in cases when the contract was rewarded to too low a bidder.

- The specifications should be written so that the difference of the rock mass as explored and the rock mass as-is after construction, e.g., excavation, is minimized unless the design provides for it otherwise. Control of excavation procedures, even if not very popular with many heavy construction contractors, is not believed to represent excessive and unreasonable interference with construction. After all, placement of concrete is very well controlled indeed in order to assure its ultimate quality. And in structures in rock, concrete and rock are both integral components of the final structure.

3.7.2 Stabilization of high cut slopes. It is always a problem to coordinate the progress rate of large-scale, even well controlled excavation with the time-consuming installation of slope support believed to be required. Specification calling for installation of e.g., rock anchors/tendons "as soon as possible" after excavation is not sufficient. Run-away excavations, with installation of rock support lagging behind are often observed if work is not based on strong specifications comparable to those now customary in tunneling in many parts of the world. Of course, this can lead to the demonstration that the rock support specified was not really required for immediate stabilization (but still would have nevertheless produced some increased margin of safety). This reporter has on file several well documented cases in which largely unsupported rock slopes were anticipated to fail sooner or later after excavation but have performed quite satisfactorily under even adverse conditions without the rock support originally foreseen in the formal design. However, in most cases it would have been quite impossible to analytically support, by accepted methods, the stability as actually observed. In such cases, performance monitoring is resorted to. Some of the dilemmas faced then have been discussed above.

3.8 Conclusions

The present discussions of the state-of-the-art in rock slope and rock foundation engineering are concluded with the following recommendations:

- o For any rock engineering investigation the complete system should be followed through. Both gaps and excesses should be avoided. Omissions may save some money up-front but become ultimately quite costly, while excesses are generally non-productive.

- o Continuing control over the entire investigation is essential, with intermittent reviews of results obtained, with field investigation and analysis providing most important contribution to final results.

- o Any designs need to be based on realistical idealization (modeling) of the natural conditions, based on both geologic field work and analysis and verified by engineering analysis.

- o Construction aspects need to be included in the design. Design and specifications should reflect rock engineering perspectives. Simpler designs are generally favored over complex designs. Both design and specifications should be reasonably flexible.

4 RECOMMENDATIONS FOR DISCUSSIONS

4.1 General

In the preceding section of this report a wide variety, maybe too wide a variety, of topics was discussed or only touched upon believed by the reporter to be of real significance in engineering applications. In the following a catalog of questions in rock mechanics is presented which might lend themselves for more general discussion. It is well realized that many other questions raised can only be dealt with case by case, application by apapplications. In the following a catalog of questions in rock mechanics is presented which might lend themselves for more general discussion. It is well realized that many other questions raised can only be dealt with case by case, application by application.

In the introduction of this report it is stated that it is to be limited to the engineering perspectives of the Session's theme. In the course of the preceding discussions it has, however, become obvious that no well defined boundary really exists between theoretical (analytical) and engineering aspects. This is reflected in the following listing of potential topics for discussion.

In the following four complexes for discussion are presented, with only one or possibly two to be discussed during this Congress. It is hoped that discussions elsewhee will reflect on such questions which cannot be coverd at this time.

4.2 Effectiveness of rock mechanics

In order to maximize the effectiveness of rock mechanics in future engineering applications continuing internal critique of present usage is required. In this reporter's view the following sets of topics and questions should be faced in a discussion.

- o "Problem solving power" of rock mechanics in engineering applications, with:

  - o Definition of strong and particularly of weak points of the overall system for specific design tasks, here surface excavation and dam foundations. What can, or should, be done about it? And

  - o Translation of results of analysis and conclusions thereof into actual design solutions. Can we be satisfied with the present real possibilities?

- o Cost of rock mechanics contribution as related to cost of design and construction, especially considering:

  - o Geologic and rock mechanics mapping and classification;

  - o Drilling, including logging, sampling, and down-in-the-hole testing;

  - o Testing, both in-situ and that of cores and/or samples in the laboratory; and o Analysis, maybe referring to typical stability analysis for slopes in jointed rock.

o Cost effectiveness, of course, is extremely difficult to quantify, although it is fairly expected in any engineering application. The sometimes perceived notion of particularly experienced heavy construction professionals that rock mechanics contributions often do not relate to the real problems on hand while showing only limited concern for cost effectiveness and scheduling has, if true, to be counteracted most vigorously. Rock mechanics and rock engineering practitioners have to be prepared to fully justify cost and time expended for their contribution, one way or another, if interested in continuing involvement.

This reviewer believes that the cost effectiveness of the following packages and also of components of such packages should be looked into. It appears, intuitively, that neither under- nor over-studying a particular aspect is productive, with the reasonably productive range of involvement not definable in general terms.

- Overall rock mechanics contribution, up and above conventional engineering geological exploration, with the most pertinent question being asked of how much should be spent to obtain more than descriptions of regional geology, hydrogeology, and seismicity, plus complete logs of borings.
- Within a complete review the cost effectiveness of conventional geological explorations should also be discussed in order to establish reasonable cost ratios between engineering geological and rock mechanics components of an overall investigation.
- In-situ testing of geologic discontinuities, defined as potential planes or zones of weakness, particularly in rock slope design but also frequently governing the design of dam foundations, is always a potential rock mechanics contribution. Whoever has ever performed in-situ tests is well aware that limiting expenditures for such tests below reasonable limits is nonproductive, as it only produces "guess-timates" of design parameters. Overtesting, possibly unrelated to models defined and design analyses required is also nonproductive. But what is a reasonable expenditure for in-situ testing? Although most difficult to answer, this question should at least be discussed.

4.3 Weak rock engineering

Actual and potential contributions by the rock mechanics community to weak rock engineering should be discussed, which will hopefully lead to defining interfaces with soil mechanics/engineering, engineering geology, and geology. Of course, interfaces are seen as opportunities for productive collaboration and not as professional boundaries.

In slope and foundation design in weak rock the following topics appear to be worthy of discussion.

- Engineering properties of integral weak rock material, as affected by geological processes and also geologic structures, such as stratification differentiating in present and future properties.
- Failure modes in slopes in weak rock with well-defined geologic structures, with low strength of weak rock becoming additional input, as compared to most conventional rock mechanics slope stability analyses. Or to pose the key question more simply, when is the assumption of a slip-circle reasonable and when not? Or where is the interface between soil mechanics and rock mechanics in slope stability analysis?

4.4 Stabilization of rock slopes

This reviewer believes quite strongly that the following three sets of possibilities to be dealt with in rock slope stabilization design should be discussed in close context with each other.

- Use of pre-tensioned (active) vs. post-tensioned (passive) rock reinforcement in its broadest sense;
- Applied in hard vs. weak rock, or combinations of both;
- For temporary vs. permanent support.

The following topics could be covered individually.

- Analytical concepts;
- Design possibilities;
- Time element in installation of rock reinforcement;
- Performance monitoring and its full spectrum, and finally;
- Pros and contras of technically and economically feasible combinations, and design compromises, with uncertainties, such as risks involved.

4.5 Rock mechanics and construction in rock

It has been pointed out that more systematic cross-breeding between rock mechanics and actual construction in rock would be desirable for increased effectiveness of rock mechanics in the practice of heavy construction, by significantly improving designs in rock and decreasing respective construction costs. It is believed that this role and contributions to this end should be strengthened. Two key aspects appear to be most important, beyond the role of rock mechanics in design.

- Rock mechanics/engineering and specifications, with technically and economically relevant contributions to contract documents, such as:

- Type of rock reinforcement believed to be applicable and effective;
- Contractual provisions for effective field control of, e.g., excavation procedures in rock; and
- Provisions for execution of typical rock engineering designs, e.g., rock stabilizations.

- Role of rock mechanics/engineering in actual construction in the field:
  - Control and monitoring of construction techniques; and
  - Monitoring of performance of designs executed.
  - Additionally and most important, carrying out of continuing investigations during construction, of course, in close collaboration with geologists.

INGENIEURTECHNISCHE ÜBERLEGUNGEN ZU ÜBERTÄGIGEN UND OBERFLÄCHENNAHEN FELSBAUWERKEN

Engineering perspectives of surface and near-surface excavations in rock

Le point de vue de l'ingénieur sur les ouvrages au rocher, en surface ou en profondeur

Klaus W. John
Ruhr-Universität, Bochum, BR Deutschland

ZUSAMMENFASSUNG

Dieser Hauptbericht stellt derzeitige Tendenzen im obertägigen und oberflächennahen Felsbau aus bautechnischer Sicht dar: zunehmend systematische Bearbeitung, wobei Berechnungen weitgehend integriert sind. Viele Beispiele behandeln auch wenig festes Gebirge, dessen bautechnische Bedeutung zunimmt. Wirklich neuartige Beiträge sind selten. Derzeitiger Kenntnisstand wird in folgender systematischer Abfolge behandelt: Planung, Erkundung, Versuche, Idealisierung = Modellvorstellung, Berechnung, Entwurf und Bauausführung. Beträchtliche Fortschritte werden beobachtet, der Bericht weist auf zahlreiche, noch kritische Fragen hin, wie Überwachung von Erkundungen, übertriebener Bezug auf Routinen, repräsentative Versuchsergebnisse, realistische Modellvorstellungen über natürliche Verhältnisse, Behandlung alter Rutschungen, Eingangsdaten für Berechnungen, Grenzen der Rückrechnung, aktive oder passive Felsbewehrung, Anpassungsfähigkeit der Bauausführung mit entsprechenden Ausschreibungsunterlagen.

SYNOPSIS

This co-general report first exposes, from an engineering perspective, present trends in surface and near-surface construction in rock: system approach increasingly applied, with analysis being an integral part. Many case histories also cover weak rock indicating its engineering significance. Truly novel approaches appear to be rare. Present state of art is reviewed following system sequence planning/exploration/testing/idealization = modeling/analysis/design/construction. Considerable progress can be observed but report raises many still critical issues, such as supervision of exploration, over-reliance on routines, representative test results, realistic modeling of natural conditions, treatment of "ancient" landslides, input for analysis, limits of back-analysis, active vs. passive rock reinforcement, and flexibility in construction with specifications to provide for it.

RÉSUMÉ

Ce co-rapport général présente le point de vue de l'ingénieur sur les récentes perspectives concernant les ouvrages en milieu rocheux, en surface ou en profondeur: on utilise de plus en plus fréquemment une approche systématique, dont les analyses deviennent part intégrale. De nombreux cas concrêts concernent les roches tendres, d'où leur importance pour l'ingénieur. Les approches totalement nouvelles semblent être rares. On examine ici l'état actuel des connaissances, puis on présente les phases successives de l'approche systématique: plannification/exploration/essais/idéalisation-simulation/analyses/conception/construction. De considérable progrès sont discernables, mais on attire l'attention sur plusieurs points critiques, notamment la supervision des explorations, l'excès de confiance en les méthodes routinières, la simulation réaliste des conditions en place, le traitement des "anciens" glissements, les limitations des rétro-analyses, les conditions actives ou passives des soutènements au rocher, la flexibilité des méthodes de construction ainsi que les moyens de l'obtenir.

1 EINFÜHRUNG

1.1 Thema und Unterthema dieser Sitzung

Es wurden insgesamt 46 Arbeiten mit einem Umfang von über 300 Seiten für diese Technische Sitzung B angenommen, die wie folgt gegliedert war:

- Standsicherheit von Felsböschungen (20 Arbeiten),
- Gründungen auf und im Fels, einschließlich Gründungen von Talsperren (19 Arbeiten),
- Oberflächennahes Bauen, insbesondere in Stadtbereichen (6 Arbeiten).

Verschiedene Arbeiten, die für diese Sitzung angenommen und durchgesehen wurden, hätten auch entweder in Sitzung A "Erkundung und Bewertung von Projektstellen" oder Sitzung C (Tiefe untertägige Hohlräume" eingeordnet werden können. Solche Überschneidungen zwischen Sitzungen sind jedoch andererseits wieder wünschenswert, da sie der möglichen Abkapselung von verschiedenen Aktivitäten in den Bereichen Felsmechanik und Felsbau entgegenwirken.

1.2 Ziele dieses Hauptberichtes

Die Hauptberichte zu den Sitzungen dieses Kongresses sollten

- allgemeine Tendenzen, wie sie sich aus den vorgelegten Arbeiten ersehen lassen, angemessen herausstellen,
- eigene Ansichten zum Stand der Technik in den jeweiligen Themenbereichen darstellen und
- spezifische Themen für offizielle und inoffizielle Diskussionen auf dem Kongreß vorschlagen.

Das Organisationskomitee hat dieser Sitzung zwei Hauptberichterstatter zugeordnet, die sich entschieden haben, zwar alle vorgelegten Arbeiten jeweils durchzusehen, sie aber nach unterschiedlichen Gesichtspunkten zu bewerten und zu diskutieren, wobei

- R. E. Goodman mehr theoretische und analytischen Ausblicke verfolgen und
- K. W. John sich auf mehr Praxis-orientierte Überlegungen des Ingenieurs beschränken würde.

1.3 Inhalt der vorgelegten Arbeiten

Der Hauptbericht von R. E. Goodman beinhaltet Tabellen, in denen die Inhalte aller Arbeiten dieser Sitzung sowohl analysiert als auch in interessante Stichworte des Rahmenthemas und der Unterthemen gegliedert wurden.

2 DURCH DIE ARBEITEN WIDERGESPIEGELTE ENTWICKLUNG

2.1 Allgemein

Bei der ersten Durchsicht der vorgelegten Arbeiten war der Berichterstatter zunächst versucht, alle ausgewählten Arbeiten ziemlich ausführlich zu diskutieren, um dann eigene Beiträge und/oder Fallbeschreibungen anzufügen, die diesen Bericht vervollständigen. Bald entschied er jedoch, eine komprimierte Gesamtübersicht des Rahmenthemas zu geben, wobei besonders wichtig erscheinende Sachfragen und Tendenzen herausgestellt werden sollten, ob sie nun in den vorgelegten Arbeiten abgehandelt worden sind oder noch zu erörtern wären. Es wird erhofft, daß sich eine weiterführende Behandlung von ausgewählten Sachfragen aus der Diskussion während des Kongresses ergeben wird, da dieser Berichterstatter zu einigen von ihm angeschnittenen Fragen selbst keine überzeugenden Antworten hat.

2.2 Beobachtete Tendenzen

Die Durchsicht der Arbeiten offenbarte gewisse Tendenzen, von recht deutlich bis reichlich vage. In dieser Reihung werden sie nachstehend vorgestellt.

2.2.1 Stärkere Tendenzen. Dieser Berichterstatter kam zu dem Schluß, ganz sicher mit einigem Wunschdenken, daß die vorgelegten Arbeiten drei Tendenzen recht deutlich, recht überzeugend vermitteln:

- Systematische Bearbeitung zur Lösung von und als Beitrag zu Aufgaben zur Felsböschungs- und Gründungsstandsicherheit, bestehend aus der vollständigen Abfolge von Planung, Erkundung, Versuchen, Idealisierung (= Modellvorstellung), Berechnung, Entwurf (oder Beitrag zum Entwurf). Diese Tendenz spiegelt sich in mehr als der Hälfte aller vorgelegten Arbeiten sowie in fast allen Falldarstellungen und mehr allgemeinen Diskussionen wider. Während anscheinend weitgehende Übereinstimmung über die einzelnen Elemente des felsbaumechanischen (oder geotechnischen) Beitrages besteht, wird der Kostenfaktor nur in einer Arbeit behandelt; die Frage des Verhältnisses von Kosten und Nutzen bleibt immer noch mehr oder weniger ausgeklammert.
- Berechnungen im allgemeinen und geologische sowie felsmechanische Berechnungen im besonderen sind ganz eindeutig als ein Hauptbestandteil des felsbaumechanischen Beitrages eingeführt. Lediglich Bohrungen niederzubringen und/oder Versuche durchzuführen, ohne anschließende ernsthafte analytische Überlegungen, erscheint der Vergangenheit anzugehören. Die Ergebnisse der Berechnungen und ihr Beitrag zum eigentlichen Entwurf werden allerdings immer noch von den Schwierigkeiten beeinflußt, realistische und nicht mehr als angemessen konservative Gedankenmodelle (oder Idealisierungen der natürlichen Gegebenheiten) samt Kennwerten herzuleiten. Rückrechnungen sind oft sehr nützlich, häufig aber doch nicht so produktiv wie vielfach angenommen, zum Beispiel in Fällen, in denen die Umweltbedingungen des rückgerechneten Vorganges lediglich vermutet werden können.
- Festes zu wenig festem Gebirge: Mehr als 50 % der vorgestellten Fallbeschreibungen beziehen sich auf Böschungen und Gründungen in wenig festem, im allgemeinen verwitterten Gebirge, in zahlreichen Fällen in Verbindung mit kompetenterem Gebirge.

Wie zu erwarten, hat sich der Entwurf von Böschungen des Tagebergbaues öfter mit wenig festem Gebirge zu befassen als der von Talsperrengründungen. Der Rezensent ist recht befriedigt, das zunehmende Interesse an weniger festem Gebirge im Zusammenhang mit Bauaufgaben zu entdecken, woraus sich die Notwendigkeit für Felsmechaniker und Felsbauingenieure ableiten läßt, sich noch systematischer mit weniger festem Gebirge zu befassen, als dies in der Vergangenheit üblich war. Mit seinem fortgesetzten Kontakt zu Bauaufgaben im festen Fels (Aufgaben, die sich nicht immer als Probleme darstellen) hat dieser Berichterstatter im Laufe der Zeit erfahren, daß fester Fels, auch wenn weitgehend geklüftet, wird er beim Aushub von Tunneln und anderen Felsbauwerken schonend behandelt, sich oft erstaunlich gut, ja geduldiger verhält, als eigentlich aufgrund von konstitutiven Gleichungen und möglichen Bruchmechanismen zu erwarten gewesen wäre. Die wirkungsvolle Bearbeitung von Aufgaben im Zusammenhang mit weniger festem Fels wird den Gesamtbeitrag der Felsmechanik in Bauwesen und Bergbau verstärken.

2.2.2 Schwächere Tendenzen. Zwei Fragen, die schon in der Vergangenheit im einzelnen diskutiert wurden, werden auch von einigen der vorgelegten Arbeiten reflektiert:

- Systeme zur Gebirgsklassifizierung und ihre systematische Anwendung, nicht nur im Tunnelbau (eine Arbeit), sondern auch beim Entwurf von Felsböschungen, scheinen zuzunehmen. Obwohl diese Tendenz ermutigt werden sollte, da sie die Verständigung zwischen Geologen und Ingenieuren erleichtern dürfte, wird die Anwendung der Gebirgsklassifikation als primäres Entwurfsverfahren, weit außerhalb der ursprünglichen Absichten der jeweiligen Systeme, mit einiger Skepsis betrachtet.

- Andere Bruchmechanismen als Gleiten: Verschiedene Arbeiten beziehen sich auf das Kippen und damit zusammenhängende komplexe Bruchmechanismen, sowohl für Berechnung als auch Entwurf. Es muß jedoch festgestellt werden, daß überzeugende, in der letzten Dekade entwickelte analytische Verfahren noch immer nicht verbreiteten Eingang in die Felsbaupraxis gefunden haben.

2.2.3 Bestehende Tendenzen, die nicht oder kaum widergespiegelt werden; Sachfragen und/oder Tendenzen, die für Entwurf und Ausführung von Felsböschungen (und Felsgründungen) für wichtig gehalten werden, werden von der vorliegenden Sammlung von Veröffentlichungen nicht oder kaum widergespiegelt:

- Wahrscheinlichkeitsverfahren zur Abschätzung der tatsächlichen Standsicherheit von Böschungen (eine Arbeit), die, obwohl vielversprechend, sich in der Ingenieurpraxis aber nicht recht durchsetzen konnten, möglicherweise, weil zu viele Ingenieure die, wenn auch geringe, Möglichkeit des Versagens ihrer Entwürfe verdrängen wollen.

- Neuartige Konzeptionen für die Stabilisierung von Felsböschungen mittels (Stahl-) Bewehrung im weitesten Sinne, zum Beispiel die systematische Gegenüberstellung von Vor- und Nachteilen von vorgespannten und schlaffen Felsbewehrungen. Die Tendenz zur schlaffen Bewehrung, die vereinfachte Bauausführung mit offensichtlicher Wirksamkeit kombiniert, wie sie im Tunnelbau beobachtet werden kann, sollte und wird auch tatsächlich für die Stabilisierung von Felsböschungen zur Kenntnis genommen werden. Diese Frage und Folgerungen daraus, für die Erstellung und Sicherung von Felsböschungen, sollten während dieses Kongresses behandelt werden, auch vielleicht aus dem Grund, weil diese Fragen in den vorgelegten Arbeiten überhaupt nicht angeschnitten wurden.

- Sich aus der Bauausführung ergebende felsmechanische Aufgaben wurden kaum berührt, mit der Ausnahme einer Arbeit über die Bewertung des Einflusses von Sprengarbeiten auf die Standsicherheit und den Entwurf einer Felsböschung. Sprengen und/oder Reißen verdienen nach Meinung dieses Berichterstatters mehr Aufmerksamkeit in Felsmechanik und auch Geowissenschaften, als eine ständige mögliche Ursache von wirklichen Problemen auf einer Felsbaustelle.

2.3 Folgerungen

2.3.1 Frühere Beiträge. Während der vergangenen zwei Dekaden wurden auf früheren Tagungen dieser und verwandter Berufsvereinigungen neue Ideen sozusagen "en masse" vorgestellt, von denen fast alle in mehr oder weniger engem Zusammenhang mit dem Rahmenthema dieser Sitzung stehen. Einige werden nachstehend aufgeführt:

- Geologisches Trennflächengefüge (wie Klüftung), das oft die bautechnischen Eigenschaften und das tatsächliche Verhalten von Gebirgskörpern beherrscht, was ein zunehmendes Interesse von Geologen vieler Schattierungen im Bereich des Felsbaues mit sich brachte.

- Gebrauch von vorgespannten Felsankern für die Sicherung von Felsböschungen, zunächst als sehr einfache, dann immer höher entwickelte Systeme, mit der unaufhörlichen Diskussion der Dauerhaftigkeit dieser Entwürfe.

- Hochentwickelte Versuche an relativ kleinen Proben sowohl des Gesteins als auch des Gebirges, mit kritischer Diskussion von Fragen wie der repräsentativen Größe und (Entnahme-)Ort von Probekörpern und der Extrapolation von Versuchsergebnissen.

- Anwendung des Verfahrens der finiten Elemente für Berechnungen in einem sehr weiten Bereich der Felsmechanik und des Felsbaues, mit einer überaus raschen Abfolge von verbesserten Varianten, die jedoch die Anwendung in der Baupraxis nicht immer erleichterten.

- Neue Konzeptionen des Tunnelbaus und dementsprechend des Tunnelentwurfes, die hier nur erwähnt werden sollen, weil sie allen im Felsbau tätigen Berufsgruppen überzeugend bewiesen haben, daß die Schonung des Gebirges bei der Bauausführung und die Motivierung des Unternehmers hierzu sich für alle Beteiligten ausgezahlt hat.

2.3.2 Heutige Beiträge. Die durchgesehenen Arbeiten und andere von diesem Berichterstatter beobachtete Anzeichen deuten darauf hin, daß die mit Felsmechanik und Felsbau befaßten Berufsgruppen sich derzeit nicht in einer Periode dramatischer neuer Beiträge befinden. Die Zeit der anregenden Vorträge und der lebendigen Diskussionen im Gelände, im Labor und im Entwurfsbüro scheint zumindest zeitweise vorüber, während die Berufsgruppe vorwiegend dabei ist, frühere Errungenschaften zu konsolidieren und schon länger vorliegende Arbeitsansätze und -ideen endlich anzuwenden, wobei Produktivität und Kosten-Nutzen-Abschätzungen langsam aber sicher in die Überlegungen mit einbezogen werden. Es scheint, daß die Ingenieurkomponente in der Felsmechanik endlich mehr Kraft gewinnt, eine Beobachtung, die der Berichterstatter als etwas konventioneller Bauingenieur mit Befriedigung zur Kenntnis nimmt. Trotzdem, neue Konzeptionen stehen schon wieder zur Verfügung und warten auf die Entdekkung durch die Fachwelt und ihre Einfügung in die systematische Problembearbeitung: die analytische Konzeption des "Schlußsteines" soll hier nur als ein Beispiel dazu erwähnt werden.

3 ÜBERBLICK ÜBER DEN STAND DER TECHNIK

3.1 Allgemein

Dieser Gesamtüberblick über den Stand der Technik der ingenieurtechnischen Bearbeitung von Felsböschungen und Gründungen in und auf Fels wird sich über den durch die vorgelegten Arbeiten vorgegebenen Bericht hinaus erstrecken. Wie bereits weiter oben erwähnt, wird dieser Berichterstatter die Überlegungen des ingenieurtechnischen Bereiches betonen, in denen Beiträge zur Lösung von tatsächlichen Problemen und Entwurfsaufgaben gefordert werden. Im Verlauf dieses Überblicks wird der Berichterstatter es auch wagen, seinen Bedenken über den derzeitigen Stand der Technik Ausdruck zu verleihen, Erkenntnissen, die er in den vergangenen dreißig Jahren in diesem Fachgebiet gewonnen hat, und Sorgen, die sich in dieser Zeit nur zu einem kleineren Teil als grundlos erwiesen haben. Obgleich nicht im Vordergrund, so sollen doch auch Beiträge und die Grenzen von theoretischen Ansätzen im Gedächtnis gehalten werden.

Es wird erhofft, daß dieser Überblick eine zusätzliche Grundlage für die Auswahl von wirklich bedeutungsvollen Themen für die Diskussionen während dieses Kongresses liefert. Ein internationaler Kongreß ist der perfekte Anlaß, vorliegende und zukünftige Möglichkeiten vorzustellen. Er ist aber auch die beste Gelegenheit, sich auch mit den schwächeren Gliedern der verschiedenen felsbaumechanischen Systeme zu befassen, um letztlich den Beitrag dieses Spezialgebietes sowohl für Bauwesen als auch Bergbau noch weiter zu verbessern.

3.2 Fels

3.2.1 Gebirge. Die Unterscheidung zwischen Fels (oder Gestein) und Gebirge liegt seit langer Zeit fest, die bautechnischen Folgerungen daraus wurden und werden immer wieder behandelt. Die Pseudo-Gleichung "Gestein + geologisches Trennflächengefüge (von Mikrorissen über Klüftung zu Störungen) = Gebirge" stellt eine der wichtigsten Grundlagen der meisten derzeitigen Beiträge der Felsbaumechanik dar.

3.2.2 Geologisches Trennflächengefüge. Es besteht hier keinerlei Frage über die technische Bedeutung der größeren Gefügeelemente, wie Großklüften und Störungen, für bautechnische Entscheidungen im Felsbau. Demgegenüber ist die bautechnische Bewertung von weniger deutlich ausgeprägten Gefügeelementen, wie Kleinklüften und Mikrorissen, viel weniger eindeutig. Obwohl es eine Vielzahl von Verfahren zur Kartierung, Beschreibung, Darstellung und Analyse aller Kategorien von Gefügeelementen gibt, sind die tatsächlichen Eingangswerte in Ingenieuranalysen, streuende Raumdaten von Kontinuitäten (oder Diskontinuitäten) solcher Klüfte, nicht gerade sehr überzeugend. Die Auswirkung von Baumaßnahmen auf geklüftetes Gebirge erscheint ein weiterer strittiger Punkt zu sein. Ungeeignete Verfahren, wie nicht gebirgsschonendes Sprengen und exzessives Reißen, neigen dazu, den Einfluß der Klüftung überzubetonen, während gebirgsschonende Bauverfahren wie moderne Spreng- und Tunnelbauweisen den negativen Einfluß von Kleinklüften weitgehend hintanhalten. Eine objektive Bewertung ist recht schwierig, was entweder zu Über- oder Unterbewertung führt. Für die Bewertung der bautechnischen Bedeutung der Klüftung im Gebirge wird empfohlen, sich nicht nur auf vorgelegte Gefügedaten wie Winkel- oder flächentreue Poldiagramme zu beziehen, sondern auch in Beziehung mit den wirklichen Verhältnissen zu bleiben und auch zu verfolgen, wie sich diese während der Bauausführung darstellen. Dies trifft vor allem für Entwurfsingenieure zu. Zu leicht führen numerische und graphische Daten, wenn losgelöst von den tatsächlichen Verhältnissen im Gelände, zu abstrakten, unrealistischen Gedankenmodellen.

3.2.3 Fester zu wenig festem Fels. Die frühzeitige Entwicklung von Felsmechanik in den alpinen Ländern, und auch in Skandinavien, begünstigte anfänglich die Mechanik des festen Felsens und ihre Anwendung auf entsprechende Bauaufgaben. Auf der Grundlage der sich immer wiederholenden Erfahrung vor Ort kam dieser Berichterstatter schon frühzeitig zu dem Schluß, daß schonend behandeltes Gebirge, zum Beispiel mittels gebirgsschonender Löse- und Aushubverfahren, sich häufig als weitgehend gutmütig erweist, zumindest in bezug auf kleinere und mittlere Standsicherheitsprobleme. Dementsprechend hat er an sich hochinteressante Felsmechanikaufgaben in den Hintergrund schwinden sehen, sobald gebirgsschonende Verfahren durchgesetzt werden konnten. Im Gegensatz dazu erscheint weniger festes, jüngeres und/oder verwittertes Gebirge weniger gutmütig, ja sogar heimtückisch, wenn unfachgemäß behandelt. Zusätzlich, zumindest im Bereich des allgemeinen Bauwesens, wird weniger festes Gebirge weit häufiger angetroffen als wirklich kompetentes Gebirge. Es soll in diesem Zusammenhang festgestellt werden, daß die Entwicklung der Mechanik und die Bautechnik des weniger festen Gebirges eine Herausforderung darstellt, die eine enge Zusammenarbeit von Boden- und Felsmechanikern erfordert, wobei beide Gruppen der Unterstützung durch Geologen und Ingenieurgeologen bedürfen.

3.3 Gesamte Unterstützung eines Projektes im Fels

3.3.1 Systematische Bearbeitung. Die systematische Bearbeitung, die bei der Untersuchung von verschiedensten Projekten im Fels verfolgt wird, spiegelt sich in vielen Arbeiten dieser Sitzung wider. Die nachstehende Abfolge ist heute (beinahe) alltägliche Routine:

- Festlegung der Felsbauaufgabe innerhalb des gesamten Planungsverfahrens, zum Beispiel der Entwurf und die Erstellung eines Tagebergbaues oder einer Talsperrengründung (oder eines Tunnels).

- Erkundung der vorliegenden natürlichen Verhältnisse, mit geologischer Analyse.

- Versuche, in-situ und/oder im Labor.

- Idealisierung der natürlichen Bedingungen mit dem Ziel, ein "Gedankenmodell" zu entwickeln.

- Ingenieurtechnische Analyse dieses Modelles.

- Entwurf und/oder Beitrag zum Entwurf und zu Vertragsunterlagen und

- Bauausführung, mit geotechnischer Beaufsichtigung der zur Anwendung gelangenden Bauverfahren und ingenieurgeologischer Aufnahme der wirklichen Gegebenheiten.

3.3.2 Anzustrebende Ziele und Ergebnisse. Die Gesamtqualität der letztendlichen Beiträge des vorstehenden Systems hängt häufig von der Konsequenz ab, mit der die systematische Bearbeitung wirklich durchgeführt wird. Übertrieben detaillierte Behandlung von Einzelelementen entspräche dem "Reiten eines Steckenpferdes", zu dürftige Bearbeitung von Teilbereichen wären bautechnische Unterlassungssünden. Dieser weiterführende Überblick hält sich weitgehend an die oben aufgeführte Abfolge des Untersuchungssystems.

Spezifischer gesehen sollten sich die zu erwartenden Ergebnisse einer Untersuchung an den Ansprüchen der vorgegebenen Ingenieuraufgabe orientieren. Zwei oder möglicherweise drei diesbezügliche Grundaufgaben können festgehalten werden:

- Entwurf eines Tagebergbaues, mit Böschungen entweder nahe dem oder im Grenzgleichgewicht, wobei zu berücksichtigen ist, daß Bergleute gewöhnt sind, mit der Möglichkeit von Verbrüchen zu leben. So können Verbruchmechanismen häufig tatsächlich beobachtet werden, wodurch Rückrechnungen sehr sinnvoll werden und somit wirklich überzeugende Ergebnisse liefern können.

- Talsperren- (oder andere) Gründungen in und auf Fels, recht häufig in Verbindung mit Böschungsstandsicherheitsaufgaben, mit der beherrschenden Verpflichtung, "absolut sichere" Entwürfe zu entwickeln, Verbrüche müssen ausgeschlossen bleiben, können häufig nicht einmal diskutiert werden. Die somit oft beträchtlichen in den Entwurf eingebauten Sicherheitsspannen verschleiern häufig die tatsächliche Ingenieurleistung, obgleich die Frage nach der tatsächlichen Sicherheit eines solchen Felsbauentwurfes nur sehr selten überzeugend beantwortet werden kann.

Die dritte mögliche Problemklasse, zwischen Bergbauaufgabe und Gründungsentwurf gelegen, umfaßt folgenden Aufgabenbereich.

- Standsicherheit, Bauwesen oder im Bergbau, von natürlichen Böschungen, mit oder ohne beobachtete Rutschung, mit folgender Unterteilung:

  - "Potentielles" Standsicherheitsproblem, wobei der Entwurf einen Zuwachs an Sicherheit in bezug auf die naturgegebene, als sicher bewertete Situation nachzuweisen hat.

  - "Vorgeschichtliche Rutschung", wie vom Geologen kartiert, mit möglicher Wiederaktivierung durch natürliche Ursachen oder bauliche Eingriffe.

  - "Aktive Rutschung", die stabilisiert werden soll. Die reale Bearbeitung dieser Problemklasse ist trotz oft großem Aufwand nicht immer zu erfolgreich. Aus diesem Grund soll im Verlauf dieses Überblicks auf diese Sonderfrage noch einmal zurückgekommen werden.

3.3.3 Kosten der Erkundung und Kosten-Nutzen-Verhältnis. Lediglich eine Arbeit berührt die Frage der Kosten von Erkundungen für Talsperrengründungen, setzt sich aber nicht mit der realen Aussagekraft von Erkundungen und dem Kosten-Nutzen-Verhältnis auseinander. Die Kosten von Erkundung, Versuchen und Berechnungen haben sich im Verlauf der letzten Dekade beträchtlich erhöht, und das nicht unbedingt wegen komplexerer Projekte, sondern auch wegen immer komplizierter werdenden technischen, gesetzgeberischen und verwaltungstechnischen Randbedingungen, wobei einige Erschwernisse auch von der geotechnischen Berufsgruppe zu verantworten sind. Gleichzeitig erfordert die Durchführung von Erkundungen immer mehr Zeit, was auch wieder im Rahmen der Projektfinanzierung bewertet werden muß.

Nach dem heutigen Sachverhalt ist dieser Berichterstatter überzeugt, daß sich die gesamte Fachgruppe mehr mit der Frage der Kosten und des Wirkungsgrades des felsmechanischen Beitrages auseinandersetzen sollte, als dies aus den hier vorgelegten Arbeiten hervorgeht. Tatsächlich ist der Geotechniker vor zwei Alternativen gestellt: Auf der einen Seite führt der Konkurrenzdruck zum Unterbieten und damit fast unmittelbar zur "Unter"-Erkundung. In ganz seltenen Fällen bietet sich jedoch auch die Möglichkeit der "Über"-Erkundung, wobei Ergebnisse nicht unbedingt proportional zu den Aufwendungen nützlicher werden. Beide Extreme können zu leicht dazu führen, daß der Beitrag der Felsmechanik an Überzeugungs- und Aussagekraft verliert.

Im folgenden werden zwei mit der Frage der Erkundungskosten eng zusammenhängenden Problembereiche angesprochen, um möglicherweise eine entsprechende Diskussion anzuregen.

- Das Haushaltsdilemma. Vor der eigentlichen Bauphase, ja vor dem eigentlichen Entschluß zum Bau durchzuführende Arbeiten, wie Erkundungen und Untersuchungen, müssen notwendigerweise mit begrenzten Mitteln auskommen, was leicht zu engen Aufgabestellungen führen kann. Fast immer werden dazu positive (und optimistische) Aussagen erwartet. Die hiervon abgeleiteten Entwürfe sind dann nicht immer realistisch. Überwältigend sind oft die Kosten von Bauverzögerungen. Unerwartete Bedingungen erfordern leicht kostspielige Entwurfsänderungen, die zwar weniger teuer als Bauunterbrechungen, aber viel kostspieliger als bessere, vollständigere Erkundungen sind. Das Budget für die Bauausführung muß all das verkraften, was nur allzu häufig zu dramatischen Kostenüberschreitungen und damit zusammenhängenden Rechtsstreiten führt.

- Anpassungsfähige Verfahren. Ein vernünftiges Kosten-Nutzen-Verhältnis kann nur erreicht werden durch einen Kompromiß zwischen "angemessener" Erkundung vor Baubeginn, wobei die Erkundung während der Bauausführung weitergeführt wird, und "vernünftiger" Anpassungsfähigkeit während der Bauausführung, die durch Entwurf und Bauvertrag ermöglicht werden muß, um mit nunmal unvermeidlichen Änderungen der angenommenen Verhältnisse fertig werden zu können. Solche "Änderungen" werden zu leicht der Geologie zur Last gelegt, auch wenn sich zum Beispiel die gewählten Bauverfahren als ungeeignet oder unproduktiv erweisen. Selbstverständlich, "angemessene" Kompromißlösungen für tatsächliche Gegebenheiten (Ingenieuraufgabe, geologische Gegebenheiten, örtliche Umwelt und/oder Gebräuche) sind nur möglich, wenn die Risiken aus Entwurf und Bauausführung, zusammen mit den sich daraus ergebenden Kostenbelastungen, auch "angemessen" verteilt worden sind.

3.4 Erkundung im Gelände, geologische Analysen und Versuche

3.4.1 Allgemein. Die unterschiedlichen, in diesem Abschnitt zusammengefaßt behandelten Tätigkeiten führen zu einem qualitativen und teilweise bereits quantitativen Modell der natürlichen Verhältnisse in Verbindung mit der vorgesehenen Bauaufgabe. Dieses Modell, ob man es nun physikalisches oder geotechnisches Modell nennt, wird durch nachfolgende ingenieurmäßige Berechnungen und Entwürfe nur ganz selten drastisch verändert. Die Möglichkeit für tiefgreifende Änderungen besteht vielmehr während der Bauausführung durch das Aufschließen von unerwarteten geotechnischen Verhältnissen, mit oder ohne geologische Einflüsse. Es ist ganz offensichtlich geworden, daß die Qualität des hier beschriebenen Gedankenmodells, das schon frühzeitig während der Projektbearbeitung entwickelt wird, die Qualität der Ingenieurberechnungen, des Entwurfes und der Vertragsunterlagen ganz maßgeblich beeinflußt.

3.4.2 Erkundung. Bei dieser Gelegenheit sollen einige Überlegungen vorgetragen werden, die diesen Berichterstatter immer wieder beschäftigt haben:

- Die Qualität von Kernbohrungen, die immer noch das wichtigste Verfahren zur Erkundung von Gebirgskörpern sind und einen großen Teil der Erkundungsmittel verbrauchen. Probleme werden bei den folgenden Fragen gesehen.

  - Das Niveau der Bohraufsicht im Gelände ist das Ergebnis von vertraglichen und anderen Kostenüberlegungen. Es ist betrüblich, daß der alte Streit zwischen geleisteten Bohrmetern im Verhältnis zur tatsächlich erhaltenen Information immer noch zu oft zugunsten der reinen Bohrmeter entschieden wird, besonders bei starkem Wettbewerbsdruck.

  - Die Aufnahme von Daten im Gelände und Führung der Bohrprotokolle durch oft wenig qualifiziertes Personal, wobei geringer Kerngewinn oft als unvermeidlich angesehen wird, obwohl es bekannt ist, daß dieser sich häufig aus der Kombination von geologisch-geotechnischen Verhältnissen und Qualität von Bohrgerät, Bohrmannschaft und der Bohrüberwachung ergibt.

  - Datenverarbeitung und Beurteilung der Bohrergebnisse werden oft routinemäßig mittels genormter Verfahren durchgeführt, wobei zu viele, teuer erworbene Information ungenützt bleibt, für weitergehende geologische und geotechnische Analysen und bautechnische Entwürfe.

  - Übermäßiges Vertrauen in Gebirgsklassifizierungen, bei unzulässiger Extrapolation und daraus abgeleiteten Folgerungen. Diese Systeme umfassen den Bereich vom sehr einfachen, geradlinigen RQD-Wert zur allgemeinen Beschreibung der Klüftung, bis zu komplexeren, aber doch nicht sehr komplizierten Gebirgsklassifizierungssystemen. Hierbei stehen unterschiedliche Systeme zur Verfügung, die als sehr nützlich erachtet und von diesem Berichterstatter voll und ganz unterstützt werden, vorausgesetzt sie dienen dazu, die Verständigung zwischen den an einem Projekt beteiligten Gruppen zu fördern. Der Gebrauch eines einfachen oder auch nicht ganz einfachen Gebirgsklassifizierungssystems als den Schlüssel zu komplexen Entwurfsaufgaben wird dagegen nicht begrüßt. Diese Kritik bezieht sich sowohl auf den Gebrauch des RQD als den alleinigen geotechnischen Beitrag zum Tunnelentwurf, d. h. für die Bestimmung von Tunnelbauweisen und -ausbau. Es bezieht sich aber auch auf den Gebrauch der Klassen eines an sich sehr brauchbaren Klassifizierungssystems als der Grundlage für den Entwurf von Felsböschungen, wie es in einer Arbeit vorgeschlagen wird. Es ist bewiesen, daß der RQD-Wert weitgehend durch mittelmäßige Bohrverfahren beeinflußt wird, vor allem in möglicherweise anisotropen Sedimentgesteinen und metamorphen Gesteinen. Ein übertriebenes

Vertrauen in allumfassende Gebirgsklassifizierungen hat, in der Sicht dieses Berichterstatters, zu einer Unterschätzung des Nutzens von Verfahren der positiven Baugrundbeherrschung und anderen modernen Bauverfahren gerade im wenig festen und engständig geklüfteten Gebirge beigetragen.

3.4.3 Geologische Analyse. Jede Erkundung, die im allgemeinen von Geologen oder Ingenieurgeologen durchgeführt und/oder überwacht wird, wird geologische Analysen enthalten, die oft nicht sehr deutlich gemacht und auch unbefriedigend dokumentiert werden, aber trotzdem von ausschlaggebender Bedeutung für die Entwicklung des "Modelles" sind, das die Grundlage für die nachfolgenden Arbeiten ist. Für diesen Berichterstatter ist es ganz zulässig, daß geologische Analysen dazu neigen, mehr qualitativ als quantitativ zu sein, mit der Ausnahme der Bearbeitung von geologisch-geometrischen Zusammenhängen. Der Verfasser hat es am eigenen Leibe erfahren, daß eine vorbildlich dargestellte, ausgesprochen modern-numerische Analyse, d. h. Berechnung, völlig wertlos, ja gefährlich sein kann, wenn ihr lediglich vereinfachend angenommene Modelle oder Mechanismen (oft aus den Akten entnommen) zugrunde liegen und nicht das Modell, das vom erfahrenen Geologen, der sich lange mit dem gegebenen Gelände beschäftigt hatte, vorgeschlagen wurde.

Geologisch-geometrische Analysen unter Verwendung von flächentreuen (oder winkeltreuen) Lagekugeldarstellungen, in denen Kluftstellungen dargestellt und in Wechselbeziehung gebracht werden, sind heute ein fest eingeführtes Werkzeug bei Standsicherheitsuntersuchungen für Felsböschungen und anderen Aufgaben des Felsbaues. Vor fünfzehn Jahren hatte dieser Berichterstatter einiges Vergnügen, dieses geologische Werkzeug erst einmal für sich zu entdecken, es bei Aufgaben der Standsicherheitsbewertung für Felsböschungen und Talsperrengründungen anzuwenden und auch darüber zu berichten. Inzwischen hat er aber in aller Deutlichkeit die Unzulänglichkeiten, ja Gefahren dieses grundsätzlich positiven Verfahrens erkannt, wie:

- Durch den fast ausschließlichen Gebrauch von mittleren Raumstellungen, die als repräsentativ angenommen werden, wird die Streuung der Gefügedaten vernachlässigt, was möglicherweise zu unvorsichtigen Folgerungen führen kann.

- Wenn die Streuung der Gefügedaten berücksichtigt wird, dann werden die Ergebnisse in bezug auf die kinematischen Verhältnisse und sich daraus ergebende Standsicherheitsprobleme mit einiger Wahrscheinlichkeit sehr vage, was dazu führt, daß die Aussage von recht komplizierten graphischen oder auch numerischen Darstellungen nicht viel mehr als ein "nein oder vielleicht auch ja" (oder umgekehrt) gedeutet werden kann.

- Mit wirklichen Eingabewerten, die oft keinesfalls einfach überblickbare geologische Strukturen darstellen, werden die meisten der heute zur Verfügung stehenden Verfahren für alle, außer graphische und numerische Zauberkünstler, praktisch undurchführbar.

An dieser Stelle möchte der Berichterstatter raten, die tatsächlichen, im Gelände beobachteten Verhältnisse nicht von den im Büro zu entwickelnden graphischen und anderen Gedankenmodellen gedanklich zu trennen. Wenn immer möglich, sollten Folgerungen aus Berechnungen kritisch und objektiv mit den tatsächlichen Verhältnissen im Gelände verglichen werden und umgekehrt. Ein solcher fortlaufender kritischer Vergleich trägt dazu bei, "Papierkeile" rechtzeitig zu beseitigen, was erforderlich ist, da sich ein wirklichkeitsgerechter Entwurf ja nur an wirklichen "Felskeilen" orientieren sollte.

3.4.4 Versuche. Eine Abhandlung des gesamten Untersuchungsvorganges ist ganz bestimmt nicht vollständig ohne zumindest kurzen Bezug auf die Durchführung von felsmechanischen Versuchen. Für den Entwurf von Felsböschungen ist die Scherfestigkeit von geologischen Gefügeelementen, wie Großklüften und Störungen, die ausgeprägte Schwachstellen im Gebirge darstellen, von entscheidender Bedeutung. Für Felsgründungen wird die Verformbarkeit der Gebirgsmasse in die Betrachtung einbezogen. Laborversuche und in-situ Versuche zur Bestimmung dieser beiden Schlüsselkennwerte wurden in den letzten beiden Dekaden sehr ausführlich diskutiert. 23 Jahre nach den bahnbrechenden in-situ Großversuchen für die Felsgründung der Kurobe IV Bogenstaumauer in Japan, an deren Durchführung dieser Berichterstatter aktiv beteiligt war, hat er jedoch immer noch Schwierigkeiten, repräsentative Scherfestigkeitskennwerte für sich weit erstreckende mögliche Scherflächen in geklüftetem Gebirge zu bestimmen. Dabei sollten die entsprechenden Parameter realistisch, aber nicht übermäßig konservativ sein. Das Erstellen von übermäßig vorsichtigen Daten ist natürlich sehr einfach. Es scheint, daß man sich heute immer noch mit folgenden Problemen bei Versuchen im Fels herumschlagen muß:

- Auswahl von wirklich repräsentativen Proben(entnahme)stellen für die Entnahme von Bohrkernen und/oder für die immer sehr begrenzte Anzahl von in-situ Versuchen. Diese beruht fast immer auf recht begrenzten geologischen Angaben und ist immer, so oder so, sehr stark von begrenzten Zugangsmöglichkeiten beeinflußt.

- Die Umwandlung von Rohdaten aus den Versuchen in Entwurfsparameter verlangt fast immer beträchtliche Interpretationen, die nicht immer sehr objektiv sein können, oft aber recht intuitiv ausfallen. Die Versuchung besteht, einfach obere und untere Grenzwerte in die Berechnung einzugeben. Aber gerade für Untersuchungen zur Böschungsstandsicherheit löst dieser Ausweg nicht immer das vorliegende Problem, da die vorgegebene Streuung nur zu leicht bei den Rechenergebnissen den Bereich von standsicher bis zu nicht standsicher ergeben kann.

- Flächenmäßige und räumliche Extrapolation von Versuchsergebnissen: Sogar die höchstentwickelten und aufwendigsten in-situ Versuche (mit der Ausnahme von geophysikalischen Verfahren) umfassen bestenfalls einige Quadrat- bzw. Kubikmeter. Eine Extrapolation der Versuchsergebnisse auf Scherflächen von einigen 10.000 m für einen typischen Tagebergbau oder einen Gebirgskörper von einigen Millionen m^3 für die Felsgründung einer großen Talsperre sind sehr schwierig, um es optimistisch auszudrücken. Dieser Berichterstatter ist heute überzeugt, daß fast alle dieser Extrapolationen recht nahe an die unangemessen konservative Grenze des möglichen Ergebnisbereiches führen.

3.4.5 Idealisierung der natürlichen Bedingungen. Die nachfolgende Aufzählung faßt Überlegungen des Berichterstatters über diesen überaus wichtigen Festpunkt innerhalb des Untersuchungsvorganges zusammen, der praktisch den Kontakt zwischen Geologen und Ingenieuren darstellt.

- Die Idealisierung, d. h. ein im wesentlichen qualitatives physikalisches Modell der natürlichen Verhältnisse, liefert die Grundlage für die meisten, ja sogar alle nachfolgenden Berechnungen und Entwürfe.

- Das Modell muß wirklichkeitsgetreu sein, wenn auch mit einer konservativen Tendenz. Es darf weder von Erkundungsverfahren (z. B. den Vielschichten der Geophysik) noch von vorhandenen Rechenverfahren (einfacher Gleitkörper gegenüber kompliziertem wirklichen Verbruchmechanismus) beherrscht werden.

- Während der vergangenen Dekade wurde eine Vielzahl von möglichen Verbruchmechanismen vorgestellt:

  - Blöcke/Keile/Schichtpakete, die

  - gleiten, kippen, beulen, wobei die diesbezüglichen kritischen Kennwerte und Entwurfslösungen im vernünftigen Rahmen heute vorliegen.

  Trotz dieser recht überzeugenden Auswahl an Möglichkeiten sind diesem Berichterstatter erst in jüngster Vergangenheit die immer noch sehr großen Schwierigkeiten ganz deutlich geworden, die bei der angemessen-konservativen Deutung der natürlichen Verhältnisse von Geologen und Ingenieuren in der Praxis routinemäßig bewältigt werden sollen. Sowohl die Auswahl eines zu konservativen Gedankenmodelles als auch das Übersehen von verborgenen Merkmalen und/oder Mechanismen erscheinen fast leichter als das Aufzeigen des Modelles, das durch nachfolgendes Verhalten des Prototyps wirklich überzeugend bestätigt werden kann.

- Abmessungen. Für kleinere oder mittlere Standsicherheitsaufgaben erscheint die Auswahl eines angemessenen Modelles noch eine Aufgabe zu sein, die man mit wirklich gutem Gewissen lösen kann. Für wirklich großmaßstäbliche Aufgaben, eine natürliche Böschung, auf der eine Bogenstaumauer gegründet werden soll, oder ein größerer Teil eines Tagebergbaues, treten die einfachen, oben aufgezählten und in vielen Veröffentlichungen abgehandelten Mechanismen meist nicht allein, sondern als Bestandteil von vielschichtigen, möglichen bzw. tatsächlichen Verbrucharten auf, die zu oft auch für höchstentwickelte Rechenverfahren zu kompliziert sind.

- Parametrische Überlegungen. Seit einiger Zeit werden parametrische Studien oder Tendenz- bzw. Empfindlichkeitsuntersuchungen als Allheilmittel für die ingenieurmäßige Aufarbeitung von (zu) komplizierten Modellen dargestellt. Die Untersuchungen zielen dabei auf die Bestimmung der für das Verhalten des vorgesehenen Modells wirklich entscheidenden Kennwerte hin. Während immer betont wird, daß dabei die Eingangswerte verändert werden, kam dieser Berichterstatter zu dem Schluß, daß in vielen Fällen auch das Modell, das die Grundlage der parametrischen Untersuchung darstellt, variiert werden sollte. Dies würde natürlich parametrische Studien komplizieren, da Anzahl und Komplexität der Veränderlichen zunehmen würden. Das unterschiedliche Verhalten verschiedener Modelle, bei vergleichbaren Kennwerten, könnte sehr leicht genau so erschütternd sein wie das Verhalten eines tatsächlichen Einschnittes im Fels, in dem sich der angenommene Gleitkeil wegen unregelmäßiger und nicht durchstreichender Klüftung nicht realisiert, und der sich dementsprechend nur aufgrund der seitlichen Entspannung verformt. Dazu, der Vergleich von rechnerisch ermittelten Verformungen aufgrund einer Modellvorstellung mit Meßergebnissen am Prototyp, der sich in ganz anderer Art und Weise verhält, könnte zu unheilvollen Beurteilungen führen, die gegebenenfalls Geländemessungen als solche unberechtigterweise ins falsche Licht setzen würden.

3.4.6 "Vorgeschichtliche Rutschungen". Mit der Planung, dem Entwurf und der Bauausführung befaßte Ingenieure werden bei ihren Tätigkeiten unvermutet aufgehalten, wenn sie mit geologischen Karten und Plänen einer Baustelle konfrontiert werden, die

- "vorgeschichtliche Rutschungen" und/oder

- "aktive" und "möglicherweise aktive" Störungen aufzeigen.

Die Grundlage für solche Ansprachen und entsprechende Bezeichnungen kann von sehr oberflächlich bis äußerst umfassend reichen, je nach dem Bauvorhaben und den jeweiligen örtlichen Vorschriften und Gebräuchen. Welche Wahl bleibt hierbei für ein Bauvorhaben, als die vorgesehene Projektstelle aufzugeben und nach einer anderen zu suchen, ein Ausweg, der in weitentwickelten, geologisch komplexen Gebieten schwierig und für Bergbauprojekte unmöglich ist.

Es ist natürlich sehr gut bekannt, wie schwierig es ist, für frühere und vorgeschichtliche Vorgänge folgendes zu bestimmen:

- Zeitpunkt des Verbruches oder einer Bewegung, wobei Angaben über den geologischen Zeitpunkt nicht allzu nützlich sind, wenn diese nicht dazu dienen, die Verbindung mit anderen, sachdienlichen Ereignissen herzustellen.

- Umweltbedingungen zu diesem Zeitpunkt, die mit einiger Wahrscheinlichkeit unterschiedlich und möglicherweise ungünstiger als derzeitige Bedingungen waren.

Nur mit zuverlässigen Angaben über beide vorgenannten Fragen würde man überzeugende Grundlagen liefern sowohl für die ingenieurmäßige Abschätzung der derzeitigen Standsicherheit mittels Rückrechnung der ursprünglichen Vorgänge, als auch die Bewertung der derzeitigen Verhältnisse, mit entsprechendem Entwurf und Überwachung des gegenwärtigen und zukünftigen Verhaltens.

Wie sich die Lage heute darstellt, stellt ein Bauwerk auf oder nur nahe von geologischen Merkmalen, die als vorgeschichtlich und möglicherweise aktiv bezeichnet werden, unannehmbare Risiken dar, obgleich die derzeitige Standsicherheit sehr wohl höher sein könnte als die der angrenzenden, als standsicher bezeichneten Bereiche, die sich sehr wohl im Grenzgleichgewicht befinden können. Es besteht die Notwendigkeit, daß die Analytiker der Geologie und des Ingenieurbereiches zusammenwirken, Gedankenabfolgen und Verfahrensweisen zu entwickeln, um eine entsprechende Situation überzeugender, als bisher üblich, bewältigen zu können. Selbstverständlich soll hier nicht vertreten werden, eine Talsperre über einer möglicherweise aktiven Störung zu erstellen oder wissentlich auf einer prähistorischen Rutschung zu gründen. Es erscheint jedoch unangemessen, daß einfaches Ausweisen einer alten Rutschung oder einer aktiven Störung, ohne weitere Angaben, alle und jede Möglichkeit zur Entwicklung einer Projektstelle, trotz der vorliegenden geotechnischen Probleme, ausschließen kann.

3.5 Ingenieurberechnungen

3.5.1 Zur Verfügung stehende Verfahren. Für die Berechnung der Standsicherheit von Felsböschungen und Gründungen im Fels stehen heute zwei grundsätzliche Verfahrensweisen zur Verfügung:

- Grenzgleichgewichtsverfahren und

- Rechenverfahren der finiten Elemente.

Eine Kombination der beiden Verfahren wird mehr und mehr angewendet, sowohl für statische als auch dynamische Belastungen, wobei das mit Hilfe der Finiten Element Methode bestimmte Spannungsfeld die Grundlage liefert für die Ermittlung von entweder Sicherheitswert oder Verbruchwahrscheinlichkeit, jeweils aufgrund von Grenzgleichgewichtsbetrachtungen.

Beide Verfahren und vor allem die Kombination der beiden reagieren recht empfindlich auf sich ändernde Eingangsdaten, sowohl der Modelle als auch Kennwerte. Diese sehr wohl bekannte Schwäche wird zumindest teilweise durch Parameteruntersuchungen umgangen. Die Tatsache, daß sowohl Modelle als auch Kennwerte durch Bauverfahren beeinflußt werden können, wird nur selten in Entwurfsberechnungen berücksichtigt. Die Verformbarkeit einer Felsmasse, wie sie in der Erkundung und durch Versuche ermittelt wurde, kann durch Baumaßnahmen, z. B. die Erstellung eines nahegelegenen Aushubes, bei dem die Sprengarbeiten nicht gebirgsschonend durchgeführt wurden, dramatisch erhöht werden. Auf der anderen Seite können Schichtpakete, die parallel zu einer geneigten Einschnittsböschung einfallen, als eine vernünftige Entwurfsgrundlage gelten, solange die Felsplatten nicht durch wildes Sprengen und/ oder Reißen unterschnitten werden, somit in den Aushub abrutschen können und einem Verbruchmechanismus folgen, der beim Entwurf ausgeschlossen worden war.

3.5.2 Schlußstein-Ermittlung. Ein weiteres, typisch felsmechanisches Rechenverfahren steht heute zur Verfügung, die Ermittlung des Schlußsteines, ein Berechnungsverfahren, das von dem anderen Hauptberichterstatter und dessen Mitarbeitern entwickelt worden ist. Es handelt sich um ein sehr interessantes, intellektuell äußerst anregendes Verfahren zur Bestimmung des kritischen Schlußsteines eines Felssystems, aufgrund des geologischen Trennflächengefüges und der Ausbruchgeometrie, der, wenn rechtzeitig gesichert, den gesamten Felsbereich um den Aushub stabilisiert. Fairerweise sollte festgestellt werden, daß ähnliche, wenn auch weit weniger formalisierte Verfahren, seit längerer Zeit angewendet werden, z. B. bei der Sicherung einer Felsböschung durch kurze Felsanker, ohne unmittelbare Ankerung der tieferen Gebirgsbereiche. Das neue Verfahren erscheint sehr vielversprechend für einfaches geologisches Gefüge mit vollwirksamen Klüften, was möglicherweise vor allem einer durch Sprengungen erschütterten und aufgelockerten Felsmasse entspricht.

Das Erkennen der kritischen Schlußsteine während der Aushubarbeiten, bevor diese gleiten oder fallen, dürfte recht schwierig sein. Intuitiv erscheint das neue Verfahren vorzüglich geeignet zu sein, als Grundlage für den Entwurf von Systemsicherungen in geologischen Verhältnissen, in denen Schlußsteine vorkommen können, ohne daß diese dann während des Aushubes erst identifiziert werden müssen. Es erscheint sehr gut möglich, daß Entwürfe auf dieser Grundlage wirtschaftlicher als konventionelle Entwürfe werden, da damit örtliche Sicherungen gezielt dort verstärkt werden können, wo sie tendenzmäßig am meisten zur Standsicherheit beitragen.

3.5.3 Verbruchwahrscheinlichkeit. Bei der Anwendung der gewohnten Grenzgleichgewichtsverfahren steht der berechnende Ingenieur dem wohlbekannten Dilemma gegenüber, einen angemessenen Sicherheitswert als Entwurfskriterium vorzugeben; es sei denn, diese Entscheidung ist ihm vorenthalten durch Vorschriften, in denen für verschiedene Lastfälle entsprechende Werte vorgegeben sind. Und es steht immer etwas in Frage, ob ein vorgegebener Wert für einen vorliegenden Fall auch wirklich vernünftig ist. Wenn der Entwurfsingenieur die Wahl hat, muß er sich fragen, was ausreichend ist, was sicher ist, ein Sicherheitswert von 1,5 oder 1,2 oder ein Wert dazwischen. Seine entsprechende Entscheidung wird unmittelbar die Kosten der Maßnahmen beeinflussen, die erforderlich werden, um natürliche Verhältnisse (die gegebenenfalls mit einem Sicherheitswert von 1,0 bewertet werden) auf einen "sicheren" Sicherheitswert zu bringen. Die Tatsache, daß ein angenommener Wert unmittelbar die Kosten von erforderlichen Sicherungsmaßnahmen diktiert, hat diesen Berichterstatter nie befriedigt.

Die meisten Praktiker werden damit übereinstimmen, daß der vielversprechendste Ausweg aus dieser Zwangslage es sein würde, sich endlich die Verbruchswahrscheinlichkeit als das Entwurfskriterium für den Böschungsentwurf zueigen zu machen. Diese Möglichkeit wird seit Jahren diskutiert, wobei recht vielversprechende Verfahren vorgestellt wurden, ohne jedoch auf breiter Basis angenommen zu werden, am allerwenigsten von der Gruppe der Bauingenieure. Der Gedanke, daß ein Verbruch möglich, wenn auch unwahrscheinlich ist, wirkt offensichtlich für die meisten Bauingenieure zu abschreckend. Nachdem es die Bergbauingenieure in vielen Fällen gewohnt sind, mit dem drohenden Verbruch zu leben, scheint diese Konzeption eben dieser Berufsgruppe viel eher zu liegen.

Die Erfahrung des Verfassers aus eigenen entsprechenden Forschungsbemühungen anhand der Gleitkeil-Aufgabe und deren Anwendung im Ingenieurbereich, bei Annahme von statistischer Verteilung von sowohl Gefüge- als auch Festigkeitswerten, zeigte jedoch, daß das ganze entsprechende Verfahren zwar übersichtlich und grundsätzlich überzeugend ist, jedoch dazu tendiert, bei wirklichen Anwendungen ziemlich vage Ergebnisse zu liefern. Der erste Teil der Ergebnisse bezieht sich dabei auf die Wahrscheinlichkeit des Auftretens der einen Gleitbruch ermöglichenden kinematischen Bedingungen. Wenn diese besteht, dann wird abgeschätzt, ob angesichts der im angesprochenen Bereich wahrscheinlich zu erwartenden Scherfestigkeiten dieser an sich mögliche Verbruch auch wirklich eintreten würde. Die dabei erzielten Ergebnisse waren noch nicht brauchbar für routinemäßige Entwurfsaufgaben und damit zusammenhängende Genehmigungsverfahren.

3.5.4 Rückrechnungen. Rückrechnungen sind sehr nützlich, um realistische Scher- (oder andere) Kennwerte aus einem beobachteten Verbruch abzuleiten, der unter bekannten und quantitativ dokumentierten Bedingungen stattgefunden hat. In den meisten Fällen, die sich für die Rückrechnung anbieten, sind die Verhältnisse beim Verbruch allerdings nicht sehr zuverlässig feststellbar.

Für vorgeschichtliche (oder "uralte") Rutschungen, wie sie vom Geologen verzeichnet werden, wobei der Einfluß der Umweltbedingungen während des ursprünglichen Verbruches nur sehr selten bestimmt werden kann, muß man mit folgenden Möglichkeiten rechnen.

- Sich positiv verhaltende natürliche Böschungen können sehr wohl einen Sicherheitswert von über 1,0 besitzen, wobei die auf der Rückrechnung unter Annahme eines heute bestehenden Grenzgleichgewichtes ermittelten Kennwerte recht konservativ sein können, möglicherweise unangemessen konservativ.

- Künstliche Böschungen, die sich heute auch positiv verhalten, können zu optimistischen Beurteilungen und Folgerungen für heutige Entwürfe führen, da die früheren Bauweisen nicht mit modernen Bauverfahren vergleichbar sind. Es ist einfach einzusehen, daß Böschungen, die heute mittels unkontrollierten Sprengens und/oder Reißens erstellt werden sollen, einen flacheren Neigungswinkel haben müssen als solche, die vor 100 Jahren in vergleichbaren geologischen Verhältnissen, jedoch mit handwerklichen, ausgesprochen gebirgsschonenden Aushubverfahren erstellt wurden. Es muß gefolgert werden, daß bei so unterschiedlichen Bedingungen Rückrechnungen nicht sehr produktiv sein können.

3.6 Entwurf von Böschungen und Gründungen, mit entsprechenden Sicherungen

3.6.1 Allgemein. In den vorgelegten Arbeiten und auch in dieser Diskussion wird eine Vielfalt von "Bauwerken im Fels" behandelt, wobei Felsböschungen und Gründungen im Fels die beiden Extreme darstellen, und die Gründung einer in einem engen Tal zu erstellenden Bogenstaumauer sehr oft Böschungsstandsicherheits- und Gründungsaufgaben kombiniert.

Der Entwurf hat sich auf die Ergebnisse und Folgerungen aus der Erkundungs-Berechnungs-Abfolge zu stützen, wobei zusätzliche Gesichtspunkte betrachtet werden müssen, wie Kostenwirksamkeit, Bauausführungsfragen und allgemeine Felsbauerfahrung, oft vereinigt mit der Intuition des dienstälteren Entwurfsingenieurs. Der tatsächliche Beitrag des Untersuchungsvorganges zur Lösung der vorliegenden Entwurfsfragen ist nur schwer feststellbar. In einigen der durchgesehenen Arbeiten erscheinen die Folgerungen für den Entwurf etwas losgelöst zu sein von den tatsächlich durchgeführten Untersuchungen.

Dieser Berichterstatter hat Enttäuschungen erlebt in Fällen, in denen Böschungs- und Gründungsentwürfe entwickelt wurden, die im Gegensatz zu den geotechnischen Bemühungen standen, die für diese die Grundlage hätten liefern sollen. In anderen Fällen war dagegen der Beitrag der vorhergehenden Erkundungs-Versuche-Berechnungen-Abfolge zum Entwurf überzeugend. Wie weiter oben ausgeführt erscheint es, daß die Kombination von Problemlösekraft und Kostenwirksamkeit von felsmechanischen Untersuchungen erhöhte Aufmerksamkeit der Felsmechanik-Gemeinde verdient, um die langfristige Daseinsberechtigung dieses Berufszweiges zu stärken.

3.6.2 Entwässerung. Den meisten Mitgliedern unserer Berufsgruppe wird die folgende Bemerkung völlig überflüssig vorkommen, da es wohlbekannt ist oder sein sollte, obwohl bei der tatsächlichen Berufsausübung manchmal andere Erfahrungen gemacht werden: Die systematische Entwässerung von Felsmassen stellt die wirksamste, zuverlässigste und auch preisgünstigste Felssicherungsmaßnahme dar. Sicherungsmaßnahmen mit anderen Mitteln, wie durch Felsbewehrung und Felsankerungen, ohne voll integrierte Entwässerung, wird als sehr unbefriedigender Beitrag zum Felsbau gesehen.

3.6.3 Böschungssicherung. Die folgenden Fragen sind für den Entwurf von Sicherungsmaßnahmen an Felsböschungen von beträchtlicher Bedeutung.

- Vorgesapnnte oder schlaffe Felssicherung: die Frage, ob eine Felsbewehrung zum Zeitpunkt des Einbaues vorgespannt (aktiv) oder schlaff (passiv) sein soll, wird in zunehmendem Umfang diskutiert, und nicht nur im Tunnelbau, sondern auch für Böschungssicherungen. Diese Diskussion spiegelt sich in den hier vorgelegten Arbeiten noch nicht wider.

Vor zwanzig Jahren wurde die aktive Sicherung durch vorgespannte Felsanker in den Vordergrund gestellt, was zu hochentwickelten Entwürfen für Felsanker verschiedenster Bauweisen führte. In jüngerer Zeit wird Felsbewehrung, die beim Einbau schlaff bleibt und erst durch Bewegungen der Felsmasse aktiviert wird, ernsthaft für einen weiteren Anwendungsbereich betrachtet, nicht zuletzt wegen überraschend guter Erfahrungen im modernen Tunnelbau. Zusätzlich ist einfacher Einbau, mit Einsparungen an Kosten und Zeit, sehr zugkräftig für Entwurfs- und Baupraxis. Parallel dazu behandelt die Diskussion Fragen wie die der Möglichkeit des Verlustes der Felsstützung im Laufe der Zeit infolge von Korrosion, mit der sich daraus ergebenden Notwendigkeit für langfristige Beobachtung, Unterhalt und Reparatur und/oder Ersatz.

- Die rechnerische Behandlung von Felsstützungen durch aktive (vorgespannte) oder passive (erst durch Verformung aktivierte) Elemente beruht ausschließlich auf der Grenzgleichgewichtsüberlegung, mit oder ohne Eingaben aus FEM Berechnungen. Im allgemeinen gibt es noch keine wirklich überzeugende Unterscheidung der rechnerischen Behandlung von aktiver und passiver Felssicherung. Kovaris beharrliche Diskussion der Definition des Sicherheitswertes für Felsböschungen trug jedoch zu einer Klärung dieser Frage beträchtlich bei. Allmählich entwickelt sich allgemeine Übereinstimmung, daß durch vorgespannte Stützung die aktiven Kräfte im Zähler des den Sicherheitswert definierenden Verhältnisses vermindert würden. In Fällen, in denen die passive Stützung erst allmählich durch Verformungen in der Felsmasse erzeugt wird, wird diese dazu tendieren, die passiven Kräfte im Zähler des Sicherheitsverhältnisses zu erhöhen.

- Die tatsächliche Beanspruchung von auf der ganzen Länge verpreßter Felsbewehrung, über reinen Zug hinaus, wird ebenfalls diskutiert, wobei zusätzliche Scherbeanspruchungen zur Kenntnis genommen werden, die zu einer wirksamen Verdübelung des geklüfteten Gebirges führt, die sich im Tunnelbau als besonders wirksam herausgestellt hat. Eine Verdübelung von einzelnen potentiellen Scherflächen mittels Bewehrungseisen normal zu diesen Flächen ist auf der anderen Seite jedoch als recht unwirksam bekannt.

3.6.4 Überwachung. Die Rolle der Überwachung, sowohl der Verformung der Felsmasse als auch der Beanspruchung der Felssicherungselemente, als vollintegriertem Teil des Entwurfes von Felsböschungen hat sich bei bergbaulichen und bautechnischen Anwendungen eindeutig durchgesetzt. Dies bestätigt die Tatsache, daß die Ergebnisse der höchstentwickelten rechnerischen Standsicherheitsnachweise doch nicht die Gewißheit vermitteln, die für den tatsächlichen Entwurf gefordert werden muß.

Obgleich immer wieder im einzelnen behandelt, muß man sich bei der Überwachung des Verhaltens von Felsböschungen immer noch mit folgenden Engpässen auseinandersetzen:

- Welche Verformungen sind für vorliegende Verhältnisse und einen gegebenen Entwurf tatsächlich zulässig, wobei die Verformungsgeschwindigkeit und deren Änderung nicht immer ausreichende Leitlinien für die Beobachtung liefern.

- Dauer der Beobachtung, während der Aushubarbeiten und, wenn darüber hinaus, um wieviel länger, wobei aufwendige Felssicherungen mit großer Wahrscheinlichkeit Beobachtung ihres Verhaltens über die volle Nutzungszeit der Bauwerke, und darüber hinaus, erfordern dürften.

- Reaktion auf die Ergebnisse der Überwachung, wobei Entscheidungen oft längere Zeit nach Abschluß der Baumaßnahmen erforderlich werden, mit oder ohne Verteilung der Haftung für das Verhalten des ganzen Entwurfes bzw. Bauwerkes oder nur der Felssicherungselemente.

3.7 Bauausführung

3.7.1 Vertragsunterlagen. Der Entwurf von allen Felsbauwerken und die Vertragsunterlagen für deren Erstellung sollten die geotechnischen Vorgaben wirklichkeitsgetreu widerspiegeln. Inwieweit das geschieht, entspricht der Stärke oder Schwäche des jeweiligen Beitrages. Die nachfolgenden, mit den Vertragsunterlagen zusammenhängenden Fragen erscheinen für erfolgreiche Bauausführung im Felsbau sehr wichtig.

- Anpassungsfähige Vertragsunterlagen: Nicht einmal die umfassendsten Untersuchungen können für völlig wirklichkeitsentsprechende und somit endgültige Modellvorstellungen und Kennwerte bürgen. Somit können "Änderungen der geologischen Verhältnisse" niemals ganz vermieden werden. Auf solche Änderungen muß durch Vorsorge im Bauvertrag und durch angemessene Anpassungsfähigkeit der Bauverfahren vorbereitet werden, damit es möglich bleibt, diese ohne Änderungen der ursprünglichen vertraglichen Abmachungen und hoffentlich ohne Rechtsstreitigkeiten zu bewältigen. Eine solche vernünftige Anpassungsfähigkeit bei der Bauausführung ist im Bergbau leichter zu erreichen als im Bauwesen. Unvernünftige Anpassungsfähigkeit kann allerdings nie erwartet werden. Die Anpassung eines vorliegenden Entwurfes an örtliche geologische Verhältnisse, wie sie während der Bauarbeiten erst aufgeschlossen werden, wird für die meisten Baubereiche als weitgehend unrealistische Illusion betrachtet.

- Einfache oder komplexe Vertragsunterlagen: Zu viele Felsbauvorhaben werden in der Hoffnung ausgeschrieben, daß möglichst einfache Vertragsunterlagen zu niedrigsten Angeboten führen, auch wenn diese keinesfalls realistische Kosten darstellen. Das erscheint nicht sehr vernünftig. Übermäßige Nachforderungen, die sich aus "geänderten Verhältnissen" ableiten lassen, die nicht durch vereinfachte Vertragsunterlagen abgedeckt waren, mit sich daraus ergebenden Bauverzögerungen und/oder Rechtsstreitigkeiten, führen sehr leicht zu Gesamtkosten, die weit über das hinaus-gehen, was mit einem wirklichkeitsgerechteren Vertrag hätte erreicht werden können, der für Änderungen während der Bauausführung Sorge getragen hätte, ohne Nachforderungen zu erzwingen. Selbstverständlich kann die Frage von möglichst vernünftigen und wirklichkeitsgerechten Vertragsunterlagen nicht losgelöst werden von der Frage der einseitigen vollen oder getrennten Teilverantwortungen. Dieser Berichterstatter hält daran fest, daß das geologische Risiko von größeren Felsarbeiten beim Bauherrn verbleiben muß. Wird dieses Risiko dem Unternehmer zugeschoben, führt das unmittelbar zu Nachforderungen, die darauf abzielen, die Kosten dieses Risikos zuverlässig abzudecken, ganz besonders in den Fällen, in denen ein Auftrag an Unterbieter vergeben wurde.

- Die Vertragsunterlagen sollten so abgefaßt werden, daß der Unterschied zwischen Gebirge vor und nach der Bauausführung, d. h. der negative Einfluß von Baumaßnahmen auf die Gebirgseigenschaften, möglichst klein gehalten wird, außer der Entwurf sieht diesen Unterschied wirklich vor. Die Überwachung der Bauverfahren, wie Sprengen und Reißen, obgleich von vielen Tiefbauunternehmen nicht gerade willkommen geheißen, wird nicht als unangemessene Einflußnahme auf die Bauausführung angesehen. Danach wird schließlich der Betoniervorgang sehr gründlich überwacht, um die letztendliche Betongüte sicherzustellen. Und bei Bauvorhaben im Fels sind Beton und Gebirge die wesentlichen Bestandteile des Gesamtbauwerkes.

3.7.2 Stabilisierungsmaßnahmen für hohe Einschnittböschungen. Es ist immer schwierig, den Ablauf von großräumigen Felsabträgen, auch wenn an sich angemessen überwacht, mit dem zeitraubenden Einbau von erforderlich erachteten Sicherungsmitteln abzustimmen. Technische Vorschriften, die den Einbau von Felsankern "sobald als möglich" verlangen, sind nicht wirklichkeitsgerecht. Rasch fortschreitender Aushub, bei zurückbleibendem Einbau der Felssicherung, wird häufig beobachtet, wenn die Bauleistungen nicht so streng wie im modernen Tunnelbau in vielen Teilen der Welt festgelegt sind. Das kann natürlich dazu führen, daß es sich herausstellt, daß die Sicherungen für die kurzfristige Standsicherheit nicht unbedingt erforderlich sind (aber trotzdem eine zusätzliche Sicherheitsspanne erzeugt hätten). Dieser Berichterstatter hat in seinen Unterlagen einige gut beschriebene Fälle, in denen weitgehend ungesichert gebliebene Felsböschungen sich auch unter kritischen Bedingungen dann doch recht befriedigend verhalten haben, obwohl erwartet worden war, daß sie früher oder später nach dem Felsabtrag versagen würden. In den meisten Fällen wäre es jedoch unmöglich gewesen, die so beobachtete offensichtliche Standsicherheit mit normalen Verfahren auch rechnerisch nachzuweisen. Dann muß auf die systematische Überwachung des Verhaltens dieser Böschungen zurückgegriffen werden, wobei man sich dann allerdings mit einigen der weiter oben behandelten Zwangslagen auseinandersetzen muß.

3.8 Schlußfolgerungen

Die vorliegende Diskussion des Standes der Technik in bezug auf Felsböschungen und Gründungen im Fels soll mit folgenden Empfehlungen abgeschlossen werden.

- Für jede felsbautechnische Untersuchung sollte die gesamte Systemabfolge eingehalten werden. Sowohl Lücken als auch Übermaße in der Bearbeitung sollten vermieden werden. Auslassungen mögen anfangs einige Mittel einsparen, können sich aber letzthin als recht kostspielige Unterlassungen herausstellen, während über die Aufgabe hinausgehende Bearbeitungen im allgemeinen nicht produktiv sind.

- Fortlaufende Kontrolle über den gesamten Untersuchungsablauf ist wesentlich, mit zwischengeschalteten Überprüfungen der Teilergebnisse, wobei die Geländearbeiten und die Berechnungen die für den endgültigen Entwurf bedeutendsten Ergebnisse liefern.

- Jeder Entwurf bedarf der Grundlage einer wirklichkeitsgetreuen Idealisierung (des Gedankenmodelles) der natürlichen Gegebenheiten, die sich aus der geologischen Erkundung im Gelände und der geologischen Analyse ergibt, wobei Ergebnisse durch ingenieurmäßige Berechnungen bestätigt werden müssen.

- Ausführungstechnische Gesichtspunkte müssen im Entwurf berücksichtigt werden. Entwurf und Vertragsunterlagen sollten felsbaumechanische Überlegung berücksichtigen. Einfache Entwürfe sind im allgemeinen komplizierten Entwürfen vorzuziehen. Sowohl Entwurf als auch Vertragsunterlagen sollten im angemessenen Rahmen anpassungsfähig sein.

4 EMPFEHLUNGEN FÜR DIE DISKUSSION

4.1 Allgemein

Im vorhergehenden Teil dieses Berichtes wurde vielleicht eine zu große Vielzahl von Themen abgehandelt oder auch nur angesprochen, die nach Meinung dieses Berichterstatters von wirklicher Bedeutung bei der ingenieurmäßigen Bearbeitung sind. Im folgenden wird ein Katalog von Fragen der Felsbaumechanik vorgestellt, die sich für eine allgemeine Diskussion eignen. Es wird aber auch anerkannt, daß viele andere der angesprochenen Fragen nur Fall für Fall, Anwendung für Anwendung, behandelt werden können.

In der Einführung wurde festgestellt, daß sich dieser Bericht auf ingenieurmäßige Überlegungen zum Leitthema dieser Sitzung beschränken sollte. Im Laufe der vorhergehenden Diskussion wurde es jedoch offensichtlich, daß es keine eindeutige Grenze gibt zwischen theoretischen (analytischen) und ingenieurtechnischen Seiten des vorliegenden Themenbereiches. Diese Erkenntnis spiegelt sich in der folgenden Aufzählung von möglichen Diskussionsthemen wider.

Nachfolgend werden vier Diskussionsbereiche vorgestellt, von denen einer oder möglicherweise zwei während dieser Tagung behandelt werden könnten. Es steht zu hoffen, daß Diskussionen bei anderen Gelegenheiten sich mit Fragen befassen, die hier vorgeschlagen sind, aber nicht beim vorliegenden Anlaß behandelt werden konnten.

4.2 Wirkungsgrad der Felsmechanik

Um die Wirksamkeit der Fels(bau)mechanik in zukünftigen ingenieurmäßigen Anwendungen weiter zu verbessern, ist eine fortlaufende, interne Kritik der heute üblichen Anwendungen erforderlich. Nach Meinung dieses Berichterstatters sollten die folgenden Themen- und Fragegruppen in einer grundsätzlichen Diskussion behandelt werden.

- Beitrag zur Lösung von Felsmechanikaufgaben des Ingenieurbereichs, mit
    - Bestimmung von starken und besonders schwachen Stellen im Gesamtsystem für bestimmte Entwurfsaufgaben, im vorliegenden Fall für übertägige Felsarbeiten und Talsperrengründungen. Was kann oder was sollte zur Verbesserung getan werden? Und
    - Übertragung der Ergebnisse von Berechnungen und Rückschlüssen daraus in wirkliche Entwurfslösungen. Kann man mit den heutigen wirklichen Möglichkeiten zufrieden sein?
- Kosten des felsmechanischen Beitrages, im Vergleich mit den Kosten von Entwurf und Bauausführung, unter besonderer Betrachtung der
    - geologischen und felsmechanischen Kartierung und Klassifizierung,
    - Bohrarbeiten, einschließlich der Protokollierung, Probenentnahme und in-situ Versuchen in Bohrungen,
    - Versuchswesen, sowohl in-situ als auch an Bohrkernen und/oder sonstigen Proben im Labor und zu
    - Berechnungen, möglicherweise in bezug auf typische Standsicherheitsnachweise für Böschungen in geklüftetem Gebirge.
- Kosten-Wirkungsgrad. Dieser ist natürlich sehr schwer zu quantifizieren, obgleich dies fairerweise für alle Beiträge im Ingenieurbereich erwartet werden muß. Sollte die manchmal besonders von erfahrenen Tiefbaufachleuten vertretene Vorstellung auch nur einen Schein der Berechtigung haben, nämlich daß die Beiträge der Felsmechanik sich oft nicht auf die wirklich anstehenden kritischen Fragen bezögen, wobei auch noch begrenztes Verständnis für Kosten und Zeitplan aufgebracht wird, dann müßte diesem entschlossen entgegengetreten werden. Felsmechanikfachleute müssen darauf vorbereitet sein, für den Kosten- und Zeitaufwand ihrer Beiträge Rechenschaft abzulegen, so oder so, falls sie an einer fortdauernden Mitarbeit im Felsbau interessiert sind.

  Dieser Berichterstatter glaubt, daß der Kosten-Wirkungsgrad der nachstehenden Beiträge und Teilbeiträge unter die Lupe genommen werden sollte. Intuitiv erscheint es, daß weder untertriebener noch übertriebener Aufwand für eine Teilleistung fruchtbar ist, wobei der besonders wirksame Bereich eines Beitrages nicht allgemein festgelegt werden kann:

    - Beitrag der Felsmechanik über herkömmliche ingenieurgeologische Erkundungen hinaus, wobei die immer wieder gestellte Schlüsselfrage die nach dem Aufwand ist, der erforderlich wird, wenn mehr als Beschreibungen der Regionalen Geologie, der Hydrogeologie und der Seismizität, mit entsprechender Dokumentation, erforderlich erachtet werden.
    - Innerhalb einer vollständigen Überprüfung sollte auch die Kostenwirksamkeit der üblichen geologischen Erkundung beleuchtet werden, um zu angemessenen Kostenverhältnissen zwischen ingenieurgeologischen und felsmechanischen Teilbeiträgen einer gesamten geotechnischen Untersuchung zu gelangen.
    - In-situ Versuche an geologischen Trennflächen, die als mögliche Verbruchflächen und -zonen erkannt sind, stellen immer einen möglichen felsmechanischen Beitrag dar, sowohl beim Entwurf von Felsböschungen als auch besonders bedeutungsvoll beim Entwurf von Talsperrengründungen. Wer immer selbst in-situ Versuche durchgeführt hat, weiß sehr genau, daß eine Verminderung des Aufwandes für solche Versuche unter eine vernünftige Grenze nicht sehr produktiv ist, da sie bestenfalls zu vagen Schätzungen von Entwurfskennwerten führt. Zu aufwendige Versuche, möglicherweise ohne Abstimmung auf vorliegende Modellvorstellungen und erforderliche Entwurfsberechnungen sind ebenfalls wenig ertragreich. Aber welche Aufwendungen für in-situ Versuche sind vernünftig? Obgleich wohl schwierig zu beantworten, so sollte diese Frage doch einmal angesprochen werden.

4.3 Technik des wenig festen Gebirges

Tatsächliche und mögliche Beiträge durch die Felsmechanikfachwelt zu ingenieurtechnischen Fragen zu wenig festem Fels sollten behandelt werden. Das könnte, hoffentlich, zu einer Klärung der Kontaktstellen mit Bodenmechanik und Grundbau, Ingenieurgeologie und Geologie führen. Selbstverständlich werden Berührungsstellen als Gelegenheit zur produktiven Zusammenarbeit und nicht als Grenze zwischen Fachbereichen gesehen.

Für den Entwurf von Böschungen und Gründungen in wenig festem Gestein erscheinen folgende Fragen eine Diskussion zu rechtfertigen.

- Bautechnische Eigenschaften von wenig festem Gestein und der Einfluß von geologischen Vorgängen und geologischen Strukturen, wie Schichtung und Schieferung, wobei in gegenwärtige und zukünftige Eigenschaften unterschieden werden muß.
- Verbruchformen und -arten von Böschungen

in wenig festem Gebirge mit ausgeprägtem geologischen Gefüge, wobei die geringe Gesteinsfestigkeit im Vergleich zu üblichen Standsicherheitsuntersuchungen für Felsböschungen zusätzlich berücksichtigt werden muß. Oder um die Schlüsselfrage unmittelbarer zu stellen, wann ist die Annahme eines Gleitkreises in wenig festem geklüfteten Gebirge noch zulässig, oder wo liegt die Grenze zwischen Boden- und Felsmechanik in bezug auf Standsicherheitsnachweise für Böschungen?

4.4 Sicherungen von Felsböschungen

Dieser Rezensent ist der Meinung, daß die nachfolgenden drei unterschiedlichen Möglichkeiten, die den Entwurf von Böschungssicherungen fast immer beeinflussen, in engem gegenseitigen Zusammenhang diskutiert werden sollten:

- Verwendung von vorgespannter (aktiver) oder schlaffer (passiver) Felsbewehrung im weitesten Sinne.
- Anwendung im festen oder wenig festen Gebirge oder in Kombination von beiden.
- Für zeitweilige oder (langzeitige) Sicherung.

Die folgenden Fragen sollten im einzelnen behandelt werden:

- Möglichkeiten der Berechnung,
- mögliche Entwurfsvarianten,
- Einfluß des Zeitfaktors beim Einbau der Felssicherung,
- Überwachung des Verhaltens einer Felsböschung, mit allen sich daraus ergebenden Fragen, und abschließend
- Vor- und Nachteile von technisch und wirtschaftlich möglichen Kombinationen und Entwurfskompromissen, unter Berücksichtigung der jeweiligen Unsicherheiten und Risiken.

4.5 Felsmechanik und Felsbau

Es wurde darauf hingewiesen, daß ein wirkungsvollerer Austausch zwischen Felsmechanik und dem tatsächlichen Bauen im Fels wünschenswert wäre, um den Wirkungsgrad der Felsmechanik in der Tiefbaupraxis zu erhöhen, was durch nachweisbar bessere Entwürfe für Felsbauwerke und Verringerung der entsprechenden Kosten erreicht werden könnte. Es erscheint deutlich, daß die Beiträge in dieser Richtung verstärkt werden sollten. Zwei Schlüsselfragen erscheinen dabei besonders wichtig, über den Beitrag der Felsmechanik zum eigentlichen Entwurf hinaus.

- Beitrag der Felsmechanik zu den technischen Vertragsunterlagen, mit technischer und wirtschaftlicher Bedeutung, wie
  - Arten der Felssicherung, die für anwendbar und wirksam erachtet werden,
  - vertragliche Absicherung einer wirksamen Bauüberwachung, z. B. von Aushubverfahren im Fels und
  - technische Vorschriften für die Ausführung von typischen felsbaulichen Entwürfen, wie Felssicherungen.
- Rolle der Felsmechanik auf Felsbaustellen:
  - tatsächliche Kontrolle und Beobachtung von Bauverfahren,
  - Überwachung des Verhaltens von ausgeführten Felsbauwerken und,
  - zusätzlich und besonders wichtig, Durchführung von fortlaufenden Untersuchungen während der Bauausführung, selbstverständlich in enger Zusammenarbeit mit Geologen.

DISCUSSIONS / DISCUSSIONS / DISKUSSIONEN

STABILITY OF ROCK SLOPES

Stabilité des pentes rocheuses

Standsicherheit von Felsabhängen

R. Oliveira
Chairman

GENERAL REPORTS: Presented by Professor R. Goodman and Professor K. John

PRESENTED BY: Professor Richard E Goodman, Berkeley, U.S.A.

DISCUSSION

Question: Dr Chappell, Australia

Mechanisms in blocky material cause load redistributions and formation of hinge mechanisms. Does the lower bound take this into account? What is a keystone can become a hinge very readily due to slip of joints, development of cracks or failure of material. This can account for progressive soft or hard rock mass failures.

Answer:

The block theory is a lower bound theory. At the stability analysis level we can discuss stresses and these are transformed to forces on the faces of the key block, (tractions in other words). There is no attempt at this point to discuss the stresses within the blocks. We are now working on a 3 dimensional discrete portion of the analysis, and at that point we would be able to do in 3 dimensions perhaps what you are discussing. There are existing analyses for 2 dimensions that do this including your own photoelastic analysis, discrete element analysis of Cundall and Voegele, Burman and Burn and others and of course finite element analysis and now boundary element analysis, but we are looking at an easier solution, a lower bound limit solution based first on geometry, second on statistics.

Question: Prof K Zhu, China

Prof. Goodman and Prof. John have made a splendid general report, and summarized all the papers presented in this Congress. I would like to know how would you differentiate near surface excavations from the deep mining. The keyblock theory is said to be restricted at present only to hard rocks. It seems to me the important thing is that the relative stiffnesses and strength of the block and the interfaces that are important. So I would like to know how could you define hard rocks from soft or weak rocks? The final point is that both the in-situ stresses or more properly I should say the rebound deformation after excavation and also the seepage pressure are important items to be dealt with if support is to be provided. So would that be proper to say stress and deformation data are not essential to the performance of a key block analysis in such cases?

Answer: Dr. Goodman

Thank you Dr Zhu for your questions. The first about near-surface and deep underground: the method of Hoek and Bray has been used for slopes and extended by Hoek and Brown to underground without perhaps consideration of the forces of stresses, the initial stresses and the concentrated stresses around an underground excavation as they affect rigid block motions.

This is a difficult problem and is presently being studied by a number of people, but I think that is a key to the answer to your question. In the underground at depth the stress flow around the excavation reinforces some blocks and forces others to move and this cannot be neglected, whereas in a near surface excavation at least it is a different statement of the problem. The stresses may be important or they may not be important depending on the way the work is being carried out. Certainly a surface excavation has less importance of the flow of stresses than an underground excavation because there isn't a complete path for stress to envelope the excavation. The second question concerned hard versus soft rock. When we made this theory we took the very easiest set of assumptions that there was no new rock cracking, no new fractures. If the rock is weak one can expect new fractures and then that theory is not complete. It could have on top of it some analysis of new fractures but until you do that you haven't addressed weak rock. Therefore the hard versus soft rock is a question of some kind of dimensionless strength or dimensionless stress to strength ratio. Very soft rock could still be considered hard and we saw that in the picture from Bougainville, where in terribly broken and decomposed rocks still there was an influence of faulting and a wedge moved. The third question I think I answered previously in response to Dr. Chappell, about water. The water is taken as a force in the stability analysis it is very important but less important in the initial lower bound analysis of geometry.

GENERAL REPORTS: Presented by Professor R. Goodman and Professor K. John

PRESENTED BY: Professor Klaus W. John, West Germany

DISCUSSION

Question: Professor Hans Grob, Switzerland

I should like to come back to the question of active or passive rock anchors which Mr John was raising. He was actually stating in his report that he felt that more passive anchors should be proven or installed in slopes as they do very well in tunnels, but I feel that these are two different questions. Whenever we have a failure in slopes the movement is actually destroying the cohesion, and sliding friction is less than stable friction so we should prevent the movement and I mean we should pre-stress the rock anchors in slope. In tunnelling the task of the bolts is different. They just should keep in shape the vault which is formed by the rock above the opening and whenever a movement occurs there is some sort of squeezing in and that adds to the stability. So there in that case unstressed bolts will do.

Answer: Prof. John

I agree 100% with your considerations. On the other hand I see a trend that people shy away from the very complicated technology of high capacity pre-stressed anchors and rather go for rock reinforcement. I know of some rather extensive installations in Vancouver by B.C. Power and they felt there was not time enough to go up there, drill, install tendons, wait till they set, then come back pre-stress them, and they just went ahead and installed unpre-stressed rebars and they worked well. The slopes stood up and I'm really torn. I see the theoretical arguements that we should really have active anchors, but on the other hand I'm also aware of problems with active anchors, and see many people just rather do non pre-stress anchor than nothing or avoiding in all too complicated procedures.

GENERAL REPORTERS: Prof. R. Goodman and Prof. K. John

PAPER: Rock slope engineering in Hong Kong

AUTHORS: E. Brand, S. Hencher and D. Youdan

PRESENTED BY: S. Hencher

DISCUSSION

Question: Dr. R. Benson, Canada

This question of corrosion of anchors, either active or passive is one that concerns me considerably for we know very little really about the performance of anchors in-situ, either single or in company with others, and I note you say that someone in Hong Kong is beginning a guide on the question of corrosion of anchors. Of course I put this to the general audience, but will this guide be available soon, and does it deal with the long term performance and the corrosion of anchors at depth?

Answer: S. Hencher

The guide has been under preparation now for about 2 years and it's certainly in the final stages. It's been very heavily researched on the basis of all existing codes, Swiss code and the draft British code especially I think, but it is concentrating on corrosion aspects rather than the design of anchors to hold up slopes. I don't know very many details about it, it will be available probably over the next few months and will be on sale in Hong Kong.

GENERAL REPORT: Presented by Prof. R. Goodman and Prof. K. John.

PAPER: Pit slope design methods; Bougainville Copper Limited Open Cut

AUTHORS: J.R.L. Read and G.N. Lye

PRESENTED BY: J.R.L. Read

DISCUSSION

Question: Dr. McMahon, Australia

As the consultant involved in the probabilistic aspects of the Bougainville work I think I would like to make one point, and that is that people who know very little about statistics and probability will tell you that safety factors are equivalent to probabilistic criteria and that the two methods will give you essentially the same thing. Well the very very best of deterministic work was done at Bougainville. It came up with random joint orientations because without statistics you can't prove anything else once you get more than a few. It was unable to look at the joint length characteristics because without statistics you cannot study the probability distributions of joint lengths, and it gave slope angles between haul road segments of around 38 degrees. Now by looking at the same data in a different way it was possible to take into account the orientations of the fractures, the fact that most of their lengths were small and we found an increase in slope angle of 9 degrees to 47 degrees between haul roads, and as Mr. Read has said he doesn't think he has hit the optimum yet, and as he pointed out this has increased the ore reserves of the mine by 30% and as a value in excess of several hundred million dollars.

Comment: John Trudinger, Australia

I thought that John Read's paper highlights an interesting problem and I've seen it at other open pit mines. When a design has been arrived at, based on sound data hopefully, quite often it is very difficult to achieve in practice because of the shovel system of excavating the faces. That's something that is not always foreseen in the early stages of design when maybe the equipment that is ultimately going to be used hasn't even been decided on.

Comment: John Read:

If I can just comment on what John Trudinger said. That is very true and I think it should be recognized by the Mining Industry that we only use shovels and trucks because that's the way it has always been done, and that planners in the future should perhaps look to other methods of excavation when they are considering their slope design methods. We don't necessarily have to be bound to shovel excavation if there are better methods available to us, and we should not look to the past, we should be looking forward.

Answer: John Trudinger, Australia

This relates of course to the final slopes. Obviously the mining engineers will still want to use shovels for the bulk of the ore excavation.

GENERAL REPORT: Prof. R. Goodman and Prof. K. John

PAPER: Open cut slope design using probabilistic methods

AUTHORS: P. Morriss and H.J. Stoter

PRESENTED BY: P. Morriss

DISCUSSION

Question: Dr. McMahon, Australia

I would like to ask Mr. Morriss if he thinks it would have been possible to have designed that slope on a deterministic basis and what answer would he have got and how it would have compared to the one he has?

Answer: Mr. P. Morriss

I think that because we actually used deterministic analysis on a particular analysis. What was important was that we endeavoured to model all the measured variability and therefore end up with a distribution of safety factors from which we could derive a probability of failure and from that point use an economic analysis. I think the major limitation with using a purely deterministic approach is that you have not got this ability of using an economic approach to the slope design and you also have no idea of how conservative your resulting safety factor is, in other words is 1.2 the magic number or is 1.1 adequate and you can't determine those answers without understanding or measuring the variability in the parameters that form the safety factor.

Question: P. A. Gray, Australia

In your paper you mentioned you conducted 240 shear box tests using the Hoek shear box. Were these tests performed on samples from borecore or lumps? It is our experience that Hoek shear box tests are very variable since there is no control on strain rate and little control on shear direction. You also mention that shear tests were performed on each surface type. Your Fig. 9 shows over 100 data points for such tests on one surface type which would indicate a small number of different surface types tested. Is the geology such that only a small number of surface types exist or did you selectively choose a small number of surface types to test? If the latter, how does this affect your probability calculations?

Finally the slide you showed in your presentation indicated a failure which you said represented the 10% probability of failure. Did you back analyse this failure and if so would these results then influence your stability analyses and overall optimal slope angle?

Answer: Mr. P. Morris

The Hoek shear box tests were carried out entirely on borehole core, mainly of HQ size. The shearing direction was chosen by inspection of the core to endeavour to minimise shearing resistance (e.g. parallel to any striations). Because the rock was weak, vertical loads were chosen to avoid crushing on the contact surface if possible. The pressure cell was then used to maintain vertical stress levels during the vertical dilation which occurred with the rougher surfaces. Whilst I agree that there is no control on strain rate, we did not get much uncontrolled sliding occurring, and usually observed a residual type of stress-strain curve which was reproducible.

Regarding the number of surface types, we allocated each surface a coating and roughness. The three main coating types were:

B - clear
A - clay coated
F - Goethite/Iron coated

and the roughness was designated 1 - 5 on the core scale, as per Piteau (see references). Whilst there was a scatter of results throughout the surface types, the B2, B3, A3 and A4 surfaces provided a high percentage of the total, and these were the ones used by Lilly to do the regressions on. For the lesser surfaces the statistics relating to their group were used for sampling in analyses. The frequency of occurrence in the ground of each surface type was analysed from the drill core data sheets, representing a sample of over 15,000 surfaces. No statistical bias by location in the pit walls was observed.

FOUNDATIONS ON AND IN ROCK, INCLUDING DAM FOUNDATIONS

Fondations dans et sur les rochers, y compris les fondations de barrage

Fundamente auf und im Fels, einschliesslich Stauwerkfundamente

O. Moretto
Chairman

PAPER: Pre-failure dilatancy and the stress distribution in a closely jointed rock mass

AUTHORS: M. J. Pender, C. J. Graham and W. J. Gray

PRESENTED BY: M. J. Pender

DISCUSSION

Question: Prof. Tan Tjong Kie, China

Dilatancy is a very important phenomenon which works frequently of course in underground openings, slopes and even prior to earthquakes. We know now that earthquakes are preluded by dilatancy effects which may take place in an area of 10,000 square kilometres. At the moment only a few papers discuss such an important problem and therefore I wish to congratulate Dr Pender with his paper. And now may I ask some questions. Have you analysed your test results with equation No. 10, where you take the dilatancy D as a constant, and you have some solutions with the FEM. It seems to me that D cannot be a constant, as you have used for your computations. In my opinion dilatancy is due to micro-fissuring and micro-fissuring is always non linear, but I think you may not introduce further difficulties if you take the dilatancy as a function of the current state of stress and not only as a constant. (The incremental stress-strain relationships are usually non-linear and dilatancy is suppressed by increasing hydrostatic stress).

Answer: Dr. Pender

When I first got interested in this phenomenon of dilatancy I looked through the literature and I agree with your statement that it does seem to be a phenomenon that has not been thoroughly investigated, or certainly thoroughly presented in the rock mechanics literature. So the idea we had in setting out on these analyses was to try and reduce the phenomenon to the simplest possible terms, so that we might be able to reach some conclusions and maybe make some statement to the rock mechanics community illustrating the significance of the phenomenon. So hence our starting point of the idealisation that it is a simple linear dilatancy. But I quite agree with your comment about non-linearity, that the stress strain curves that I show have a fairly definite tendency for the dilatant volume change to be non-linear and furthermore there is another element of non-linearity that needs to be considered and that is the effect that with increasing normal stress there is a tendency to suppress the dilatancy.

It seemed to me that if we tried as a first step though to incorporate all of those non-linear effects we had a fair chance of getting rather lost in the analysis, so I regard it as a worthwhile first step to reduce it to these very simple linear terms and explore the consequences of that linear behaviour as a prelude to some later analysis, where we would attempt a more formal modelling of the non-linear behaviour.

Question: Dr. Chappell, Australia

Dilatancy in blocky material is of two types, namely, that from joint roughness and that from geometrical block movements. What dilatancy is being measured? That is joint dilatancy or block dilatancy? Mechanisms come into block dilatancy more than they come into joint dilatancy. What mechanisms did you observe?

Geometric dilatancy mechanisms readily induce tension. If you keep the geometric dilatancy down you create an interlock whereas for the joint dilatancy you have a strength enhancement all dependent on the constraint conditions. In your diagrams or photographs you had a system of blocky material. Did you consider these two types of dilatancy?

Answer: Dr. Pender

I can't say that I have. The 2 slides I showed were really just illustrations of the type of rock conditions that we have in New Zealand and once again a further step from the analysis we have done would be to do a modelling of those situations taking account of the physical size of the blocks and the characteristics of the joints. In the second example I showed of the jointed ignimbrite that would certainly be a reasonably attractive analysis because the shapes of the blocks over quite a large area of that ignimbrite were fairly well controlled into those rectangular block shapes I showed. I would still regard that as a further step in the analysis, reiterating my earlier point that what we wanted to do was to do the simplest analysis one could imagine and if one does that it's hard to beat taking the material as a homogeneous isotropic continuum.

Question: From Chairman, Prof. O. Moretto

I would like simply to mention the question that this phenomenon of dilatancy has been known to the reinforced concrete engineering industry since about the beginning of the century. As a matter of fact a well known French engineer by the name of Concider invented the spiral reinforced columns. The spiral reinforced columns are based exclusively on the phenomenon of dilatancy, so I would suggest to those people who are working on this problem that they go also to the reinforced concrete field and probably they will find quite a number of investigations made about behaviour of concrete and the triaxial strength of such discontinuous or non-homogeneous systems.

PAPER: Dynamic stability of concrete dam foundations

AUTHORS: G.A. Scott and K. J. Dreher

PRESENTED BY: G.A. Scott

DISCUSSION

Question: Dr. F. H. Cornet, France

In your analysis, an important input datum is the accelerogram. How did you obtain yours - are they synthetic or natural ones? What are the normalisation parameters: acceleration, velocity or displacement? How did you choose those input data?

Answer: Dr. Scott

The accelerograms are synthetic. The procedure we have adopted for developing them involves first establishing a site response spectrum based on the magnitude and distance of the design earthquake. This is done by statistically evaluating spectra from historical earthquakes, and normalizing them by a parameter called spectrum intensity, which is the area under the velocity response spectrum curve between periods of 0.1 and 2.5 seconds. We feel that spectrum intensity is a better indication of structural response in terms of the energy in an earthquake than simple parameters such as peak acceleration. Three independent accelerograms are then generated to closely match the site response spectrum.

Question: Dr. B. McMahon, Australia

For a long time now I have been very puzzled by the fact that although there are many examples of natural rock slopes that have failed during earthquakes there has been no example of an open pit mine slope that has failed during an earthquake. We have seen cases where slopes have failed later so we know they were very close to critical equilibrium at the time of quite large earthquakes, say 7 on the Richter scale at a distance of about 200 kilometres. We have come to the theory that this is because a rock slope takes a very long time to fail and that an earthquake only acts for a short duration, so when a natural slope fails it is because it is the end result of maybe thousands of earthquakes whereas when an open pit slope is subject to an earthquake it may be the first one in its life, and therefore it doesn't have time to dilate and fail.
Now this is very important to us, because if this is so, then we don't design for the worst earthquake that could happen in the life of the pit, but for the worst earthquake that is going to happen many times in the life of the pit. Would you like to comment on this?

Answer: Dr. Scott

I think your point is well taken. The demand for public safety in the United States has forced us into very conservative design earthquake parameters for public works, and I think this will continue to be the case. Certainly one should also consider the possibility of smaller earthquakes that could occur many times during the life of a structure. An open pit slope may also present a different sort of design problem.

Question: Dr. S. R. Hencher, Hong Kong

I would like to make a comment which I think is relevant to Dr. McMahon's statement. I carried out some research on the dynamic shear strength of rock joints in 1973 to 1976 and one of the conclusions I came to was that the shear strength under dynamic loading was in fact higher than under static loading. So in fact it needed a higher pulse to get the thing moving than you would calculate from static experiments.

Once the thing was moving then in fact the shear strength was lower than you would estimate from static experiments. So I think this observation that slopes don't fail during earthquakes and also don't fail during blasts is related to this increased shear strength because the stresses just don't have time to effect the joint. Another is, that Sarma in 1975 wrote a paper on slope stability and there he considered various accelograms and he worked out the peak displacements due to all of these various accelograms and he came down to the conclusion that instead of using a very complex analysis method, you could just put in a half sine pulse or a triangular pulse or rectangular pulse to simulate all sorts of earthquakes.

Answer: Dr. Scott

I am aware of your research and it is quite interesting. The response of interstitial water pressure is also of great importance, and we hope to address this to some extent in papers to be presented at the 24th U.S. Symposium on Rock Mechanics and the 8th World Conference on Earthquake Engineering. Certainly research in this area is just getting off the ground and I hope that it continues.

Question: By Co-Chairman, Professor Poulos

I would like to inject a question of my own at this stage. Mr. Scott has done a fairly sophisticated sort of analysis, a three dimensional dynamic analysis, of his dam, and yet he's perhaps ignored the effect of the foundation beneath and adjacent to the dam. I wonder whether he might like to comment on what influence the foundation compliance might have on the forces that he would compute and consequently the stability assessment for his dam.

Answer: Dr. Scott

A foundation was modelled under the dam in the dynamic finite element analysis from which the forces were computed. The deformation properties of the foundation must certainly be considered when modelling this part. Therefore, to a certain extent the effects of foundation interaction are actually included in the calculation of the forces.

Question: Prof. A. Bello, Mexico

For foundation in and on rock; do we have to care about concepts like bearing capacity and settlement analysis or not? These concepts were missed in this session on foundations.

Answer Dr. Scott

The strength and deformation properties of a hard rock mass are to a large extent controlled by the discontinuities present. Certainly the deformation and strength of a foundation should be assessed, and from this point of view perhaps settlement and bearing capacity take on a different meaning. The type of structure should also be considered. For example, a concrete arch dam abutment may require more assessment than a building foundation. The point at which a soft rock should be analyzed as a soil is not well defined, and should also depend on the situation.

Question: Dr. G. Truscott, Australia

I have recently been involved in the stability of a 60m high mass concrete dam structure on low angle shears under dynamic earthquake loading in the Phillippines. Due to time and financial limitations displacements under earthquake loading were calculated using the relationships developed by Makdisi and Seed, (ASCE Geot Eng) and Sarma (Geotechnique) which were developed from analysis of a large number of earthquake accelograms for stability determinations of earth and rockfill dams. Have the authors considered the use of this data and could they comment please?

Answer: Dr. Scott

We have not considered the use of these relationships for

concrete dams. We currently do not allow permanent deformations in the foundations of our concrete dams calculated from dynamic loading. The use of these simplified procedures requires some knowledge of the natural frequencies, damping, and other properties of the structures involved. I would be careful when extrapolating these parameters for concrete structures, which may behave quite differently than the earth structures for which the relationships were developed. In particular, the distribution of acceleration throughout the structure may be significantly different.

PAPER: Horizontal behaviour of pier foundation of a soft rock slope

AUTHOR: H. Maeda

PRESENTED BY: H. Maeda

DISCUSSION

Question: Prof. Romana, Spain

You have explained and analysed what happened to pier B (ground angle 30 degrees), but you have tested two piers, pier A and pier B. Could you tell me what happened to pier A. (ground angle 20 degrees)?

Answer: Dr. Maeda

Through the load test, the displacement of pier A was small compared with the displacement of pier B.

As a result, the failure of surrounding ground of pier B occurred prior to the failure of surrounding ground of pier A.

Therefore, pier A was assumed to act like an anchored pier or reaction pier against pier B in this loading experiment.

At the final stage of loading, which load was 9.81 MN, the pierhead displacement of pier A was 35 mm and tensile cracks of surrounding ground of pier A were observed.

Question: Prof. A. Bello, Mexico.

To all presenting papers on Foundations On And In Rock.

In this session related to Foundations On And In Rock we have listened to very important and interesting aspects of the shear strength related to the jointing of rocks, even on dynamic efforts. I would like to ask any of the authors of the papers presented how was that information applied to mechanism of failure, or potential failure for the concepts we are used to having in soil mechanics: the bearing capacity and the settlement analysis of foundation. Do we have to be careful about this concept in rock or not?

Answer: Dr. Pender

One comment that was going through my mind when we were discussing the second paper about the foundation for the dam and that perhaps does take up the point that the discussor has raised relates to the concept of designing slopes under earthquakes for a performance criterion based on displacement rather than a factor of safety.

I think that idea was first proposed in relation to slopes in soil and in that situation for many soils we have a behaviour where the peak strength of the material can be mobilized and then after that is mobilized, with further displacement, there is a drop in the strength that is available. But quite often this is a relatively slow drop off in strength, and so one can argue that one could design a slope or an embankment so that there will be some displacement under an earthquake and one also has the feeling then that subsequent to the earthquake there is a chance for the soil to undergo a self healing process so it will consolidate, pore pressures will be dissipated etc. and some of its former strength will be gained. When one comes to thinking about a rock slope or an element underneath a foundation, a joint, one can similarly apply the same reasoning, but I'm not sure that in that case the argument is quite so convincing, because the peak strength of a joint surface is generated at a very small displacement and then if it is a rough joint surface the post peak decrease in strength occurs fairly rapidly, and so it seems that after that's occurred there is no further healing process that can take place. Once the joint has been displaced through its peak strength the strength that one has available is then the ultimate or residual strength of the surface itself.

So in designing slopes in rock masses considering this type of behaviour it seems to be that it is not quite so convincing that one can rely on that mechanism and maybe one has to think about it from the point of view that if the joint surfaces are rough and one wants to capitalize on the joint roughness angle one has to try and design the situation so that this peak strength is not dissipated and so one has to think about it in terms of preventing displacement rather than letting that displacement take place.

Question: Dr. Barry McMahon, Australia

I was concerned about the discussion about peak strength and residual strength because my observation is that in large rock slopes peak strength is never mobilized. It's something that we only see in the laboratory when we carry out these tests on small specimens, and it appears to be a very scale dependent quantity. We get good peak strengths in the lab because we have asperities which are very rough and of a scale about the size of our test, but in the field it appears that because the rock mass or the sliding block is probably only making contact in a very small percentage of its surface area it is tending to crush those very small asperities and that our shear strength in fact is equal to the fundamental or residual laboratory strength, plus the field angle of roughness which is something like 3-10% wavelength of the height of the slope or a roughness which has a very large wavelength, and that all the other small roughnesses are simply destroyed. Would you like to comment on that?

Answer: Dr. Pender

I think there is much that I agree with in that comment.

When one looks at the roughness of a joint surface quite often the measurements are made looking at an exposed half of a joint so we don't in fact know what the condition of the mateing surface was, and as you say in some situations there may be a joint surface that in a small scale looks to be very rough indeed, but when you examine it on the large scale the roughness isn't quite so great. But what one also has to consider is not only the roughness of the joint surface itself but how well the two halves of the joint fit together and your comment that in many situations the two halves don't fit particularly well at all so they are only in contact at a few asperities is a very valid one. However, going to my own experience in an environment that I am used to looking at I always come back to thinking about the slopes in the Wellington Greywacke. (One of the examples which I showed in my presentation earlier). In that material one expects a basic friction angle in the material to be typically in the middle 30s range that many people have reported for sandstone type materials, but yet it is very apparent from examining the behaviour of those slopes that the friction angle that one can rely on is in fact very much greater than that base friction angle, so the combined effect of the close jointing and the way in which all the blocks interlock gives one a rather larger available friction angle for the mass.

NEAR-SURFACE CONSTRUCTION ESPECIALLY IN CITIES

Construction à faible profondeur surtout dans les villes

Oberflächennahe Konstruktionen, insbesondere in Städten

S. Bjurstrom
Chairman

PAPER: Stability of a deep excavation in diagenetically consolidated mudstone

AUTHORS: R. Gruter and W. Wittke

PRESENTED BY: W. Wittke

DISCUSSION

Written Question: Prof. Tan, China

Side wall displacements and dislocations between rocklayers and bottom upheavals of pits are a general phenomena, which in rheology are known as due to elastic recovery. In the Ghe Zhou dam site these displacements exceeded 10cm within one month and proceeded logarithmic with time for many years.

Verbal Question: Prof. Tan, China

I am of the opinion that one of the most phenomenal and inherent features of rock masses is that it contains initial stresses. So Prof. Wittke has estimated the initial stress to be in the order of 10 to 20 kilograms per square centimetre. That is fairly common for mud rocks. I will tell you a story about the most important dam in China. It is work on multilayered mudrocks and sandstone. We have excavated a pipe for a pit for the power house and so I have expected that lateral deformations will occur, but at that moment nobody believed me, because nobody could believe that mudrocks can have such large initial stresses. After two months we have measured dislocation of over 8cm and these dislocations are increasing logarithmically with time. So we have measured the initial stresses, (not calculated but also measured) and the initial stresses in mudrock are in the order of 10-30 kilogram per square cm. I was asked by the authorities to give my attention to this important problem, as to in what motion of pits maybe of dangerous character. If you don't be prudent and we put the powerhouse inside the moving walls then the powerhouse will be distorted and maybe crushed.

Verbal reply: Prof. Wittke, Germany

I do not quite agree with your comment Prof. Tan. I do believe that one has to look at the reasons for the stresses and we did this and it is briefly reported in the paper. Actually this is a diagenetically compacted mudstone and so this means that it was originally sedimented as a clay and then it was subjected to overburden in geological times, probably (as I got information from geologists) up to a magnitude of 600 metres, and as a consequence of that horizontal stresses must have developed due to a high Poisson's ratio at the beginning because the material was soft at the beginning, and then it was compacted and then the overburden was removed to the present level and only part of the stresses were relieved, only that part according to the new Poisson's ratio, and there was a large magnitude left. That is the explanation I believe and the reason why we did not predict it is first that the stress measurement techniques actually fail in this type of rock because it is slaking, and we couldn't find a method which worked and secondly the whole area was actually on a so-called platform, with a width of about 800 metres with slopes to all sides. We thought there might have been stress relief before, due to the situation and also we had a soft layer underneath so we thought there might have been sufficient stress relief, but obviously the soft layer did not extend over the whole area and so probably the stresses were still retained. There was also some explanation for the geological structure of the area which could explain practically everything. For example the horizontal or flatly inclined slicken-sides in the brown mudstone were explained as being due to shear as a consequence of horizontal stresses. So in my experience we do not always have to expect high stresses in mudstones. You shouldn't draw this general conclusion: you have to look at the geological history and then you might find reason for it.

Written Reply: Prof. Wittke, Germany

The displacements of the side walls in the presented case were caused by a release of high horizontal stresses. The heavings of the invert were due to these stresses and a comparatively low shear strength of the rock mass at the elevation of the floor of the open excavation. The existence of the horizontal stresses as described in the paper was due to the removal of 600m of overburden from the diagenetically compacted rock in geological times. Side wall displacements and floor heavings are therefore not to be considered as general phenomena. One must look at the conditions of every case, whether they are likely to occur or not.

PAPER: Roof stability of shallow tunnels in isotropic and jointed rock

AUTHOR: P. Egger

PRESENTED BY: P. Egger

DISCUSSION

Question: Prof. A. Bello, Mexico

1. It is interesting to note the author has taken out of consideration the potential plastic zones assumed by Terzaghi at the walls; however, this condition calls for a certain amount of compressive strength: Have you had an indication for this parameter?

2. Do you have any indication for the ratio of horizontal to vertical stresses in the vertical limits of the body taken for limit stress equilibrium?

Answer: Prof Egger

The first question was about potential plastic zones above the roof like Terzaghi considered. In fact all the zone which is limited by these two vertical failure planes is considered as plastified and at every point the Mohr-Coulomb criterion is assumed to be just attained. That is the reason why our circles are always touching the Mohr-Coulomb envelope.

The second question was whether or not I assumed the material to have a compressive strength greater than zero. I worked with equivalent stresses which are equal to sigma plus the intercept H and according to this intercept H in fact the material has compressive strength. And the third question was about the consideration on horizontal stresses (initial stresses) in the ground. In fact this is a method based on limit equilibrium and we pose in fact the conditions of the limits which are assumed to be vertical and we have at this limit horizontal and tangential stresses at limit equilibrium and the horizontal stresses given by the failure criterion applied to the tunnel axis.

Question: Dr.R. Benson, Canada

This is not a question, but just a quick comment. I have been involved in two cases recently where failure occurred and the failure surface reached the ground surface. The cases were in soft weak rock, with saprolite.

The failure surfaces developed were similar to those predicted by the author but did involve the sides of the tunnel. Failure was preceded by lateral yielding of the sides, followed by excessive downward yielding of the arch crushing and buckling the steel ribs, then development of a log-spiral failure surface reaching up to the surface. The key is as always, to prevent excessive lateral yielding from occurring.

PAPER: Comparison Of Five Empirical Tunnel Classification Methods - Accuracy, Effect Of Subjectivity And Available Information

AUTHORS: H.H. Einstein, D.E. Thompson, A.S. Azzouz, K.P. O'Reilly, M.S. Schultz and S. Ordun

PRESENTED BY: H.H. Einstein

DISCUSSION

Question: Prof. Fairhurst

How was the support designed i.e. what was the basis against which the 5 methods were compared?

Answer: Prof. Einstein

Prof. Fairhurst asked me how was the original support designed. This was a combination of the rock load approach by Cording, and adaptation of the Terzaghi method and combined with R.Q.D. and a linear frame analysis

Question: Dr. B. McMahon,Australia

I once had the experience of working in two almost identical tunnels at much the same time with two different contractors with two different unbalanced bids. One fellow had a high price for steel and put it in at every opportunity, the other fellow had a low price for steel and we had to plead with him to put it in. Now with these empirical classifications which are correlated with actual practice, how do you consider these quite substantial psychological and economic effects. The difference was between 5 and 30% steel support in these two tunnels. In my work I have tried to correlate R.Q.D. with Geomechanics Classification and Q. I have found that there is a very high degree of correlation between these things, particularly in Rock Mass Rating which isn't surprising because the RQD and things that correlate highly with RQD make up some 70 to 80% of what goes into the Rock Mass Rating. So if these things all correlate very closely together why don't we just use the simplest one?

Answer: Prof. Einstein

I want to answer the two things separately. I agree with your first statement and that was basically what I wanted to say in my last comment, that these empirical methods by definition are based on underlying cases and whatever happened in that underlying case, be that geology or what the contractor or the designer did enters into the empirical method, so that is reflected and may give you over-conservative supports in some cases or the opposite in others. The other point: I agree that RQD expressing fracture spacing, in many instances the fracture spacing is a predominant factor. Actually if you look at what Barton does he has another factor adding to that in the same way as Bieniawski so they have over emphasized this fracture spacing together with some other parameters. In my experience I had some cases where they gave us not the good correlations that you evidently encountered so I am still open to see more cases before I personally form an opinion on that.

Comment by Prof. Wittke, Germany:

I would like to comment on the remark of Dr. McMahon, that the application of classification systems could have been limited to the RQD system only. According to my experience all the parameters describing the geometry of the families of discontinuities, as e.g. orientation, extent, asperity, unevenness, fillings etc., may have an influence on the stability of an underground structure and thus on its design in a practical case. As far as I know the RQD system for example does not account for the orientation. Therefore in practical cases, if stability analyses are not being performed, I rather prefer to rely on an interpretation of the results of boreholes with regards to the geometry of the discontinuities - as extensive as possible - and on my experience and disregard classification systems.

ROCK MECHANICS ISSUES IN UNDERGROUND MINING, LARGE PERMANENT EXCAVATIONS AND THERMOMECHANICAL PROBLEMS

Questions de mécanique des roches dans les mines souterraines, des excavations permanentes de grandes dimensions et des problèmes thermo-mécaniques

Felsmechanische Probleme im Untertagebau und in großen, langlebigen Hohlräumen, und Probleme der Thermomechanik

Charles Fairhurst
University of Minnesota, Minnesota, USA
Barry H. G. Brady
University of Minnesota, Minnesota, USA

SYNOPSIS

This component of the report on Theme C is concerned with rock mechanics issues in underground mining, large permanent excavations and thermomechanical problems. Contributions related to mining rock mechanics (theme segment C1) consider most of the topics of interest in mine design and mining practice. Mine global stability is considered from a pragmatic point of view. The mechanics of caving methods of mining are not represented. Three types of papers are represented under the theme segment C2 concerned with large permanent excavations. One group reflects attempts to improve analytical procedures for excavation design. The second group provides case studies of site characterization, design analysis and rock mass performance during construction. In the third group, rock mechanics concerns are of a more qualitative nature. Under theme segment C5, the topics represented include thermal properties of rocks, engineering design and performance of heat storage schemes, and sub-surface isolation of nuclear waste.

Summary

Contributions to Theme C of the Congress have been related to various geomechanics aspects of deep subsurface engineering in rock. The range of types and duty roles for excavations, from free-standing openings generated in mine production activities to comprehensively engineered structures for the long term containment of radioactive wastes, justifies separate consideration of the geomechanics issues raised and principles to be applied in the various cases of excavation design.

The topical area C1 related to ''Mining Excavations and Mining Methods'' produced a set of papers that is reasonably representative of the current state of Rock Mechanics practice in mining engineering. Many papers demonstrated that a philosophically sound, observational methodology linked to computational methods of analysis is well embedded in industrial practice. A number of papers addressed the interesting issue of design to assure mine global stability, using ideas which have evolved from analysis of rock performance under idealized loading conditions. Two papers related to this topic which describe control and location of rock bursts, confirm that more fundamental analysis of the mine global stability problem is justified. The geomechanics of caving methods of mining were not addressed in any of the papers in this area, although the topic remains a potentially fruitful area for basic and applied research.

The task of reviewing topical area C2 was divided; Dr. Hiramatsu considered the papers on tunnels, while these reviewers examined those dealing with large permanent underground openings - power stations and oil storage caverns. It was noteworthy that all of the papers on large openings were motivated by practical examples; some sought to develop an improved basis in mechanics for observed field behavior, while others described, in varying detail, the design procedures and practical constraints for specific case examples. It seems generally recognized that advance in the design of large caverns depends on improved understanding of the response of the rock mass to excavation - induced changes in the force and displacement fields. This, in turn, will depend on knowledge of the mechanical properties of and interaction between the constituent elements (rock, joints, fractures etc.) of the rock mass. Since most of the design and strategy for excavation and support must be developed before the start of construction, much of this understanding must be of a general ''generic'' variety, derived from carefully validated numerical model studies. Generic studies are intended to reveal critical combinations of practical conditions (joint friction and orientations, excavation sequence, shape, orientation, principal stress directions etc), and measures to ameliorate such situations should they arise.

Relatively few papers were presented in the topical area C5 entitled ''Nuclear Wast Isolation and Thermal Behavior of Rocks''. This is not taken to represent general satisfaction with the current state of waste repository engineering. It may represent an hiatus between the preliminary phase of establishing the technical feasibility of subsurface waste isolation, and the executive phase involving detailed design and construction of a repository. An important issue identified in one of the papers is related to the impossibility of a comprehensive, pre-development characterization of rock mass conditions at a selected repository site. Thus, although various studies indicate the general technical feasibility of constructing a geomechanically sound repository, the design and construction philosophy applied in practice must take due account of uncertainties about the integrity of a prospective repository site. This philosophy must accept the possibility of exposure of adverse structural or hydrogelogic conditions during excavation in the repository domain, and lead to formulation of appropriate engineering responses to these conditions when they are encountered. The general ramifications in repository construction of such a methodology do not appear to be appreciated in either the geomechanics community or bodies responsible for nuclear waste management.

Introduction

The theme considered in this report, ''Deep Underground Excavations'', embraces a broad range of interests in engineering rock mechanics. It varies from a discipline with a long history, exemplified by mining engineering, to one with topical currency and significant social and economic ramifications, represented by nuclear waste isolation. There is clearly a problem in attempting to unify the engineering principles educed in the range of papers related to the general theme, due to the diverse engineering notions involved in the various areas of geomechanics practice. However, some specific issues can be identified in the papers which deserve current attention in their own right, and which may be appropriate topics for research in the future.

Prior to general discussion of the various topical areas considered under this theme, it is reasonable to define the scope of technical interest represented by ''Deep Underground Excavations''. The reporters interpret the theme to apply to excavations in two general categories: those at such depth that the ground surface can be neglected, so that one is dealing notionally with infinite-body problems; or, alternatively, the theme may apply to Rock Mechanics problems in which the states of stress generated in a rock structure are sufficient to cause extensive failure in the rock mass. In both cases, the implication is that the ambient state of stress prior to excavation will exercise a significant role in the subsequent engineering performance of the rock medium.

The problem posed in the design, development and operation of deep underground excavations is prediction and control of induced displacements in the rock mass. The role of Rock Mechanics in excavation design is now properly established in engineering practice. The reporters have sought, in their review of papers, to identify contributions by authors to evolution of concepts of excavation design in various geomechanical and engineering environments. Their observations are made with reference to the distinct methodologies which may be employed in Rock Mechanics design practice.

There are three possible approaches to underground excavation design, reflecting various levels of reliance on the principles of engineering mechanics. The most elementary design approach is the application of a selected rock mass classification to a particular design exercise. A rock mass classification scheme purports to represent, by some simple index, a capacity to predict the local and global response of a rock mass to some perturbation imposed by excavation activity. Such schemes represent no more than an empirical interpolation, and sometimes an extrapolation, of observed rock mass performance, from a limited and questionable data base. The reporters are pleased to observe that no new, unique, universal classification scheme has been proposed in the current session, and little reference has been made to their application. Instead, their use as ''baseline'' design aids now seems properly appreciated in the Rock Mechanics profession. The approach recommended by Starfield and Detournay (1981) wherein rock mass/excavation situations are subdivided into several categories (near-surface plastic (continuum) behavior, deep plastic (continuum) behavior, jointed (discontinuum) deformation ...etc), each warranting a different analytical procedure, seems a more appropriate basis on which to develop a classification scheme.

The second design methodology involves the development of a simple conceptual model of a rock mass from the site characterization data. A design is then undertaken using some analytical or computational routine which captures the key modes of rock mass response defined by the conceptual model. This approach is most useful when linked to an observational scheme, which involves the progressive re-definition of the properties of elements of the conceptual model. Most of the papers presented in this thematic component of the congress are concerned with either analytical case studies or design exercises implicitly, if not intentionally, exploiting this approach.

The third design methodology aims to predict the performance of a rock mass by comprehensive understanding of the global constitution of the medium. This presumes that it is possible to formulate a constitutive model for a particular rock mass, and to determine the various parameters which characterize the medium. The prospective response of the medium, under the conditions imposed by excavation, is then allowed to evolve naturally in an analysis incorporating the governing equations for the system, the constitutive equations, and the initial and boundary conditions for the problem domain. A few of the papers adopt this approach. As will be apparent in the following discussion of the various topics covered in this thematic session, there are specific problems in Rock Mechanics where this rigorous engineering mechanics formulation appears to be required for their effective solution. Research related to these particular problems presages new initiatives in the areas of theoretical mechanics, constitutive equations for frictional media, and computational mechanics. However, it is doubtful if this approach can ever provide a definitive answer to a particular design problem. Instead, the objective is to recognize unfavorable modes of rock mass behavior, and to formulate appropriate engineering responses to such behavior.

In setting out to report on the technical submissions of the various authors, the reporters' objectives have been to note significant contributions, identify points of contention, and, in some cases, elaborate on issues of more general geomechanics interest. Not all submitted papers have been discussed in this report, for reasons of space or lack of uniqueness of the contribution to the development of the theme.

Theme Segment C1: Mining Excavations and Mining Methods.

Underground mining excavations, and the rock structures produced by ore extraction from an orebody, occupy a fairly unique position in rock mechanics practice. Almost as a matter of course, states of stress sufficient to generate relatively large zones of rock mass failure, or slip or separation on planes of weakness, are generated in the orebody or the adjacent country rock, at some stage in the mining program. The mining objective is to control rock mass displacements in a way which will not compromise current or future mining activity. Control of displacements may be effected on a local scale, where various support and reinforcement measures may be applied in excavation of peripheral rock, or through development of pillar support. Control of the global stability of a mine structure presents a more formidable problem. In supported mine structures, backfill emplaced in mined voids is the prime practical method of preventing large-scale catastrophic displacements in the system. For mining methods unsuited to the use of backfill, techniques for identification of potential global instability, and mitigating measures to be applied in the rock structure, assume particular importance. This issue is considered in some detail in later discussion.

A feature of mining geomechanics practice is that, in comparison with civil engineering subsurface construction projects, the density of rock mass characterization data obtained in the pre-production phase of mining is low. This imposes a particular burden on mine site characterization, the problem being to generate geomechanically significant data from a limited set of observations. This assertion applies to determinations of the pre-mining state of stress in a rock mass, as well as to other components of the site investigation program.

The significance of the in-situ stress problem is demonstrated in the paper by Chunting and Zhaoxian. Their observations of the in-situ state of stress at the Jinchuan mine indicate that the in-situ distribution of the field stress tensor is highly non-homogeneous, being influenced by the location of measurement points relative to the dominant structural features. A usable imput to any subsequent analytical scheme is a volume average of the field stress tensor throughout the problem domain. Their work confirms that a satisfactory determination of the pre-mining field stress tensor requires consideration of the local structural geology, and of the location of the measurement points relative to the structural features.

The role of Rock Mechanics in a mining pre-feasibility study is reflected in the paper by Page et al. The paper is also a demonstration of a valid application of a rock mass classification scheme in an initial mine evaluation exercise. In this general type of study, geomechanical data is sparse, and the requirement is to examine such parameters as support needs, sustainable excavation spans and feasible mining layouts.

Extensive reference to application of classification schemes is made in the paper by Pagamehmetoglu et al. Earlier observations in this report, related to intrinsic deficiencies in the notion of a universal classification scheme for prediction of rock mass response to excavation, are confirmed in this paper. In particular, it is noted that one commonly-used classification scheme overestimated support and reinforcement requirements for access excavations, at the subject mine site.

As was stated in the introduction, there are three general procedures for design of underground excavations. Mining Rock Mechanics practice, with limited capacity for detailed rock mass exploration and characterization, typically follows the second of the prescribed procedures. Excavation design is thus based on some tentative conceptual model of the rock mass and its constitutive properties. Re-evaluation of the properties of the model, or revision of the model itself, proceeds as the rock mass is explored, or responds to excavation activity, during the progress of mining. Several papers represent excellent case studies in this mine geomechanics methodology. These entail detailed description of preliminary site data collection, proposal of a geomechanical model of the subject rock mass, description of the selected method for conducting design analyses, and conclusions from such analyses. In some cases, the case studies are supported by field observations of rock performance reflecting the adequacy of the mine geomechanical model, or the design methodology itself. The papers by Borm, Wallner, Ramirez Oyanguren et al. and Vujec are representative of this class of paper. The selected constitutive behavior for the rock mass may be quite complex. For example, both Borm and Wallner invoke models to accommodate the creep deformation behavior of salt.

The study by Crea et al. of geomechanics activity for extraction design at the Campiano mine includes a description of computational modelling of backfill placed in mine voids. In view of the large difference in moduli of mine rock and backfill, the stresses developed in the backfill appear excessive. This paper may have benefitted from brief comparison of the results of these analyses with those conducted in various Australian and North American mine backfill studies.

In their description of geomechanics conditions at the Soldado mine, Bolivar and Krstulovic introduce some rather misleading notions related to the pre-mining state of stress in a rock mass. The postulate that all non-gravity contributions to the pre-mining state of stress can be ascribed to tectonism is incorrect. Issues such as erosion and isostasy, thermomechanical residual stresses, and topography are also important. The reported measurements of pre-mining stresses also appear inadequate, since the measurement sites are clearly within the zone of influence of the major mining excavation.

Common modes of response of rock around mine excavations include:

- o rigid body displacement of structurally defined blocks from stope walls and crown;
- o slip on penetrative planes of weakness;
- o separation on planes of weakness in low stress environments;
- o spalling at pillar and abutment surfaces;
- o penetrative failure, or crushing, in highly stressed components of a mine structure.

Papers concerned with each of these modes of response, and with design and operating practices to address them, have been presented for discussion.

In their review of the mining of a series of tabular, inclined orebodies, Bywater et al. refer to the use of a Boundary Element Method of stress analysis, describing its use in developing an extraction sequence for the orebodies. At the high extraction ratios developed in the subject rock mass, pillar stresses sufficient to cause penetrative failure were generated. The unstated assumption in their computational scheme is that mine pillars are subject to homogeneous, uniaxial compression. This seems at variance with common experience of the states of stress that develop in pillars, and to contradict the results of measured, in-situ pillar stresses. This may account for some apparent differences noted in the paper in rock mass strengths determined by various retrospective analyses of observed pillar failures. However, the paper demonstrates a useful principle. That is, that a computational scheme, based on even the most simple constitutive behavior, can be used to devise successful mining sequences for a complex mining configuration, if formally linked in application to an observational scheme.

Mining conditions in the inclined tabular orebody described by Kotze are in direct contrast with those described by Bywater. Pre-mining states of stress were low. Problems of local ground control arose from adversely oriented fractures generating discrete blocks, subject to gravity load. These could not be supported in the stope hangingwall by the shear resistance mobilized on fracture surfaces by the prevailing normal stresses. Perhaps the main conclusion from this paper is the necessity, in mining practice, to proceed expeditiously with comprehensive rock mass characterization as soon as access to and exposures of orebody rock are developed after commencement of mining. Conditions leading to instablity, or other intolerable modes of rock mass response, may then be recognized prior to their large-scale development, and control and mitigative measures planned. Such a geomechanics methodology might now be accepted as the normal mode of action in mining rock mechanics practice.

The particular mode of local rock mass response considered by Beer et al. is hangingwall deflection, under the combined effects of gravity and the in-situ stress field, around stopes in a jointed, stratiform orebody. The paper discusses the development of a hybrid Boundary Element-Finite Element code. The distinguishing feature of the Finite Element formulation is that it allows separation to occur between element surfaces, suggesting a capacity for rigid body translation of the blocks which constitute the stope hangingwall span. However, the conceptual model still exploits only small displacements in the jointed rock mass, which are not usually observed in practice. In their discussion of the mechanics and modelling of excavation near-field behavior in stratified rock, reference could usefully have been made to the experimental observations of Sterling (1980) and Sterling and Nelson (1978). These may have provided a data set for verification of the computational scheme. Similarly, the work by Voegele (1982) and Lorig (1982) on alternative computational schemes for analysis of stope roof performance, based on various versions of the Distinct Element Method of Cundall (1971, 1974, 1980), complements that presented in this paper. These comments notwithstanding, the solution procedure is an interesting one. The reporters will await with interest the execution of properly designed verification studies for the scheme, to determine the extent to which it can predict correspondence with mechanical reality.

Analytical case studies are frequently productive exercises in mining rock mechanics, since the elapsed time between design of a mine component, and field observation of its performance, is relatively small. Continuity of observation of rock mass behavior then presents no great difficulty. Additionally, the rock mass is frequently driven well into the large strain, non-linear range of deformation. An interesting mine-based case study of this type is presented by Coulthard et al. The particular exercise involved comparison of the field performance of a mine pillar with performance predictions from Finite Element and Boundary Element studies of the pillar. Both computational schemes were capable of modelling slip on the faults transgressing the pillar. The study suggested that the Finite Element scheme was most effective in modelling pillar performance, probably because of its additional capacity to accommodate elastoplastic response in the rock medium adjacent to fault surfaces subject to slip. This conclusion leads directly to the notion that the most useful computational schemes in Rock Mechanics may be hybrid methods, capable of modelling different modes of rock behavior in different domains. They may be developed by coupling, for example, a Boundary Element routine, to represent far-field and linear near-field rock, and a differential scheme (such as Finite Elements or Distinct Elements) to represent the comparatively small domains of non-linear response. The recent development of codes based on such hybrid formulations has been reported by Brady and Wassyng (1981), Beer and Meek (1981), and Yeung (1982), among others.

In the general area of analytical methods for excavation design, Filatov et al. report what appears to be an enhanced method of photoelastic analysis. It is claimed that the technique can yield directly distributions of the principal stresses in a model, presumably by some form of post-processing. The fundamental principles of the method are not described in the paper, and are available only in the Russian literature.

Two papers are concerned with techniques for relatively large scale, but local, control of rock displacements around stoping excavations, by support and reinforcement. As opposed to the superficial reinforcement provided by rockbolts, tendons, cables, or rock anchors are sufficiently long to provide, potentially, more coherent integration of stope wall rock with the country rock. The construction and application of rock anchors are described by Herbst. The studies by Fuller suggest that extensive reinforcement of stope walls and crowns is not, at this stage, a sound mechanical or economic proposition for non-entry methods of mining.

In the preceding discussion, it was observed that mining Rock Mechanics is concerned with control of both the local and the global performance of the host rock medium for mining. Papers discussed to this stage have been concerned purely with local stability. Analysis of mine global stability is not feasible by application of the techniques described earlier.

Mine global instability results when small perturbations in the state of equilibrium of a mine structure can provoke large displacements in the rock medium. Bumps and bursts are expressions of mine global instability. Contributions to the understanding of the mechanics of mine instability have been made by Cook (1965), Salamon (1970), and Starfield and Fairhurst (1968). From their work, useful design notions have evolved, in which mine stability is sought through consideration of the relative stiffnesses of mine support elements and adjacent country rock. An equivalent approach is based on determination of energy available in a system, and the energy required to crush the support system. Application of these principles in mine design is described in interesting papers at this meeting, by Saarka and Weber. The paper by Petukov presents a general review of rock burst research in the U.S.S.R., and describes the development of mining strategies to mitigate instability problems in mining practice.

Mine global instability is a pervasive problem, and is not restricted to deep mining. Any method of mining which results in states of stress approaching the in-situ strength of the rock mass throughout an extensive volume of rock poses a potential instability problem. Current notions of prevention of instability are based on observations of the performance of rock specimens in various types of laboratory test rigs. The results of such tests, conducted under conditions of one degree of freedom, are usually interpreted in terms of a macroscopic strain-softening constitutive model for the rock medium. Few theoretical mechanicians would support such interpretations. In fact, test specimens are subjected to poorly prescribed boundary conditions, which bear little relationship to the in-situ boundary conditions on an element of rock, and highly non-homogeneous conditions of stress and strain develop in a test specimen. These facts allow little to be said about the post-elastic performance of rock in general, or to draw any valid comparison between the laboratory and in-situ performances of elements of rock. The most promising approaches to elucidation of the mine stability problem may be developed from fundamental engineering mechanics. For example, the general principles of elastic stability, as presented by Thompson and Hunt (1973), may be elaborated into a more specific treatment, allowing the development of unstable bifurcation paths in the mining induced loading of the interior of an infinite or semi-infinite body.

Alternative formulations of procedures for global instability analysis may be based, in the future, on large strain elasto-plastic analysis of hardening, frictional media. Bifurcation and localization phenomena in such systems have been the subject of rigorous experimental and analytical study for some time. For example, Rudnicki and Rice (1975) discuss conditions for localization of deformation, and the generation of shear bands in a body of rock. Vardoulakis (1979) describes bifurcation in the deformation response of a granular material under simple loading conditions. The constitutive equations used in these analyses, involving work hardening and non-associated flow, appear capable of capturing such modes of response as unstable bifurcation, without the necessity to postulate strain-softening behavior.

Having observed the significance of the problem in mining Rock Mechanics, it is appropriate that one paper describes a system for practical management of the rock burst phenomenon. The seismic network discussed by de Jongh represents an attempt to identify burst-prone areas of a mine, so that practical mitigative measures can be implemented. The necessity for such a system represents direct confirmation that a significant commitment of research effort is still required, for solution of a problem whose frequency is bound to increase in future mining practice.

In concluding this report on session C1- ''Mining Excavation and Mining Mechods'', it is noted that the distribution of topics for papers represents a fair sampling of mining Rock Mechanics practice. Activities such as mine site characterization, excavation and pillar design analyses, field observations of rock mass performance, and retrospective analysis to establish mine design criteria, are given due attention. Several papers also consider the evolution of site-specific design strategies to accommodate particular local conditions, and these are a natural part of mining geomechanics activity.

One aspect of underground mining that is distinguished by its absence from the session is the geomechanics of caving methods of mining. In fact, a reasonable observation is that, apart from the exploratory work by such people as MacMahon (1969), Mahtab (1976), and Kendorski (1978), there has been little real effort devoted to the fundamental mechanics of caving. Issues of interest include cave propagation, draw, and interaction of the mobilized mass with the underlying medium in which extraction and service openings are located. The difficulties in analysis of these issues are not inconsiderable. The emergence of powerful and more versatile computational methods of analysis suggests that re-examination of these topics may now be worthwhile.

Theme Segment C2: Permanent Excavations Including Tunnels, Power Stations and Storage Caverns

The papers discussing large permanent underground excavations may be considered in three categories:
1) those seeking to develop an improved analytical framework, albeit still idealized, to provide greater agreement between the predicted and observed deformation behavior of the large excavations.
2) case histories in which detailed rock mechanics investigations were conducted in advance of excavation and used in design. In some cases observations of the actual deformation behavior of the cavern allow valuable comparisons to be drawn
3) case histories where the rock mechanics questions were of a more qualitative nature, introduced as a consequence of difficulties encountered during the project.

The papers by Berest, Monsseau et al; Minh, Berest and Bergues; Berest and Nguyen comprise the first category. The first discusses containment of liquefied (compressed) petroleum gas, butane, and propane in unlined cavities below the water table. In analyzing the water pressures required to ensure no leakage of gas from the cavity the authors draw attention to the phenomena of ''gas entry pressure'' whereby gas is to a limited degree, restricted from leaking into the rock formation, even in the absence of a 'suitable' hydraulic gradient, by the formation of menisci between the grains. It is also shown that, taking into account the shape (i.e. finite size) of the cavity, it is possible for a recycling flow to develop, whereby gas may leave the cavity at the top, recirculating to enter again towards the cavity base. Such considerations could be significant in determining cavity depths and conditions of containment for Liquid Petroleum and other gases.

The papers by Minh et al. and Berest and Nguyen both deal with time dependent behavior. The first examines a strain softening elasto-visco-plastic model of rock deformation and shows that the model can satisfactorily explain both the behavior of brittle rock and softer 'plastic' rock as observed by loads imposed on tunnel supports. The second involves a comprehensive analysis of closures (i.e. volume reductions) in salt cavities for oil storage, and demonstrates that a visco-plastic model (i.e. with essentially a zero friction angle in $\tau - \sigma$ space) more satisfactorily accounts for observed closure effects than an elasto-plastic model - which neglects time dependence.

The remaining papers are all concerned with case-histories.

Hibino, Motojima and Kanagawa's contribution is an important one, reporting the results of numerical analyses, and determinations of stresses and displacements for 12 underground caverns. Agreement between computed and observed behavior is good, and the authors offer useful general conclusions, including the need to consider rock jointing in numerical analysis of the excavation, and the observation that stresses in the concrete lining ''do not depend on the horizontal component of ground pressure, but on the convergence value of the cavern walls''. This suggests that direct observation (and control) of wall displacements is in such cases a more useful practical measure for this purpose of support design than is stress determination.

The contribution by Kondo and Yamashita is a very good example of rock mechanics applied to design - with careful studies conducted in advance of excavation for purposes of prediction, measurements made during construction for safety and control, and a discussion of the causes for occasional differences between prediction and observation.

Wittke and Soria present a comprehensive discussion of extensive rock mechanics studies of a large underground power house. This paper can serve as a useful reference for such studies. The comparison between overcoring and flat-jack results of stress determinations tends to illustrate the remaining lack of reliability in stress determination procedures. Careful examination seems warranted of the effects of geological heterogeneity (joints, folding etc.) on in-situ stress distributions, and of the simplifications (e.g. linear elasticity, isotropy) invoked in interpreting overcoring (and doorstopper) techniques on the validity of such methods.

Bonapace presents a valuable example of a pressure tunnel design in which recognition that the rock mass could contribute significantly to the lining resistance to internal pressure resulted in substantial reduction of the size of both concrete and steel (pressure) linings.

Bergman and Stille describe a careful analysis and sound practical approach to a problem of rock bursting in a large oil storage excavation. Wire mesh bolted to the rock was found to be a satisfactory means of minimizing the detachment of rock slabs produced by minor bursts. The apparent variability in ''hardness'' of the rock - from the point load index - apparently produced correspondingly variable stresses - concentrated in the harder (and more brittle) sections. This again cautions against extrapolation of point determinations of in-situ stresses beyond the immediate vicinity (several borehole diameters) of the determination, especially in jointed and geologically variable rock formations.

The papers by Saito, Tsukada et al., and by Bai, Zhu, and Wang both discuss the problem of rock bursts - in a tunnel and in a power station, and the manifestation of burst-prone regions by discing of cores in core drilling. The authors recognize that the combination of high strength (brittle) zones of rock and the associated high stresses in these zones, is conducive to bursting. The apparent increase in such rock burst occurrence in civil engineering suggests that a discussion of procedures and techniques to minimize the damaging consequences, some perhaps borrowed from mining (e.g. destressing or water infusion), could be valuable.

The case study by Ceriani and Nord of the Majes Project, although containing little detailed discussion of rock mechanics, does include at least one valuable comment for rock mechanics specialists: ''One of the major experiences gained is that is is most important to focus the investigation efforts on the difficult rock conditions. In this case difficult rock was 2.3 percent of the tunnel length but was 5-10 times more cost- and time-consuming. The effort to localize the difficult rock conditions should match that ratio''.

This suggests the value of a coordinated effort to gather details of how difficult rock conditions have been addressed, successfully and unsuccessfully, in various situations; and suggestions as to practical procedures for dealing most effectively with such occurrences.

Overall, the papers presented to the Congress on large underground openings represent a reasonable cross-section of current applications of rock mechanics principles to such projects, viz numerical procedures, especially the finite element method, are used to provide a working basis on which to judge the stability of the excavation. In-situ measurements help to provide input on mechanical properties and in-situ conditions to the numerical model. Recognizing the limitations of such analyses and in-situ measurements, care is taken to monitor the actual deformation behavior and stability of the excavation, with ''sound'' judgment always required to interpret the usual differences observed, due mainly to variations of the local geological and rock-mass details.

The need for greater physical validation of the powerful numerical models should be stressed. Such models are now widely used as the basis for design studies - but there has been relatively little direct physical verification of them, especially the interactions between discontinuities and the resultant overall deformation behavior. Realistic incorporation of 'discontinuities' in physical models implies test 'specimens' (or assemblages) of the order of one to several meters in linear dimension. Tests on such a scale are costly, and must be carefully selected to test combinations of conditions that provide a critical, or especially informative, test of the numerical model. Such validation is the key missing link between the numerical model that allows consideration of discontinuities, and the confident use of such models in full-scale practical design.

The large size of permanent underground excavations, up to 40 meters wide and 100 m high for some salt cavities, and 20 m wide by 30-40 m high in ''hard-rock'' excavations - and the increasing use of long cable supports, suggest the question - how large a ''permanent'' underground excavation is feasible in rock of various types? Conversely, to what extent can equipment and systems (e.g. power stations) be economically re-designed so as to fit into smaller, perhaps multiple, excavations where the rock type and quality and geological conditions dictate a smaller than ''standard'' size of cavern?

Theme C5: Nuclear Waste Isolation and Thermal Behavior of Rocks

Emplacement of spent reactor fuel, packaged in various forms, in mined subsurface excavations is currently regarded as the preferred terminal storage scheme for these materials. In concept, this scheme involves the development of galleries or repository rooms in rock formations, at sufficient depth and with suitable geomechanical and hydrogeological properties, to assure waste containment at the repository horizon. The site properties and operating depth are to be chosen to eliminate the possibility of radionucleide release into the biosphere from waste packages emplaced in the repository gallery.

The subsurface storage of certain nuclear wastes is complicated by their long-term heat emission properties. Emplacement of heat sources in a rock mass generates thermal stresses, with the attendant possibility of fracturing of the rock mass, or separation and slip on planes of weakness. In the usual case where a subsurface rock mass is saturated with water, the emplaced heat sources will produce convective flow of groundwater in any natural and induced fissures in the medium.

Credible mechanisms for the migration of radionucleides from a subsurface repository to the biosphere involve their transport by solution in the natural or thermally-induced groundwater flow. Groundwater flow is governed by fissure aperture, connectivity, and temperature gradients, and these are related to the state of stress in the medium (including thermal stress components), and rock mass properties. Development of effective containment strategies requires comprehensive understanding of the thermo-hydro-mechanical response of a rock mass to excavation development and emplaced heat sources.

The containment options presented in the notional performance specifications of a waste repository, given a particular set of site conditions and for an imposed thermal load in the rock medium, include:
(a) transport of any subsurface release of radionucleide to the biosphere can occur so slowly that these materials can constitute no long-term biological hazard; i.e., only harmless decay products could conceivably appear at the ground surface after an indefinite elapsed time;
(b) the repository and its near-field elements are designed such that migration of contaminated groundwater from the repository to the far-field and biosphere is effectively prevented.

Option (a) uses a natural geologic barrier for waste containment. Option (b) is based on the design and emplacement of engineered barriers to control both groundwater contamination, and its dispersion is the rock medium. Either of these two principles, or a combination of them, may be employed in the formulation of the primary operating mode for a repository. The obvious requirement is that the selected operating mode and repository design be compatible with the hydro-mechanical properties of the rock mass, or that a site be chosen specifically to implement a selected operating mode.

The nature of the nuclear waste isolation problem, particularly the need to predict the thermo-hydro-mechanical performance of a rock mass for hundreds or thousands of years, imposes great demands on geomechanics as a field of engineering science. It is clear that the design methodology applied in repository engineering must be of the third type mentioned in the introduction to this report; i.e., it should evolve from a comprehensive understanding of the global constitution of the geologic medium. With some critical exceptions to be considered later, the scope for applications of an observational methodology is more limited than in current, conventional geomechanics practice.

Among the papers presented at this meeting, that by Hudson is a concise specification of the problems involved in subsurface containment of low-level (non-heat emitting) nuclear wastes, as will apply in the U.K. terminal isolation programme. The author proposes general principles in line with those discussed above. The crucial issue raised by Hudson is that of rock mass characterization. In the domain where most detailed information is required on the state of the rock mass, extensive drilling is precluded. The problem is, of course, that extensive site exploration by core drilling destroys the hydraulic integrity of the natural geologic barrier. Only limited penetration, by drill holes, of a candidate rock mass for a repository can be tolerated. A solution to this problem is proposed later.

The need to understand the thermomechanical properties of rock materials and rock masses has stimulated, in recent years, a number of laboratory and field investigations devoted to this topic. Ehara et al. report the results of laboratory studies of the thermal response of crystalline rocks. The microscopic response of the granite and basalt involved the development of micro-cracks in the rock fabric. The occurrence of such irreversible effects may have significant implications in the long term performance of a crystalline rock mass as a containment medium. The problem of in-situ determination of the thermal properties of rock is described by Kuriyagawa et al. The temperature variation of the thermal conductivity of a granite rock mass has been determined by retrospective analysis of in-situ temperature observations around a heat source, using a Finite Element code for the inversion.

In their discussion of rock thermal investigations related to the ventilation of Campiano mine, Barla et al. describe measurements of rock thermal properties, and development of a computational model of heat transfer to mine air. In practice, the analysis is much more complicated than is suggested by the treatment in this paper. The extensive work by Starfield (1966, 1967) could now be regarded as the base from which studies of heat flow into ventilation air streams should develop.

The paper by Bjurstrom et al. is related to the large scale storage of low quality heat, in hot water, in subsurface chambers in granite. The nature of the heat storage-recovery process means that the chamber peripheral rock will be subject to temperature cycling, over a long period, in the range 40-115 degrees C. The authors conclude that hysteresis effects associated with the thermal cycling will lead to an asymptotically decreasing accumulation of deformation in the excavation periphery, and mechanical stabilization of the boundary rock.

While no terminal repository has yet to be constructed for the storage of high-level wastes, a temporary subsurface repository, described by Roshoff et.al., has been designed and excavated. The repository is at shallow depth in crystalline rock. Waste containment is provided by the waste form itself, and various engineered barriers. Since no thermal loads are applied in the interior of the rock medium, the performance of this storage system is not highly relevant to the long term containment problem.

Some insight into one of the options for high level waste isolation being considered in the U.S.A. is provided in the paper by Wagner et al. The four candidate sites discussed are in dome and stratiform salt formations, at depths between 500m and 1000m below ground surface. The computational exercise described in the paper involved subjecting a prototypical repository gallery, at each site, to a prototypical thermal load imposed by emplaced waste containers. Due account was taken of site specific conditions, including in-situ temperature field, stress field, and salt properties, and their particular creep characteristics. An unspecified ''baseline'' creep law, with appropriate empirical parameters for each salt formation, was employed in a Finite Element computational scheme, to model repository room performance at each site. The results of these studies demonstrate significant differences in response at the four sites, associated with their different creep properties, ambient stress fields and thermal conditions. However, the results appear to imply that the technical requirements of repository operation, involving retrievability, gallery local stability, and ambient temperature conditions in workspaces, can be achieved in a practical repository design.

It was observed, in the preceeding discussion, that a major problem in repository design is that comprehensive rock mass characterization at the prospective site is precluded. A reasonable response to this restraint is to select a site in an area with extensive, uniform geology, characterize the far field domain, and use some geostatistical principle to interpolate rock mass properties at the repository site and in its near-field. Virtually all detailed site characterization will then occur as openings are developed in the repository domain. Such a procedure is adequate only if provision is made to accommodate, in the repository construction phase, adverse geological features which will undoubtedly be exposed by development operations. The existence of such features, and the absence of prior planning and specification of measures to reconstitute the hydraulic integrity of the repository domain, could lead to invalidation of an otherwise acceptable and sound candidate site. Satisfactory handling of this aspect of repository design and construction would then appear to be a key element in the engineering geomechanics of nuclear waste isolation.

REFERENCES

1. Beer, G. and Meek, J.L. (1981) ''The Coupling of Boundary and Finite Element Methods for Infinite Domain Problems in Elasto-Plasticity'', In ''Boundary Element Methods'', C.A. Brebbia, ed., Springer-Verlag, New York, pp 575-592.

2. Brady, B.H.G. and Wassyng A. (1981) ''A Coupled Finite Element-Boundary Element Method of Stress Analysis''. Int. J. Rock Mech. Min. Sci. and Geomech. Abstr., Vol 16, pp 235-244.

3. Cook, N.G.W. (1965) ''The Basic Mechanics of Rockbursts''. In ''Rock Mechanics and Strata Control in Mines'', S. African Inst. Min. Metall., pp 56-66.

4. Cundall, P.A. (1971) ''A Computer Model for Simulating Progressive Large Scale Movements in Blocky Rock Systems''. Proc. Int. Symp. on Rock Fracture, ISRM, Nancy, Paper II-8.

5. Cundall, P.A. (1974) ''A Computer Model for Rock Mass Behavior using Interactive Graphics''. U.S. Army Corps of Engineers, Technical Report MRD 2-74 (Missouri River Division).

6. Cundall, P.A. (1980)'' UDEC - A Generalized Distinct Element Code for Modelling Jointed Rock''. Technical Report to United States Army (European Research Office).

7. Gumbel, E.J. (1958) ''Statistics of Extremes''. Columbia University Press, New York.

8. Kendorski, F.S. (1978) ''The Cavability of Ore Deposits''. Mining Engineering, June, pp 628-631.

9. Lorig, L.J. and Brady, B.H.G. (1982) ''A Hybrid Discrete Element Boundary Element Method of Stress Analysis'', Proc. 23rd U.S. Rock Mechanics Symposium, pp 628-636.

10. Mac Mahon, B.K., and Kendrick, R.F. (1969) ''Predicting the Block Caving Behavior of Orebodies'', S.M.E. of A.I.M.E. Preprint No. 69-AU-51, S.M.E. Annual Meeting.

11. Mahtab, M.A. and Dixon, J.D.(1976) ''Influence of Rock Fractures and Block Boundary Weakening on Cavability''. S.M.E. of A.I.M.E. Transactions, March, pp 6-12.

12. Mahtab, M.A. and Yegulalp, T.M. (1982) '' A Rejection Criterion for Definition of Clusters in Orientation Data''. Proc. 23rd U.S. Rock Mechanics Symposium, pp 116-123.

13. Rudnicki, J.W. and Rice, J.R. (1975) ''Conditions for the Localization of Deformation in Pressure-Sensitive Dilatant Materials'', J. Mech. Phys. Solids, Vol. 23, pp 371-394.

14. Salamon, M.D.G. (1979) ''Stability, Instability, and the Design of Pillar Workings.'' Int. J. Rock

Mech. Min. Sci. and Geomech. Abstr., Vol 7, pp 613-631.

15. Starfield, A.M. and Fairhurst, C. (1968) ''How High-Speed Computers Advance Design of Practical Mine Pillar Systems''. Eng. Min. Journal, May, pp 78-84.

16. Starfield, A.M. and Dickson, A.J. (1967) ''A Study of Heat Transfer and Moisture pick-up in mine airways''. S. African Inst. Min. and Met., Dec. 1967, pp 211-234.

17. Starfield, A.M. (1966) ''The computation of Temperature Increases in Wet and Dry Airways''. Mine Ventilation Soc. of S. Africa. Vol. 19, pp 157-165.

18. Starfield, A.M. and Detournay E. (1981) ''Conceptual Models as a Link Between Theory and Practice: A General Approach With A Particular Application to Tunnelling''. Proc 22nd U.S. Rock Mechanics Symposium, pp 398-401.

19. Sterling, R.L. (1980) ''The Ultimate Load Behavior of Laterally Constrained Rock Beams''. Proc. 21st U.S. Rock Mechanics Symposium, pp 533-542.

20. Sterling, R.L. and Nelson, C. (1978) ''An Experimantal Investigation of the Strength of Laterally Constrained Rock Beams''. Experimental Mechanics, July 1978, pp 261-268.

21. Thompson, J.M. and Hunt, G.W. (1973) ''A General Theory of Elastic Stablity'', Wiley, London.

22. Vardoulakis, I. (1979) ''Bifurcation Analysis of the Triaxial Test on Sand Samples'', Acta Mechanica, Vol. 32, pp 35-54.

23. Voegele, M.D. and Fairhurst, C. (1982) ''A Numerical Study of Excavation Support Loads in Jointed Rock Masses''. Proc. 23d U.S. Rock Mechanics Symposium, pp 673-683.

24. Yeung, D., and Brady, B.H.G. (1982) ''A Hybrid Quadratic Isoparametric Finite Element-Boundary Element Code for Underground Excavation Analysis'', Proc 23rd U.S. Rock Mechanics Symposium, pp 692-703.

QUESTIONS DE MÉCANIQUE DES ROCHES DANS LES MINES SOUTERRAINES, DES EXCAVATIONS PERMANENTES DE GRANDES DIMENSIONS ET DES PROBLÈMES THERMO-MÉCANIQUES

Rock mechanics issues in underground mining, large permanent excavations and thermomechanical problems

Felsmechanische Probleme im Untertagebau und in großen, langlebigen Hohlräumen, und Probleme der Thermomechanik

Charles Fairhurst
Université de Minnesota, Minnesota, EU

Barry H. G. Brady
Université de Minnesota, Minnesota, EU

SOMMAIRE

Cette partie du rapport sur le thème C traite des questions de mècanique des roches dans les mines souterraines, des excavations permanentes de grandes dimensions et des problèmes thermo-mécaniques. Les contributions se rapportant à la mécanique des roches minière (segment C1 du thème) couvrent la plupart des sujets d'intérêt dans l'étude des mines et la pratique minière. On considère la stabilité globale de la mine d'un point de vue pragmatique. La mécanique des techniques d'exploitation des mines par foudroyage n'est pas représentée. Trois genres de communications sont représentées dans le segment C2 du thème qui s'adresse aux excavations permanentes de grandes dimensions. Un ensemble de communications se rattache à des tentatives d'amélioration des procédés analytiques pour les études d'excavations. Le second ensemble présente des études de cas sur la caractérisation des sites, l'analyse de la conception du système et le comportement de la masse rocheuse pendant la construction. Les développements en mécanique des roches sont de nature plus qualitative dans le troisième groupe de communications. Dans le segment C5 du thème, les questions représentées comprennent les propriétés thermiques des roches, la conception de construction et le comportement en fonctionnement des systèmes d'entrepôt de chaleur ainsi que l'isolement souterrain des déchets nucléaires.

RESUME

Les contributions au thème C du congrès ont concerné différents aspects géomécaniques des techniques souterraines dans les roches en profondeur. La variété des types d'excavations et leurs applications, depuis les ouvertures non consolidées obtenues lors d'activités de production minière, jusqu'aux structures très élaborées utilisées pour la rétention de déchets radioactifs, justifie une étude particulière de chacune des questions de géomécanique évoquées et des principes à appliquer dans les différent types d'excavations.

L'ensemble des communications du groupe C1: "Excavations Minières et Méthodes Minières" est à peu près représentatif de l'état actuel de la pratique de Mécanique des Roches dans les techniques minières. De nombreuses communications ont démontré qu'une méthodologie d'observation basée sur des principes solides et liée à des méthodes d'analyse sur ordinateurs est fermement établie dans la pratique industrielle. Un certain nombre de communications se sont attaquées à la question intéressante des études de projets destinées à assurer la stabilité globale de la mine, utilisant des idées développées à partir de l'analyse du comportement des roches sous des conditions de charge idéalisées. Deux études concernant ce sujet et décrivant le contrôle et la localisation des écaillements rocheux confirment l'intérêt d'approfondir l'analyse fondamentale de la stabilité globale des mines. La géomécanique des méthodes d'exploitation minière par foudroyage n'a été traitée par aucune des communications de ce groupe, bien que ce sujet soit encore un domaine potentiellement riche pour les recherches fondamentales et d'application.

La tâche de la revue des sujets du groupe C2 a été divisée; le Dr. Hiramatsu a considéré les communications sur les tunnels, tandis que les présents rapporteurs ont examiné celles concernant les cavités souterraines permanentes de grande taille - centrales génératrices d'électricité et cavités pour entreposer le pétrole. Il était intéressant de remarquer que toutes les communications sur les cavités de grande taille étaient motivées par des exemples pratiques; certains cherchaient à affiner les principes de mécanique de façon à pouvoir rendre compte des comportements observés sur le terrain, tandis que d'autres décrivaient avec plus ou moins de détails les procédés d'étude de projets et les contraintes pratiques dans des cas d'exemples spécifiques. Il semble qu' il soit généralement reconnu que les progrès dans les études de projets de cavités de grande taille dépendent d'une meilleure compréhension de la réponse de la masse rocheuse au processus d'excavation - changements induits dans les champs de force et de déplacement. Cette meilleure compréhension dépend à son tour de la connaissance des propriétés mécaniques et de l'interaction des composants de la masse rocheuse (roches, diaclases, fractures etc.). Puisque l'étude du projet et de la stratégie d'excavation et de consolidation doit être en grande partie menée à bien avant le début de la construction, cette compréhension doit pour la plupart être de type général, "générique" et doit être dérivée d'études de modèles numériques soigneusement validées. le rôle des études génériques est de mettre en relief des combinaisons critiques des conditions pratiques (orientation et friction des diaclases, séquence d'excavation, forme, orientation, directions principales des contraintes etc.) et des mesures pour l'amélioration de ces situations lorsqu'elles surviennent.

Relativement peu de communications on été présentées sur les sujets du groupe C5, intitulé: "isolement des Déchets Nucléaires et comportement Thermique des Roches".

Cela ne sous-entend pas une satisfaction générale en ce qui concerne l'état actuel des techniques d'entrepôt des déchets. Cela peut représenter un hiatus entre la phase préliminaire de l'établissement de la possibilité technique de l'isolement souterrain des déchets et la phase d'exécution qui implique l'étude détaillée d'un projet et la construction d'un dépôt. Une question importante, identifiée dans une des communications,se rapporte à l'impossibilité, avant construction, de la détermination complète des conditions de la masse rocheuse à un site de dépôt choisi. Ainsi donc, bien que des études variées indiquent que la construction d'un dépôt sain du point de vue mécanique est techniquement possible, l'étude d'un projet et les principes de construction appliqués dans la pratique doivent tenir compte des incertitudes en ce qui concerne l'intégrité d'un site d'entrepôt éventuel. Ces principes doivent accepter l'éventualité où des conditions structurales ou hydrologiques défavorables seraient révélées par l'excavation dans la région de l'entrepôt et ils doivent conduire à la formulation de réponses techniques appropriées à ces conditions lorsqu'elles sont rencontrées. Les ramifications générales d'une telle méthodologie dans la construction d'entrepôts ne semblent pas être considérées à leur juste importance, ni par la communauté géomécanique, ni par les autorités responsables de la question des déchets radioactifs.

INTRODUCTION

Le thème considéré dans ce rapport: " Les Excavations Souterraines Profondes ", inclut une grande variété de zones d'intérêt dans les techniques de mécanique des roches, depuis une ancienne discipline comme celle des techniques minières, jusqu'à un sujet d'actualité aux importantes ramifications sociales et économiques, comme l'isolation des déchets radioactifs. Il y a un problème évident lorsqu'on s'efforce d'unifier les principes techniques de construction, présentés dans un ensemble varié de communications se rapportant au thème général; ceci est dû aux diverses notions techniques impliquées dans les différents domaines de la pratique géomécanique. Cependant on peut identifier dans les communications des questions spécifiques qui méritent une attention particulière en ce moment et qui peuvent devenir des sujets de recherche appropriés dans le futur.

Avant la discussion générale des divers domaines d'intérêt considérés en rapport avec ce thème, il faut tenter ici de définir la portée de l'intérêt technique que présentent les "excavations Souterraines Profondes". Les rapporteurs interprètent ce thème comme s'appliquant aux excavations de deux catégories générales: les excavations à une telle profondeur que la surface du sol peut être négligée, de telle sorte que l'on a à faire à la notion des problèmes de type masse-infinie; ou, alternativement, le thème peut s'appliquer aux problèmes de mécanique des roches dans lesquels les états de contrainte engendrés dans une structure rocheuse suffisent à causer des défaillances importantes de la masse rocheuse. Dans les deux cas il est impliqué que l'état de contrainte ambiante avant l'excavation exercera un rôle important dans la performance ultérieure du milieu rocheux.

Le problème posé par l'étude, le développement et l'utilisation d'excavations souterraines profondes est celui de la prédiction et du contrôle des déplacements induits dans la masse rocheuse. Le rôle de la Mécanique des Roches dans les études de projets d'excavations est maintenant fermement établi dans les pratiques de l'industrie. Les rapporteurs ont cherché dans leur revue des communications à identifier les contributions des différents auteurs à l'évovution des concepts d'étude d'excavation dans divers environnements géomécaniques et techniques. Leurs observations se réfèrent aux différentes méthodologies qui peuvent être utilisées dans la pratique des études de projets en mécanique des roches.

Il y a trois approches possibles aux problèmes de l'étude des excavations souterraines, qui font appel à des niveaux variés aux principes de mécanique. L'approche d'étude la plus élémentaire est l'application d'un choix de classification de la masse rocheuse à un exercice d'étude de projet particulier. un système de classification des masses rocheuses aspire à représenter, par quelque indice simple, la capacité de prédire la réponse locale et globale d'une masse rocheuse à une perturbation imposée par une activité d'excavation. De tels systèmes ne représentent guère qu'une interpolation empirique, et , quelquefois une extrapolation, de la performance des masses rocheuses observées, à partir d'une base de données limitée et contestable. Les rapporteurs sont heureux d'observer qu'aucun nouveau et unique système de classification universelle n'a été proposé lors de cette session et qu'il n'y a guère eu de référence à leur application. Par ailleurs leur utilisation comme outils d'étude "primaires" semble maintenant être bien comprise par le milieu professionnel de la Mécanique des Roches. L'approche recommandée par Starfield et Detournay (1981) qui subdivise les excavations en plusieurs catégories selon les conditions de la masse rocheuse (à comportement plastique (continuum) proche de la surface, à comportement plastique (continuum) en profondeur, à déformation en diaclases (discontinuum) ...etc.), réclamant chacune une procédure analytique différente, semble être une base mieux appropriée sur laquelle développer un système de classification.

La deuxième méthodologie d'étude inclut le développement d'un modèle conceptuel simple de la masse rocheuse à partir des données de caractèrisation du site. Une étude de projet est alors mise en train, utilisant une routine d'analyse ou de calcul qui comprend les modes clef de la réponse de la masse rocheuse définie par le modèle conceptuel. Cette approche est particulièrement utile quand elle est liée à un système d'observation qui inclut la re-définition progressive des propriétés du modèle conceptuel. La plupart des communications présentées dans ce composant thématique du congrès concernent soit des études analytiques de cas soit des exercices d'étude qui utilisent implicitement, ou même délibérément cette approche.

La troisième méthodologie d'étude a pour but la prévision du comportement de la masse rocheuse en se basant sur une compréhension approfondie de la constitution du milieu. Ceci présume qu'il est possible de formuler un modèle constitutif pour une masse rocheuse particulière, et de définir les différents paramètres qui caractèrisent le milieu. La réponse prédite du milieu peut ensuite évoluer naturellement dans les conditions imposées par l'excavation selon une analyse qui incorpore les équations directrices du système, les équations consécutives et les conditions initiales et de limites pour le domaine du problème. Un petit nombre de communications adoptent cette approche. Comme il sera apparent dans la discussion qui suit des différents sujets couverts dans cette session thématique, il y a des problèmes spécifiques de Mécanique des Roches qui semblent réclamer pour leur solution effective cette formulation mécanique rigoureuse. La recherche qui traite de ces problèmes particuliers fait présager de nouvelles initiatives dans les domaines de la mécanique théorique, des équations consécutives pour les milieux en friction et le calcul mécanique. Cependant on peut douter que cette approche puisse jamais fournir une réponse définitive à un problème d'étude particulier. D'un autre côté le but est de reconnaître les modes de comportement défavorables de la masse rocheuse et de formuler des réponses techniques appropriées à de tels comportements.

En entreprenant ce rapport sur les communications techniques des différents auteurs, les rapporteurs ont cherché à prendre note des contributions importantes, à identifier les points contentieux et, dans certains cas, à commenter sur les questions d'un intérêt plus généralement géomécanique. Tous les papiers présentés ne sont pas discutés dans ce rapport. Pour des questions de manque de place on a dû omettre les communications n'offrant pas une contribution unique au développement du thème.

SEGMENT C1 DU THEME: Excavations Minières et Méthodes Minières.

Les excavations minières souterraines et les structures rocheuses produites par l'excavation du minerai occupent une place plutôt unique dans la pratique de la mécanique des roches. Presque tout naturellement, à un stade quelconque du programme minier, des états de contrainte se produisent dans la masse rocheuse qui sont suffisants pour créer des zones de défaillance relativement grandes de la masse rocheuse, des glissements ou des séparations le long des plans de faiblesse. L'objectif des mineurs est de contrôler les déplacements de la masse rocheuse de façon à ne pas compromettre l'activité minière du moment ou dans le futur. Le contrôle des déplacements peut se faire à l'échelle locale où l'on peut appliquer des mesures variées de soutènement ou de renforcement lors de l'excavation des roches encaissantes, ou construire des piliers de soutien. Le contrôle de la stabilité globale de la mine présente un problème beaucoup plus considérable. Dans les structures de mines soutenues, le remblayage dans les interstices miniers est un des principaux moyens pratiques pour empêcher des déplacements catastrophiques à grande échelle dans le système. Dans les méthodes minières qui ne se prêtent pas à l'utilisation du remblayage, les techniques permettant l'identification d'une instabilité globale potentielle et des mesures d'atténuation à appliquer à la structure rocheuse prennent une importance particulière. Cette question est discutée en détail plus loin dans la discution.

Une caractéristique de la pratique géomécanique minière est, qu'en comparaison avec les projets de constructions souterraines du génie civil, la densité des données de caractérisation de la masse rocheuse obtenues dans la phase de pré-production des mines est faible. Ceci rend la caractérisation du site minier particulièrement difficile, car il s'agit de produire des données significatives à partir d'un ensemble limité d'observations. Cela s'applique à la détermination de l'état de contrainte de la masse rocheuse avant les opérations de mine aussi bien qu'aux autres composants du programme d'investigation du site.

L'importance du problème de la contrainte in-situ est démontrée dans la communication de Chunting et Zhaoxian. Leurs observations de l'état de contrainte in-situ à la mine de Jinchuan indique que la distribution in-situ du tenseur de contrainte du champ est extrèmement non-homogène, car elle est influencée par la position des points de mesure par rapport aux éléments structuraux caractéristiques dominants. Une donnée utilisable dans un système analytique dans le futur serait la moyenne de volume du tenseur de contrainte du champ dans tout le domaine du problème. Leur travail confirme qu'une détermination satisfaisante du tenseur de contrainte du champ, avant l'exploitation de la mine, exige que l'on considère la structure géologique locale et la localisation des points de mesure par rapport aux éléments structuraux.

Le rôle de la Mécanique des Roches lorsqu'on étudie la possibilité de miner est exposé dans la communication de Page et al.. Cette communication est aussi une démonstration de l'application valable d'un système de classification de la masse rocheuse à un exercice d'évaluation de mine initial. Généralement dans ce type d'étude, les données géomécaniques sont rares et il est nécessaire d'examiner des paramètres comme les besoins de consolidation, les dimensions maximum des cavités et les dispositions possibles du plan de mine.

La communication de Pagamehmetoglu et al. se réfère considérablement à l'application des systèmes de classification. Cette communication confirme les remarques faites précédemment dans ce rapport concernant les limitations du concept de système de classification universel pour la prédiction de la réponse de la masse rocheuse à l'excavation. En particulier il est remarqué qu'un système de classification d'usage courant surestimait les besoins de renfort et de soutènement pour les excavations d'accès au site minier en question.

Comme il a été dit dans l'introduction, il y a trois procédés généraux d'étude de projets d'excavations souterraines. La pratique minière de la Mécanique des Roches, offrant une capacité limitée d'exploration et de caractérisation détaillée de la masse rocheuse, suit typiquement le deuxième des procédés prescrits. L'étude de l'excavation est donc basée sur un modèle conceptuel préliminaire de la masse rocheuse et de ses propriétés constitutives. La ré-évaluation des propriétés du modèle, ou la révision du modèle lui-même, progresse avec l'exploration de la masse rocheuse, ou en réponse à l'activité d'excavation pendant la progression de la mine. Plusieurs communications représentent des études de cas excellentes de cette méthodologie de géomécanique des mines. Celles-ci impliquent une description détaillée du rassemblement des données préliminaires sur le site, la proposition d'un modèle géomécanique de la masse rocheuse en question, la description de la méthode choisie pour les analyses de l'étude du projet et les conclusions de ces analyses. Dans certains cas les études sont appuyées sur des observations des performances des roches sur le terrain, qui reflètent la justesse du modèle géomécanique de la mine, ou de la méthodologie d'étude elle-même. Les communications de Borm, Wallner, Ramirez, Oyanguren et al. et de Vujec sont représentatives de cette catégorie de communications. Le comportement constitutif choisi pour la masse rocheuse peut être très complexe. Par exemple, Borm et Wallner invoquent tous deux des modèles qui prennent en compte le comportement de déformation par reptation des sels.

L'examen par Crea et al., de l'activité géomécanique pour l'étude du projet d'extraction à la mine de Campiano inclut la description d'un modèle de calcul pour le remblayage placé dans les interstices miniers. Vue la grande différences entre les caractéristiques de la roche de la mine et de remblayage, les contraintes développées dans le remblayage paraissent excessives. cette communication aurait tiré profit d'une brève comparaison des résultats de ces analyses avec celles conduites dans les différentes études de remblayage dans les mines australiennes et nord-américaines.

Dans leur description des conditions géomécaniques à la mine de Solado, Bolivar et Krstulovic ont introduit des concepts plutôt fallacieux ayant rapport à l'état de contrainte dans une masse rocheuse avant les opérations minières. Le postulat que toutes les contributions à l'état de contrainte avant les opérations minières qui ne proviennent pas de l'action de la gravité peuvent être attribuées à l'activité tectonique est incorrect. Les problèmes d'érosion, de contraintes résiduelles thermodynamiques d'isostasie et de topographie ont aussi leur importance. Les mesures de contrainte précédant les opérations minières paraissent aussi inadéquates, puisque les sites des mesures sont clairement à l'intérieur de la zone d'influence de l'excavation minière principale.

Les modes communs de réponse des roches autour des excavations minières comprennent:

- o Déplacements en corps rigides de blocs structurellement définis à partir des parois d'abattage et des clefs de voûte.
- o Glissements le long des plans de faiblesse pénétratifs
- o Séparation le long des plans de faiblesse dans des environnements de faible contrainte.

- o Ecaillement aux surfaces de contact des èpaulements et des piliers.
- o Défaillances pénétrantes, ou écrasement dans les composants soumis à de fortes contraintes dans la structure minière.

Des communications se rapportant à chacun de ces modes de réponse et aux pratiques d'étude et d'opération qui s'y rattachent ont été présentées pour discussion. Dans leur revue de l'exploitation de plusieurs gisements miniers tabulaires inclinés, Bywater et al. se réfèrent à l'utilisation d'une méthode d'élément limite d'analyse des contraintes et décrivent son emploi dans le développement d'une séquence d'extraction des gisements de minerai. A cause des rapports d'extraction élevés dans la masse rocheuse en question, les contraintes au niveau des piliers ont été suffisantes pour produire des défaillances pénétrantes. Dans leur système de calcul, il est impliqué que les piliers de mine sont sujets à une compression homogène uniaxiale. ceci semble être en désaccord avec l'expérience commune des états de contrainte qui se développent dans les piliers et en contradiction avec les résultats de mesures in-situ des contraintes dans les piliers. Cela peut expliquer certaines différences apparentes notées dans cette communication entre les mesures de solidité des piliers dans différentes analyses rétrospectives des défaillances observées des piliers. Cependant la communication démontre un principe utile. Ce principe est qu'un système de calcul, même basé sur un comportement constitutif des plus simplifiés, peut être utilisé pour la mise en œuvre de séquences d'exploitation de mine efficaces dans le cas d'une configuration minière complexe, s'il est formellement lié à un système d'observations pendant l'application.

Les conditions de l'exploitation minière dans le gisement tabulaire incliné décrit par Kotze sont en contraste direct avec celles décrites par Bywater. Les états de contrainte avant l'exploitation étaient faibles. les problèmes locaux du contrôle de la masse rocheuse étaient causés par la présence de fractures orientées selon des plans tels qu'ils produisaient des blocs séparés, soumis à l'action de la gravité. Ceux-ci ne pouvaient pas être soutenus dans le toit par la résistance au cisaillement développée sur les surfaces de fracture par les contraintes normales prédominantes. La conclusion principale que l'on pourrait peut-être tirer de cette communication est la nécessité, dans la pratique minière, de procéder rapidement à une caractèrisation approfondie de la masse rocheuse dès que le gisement est atteint et le minerai exposé au début des opérations de mine. Les conditions conduisant à une instabilité, ou à d'autres modes de réponse inacceptables de la masse rocheuse, peuvent être repérées avant leur développement sur une grande échelle et des mesures d'atténuation et de contrôle peuvent être prévues. Une telle méthodologie géomécanique pourrait être maintenant acceptée comme le mode d'action normal dans la pratique minière de la mécanique des roches.

Le mode particulier de la réponse de la masse rocheuse locale, considérée par Beer et al., est la déflection du toit sous l'effet combiné de la gravité et du champ de contrainte in-situ, près de la paroi d'abattage dans un gisement stratifôrme avec diaclases. cette communication traite du développement d'un code hybride: Elément Limite-Elément Fini. Ce qui caractérise la formulation de l'Elément limite est qu'elle permet la séparation entre les surfaces élémentaires, ce qui suggère la possibilité de translation en corps rigides des blocs qui forment la voûte des toits, cependant, le modèle conceptuel n'envisage encore que des petits déplacements dans la masse rocheuse à diaclases que l'on n'observe pas d'habitude dans la pratique. Dans leur discussion de la mécanique et de l'élaboration du modèle du comportement, dans le domaine proche, des excavations dans les roches stratifiées, une référence aux observations de Sterling (1980) et de Sterling et Nelson (1978) aurait été bénéfique.

Celles-ci auraient pu fournir un ensemble de données pour la vérification du système de calcul. De même les travaux de Voegele (1982) et de Lorig (1982) concernant des systèmes de calcul de remplacement, pour l'analyse de la performance des toits, basés sur différentes versions de la méthode de l'Elément distinct de Cundall (1971, 1974, 1980) sont complémentaires de ceux présentés dans cette communication. Mais malgré ces commentaires le procédé de solution est intéressant. Les rapporteurs attendront avec intérêt l'exécution d'une étude de vérification proprement conduite du système, afin de déterminer jusqu'où ses prédictions correspondent avec la réalité mécanique.

Les études de cas analytiques sont fréquemment des exercices rentables en mécanique des roches minière, car la période entre l'étude d'un composant de la mine et l'observation sur le terrain de son comportement est relativement courte. La continuité de l'observation du comportement de la masse rocheuse ne présente alors pas grande difficulté. De plus, la masse rocheuse est souvent poussée bien avant dans le domaine des fortes déformations non linéaires. Une intéressante étude de ce genre de cas dans une mine a été présentée par Coulthard et al.. Cet exercice avait pour objet la comparaison entre le comportement sur le terrain d'un pilier de mine et celui prévu par les méthodes de l'Elément Fini et de l'Elément Limite. Les deux systèmes de calcul étaient capables de fournir un modèle du glissement le long des failles affectant le pilier. L'étude suggérait que le système de l'Elément Fini offrait un modèle plus efficace du comportement du pilier, probablement à cause de la capacité supplémentaire d'accomoder la réponse élasto-plastique du milieu adjacent aux surfaces des failles sujettes au glissement. Cette conclusion conduit directement à la notion que les systèmes les plus utiles en Mécanique des Roches pourraient être les méthodes hybrides de calcul, capables de fournir des modèles des différents modes de comportement des roches dans différents domaines. On pourrait les développer en couplant, par exemple, une routine d'Elément Limite pour représenter les roches du domaine éloigné et les roches à comportement linéaire du domaine proche, et un système différentiel (tel qu'Elément Fini et Element Distinct) pour représenter les domaines relativement restraints de la réponse non linéaire. le développement récent de codes basés sur de telles formulations hybrides a été rapporté par Brady et Wassyng (1981), Beer et Meek (1981) et Yeung (1982) entre autres.

Dans le domaine général des méthodes analytiques pour les études d'excavations, Filator et al. rapportent une méthode apparemment améliorée d'analyse photo-élastique. Ils revendiquent pour cette technique la capacité de donner directement la distribution des contraintes principales dans un modèle; nous supposons qu'ils utilisent pour cela une méthode de traitement a-postériori. Les principes fondamentaux de la méthode ne sont pas décrits dans la communication et ne sont accessibles que dans la littératire russe.

Deux communications traitent des techniques de contrôle des déplacements rocheux à une échelle relativement grande, mais locale, dans l'environnement des excavations en gradins par le moyen de soutènement et de consolidation. A l'opposé du renforcement superficiel ménagé par le boulonnage court, le boulonnage long, tendons, cables et ancrage profond des roches , a suffisament de profondeur pour offrir, potentiellement, une intègration plus cohérente de la roche de la paroi avec les roches encaissantes. La construction et l'application de l'ancrage des roches sont décrites par Herbst. Les études de Fuller suggèrent qu'un renforcement considérable des parois et des voûtes n'est pas, à ce point, une proposition économique ou mécanique raisonnable pour les méthodes minières sans descente.

Dans la discussion précédente, nous avons observé que la mécanique des roches minière s'intéresse au contrôle du comportement local et global du milieu de la roche encaissante du point de vue de la mine. Les communications que nous avons discutées jusqu'ici se sont adressées purement à la stabilité locale. L'analyse de la stabilité globale n'est pas possible en utilisant les techniques précédemment décrites.

Une instabilité globale de la mine se produit lorsque de petites perturbations dans l'état d'équilibre de la structure de la mine peuvent provoquer de grande déplacements dans le milieu rocheux. Les renflements et les écaillements sont l'expression d'une instabilité globale de la mine.Cook (1965), Salamon (1970) et Starfield et Fairhurst (1968) ont apporté leur contribution à la compréhension de la mécanique de l'instabilité minière. A partir de leurs travaux se sont développées des notions utiles pour les études de projets dans lesquels on recherche la stabilité de la mine en considérant la solidité relative des éléments de support de mine et des roches encaissantes. Une approche équivalente se base sur la détermination de l'énergie disponible dans un système et de l'énergie nécessaire à l'écrasement du système de soutènement. L'application de ces principes dans l'étude d'une mine a été décrite dans d'intéressantes communications à ce congrès par Saarka et Weber. La communication de Petukov présente une revue générale de la recherche sur les écaillements rocheux en U.R.S.S. et décrit le développement de stratégies destinées à minimiser les problèmes d'instabilité dans la pratique minière.

L'instabilité globale est un problème largement répandu et qui ne concerne pas seulement les mines profondes. Une méthode minière qui crée des états de contrainte approchant la limite de solidité in-situ de la masse rocheuse dans un volume étendu de roches pose un problème d'instabilité potentielle. Les concepts courants de prévention des instabilités sont fondés sur des observations du comportement de spécimens de roches dans différents équipements expérimentaux au laboratoire. Les résultats de ces tests à un degré de liberté sont habituellement interprétés en termes d'un modèle constitutif macroscopique de l'adoucissement des matériaux en déformation dans le milieu rocheux. On ne trouve guère de spécialiste de mécanique théorique pour soutenir de telles interprétations. En fait, les spécimens d'expérience sont soumis à des conditions limites mal définies, qui ont peu de rapport avec les conditions limites in-situ d'un élément de roche, et des conditions de contrainte et de déformation fortement non-homogènes se développent dans le spécimen expérimental. Ces faits ne permettent pas de dire grand-chose du comportement général post-élastique des roches, ni d'obtenir une comparaison valide entre les comportement en laboratoire et in-situ des éléments de roches. Les approches les plus prometteuses pour la solution du problème de la stabilité de la mine, peuvent être développées à partir des techniques de mécanique fondamentale. Par exemple, les principes généraux de la stabilité élastique, présentés par Thompson et Hunt (1973) peuvent être utilisés pour l'élaboration d'un traitement plus spécifique permettant le développement de voies de bifurcation instables dans les forces induites par les opérations minières dans l'intérieur d'un corps infini ou semi- infini.

Des formulations de remplacement pour l'analyse de l'instabilité globale peuvent être fondées, dans le futur, sur l'analyse de grandes déformations élasto-plastiques d'un milieu en friction se durcissant. les phénomènes de localisation et de bifurcation de tels systèmes sont le sujet depuis quelques temps d'études analytiques et expérimentales rigoureuses. Par exemple, Rudnicki et Rice (1975) ont discuté des conditions pour la localisation de la déformation, et la création de bandes de cisaillement dans un corps rocheux. Vardoulakis (1979) décrit la bifurcation dans la réponse de déformation d'un matériau granuleux sous des conditions de charge simples. les équations constitutives utilisées dans ces analyses, qui comprennent le durcissement en déformation et l'écoulement non associé, semblent pouvoir inclure des modes de réponse comme la bifurcation instable sans devoir postuler nécessairement un comportement d'adoucissement du matériau en déformation.

L'importance du problème ayant été observée en Mécanique des Roches minière, il semble approprié qu'une communication décrive un système de conduite à observer en ce qui concerne le phénomène d'écaillement des roches. Le réseau sismique traité par de Jongh représente une tentative pour identifier les zones de la mine sujettes à des écaillements afin que des mesures de minimisation puissent être mises en œuvre. Le besoin d'un tel système est une confirmation directe qu'un effort de recherche important est encore nécessaire afin de trouver la solution d'un problème dont la fréquence augmentera inexorablement dans le futur de la pratique minière.

Il est à remarquer, en concluant ce rapport sur la session C1: "les Excavations Minières et les Méthodes Minières" que la répartition des sujets des communications présente un échantillon raisonnablement représentatif de la pratique minière de la Mécanique des Roches. La caractérisation du site minier, les analyses d'excavations et d'études de piliers, les observations sur le terrain du comportement de la masse rocheuse et l'analyse rétrospective pour l'établissement de critères d'étude de projets miniers reçoivent l'attention qui leur est due. Plusieurs communications considèrent aussi l'évolution de stratégies d'études spécifiquement liées au site pour tenir compte de conditions locales particulières, ces stratégies faisant normalement partie de l'activité géomécanique minière.

Un aspect de l'activité minière souterraine qui se distingue par son absence de cette session du congrès est la géomécanique des méthodes d'exploitation minière par foudroyage. En effet, on peut constater, que, mis à part les travaux préliminaires de MacMahon (1969), Mahtab (1976) et Kendorski (1978), peut d'efforts se sont adressés à la mécanique fondamentale du foudroyage. Parmi les questions d'intérêt, on peut citer: la propagation dans les effondrements, l'extraction et l'interaction de la masse mobilisée et du milieu sous-jacent dans lequel l'extraction a lieu et où les ouvertures de service sont pratiquées. La difficulté de l'analyse de ces questions est loin d'être négligeable. l'apparition de méthodes de calcul puissantes et plus souples pour l'analyse suggère que le ré-examen de ces sujets peut être à l'heure actuelle de grande valeur.

SEGMENT C2 DU THEME: Les Excavations Permanentes Comprenant les Tunnels, les Centrales Electriques et les Cavernes d'Entrepôt.

Les communications qui traitent des excavations souterraines profondes de grandes dimensions peuvent être séparées en trois catégories:

1) Celles qui cherchent à développer un cadre analytique amélioré, quoiqu'encore idéalisé, ayant pour but un meilleur accord entre le comportement de déformation prévu et le comportement observé dans les excavations de grandes dimensions.
2) Des rapports de cas où des investigations en mécanique des roches ont été conduites de manière détaillée avant l'excavation et ont été utilisées dans l'étude du projet. Dans certains cas, l'observation du comportement de déformation de la caverne dans la réalité, permet de faire des comparaisons de grande valeur.
3) Des rapports de cas où des questions de mécanique des roches d'une nature plus qualitative étaient introduites à cause de difficultées rencontrées durant l'exécution du projet.

Les communications de Berest, Monsseau et al; de Minh,

Berest et Bergues; de Berest et Nguyen font partie de la première catégorie. La première traite de la rétention du gaz naturel liquéfié (comprimé), du butane et du propane dans des cavités sans revêtement situées au dessous de la surface de l'aquifère. En analysant les pressions d'eau nécessaires pour que la cavité ne laisse pas fuir le gaz, les auteurs attirent notre attention sur le phénomène de "pression d'entrée des gaz" par lequel la formation de ménisques entre les grains empêche jusqu'à un certain point le gaz de s'échapper dans la formation rocheuse, même en l'absence d'un gradient hydraulique "convenable". On nous montre aussi qu'en tenant compte de la forme (c'est à dire de la dimension finie) de la cavité, il est possible qu'une circulation de recyclage se développe, où le gaz peut s'échapper par le sommet de la cavité, circuler, puis pénétrer de nouveau dans la cavité par la base. Ce genre de considérations pourrait être important pour la détermination de la profondeur des cavités et des conditions de rétention du gaz naturel liquide et d'autres gaz.

Les communications de Minh et al. et de Berest et Nguyen traitent toutes deux de comportements indépendants du temps. La première examine un modèle élasto-visco-plastique, avec adoucissement des matériaux, de la déformation rocheuse et montre que ce modèle peut expliquer de manière satisfaisante, à la fois le comportement cassant des roches, et le comportement plastique observé lors de l'imposition de charges sur les supports des tunnels. La seconde comporte une analyse approfondie de la fermeture (c'est à dire des réductions de volume) des cavités dans les sels destinées à l'entrepôt du pétrole et démontre qu'un modèle visco-plastique (c'est à dire avec un angle de friction zéro dans un espace $\tau - \sigma$) rend compte de manière plus satisfaisante des effets de fermeture observés qu'un modèle élasto-plastique qui néglige l'influence du temps. Les autres communications traitent toutes d'exemples particuliers.

La communication de Hibino, Motojima et Kanagawa est importante. Elle rapporte les résultats d'analyses numériques et des déterminations des contraintes et des déplacements dans douze cavernes souterraines. L'accord entre les comportements calculé et observé est bon et les auteurs présentent des conclusions générales utiles, y - compris la nécessité de tenir compte des diaclases dans les roches pour l'analyse de l'excavation, et ils notent que"les contraintes dans le revêtement en béton sont indépendantes de la composante horizontale de la pression du terrain, mais dépendent de la magnitude de la convergence des parois de la caverne." Ceci suggère que l'observation directe (et le contrôle) des déplacements des parois est, dans de tels cas, une mesure plus pratiquement utile pour le propos de l'étude de la consolidation que la détermination de la contrainte.

La contribution de Kondo et Yamashita est un très bon exemple d'application de la mécanique des roches à l'étude projets - avec des études sérieuses conduites avant l'excavation et ayant pour but la prédiction, avec des mesures effectuées pendant la construction pour la sécurité et le contrôle, et avec une discussion des causes des différences occasionnelles entre les prévisions et l'observation.

Wittke et Soria présentent une discussion approfondie d'études détaillées de mécanique des roches dans le cas d'une station génératrice souterraine de grandes dimensions. Cette communication peut servir de référence utile pour les études de ce type. La comparaison des résultats des déterminations de la contrainte par la méthode de vérin plat et de sur-carottage, tend à illustrer le manque de fiabilité qui existe encore dans les procédés de détermination de la contrainte. Un examen sérieux des effets de l'hétérogénéité géologique (diaclases, plissements, etc.) sur la distribution in-situ de la contrainte et des simplifications (par exemple l'élasticité linéaire, l'isotropie) invoquées dans l'interprétation des techniques de sur-carottage (et de "doorstopper") sur la validité de ces méthodes semble s'imposer.

Bonapace présente l'exemple de grande valeur d'une étude d'un tunnel en pression où le fait de reconnaître que la masse rocheuse peut offrir une contribution importante à la résistance du revêtement à la pression interne, a eu pour résultat une réduction substantielle de l'épaisseur du chemisage en acier (pression) et du revêtement en béton.

Bergman et Stille décrivent une analyse soigneuse et une approche pratique valable d'un problème d'écaillement des roches dans une excavation de grandes dimensions destinée à l'entrepôt du pétrole. Du grillage, boulonné au rocher s'est révélé un moyen satisfaisant pour minimiser la séparation des plaques rocheuses produite par des écaillements mineurs. La variabilité apparente de la "dureté" des roches autour de l'indice de charge ponctuelle semble avoir produit des contraintes qui varient en rapport, concentrées dans les sections plus dures (et plus cassantes). Ceci nous avertit encore une fois des dangers de l'extrapolation des déterminations ponctuelles des contraintes in-situ au delà de la vicinité immédiate (plusieurs diamètres de forage) de la détermination, spécialement dans les formations géologiques variables ou à diaclases.

Les communications de Saito, Tsukada et al.,et de Bai, Zhu et Wang traitent toutes deux des problèmes d'écaillement des roches dans un tunnel et dans une centrale génératrice, et de la manière dont les zones sujettes aux écaillements se manifestent par la séparation en disques des carottes lors du carottage. Les auteurs remarquent que la combinaison de zones de roches de grande solidité (cassantes) et des grandes contraintes associées à ces zones, conduit aux écaillements. L'augmentation apparente de la fréquence de ces écaillements dans les travaux de génie civil suggère qu'une discussion des procédés et des techniques destinées à en minimiser les conséquences néfastes pourrait être de grand intérêt: certains de ces procédés, comme le soulagement des contraintes et l'injection d'eau peuvent être empruntés aux techniques minières.

L'étude de cas de Ceriani et Nord du projet Majes, quoique traitant peu des problèmes de mécanique des roches, inclut du moins un commentaire de prix pour les spécialistes de mécanique des roches: "Une des leçons les plus utiles est qu'il est trés important d'orienter les efforts d'investigations vers les difficultés crées par certaines conditions particulièrement délicates des roches. Dans le cas présent, les roches à problème constituaient 2,3 pour cent de la longueur du tunnel mais elles ont coûté de 5 à 10 fois plus de temps et d'argent. L'effort fait pour localiser les conditions rocheuses particulièrement délicates devrait être en rapport."

Ceci suggère la valeur d'un effort coordonné pour rassembler des détails sur la manière dont on s'est adressé à la question des conditions rocheuses difficiles, avec ou sans succès, dans des conditions variées, et des suggestions sur les procédés pratiques pour s'occuper de ces conditions lorsqu'elles se produisent.

Dans l'ensemble, les communications concernant les grandes cavités souterraines forment un échantillon représentatif des applications actuelles des principes de mécanique des roches à de tels projets, à savoir, les méthodes numériques, spécialement la méthode de l'élément fini, qui sont utilisées dans le but de fournir une base de travail sur laquelle on juge de la stabilité de l'excavation. Les mesures in-situ aident à fournir des données sur les propriétés mécaniques et les conditions in-situ pour le modèle numérique. En gardant à l'esprit les limitations de ces analyses et des mesures in-situ, on surveille avec soin le comportement de déformation réel et la stabilité de l'excavation, et il est toujours nécessaire d'utiliser son "bon sens" dans l'interprètation des différences habituellement observées, qui sont principalement dues à des variations locales de la géologie et des caractéristiques de la masse rocheuse.

Nous devons insister sur la nécessité d'une confirmation physique plus grande des puissants modèles numériques. L'utilisation de tels modèles comme bases d'étude de projets est maintenant largement répandue - mais ils n'ont été que relativement peu vérifiés dans la réalité physique, en particulier les interactions entre les discontinuités et le comportement général de déformation résultant. L'incorporation réaliste des "discontinuités" dans les modèles physiques implique le prélèvement d'échantillons (ou d'ensembles) de dimensions linéaires de l'ordre d'un à plusieurs mètres. Des essais à une telle échelle coûtent cher et doivent être choisis soigneusement pour la vérification des combinations de conditions qui fournissent une mesure critique ou spécialement féconde en informations du modèle numérique. Une telle vérification est le maillon crucial qui manque pour relier le modèle numérique, qui permet la considération des discontinuités, et l'utilisation en confiance de tels modèles dans les études de projets pratiques dans la réalité.

Les grandes dimensions des excavations souterraines profondes, jusqu'à 40m de large et 100m de haut pour certaines cavités dans les sels, et 20m de large et 30 à 40m de haut dans les excavations en roches dures et l'augmentation de la pratique du renforcement par longs cables suggèrent la question suivante: jusqu'à quelles dimensions peut-on pratiquer une excavation souterraine"permanente" dans des roches de types différents? D'autre part jusqu'où peut-on changer les équipements et les systèmes (par exemple des centrales génératrices) pour les incorporer dans des excavations plus petites, parfois multiples, lorsque le type de roches et la qualité des conditions géologiques obligent à des dimensions de cavernes plus petites que les dimensions standard?

SEGMENT C5 DU THEME: Isolement des Déchets Nucléaires et Comportement Thermique des Roches.

Le placement, après usage,du combustible nucléaire épuisé dans des excavations souterraines, est actuellement considéré comme le système d'entrepôt final préféré pour ces matériaux. La conception de ce système implique le développement de galeries ou de chambres d'entrepôt dans des formations rocheuses, à une profondeur suffisante et avec des propriétés géomécaniques et hydrologiques telles que la rétention des déchets dans l'horizon du dépôt soit assurée. On doit choisir les propriétés du site et la profondeur d'opération de façon à éliminer la possibilité d'échappement des éléments radioactifs dans la biosphère à partir des déchets placés dans la galerie du dépôt.

Le dépôt souterrain de certains déchets radioactifs est rendu plus compliqué par leurs caractéristiques d'émission thermique à long terme. Le placement de sources de chaleur dans la masse rocheuse crée des contraintes thermiques, avec possibilité de fracturation de la masse rocheuse ou de séparation et de glissements le long des plans de faiblesse. Dans le cas habituel où une masse rocheuse souterraine est saturée d'eau, les sources de chaleur produiront un courant de convection des eaux souterraines dans toutes les fissures naturelles ou induites du milieu.

Des mécanismes possibles pour la migration des éléments radioactifs à partir d'un dépôt souterrain jusqu'à la biosphère incluent leur transfer en solution dans le courant naturel ou induit par la chaleur des eaux souterraines. Le déplacement des eaux souterraines est contrôlé par la largeur des fissures, la connectivité, et les gradients de température; ceux-ci comme celle-là étant liés à l'état de contrainte du milieu (y compris la composante de contrainte d'origine thermique) et aux propriétés de la masse rocheuse. Le développement de stratégies efficaces pour la rétention exige une compréhension approfondie de la réponse thermo-hydro- mécanique de la masse rocheuse à l'excavation et aux sources de chaleur entreposées.

Les options pour la rétention, présentées dans les spécifications d'opération pour un dépôt de déchets nucléaires étant donné un ensemble particulier de conditions du site et une charge thermique imposée dans le milieu rocheux, comprennent:

(a) Le transport d'éléments radioactifs relâchés sous la surface jusqu'à la biosphère peut se produire si lentement que ces matériaux peuvent ne pas présenter de danger biologique à long terme; c'est à dire que seuls des produits de désintégration radioactive sans danger pourraient, dans notre conception, apparaître à la surface du sol après une période de temps indéfinie.
(b) La zone d'entrepôt et ses éléments du domaine proche sont conçus de telle manière que la migration des eaux souterraines contaminées du dépôt jusqu'au domaine éloigné et à la biosphère est effectivement éliminée.

L'option (a) utilise une barrière géologique naturelle pour la rétention des déchets. L'opion (b) est basée sur l'étude et le placement de barrières artificielles pour contrôler en même temps la contamination des eaux souterraines et la dispersion dans le milieu rocheux. chacun de ces deux principes, ou leur combinaison, peut être employé pour formuler le mode opératoire primaire d'un dépôt. Une condition requise évidente est que le mode opératoire choisi et la conception du dépôt soit compatible avec les propriétés hydro-mécaniques de la masse rocheuse ou qu'un site soit choisi spécifiquement pour l'exécution du mode opératoire choisi.

La nature du problème de l'isolement des déchets radioactifs, en particulier la nécessité de prédire le comportement thermo-hydro-mécanique d'une masse rocheuse pour des centaines ou des milliers d'années,impose de grandes exigences à la géomécanique en tant que science de l'ingénieur. Il est clair que la méthodologie d'étude appliquée à la conception des dépôts doit être du troisième type mentionné dans l'introduction à ce rapport; c'est à dire qu'elle doit se développer à partir d'une compréhension approfondie de la constitution globale du milieu géologique. A part certaines exceptions critiques que nous considèrerons plus loin, la portée de l'application d'une méthodologie d'observation est plus limitée que dans la pratique géomécanique conventionnelle actuelle.

Parmi les communications présentées à cette session, celle de Hudson est une description concise des problèmes rencontrés dans la rétention souterraine des déchets nucléaires de bas niveau (n'émettant pas de chaleur), qui sera applicable dans le programme d'isolement final du Royaume-Uni. Les auteurs proposent des principes généraux en accord avec ceux discutés ci- dessus. La question cruciale posée par Hudson est celle de la détermination des caractères de la masse rocheuse. Dans un domaine où on exige principalement une information détaillée sur l'état de la masse rocheuse, un forage intensif est exclu. Le problème est, bien sûr, qu'une exploration détaillée du site par carottage détruit l'intégrité hydraulique de la barrière géologique naturelle. Seule une pénétration limitée de la masse rocheuse sélectionnée pour le dépôt, par des forages, peut être tolérée. on propose plus loin une solution à ce problème.

La nécessité de comprendre les propriétés thermomécaniques des matériaux rocheux et des masses rocheuses a stimulé, dans les années récentes, un certain nombre de laboratoires et a motivé des investigations sur le terrain s'adressant à cette question. Ehara et al. rapportent les résultats d'études de laboratoire de la réponse thermique des roches cristallines. La réponse au niveau microscopique du granite et du basalte tient compte de la formation de micro-fissures dans la structure rocheuse. L'existence de ces effets irréversibles peut avoir des implications importantes dans le comportement à long terme de la masse cristalline comme milieu de rétention. Le problème de la détermination in-situ des propriétés thermiques des roches est décrit par Kuriyagawa et al.. La variation, selon la température, de la conductivité thermique d'une masse rocheuse granitique a été déterminée par une analyse rétrospective d'observations de températures in-situ autour d'une source, en utilisant un code d'Elément Fini pour l'inversion.

Dans leur discussion des investigations thermiques des roches concernant la ventilation de la mine de Campiano, Barla et al. décrivent les mesures des propriétés thermiques des roches et l'élaboration d'un modèle de calcul du transfert de la chaleur à l'air de la mine. En pratique, cette analyse est beaucoup plus compliquée que cela n'est suggéré par le traitement de la question dans cette communication. Les travaux détaillés de Starfield (1966, 1967) pourraient être maintenant considérés comme la base à partir de laquelle les études de transmission de la chaleur aux courants d'air de ventilation devraient se développer.

La communication de Bjurstrom et al. concerne l'emmagasinage à une grande échelle de chaleur de basse qualité (basses températures), par l'intermédiaire d'eau chaude, dans des chambres souterraines dans le granite. La nature du processus d'emmagasinage-récupération de la chaleur implique que les roches adjacentes à la chambre seront sujettes à un cycle de température, sur une longue période, dans la région de 40 à 115 degrés C.. Les auteurs concluent que les effets d'hystérésis associés au cycle thermique conduiront à une accumulation qui décroître asymptotiquement des déformations à la périphérie de l'excavation et à une stabilisation mécanique des roches périphériques.

Quoiqu'aucun dépôt final n'ait été construit pour la rétention de déchets nucléaires de haut niveau, un entrepôt transitoire, décrit par Roshoff et al. a été étudié et creusé. L'entrepôt est à faible profondeur dans des roches cristallines. La rétention des déchets est effectuée par la forme même des déchets et par des barrières artificielles variées. Puisqu'aucune charge thermique ne s'applique à l'intérieur du milieu rocheux, la considération du comportement de ce système de dépôt ne peut guère s'appliquer au problème de la rétention à long terme.

La communication de Wagner et al. offre quelque lumière sur une des options pour l'isolement des déchets radioactifs de haut niveau qui est actuellement examinée aux U.S.A.. Les quatre sites possibles se trouvent dans des formations salines en dôme et stratiformes, à des profondeurs s'étendant de 500m à 1000m sous la surface. Dans l'exercice de calcul décrit dans la communication, on soumet une galerie d'entrepôt prototype, à chaque site, à une charge thermique prototype imposée par le placement de récipients contenant des déchets. Il a été dûment tenu compte des conditions spécifiques des sites, y compris le champ de température in-situ, le champ de contraintes, les propriétés des sels et leurs caractéristiques de reptation. Une loi de reptation de base non spécifiée, avec des paramètres empiriques appropriés à chacune des formations salines, a été employée dans un système de calcul à Elément Fini en vue d'établir un modèle du comportement de la chambre de dépôt à chaque site. les résultats de ces études montrent des différences importantes de la réponse dans les quatre sites, correspondant à leurs différentes propriétés de reptation, de champ de contraintes ambiantes et aux conditions thermiques. Cependant la communication semble impliquer que les caractéristiques requises pour l'opération de l'entrepôt, y compris la possibilité de récupération des déchets, la stabilité locale de la galerie et les conditions de température ambiante dans les espaces de travail, sont réalisables dans un projet pratique d'entrepôt.

Il a été observé, dans la discussion précédente, qu'un problème majeur de l'étude des entrepôts est que la caractérisation approfondie de la masse rocheuse dans le site étudié est exclue. Une réponse raisonnable à cette restriction est la sélection d'un site dans une zone étendue de géologie uniforme, la caractérisation du domaine éloigné et l'utilisation d'un principe géostatistique pour l'interpolation des propriétés de la masse rocheuse au site du dépôt et dans le domaine proche. Virtuellement donc, toute la caractérisation détaillée du site aura lieu lors de la perforation des ouvertures dans le domaine de l'entrepôt. Une telle procédure n'est adéquate que s'il est fait provision de tenir compte pendant la phase de construction de l'entrepôt des structures géologiques défavorables que l'on ne manquera pas de rencontrer au cours du développement des opérations. L'existence de telles structures, liée à l'absence de planification et de spécification des mesures à prendre pour reconstituer l'intégrité hydraulique du domaine du dépôt, pourrait conduire à l'invalidation de la candidature d'un site par ailleurs sain et acceptable. Une approche satisfaisante de cet aspect de l'étude et de la construction d'un projet d'entrepôt semblerait donc être un élément clef dans la géomécanique de l'ingénieur pour l'isolement des déchets nucléaires.

SOURCES

1. Beer, G. and Meek, J.L. (1981) "The Coupling of Boundary and Finite Element Methods for Infinite Domain Problems in Elasto Plasticity", in "Boundary Element Methods", C.A. Brebbis, ed., Springer-Verlag, New York, pp 575-592.
2. Brady, B.H.G. and Wassyng A. (1981) "A Coupled Finite Element-Boundary Element Method of Stress Analysis". Int. J. Rock Mech. Min. Sci. and Geomech. Abstr., Vol 16, pp 235-244.
3. Cook, N. G. W. (1965) "The Basic Mechanics of Rockbursts". In "Rock Mechanics and Strata Control in Mines", S. African Inst. Min. Metall., pp 56-66.
4. Cundall, P.A. (1971) "A Computer Model for Simulating Progressive Large Scale Movements in Blocky Rock Systems". Proc.Int. Symp. on Rock Fracture, ISRM, Nancy, Paper II-8.
5. Cundall, P. A. (1974) " A Computer Model for Rock Mass Behaviour using Interactive Graphics". U.S. Army Corps of Engineers, Technical Report MRD 2-74 (Missouri River Division).
6. Cundall, P.A. (1980) " UCEC - A Generalised Distinct Element Code for Modelling Jointed Rock". Technical report to United States Army (European Research Office).
7. Gumbel, E.J. (1958) "Statistics of Extremes". Columbia University Press, New York.
8. Kendorski, F.S. (1978) "The Cavability of Ore Deposits". Mining Engineering, June, pp 628-631.
9. Lorig, L.J. and Brady, B.H.G. (1982) "A Hybrid Discrete Element Boundary Element Method of Stress Analysis", Proc. 23rd U.S. Rock Mechanics Symposium, pp 628-636.
10. Mac Mahon, B.K., and Kendrick, R.F. (1969) "Predicting the Block Caving Behavior of Orebodies", S.M.E. of A.I.M.E. Preprint No 69-AU-51, S.M.E. Annual Meeting.
11. Mahtab, M.A. and Dixon, J.D. (1976) "Influence of Rock Fracture and Block Boundary Weakening on

Cavability". S.M.E. of A.I.M.E. Transactions, March, pp 6-12.

12. Mahtab, M.A. and Yegulalp, T.M. (1982) " A Rejection Criterion for Definition of Clusters in Orientation Data". Proc. 23rd U.S. Rock Mechanics Symposium, pp 116-123.

13. Rudnicki, J.W. and Rice, J.R. (1975) " Conditions for the Localization of Deformation in Pressure-Sensitive Dilatant Material", J. Mech. Phys. Solids Vol. 23, pp 371 394.

14. Salamon, M.D.G. (1979) "Stability, Instability, and the Design of Pillar Workings." Int. J. Rock Mech. Min. Sci. and Geomech. Abstr., Vol 7, pp 613-631.

15. Starfield, A.M. and Fairhurst, C. (1968) "How High-Speed Computers Advance Design of Practical Mine Pillar Systems". Eng. Min. Journal, May, pp78-84.

16. Starfield, A.M. and Dickson, A.J. (1967) "A Study of Heat Transfer and Moisture Pick-up in Mine Airways". S.African Inst. Min. and Met., Dec. 1967. pp 211-234.

17. Starfield, A.M. (1966) "The Computation of Temperature Increases in Wet and Dry Airways". Mine Ventilation Soc. of S. Africa. Vol. 19,pp 157- 165.

18. Starfield, A.M. and Detournay E. (1981) "Conceptual Models as a Link Between Theory and Practice: A General Approach with a Particular Application to Tunnelling". Proc. 22nd U.S. Rock Mechanics Symposium, pp 398-401.

19. Sterling, R.L. (1980) "The Ultimate Load Behaviour of Laterally Constrained Rock Beams". Proc. 21st U.S. Rock Mechanics Symposium, pp533-542.

20. Sterling, R.L. and Nelson, C. (1978) " An Experimental Investigation of the Strength of Laterally Constrained Rock Beams". Experimental Mechanics, July 1978, pp 261-268.

21. Thompson, J.M. And Hunt, G.W. (1973) " A General Theory of Elastic Stability", Wiley, London.

22. Vardoulakis, I. (1979) "Bifurcation Analysis of the Triaxial Test on Sand Samples", Acta Mechanica, Vol.32, pp 35-54.

23. Voegele, M.D. and Fairhurst, C. (1982) " A Numerical Study of Excavation Support Loads in Jointed Rock Masses". Proc. 23rd U.S. Rock Mechanics Symposium, pp 673-683.

24. Yeung, D., and Brady, B.H.G. (1982) "A Hybrid Quadratic Isoparametric Finite Element - Boundary Element code for Underground Excavation Analysis", Proc. 23rd U.S. Rock Mechanics Symposium, pp 682-703.

FELSMECHANISCHE PROBLEME IM UNTERTAGEBAU UND IN GROSSEN, LANGLEBIGEN HOHLRÄUMEN, UND PROBLEME DER THERMOMECHANIK

Rock mechanics issues in underground mining, large permanent excavations and thermomechanical problems

Questions de mécanique des roches dans les mines souterraines, des excavations permanentes de grandes dimensions et des problèmes thermo-mécaniques

Charles Fairhurst
Universität Minnesota, Minnesota, USA

Barry H. G. Brady
Universität Minnesota, Minnesota, USA

ÜBERBLICK

Dieser Teil des Berichts über Thema C befaßt sich mit felsmechanischen Problemen im Untertagebau und in großen, langlebigen Hohlräumen. Ferner werden Probleme der Thermomechanik erörtert. Die Beiträge über felsmechanische Probleme des Bergbaus (Teil C1) erörtern die meisten aktuellen Themen im bergbaulichen Gewerbe und der Grubenplanung. Das Thema der Gesamtstandfeste eines Bergwerks wird von einem pragmatischen Gesichtspunkt her betrachtet. Die Beiträge über große, langlebige Hohlräume (Teil C2) lassen sich in drei Gruppen unterteilen: Diejenigen der ersten Gruppe spiegeln das Bemühen wider, analytische Verfahren der Grubenplanung zu verfeinern. Die der zweiten stellen Berichte über Standortvoruntersuchungen, Entwurfsanalysen und das Verhalten des Gebirges während des Ausbruchs dar. Die Beiträge der dritten Gruppe befassen sich mit felsmechanischen Anliegen überwiegend qualitativer Art. Die Themen des Teils C5 umfassen thermische Gebirgseigenschaften, Fragen der Planung und Leistung von Wärmespeicherprojekten, sowie die unterirdische Endlagerung radioaktiver Abfallstoffe.

ZUSAMMENFASSUNG

Die Beiträge zum Thema C des Kongresses behandeln verschiedene geomechanische Aspekte des Ausbruchs in tiefliegendem Gestein. Die Vielfalt der Arten und Zwecke von Untertageaufschlüssen, vom unbefestigten Grubengebäude im Untertagebergbau zu detailliert vorgeplanten Hohlräumen zur Endlagerung radioaktiven Abfalls, rechtfertigt die gesonderte Betrachtung der angesprochenen geomechanischen Probleme sowie der jeweils anzuwendenden Entwurfsgrundsätze.

Das Thema C1 "Bergbauliche Hohlräume und Bergbaumethoden" brachte eine Anzahl von Beiträgen hervor, die den gegenwärtigen Stand der angewandten Felsmechanik im Bergbau ziemlich genau widerspiegeln. Viele Beiträge bestätigen, daß eine philosophisch fundierte, auf der Beobachtung fußende, und durch die Rechenanalyse gestützte Methodologie im Bergbaugewerbe breite Anwendung findet. Eine Reihe von Beiträgen sprechen das interessante Thema der Verbindung zwischen Grubenplanung und -gesamtstandfeste an, unter Anwendung von Ideen, die man bei der Analyse des Gebirgsverhaltens im Zustand idealisierter Belastung entwickelt hat. Zwei Beiträge, die sich mit diesem Thema befassen und die Ursachen und Vorkommen von Gebirgsschlägen erörtern, bestätigen, daß eine grundlegendere Analyse des Grubenfestigkeitsproblems gerechtfertigt ist. Geomechanische Probleme im Bruchbau wurden nirgendwo in diesen Beiträgen erörtert, obwohl dieses Thema nach wie vor ein potentiell fruchtbares Gebiet der Grundlagen- und Angewandten Forschung ist.

Die Aufgabe, über die Beiträge des Thema C2 zu berichten, wurde geteilt: Diejenigen über Tunnelbau wurden von Dr. Hiramatsu gesichtet, während die über große, langlebige untertägige Hohlräume (Kraftwerke und Ölspeicherkavernen) von den Verfassern geprüft wurden. Es war bemerkenswert, daß alle Beiträge über große Hohlräume von Beispielen aus der Praxis ausgingen: Die einen bemühten sich um Verbesserung der bergbaumechanischen Basis der Beobachtung vom Gebirgsverhalten; die anderen beschrieben, mehr oder weniger ausführlich, den Planungsvorgang spezifischer Projekte sowie deren praxisbedingten Einschränkungen. Es scheint allgemein akzeptiert zu sein, daß Fortschritte in der Planung großer Kavernen vom besseren Verständnis der Reaktionen des Gebirges auf Ausbrüche abhängig sind - künstlich veranlaßte Veränderungen im Kraft- und Verschiebungsfeld. Dies hängt wiederum von unserer Kenntnis der mechanischen Eigenschaften der Gebirgskomponenten (Gestein, Störungen, Klüftung), sowie deren gegenseitigen Abhängigkeit ab. Da der Großteil der Planung, die Ausbruchs- und die Ausbaustrategie bereits vor Konstruktionsbeginn fertig sein muß, wird diese Kenntnis in großem Maß zwangsläufig allgemeiner, "generischer" Art sein, die man aus sorgfältig überprüften, numerischen Modellen entwickelt hat. Generische Untersuchungen sollen kritische

Kombinationen auftretender Verhältnisse, wie Kluftreibung und -richtungen, Reihenfolge der Ausbruchsphasen, Gestalt und Richtung des Aufschlusses, Hauptspannungsrichtungen usw. enthüllen, sowie Maßnahmen vorschlagen, solche kritischen Verhältnisse zu überkommen, sollten sie tatsächlich auftreten.

Es gab verhältnismäßig wenige Beiträge zum Thema C5 "Beseitigung radioaktiven Abfalls und thermisches Verhalten von Felsen". Dies wird nicht als Zeichen allgemeiner Zufriedenheit mit dem gegenwärtigen Stand der Planung und Durchführung von Lagern verstanden. Es dürfte eher eine Lücke zwischen der Frühphase der Diskussion über die technische Durchführbarkeit unterirdischer Endlagerung nuklearen Abfalls und der Durchführungsphase der Lagerplanung und des Lagerbaus sein. Ein schwerwiegendes Problem, das einer der Beiträge aufzeigt, betrifft die Unmöglichkeit einer umfassenden geologischen und felsmechanischen Voruntersuchung an geplanten Standorten für Lager. Daher muß, obwohl verschiedene Studien die allgemeine technische Möglichkeit eines geomechanisch stabilen Lagers bestätigen, die Planungs- und Bauphilosophie Ungewißheiten in bezug auf die geomechanische Integrität eines prospektiven Lagerstandorts gebührend berücksichtigen. Die Philosophie muß die Möglichkeit akzeptieren, daß während der Aufschlußarbeiten konstruktionswidrige oder widrige tektonische oder hydrogeologische Bedingungen im Bereich des Lagers angetroffen werden, und zur Entwicklung geeigneter Abhilfemaßnahmen auf solche Bedingungen führen, wenn sie auftreten. Die weitreichende Bedeutung einer solchen Methodik im Bau von Abfalllagern scheint bisher weder von den Fachleuten der Felsmechanik noch von den für die Beseitigung nuklearen Abfalls verantwortlichen Instanzen erkannt zu sein.

EINLEITUNG

Das Thema dieses Berichts über Teil C2, "Tiefe unterirdische Hohlräume und Abbaue", umfaßt ein breites Spektrum von Richtungen der angewandten Felsmechanik. Es erstreckt sich von alten, etablierten Disziplinen wie der des Erzbergbaus bis zu jungen, aktuellen Anwendungen mit beachtlicher gesellschaftlicher und wirtschaftlicher Bedeutung, wie der Beseitigung nuklearen Abfalls. Wegen der Vielfalt der Themen in den verschiedenen Bereichen der Geomechanik lassen sich die technischen Prinzipien der Beiträge nicht leicht auf einen gemeinsamen Nenner bringen. Man kann jedoch in den Beiträgen einige Probleme identifizieren, die nicht nur an sich besondere Aufmerksamkeit verdienen, sondern auch geeignete Gegenstände der zukünftigen Forschung sein dürften.

Vor einer allgemeinen Erörterung der Themen des Teils C2 scheint es angemessen, den Begriff "tiefe unterirdische Aufschlüsse" sowie seine Relevanz zur Felsmechanik genauer zu beschreiben. Das Thema wurde von den Beiträgen hinsichtlich seiner Anwendbarkeit auf Untertageaufschlüsse in zweierlei Weise aufgefaßt: Einmal diejenige, daß sie sich in so großer Teufe befinden, daß die Erdoberfläche außer Betracht bleiben darf. Man hat also theoretisch mit Geometrie "unendlicher Körper" zu tun. Andererseits läßt sich das Thema auf felsmechanische Probleme beziehen, wo der Spannungszustand, der sich in Felsstrukturen aufbaut, ausreicht um zu erheblichen Brucherscheinungen zu führen. In beiden Fällen muß man davon ausgehen, daß der herrschende Spannungszustand vor dem Ausbruch eine bedeutende Rolle im darauffolgenden Verhalten des Gebirges spielen wird.

Das Problem, das sich für die Planung, den Ausbruch und den Betrieb tiefer unterirdischer Aufschlüsse stellt, ist die Voraussage und Beherrschung von baubedingten Gebirgsbewegungen. Das Mitwirken der Felsmechanik bei der Planung von Hohlräumen wird überall im Ingenieurwesen als unabdingbar angesehen. Die Verfasser dieses Berichts haben sich bemüht, wichtige Beiträge zur Entwicklung von neuen Konzepten in der Planung von Untertageaufschlüssen unter verschiedenen geomechanischen und technischen Bedingungen zu identifizieren. Ihre Bemerkungen beziehen sich auf die ausgeprägten Methodologien, die in der felsmechanischen Planungspraxis Anwendung finden.

Es gibt drei mögliche Ansätze zur Grubenplanung, die verschiedenen Stufen der Abhängigkeit von bautechnischen Grundlagen entsprechen. Der elementarste Ansatz beinhaltet lediglich die Verwendung eines gewählten Gesteinsklassifikationsschemas bei einer gegebenen Planungsaufgabe. Es wird behauptet, ein Gesteinsklassifikationsschema besitze, mittels eines einfachen Indexes, die Fähigkeit, das durch den Ausbruch bedingte örtliche und großräumige Gebirgsverhalten vorauszusagen. Solche Schemata stellen nichts weiter dar als eine empirische Interpolation, zuweilen auch eine Extrapolation, des beobachteten Gebirgsverhaltens, auf der Basis von begrenzten und fraglichen Daten. Die Verfasser freuen sich, daß bei dieser Tagung kein neues, einzigartiges Klassifikationsschema vorgelegt wurde, und daß von der Anwendung solcher Klassifikationsschemata kaum die Rede war. Vielmehr scheint heute unter Geomechanikern ihr Wert als "Ausgangshilfsmittel" realistischer angesehen zu werden. Die Vorschläge von Starfield und Detournay (1981), wobei Gebirgstypen und Ausbruchsbedingungen in mehrere Kategorien unterteilt werden (oberflächennahes, plastisches (ununterbrochenes) Gebirgsverhalten; tiefliegendes, plastisches (ununterbrochenes) Gebirgsverhalten; (unterbrochene) Bruchverformung usw.), und wobei jeder Fall ein anderes analytisches Verfahren bedingt, scheinen eine geeignetere Grundlage für Klassifikationsschemata zu sein.

Beim zweiten Planungsansatz wird von den Daten der Standortsvoruntersuchung ein abstraktes Verhaltensmodell entwickelt. Mittels analytischer oder rechnerischer Verfahren wird dann ein Entwurf erstellt, der den wichtigsten Arten von Gesteinsverhalten, wie sie im Verhaltensmodell definiert sind, Rechnung trägt, wenn gleichzeitig die Parameter des Verhaltensmodells durch Beobachtungen fortlaufend revidiert werden können. Die meisten Beiträge des Themas C befassen sich entweder mit analytischen Projektstudien oder Projektplanungen, und verwenden, wenn nicht absichtlich dann implizit diesen Ansatz.

Der dritte Ansatz versucht, das Gebirgsverhalten mittels umfassender Kenntnisse der großräumlichen Zusammensetzung des Gesteins vorauszusehen. Dies setzt voraus, daß es möglich ist, für ein beliebiges Gebirge ein konstitutives Modell zu entwickeln und ferner die verschiedenen gebirgstypischen Parameter festzustellen. Das zu erwartende ausbruchsbedingte Gebirgsverhalten entwickelt sich dann auf natürliche Weise in einer Analyse, die die Bestimmungs- und Stoffgleichungen des Systems, sowie die Ausgangs- und Grenzbedingungen des Aufgabenbereichs mit einbeziehen. Einige der Beiträge verwenden diesen Ansatz. Wie wir unten bei der Diskussion der Beiträge zu diesem Thema sehen werden, gibt es in der Felsmechanik bestimmte Probleme, bei deren Lösung dieses rigorose Entwurfsverfahren unabdingbar erscheint. Die Forschung auf diesem Gebiet kündigt neue Initiativen in den Bereichen der theoretischen Mechanik, konstitutiver Formeln für Reibungsmedia, und in der mathematischtheoretischen Mechanik an. Es ist jedoch zu bezweifeln, ob dieser Ansatz jemals die Lösung eines gegebenen Entwurfsproblems ermöglicht. Das Ziel ist vielmehr, ungünstige Gebirgsverhaltensformen zu erkennen, und dementsprechende bergbautechnische Verfahrensweisen zu formulieren.

In diesem Bericht über die Referate des Thema C haben sich die Verfasser vorgenommen, auf wichtige Beiträge sowie auf Meinungsverschiedenheiten hinzuweisen und, in einigen Fällen, Themen von allgemeiner felsmechanischer Relevanz ausführlich zu erörtern. Aus Platzgründen haben sich die Verfasser auf diejenigen Beiträge beschränkt, die für die Fachdiskussion von größtem Interesse sind.

THEMA C1: BERGBAULICHE HOHLRÄUME UND BERGBAUMETHODEN

Untertägige Aufschlüsse sowie die durch Abschlag von Erz in einem Erzkörper bedingten Gebirgsstrukturen nehmen einen besonderen Platz in der Anwendung der Felsmechanik ein. Irgendwann im Laufe eines Erzbergbauprojekts ergeben sich fast zwangsläufig Spannungszustände, die im Erzkörper oder im Nebengestein relativ große Bereiche von Gesteinsbrüchen, Rutschungen oder Absonderungen entlang von Schwächezonen verursachen. Das bergbauliche Ziel ist die Beherrschung von Gebirgsbewegungen auf eine solche Art und Weise, die weder den gegenwärtigen noch einen zukünftigen Abbau einschränkt. Gesteinsbewegungen kann örtlich entweder durch diverse Stütz- oder Ausbaumaßnahmen im Ausbruch des Nebengesteins vorgebeugt werden, oder durch Stützpfeiler im Erz. Die Beherrschung der Gesamtstandfeste eines Grubengebäudes ist problematischer. In ausbaubedürftigen Gruben ist der Versatz der Abbaue mit "backfill" am meisten praktiziert, um Einbrüche katastrophalen Ausmaßes im System zu vermeiden. Bei Abbaumethoden, wo Versatz nicht anwendbar ist, gewinnen Methoden zur Feststellung potentieller Instabilität und der im Abbau anzuwendenden Abhilfemaßnahmen stark an Bedeutung. Dieses Thema wird weiter unten ausführlicher erörtert.

Ein Merkmal geomechanischer Tätigkeit im Erzbergbau im Vergleich zum Tiefbau ist, daß die während der Explorations- und Developmentphase gewonnenen Daten über Gebirgs- und Erzcharakteristika relativ spärlich sind. Die Voruntersuchung eines Bergwerksstandorts wird daher durch die besonders wichtige Aufgabe gekennzeichnet, geomechanisch zuverlässige Daten aus einer begrenzten Anzahl von verfügbaren Beobachtungen zu gewinnen. Diese Behauptung bezieht sich sowohl auf die Bestimmung von Spannungszuständen im Gebirge vor dem Ausbau, als auch auf die übrigen Teilbereiche von Standortsuntersuchungen.

Die Bedeutung des Problems der in-situ Spannungen wird in dem Beitrag von Chunting und Zhaoxian aufgezeigt. Ihre Beobachtungen des in-situ Spannungszustands im Bereich der Jinchuan-Grube weisen darauf hin, daß die in-situ Verteilung des Feldspannungstensors in hohem Maße inhomogen ist und von der auf die Hauptstrukturelemente bezogenen Lage der Meßpunkte beeinflüßt wird. Eine brauchbare Eingabe in ein darauffolgendes Gebirgsschema ist der Volumendurchschnitt des Feldspannungstensors im Projektbereich. Ihre Untersuchung bestätigt, daß, um den Feldspannungstensor vor dem Ausbau zufriedenstellend zu bestimmen, die Geologie und Tektonik um den Standort herum berücksichtigt werden muß, und daß die Meßpunkte den strukturellen Elementen entsprechend plaziert werden.

Die Rolle der Felsmechanik in Wirtschaftlichkeitsvoruntersuchungen im Bergbau wird in dem Beitrag von Page et al. deutlich. Der Beitrag zeigt auch eine berechtigte Anwendung eines Gebirgsklassifikationsschemas in Bergbauvoruntersuchungen. Bei allgemeinen Studien dieser Art liegen nur wenige geomechanische Daten vor und es besteht die Notwendigkeit, solche Parameter wie Ausbaubedarf, maximale Grubenspannweiten und optimale Auslage der Abbaue zu untersuchen.

In dem Beitrag von Paşamehmetoglu et al. wurde eingehend auf die Anwendung von Gebirgsklassifikationsschemata verwiesen. Die oben angeführten Bemerkungen in bezug auf die wesentlichen Schwächen im Konzept eines universellen Klassifikationsschemas zwecks Voraussage der Reaktion des Gebirges auf den Ausbruch werden in diesem Beitrag bestätigt. Insbesonders wird erwähnt, daß ein oft verwendetes Klassifikationsschema den Stützungs- und Verstärkungsbedarf beim Zugangsausbruch einer Grube überschätzte.

Wie in der Einleitung dieses Berichts erwähnt wurde, gibt es, allgemein gesagt, drei Planungsansätze im Grubenbau. Angewandte Felsmechanik im Bergbau, mit beschränkten Möglichkeiten einer detaillierten Exploration und Bestimmung der Gesteinsparameter, folgt normalerweise auf den zweiten der beschriebenen Ansätze. Der Grubenentwurf basiert daher auf einem abstrakten Modell des Gebirges und seiner konstitutiven Eigenschaften. Vorstellungen über die Parameter des Modells, ja das Modell selbst, werden bei fortschreitender Exploration und aufgrund von Beobachtungen des Gebirgsverhaltens während des Ausbruchs ständig revidiert. Mehrere Beiträge stellen ausgezeichnete Abrisse der Anwendung dieser Methodik dar. Sie umfassen die detaillierte Beschreibung der ursprünglichen Datensammlung am Standort, des Vorschlages eines geomechanischen Modells für das betreffende Gebirge, die Beschreibung des letztlich gewählten Verfahrens, nach dem Entwurfsanalysen durchgeführt wurden, und die den Analysen entspringenden Schlußfolgerungen. In manchen Fällen werden die Studien durch Feldbeobachtungen vom Gebirgsverhalten untermauert, die die Zulänglichkeit des geomechanischen Grubenmodells bzw. die Entwurfsmethodik selbst bestätigen. Typisch für solche Studien sind die von Borm, Wallner, Ramirez, Oyanguren et al. und Vujec et al. Das gewählte konstitutive Verhalten des Gebirges kann sehr komplex sein. Sowohl Borm als auch Wallner zum Beispiel konstruieren Modelle, die das Kriechverhalten von Salzen mit einbeziehen.

Die geomechanischen Untersuchungen in der Grubenplanung von Campiano von Crea et al. beschreiben u.a. die Modellrechnungen des in ausgeerzten Grubenteilen eingebrachten Fremdversatzes. Angesichts der Unterschiede in der Beschaffenheit des Gebirges und des Versatzes scheinen die Spannungen im Versatz übermäßig groß zu sein. In diesem Beitrag wäre es vielleicht nützlich gewesen, die Ergebnisse der Campiano-Analysen mit denen von Analysen australischer und nordamerikanischer Bergbauversatzstudien zu vergleichen.

Bei der Beschreibung geomechanischer Bedingungen der Soldado-Grube (Bolivar und Krstulovic) gibt es einige irreführende Behauptungen bezüglich des Gebirgsspannungszustands vor dem Ausbau. Die Behauptung, daß alle Spannungsfaktoren außer dem der Schwerkraft der Tektonik zuzuschreiben sind, ist falsch. Faktoren wie Abtragung und Isostasie, bleibende thermomechanische Spannungen und Oberflächenformen spielen auch eine Rolle. Ferner scheinen die zitierten Meßwerte ursprünglicher Spannungen unannehmbar zu sein, da die Messungspunkte eindeutig innderhalb des Einflußbereichs des Hauptgrubengebäudes liegen.

Häufige Gebirgsbewegungsarten im Bereich von Grubengebäuden sind, u.a.:

- Dislokation kompakter kleintektonisch begrenzter Blöcke an den Seiten und am First von Kavernen;
- Rutschungen entlang durchziehender Schwächeflächen;
- Abbruch an Schwächeflächen in den Bereichen niedriger Spannung;
- Abplatzungen an Pfeilern und Widerlagern;
- Einstürze oder Brüche starkbelasteter Komponenten eines Grubengebäudes.

Beiträge, die sich mit jeder dieser Verhaltensweisen, sowie mit geeigneten Entwurfs- oder Vorbeugungsmaßnahmen befassen, wurden dem Kongreß vorgelegt.

In ihrem Überblick über den Ausbau einer Anzahl von tafelförmigen, geneigten Erzkörpern verweisen Bywater et al. auf die Verwendung einer Grenz-Element-Methode der Belastungsanalyse und deren Wert bei der Ausbauplanung in den Erzkörpern. Bei den im Gesteinsverband entwickelten hohen Abbauraten ergaben sich Pfeilerspannungen, die stark genug waren, Bruchversagen zu verursachen. Die nicht ausgesprochene Annahme ihrer Berechnung ist die, daß die Grubenpfeiler einer homogenen, einachsigen Kompression ausgesetzt sind. Dies scheint einmal mit der allgemeinen Erfahrung in bezug auf die sich in Pfeilern entwickelnden Spannungszustände nicht übereinzustimmen, zum anderen den Ergebnissen von Messungen von in-situ Pfeilerspannungen zu widersprechen. Vielleicht erklärt dies einige im Beitrag erwähnte Unterschiede der Gebirgsfeste, die mittels verschiedener nachträglicher Analysen beobachteter Pfeilerbrüche ermittelt wurden. Der Beitrag zeigt jedoch ein nützliches Prinzip auf: Selbst ein Rechenmodell, das nur von einfachsten Gesteinsverhaltungsweisen und -parametern ausgeht, kann in der Abbauplanung komplexer Erzkörper erfolgreich angewendet werden, sofern es nur in engem Zusammenspiel mit fortlaufenden Beobachtungen vor Ort steht.

Die Abbaubedingungen des von Kotze beschriebenen tafelförmigen Erzkörpers stehen in direktem Kontrast zu denen des von Bywater beschriebenen Erzkörpers. Ursprüngliche Spannungszustände waren niedrig. Schwierigkeiten im Abbau wurden durch Blöcke verursacht, die entlang von ungünstig orientierten Kluftschwärmen unter dem Einfluß der Schwerkraft gelockert waren. Die Scherwiderstände auf den Kluftflächen, die sich aus den herrschenden Normalspannungen ergaben, reichten nicht aus, um die Blöcke im Hangendem der Abbaue im Verband zu halten. Die Hauptschlußfolgerung aus diesem Beitrag dürfte die sein, daß es im Bergbau notwendig ist, umgehend nach Beginn des Bergbaues mit umfassenden Gebirgsanalysen zu beginnen, sobald Erz und Nebengestein angefahren sind. Zustände, die zur Instabilität bzw. zu anderen unannehmbaren Reaktionsarten des Gebirges führen, dürften dann erkannt werden, bevor sie eine ernste Gefahr werden, und Kontroll- und Vorbeugungsmaßnahmen können getroffen werden. Eine solche geomechanische Methodik dürfte nunmehr als normales Verfahren der Felsmechanik im Bergbau akzeptiert sein.

Beer et al. untersuchen die Durchbiegung des Hangenden unter dem gleichzeitigen Einwirken der Schwerkraft und des lokalen Spannungsfelds im Grubengebäude eines geklüfteten, schichtgebundenen Erzkörpers. In ihrem Beitrag wird die Entwicklung eines hybriden Boundary-Element-Finite-Element-Programms erörtert. Das kennzeichnende Merkmal der Methode der Finiten Elemente ist, daß sie die Möglichkeit der Absonderung an Gesteinsflächen mitberücksichtigt und damit die Möglichkeit andeutet, daß alle Blöcke, die in der Spannweite des Ausbruchsfirstes liegen, als ein einheitlicher, starrer Block reagieren könnten. Das theoretische Modell geht jedoch lediglich von kleinen Versetzungen im geklüfteten Gestein aus, die in der Praxis normalerweise nicht vorkommen. Bei ihrer Erörterung der Technik und der Modellrechnung vom abbaubedingten standortsnahen Verhalten von Schichtgesteinen wäre es nützlich gewesen, auf die Versuche von Sterling (1980) und Sterling und Nelson (1978) zu verweisen. Diese hätten Daten zur Verifizierung des Rechenschemas liefern können. Ferner schließen die Arbeiten von Voegele (1982) und Lorig (1982), die sich mit alternativen Rechenschematen zur Analyse des Grubenfirstverhaltens befassen und auf verschiedenen Versionen der Distinct-Element-Methode von Cundall (1971, 1974, 1980) fußen, diesem Beitrag an. Ungeachtet dieser Bemerkungen bleibt das Lösungsverfahren von Beer et al. interessant. Die Verfasser sehen der Durchführung fachlich geplanter Verifizierungsuntersuchungen mit Interesse entgegen, um zu erfahren, inwiefern Vorhersagen entsprechend dem Verfahren mit den Realitäten der Mechanik übereinstimmen.

Analytische Fallstudien sind oft nützliche Übungen der Felsmechanik im Bergbau, da die Zeit zwischen dem Entwurf einer Grubenkomponente und der Beobachtung ihres Verhaltens vor Ort relativ kurz ist. Daher stellt die Kontinuität der Beobachtung des Gebirgsverhaltens kein wesentliches Problem dar. Ferner wird das Gebirge oft weit in die Phase des hochbelasteten nichtlinearen Verformungsverhaltens getrieben. Eine interessante Untersuchung dieser Art wurde von Coulthard et al. vorgelegt. Darin wurde das Verhalten eines Pfeilers mit auf den Finite-Element- und Boundary-Element-Methoden basierten Voraussagen verglichen. Beide Methoden besaßen die Fähigkeit, Gleiten auf Klüften im Pfeilerbereich im Modell darzustellen. Die Untersuchung deutete an, daß die Methode der Finiten Elemente am effektivsten bei der Darstellung des Pfeilerverhaltens im Modell ist, wahrscheinlich wegen ihrer zusätzlichen Fähigkeit, das elastoplastische Verhalten von Gestein in der Nähe von Störflächen, an denen Rutschungen auftraten, mit zu berücksichtigen. Dieser Schluß führt direkt zu der Annahme, daß das der Felsmechanik nützlichste Rechenschema ein Hybrid sein darf, das zur Darstellung verschiedener Gebirgsverhaltensarten in verschiedenen Bereichen fähig ist. Es kann zum Beispiel durch eine Kombination der Boundary-Element-Methode (die ortsfernes und lineares ortsnahes Gebirge darstellt) und einer anderen Methode wie zum Beispiel der Finite-Element- oder Distinct-Element-Methode (die verhältnismäßig kleine Bereiche nichtlinearen Verhaltens darstellen) zustande gebracht werden. Unter anderen berichten Brady und Wassung (1981), Beer und Meek (1981), und Yeung (1982) über die Entwicklung solcher hybriden Verfahren.

Auf dem Gebiet der analytischen Verfahren bei der Grubenplanung berichten Filatov et al. über eine anscheinend verbesserte Methode der photoelastischen Analyse. Es wird behauptet, die Methode bringe unmittelbar die räumliche Verteilung der Hauptspannungen im Modell hervor, vermutlich durch irgendeine Art der Nachverarbeitung. Grundlegendes zur Methode wurde im Beitrag nicht erörtert und ist nur in der russischen Literatur zu lesen.

Zwei Beiträge erörtern Methoden der relativ umfangreichen, jedoch örtlich begrenzten Kontrolle von Felsverschiebungen in Kavernen durch Stütz- und Bewehrungsmaßnahmen. Im Gegensatz zur oberflächlichen Bewehrung durch Felsnägeln stellt die langfristigere Bewehrung durch Drahtflechsen, Kabel oder Felsanker eine potentiell effektivere Integrierung der Grubenwandung mit dem Nebengebirge dar. Die Konstruktion und Anwendung von Felsankern wird von Herbst erörtert. Die Untersuchungen von Fuller deuten an, daß die umfangreiche Bewehrung von Grubenwänden und -firsten zur Zeit weder technisch noch wirtschaftlich für offenen Bruchbau zu rechtfertigen ist.

Die Verfasser haben oben erwähnt, daß im Bereich des Bergbaus sich die Felsmechanik mit der Beherrschung sowohl des örtlichen als auch des Gesamtverhaltens des umliegenden Gebirges während des Abbaues befaßt. Die bisher erörterten Beiträge haben sich ausschließlich mit lokaler Standfeste befaßt. Die Analyse der Gesamtstandfeste anhand der oben beschriebenen Methoden ist undenkbar.

Die Unfestigkeit im Gesamtbereich eines Bergwerks entsteht dann, wenn kleine Störungen im Gleichgewichtszustand der Grube erhebliche Gesteinsverschiebungen verursachen können. Wölbungen und Gebirgsschläge sind Ausdruck dieser Unfestigkeit. Beiträge zur Erfassung des Unfestigkeitsproblems kommen von Cook (1965), Salamon (1970), und Starfield und Fairhurst (1968). In ihren Arbeiten sind nützliche Planungsbegriffe entwickelt worden, anhand derer man Bergwerksstandfeste durch die Betrachtung der relativen Bruchfestigkeit von Stützelementen und vom Nebengestein zu bewerten versucht. Ein gleichwertiger Ansatz beruht auf der Bestimmung der Energie eines gegebenen Systems sowie der Energie, die zum Zusammenbruch des Systems notwendig ist. Die interessanten Beiträge von Saarka und Weber erörtern die Anwendung dieser Grundsätze bei der Bergwerksplanung. Der Beitrag von Petukov gibt einen allgemeinen Überblick über die Gebirgsschlagforschung in der U.d.S.S.R. und beschreibt die Entwicklung von Richtlinien zur Vorbeugung von Unfestigkeitsproblemen im Bergbau.

Das Problem der Gesamtunfestigkeit eines Bergwerks ist immer gegenwärtig, und zwar nicht nur in großen Tiefen. Alle Abbaumethoden, die Spannungszustände in einer ausreichend großen Gesteinmasse verursachen, schaffen ein potentielles Unfestigkeitsproblem. Gegenwärtige Vorstellungen in bezug auf das Verhindern von Unfestigkeitsbedingungen fußen auf der Beobachtung des Verhaltens von Gesteinsproben in verschiedenen Laborversuchen. Die Ergebnisse solcher unter den jeweiligen Beschränkungen durchgeführten Untersuchungen werden meistens je nach dem makroskopischen belastungslindernden konstitutiven Gebirgsmodell interpretiert. Nur wenige Theoretiker würden solche Interpretationen unterstützen. In der Tat werden Gesteinsproben ungenügend vorgeschriebenen Grenzbedingungen unterstellt, die wenig mit den in-situ Grenzbedingungen eines Gebirgselements gemeinsam haben, und es entwickeln sich in einer Gesteinsprobe sehr unhomogene Spannungs- und Belastungszustände. Angesichts dieser Tatbestände läßt sich wenig zum allgemeinen postelastischen Gebirgsverhalten sagen, und ein gültiger Vergleich zwischen dem Labor- und dem in-situ Verhalten des Gesteins ist unmöglich. Die hoffnungsvollsten Ansätze zur Beleuchtung des Unfestigkeitsproblems dürften aus dem Bereich der grundlegenden Ingenieurtechnik kommen. Die Grundsätze der elastischen Stabilität von Thompson und Hunt (1973) zum Beispiel könnten zu einer detaillierteren Abhandlung ausgebaut werden, die die Entwicklung instabiler Bifurkationswege im abbaubedingten Belastungsfeld im Inneren eines unendlichen bzw. halbinfinitiven Körpers ermöglicht.

Alternative Formulierungen von Verfahren der Gesamtunfestigkeitsanalyse dürften künftig auf der Hochbelastungs-, elastoplastischen Analyse von sich erhärtenden Reibungsmitteln beruhen. Bifurkations- und Lokalisierungserscheinungen solcher Systeme sind schon lange Gegenstand rigoroser experimenteller und analytischer Untersuchungen. Rudnicki und Rice (1975) zum Beispiel erörtern Bedingungen zur örtlichen Festlegung von Verformungserscheinungen sowie das Zustandebringen von Scherzonen in einem Gesteinskörper. Vardoulakis (1979) beschreibt Bifurkation im Verformungsverhalten eines körnigen Stoffes unter einfachen Belastungszuständen. Die konstitutiven Gleichungen, die in diesen Analysen verwendet wurden, einschließlich Beanspruchungshärtung und nichtassoziierten Fließens, scheinen die Fähigkeit zu besitzen, solche Verhaltensformen wie instabile Bifurkation zu behandeln, und zwar ohne die Notwendigkeit, belastungslinderndes Verhalten postulieren zu müssen.

Da die Bedeutung dieses Problems in der Felsmechanik des Bergbaus erwähnt wurde, trifft es sich gut, daß einer der Beiträge ein System zur praktischen Beherrschung von Gebirgsschlägen beschreibt. Das von de Jongh erörterte seismische Netzwerk stellt den Versuch dar, einschlagsgefährdete Bereiche eines Bergwerks zu identifizieren, damit Vorbeugungsmaßnahmen getroffen werden können. Die Notwendigkeit eines solchen Systems bestätigt den immer noch bestehenden Forschungsbedarf auf diesem Gebiet, das mit Sicherheit im künftigen Bergbau wegen der zunehmenden Anzahl von Gebirgsschlägen an Bedeutung gewinnen wird.

Abschließend sei zum Thema C1 "Bergbauliche Hohlräume und Bergbaumethoden" erwähnt, daß die Beitragsthemen einen guten Querschnitt von Problemen der Felsmechanik im Bergbau präsentieren. Aufgaben wie denen der Standortvoruntersuchung, der Abbauanalyse und der Analyse der Pfeilerdimensionierung, der Beobachtung des Gebirgsverhaltens vor Ort, und der nachträglichen Analyse zwecks Bestimmung bergbaulicher Planungskriterien, wird gebührliche Aufmerksamkeit geschenkt. Einige Beiträge befassen sich auch mit der Entwicklung ortsspezifischer Planungsrichtlinien, die bestimmte ortsrelevante Bedingungen mit einbeziehen. Solche Richtlinien bilden einen unabdingbaren Teil der Geomechanik des Bergbaus.

Ein Thema zum Untertagebau allerdings, das im Kongreß nirgendwo zur Sprache kam, betrifft die Geomechanik des Bruchbaus. In der Tat darf man sich die Bemerkung erlauben, daß - abgesehen von Vorarbeiten wie denen von MacMahon (1969), Mahtab (1976) und Kendorski (1978) - die grundlegende Technik des Bruchbaus recht wenig Aufmerksamkeit auf sich gezogen hat. Themen von Interesse sind in diesem Zusammenhang u.a. die Bruchkammervorbereitung, Einzugsbereich und Einfluß des Abbruches auf das darunterliegende Medium, in dem sich Förder- und Betriebsstrecken befinden. Die mit der Analyse dieser Probleme verbundenen Schwierigkeiten sind beträchtlich. Die Entwicklung von leistungsfähigeren, vielseitig verwendbaren Methoden der Rechenanalyse läßt einen erneuten Ansatz zur Untersuchung dieser Probleme heute lohnenswert erscheinen.

THEMA C2: LANGLEBIGE GRUBENBAUE EINSCHLIEßLICH TUNNEL, KRAFTWERKE UND UNTERIRDISCHE SPEICHERHOHLRÄUME

Die Beiträge über große permanente Untertageaufschlüsse dürfen in drei Gruppen unterteilt werden: Erstens diejenigen, in denen der Versuch unternommen wird, einen besseren, wenn nach wie vor idealisierten, analytischen Rahmen zu entwickeln, um ein größeres Maß an Übereinstimmung zwischen dem vorausgesagten und dem vor Ort beobachteten Verformungsverhalten der großen Grubenbaue herbeizuführen. Zweitens diejenigen der Fallstudien, bei denen vor dem Ausbruch detaillierte felsmechanische Untersuchungen durchgeführt wurden und dann bei der Grubenplanung verwendet wurden. In einigen Fällen ermöglichen Beobachtungen des tatsächlichen Verformungsverhaltens der Kavernen wertvolle Vergleiche. Drittens diejenigen der Fallstudien, deren felsmechanische Anliegen qualitativerer Art sind. Diese Untersuchungen wurden eingeleitet aufgrund von Schwierigkeiten, die im Lauf der Projekte zutage traten.

Die Beiträge von Berest, Morisseau et al.; Minh, Berest und Bergues; und von Berest und Nguyen bilden die erste Gruppe. Der erste Beitrag befaßt sich mit der Speicherung von Flüssiggas (LPG), Butan und Propan in Kavernenspeichern ohne Innenschale unter dem Grundwasserspiegel. Bei der Analyse des zum Vermeiden von Gasaustretung notwendigen Wasserdrucks verwiesen die Verfasser auf das Phänomen des "gas entry pressure", durch den das Gas selbst beim Fehlen eines 'geeigneten' hydraulischen Gefälles bis zu einem gewissen Grad daran gehindert wird, in das Nebengestein einzudringen, und zwar durch das Zustandekommen von Menisken zwischen den Partikeln. Es wird ferner gezeigt, daß abhängig von der Gestalt (d.h. der endlichen Größe) des Hohlraums ein Recycling-Flow entstehen kann: Das Gas diffundiert oben aus dem Hohlraum hinaus und wandert nach unten, um unten wieder dem Hohlraum zuzufließen. Solche Erscheinungen können bei der Bestimmung der Kavernenteufe von Bedeutung sein, sowie bei der Bestimmung der zur erfolgreichen Speicherung von LPG und anderen Gasen notwendigen Voraussetzungen.

Die Beiträge von Minh et al. und Berest und Nguyen befassen sich beide mit zeitabhängigem Verhalten. Der erste untersucht ein belastungslinderndes elasto-visko-plastisches Gesteinverformungsmodell und zeigt, daß das Modell das Verhalten von sowohl sprödem als auch weicherem 'plastischen' Gestein zufriedenstellend erklären kann, wie bei der Belastung von Tunnelstützen beobachtet wurde. Der zweite Beitrag berichtet über eine umfangreiche Analyse von Zuwachsungen (Volumenverminderung) bei Ölspeicherkavernen im Salz und zeigt, daß ein durch einen im wesentlichen Null-Reibungswinkel der Scher- und Normalspannungen gekennzeichnetes viskoplastisches Modell die beobachteten Zuwachsungserscheinungen zufriedenstellender erklärt, als ein elastoplastisches Modell, das die Zeitabhängigkeit nicht berücksichtigt.

Die restlichen Beiträge befassen sich mit Fallstudien.

Der Beitrag von Hibino, Motojima und Kanagawa ist wichtig. Er berichtet über die Ergebnisse von numerischen Analysen und ferner die Spannungs- und Verschiebungsbestimmungen bei zwölf unterirdischen Kavernen. Das berechnete und das beobachtete Gesteinsverhalten stimmen gut überein, und die Verfasser bieten nützliche allgemeine Schlüsse an, darunter die Notwendigkeit, bei numerischen Analysen des Aufbruchs Kluftfugen zu berücksichtigen, sowie die Beobachtung, daß Spannungen im Betonausbau nicht von der nicht von der horizontalen Spannung des primären Spannungsfelds abhängen, sondern von der horizontalen Verdrängung der Kavernenwände. Damit wird darauf hingewiesen, daß die direkte Beobachtung (und Beherrschung) von Wandverdrängungen für Ausbauentwürfe nützlicher ist als Spannungsmessungen.

Der Beitrag von Kondo und Yamashita ist ein sehr gutes Beispiel für die Anwendung der Felsmechanik bei der Grubenplanung, von sorgfältigen Untersuchungen vor dem Aufschluß zum Zweck der Vorhersage, zur Durchführung von Messungen während des Vortriebs zur Bestätigung und als Sicherheitsmaßnahme. Die Ursachen gelegentlicher Diskrepanzen zwischen den Voraussagen und den Beobachtungen vor Ort wurden sorgfältig untersucht.

Wittke und Soria legen eine detaillierte Diskussion über umfangreiche felsmechanische Untersuchungen an einem großen unterirdischen Kraftwerk vor. Diesen Beitrag kann man bei ähnlichen Untersuchungen als nützliche Bezugsquelle zur Hand nehmen. Ein Vergleich der durch "Overcoring" gewonnenen Spannungswerte mit denen, die durch Anwendung der Druckkissen-Methode gewonnen wurden, unterstreicht die noch immer bestehende Unzuverlässigkeit der Verfahren zur Spannungsbestimmung. Eine sorgfältige Untersuchung des Einflusses geologischer Heterogenität (Klüfte, Falten usw.) auf in-situ Spannungsverteilungen scheint angebracht zu sein, sowie des Einflusses der vereinfachten Interpretation (z.B. lineare Elastizität, Isotropie) der "Overcoring-" und "Doorstopper-"Testmethoden auf die Gültigkeit solcher Verfahren.

Bonaface legt ein wertvolles Beispiel eines Druckstollenentwurfs vor. Die Erkenntinis, daß das Gebirge bedeutend zur Widerstandsfähigkeit der Betonauskleidung gegen Innendrücke beitragen könnte, führte zu einer erheblichen Reduktion der erforderlichen Betonauskleidung als auch der erforderlichen Panzerung.

Bergman und Stille bringen eine sorgfältige Analyse eines Gebirgsschlagsproblems einer großen unterirdischen Ölspeicherkaverne, sowie den praktischen Ansatz zu seiner Lösung. Es wurde festgestellt, daß ein mit Felsnägeln im Gestein befestigtes Drahtgeflecht als befriedigendes Mittel zur Verhinderung von Abplatzungen diente. Die augenscheinliche Unterschiede in der Gesteins"härte" - vom Index der Punktlast - führten angeblich zu entsprechenden unterschiedlichen Spannungen, die sich in den härteren (und spröderen) Teilen konzentrierten. An dieser Stelle sollte wieder vor der Extrapolation der in-situ Spannungsmessungen über die unmittelbare Umgebung der Messungspunkte (einige Bohrkerndurchmesser) hinaus gewarnt werden, vor allem in geklüfteten und geologisch vielfältigen Gesteinsmassen.

Die Beiträge von Saito et al. und von Bai et al. befassen sich mit dem Problem des Gebirgsschlags sowohl in einem Tunnel als auch in einer Kraftwerkskaverne, und ferner mit den Zonen, die durch den scheibenförmigen Zerfall von Bohrkernen als schlagsgefährdet erkennbar sind. Die Verfasser erkennen an, daß das Zusammentreffen von Gesteinen mit großer Bruchhärte (spröde) mit hohen Spannungen in diesen Bereichen das Auftreten von Gebirgsschlägen begünstigt. Die wachsende Anzahl von Gebirgsschlägen im Tiefbauwesen weist darauf hin, daß die Erörterung von Verfahren, die die schädlichen Folgen minimieren, und die zum Teil vielleicht vom Erzbergbau entlehnt sind (z.B. Entlastungsverfahren, Wasserinfusion), wertvoll sein könnte.

Obwohl in der Fallstudie des Majes-Projekts von Ceriani und Nord wenig zum Thema Felsmechanik gesagt wird, enthält sie doch zumindest eine für Felsmechaniker wertvolle Bemerkung: "Eine der wichtigsten Erfahrungen aus diesem Projekt ist die, daß sich die Untersuchungen auf die schwierigen Abschnitte konzentrieren sollten. In diesem Falle machten schwierige Gesteinsabschnitte nur 2,3% der Tunnelstrecke aus, die Ausbruchskosten und -dauer waren jedoch das 5- bis 10fache. Die Bemühung, schwierige Gesteinsbedingungen zu orten, sollte diesem Verhältnis entsprechen."

Es scheint daher wert, gezielt Einzelheiten darüber zusammenzutragen, wie schwierige Gesteinsbedingungen behandelt worden sind, mit oder ohne Erfolg und in verschiedenen Situationen, sowie Vorschläge zur wirksamsten Bewältigung solcher Vorfälle.

Im Großen und Ganzen stellen die dem Kongreß vorgelegten Beiträge über große untertägige Hohlräume einen vernünftigen Querschnitt gegenwärtig angewandter felsmechanischer Grundlagen bei solchen Projekten dar. Numerische Verfahren, insbesonders die Methode der Finiten Elemente, dienen als Grundlage zur Beurteilung der Standfeste der Kavernen. In-situ Messungen bestätigen oder korrigieren die mechanischen Eigenschaften und in-situ Bedingungen für das numerische Modell. Die Grenzen solcher Analysen und in-situ Messungen werden wohl erkannt, und man bemüht sich, das tatsächliche Verformungsverhalten sowie die Standsicherheit der Ausbrüche zu überwachen. Die "gesunde" Urteilsfähigkeit spielt immer eine wichtige Rolle bei der Interpretation der üblichen Unterschiede zwischen Modell und Wirklichkeit, die sich meistens aus Variationen der geologischen und felsmechanischen Daten ergeben.

Die Notwendigkeit weiterer Überprüfung der leistungsfähigen numerischen Modelle sollte betont werden. Solche Modelle finden breite Anwendung als Grundlage von Dimensionierungsuntersuchungen, sind jedoch in relativ geringem Ausmaß physisch verifiziert worden, insbesonders was die Interaktionen zwischen Diskontinuitäten und dem darausfolgenden Gesamtverformungsverhalten anbelangt. Die realistische Darstellung von 'Diskontinuitäten' in physischen Modellen bedingt die Untersuchung von Testproben (bzw. Montagen) zwischen einem und mehreren Metern linearer Größenordnung. Versuche solchen Ausmaßes sind kostspielig und müssen sorgfältig ausgewählt werden, um Kombinationen von Bedingungen zu testen, die eine kritische bzw. besonders informative Überprüfung des numerischen Modells ermöglichen. Solche Überprüfungen sind das fehlende Glied zwischen dem Modell, das die Berücksichtigung von Diskontinuitäten ermöglicht und der vertrauensvollen Anwendung solcher Modelle bei der praktischen Planung.

Die beachtliche Größe permanenter Aufschlüsse unter Tage (manche Ausbrüche im Salz sind bis zu 20m breit und 100m hoch; Ausbrüche im festen Gestein sind bis zu 20m breit und 30-40m hoch), sowie die zunehmende Verwendung langer Kabelanker, werfen die Frage auf, wie groß ein 'permanenter' Ausbruch unter Tage sein darf, in welchen Gesteinstypen. Oder, andersherum, inwieweit kann man Geräte und Systeme (z.B. Kraftwerke) kostengünstig neu entwerfen, um sie dort, wo die Gebirgsart und -beschaffenheit kleinere Hohlräume als die gewöhnlichen erfordern, in kleineren, vielleicht mehreren Hohlräumen unterzubringen?

THEMA C5: BESEITIGUNG NUKLEAREN ABFALLS UND THERMISCHES VERHALTEN VON FELSEN

Die Einlagerung nuklearen Abfalls, auf verschiedene Art und Weise verpackt, in Aufschlüssen unter Tage, wird derzeit als Lösung des Endlagerungsproblems vorgezogen. Im Prinzip bedeutet dieser Plan den Vortrieb von Gallerien oder Lagern im festen Gestein, dessen Teufe, sowie geomechanische und hydrologische Eigenschaften gewährleisten müssen, daß die eingelagerten Abfallstoffe im Lagerhorizont eingeschlossen sind. Die Eigenschaften und die Teufe des Lagers sollen das Austreten von Radionukleiden in die Biosphäre unmöglich machen.

Die Endlagerung nuklearer Abfallstoffe unter Tage wird durch die langfristigen Wärmeausstrahlungseigenschaften dieser Stoffe erschwert. Die Einlagerung von Wärmequellen im Gebirge erzeugt thermische Spannungen und es entsteht die Möglichkeit von Bruch-, Absonderungs- und Rutscherscheinungen an Schwächeflächen. Normalerweise, wo Gesteine in der Teufe mit Formationswasser saturiert sind, werden die Wärmequellen Konvektionsströmungen im Grundwasser in allen natürlichen und baubedingten Klüften im Gestein verursachen.

Radionukleide können vom Lager in die Biosphäre gelangen, indem sie im Grundwasser gelöst entweder durch natürliche oder thermische Strömung hinaustransportiert werden. Die Strömung des Grundwassers wird durch das Vorhandensein, die Größe und die Verbindung von Spalten, sowie von Wärmegefällen bedingt. Diese wiederum werden durch den Spannungszustand des Gebirges (einschließlich der thermischen und Spannungskomponenten) sowie die Gebirgseigenschaften bedingt. Die Entwicklung effektiver Lagerungsstrategien setzt umfassende Kenntnisse der thermo-hydromechanischen Gebirgsreaktionen auf den Ausbruch und auf die eingelagerten Wärmequellen voraus.

Die Grundalternativen eines hypothetischen Lagers unter gegebenen Standortsbedingungen und unter einer gegebenen Wärmebelastung sind:
(a) Der Transport untertage ausgekommener Radionukleide zur Biosphäre geschieht dermaßen langsam, daß diese Stoffe keine langfristige biologische Gefahr darstellen. D.h., nur unschädliche Zerfallsprodukte treten an der Oberfläche auf nach einer unbestimmten Zeit.
(b) Das Lager und seine ortsnahen Komponenten sind so beschaffen, daß das Austreten radioaktiven Grundwassers vom Lager in die weitere Umgebung oder gar die Biosphäre effektiv ausgeschlossen ist.

Alternative (a) verwendet bei der Lagerung eine natürliche geologische Barriere. Alternative (b) basiert auf dem Entwurf und der Konstruktion von technischen Barrieren, um sowohl die Verseuchung des Grundwassers als auch dessen Bewegung im Gebirge zu beherrschen. Beide Alternativen, oder auch eine Kombination davon, können bei der Bestimmung der primären Betriebsart eines Lagers eingesetzt werden. Man braucht kaum zu erwähnen, daß die gewählte Betriebsart und der Lagerentwurf den hydromechanischen Eigenschaften des Gebirges entsprechen muß, oder daß ein gewählter Standort der gewählten Betriebsart genau entsprechen muß.

Das Problem der Endlagerung nuklearen Abfalls, insbesonders die Notwendigkeit, das thermo-hydro-mechanische Verhalten des Gebirges auf Jahrhunderte oder Jahrtausende vorauszusagen, stellt hohe Anforderungen and die Geomechanik als Bergbauwissenschaft. Es liegt auf der Hand, daß die im Lagerbau angewandte Entwurfsmethodik der in der Einleitung erwähnten dritten Kategorie angehören muß. D.h., sie soll einer umfassenden Kenntnis der großräumigen Zusammensetzung des geologischen Mediums entspringen. Die Möglichkeiten, eine auf der Beobachtung basierte Methodik anzuwenden, sind hier geringer als anderswo in der gegenwärtigen konventionellen Praxis der Geomechanik (einige Ausnahmen werden unten erörtert).

Unter den vorgelegten Beiträgen beschreibt der von Hudson genau die Probleme, die mit der Beseitigung schwach radioaktiven Abfalls (nicht wärmeausstrahlend) im Endlagerungsprogramm von England verbunden sind. Der Verfasser schlägt allgemeine Grundsätze vor, die mit den oben erörterten übereinstimmen. Das von Hudson erwähnte Hauptproblem ist das der Gebirgsvoruntersuchung: Ausgerechnet in dem Bereich, wo die ausführlichsten Informationen über die Beschaffenheit des Gebirges notwendig sind, sind detaillierte Bohrungen ausgeschlossen. Das Problem besteht natürlich darin, daß die umfangreiche Standortuntersuchung durch Kernbohren die hydraulische Integrität des Gebirges zerstört. Nur eine begrenzte Perforation des potentiellen Lagergebirges durch Bohren kann toleriert werden. Eine Lösung dieses Problems wird unten vorgeschlagen.

In den letzten Jahren hat die Notwendigkeit, die thermomechanischen Eigenschaften von Gebirgen und Gesteinsarten zu verstehen, eine Anzahl von Labor- und Felduntersuchungen angeregt. Ehara et al. legen die Ergebnisse von Laborversuchen zum thermischen Verhalten kristallinen Gesteins vor. Granit und Basalt reagierten im mikroskopischen Bereich mit der Anlage von Haarrissen im Mineralverband . Solche irreversiblen Erscheinungen dürften von ausschlaggebender Bedeutung für das langfristige Verhalten eines Kristallingesteins als Endlagerstätte sein. Das Problem der in-situ Bestimmung der thermischen Eigenschaften eines Gebirges wird von Kuriyagawa et al. erörtert. Die Wärmeleitfähigkeit eines Granitstockes sowie dessen Temperaturvariationen wurden durch Rückanalysen der um eine Wärmequelle herum vor Ort beobachteten Temperaturmessungen bestimmt. Zur Inversion wurde ein Finite-Element-Programm verwendet.

Bei ihrer Erörterung der Untersuchungen zur Thermik des Gebirges im Zusammenhang mit der Bewetterung im Campiano-Bergwerk beschrieben Barla et al. die Messung gebirgsthermischer Eigenschaften sowie die Entwicklung eines Rechenmodells, das den Wärmetransfer zur Grubenluft zu erfassen versucht. Diese Analyse ist in der Praxis viel komplizierter als die Erörterung in diesem Beitrag andeutet. Die umfangreichen Arbeiten von Starfield (1966, 1967) dürften als der geeignetste Ausgangspunkt von Untersuchungen zum Wärmetransfer in die Bewetterungsströme von Bergwerken betrachtet werden.

Der Beitrag von Bjurstrom et al. behandelt die großangelegte Speicherung von Wärme niedrigen Grades in erhitztem Wasser in untertägigen Hohlräumen im Granit. Die Art des Wärmespeicherungs- und rückgewinnungsprozesses hat zur Folge, daß das umgebende Gestein langfristigen Temperaturzyklen zwischen 40 und 115 Grad Celsius ausgesetzt wird. Die Verfasser kommen zum Schluß, daß die mit thermischen Stömungen verbundenen Hystereseerscheinungen zur asymptotischen Abnahme der gesamten Gebirgsverformung innerhalb der unmittelbaren Umgebung des Ausbruchs führen, und damit zur mechanischen Stabilisierung des Grenzgesteins.

Während bisher kein Endlager für hochradioaktive Abfallstoffe konstruiert werden mußte, hat man ein vorläufiges Untertagelager entworfen und ausgelegt. Es wird von Roshoff et al. beschrieben. Es befindet sich in geringer Teufe im Kristallgestein. Die Abfallagerung wird von dem Stoff selbst sowie von verschiedenen eingebauten technischen Barrieren gesichert. Da keine thermischen Belastungen im Inneren des Gebirges vorhanden sind, ist das langfristige Verhalten dieses Lagersystems dem Endlagerungsproblem nicht sehr relevant.

Einen Einblick in eine der Alternativen zur Endlagerung hochradioaktiver Abfälle, die in den U.S.A. in Betracht gezogen wird, gibt der Beitrag von Wagner et al. Die vier potentiellen Lagerstandorte befinden sich in Salzdomen und schichtigen Salzlagern, in Teufen zwischen 500m und 1000m unter der Oberfläche. Die im Beitrag beschriebene Rechenaufgabe bestand darin, auf jedem der vier Standorte eine prototypische Lagergallerie einer durch eingelagerte Abfallbehälter bedingten prototypischen Wärmebelastung auszusetzen. Standortspezifische Bedingungen, u.a. die in-situ Temperatur- und Spannungsverteilung, die Salzeigenschaften sowie deren Kriechverhalten, wurden gebührend berücksichtigt. Ein nicht näher beschriebenes "Ausgangs"-Kriechgesetz mit entsprechenden empirischen Parametern für jede Salzformation wurde im Rahmen eines Finite-Element-Rechenprogramms verwendet, um das räumliche Verhalten der Endlagerhallen jedes Standorts modellhaft darzustellen. Die Ergebnisse der Untersuchungen weisen erheblich unterschiedliche Verhaltensweisen in den vier Standorten auf, die sich aus den unterschiedlichen Kriecheigenschaften, herrschenden Spannungsfeldern und thermischen Bedingungen ergeben. Die Ergebnisse weisen jedoch darauf hin, daß die technischen Voraussetzungen für den Betrieb eines Abfallagers, d.h. die Faktoren Rückholbarkeit, örtliche Standsicherheit und herrschende Temperaturbedingungen im Betriebsraum, bei einer praktischen Lagerplanung erfüllt werden können.

Es wurde oben behauptet, ein Hauptproblem der Lagerplanung darin liegt, daß detaillierte Untersuchungen der Gesteinsparameter am gewählten Standort ausgeschlossen sind. Eine vernünftige Lösung dieses Problems besteht darin, ein Gebiet mit einheitlicher Geologie über größere Erstreckung hin auszuwählen, die Formationen standortsfern zu untersuchen, und dann mit geostatistischen Verfahren die Formationseigenschaften am und um den Lagerstandort herum zu interpolieren. Die Standortsuntersuchung selbst wird dann gleichzeitig mit dem Aufschluß des Lagers durchgeführt. Ein solches Verfahren ist nur dann vertretbar, wenn für die Konstruktionsphase ausreichend Vorsorge und Spielraum eingeplant ist, um widrige geologische Verhältnisse, wie sie mit Sicherheit beim Auffahren des Grubengebäudes erwartet werden können, technisch zu bewältigen. Das Vorhandensein widriger Erscheinungen und das Fehlen früher Planung und Festlegung von Maßnahmen, die hydraulische Integrität des Lagerbereichs während der Konstruktion wiederherzustellen, könnte einen sonst akzeptablen und sicheren potentiellen Lagerstandort unbrauchbar machen. Eine zufriedenstellende Handhabung dieses Aspektes der Lagervorplanung und des Lagerbaus scheint daher ein Schlüsselbestandteil der geomechanischen Planung für das Endlagern nuklearen Abfalls zu sein.

LITERATUR

1. Beer, G. and Meek, J.L. (1981) "The Coupling of Boundary and Finite Element Methods for Infinite Domain Problems in Elasto-Plasticity". In "Boundary Element Methods", C.A. Brebbia, ed., Springer-Verlag, New York, pp 575-592.

2. Brady, B.H.G., and Wassyng, A. (1981) "A Coupled Finite Element-Boundary Element Method of Stress Analysis". Int. J. Rock Mech. Min. Sci. and Geomech. Abstr., Vol. 16, pp 235-244.

3. Cook, N.J.W. (1965) "The Basic Mechanics of Rockbursts". In "Rock Mechanics and Strata Control in Mines", S. African Inst. Min. Metall., pp 56-66.

4. Cundall, P.A. (1971) "A Computer Modell for Simulating Progressive Large Scale Movements in Blocky Rock Systems". Proc. Int. Symp. on Rock Fracture, ISRM, Nancy, Paper II-8.

5. Cundall, P.A. (1974) "A Computer Model for Rock Mass Behavior Using Interactive Graphics". U.S. Army Corps of Engineers, Technical Report MRD 2-74 (Missouri River Division).

6. Cundall, P.A. (1980) "UDEC - A Generalized Distinct Element Code for Modelling Jointed Rock". Technical Report to United States Army (European Research Office).

7. Gumbel, E.J. (1958) "Statistics of Extremes". Columbia University Press, New York.

8. Kendorski, F.S. (1978) "The Cavability of Ore Deposits". Mining Engineering, June, pp 628-631.

9. Lorig, L.J. and Brady, B.H.G. (1982) "A Hybrid Discrete Element Boundary Element Method of Stress Analysis". Proc. 23rd U.S. Rock Mechanics Symposium, pp 628-636.

10. MacMahon, B.K., and Kendrick, R.F. (1969) "Predicting the Block Caving Behavior of Orebodies". S.M.E. of A.I.M.E. Preprint No. 69-AU-51, S.M.E. Annual Meeting.

11. Mahtab, M.A. and Dixon, J.D. (1976) "Influence of Rock Fractures and Block Boundary Weakening on Cavability". S.M.E. of A.I.M.E. Transactions, March, pp 6-12.

12. Mahtab, M.A. and Yegulalp, T.M. (1982) "A Rejection Criterion for Definition of Clusters in Orientation Data". Proc. 23rd U.S. Rock Mechanics Symposium, pp 116-123.

13. Rudnicki, J.W. and Rice, J.R. (1975) "Conditions for the Localization of Deformation in Pressure-Sensitive Dilatant Materials". J. Mech. Phys. Solids, Vol. 23, pp 371-394.

14. Salamon, M.D.G. (1979) "Stability, Instability and the Design of Pillar Workings". Int. J. Rock Mech. Min. Sci. and Geomech. Abstr., Vol. 7 pp 613-631.

15. Starfield, A.M. and Fairhurst, C. (1968) "How High Speed Computers Advance Design of Practical Mine Pillar Systems". Eng. Min. Journal, May, pp 78-84.

16. Starfield, A.M. and Dickson, A.J. (1967) "A Study of Heat Transfer and Moisture Pick-Up in Mine Airways". S. African Inst. Min. and Met., Dec. 1967, pp 211-234.

17. Starfield, A.M. (1966) "The Computation of Temperature Increases in Wet and Dry Airways". Mine Ventilation Soc. of S. Africa, Vol. 19, pp 157-165.

18. Starfield, A.M. and Detournay, E. (1981) "Conceptual Models as a Link Between Theory and Practice: A General Approach with a Particular Application to Tunnelling". Proc. 22nd U.S. Rock Mechanics Symposium, pp 398-401.

19. Sterling, R.L. (1980) "The Ultimate Load Behavior of Laterally Constrained Rock Beams". Proc. 21st U.S. Rock Mechanics Symposium, pp 533-542.

20. Sterling, R.L. and Nelson, C. (1968) "An Experimental Investigation of the Strength of Laterally Constrained Rock Beams". Experimental Mechanics, July 1978, pp 261-268

21. Thompson, J.M. and Hunt, G.W. (1973) "A General Theory of Elastic Stability". Wiley, London.

22. Vardoulakis, I. (1979) "Bifurcation Analysis of the Triaxial Test on Sand Samples". Acta Mechanica, Vol. 32, pp. 35-54.

23. Voegele, M.D. and Fairhurst, C. (1982) "A Numerical Study of Excavation Support Loads in Jointed Rock Masses". Proc. 23rd U.S. Rock Mechanics Symp., pp 673-683.

24. Yeung, D. and Brady, B.H.G. (1982) "A Hybrid Quadratic Isoparametric Finite Element - Boundary Element Code for Underground Excavation Analysis." Proc. 23rd U.S. Rock Mechanics Symposium, pp 692-703.

DEEP MINING EXCAVATIONS

Excavations minières à grande profondeur

Tief unterirdischer Bergbau

Barry H. G. Brady
University of Minnesota, Minnesota, USA

Mr Chairman, ladies and gentlemen, the first point I would like to address is the absence of Professor Fairhurst, he has asked me to convey his apologies and to note that the fact that he is now in Hawaii and sitting in the sun drinking beer was in no way intentional, but it was strictly an error that caused that.

This report is on Theme C of the Congress, Deep Underground Excavations, and my contribution is related to deep mining excavations. Originally I was told that I should be profound and provocative in 12 to 15 minutes, yesterday I was told I had to be profound and provocative for 30 minutes. Now the latter charter strikes me as being profoundly boring on the one hand or quite dangerous on the other, so I assure you I won't try to stick to that commission.

It is appropriate that the Technical Proceedings of this Congress open with the discussion of mining rock mechanics. Australia is a great mining nation, is one of the leading mineral producers and exporters in the world, has made significant contributions to the development and evolution of mining practices, and this has been achieved by highly talented individuals working within the mining companies and by commitments by the mining companies themselves: particularly, it was noted earlier this morning by companies such as Mount Isa Mines Limited and The Broken Hill Mines; also because of a commitment by the Federal Government through the Commonwealth Scientific and Industrial Research Organization and the Division of Applied Geomechanics, and I am pleased to note that I now work with that group which has been a productive and innovative group in geomechanics research in Australia.

Just a general comment before we start on the core of this presentation, there were about 24 papers presented in the Section related to Deep Mining Excavations but in that set of papers none were related to caving, and I think that should be of particular concern to the geomechanics business because rock caving or sublevel caving but particularly rock caving is one of the least understood methods of mining geomechanically and is a potentially fruitful area for research. With the emergence of very powerful computational methods its a method which could be researched in a way much more efficiently than was possible previously through these computational models. Also there is the possibility of physical modelling of those processes.

Considering a schematic cross-section of a set of excavations well below the ground surface, so we are dealing with a supported mine structure. For the rest of this presentation I will only be concerned with supported mining methods, having noted the absence of papers on caving. The problem we are dealing with is the design of a set of mining excavations with an intermedium, with a traction free surface. The objective is to control the displacements in the mine near field domain. The problem in mining practice in general is that the rock mass, that is the mine near field rock, is subjected to stresses greater than the in situ strength of the rock, or slip and separation occurs on pieces immediately around the mine excavations. Now that is in distinct contrast to civil engineering rock mechanics and as far as I can see is the core of the mining engineering or the mining rock mechanics problem. That is to be able to achieve efficient extraction in the medium, over-stress the rock or induce significant slip and separation and yet maintain some level of overall control. So the rock mechanics, and here I disagree with some of the earlier speakers this morning, mining rock mechanics as far as I am concerned is the medium, is the situation, in which one is likely to learn most about the most interesting properties of rock masses under load. So in the supported methods of mining the objective is to maintain displacements immediately around these excavations to small orders, typically elastic orders of magnitudes. One should be designing to preclude large rigid body displacements. Now although the geomechanics principles of caving and supported mining methods are quite distinctively similar in fact there is a common methodology from which mining rock mechanics would proceed.

These principles, these ideas are illustrated by some general notions on how mining rock mechanics proceeds in some sort of operational environment. You will note that the methods available are designed by precedent, and that would be the way in which underground mining excavations would have proceeded for many years, probably up until the mid 60's for example. There would have been very little real engineering mechanics input into design of underground excavations. This design by precedent is represented in modern mining rock mechanics practice by these rock mass classification schemes. The question of classification schemes is a rather contentious one, and in general engineering geologists would tend to favour them, but anybody with a strong background in engineering mechanics would treat them with a certain amount of suspicion. Now I would make one reservation about that statement, that is that there is a real role for these classification schemes in certain aspects of mining rock mechanics. The particular area is of course in the business of pre-feasibility studies where data on the state of the rock mass is sparse and one is looking for some fair assessments of the way the rock mass is going to behave or to make some initial estimates of the stope spans, pillar spans and so on, which might be attainable in the particular mining environment and the prospective mining environment.

The paper by Page for example is a useful indication of the way in which those schemes would be used. The paper by

Pasamehmetoglu demonstrates the deficiencies of the pervasive application of these geomechanics classifications schemes and mining rock mechanics practice and what the authors of that paper noted was a distinct divergence between the predicted performance of mining excavations on the basis of the classification schemes and the actual performances observed as mining progressed. I will just note at this stage two other schemes: the observational principle, which has always been exploited in geomechanics practice; and a more rigorous principle where one would be appealing to the basic mechanics of the constitutive behaviour of the material, the governing equations for the medium, and then one would look at the initial conditions and final conditions of the medium such as associated with excavation. These are the three methods, there is an increasing emphasis on engineering mechanics, it also represents an increased effort in the formal implementation of these schemes in rock mechanics practice. I am going to spend a fair bit of time, probably the majority of this talk, on the observational principle, at the end of it we will return to this question of rigorous principle in geomechanics practice.

There are six phases, six components of the methodology of the method of application of the observational method of mining excavation design. The main point about this system is that is recognizes the inherent sparseness of geomechanical data which can be collected in initial site investigation of the mine, and the need for formulation of a site geotechnical model which is going to be modified as pre-production development proceeds and also as production proceeds. Of the 21 papers which have been presented in this session at least 12 of them would have been implicitly related to this observational principle. One of the outstanding cases is provided in the paper by Bywater, Cowling and Black from Mount Isa. It represents the application of the observational method, to a very complex mining situation in which a very simple model is used in the design basis, in that case the boundary element method. What this shows us is that excavation sequences and stope and pillar dimensions can be developed to quite a rigorous engineering standard providing this closed loop arrangement is proceeded with in a rational way. The other point I would make about this method is that at this stage computational methods are firmly embedded in computational schemes of stress analysis, firmly embedded in this methodology. For example, of the papers implicitly related to this methodology at least half of them, maybe three quarters of them exploited either the boundary element methods or finite element methods of stress analysis. The methods which were used were not in fact always exploiting these simple constitutive models of rock mass and in many cases more complex behaviour was assumed. For example the German workers were dealing with supposedly softening materials and also with materials subject to creep.

I would like to comment on a few particular aspects of this method, particularly related to the question of site exploration and characterization. I particularly want to talk about the state of stress in a fractured medium. A typical assumption in geomechanics practice is that it is a straightforward matter to go into a mine or a mining area and measure the pre-mining state of stress. All mine environments are fractured, the minerals are there because of the fracturing of the mass, that would be a reasonable geological interpretation at least, and the question is how does one measure or how does one determine a representative value for the pre-mining field stress tensor in this sort of medium. What I have set up here is a simple model of a jointed medium transgressed by a continuous feature and that represents just a body of rock embedded in this domain which is an infinite elastic body. If we look at the next slide, we look at the stress distribution in that medium. In this case we are looking at horizontal far field stresses which represent the average state of stress in the medium, we say 75 horizontal and 25 vertical and you see the stress distribution in the medium. If we were to determine the state of stress satisfactorily in this system, then one would need to sample the state of stress, there are a number of points in there, and the average value would represent the quantities which one would use in some subsequent computational analysis. The point I am making here is that it is not a straightforward matter to determine the state of stress in a jointed or faulted domain, but one should develop some sampling strategy, measure the state of stress at a certain number of points and have some volume averaging scheme which is going to allow you to establish the average far field state of stress for that particular domain.

In a case with a high deviator stress component and therefore you would expect to see something odd. This is the same case where we were looking at contours of major principal stress, contours of minor principal stress contours in a hydrostatic stress field. What was done there was to relax this computational model, allowing it to proceed through a series of relaxations, relaxing the far field stresses. What you see is the same case. There should have been 25 horizontal 25 vertical in this case, but you see these large divergences from that hydrostatic state. The point is then, that if one is going to measure the pre-mining state of stress and apply it in this observational methodology, in fact one should be making a large number of observations through a large volume of rock. Now the general point I am making here, of the complex tasks presented by measuring the state of stress in a fractured and jointed rock mass is a fact borne out by the paper by Chunting and Zhaoxian which will be presented shortly.

The third component of the methodology which I presented was design analyses, and as I have noted already the computational methods of analysis are well embedded in mining practice. In fact the papers which are presented display a very general acceptance of the role of computational mechanics in mining rock mechanics practice. About half the papers had this strong computational mechanics base. More recently there have been computational schemes developed which aim to exploit the advantages of the different methods of analysis, for example the finite element method is a very powerful method of analysis because it can accept quite complex constitutive behaviour, but it is inefficient in execution because you have to discretize the complete problem domain. The boundary element method on the other hand allows you to construct a super element which can then be linked up with these more complex methods of analysis.

The general principle can be demonstrated by a tunnel, but it could be any mining excavation and then one would define the near field of that mining excavation by a set of elements in which complex behaviour is going to be allowed to occur. It might be just slip and separation, it might be a sort of highly non-linear deformation of the system and then that can be embedded in an infinite domain of elastic material so that this surface represents the interface between a boundary element domain and a finite element domain. You can actually construct a stiffness matrix for that surface and treat the far field as an elastic super element. And that represents, as far as I can see, a major advance which is possible in mining geomechanics practice in the near future and I know there are people in the Division of Geomechanics, and also in universities in Australia, who are active in that area. For example there is paper in this meeting presented by Beer and Meek. Now the advantage of this system is that the power of analytical method is concentrated in the area where it is needed, that is immediately around the mining excavations. Out in the far field small perturbations occur and an elastic model is quite appropriate. The near field is where the action is and where one needs the computational capacity. Now mining excavation design problems typically involve that sort of problem. In the near field we noted that the rock is typically overstressed or it is going to be subject to slip and separation. One needs the capacity to handle that in any rigorous computational approach and these methods as I see it provide the way forward in that area.

Now this is a bit of a digression, but I was hoping to discuss or see the presentation of the paper by Drs. Beer and Meek, and I would like to illustrate a bit of work which one of my students has done at the University of Minnesota related to this particular scheme. We see a typical jointed medium and it demonstrates the major response. There is a block jointed medium in which one is going to generate an excavation, there are some notions which float around related to arching effects and so on, they are arm waving notions; they are not rigorous in a mechanical sense, but the terminology which is used is a ground arch above the excavation. In the immediate roof of the excavation there is linear arching action also called a voussoir arch. You can look at the way in which one can approach the analysis of this sort of problem. So one would have a blocky domain in which one is going to create an excavation and that is going to be embedded in the elastic medium which constitutes the far field for the mine excavation. So when these blocks are removed there will be the possibility of slip and separation, in this case we are dealing only with that mode of response, slip and separation between these blocks which constitute the excavation periphery. If we look at the next slide we see the effect of this. This is in the field of Pxx the horizontal field stress of 0.34 MPa, Qyy. The units don't make a difference because it is dealing with friction only so you can take that off. So the horizontal stress is half the vertical stress. What one sees in this sort of model is zones of separation, so that represents separation on the horizontal joint, that represents horizontal separation on the vertical joint and then one sees a slip in the abutment area. You also see this opening of cracks in the floor which is of no mining consequence but at least the model does the right sort of thing there. This is on the effect of combined field stresses and gravity loading. Now that is an entirely reasonable conceptual response of the medium to the generation of the opening. If we look at the state of stress around that excavation what one sees is that if slip and separation were prevented this represents the geometry of the opening and this represents the state of stress around the medium, this would represent the horizontal stress acting on features above the haunch of the opening if the joints were locked that is high friction if you introduce a low angle of friction then in fact you do see that slip occurs, so slip and separation can be readily modelled in these sorts of cases. A similar consideration applies to the paper presented by Drs. Beer and Meek except in their case rigid body translations and rotations to blocks are not accepted. I am sure Dr. Meek will want to discuss that point later.

There was a paper presented in this Congress by Crea et.al. related to the Campiano mine in Italy that reports some observations or some analysis of the performance of mine backfill in the support of those mine excavations. The comment I would make about that is that the authors could well have taken some note of the work that has been done in the last 10 to 15 years in Australia and Canada on mine backfill. In straightforward terms that paper seems to be a long way off the pace, and I think it gave a rather misleading view of how backfill performs. In fact there seems to be, although again I stand to be corrected, some sort of consensus that the mechanics of mine backfill is well understood.

Now I would like to suggest that is not the case. There are at least three modes of behaviour in typical mining application practices, there are at least three modes of performance of the fill as a support element. It can act as purely a local support at the surface of the de-stressed element of rock. So for example, if one had a thin remnant separating two stopes which had been destressed and one would normally emplace fill in the excavations, just to maintain the mechanical integrity of that system. Very little effect on the stress distribution in the medium would merely provide a kinematic restraint to displacement of those blocks. I have suspected that in many mining applications that mode, or this other one where the role of the fill is to generate a support pressure which then modifies the stress distribution of the medium, where in fact it acts as a local support system so any displacements of these units against the fill mobilizes the passive resistance and that acts to allow the maintenance of a high stress gradient in the fractured domain. They are what I call the most common modes of behavior of backfill in mining practice. The method which was analysed by Crea and I think it tends to be the method which is used generally in practice is to assume that the rock mass demonstrates pseudo-continuous behaviour and that the fill acts as a local support system. Any displacement of the stope boundary mobilizes a state of stress in the body of the fill. I do not think that represents the ways in which fill is currently being used with a realistic model of its performance.

Now I would like to make another comment about backfill, and that is that its mode of response or its performance and significance is definitely going to increase as mining depths increase. For example, one could postulate that as the mining depth increases the mode of the fill will be first of all by compression and consolidation to absorb the energy which would otherwise be propagated through the mining domain and that represents an energy approach to fill design in deep mine excavations. The second point is, I think quite possible that fill would act as a sink or absorber for the transient energy which is released. The gist that I am making here is that there is plenty of scope for investigations into the performance of fill in deep mining excavations. Up until now we have been talking about problems of local stability, that is the rock in the immediate periphery of the mine excavation. I would now like to turn to the problem of mine global stability that is the potential for the mine structure consisting of a set of stopes and pillars, abutments in the near field rock to undergo a catastrophic change in geometry and to some small imposed perturbation. Now this is an issue which is addressed in four or five papers in the course of this Congress, bursts in South Africa bumps in other places and post peak performance of rock is to be considered by Dr Weber from France. In all cases I would suggest this is an area which is probably one of the most critical which remains in mining rock mechanics practice. I make this point because it is wrong to assume that bursts or mine global instability or bumps are restricted to deep level mining. The problem is a pervasive one, it will be prevalent in any areas where high extraction ratios exist and the near field rock mass, is capable of storing energy, which can be released in an unstable way to crush pillars and cause catastrophic displacements immediately around excavations.

The concept produced by Dr Salomon in a paper around about 1970 and it is the basis on which the understanding of mine instability has proceeded. What we are dealing with here is a specimen which could represent a mine pillar which is being loaded through a spring which will represent the near field rock. This line represents the performance characteristic of the specimen, and shows the so called strain softening behaviour. Whereas the performance of the country rock or the spring is represented linearly so it is assuming that it is a linear medium loading a specimen which goes through a peak in strength into this softening domain. What we are looking at here is a plot of the variation of the stiffness of the pillar as a function of the displacement or the degree of axial compression of that specimen. About two or three years ago Professor Brown and I went and looked at this problem of instability and so what we did was set up a series of hypothetical mining layouts to consider this problem. This is represented by idealisation of the specimen being loaded where a pillar is represented and can be extracted. We can measure the deformability or the stiffness of the country rock merely by applying loads at the pillar position and using some boundary element formulation for example to calculate the mine local stiffness. We considered a series of hypothetical mining layouts, all designed to achieve an extraction ratio of 75% and we looked at the stiffness of the pillars and the local stiffness of the country rock and collected information on λ to λ'. λ' represents a post peak stiffness, assuming that you can represent the post failure performance of a specimen in terms of the elastic response. What we were looking at was this ratio λ over λ' as a function of width to height ratio and we achieved these sorts of results. From these sources, this was a comprehensive survey on literature, we obtained bounds for the possible post peak response of mine pillars.

The resulting estimate, the so called stability index, is the sum of mine local stiffness and the post peak stiffness. Now the requirement for the stability of the system would be for $kl + \lambda'$ to be positive and what this showed is that for a homogeneous isotropic medium that in fact one does not according to the analysis attain stability. But for a medium with pre-mining factors in it in fact one can achieve a level of stability. The function of these natural factors is to in fact introduce a dissipative mechanism, a yielding mechanism in the pillar.

The method we used for that analysis was based on very simple geometry, and would not be applicable with more complex geometry. If you want to assess the stability of this mine structure we would have to undertake a more comprehensive analysis and the way in which that might be conducted is shown on the next slide where one would hope to set up a method of analysis where the constitutive equations, the governing equations for the system and the initial conditions would allow an analysis of the evolution of any instability in the system naturally. So this would represent a complication for the philosophy of a computational scheme which would allow one to pursue analysis of mine global instability. I will stop there, there is scope later in this session for more comprehensive discussion of the question of mine global instability.

DEEP UNDERGROUND EXCAVATIONS ESPECIALLY TUNNELS, COAL MINING AND SUBSIDENCE CAUSED BY THEM

Tiefunterirdische Exkavationen besonders Tunnel- und Kohlenbergbau und die dadurch verursachten Bodensenkungen

Excavations souterraines et profondes, notamment des tunnels, l'exploitation minière des charbons et des affaissements du terrain causés par eux

Yoshio Hiramatsu
Kyoto University, Kyoto, Japan

SUMMARY:
This general report describes shortly the recent researches carried out so far in the geomechanics and geotechnology concerning deep underground excavations, especially tunnels, coal mining and ground subsidence caused by them, which are the topics under the Sub-Theme C2, C3 and C4 in this Congress, and introduces the important part of the papers submitted to this Congress under these Sub-Themes to suggest particular area for discussion.

ZUSAMMENFASSUNG:
Dieser allgemeine Bericht beshreibt kurz die neuesten geomechanischen und geotechnischen Untersuchungen über tiefunterirdischen Ausschachtungen, besonders bei Tunnel- und Kohlenbergbau und die dadurch verursachten Bodensenkungen, welche Gesprächstoffe unter Sub-Thema C2, C3 und C4 dieses Kongresses sind, und leitet den wichtigen Teil der an diesen Kongress unter den vorgenannten Nebenthemen vorzutragenden Referate ein, um besondere Diskussionsbereiche vorzuschlagen.

RÉSUMÉ:
Ce rapport général déscrit brièvement des recherches récentes qui ont été éffectuées dans les domaines de géomécanique et de géotéchnique, sur des excavations souterraines et profondes, notamment, des tunnels, l'exploitation minière des charbons et des affaissements du terrain causés par ces premiers, ceus qui sont les sujets des sous-thèmes C2, C3 et C4 dans ce congrés, et introduit des parties importantes des études qui ont été présentées à ce congrés sous ces sous-thèmes pour but de suggèrer des zones particulières pour discussion.

1. TUNNEL

1.1 Introduction

Tunnel engineering techniques are very closely related to rock mechanics. It is not possible to summarize here the vast amount of informations now available on this subject, therefore only a few comments will be presented.

Research works on the tunnel engineering may be classified into two broad categories: one, of theoretical nature, which aims at explaining the process of deformation and failure of tunnels, and the other, of more practical nature, which deals with the design and construction of tunnels. Researches on rock mechanics related to these two categories of tunnel engineering are summarized as shown in Fig. 1.1.

Studies belonging to the first category do not have direct relationship with design and construction at the present time. They are, however, works of basic research and highly important for the development of tunnel engineering in the future. The basic research of the tunnel engineering contains many different approaches. In the first approach the natural ground is studied by transforming it into a mechanical model, followed by analysing stress, strain, displacement and stability. If the ground is presumed to be a continuous body, the analysis can be carried out by applying the continuum mechanics, i.e. the theory of elasticity, of plasticity and of visco-elasto-plasticity. An analytical solution may be possible in this case, provided that the boundary conditions are simple. If the boundary conditions are complex, numerical analysis by the newly developed digital computer must be carried out. The same goes true when complicated material properties must be taken into consideration. This approach, however, cannot be applied unless the crack spacing is much smaller than the size of the tunnel so that the natural ground can be considered as being a homogeneous body.

The continuum mechanics approach can also be applied, with some effort and modification, to the grounds where the rock bolts or rock anchors exist. It cannot, however, be employed if the crack spacing is of similar order as the size of the tunnel opening. In such a case, the other approach which takes into account the existence of discontinuity is necessary. It is not possible to carry out these analyses without numerical techniques. Cundall (1974) and also Kawai (1977) already put forward a method of numerical analysis of discontinuity by simulating a rock mass with rigid or deformable blocks. For example the analysis of stability for the ground where joints appear at the crown of a tunnel was conducted by Shi (1981) considering the equilibrium of the forces acting on each block.

Laboratory experiments act important roles to investigate the stability of rock masses with discontinuities. Base-friction modeling can provide an qualitative answer to the question of tunnel stability (Egger, 1979; Bray, 1981). Continuum model experiments have been conducted in laboratories (Kaiser, 1981; Konda 1981), but these model tests suffer from fulfilling the condition of similitude.

As for the researches belonging to the second category, the majority deals with field measurements. A great deal of measuring results, including the development of new measuring methods and their evaluation, have already been reported (Sakurai, 1981). Many of these reports are concerned with the NATM technique. In addition, ambitious studies dealing with the decision-making in designing stage have been done. These studies collect the results of previous measurements and apply them to the decision-making with the help of the procedure based on statistics and the probability theory (Einstein, 1978; Ashley, 1981).

In the tunnel engineering, the problems concerned with underground water can not be forgotten. The underground water flow will influence

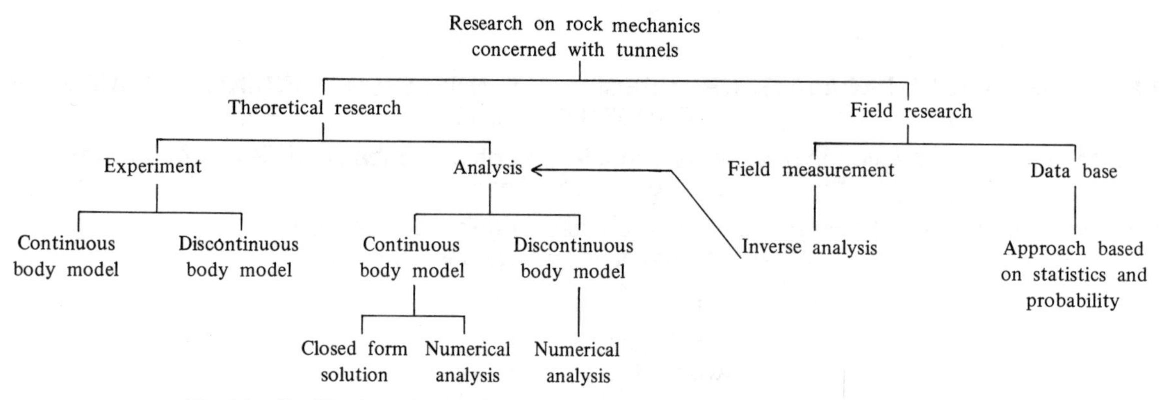

Fig. 1.1 Classification of research works on rock mechanics concerned with tunnels.

the stability of openings especially when the rock mass is fractured and sheared. Recently the hydro-mechanical analysis has been introduced to develope new numerical techniques.

Having surveyed the state of research on the tunnel engineering, now several topics in relation to tunnelling will be introduced including the Seikan tunnel, 53.85 km undersea tunnel of which pilot tunnel was completed in January 1983.

1.2 Analysis

There are two major ways to analyse the behaviors of ground and tunnel constructions; the mathematical analysis affording closed form solutions and the numerical analyses, such as the finite element method (FEM) analysis. The FEM analysis makes possibles to take into account the complex mechanical characteristics of the natural ground and the excavation sequences. Such factors, on the other hand, cannot be dealt with in the mathematical analysis. Consequently, they have to analyse on simple boundary conditions, e. g. a circular tunnel based upon the hydrostatic initial stress condition. Nevertheless, results of analyses by this method considering realistic mechanical characteristics, e. g. visco-elasto-plastic body and strain-softening characteristics, have been reported (Tanimoto, 1981; Brown, 1982).

In analysing underground structures, e. g. tunnels, it is important to evaluate correctly the parameters representing the mechanical properties of the ground. If failed in that, the result of analysis will bear little resemblance to the real-life behaviour, no matter how sophisticated the mechanical model may be. Futhermore, the estimation of the initial earth pressure is a major problem for the analysis. As we cannot give a fixed value for each parameter as the input data needed in the analysis, the parametrical analysis, which utilizes the upper and lower limit values, may be more effective. In analysing with various input data, the closed form solution, if any, may be convenient. For this purpose, the derivation of the analytical results and the development of various kinds of design charts have become important parts of tunnel engineering. Much research has already been done along this line. The parametrical analysis is also possible in case of the numerical analysis, and many methods have been devised. It is, however, desirable to have a simple analytical model in this case.

As mentioned previously, complex mechanical characteristics, even the tri-dimentional excavation process, can be taken into account in the analysis by employing such numerical methods as FEM analysis or the boundary element method analysis (Brady, 1979; Kobayashi, 1980). Nevertheless, determination of appropriate input data may be difficult in the present stage even if rigorous analytical results are obtained. On the other hand, considerable effort is being made in simplifying the analytical method, for example, the behaviour of the ground around the face of a tunnel, which requires, strictly speaking, tri-dimentional analysis, is made to be analysed bi-dimentionally with various efforts (Sakurai, 1978; Dillen, 1979; Kaiser, 1981). The analytical method obtained will yield, of course, only the approximate values, and may be possible to be applied to similar problems concerned with a tunnel.

The following steps are involved in designing a tunnel with the results of analysis explained so far. First, the initial stress as the load is estimated, in some cases by field measurements such as overcoring method. This is followed by the assumption of a mathematical model which embodies mechanical characteristics of the ground, e.g. elasticity, elasto-plasticity, isotropy, or anisotropy. The parameters representing the mechanical properties of the model are then determined by either laboratory experiments or in-situ measurements. Next, analysis is carried out using the parameters obtained to find the distributions of stress and displacement, and loads acting on timberings, linings, and rock bolts. Finally, the supporting method and the excavation method are designed so as to attain stability of the ground as well as the constructions for support.

The above-outlined approaches could work well theoretically, but in practice there are many problems because we have in general no sufficient knowledge of the initial stress and the mechanical characteristics of the ground.

Now the author should like to review some papers relating to the present subject submitted to this Congress. Romana et al. have proposed their rock classiffication system to provide fundamental data for planning construction of tunnels by means of field inventory of existing tunnels.

Doucerain has investigated the elastic moduli of the rock mass by an experimental study in which very high hydraulic pressure is applied to an underground penstock. Tan has proposed two dimensional dilatancy constitutive equations for swelling rock derived from some rheological tests, and calculated the dilatant zone around a tunnel based on the equations obtained. Saari et al. have analyzed squeezing of seams by driving a tunnel across them by means of theoretical analysis and numerical analyses on non-dilatant, cohesive and dilatant, frictional material models, and the results obtained have been discussed by comparing them with the laboratory model test results.

1.3 Design Method

As the design method for tunnels, the convergence-confinement method or the new Austrian tunnelling method (NATM) has recently attracted a good deal of attentions. For the sake of simplicity, this method will be described hereafter as NATM. This method is worth notice, because of the idea based on rock mechanics widely introduced and extensive practical use, while the conventional methods are also practiced worldwide which are based on experiences over a long period of time. The former procedure is expected to be applied to other underground structures. Since tunnel excavation consists of many repeated processes compared with those structures, it is more likely to be standardized.

It can be pointed out that this method has the following two major features. First, the inherent strength of the surrounding ground is tried to be utilized to support the opening to the greatest degree, while the support plays rather a supplementary role. It is commonly believed that the conventional supports should be designed to bear the load acting on them. Therefore, the NATM procedure is expected to de-

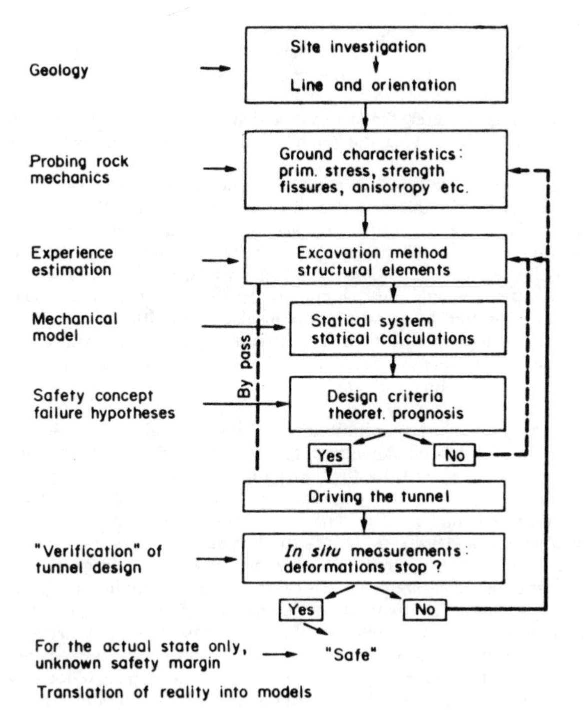

Fig. 1.2 Example of design procedure for tunnels proposed by Duddeck (1980)

crease the load to be borne by the support, which results in the economical support system.

It is almost impossible to predict the behavior of the ground around a tunnel in the stage of design. The second feature is in that in-situ measurements are carried out to provide the newest data about the characteristics and behaviours of the ground which are fed back to the analysis to contribute to complete a rational design of a tunnel. Fig. 1.2 shows the example of design procedure for tunnels proposed by Duddeck (1980).

The support theory for NATM is explained convincingly by the two characteristic lines such as those shown in Fig. 1.3 (Fairhurst, 1980). The one of the curves expresses the relationship between the radial pressure required to be applied to the ground to prevent the inward

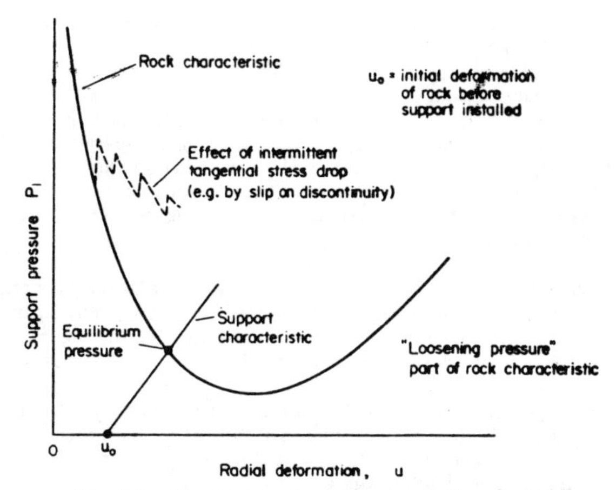

Fig. 1.3 Idealized rock and support characteristics and their interaction. (Fairhurst, 1980)

deformation and the inward deformation that has occurred at that point. The other curve is the confinement curve for the support representing the relationship between the outward pressure acting on the rock surface and the radial inward deformation of the support. The final equilibrium that enables determination of the support pressure is given by the intersection of these two curves. Provided that these two curves for the tunnel are given, this conception is very useful to design a tunnel support reasonably. Many investigators have made efforts to obtain these two curves analytically or experimentally, but these are not determined in practice at present (Kerisel, Lombardi, Egger, 1980). It is hoped to investigate the present problem by the aid of in-situ measurements.

It will be important to take into account the time dependent or three-dimentional factors for designing the tunnel support according to this theory. Fig. 1.4 shows an example of the variation of the characteristic lines which is obtained in the several stages of excavation and supporting of a tunnel (Duddeck, 1980).

Two papers have been submitted to this Congress concerning NATM procedure. Gartung et al. have reported their experiences in constructing twin tunnels, spaced by a narrow pillar, 36 cm thick, in weak sedimentary rocks. The success in economical and safe construction of the twin tunnels may be due to the refined techniques related to the procedures of NATM. Maidl has investigated an improved supporting method and materials for support of shafts devised from the experiences of modern techniques for tunnel supports.

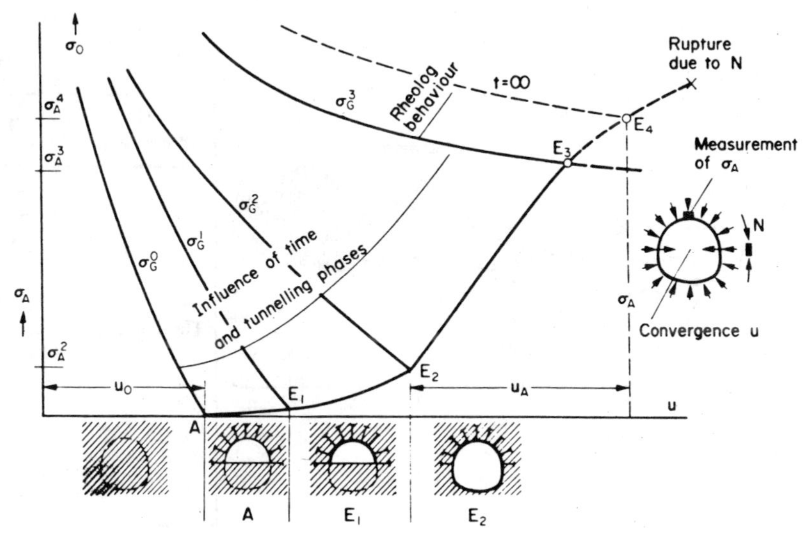

Fig. 1.4 Development of the characteristic lines. (Duddeck, 1980)

A number of papers treat the problem of tunnel linings. Romana et al. have tested the mechanical behaviours of swelling marls to determine the parameters representing them, which have been used as the basic data for analysis of the forces acting on the lining of tunnels. Altounyan et al. have studied the pressure acting on the lining of a shaft during sinking through deep aquifer rocks, and have obtained several instructive informations about it. Kobayashi et al. have studied the support systems for tunnels driven in a ground of extremely soft nature. They have obtained the results that very rigid and strong supports would be necessary for such soft grounds.

Holmgren has reported the experimental investigations on the strength and deformability of steel fibre reinforced shotcrete lining, with the results that the lining behaves more plastically and the combination with rock bolts much simplifies the support system. Ortlepp has studied the design of supports for deep hard-rock tunnels based on the experiences of high stress in rock and rock burst damage, and has found that yielding of the support elements is closely related to damage due to high stress, and proposed the concept of critical bond length of rock bolts as the determinant of their rupture.

The Seikan Undersea Tunnel is under construction in Japan for the purpose of connecting Honshu and Hokkaido by a railway. The construction of this tunnel is characterized in that (1) it is an undersea tunnel, (2) it is long (53830m in total length, 23300m from shore to shore), (3) it is deep (the maximum depth being 240m from the sea level, 140m under the sea-bottom), (4) the greater part of ground is composed of loose sandy mudstone, and (5) the sectional area is large (the interior area being $88m^2$). These characteristics will be interesting from a standpoint of rock mechanics, so that the important points will be introduced here. Fig. 1.5 illustrates the geological profile (Nisugi, 1981).

The Seikan Tunnel project, though confronting with many difficulties has conquered them through the accumulated experiences and the state of arts, and advanced to the present status where the pilot tunnel was completed on 27th January 1983, and the unexcavated length of the main tunnel remains no more than 2.7 km.

The new techniques for tunnelling through undersea weak strata have been developed through the Seikan tunnel project, which are summarized as:

(1) horizontal exploratory long boring system,
(2) grouting system for impeding water inflow,
(3) shotcrete system for stabilizing the excavated tunnel surfaces,
(4) support system to resist high earth pressures in soft rock masses.

1.4 Field Measurements and Inverse Analysis

Various behaviours of the ground and supports during the excavation of a tunnel will not necessarily coincide with those predicted in the designing stage, and in most cases, they are different. The input data obtained by in-situ tests at a particular site in a long tunnel cannot be applied to the entire length of the tunnel, even if they are adequately evaluated. Under these circumstances, it is almost impossible to analyse correctly the behaviour of a tunnel in the stage of design, and it will be desirable to alter the original design, if necessary, by considering the results of observing the behaviour of the tunnel during excavation.

It may be possible to examine whether the input data used in designing the tunnel are suitable or not, and to examine whether the construction of a tunnel can be carried out according to the original plan or not by comparing the behaviours of the tunnel observed during construction with the original predictions. In other words, we can assess the initial ground pressure and the parameters representing the mechanical properties of ground by measuring the displacement, stress or strain. Such an analysis is termed as an inverse analysis, and the problem dealt with by the inverse analysis is called either the identification problem or the characterization problem.

Many studies have been reported on the inverse analysis. Kovari (1977), for example, devised a method to calculate earth pressure from the relative displacements in the axial direction of tunnel supports and the changes in curvature by inverse analysis. He named it the integrated measuring technique. The same idea, however, had been put forward by Murayama (1968), who measured strain with electrical resistance strain gages, while Kovari adopted mechanical measuring techniques to make the measurements easier so as to be applicable to tunnel measurement. The method of determining earth pressure acting on an underground structure from the displacements of the structure by inverse analysis has occasionally been proposed in the engineering fields other than tunnels, e.g. the piles or the sheet piles (Gioda 1981). In addition, there have been methods aiming at obtaining the initial earth pressure or the material constants of the ground from the displacement and strain of the ground at the time of the excavation by inverse analysis. Gioda and others (1980, 1981) published many papers on the studies on the inverse analysis in the field of geomechanics. Their research works include the inverse analysis of non-linear problems. It is expected that the inverse analysis of the results of field measurements will act an increasingly important role in geotechnical engineering.

Sakurai (1981, 1982) attempted to obtain the strain of the ground caused by excavation of a tunnel from the measured displacement, and proposed a method of administering the construction of a tunnel by comparing the result obtained with the limit strain of the ground provided that the number of measuring points both on the wall surface and in the ground are sufficiently large. Even when the number of measuring points is not large, it may be possible to evaluate the strain by conducting the inverse analysis to find the initial earth pressure and the mechanical constants, followed by calculating the strain distribution around the tunnel by using such methods as the FEM. This method can be applied to design as well as to construction management. According to this method, the numerical simulation makes it clear that even though there comes into being the plastic region

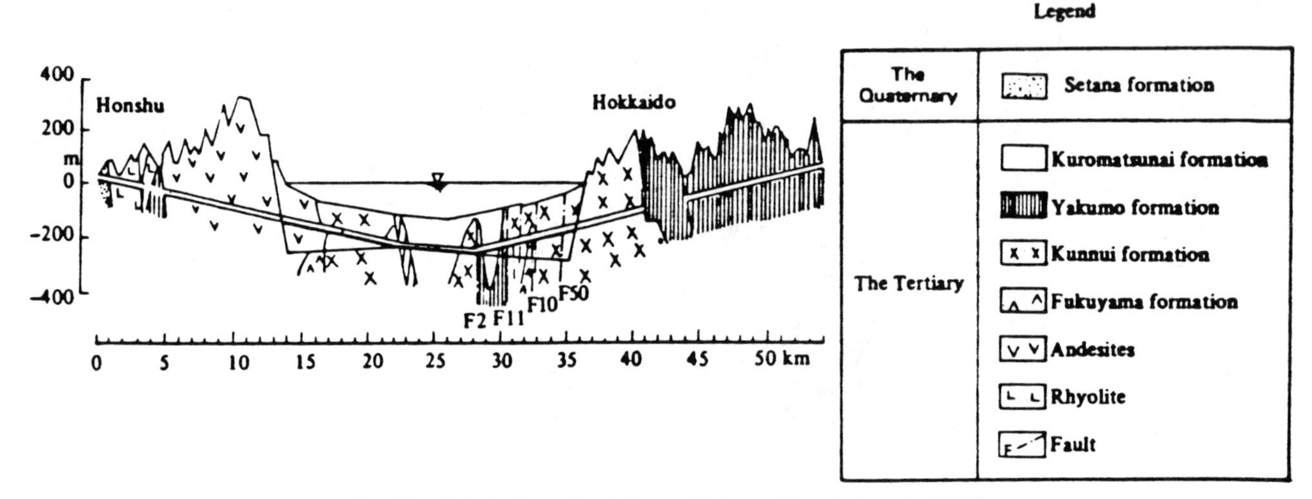

Fig. 1.5 Geological profile of Seikan Undersea Tunnel. (Nisugi, 1981)

the analysis can be carried out with sufficient accuracy by using the equivalent elastic modulus and the initial earth pressure. Since this method relies on the strain and not the stress, it has the additional advantage of not having to go through the stress analysis.

On the basis of the Kaiser et al.'s study on the rate of convergence function by model experiments in a highly stressed medium, it has been found that a tunnel behaviour assumes one of the three modes, i.e. prefailure, stable yield zone propagation and unstable or rupture mode. Based on these convergence rate functions, the performance of tunnels excavated in the same rock materials can be predicted from convergence measurements.

Saito et al. have discussed the mechanism of rock bursts occurring at the face of the deep tunnel, deeper than 1000m from the earth surface, driven in hard igneous rocks. Collecting the data about the original stress state, the change in stress caused by progressing the tunnel face and the mode of core disking along the tunnel line, they have found that the stress state has been highly concentrated elastically and the mechanism of rock bursts has been very similar to that of core disking.

2. COAL MINING

2.1 Room and Pillar Mining

Introduction
In the room and pillar mining, coal pillars are left to support the superincumbent strata. Because the stress levels are generally low, local support requirements are modest, and accordingly the support costs are low. Furthermore, surface damage is virtually eliminated if room and pillar workings are properly designed. The low capital cost, flexibility in production and easiness of operation make the room and pillar mining one of the favourable methods of coal mining. However, because the size of pillars increases with increase in the depth of mining and in seam thickness, this method is confined to the mining of relatively shallow and moderately thick coal seams. In recent years, two basic pillar extraction techniques have been employed, namely conventional pillar extraction and rib pillar extraction.

Designing coal pillars
The pillar sizing is fundamental in the design of room and pillar mining. The first step is to determine the stress distribution in the coal pillar. This will be done with the aid of either finite element or boundary element stress analysis techniques. The numerical method using finite difference displacement equations of elastically supported plates will be also available, which was adopted by Sheorey et al. (1982) to investigate the pillar load in Indian coal mines. However, the simplest approach to determine the pillar load, or more correctly the average pillar stress, is by the tributary area theory, which will present the upper limit of the average pillar stress.

The next step in the design process is to determine the critical stress level, or the strength of coal pillars. Efforts have been made to estimate the strength of coal pillars on the basis of unconfined compression tests (Salamon (1967), Bieniawski (1968) and many others). Huatrulid (1976) assorted these proposed formulas into two types, namely the compressive strength-size relationship and the compressive strength-shape relationship. Basing on these works and the tributary area theory, Bieniawski has presented new two approaches to coal pillar design. The ultimate strength approach contends that pillars will fail when the applied load reaches the compressive strength of the pillars. It presumes that the load bearing capacity of a pillar reduces to zero at the moment the stress reaches its ultimate strength. The progressive failure approach emphasizes the non-uniform stress distribution in the pillar. Failure is initiated at the most crucial point and propagates gradually to the ultimate failure. The new pillar sizing method proposed by Bieniawski adopts failure initiation as the criterion rather than the ultimate failure, taking the safety factors ranging from 1.5 to 2.0.

Evaluation of safe roof spans
The roof stability is a basic problem in room and pillar workings and has been studied by a number of investigators. The room and pillar mining using conventional or continuous mining equipments is a cyclic process composed of the following principal steps:
1) coal winning and loading,
2) setting of temporary supports,
3) installation of permanent supports.

Wade et al. (1982) reported that, in the room and pillar workings in United States of America, approximately 51 percent of the roof fall fatalities occurred while the roof was in an unsupported condition between steps 1 and 2 listed above, approximately 19 percent of the fatalities occurred while the roof is in a temporarily supported condition between steps 2 and 3, and approximately 30 percent of the fatalities took place under permanently supported roof. These statistics show that the time effect has to be taken into consideration for assessing the safe roof spans and designing the support system in room and pillar mining.
As a means of roof stability appraisal, rock classifications have received increasing attention in recent years. Rock classifications are well known empirical index for assessing the stability of underground openings in rock. Sheorey et al. (1982) presented the roof bolting system with wire rope stitching in the Indian room and pillar workings based on a rock classification and concluded that this procedure is very effective. In this classification, six parameters, i.e. rock quality designation, spacing of bedding planes, rock strength, hazardous structural features, susceptibilities to weathering and water are taken into consideration. Importance ratings are allocated for each parameter, a higher value indicating a better rock condition. The proposed scheme of rating the parameters is modified from the approach recommended by Bieniawski (1976), and takes into cognizance of the influences of the depth of workings and the abutment stresses due to workings in close proximity.

Bieniawski has proposed a method for assessing safe roof spans by means of a rock mass classification system, specifically the geomechanics classification, as modified for application to coal mining. Bieniawski and the researchers of Pennsylvania State University have studied the rock classification approach systematically since 1978. Analyzing 58 case histories of roof falls in the room and pillar mining in United States of America they have presented the relationship between the stand-up time and the roof span for various rock mass ratings. This geomechanics classification will provide the guidelines for assessing safe roof spans and selection of proper roof supports taking into account of the geomechanical properties of an individual stratum.

2.2 Longwall Coal Mining

Introduction
In the deep underground coal mining, longwall mining is superior to traditional room and pillar mining from the standpoints of productivity, resource recovery and safety. The modern mechanized longwall mining is equipped with shearers or plows for winning coal, and shield or chock selfadvancing supports, sometimes frame-type supports for face support to achieve high productivity and safety.
Nearly all longwalls in Japan and United States of America are of retreating system, while the majority of longwalls in Europe are of advancing system. The main differences between these systems may be in that the advancing longwall faces can begin production immediately, and the gate entry can be developed simultaneously with the longwall panel extraction. Conversely, gate entries have to be predeveloped in the retreat longwall faces, and the longwall panel extraction is delayed until the development work is complete. Although each method has inherent advantages and disadvantages, the choice between these two systems may usually depends on governmental regulations or on mining and geological conditions. In Germany, F. R., coal mining has steadily advanced to deeper levels, at a rate of 14 m per year, and the average working depth at the moment is about 900 m. Some of the collieries are already extracting coal from the level of 1400 m (Grotowsky (1982)). Japan and France have the similar situation in the working depth. The increase in working depth and intensive concentration of workings have forced the engineers to have a close look at the rock pressure and its effects. Prevention of coal bursts, gas-coal outbursts and gas-sandstone outbursts has become one of the most serious problems in these countries. Moreover, certain shale strata have a tendency to deteriorate when exposed to humid atmosphere, and cause excessive floor heave (Singh et al.). In the deep underground longwall mining, shearing of the roof strata at the face line, rib sloughing and general deterioration of the immediate roof and floor strata are frequently encountered.
Strata control problems were experienced primarily in the main gates and tailgates, especially at the gate-face junctions. Also in the area along the face there are several strata control problems associated with roof

falls, weak strata and faulting. Depending on the severity of these problems, face operation will be seriously hampered, resulting in a significant drop in production.

Designing longwall panel widths and interpanel pillars

The longwall panel width is a fundamental factor for strata control in the longwall mining to achieve proper caving of the roof of goaf and to minimize the face support requirements. Okamura et al. (1979) discussed the stability of the immediate roof stratum, which is separated from the upper strata and spans the goaf cavity and is capable to bend by its weight, by barodynamics model tests. They pointed out that the roof fall did not occur unless compressive yielding was initiated at the bearing points, and that it depended on the deflection of roof in its center part as well as the longwall panel width.

Wagner et al. (1982) proposed a designing method adopted in South Africa, where longwall mining is conducted in the regions overlain by one or more massive dolerite sills. These sills can span several hundred metres before failure. If the panel dimension is small a dolerite sill may span an entire panel without failing and the weight of the sill and the overlying strate may be transferred onto the panel abutments, including the face. In South African collieries the panel width has been determined by the need to induce failure of massive dolerite sills as soon as possible after the commencement of longwall operations. Equations were proposed by Salamon et al. (1972) and refined by Galvin (1982) for determining the minimum panel dimension.

The designing of interpanel pillars is connected with the determination of the longwall panel width. Two opposing criteria have been met. First, the interpanel pillar should be as narrow as possible to minimize coal loss and second, the pillar should be sufficiently wide to protect the gate roads from the effects of high abutment stresses. To enhance the extraction percentage, the concept of yield pillars should be examined. This concept is to make the coal pillars fail in a stable manner. The coal bumps occurred on interpanel pillars in three Japanese coal mines reported by Fukuda et al. were considered as failure of pillars in an unstable manner. To ensure stable failure of pillars the stiffness of the underground construction around the pillar must be greater than the post-failure slope of the stress-strain curve of yielding coal pillars.

For the stability evaluation of the active pillars, the integrity factor, which decreases as mining progresses, may be a rational parameter. Barron (1978) proposed the pillar integrity index which is defined as the ratio of the pillar core area to the total pillar cross-sectional area. This index may be a useful quantitative measure to estimate the relative structural integrity. He determined the core area on the basis of the fracture index across a pillar, which was derived from the air injection tests.

Lu has proposed a new approach to the same problem by defining an integrity factor as the ratio of the integrated residual strength to the integrated load. He has presented a set of vertical and horizontal pillar pressures measured with hydraulic borehole pressure cells. The residual strength of a full-size pillar is estimated from the horizontal pressure measured in the pillar, on the basis of laboratory-determined triaxial compressive strength, in which the in situ measured horizontal pressure is considered as the constraint, while the integrated total actual load on the coal pillar is estimated from the in situ measurements, instead of the predicted average load based on the tributary area theory or the hypothetical load postulated from the stress balance principle recommended by Wilson (1972, 1977 and 1981). Although the safety factor, for instance, which was recommended by Bieniawski, may be rational for the permanent or semipermanent structures, the integrity factor will better fit to temporary structures such as interpanel pillars.

Designing gate roads and supports

The evaluation of closure is basically important for planning the support system in gate roads. The application of the geomechanics classification, reported by Abad et al., has proved that the convergence of gate entries is concerned with the coal seam thickness, the type of entry, the working depth, the strata inclination, the supports density (steel weight per unit cubic metres of entry cavity: kg/m^3) and RMR index recommended by Bieniawski (1976). They have analyzed the convergence measured at 7500 different points in a total of 187 roadways, and have presented a formula to assess the final convergence, represented in percentage of the initial height of the road. In this formula, the convergence is related

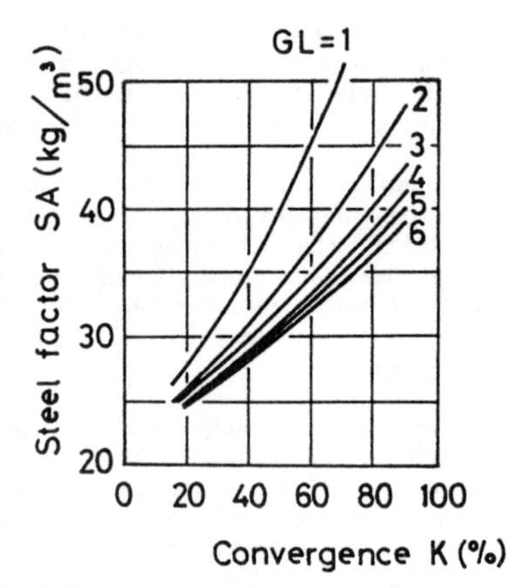

Fig. 2.1 Steel factor versus convergence and floor index GL

to the supports density and a new index, IGME 82, depending on the joint state, the strata inclination and the coal seam thickness. The support density, proposed by Abad et al., indicates the strength of the support and this approach is useful to determine the dimension of steel supports, namely the cross-section, the section weight and the centre distance of steel supports.

The approach using the support density was previously presented by Irresberger (1982) and Grotowsky (1982). They called it the steel factor in their papers and concluded from the statistical surveys that the suitable steel factor depends on the convergence and on the floor index which indicates the geomechanical condition of immediate floor strata. Fig. 2.1 shows the proposed relationship among the suitable steel factor, the convergence and the floor index. They discussed the influence of the rock pressure upon the convergence and the importance of the geometrical relation between the gate road and the longwall panel by analyzing a number of case histories in Germany, F. R. They proposed the planning procedures, developed by the Mine Support and Rock Mechanics Research Station of Bergbau-Forschung GmbH in Essen. These procedures are very systematic and practical, and are backed by the statistical survery and the compatibility tests at a series of Ruhr collieries.

The planning procedures proposed by Irresbergers (1982) are as follows. First the gate road is divided into the regions of equal load and then the convergence in each regions is calculated from the effective convergence, K_{EV}, given by:

$$K_{EV} = -78 + 0.066\,T + 4.3\,M \cdot SV + 24.3\sqrt{GL} \pm 3\%$$

where T is the working depth, M the seam thickness, SV the width of roadside packing and GL the floor index. The effective convergence, K_{EV}, is the convergence which would arise in the gate road if it were driven ahead of the longwall face line and the coal had been mined on one side and the rock pressure were unaffected by any other ribs. When other types of gate roads are planned, the convergence in each case can be estimated from K_{EV}, the method for which was proposed in his paper. Retreat gate roads, abandoned at the longwall face, only display half the convergence. When ribs are present, convergence must be corrected using the pressure calculated by the method presented by Everling et al. (1972). Once the convergence has been assessed, it is possible to find the suitable type of supports and steel factor. Rigid arches can be used in case that the convergence is less than 15 percent of the original height, and in case it is greater than 15 percent sliding arches should be employed.

Ground control at longwall faces and effects of supports

In the deep longwall faces, high abutment pressure results in severe fracturing of the immediate roof strata and slabbing of the coal face, the latter causing loss in its support potential. Furthermore, slabbing and sloping of a coal face increase the unsupported roof span and encourage local roof falls. The lateral constraint provided by the caved materials to the fractured immediate roof is also restricted by the phenomenon of discontinuous subsidence. The lack of lateral constraint in conjunction with high abutment pressures favours the development of near-vertical fractures. Because of the lack of frictional resistance, wedges of rock formed by the steeply dipping fractures can slide out of the roof close to, or at the face. This failure can lead to a complete loss of control of the roof strata and results in a major collapse. The most dangerous moment is during or immediately after the advance of the supports. Wagner et al. (1982) pointed out that support systems need to have such characteristics that the unsupported span should be as small as possible, that the support should be capable of generating high tip loads under most operational conditions and that the total supported roof area should be as small as practical.

The evaluation of suitable setting load density of face supports is the main subject in designing the face control equipments. In recent years, the majority of mechanized longwall faces are equipped with hydraulic face supports. The main functions of these supports are to support the roof strata of the working area. This is achieved firstly by applying a positive resistance normal to the plane of the roof, and secondly by displaying the capacity to resist roof lowering as the supported span increases during mining. Therefore, the support resistance is generally represented by the mean setting load density which is the force exerted by the support per unit area of the exposed roof from the rear of the canopy to the coal face.

To make clear the support effects, the roof-to-floor convergence and the pattern of roof fracturing have been researched by a number of investigators (Gupta et al. and many others). Christiaens (1982) summarized the pattern of roof fracturing observed in Belgian collieries as shown in Fig. 2.2. In shallow coal measures, inclined brittle fractures, R3, are primarily found, of which inclination will be due to the friction forces induced by the expansion of the coal towards the working area. This primary pattern can be overlaid by a more inclined, often up to 45 degrees, secondary pattern of cracks which is caused by a subsequent expansion of the coal. On the other hand, in deep coal measures, the vertical brittle fractures, R2, are found primarily in sandy strata of sandstone or dense shale. These vertical fractures may be intersected by inclined secondary cracks, R3, and the transverse shearing is encountered in the form of R4 fractures as shown in Fig. 2.2, which are mainly found in the immediate roof consisting of shale. The wedge ABC is intersected by a subsequent brittle fractures, R3, and sometimes by a horizontal brittle fracture, R1, due to horizontal compression in that wedge.

The roof behaviour has been generally considered from two different aspects, namely the main roof and the immediate roof. Christiaens (1982) has described that the arch action is borne by the coal on one side and by the caved debris on the other, and the immediate roof is fractured by brittle breaks and sometimes by final shearing. However the main roof is fissured either by brittle breaks or by the bending of the stronger strata, or because of old working. The stability of the successive main roof strata may be ensured by the horizontal forces which are active there. The shear stress is at its maximum near the abutments and slips may occur in the vertical planes, particularly if the face supports are weak.

An approach to assess the safety factor of face supports was made by Christiaens (1982), assuming the vertical fracturing in the immediate roof. He analyzed the various cases of support position and of the interval of roof fracturing, and presented the following conclusion. The stability of roof is more easily achieved if the props are brought nearer to the coal face, and the setting load density must be as high as possible. The best practical results have always been obtained with

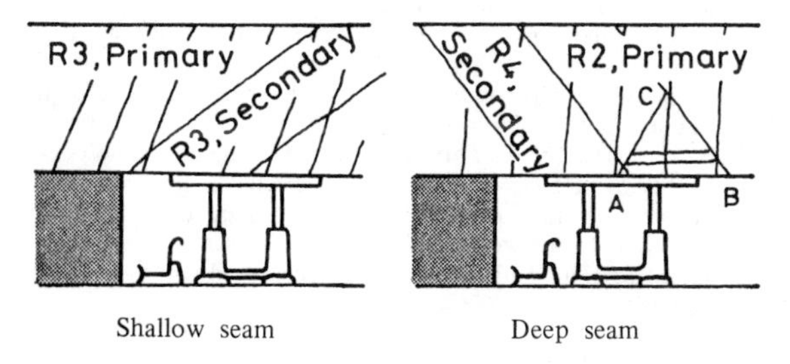

Fig. 2.2 Two types of fracture system on a face (Christiaens, 1982).

the types of powered support which have the best safety factor, k, assessed by the equation derived from the moment of the force exerted by a support against the roof, as follows:

$$k = \left\{F \cdot y + 0.15L\,(F - 100L \cdot M)\right\} / 50ML^2$$

where F is the vertical force exerted by a powered support per unit face length in kN/m, y the arm-length of F in m in relation to the rear of the canopy, L the width in m of the immediate roof to be supported, M the height of the face in m, and the hight of the roof to be supported is assumed as four times the height of the face.

Irresberger (1982) presented the criteria to determine whether the roof is controlled well or badly as follows. A roof is considered to be under good control in cases that during winning shifts, the area of local roof falls ahead of the unsupported roof area is on average less than 10% of the face length, the area of roof ahead of the supports breaking up deeper than 30 cm is less than 5% of the face length and the area of roof ahead of the supports with steps, greater than 10 cm in height, is less than 5% of the face length. Conversely, a roof is considered under poor control in cases that these three percentages are respectively greater than 30%, 20% and 10%. The fall frequency is perhaps the most important in these criteria, because increase in roof falls may necessarily slow down the rate of face advance and thus lower the production from the working face.

The rock pressure on the face, assessed by the calculation method recommended by Everling et al. (1972), is pointed out as the critical factor in planning face supports by Irresberger (1982). The relationships among the roof consistency, the rock pressure and the technical support requirements are summarized by Irresberger (1982) on the basis of experiences in Germany, F. R. Where the roof contains firm beds with a thickness of greater than 10 cm, there is a danger that the roof will settle in steps when the rock pressure is low. To prevent this, the support resistance has to be relatively high and a minimum suggested value is 400 kN/m². However, the tip-to-face interval can be as much as 1 m due to the low fall-sensitivity under firm roofs. In cases that the rock pressure is high and a clay shale in the immediate roof is overlain by firm beds, the roof will tend to break up frequently when the clay shale is less than 2 m in thickness. Support resistance should therefore be greater than 300 kN/m² and the tip-to-face interval should be less than 30 cm. When the immediate clay shale roof is thicker than 2 m, then the demands placed on face supports are lower. Depending on the particular rock pressure, support resistance should be between 250 and 300 kN/m² and the tip-to-face interval should be less than 50 cm so as to keep down frequency and area of falls from the roof.

A series of detailed observations on three retreating longwall faces in England, presented by Gupta et al., have also clarified the effect of setting pressure on support performance. Their detailed observations of roof deformation have shown that the increases in support setting load density have changed the roof strata deformation from an uneven compression zone with a tension zone ahead of the supports, to an evenly distributed compression zone over the face area. The increases in support setting load density, from 220 to 420 kN/m², have also conclusively reduced the average roof-to-floor convergence over the supported face area, the lateral expansion of the coal face into the working area, and face spalling and roof flaking.

2.3 Monitoring for Strata Control

Monitoring is a useful means to obtain important materials for strata control in many collieries. Roof subsidence, floor heave and seam expansion are monitored using various devices, and multiple extensometers are used in boreholes drilled horizontally in the seam and vertically in the roof strata. Hydraulic pressure cells and vibrating wire borehole stressmeters are used to monitor stress changes in the seam and strata. Sometimes, the initial stresses are investigated by applying the stress relief technique. In this section, the applications of these standard field monitoring techniques will be reviewed and some topics will be introduced.

The change of roof subsidence profile in the gate roads with advance of longwall mining faces was presented by Kimura et al. (1982) and Johnson et al. (1982). Johnson et al. discussed quantitatively the effect of packing behind the longwall face by analyzing the relation between pack loads and roof-to-floor convergences, measured in the coal measures in United Kingdom. A similar measurement in Japan was presented by Sugawara et al. (1981). Hebblewhite et al. (1982) presented a successful monitoring system which consisted of roof-to-floor convergence monitoring in the gate roads, plus deformation and induced stress monitoring in the seam and roof strata. They used hand-held telescopic rods between pins grouted into rock, multiple wire extensometers with expansion shell anchors and vibrating wire borehole stressmeters. Fisekci et al. (1982) presented a monitoring system in hydraulic mining of thick and steep seams, and discussed the time dependent characteristics of support load and entry deformations. Tinchon et al. (1982) reported an application of flat jack method to measure the initial stresses in French coal mines, and showed that the vertical stress increased with depth, but it was highly sensitive to the geological structures.

In this Congress, improved field monitoring techniques have been presented. Gupta et al. have presented the measurements of strata deformation and strain in the immediate roof on the face, in which strain has been computed from the relative deformation measured by anchor wire extensometers and magnetic extensometers. Lu has proposed an improved pressure measurements with hydraulic borehole pressure cells to evaluate the stability of interpanel pillars. Lu (1981) presented a technique to monitor the biaxial ground pressures using a combination of one cylindrical and two flat hydraulic pressure cells installed in a single hole. The response ratio between cell pressure and ground pressure was confirmed to be approximately one-to-one for coal.

Face support monitoring was reported by Hebblewhite et al. (1982). They installed four continuous clockwork chart recorders of the load acting on supports on the selected ones of the 89 chock shield supports used on the face, and spot readings were taken along the entire face line at regular intervals. The rapid load build-up was confirmed on both front and rear legs after the shearer passed, and it was clarified that frequent yielding of supports occurred particularly when the face was under surface escarpments and deep covers. In this Congress, Carrasco has presented a new system to monitor the dynamic stress changes of load acting on the powered supports due to roof falls by blasting in the goaf.

Remote data transmission is becoming more popular in coal mining field monitoring in connection with the development in the field monitoring system by means of acoustic emission. Applications of the acoustic emission technique were presented by Watanabe et al. (1981), Nakajima et al. (1981), Josien et al. (1982) and many others. Watanabe et al. emphasized that the impending gas outburst in a cross-measure drivage can be forecasted with a high probability by monitoring the degree of increase in acoustic emission activity after blasting. In this Congress, Schuermann has discussed the cost-effectiveness of the remote data transmission and has presented a manless processor-controlled monitoring system.

2.4 Mathematical Modeling of Coal Mining for Designing

Methematical modeling by finite element and displacement discontinuity methods has been tried by a number of investigators to analyze the rock pressure effects in the deep underground coal mining. Complicated geometrical features of total extraction systems, for instance near the gate end of a longwall face, cannot be adequately modeled by two dimensional methods. Although in principle three dimensional finite element or general boundary element models can be applied for such geometries, the high cost and effort required for such analyses may seldom be justified for the time being. On the other hand, it may be impossible to obtain the suitable mathematical description of the continuing crushing and relatively extensive fracture movements. Therefore the calculation of the rock pressure has been largely simplified.

Everling et al. (1972) presented a simple three dimensional calculation method of rock pressure, which is based on the homogeneous elasticity. In this method, it is assumed that the horizontal displacements are negligibly small and the displacement takes place only in the vertical direction at any point. From the wide application of this calculation method in Germany, F. R., it was found that the in-situ fracture movements are proportional to the deformation in a body of rock when this is assumed to be elastic (Grotowsky (1982)).

Yeates et al. have applied the displacement discontinuity method to investigate the stress states in chain pillars and used the two dimensional non-linear elasto-plastic finite element method to assess the potential stability of developement roadways in coal seams of different horizons. Similarly, Fukuda et al. have applied the two dimensional displacement discontinuity method (Crouch (1976)) to investigate the stress change in front of the longwall face, associated with the face advancing. These works have confirmed the availability of the displacement discontinuity method. This method is based on replacing actual extraction geometries by thin slits of the same plane area. This approach is promising and allows three dimensional stress analysis at relatively low cost.

Displacement discontinuity solutions were primarily used by Berry (1960) to model surface subsidence profiles due to coal mining. It was assumed that the mined panel could be modeled by a single uniform displacement discontinuity. The extension to allow more realistic boundary conditions and excavation geometries was made possible by dividing the plan area of the excavation into elements, each representing a uniform displacement discontinuity (Plewman et al. (1969), Starfield and Crouch (1973)). Diering (1980) treated piecewise homogeneous isotropic continuum properties by using elements along the geological interfaces.

Wardle et al. have extended the displacement discontinuity method to a layered anisotropic rock mass. It is assumed that a rock mass consists of an arbitrary number of layers parallel to the ground surface. The elastic properties in any layer are transversely isotropic with a vertical symmetry axis. The interfaces between the layers can be assumed as fully continuous or fully frictionless. The numerical solution procedure involves subdividing the plan area of interest into rectangular elements and each element is assigned a property code indicating whether it is mined out, intact coal or caved waste material. For elements transmitting load between roof and floor, one dimensional stress-strain properties are used. The resulting system of equation is solved iteratively, allowing non-linear coal and waste properties to be used.

The reliability of predictions obtained from a numerical model is only commensurate with the reliability of the input parameters used, so that the method used to determine the relevent input parameters is important. Although Wardle et al. have presented a number of hopeful experimental approaches including measurement of the deformability of the caved waste by in-situ plate loading, a number of problems still remain and the future works are expected in this direction.

2.5 Coal Bursts

In deep coal mines, coal bursts, i.e. brittle fractures of a part of a coal seam accompanied by seismic activity, have frequently been experienced. Coal or rock bursts are the events of dangerous dynamic fractures and occur in various measures associated with various types of workings. Some symptoms seem to be present, but since there are considerable differences in the states of affairs, it may be inadvisable to generalize the knowledge obtained from the experiences in a particular area.

The case histories in the Ruhr coal field show that the following geometrical situations are favourable to the coal burst occurrence (Bräuner, 1979). (1) Face commencing, where the seam is placed under increasing load as long as the main roof has not yet fallen. (2) Corner of a longwall panel adjacent to a former panel, where the pressure zone in front of the face crosses the pressure zone of the former panel. (3) Formation of a pillar, or drivage in the abutment pressure zone of a former panel subjected to additional loads. (4) Working under or above residual pillars in other seams, which are usually subjected to concentrated pressure by mining the higher or lower seam. (5) Driving additional roadways which cause stress increase in the seam or in the rock. Bräuner (1979) described that the most important feature of coal bursts is almost always a dynamic thrust in the coal seam occurring with great force. However, there are variations in the extent of this thrust, the mass of fractured material and consequently the degree of destruction in the mine-workings. Severe coal bursts thrust the seam several meters forward and completely filled a roadway over a distance of 100 meters.

Sugawara et al. (1981) and Kimura et al. (1982) investigated the coal bursts in the longwall faces occurred in the Miike coal mine. Heavy

Classification of coal bursts by the length of collapsed faces, L^*.

Class	L^* (m)
A	>26.1
B	13.1~26.0
C	6.6~13.0
D	0.1~ 6.5
E	Only rock noises

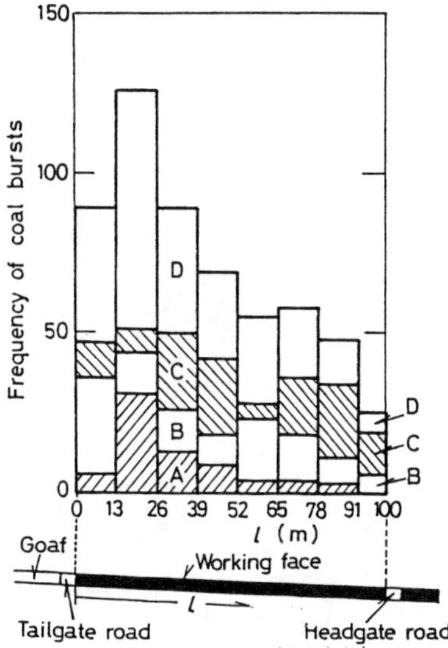

Fig. 2.3 Classification of coal bursts and histogram of each class of coal bursts in relation to the position of occurrence. (Kimura et al., 1982).

coal bursts were apt to occur near the tailgates as shown in Fig. 2.3, in which the events were classified into four classes depending on their extents. Kimura et al. (1982) pointed out that the location of coal bursts in the face migrated somewhat regularly with the face advance, namely the coal burst occurred initially near the headgate and then the location of occurrence migrated toward the tailgate and subsequently toward the headgate as the face advanced, and that coal bursts are likely to occur when the roof-to-floor convergence on the face is not great and within a certain range. The reason for the latter fact may be that an excessive convergence eliminates the hazard of coal burst by loosening the coal seam near the face.

Many investigators have emphasized that the most important geological condition for coal bursts is existence of a thick bed in the roof, consisting usually of sandstone. Bräuner (1979) described that such beds require to be at least 5 m thick and no farther than 10 m above the seam. Fukuda et al., in this Congress, has pointed out that the coal bursts in pillars occurred when a coal seam is overlain by a thick sandstone roof, thicker than four times the thickness of the coal seam. Josien et al. (1982) discussed a method to assess the hazard of coal burst, by analyzing the deformability of rock, and proposed the W_{et} index which is defined as the proportion of elastically stored energy to the disappeared energy with plastic deformation in rock, at the stage just before the strength failure in the uniaxial compression test of rock. The increase in W_{et} index, especially in thick beds, predicts the hazard due to the dynamic energy release in French collieries.

In many cases, the hazard of coal burst is removed when a floor begins to fracture or to swell. Kimura et al. (1982) confirmed that the thickness of the immediately underlying sandstone closely concerns the coal burst occurrence on the face. As shown in Fig. 2.4, the floor heaves occurred in a limited range where the underlying sandstone stratum is thinner than 2 m, in each tailgate before the dip side face arrived, and the coal bursts occurred in the area where the underlying sandstone is from 2 m to 3 m thick. Other geological factors which may concerned with the occurrence of coal bursts are faults, fluctuations in the seam thickness and changes in dip angle (Bräuner (1979) and Kimura et al. (1982)), and irregularities in strength (Fukuda et al.).

To detect the existence of high stress concentrations in the seam, test boring may be effective. Sugawara et al. (1981) proposed a procedure for the evaluation of the allowable rate of face advance, A, in m/shift, based on the test drilling. Let the thickness of a coal seam be M, the width of the zone along a coal seam into which boreholes can be drilled smoothly be l^* then A is given by:

$$A = l^* - \alpha M$$

where α is a factor to be estimated from field observations. In Miike

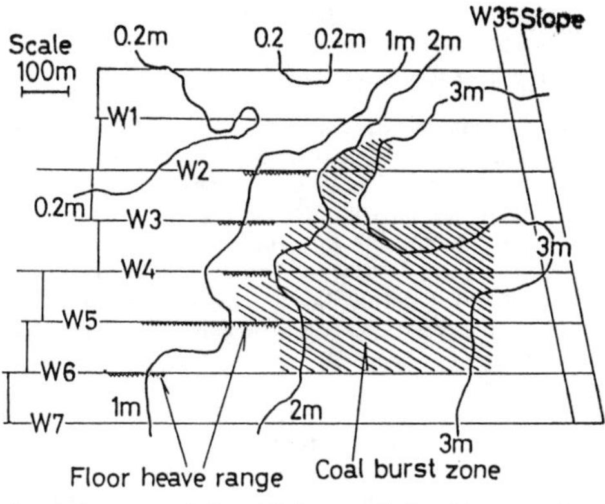

Fig. 2.4 Isopach map of the thickness of the floor sandstone with the coal burst zone and floor heave ranges in gateroads. (Kimura et al., 1982).

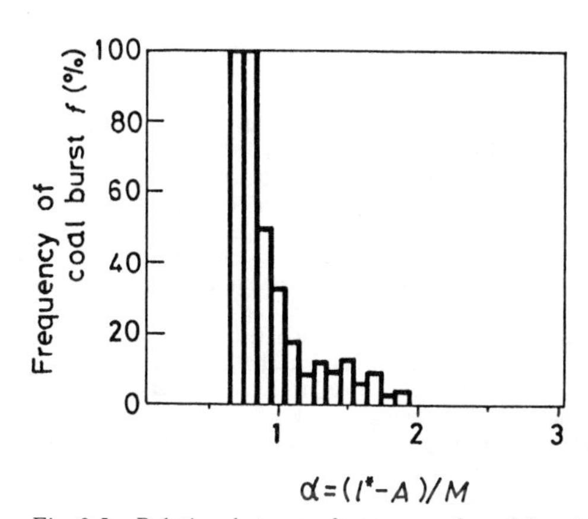

Fig. 2.5 Relation between frequency of coal bursts and the factor α. (Sugawara et al., 1981).

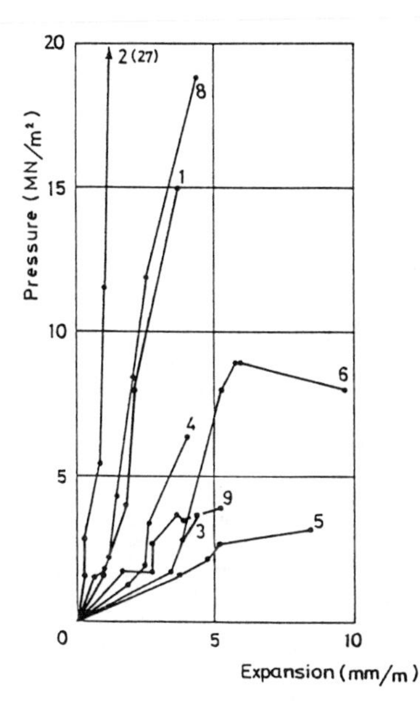

Fig. 2.6 Simultaneous measurements of the pressure acting on the pressure cells set in boreholes, and the extension of the same boreholes. (Braüner, 1979).

coal mine the frequency of coal bursts was studied in relation to the rate of face advance with the result as shown in Fig. 2.5, and it was found that the hazard of rock bursts may be removed by taking $\alpha \geqslant 2$.

Bräuner (1979) proposed the check measurements for the geomechanical state of a coal seam using the existing test boreholes as follows. The pressure in boreholes is measured with hydraulic pressure cells, and at the same time the longitudinal extension in boreholes is measured with extensometors covering the whole length of the boreholes. Fig. 2.6 shows some of the results obtained in a pillar between an entry and an approaching longwall face. The lines near straight lines, like 1, 2 and 8 in figure suggest elastic behaviour of the seam before its strength failure.

Bräuner (1975), Josien et al. (1982) and Sugawara et al. (1981) tried to monitor the state of coal seams and to predict the coal burst hazard by investigating continuously the seismo-acoustic activity due to the brittle fracture or / and shearing in the abutment in front of the face. The number of events and the accumulative energy of these events in a period of one hour or one shift have been taken as the parameters for the intensity of seismo-acoustic activities. To distinguish the period or zone of working, in which the strata is subjected to high stress, many investigators have analyzed the overall emission level, the changes in seismo-acoustic emission and the delay in release of energy.

To prevent the coal bursts the systematic stress relief boring, water infusion, loosening or stress relief blasting and their combinations have been practiced. Josien et al. (1982) reported that the drilling long relief parallel holes, 10 cm in diameter, 24 m long, at a regular interval of 2 m, proved very effective in the corner of a longwall working adjacent to an old panel. The effectiveness was monitored by measuring pressure using a pressure cell. It was found that on the effective relieving, a sudden pressure increase occurs, when a substantial volume of fine coal is recovered, and a small coal burst occurs in the hole. Nakajima et al. (1981) investigated the fracturing activity in the seam during relief boring, applying the acoustic emission technique, to ascertain its preventive effect for coal and gas outbursts. On the other hand, the effectiveness of the water injection into the hard and thick sandstone or conglomerate roof was discussed by Niu et al. (1982). This method is considered to be effective for controlling sudden roof falls of a large scale in the goaf produced by the longwall working and for relieving the high abutment pressure.

2.6 Gas Outburst

Gas outburst may be defined as a phenomenon of sudden eruption of gas accumulated in strata together with crushed coal or rock into an underground opening. Either the quantity of burst gas or the crushed material is varied, so that it will be strict to designate the phenomena as gas ejection, gas and coal outburst, and gas and rock outburst according to the quantity and kind of crushed material. But in this report they are called simply gas outbursts. Four papers are submitted to the present Congress concerning this problem.

Gas outbursts have frequently occurred in some specified coal mines in many countries producing large amount of coal in the world. A fatal accident of gas outburst erupting about 600 thousand m^3 of methane gas and 4000 tons of crushed coal, occurred in a Japanese coal mine in 1981. The prevention for occurrence of gas outbursts and disasters caused by them is one of the most important problems in these coal mines. Scientific researches have been carried out since 1920's, and several hypotheses concerning the cause for occurrence of gas outbursts are presented.

Cause for occurrence of gas outburst

The hypotheses concerning the cause for occurrence of gas outburst may be classified into three categories according to the main cause, namely the gas pressure theory, the rock pressure theory and the gas and rock pressure theory. Although these hypotheses are not coincident, the differences among them may be attributed to the cause and manner of crushing of coal, and they seem to be much the same in other points. By summarizing these hypotheses, the cause for occurrence of gas outburst is outlined by the diagram in Fig. 2.7. The gas barrier appearing in this diagram is an imaginary zone in a coal seam which has become compact due to stress concentration and plays a role to dam up outflow of gas from the inner part of the coal seam. Some researches do not

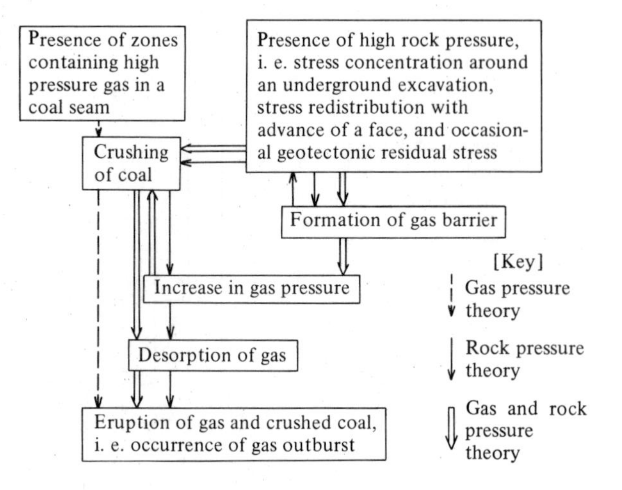

Fig. 2.7 Diagram illustrating the theories on the cause for occurrence of gas outbursts

consent to the existence of gas barriers.

Hanes et al. have reported the features and causes of gas outbursts in two Australian collieries, and suggested that though both stress and gas play a role for occurrence of gas outbursts, one of them may be more dominant than the other depending upon the local conditions.

Hiramatsu et al. have attempted, in order to study the cause for gas outburst, to explain several aspects associated with gas outbursts of large-scale having occurred in the specified coal mines in Japan. For this purpose they have investigated the volume of gas contained in coal seams, the gas pressure in pores and the rate of gas release via several routes from underground, and presented from a theoretical study a tentative criterion on the fracture of coal containing high pressure gas.

Premonition of gas outburst

It is believed that prediction of occurrence of gas outburst is generally very difficult because of poor premonition of gas outburst. The following is some results of measurement to explain this matter as an example. Oda et al. (1970) measured the increasing convergence with advance of a face in a gate road in Utashinai Coal Mine, Hokkaido, driven in disturbed strata. They used, for this measurement, five to seven sets of convergence-meter, set up at intervals of about 2 m, the front set being newly established at a point, about 2 m from the face by shifting the withdrawn last set. Fig. 2.8 shows a part of the results obtained. During excavation there appeared faults, disturbed zones and gas outbursts in connection with blasting. Each crooked line in this figure shows the profile of convergence within a short range near the face at the time when the front set of convergence-meter is newly set.

It is noted from this figure that the convergence on the occasion of gas outburst is small while large convergence is recorded either before or after the gas outburst, and that very irregular profile of convergence is obtained in the fault or disturbed zone. In the same gate road the volume of gas outflow from the range of 2 to 3 m of three boreholes drilled into the coal seam from the face was measured for 5 minutes after each blasting. The average rate of gas outflow per one borehole is plotted in Fig. 2.9. The average rate of convergence at the point, 2 m from the face, for 15 minutes measured from the time just before each blasting and the rate of gas exhaustion by ventilation after each blasting are also shown in Fig. 2.9. From this figure we cannot find any anomaly in these measured values at the time of occurrence of gas outbursts. It is noticed further that the measurements concerning gas behavior and rock pressure yield results with the same tendency, which may imply that both the gas and rock pressures participate in the geomechanical behaviour of burst-prone strata.

Prediction of gas outburst

Many kinds of measurement have been conducted for the purpose of

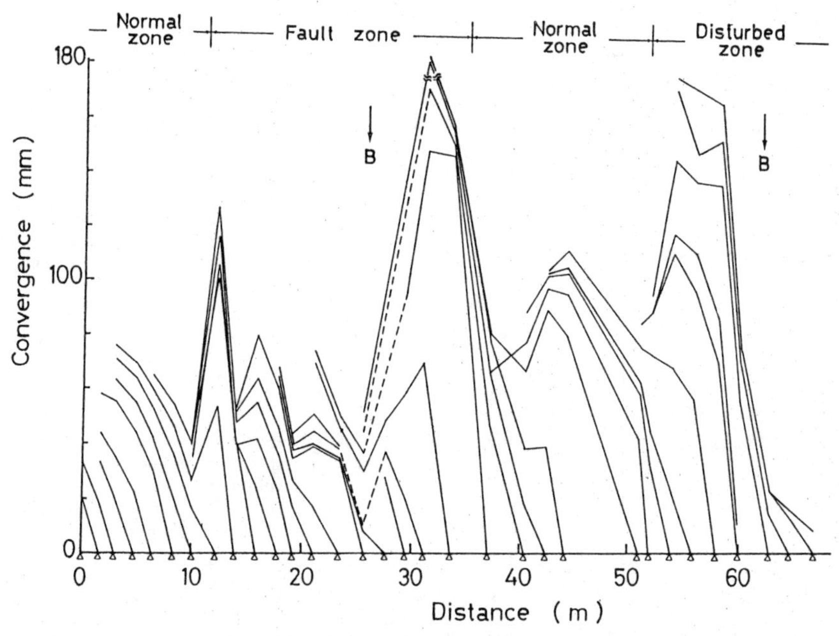

Fig. 2.8 Variation in the convergence profile with advance of a face of a gate road driven in disturbed strata. Small triangular marks indicate the position of convergence-meters, and the arrows B the occurrence of gas outbursts. (Oda, 1970).

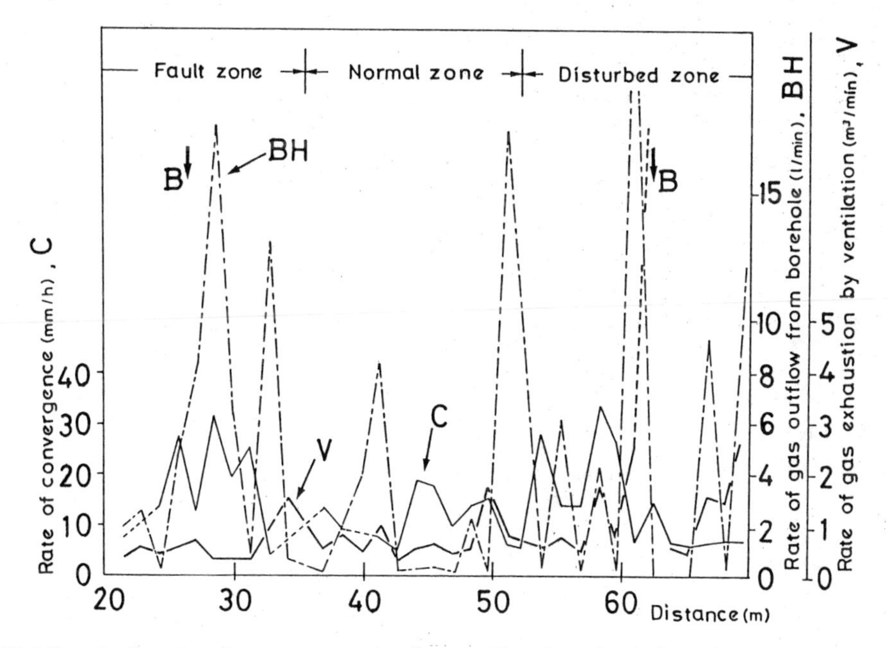

Fig. 2.9 Variations in the rate of convergence, rate of gas outflow from boreholes and rate of gas exhaustion by ventilation with advance of a face. (Oda, 1970).

predicting gas outburst. However, the phenomena occurring around a working face are much complicated. For example it seems that the rate of gas outflow from a coal seam or the gas pressure distribution in a coal seam may be influenced by the rock pressure, and that the strength and the mechanical properties of a coal seam may be influenced by its stress history. These phenomena depend largely upon the geological, geomechanical and mining technological conditions.

Moreover, occurrence of gas outbursts seems to be related to variation in measured values rather than their absolute values, which suggests that there is a room for re-examining whether the origin of gas outburst is in a specified area or not.

Under the circumstances such measurements that afford exact and reliable materials for prediction of gas outburst in all cases have not yet been found. Therefore further investigations are desired to find better measurements and to improve measuring technology for reliable prediction of gas outburst.

The major measurements which have been carried out hitherto are tabulated in Table 2.1, with the author's comments.

In specified cases, however, some reliable means may be contrived. Hanes et al. have reported that at West Cliff Colliery, where gas outbursts are associated with strike-slip faults. a technique of geological mapping

Table 2.1 Measurements for the purpose of predicting gas outbursts

Subjects	Items of measurement	Methods of measurement	Remarks	Authors of references
Nature of coal	Microcrack density	Observing coal cuttings under a microscope	Effective to some extent.	Vandeloise (1966)
	Δp index	Measuring the rate of initial desorption of gas	Effective to some extent, but in Japanese coal mines the effectiveness is indistinct.	ditto
	Brittleness	Applying the drum test for cokes	Effectiveness is indistinct.	Oda (1970)
	Strength	Testing the uniaxial compressive strength	ditto	Patching et al. (1966)
	Volume of cuttings	Measuring the volume of cuttings produced by boring	In some coal seams caution should be taken when the volume amounts to 3–4 times the ordinary volume or greater than it.	Gebirgeschlagrichtlinien (1976)
Gas contained in coal seams and its outflow	Gas concentration in air and volume of gas exhausted by air current	Measuring gas concentration and rate of air flow, and calculating the volume of gas exhausted by air current in 30 minutes after blasting	Effective to some extent. The critical values seem respectively to be 2% CH_4 in air, and 4 m^3 CH_4 exhausted by air current in 30 minutes after blasting per ton of mined coal.	Belin (1971)
	Volume of gas contained in coal	Measuring the volume of gas released from coal on grinding in a ball mill	Effectiveness is still being investigated.	ditto
	Rate of gas outflow from a borehole and rate of gas pressure rise in a borehole	Measuring the rate of free outflow of gas from a borehole and measuring the rate of gas pressure rise in a borehole after sealing the borehole mouth	Effectiveness is indistinct.	Cybulski et al. (1966)
	Temperature drop in a borehole	Measuring the temperature drop in a borehole after boring is finished	Temperature drop of 1–2°C seems to require cautions.	Kuroiwa (1964)
Rock stress and phenomena caused by excavation	Initial rock stress	Determining the rock stress by an optional method, for example the stress relief technique	Effectiveness is indistinct.	Takeuchi et al. (1976)
	Change in stress with advance of the face	Several instruments are available.	ditto	НОВИЧИХИН et al. (1975)
	Acoustic emission (AE) from coal and rock	Detecting and recording AE around a face with a suitable set of instruments	Effectiveness is being investigated.	Watanabe et al. (1981)
	Convergence	Measuring the convergence in a roadway, especially near the face	Probably effective.	Oda (1967)
	Swelling of coal face	Measuring the volume of swelling	Effectiveness is indistinct.	Szirtes (1966)
	Disking phenomena	Observing occurrence of, or the degree of disking of bored cores	Effectiveness is being investigated.	Eckart et al. (1966)

Table 2.2 Preventive measures for gas outbursts

Preventive measures	Remarks	Authors of references
Pre-mining a coal seam which is not prone to gas outburst	Very effective.	Berlin (1971)
Limiting the rate of face advance	Effective to some estent.	ditto
Suitable design of mining so as to avoid stress concentration	ditto	ditto
Vibration blasting to induce gas outburst	Effective to prevent disaster in most cases, but there remains possibility of occurring delayed gas outbursts.	The National Coal Board (1980)
Stress release by drilling boreholes with large diameter	Effective in certain collieries	Brouat (1966)
Gas drainage from coal seams in advance of mining	Effective to some extent. But it is very difficult to drain gas in virgin fields.	ГАРКАВИ et al. (1956)
Washing out of coal	Effectiveness is indistinct.	Eckart et al. (1966)
Wetting of coal seam	Effective to some extent.	Nedwiga (1966)

of coal joints has been developed which is effective to predict strike-slip faults within 45 m of the outburst threshold defined by the colliery. They have shown that the gas pressure measurement technique is also useful to predict outbursts among other various techniques.

Preventive measures for gas outburst

The major preventive measures for gas outburst which have been practiced or tested are collected in Table 2.2.

These measures may be effective to varied degrees. The effectiveness of stress relief boring may be different according to the natures of strata, and gas drainage from disturbed areas is frequently very difficult. The vibration blasting to induce gas outburst is indeed effective in most cases in avoiding disasters, but this procedure is subject to the hazard of delayed or untimely gas outburst. It should be emphasized that no satisfactory measures have yet been obtained.

The most significant themes to be investigated hereafter may be development of methods of preventing gas outbursts in disturbed areas, preventing delayed or untimely gas outbursts and so on. Pre-mining a safe coal seam would be a very promising measure. In case that no safe coal seam is found, such a technology may deserve investigation as excavating rock strata in the vicinity of apprehended coal seams.

Khristianovich et al. have reviewed the recent investigations of a wide scope concerning gas outbursts performed at the Institute for Problems of Mechanics, Academy of Sciences, USSR. The important points in them are as follows. A zone of oriented cracks with a certain width may be formed near the face when a coal seam prone to gas outbursts is excavated by virtue of acting stress and gas pressure. This zone is situated between a plastic zone which is adjacent to the face and an elastic zone subjected to concentrated stress, and is characterized by containing many cracks parallel to the face. These cracks may contain free gas of high pressure, which is apt to cause gas outbursts in case that the width of a plastic zone is small. Previous underhand or overhand mining of a safe coal seam lying near the gas outburst prone seam is effective to prevent gas outbursts, because many cracks parallel to the coal seam are already produced in the gas outburst prone seam during the previous mining. The free gas contained in these cracks will be able to flow out when this coal seam is mined, so that gas outburst will be prevented.

3. SUBSIDENCE

3.1 Introduction

The phenomena of ground movement caused by deep underground excavations are governed by several conditions such as the geological conditions, geometry of excavations and geotechnology for the excavations. Therefore the surface subsidence may assume various aspects according to the individual conditions. Deep underground excavations that cause surface subsidence may mostly be extensive underground openings produced by mining ore deposits or coal seams.

Surface subsidence is apprehended to cause damages to several surface installations, such as buildings and other constructions, railways, highways, rivers, paddy fields and so on. Surface subsidence will be prevented or restricted to a negligible extent by leaving adequately many large pillars underground, which however results in loss of mineral resources.

It would be ideal if we could realize a maximum recovery of mineral resources with a minimum damages related to surface subsidence, but in many cases we are forced, as compromise, to limit the extent of extraction of ore or to take some possible measures to control subsidences such as filling goafs. Accordingly for planning a rational mining, the prediction, control and measurement of subsidence have become one of the most important problems.

A large number of papers have been published for the problems concerning surface subsidence. In the present Congress seven papers have been submitted under this subtheme, and four of them are concerned with the subsidence caused by coal mining, two of them with that caused by mining ore deposits, while the remaining one paper treats subsidence involved in a shallow excavation.

3.2 Subsidence Caused by Mining Coal Seams

To outline the general features of surface subsidence caused by a wide underground excavation, for the sake of simplicity, the subsidence of flat earth surface caused by longwall mining of a horizontal coal seam will first be described. By extracting an area of a coal seam, the nether roof over the mined area will cave into the goaf, followed by sagging of the overlying strata, one by one successively from the lower strata to the upper ones, and at last a subsidence trough will be formed on the earth surface.

It was known already in the early years of this century that the area of the subsidence trough is larger than the mined area and that every point in this trough not only subsides but also displaces horizontally a little except the margin and the center point or the central part of the trough. Therefore the strata between the goaf and the surface are considered to move toward the mined space.

Accordingly there appear tilt and strain at every point on the trough surface. Let us denote the components of vertical and horizontal displacement at an arbitrary point on the surface by s, u_x and u_y, where x and y are horizontal coordinates, then the tilt components, i_x and i_y, and strain components, ϵ_x, ϵ_y and γ_{xy}, are given by:

$$i_x = \frac{\partial s}{\partial x}, \quad i_y = \frac{\partial s}{\partial y},$$

$$\epsilon_x = \frac{\partial u_x}{\partial x}, \quad \epsilon_y = \frac{\partial u_y}{\partial y}, \quad \gamma_{xy} = \frac{\partial u_x}{\partial y} + \frac{\partial u_y}{\partial x}.$$

The true tilt, i, and the angle ϕ between the x-axis and the direction of i are given by:

$$i = (i_x^2 + i_y^2)^{\frac{1}{2}},$$

$$\phi = \tan^{-1}\frac{i_y}{i_x}.$$

The principal strains and their directions can be found from ϵ_x, ϵ_y and γ_{xy} by the conventional equations. Let us assume that the upper diagram in Fig. 3.1 illustrate the profiles of the final surface subsidence and horizontal displacement along the line on the surface perpendicular to the longer axis of a rectangular mined out longwall panel, then the distributions of tilt and strain can be evaluated by the equations above mentioned as shown by the lower diagram in the same figure. On the narrone area on the surface where large tensile strains appear, there frequently take place tension cracks.

The substantial damages for surface buildings and other constructions induced by subsidence will be due to the tilt or strain greater than certain critical values rather than the amount of subsidence or horizontal displacement. It seems that the critical value of strain for structures may be something like 0.0005 for tension and 0.001 for compression, and the critical value of tilt for foundations of machinery may be 0.0005 and that for the ground of the new super-express railway of Japan (Shinkansen) 0.00025.

The distributions of surface subsidence and horizontal displacement will be symmetrical with respect to the center of the trough in an imaginary simple case where the earth surface and the coal seam are flat and horizontal, the geological structure is simple and homogeneous, and the mined panel is rectangular. Consequently the distributions of tilt and strain will also be symmetrical. However, in practical cases these conditions do not appear, so that the distributions of subsidence, horizontal displacement, tilt and strain will deviate from the symmetrical in varied degrees.

The seam inclination influences greatly the subsidence profile. The subsidence trough moves to the dip side with increase in dip angle of a seam, the angle of draw decreases on the rise side while increases on the dip side, and the location of the maximum subsidence point moves to the dip side from the surface point above the center of the mined panel. U.K. National Coal Board (1975) presented these relations

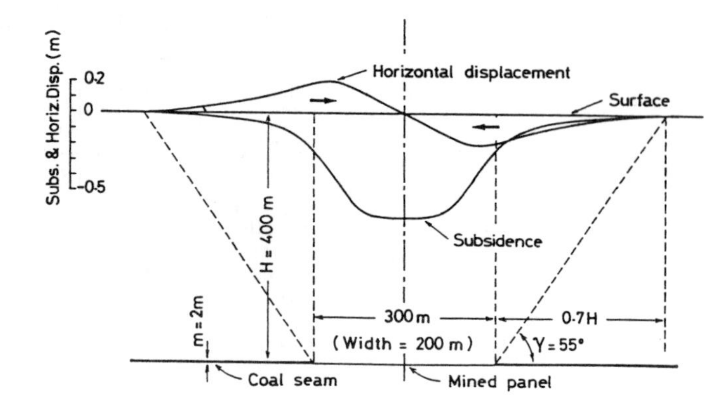

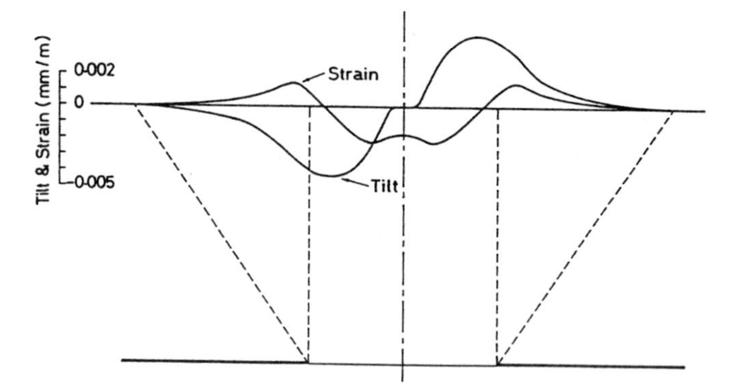

Fig. 3.1 Hypothetical profiles of subsidence and horizontal displacement (the upper diagram), and distributions of tilt and strain calculated (the lower diagram).

for whole range of dip angle based on the measurements in the British coal fields. Hiramatsu et al. (1979) reported the subsidence for moderately inclined seams investigated in Japanese coal mines. As a matter of course the horizontal displacement is also much influenced by the seam inclination.

If the ground were a continuum of which deformation obeys certain laws, the subsidence and horizontal displacement, accordingly the tilt and strain, would be related with each other. However the ground is subjected to complicated deformation, failure and movement by mining a coal seam, the subsidence and horizontal displacement may be independent. Nevertheless in practice some similarity or relationship is noticed among them. For example, horizontal displacement is approximately proportional to tilt; the analysis of subsidence by the theory of elasticity yields a subsidence profile something like an actual one. It is considered, therefore, that precise prediction of subsidence will be very difficult, but approximate predicton may be possible by several approaches.

3.3 Prediction of Subsidence Caused by Mining a Coal Seam

Prediction based on influence factors

In the early years of this century, it was already assumed that the subsidence, s, at a certain point, P, on the surface can be calculated as the continual product of the thickness of mined coal, m, the subsidence factor, a, that is the ratio of the maximum subsidence to the coal thickness, the time factor, τ, related to the time elapsed from mining, and the influence factor, e, which depends upon the mined area and its relative position to the surface point P. The factors a, τ and e range from 0 to 1, and depend on the geological condition. Besides the factor *a* depends on the mining method including the method of packing goaf.

Another assumption that is simple but implies an important meaning is the principle of superposition, which postulates that the subsidence of a certain surface point caused by mining an area of a coal seam can be obtained by summing up all the hypothetical subsidences caused by mining small areas of the coal seam which constitute the whole mined area.

Practical prediction methods based on these assumptions have been studied by many investigators, such as Keinhorst (1925), Bals (1931), Flächenträger (1938), Perz (1948), Beyer (1945), Sann (1949), Hiramatsu et al. (1956, 1967) and Salamon (1963). In fact this kind of predicting method for subsidence is convenient and yields results tolerably acceptable, but seems to be unable to give an exact subsidence profile close to the actual one. For example, the subsidence at the surface point just above the edge of a mined panel of a width, equal to or greater than half the critical width, is computed as half the maximum subsidence in all cases, though the actual subsidence is smaller than it.

Empirical method

Recently, on the basis of observation of surface subsidence in large number of British coal fields, a graphical method of predicting the subsidence profile and the distribution of strain was developed by the U.K. National Coal Board (1975).

In this method the subsidence factor, that is the ratio of the maximum subsidence, S, to the seam thickness, m, is given graphically from the depth, h, and width, w, of the panel, and the relationship between s/S and d/h is tabulated for each value of w/h, where s is the subsidence at the distance of d from the center of the trough. The angle of draw is taken as 35°, and the possible maximum subsidence is 0.9m for the caving and 0.5m for full stowing system of mining in horizontal stratifications. The maximum possible tilt in a subsidence trough caused by working a super-critical panel is estimated at 2.75 (S/h).

On the other hand, maximum tensile and compressive strains are given by the product of an factor K_3, determined graphically from the ratio w/h, and the ratio S/h, and the distribution of strain is given for each value of w/h. The maximum tensile and compressive strains over a super-critical panel are taken as 0.65 (S/h) and –0.51 (S/h) respectively. This empirical method of predicting surface subsidence and strain has been applied to coal fields in some other countries, and it has been found, that this method yielded not always good results, perhaps due to difference in geological conditions and mining procedures.

Mathematical modeling approaches

The predicting methods above mentioned let us feel uncertain in that they have no theoretical basis. An empirical method of predicting subsidence is only applicable to the coal fields where the method was established. Hereupon efforts have been made to analyze the ground movement by mathematical modeling, since more than two decades ago, for the purpose of obtianing the data to be used for prediction of surface subsidence.

Berry (1964) pointed out that it is possible to predict the final deformed state of ground qualitatively and to some degree quantitatively by the displacement discontinuities analysis on the assumption that the ground is a transversely isotropic elastic body. Dahl and Choi (1972) analyzed the subsidence on a model that behaves in an elastic- elastoplastic manner with a good result.

However the geological structures are very complex, the mechanical properties of strata are various and the boundary conditions are complicated, so that the ground movement caused by excavation is not a simple phenomenon, and mathematical modeling capable of yielding a correct picture of subsidence may be very difficult. As a matter of fact investigators have been obliged to find suitable material constants by trials so as to afford the calculated subsidence close to the measured one.

Considering that simple mechanistic models are incapable of simulating the complex strata behaviour while numerical modeling using a computer may be the only practical alternative, Mikula and Holt have examined the applicability of the finite element method to simulate subsidence. For material properties of the geotechnical model, realistic values were provided based on actual measurements and computer handling system. The analysis has been carried out by the finite element computer program of the constant strain type containing the joints elements originated by Goodman et al. It has been found from the results of analysis that the finite element modeling of subsidence is, under the present situation, not capable of yielding satisfactory results, but there are prospects of success with continuing research.

Salamon has attempted to examine the possibility of simulating the surface movement induced by mining a coal seam by elastic models of which deformation is governed by linear laws that permit superposition. The ground movement caused by mining a coal seam was analyzed on three models, namely a homogeneous isotropic model, a homogeneous transversely isotropic model and a stratified elastic medium. To judge whether a model was well simulated or not, a few critical measures of surface movement obtained in British coal fields were compared with those obtained by analysis.

In order to obtain theoretical surface movement that is consistent with the critical measures, extraordinary values had to be taken for the parameters concerning the mechanical property of the medium. Then Salamon has employed semi-empirical influence functions to attain acceptable surface subsidence prediction. He has suggested that the models can simulate ground movement qualitatively and that it may be possible to find an empirical model when the depth of mining is not unreasonably small.

Another approach to predicting surface subsidence is to regard the ground as a stochastic medium which is composed of a large number

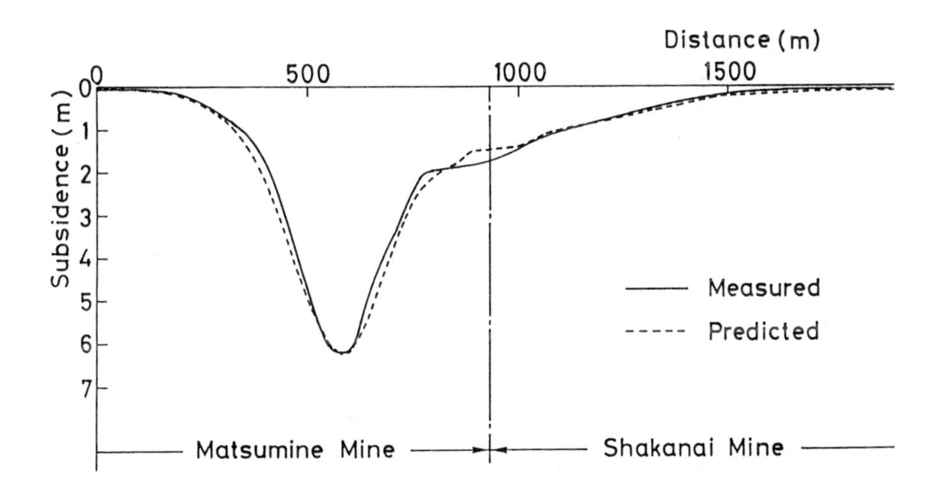

Fig. 3.2 Comparison of the predicted and measured subsidence profiles in May 1978 along No. 1 line on the surface of the Matsumine and Shakanai mines. (Hiramatsu et al., 1979).

of rock blocks of various sizes and shapes, locked together with freedom of movement at random to some extent.

Litwiniszyn (1956) suggested a new method for computing rock mass deformation induced by mining on this proposition. In the last some twenty years, this method has been improved by continual theoretical and experimental investigations. Baoshen et al. (1979) reported that the results of surface subsidence prediction by this method showed good coincidence with the measured results, and that they succeeded in solving a series of problems associated with mining under protective architectures, railways and water bodies.

3.4 Subsidence Caused by Mining Ore Bodies

Ore bodies as the object of mining are usually of irregular shape, openings produced by mining activities are diverse in shape and size, and the geological structures are more complex and irregular than those of coal fields. Therefore the surface movement caused by mining ore bodies will assume a variety of aspects, and no systematic study on this kind of subsidence has been conducted.

Goel has studied the surface subsidence in a mine working a steeply dipping tabular copper vein extending approximately 5 km along strike and to 1200 m depth. He has pointed out that the angle of draw and the angle of break are not constant, and that the location of the maximum subsidence point as well as the ratio of the volume of subsidence to the mined volume is quite different from those in mining horizontal coal seams. Particularly it should be noted that the rate of subsidence is constant for tens of years despite of non-linear mining rate. These data will show how different is the situation of subsidence in this mine from that of coal mines working flat seams.

In 1960's, the so-called "black ore" deposits were discovered in Akita Prefecture, Japan, and development works were soon started. The ground consists of mainly weak tuff and mudstone intruded by rhyolite. On the surface over the deposits there are residences, a railway, a river and paddy fields. Since the district containing these deposits were developed by two neighboring mines belonging to different companies, precise prediction of subsidence was strongly required.

This problem was studied by organizing a research committee. After investigations for several years it was found that precise prediction of subsidence by purely mathematical treatments might be impossible for the present, but prediction of subsidence in the near future could be practiced precisely by the influence function method, utilizing the results of measurements in the past two or three years as the fundamental materials for the compensation for local conditions. Fig. 3.2 shows an example of comparison of thus obtained subsidence profile and the measured one (Hiramatsu et al. 1979).

3.5 Control of Subsidence

It is clearly impossible to mine a wide area of a coal seam without surface subsidence by the longwall system. To prevent surface damages, it is necessary to restrict surface subsidence within the allowable values. The room-and-pillar method or panel-and-pillar method leaving adequate pillars will minimize the surface subsidence and damage. Stowing goafs also reduces the surface subsidence. When some specified surface structures are required to be protected from damage, it may be possible to reduce the subsidence so as to protect the structures by mining the coal seam in a special manner designed on the basis of prediction of subsidence.

Since subsidence assumes various aspects according to the type of deposits, geological conditions and mining practice, and the objects of subsidence control required are diverse, the methods of control will be various. Hereupon only the important points of the papers submitted to this Congress concerning control of subsidence will be introduced.

Galvin has dealt with a unique problem of controling failure of the massive (thicker than 30 m) dolerite sills found in the superincumbent strata in the South African coal mines. To succeed in total extraction of coal, it is necessary to induce failure of the dolerite sills soon after the commencement of mining operation. Galvin has studied the prerequisite to realize this situation. Starting from the elastic thin plate theory, he has modified it by utilizing field observations and measurements, and obtained a simple equation to calculate the minimum panel dimension required to induce dolerite failure.

Arcamone and Poirot has discussed the problem of a coal mine in South France, where both open pit development and underground extraction are carried out simultaneously.

An interesting paper has been presented by Stacey et al. describing the researches which led to success in building development over a shallow goaf in the urban area of Johannesburg. The researches are composed of investigation of an undermined site and the subsequent analysis of the stability and design requirements of the foundations and remedial measure system including the stability of deeper goaf and the surface movement. Two-dimensional displacement discontinuity stress analyses were mainly used for this purpose, and three-dimensional mining simulation techniques with displacement discontinuity elements were also used.

Acknowledgements

The author is very grateful to the many members of the Japanese Committee for ISRM for their assistance in preparing this report, and expresses especially his appreciation to Prof. S. Sakurai, Prof. S. Kobayashi, Dr. N. Oda, Dr. Y. Ohnishi Dr. T. Saito, Dr. K. Sugawara and Dr. Y. Mizuta for their valuable advices and cooperations.

References

Ashley, D. B., et al. (1981). Geological Prediction and Updating in Tunneling—A Probabilistic Approach: 22nd U.S. Sympo. Rock Mech., pp. 361–366.

Baoshen, L., Kuohuna, L. and Rougui, Y. (1979). Research in the surface ground movement due to mining: Proc. 4th Int. Cong. Rock Mech. vol. 3, ISRM.

Bals, R. (1931/32). Beitrag zur Frage der Vorausberechnung bergbaulicher Senkungen: Mitt. Markscheidew. 42/43 S, 98–111.

Barron, K. (1978). An air injection technique for investigating the integrity of pillars and ribs in coal mines: International Journal of Rock Mechanics and Mining Science and Geomechanics, Abstr. 15, p. 69–p. 76.

Belin, J. (1971). Dégagements instantanés de méthane et de charban: Rapport de synthése 1962–1699 CERCHAR. No. 41.

Berry, D. S. (1964). A theoretical elastic model of the complete Region Affected by mining a thin seam: Proc. Sixth Symp. Rock Mech., Univ. of Mo., Rolla, Mo.

Beyer, F. (1945). Über die Vorausbestimmung der beim Abban flachgelagerter Flöze auftretenden Bodenverformungen: Habil.-Schr. Marksch.-Inst. T. H. Berlin.

Bieniawski, Z. T. (1968). The effect of specimen size on the compressive strength of coal: International Journal of Rock Mechanics and Mining Science, 5, p. 325–p. 335.

Bieniawski, Z. T. (1976). The Geomechanics Classifications in rock engineering: Proceedings of Symposium on Exploration for Rock Engineering, Johanesburg.

Brady, B. H. G. (1979). A Direct Formulation of the Boundary Element Method of Stress Analysis for Complete Plane Strain: Int. J. Rock Mech. Min. Sci. & Geomech. Abst., vol. 16, pp. 235–244.

Bräuner, G. (1975). Kritische Spannungen in Kohleflözen. Glückauf 111–625.

Braüner, G. (1975). Kritische Spannungen in Kohlenflozen Glückauf, Vol. 11, No. 13, pp. 618–625.

Bräuner, G. (1979). Systematik von Gebirgsschlägen und Beispiele neuerer Über wachungs messungen: Glückauf, 115, 5, p. 196–201.

Bray, J. W. and Goodman, R. E. (1981). The Theory of Base Friction Model: Int. J. Rock Mech. Min. Sci. & Geomech. Abstr., vol. 18, pp. 453–468.

Brouat, R. (1966). The prevention of sudden outbursts in the coalpits of Cévennes Basin: Proc. of Int. Congress on Sudden Outbursts of Gas and Rock, Leipzig.

Brown, E. T. and Bray, J. W. (1982). Rock-support Interaction Calculations for Pressure Shafts and Tunnels: Proc. Rock Mechanics; Caverns and Pressure Shafts, Aachen, vol. 2, pp. 555–565.

Christiaens, P. (1982). Proposed calculation model for powered supports on a longwall face: Proceedings of 7th International Strata Control Conference, Liege.

Cividini, A., Jurina, L. and Gioda, G. (1981). Some Aspects of 'Characterization' Problems in Geomechanics: Int. J. Rock Mech. Min. Sci. Abstr., vol. 18, pp. 487–503.

Crouch, S. L. (1976). Analysis of Stresses and Displacements Around Underground Excavations: An Application of the Displacement Discontinuity Method. University of Minnesota Geomechanics Report, November 1976.

Cundall, P. A. (1974). Rational Design of Tunnel Supports; A Computer Model for Rock Mass behaviour using Interactive Graphics for the Input and Output of Geometrical Data, U.S. Army Corps of Engineers Tech: U.S. Army Corps of Engineers Technical Report MRD–2–74.

Cybulski, W., et al. (1966). Forecasting and Control of Sudden Outbursts in Polish Coal–Pits: Proc. Int. Congress on Sudden Outbursts of Gas and Rock, Leipzig.

Dahl, H. D. and Choi, D. S. (1973). Some case studies of mine subsidence and its mathematical modeling: Proc. 15th Symp. Rock Mech. Custer State Park, SD., p. 1–22.

Diering, J. A. C. (1980). Simulation of Mining in Non-Homogeneous Ground Using The Displacement Discontinuity Method. South African I.M.M Jour., Vol. 80, No. 7, pp. 159–163.

Duddeck, H. (1980). On the basic requirements for applying the convergence – confinement method: Underground Space, vol. 4, No. 4, pp. 241–247.

Eckart, D., Gimm, W. and Thoma, K. (1966). Plötliche Ausbrüche von Gestein und Gas im Bergbau: Freiberger Forschungsheft-HA409.

Egger, P. and Gindroz, C. (1979). Tunnels Ancrés A Faible Profoundeur Etude Comparative sur Modèles Physique et Mathématique: Proc. 4th Int. Cong. Rock Mech., ISRM, vol. 2, pp. 121–130.

Egger, P. (1980). Deformations at the face of the heading and deformation of the cohesion of the rock mass: Underground Space, vol. 4, No. 5, pp. 313–318.

Everling, G. (1972). Voraussage und Beurteilung des Gebirgsdruckes im Steinkohlenbergbau. 5. Internationale Gegirgsdrucktagung, London, Vortr. 18.

Einstein, H. H., et al. (1978). Decision Analysis Applied to Rock Tunnel Exploration: Engineering Geology, vol. 12, pp. 143–161.

Everling, G. and Meyer, A.–G. (1972). Ein Gebirgsdruck-Rechenmodell als Planungshilfe. Glückauf-forschungshefte, Vol. 33, No. 3, June 1972, pp. 81-88.

Fisekci, M. Y. and Parkes, D. (1982). Deformation analysis of underground openings and pillars during hydraulic mining of thick and steep seams in Western Canada: Proceedings of 7th International Strata Control Conference, Liege.

Fairhurst, C. and Damen, J. (1980). Practical inferences from research on the design of tunnel supports: Underground Space, vol. 4, No. 5, pp. 297–311.

Fläschenträger, H. (1938). Die Kostenverteilung bei gemeinsam verursachten Bergschäden im Ruhrgebiet: Mitt. Markscheidew. 49, S. 95–137.

Galvin, J. M. (1982). Total extraction of coal seams: the significance and behavior of massive dolerite sills: Chamber of Mines of South Africa Research Report No. 19/82.

ГАРКАВИ, С. М., БРАЙЦЕВ, А. В. and ВОЙНИК, И. А. (1956). Translation into Japanese from уголь.

Gebirgsschlag richtlinien (1976). Richtlinien des Landeroberbergamts Nordrhein-Westfalen für die Zulassung von Betriebsplänen für das Herstellen von Gru ben bauen in Kohlen und Nebengestein im Hinblick auf eine Gebirgsschlaggefaahr: Gebirgsschlagrichtlinien.

Gioda, G. and Maier, G. (1980). Direct Search Solution of an Inverse problem in Elastoplasticity: Identification of Cohesion, Friction angle and In Situ Stress by Pressure Tunnel Test: Int. J. Num. Meth. in Engng., vol. 15, pp. 1823–1848.

Gioda, G. and Jurina, L. (1981). Numerical Identification of Soil-Structure Interaction Pressures: Int. J. Num. Analy. Meth. in Geomech., vol. 5, pp. 33–56.

Grotowsky, U. (1982). The strata control system and its application in West German-coalmining: Proceedings of 7th International Strata Control Conference, Liege.

Hebblewhite, B. and Schaller, S. (1982). Geotechnical evaluation of the effects of shallow depth longwall extraction at Angus Place Colliery, New South Wales: Proceedings of 7th International Strata Control Conference, Liege.

Josien, J. P., et al. (1982). The Dynamic effects of Strata-Pressure

Hiramatsu, Y. and Kokado, J. (1956). On the influence factor concerning surface subsidence: J. Min. Met. Inst. Japan, vol. 72, No. 820, p. 561–564.

Hiramatsu, Y. and Oka, Y. (1967). Studies on the estimation of the ground movement due to mining: J. Min. Met. Inst. Japan, vol. 83, No. 949, p. 739–743.

Hiramatsu, Y. Okamura, H. and Sugawara, K. (1979). Surface subsidence and horizontal displacement caused by mining inclined coal seams: Proc. 4th Int. Cong. Rock Mech. vol. 1, ISRM.

Hustrulid, W. A. (1976). A review of coal pillar strength formulas: Rock Machanics, 8, p. 115–p. 145.

Irresberger, H. (1982). New planning models for winning, roadway and face supports: Proceedings of 7th International Strata Control Conference, Liege.

Johnson, G., Kellet, W. H. and Mills, P. S. (1982). Aquapak: a cementitious pack material with high water content: Proceedings of 7th International Strata Control Conference, Liege.
Rock Bursts: Proceedings of 7th Int. Strata Control Conf., Liege.

Kaiser, P. K. and Morgenstern, N. R. (1981). Time-dependent Deformation of Small Tunnels–I., II: Int. J. Rock Mech. Min. Sci. & Geomech. Abst., vol. 18, pp. 129–152.

Kaiser, P. K. (1981). A New Concept to Evaluate Tunnel Performance –Influence of Excavation Procedure–: 22nd. U.S. Sympo. Rock Mech., pp. 264–271.

Kawai, T. (1977). New discreate structural models and Generalization of the method of limit analysis: Int. Conf. of Finite Element in Non-linear Solid and Structural Mechanics, Geilo, Norway, 2, Go 4.1–Go 4.20.

Keinhorst, H. (1925). Die Berech nung von Boden senkungen im Em schergebiet. In: 25 Jahre Emschergenossenschaft 1900–1925, Essen S. 347–350.

Keriseil, J. (1980). Commentary on the general report: Underground Space, vol. 4, No. 4, pp. 233–239.

Kimura, O., Sugawara, K. and Kaneko, K. (1982). Study on the controlling of coal burst in Miike mine: Pro. of 7th Int. Strata Control Conf., Liege.

Kobayashi, S. and Nishimura, N. (1980). Elastoplastic Analysis by the Integral Equation Method: Mem. Fac. Eng. Kyoto Univ., 42, pp. 324–334.

Konda, T., Inokuma, A. and Kato, K. (1981). Three Dimensional Model Tests on Soft Ground Tunnels: Proc. Int. Sympo. on Weak Rock, vol. 3, pp. 199–204.

Kovari, K., Amstad, Ch. and Fritz, P. (1977). Integrated Measuring Technique for Rock Pressure Determination: Proc. Int. Sympo. on Field Measurements in Rock Mechanics, Zurich, pp. 289–316.

Kuroiwa, T. (1964). Gas outbursts (continuation). Journal of the mining and metallurgical Institute of Japan, vol. 80, No. 916.

Litwiniszyn, J. (1956). Application of the equation of stochastic processes to mechanics of loose bodies: Arch. Mech. Stos., 7.8 zesz. 4.

Lombardi, G. (1980). Some Comments on the convergence-confinement method: Underground space, vol. 4, No. 4, pp. 249–258.

Lu, P. H. (1981). Determination of ground pressure existing in a viscoelastic rock mass by use of hydraulic borehole pressure cells: Proceedings of International Symposium on Weak Rock, Tokyo, Japan, A. A. Bakema, Rotterdam 1, p. 459–465.

Murayama, S., Matsuoka, H. (1968). New Measuring Method of Earth Pressure on Tunnel Support, Jour. Japan Soc. Civil Engr. vol. 53, No. 3, pp. 39-42.

Murayama, S. (1978). Tunnel Support Against Increasing Earth Pressure: Proc. Int. tunnel Sympo. '78, Tokyo, b-1-1-1–8.

Nakajima, I., Watanabe, Y. and Fukai, T. (1981). Proc. of 3rd Conference on Acoustic Emission/Microseismic Activity in Geologic Structures and Materials, The Pensylvania State University, U.S.A.

The National Coal Board. (1980). Cynheidre Mine, Outbursts of Coal & Firedamp, Code of Precautions.

Nedwiga, S. N. (1966). Hydraulic Method of Controlling Sudden Outbursts of Coal and Gas in the Mines of the Donets Basin: Proc. of Int. Confression Sudden Outbursts of Gas and Rock, Leipzig.

Nisugi, I. (1981). Significance of geological features for the mechanical behaviour of rocks and rock masses: Proc. of the Int. Sym. on Weak Rock, Tokyo, pp. 1345–1353, Balkema.

Niu, X. and Gu, T. (1982). Control of the very hard roof by softening method using water injection: Proc. of 7th Int. Strata Control Conf., Liege.

НОВИЧИХИН, И. А., ЕПИШЕВ, Г. П. and УЗБЕК И. Г. (1975). Translation into Japanese from уголь.
Oda, N. (1967). The Rate of Convergence in Roadways of Coal Mine and the Influence: Journal of the Mining and Metallurgical Institute of Japan, vol. 83, No. 945.
Oda, N. (1970). Studies on prevention of gas outbursts: The thesis for Dr. degree.
Oda, N. Isobe, T. and Umezu, M. (1981). On the strata Disturbance and Gas Outburst: XIX International Conference of Research Institutes in Safety in Mines.
Okamura, H., et al. (1979). A fundamental study on roof fall phenomenon: Journal of the Mining and Metallurgical Institute of Japan, 95, 1097, p. 387–p. 392.
Patching, T. H. and Botham, J. C. (1966). Occurrence, Research and Combating Sudden Outbursts of Coal and Gas in Canada: Proc. of Int. Congress on Sudden Outbursts of Gas and Rock, Leipzig.
Pertz, F. (1948). Der Einfluß der Zeit auf die Bodenbewegungen über Abbauen: Mitt. Markscheidew. 55, S. 92–117.
Plewman, R. P., Deist, F. H., and Ortlepp, W. D. (1969). The Development and Application of A Digital Computer Method For The Solution of Strata Control Problems. South African I.M.M. Jour., Vol. 70, No. 2, Sept. 1969, pp. 33–44.
Sakurai, S. (1978). Approximate time-dependent analysis of tunnel support structure considering progress of tunnel face: Int. J. Numer. Anal. Methods in Geomech., vol. 2, pp. 159–175.
Sakurai, S. (1981). Direct Strain Evaluation Technique in Construction of Underground Opening: 22nd U.S. Sympo. Rock Mech., pp. 278–282.
Sakurai, S. (1982). Monitoring of Caverns during Construction Period: Proc. Rock Mechanics; Caverns and Pressure Shaft, ISRM sympo., pp. 433–441.
Salamon, M. G. D. (1963). Elastic analysis of displacements and stresses induced by mining of seam or reef deposits: J. S. Afr. Inst. Min. Met., p. 128–149.
Salamon, M. D. G. and Munro, A. H. (1967). A study of the strength of coal pillars: Journal S. African Institute of Mining and Metallurgy, 68, p. 55–p. 67.
Salamon, M. D. G., Oravecz, K. I. and Hardman, D. R. (1972). Rock mechanics problems associated with longwall trials in South Africa: Chamber of Mines of South Africa Research Report No. 6/72.
Saltsman, R. D., Andria, G. D. and Mayton, A. G. (1982). An overview of U.S. longwall mining systems—Productivity, mining and geological parameters: Proceedings of 7th International Strata Control Conference, Liege.
Sann, B. (1949). Betrachtungen zur Vorausberechnung von Bodensenkungen infolge Kohlenabbaues: Bergbau-Rdsch. 1, S. 163–168.
Sheorey, P. R., et al. (1982). Analysis of strata control practices in bord and pillar workings in India: Proceedings 7th International Strata Control Conference, Liege, Belgium.
Shi, G. H. and Goodman, R. E. (1981). A New Concept for Support of Underground and Surface Excavations in Discontinuous Rocks Based on Keystone Principle: 22nd U.S. Sympo. Rock Mech., pp. 290–296.
Starfield, A. M. and Crouch, S. L. (1973). Elastic Analysis of Single Seam Extraction in "New Horizons in Rock Mechanics", ed. Hardy & Stefanko, A.S.C.E., New York, pp. 421–439.
Sugawara, K., et al. (1981). Proceedings of the International Symposium on Weak Rock, Tokyo, p. 561–566.
Szirtes, L. (1966). The unusual Conditions of Pécs Coalfield, which is Endangered by Outbursts of Gas, and the Consequences which results from This: Pro. of Int. Conf. on Sudden Outbursts of Gas and Rock, Leipzig.
Takeuchi, M. and Kinoshita, S. (1976). The in-situ stress measurements in the Yubari New Colling: Journal of the Mining and Metallurgical Institute of Japan, vol. 92, No. 1063.
Tanimoto, C., Hata, S. and Kariya, K. (1981). Interaction between Fully Bonded bolts and Strain Softening Rock in Tunneling: 22nd U.S. Sympo. Rock Mech., pp. 347–352.
Tinchon, L., Daumalin, C., George, L. and Piguet, J. P. (1982). How stress varies in relation to depth and natural factors: Proceedings of 7th International Strata Control Conference, Liege.
U.K. National Coal Board (1975). Subsidence Engineer's Handbook, Mining Dept., p. 111.
Vandeloise, R. (1966). Sudden Outbursts of Coal and Methane in the Belgian Coalfield, Research Work and Methods of Prevention of Sudden Outbursts: Proc. of Int. Congress on Sudden Outbursts of Gas and Rock, Leipzig.
Van Dillen, D., et al. (1979). A Two-dimensional Finite Element Technique for Modeling Rock/Structure Interaction of a Lined Underground Opening: 20th U.S. Sympo. Rock Meck., pp. 251–258.
Wade, L. V. and Wang, C. S. (1982). Developments in ground control procedures for room and pillar coal mines in the United States: Proceedings of 7th International Strata Control Conference, Liege, Belgium.
Wagner, H. and Galvin, J. (1982). The introduction of high percentage extraction methods in South African collieries: a strata control challenge: Proceedings of 7th International Strata Control Conference, Liege.
Watanabe, Y., Nakajima, I. and Itakura, K. (1981). Proc. of Third Conference on Acoustic Emission/Microseismic Activity in Geologic Structures and Materials, The Pennsylvania State University, U.S.A.
Watanabe, Y., et al. (1981). The applications of AE techniques as a forecasting method to the rock and gas outburst in coal mine: XIX International Conference of Research Institutes in Safety in Mines.
Wilson, A. H. (1972). Research into the determination of pillar sizes, part 1: an hypothesis concerning pillar stability: Mining Engineer, 131, p. 409–p. 417.
Wilson, A. H. (1977). The effect of yield zones on the control of ground: Proceedings of 6th International Strata Control Conference, Banff, Canada.
Wilson, A. H. (1981). Stress and stability in coal ribsides and pillars: Proceedings of 1st Conference on Ground Control in Mining, West Virginia University, Morgantown, U.S.A., p. 1–p. 12.

TIEFUNTERIRDISCHE EXKAVATIONEN BESONDERS TUNNEL- UND KOHLENBERGBAU UND DIE DADURCH VERURSACHTEN BODENSENKUNGEN

Deep underground excavations especially tunnels, coal mining and subsidence caused by them

Excavations souterraines et profondes, notamment des tunnels, l'exploitation minière des charbons et des affaissements du terrain causés par eux

Yoshio Hiramatsu
Kyoto Universität, Kyoto, Japan

ZUSAMMENFASSUNG

Die bis zum gegenwärtigen Zeitpunkt durchgeführten Forschungsarbeiten auf dem Gebiet der Geomechanik und der Geotechnologie von tiefen untertägigen Hohlraumbauten unter besonderer Berücksichtigung des Tunnelbaues, des Kohlebergbaues und der dadurch verursachten Bergsenkungen (Unterthemen C2, C3 und C4) sind kurz beschrieben. Der wichtige Teil der zu diesen Unterthemen eingereichten Beiträge wird vorgestellt, um besondere Diskussionsbereiche anzuregen.

1. TUNNEL

1.1 Einführung

Tunnelbautechniken haben einen engen Bezug zur Felsmechanik. Es ist überhaupt nicht möglich, hier die Unmenge an verfügbaren Informationen über dieses Gebiet zusammenzufassen. Daher sollen nur einige wenige Bemerkungen angebracht werden.

Die Forschungsarbeiten über den Tunnelbau lassen sich in zwei breite Kategorien aufgliedern: Die eine ist von theoretischer Natur und sucht das Verformungs- und Versagensverhalten von Tunneln zu erklären. Die andere ist mehr von praktischer Natur und behandelt den Entwurf und die Ausführung von Tunneln. Felsmechanische Forschungsarbeiten zu diesen beiden Kategorien des Tunnelbaues sind in Fig. 1.1 zusammengefaßt.

Studien zur erstgenannten Kategorie haben bis heute noch keinen direkten Bezug zum Entwurf und zur Ausführung von Tunneln. Sie sind jedoch als Grundlagenforschung außerordentlich wichtig für den zukünftigen Tunnelbau. Die Grundlagenforschung des Tunnelbaues enthält viele verschiedene Näherungsmethoden. In der ersten Näherung wird der natürliche Baugrund durch ein mechanisches Modell abgebildet, in welchem die Spannungen, Verformungen, Verschiebungen sowie das Festigkeitsverhalten analysiert werden können. Unter der Annahme, daß der Baugrund ein kontinuierlicher Körper ist, kann die Untersuchung mithilfe der Kontinuumsmechanik durchgeführt werden, also nach der Theorie der Elastizität, Plastizität oder Visko-elasto-Plastizität. In diesem Fall könnte eine analytische Lösung möglich sein, wenn die Randbedingungen einfach sind. Bei komplizierten Randbedingungen lassen sich numerische Berechnungen mithilfe der neuentwickelten Digitalcomputer ausführen. Das gleiche gilt, wenn komplizierte Materialeigenschaften berücksichtigt werden müssen. Dieses Näherungsverfahren kann jedoch so lange nicht angewendet werden, wie die Kluftabstände nicht sehr viel kleiner als der Tunneldurchmesser sind und das Gebirge nicht als kontinuierlicher Körper angesehen werden kann. Das kontinuumsmechanische Konzept kann mit einiger Mühe und Modifizierung auch dann noch eingesetzt werden, wenn Felsnägel oder Felsanker vorhanden sind. Es trifft jedoch nicht zu, wenn der Kluftabstand in der Größenordnung des Tunneldurchmessers liegt. In solch einem Falle ist ein anderes Näherungsverfahren erforderlich, das die Existenz von Diskontinuitäten einbezieht. Es ist nicht möglich, diese Berechnungen ohne Zuhilfenahme numerischer Verfahren durchzuführen. Cundall (1974) wie Kawai (1977) haben bereits eine Methode der numerischen Analyse von Diskontinuitäten entwickelt, in der sie den Fels durch starre oder verformbare Blöcke nachbildeten. Beispielsweise hat Shi (1981) die Standfestigkeitsberechnung für einen Tunnel mit Klüften in der Firste durchgeführt, wobei er das Kräftegleichgewicht an jedem der Teilkörper betrachtete.

Laborversuche spielen eine wichtige Rolle bei der Untersuchung der Standsicherheit von geklüftetem Fels. Sogenannte Base-friction-Modelle können eine qualitative Antwort auf die Frage nach der Tunnelstabilität geben (Egger, 1979; Bray, 1981). Auch sind im Labor Kontinuitätsmodellversuche ausgeführt worden (Kaiser, 1981; Konda, 1981), doch leiden solche Modelle daran, daß sie die Ähnlichkeitsbedingungen nicht erfüllen.

Bei den Forschungsarbeiten der zweiten Kategorie handelt die Mehrzahl von Feldmessungen. Über einen großen Teil der Meßergebnisse einschließlich der Entwicklung neuer Meßmethoden und ihrer Auswertung wurde bereits berichtet (Sakurai, 1981). Viele dieser Arbeiten befassen sich mit der Neuen Österreichischen Tunnelbauweise (NÖT). Auch gibt es hochgestochene Unter-

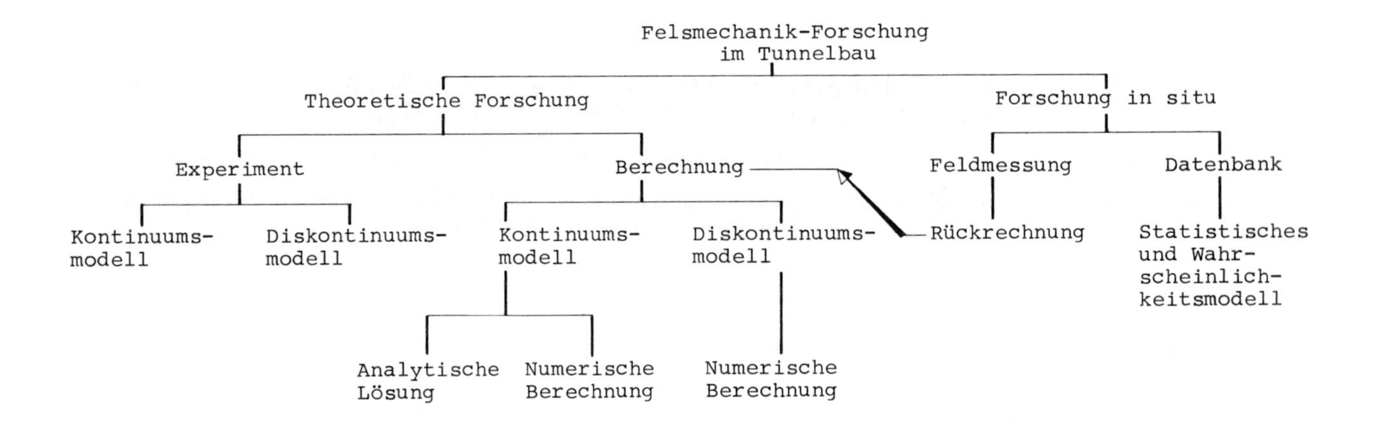

Fig. 1.1: Klassifizierung der Forschungsarbeiten zur Felsmechanik des Tunnelbaus

suchungen für den Entscheidungsprozeß beim Entwurfsstadium. Diese Studien sammeln die Ergebnisse früherer Messungen und führen sie zur Entscheidungsfindung in ein Rechenverfahren ein, das auf der Theorie der Statistik und der Wahrscheinlichkeit beruht (Einstein, 1978; Ashley, 1981).

Im Tunnelbau müssen auch die Bergwasserprobleme beachtet werden. Das unterirdische Bergwasser beeinflußt die Stabilität von Hohlräumen insbesondere dann, wenn das Gebirge geklüftet und zerschoren ist. Gegenwärtig werden hydromechanische Berechnungen angewendet, um neue numerische Verfahren für die Felshydraulik zu entwickeln.

Nach diesem Überblick über den Stand der Forschung im Tunnelbau werden im folgenden verschiedene Teilaspekte aufgezeigt einschließlich der Beschreibung des 53.85 km langen untermeerischen Seikan-Tunnels, dessen Erkundungsstollen im Januar 1983 fertiggestellt worden ist.

1.2 Berechnung

Es gibt hauptsächlich zwei Wege, das Verhalten von Gebirge und Tunneln zu berechnen: Die analytische Behandlung, wenn geschlossene Lösungen möglich sind und anderenfalls numerische Verfahren wie die Finite-Element-Methode (FEM). Die FEM kann die komplizierten mechanischen Eigenschaften des natürlichen Gebirges sowie die Ausbruchsabfolge berücksichtigen. Solche Faktoren können dagegen bei der analytischen Behandlung i.d.R. nicht einbezogen werden. Folglich muß sie sich auf einfache Randbedingungen beschränken, wie z.B. auf einen kreisförmigen Tunnel im isotropen äußeren Spannungsfeld. Ergebnisse solcher analytischer Berechnungen, die realistische mechanische Eigenschaften wie z.B. visko-elasto-plastisches Stoffverhalten und die Materialentfestigung berücksichtigen, wurden von (Tanimoto, 1981; Brown, 1982) mitgeteilt.

Bei der Berechnung untertägiger Felsbauten, z.B. bei Tunneln, ist es wichtig, die mechanischen Parameter des Gebirges richtig einzuschätzen. Wenn das nicht gelingt, hat das Rechenergebnis wenig Bezug zur Wirklichkeit, wie hochgestochen das mechanische Modell auch immer sei. Darüber hinaus ist die Einschätzung des primären Spannungszustandes ein Hauptproblem der Berechnung. Da wir keine festen Werte für jeden der Eingangsparameter angeben können, sind Parameterstudien, die die oberen und unteren Grenzwerte einsetzen, wirkungsvoller. Bei der Analyse mit verschiedenen Eingangsdaten sind geschlossene Lösungsformen - wenn es sie gibt - besonders gut geeignet. Zu diesem Zweck sind die Herleitung analytischer Ergebnisse und die Entwicklung verschiedener Arten von Entwurfsdiagrammen zu einem wichtigen Teil des Tunnelbaues geworden. Viel Forschungsarbeit wurde auf diesem Gebiet bereits geleistet. Parameterstudien sind ebenso bei den numerischen Rechenmodellen möglich, und viele Methoden wurden hierzu erdacht. Es ist jedoch wünschenswert, ein einfaches analytisches Modell für diesen Fall zu besitzen.

Wie eben erwähnt, lassen sich komplizierte mechanische Eigenschaften, selbst der räumliche Ausbruchsvorgang, bei der Anwendung numerischer Methoden wie der FEM oder der Randintegral-Element-Methode (Boundary-Element-Methode, BEM) berücksichtigen (Brady, 1979; Kobayashi, 1980). Dennoch bleibt die Bestimmung geeigneter Eingangsdaten nach wie vor schwierig, selbst wenn genaue analytische Lösungen erhalten werden. Andererseits werden beträchtliche Anstrengungen unternommen, die analytische Methode zu vereinfachen, z.B. dadurch, daß das Verhalten des Gebirges an der Ortsbrust eines Tunnels, das - streng genommen - eine räumliche Analyse erfordert, nach verschiedenartigen Vorgehensweisen anhand ebener Berechnungsmodelle untersucht werden kann (Sakurai, 1978; Dillen, 1979; Kaiser, 1981). Die erhaltene analytische Methode kann natürlich nur Näherungswerte ergeben; möglicherweise kann sie auch für ähnlich gelagerte Probleme des Tunnelbaues angewendet werden.

Die folgenden Ausführungen befassen sich mit dem Entwurf eines Tunnels auf der Basis der erwähnten Berechnungsergebnisse.

Zuerst einmal muß der primäre Spannungszustand abgeschätzt werden. In manchen Fällen geschieht dieses durch Feldmessungen wie z.B. mit der Überbohrtechnik. Alsdann folgt die Aufstellung eines mathematischen Modelles, das die mechanischen Eigenschaften des Gebirges, z.B. die

Elastizität, Elasto-Plastizität, Isotropie oder Anisotropie, verkörpert. Die Parameter der mechanischen Eigenschaften des Modelles werden durch Laborversuche oder in-situ-Messungen bestimmt. Als nächstes wird die eigentliche Berechnung mit den gefundenen Parametern durchgeführt, um die Verteilung der Spannungen und Verschiebungen sowie die angreifenden Kräfte an den Zimmerungen, Ausbauten und Felsdübeln zu ermitteln. Schließlich wird die Bauweise und Ausbaumethode so entworfen, daß die Standfestigkeit des Gebirges ebenso wie die des Ausbaues gewährleistet ist.

Die oben umrissenen Näherungsmethoden mögen theoretisch gut arbeiten, in der Praxis jedoch gibt es viele Probleme, da i.a. der primäre Spannungszustand und die mechanischen Eigenschaften des Gebirges nicht hinreichend bekannt sind.

Im folgenden möchte der Autor einige der Kongreßbeiträge zu dem hier behandelten Thema besprechen.

Romana et al. schlagen ihr Felsklassifikationssystem vor, um grundlegende Daten für die Tunnelbauplanung aus Feldaufnahmen von bestehenden Tunneln heranzuschaffen.

Doucerain untersucht die elastischen Kennwerte von Fels durch ein Feldexperiment, bei dem ein sehr hoher hydraulischer Druck auf eine untertägige Druckrohrleitung ausgeübt wurde. Tan schlägt zweidimensionale konstitutive Dilatanzgleichungen für das Schwellen von Fels als Ergebnis rheologischer Laborversuche vor und berechnet die dilatante Zone um einen Tunnel mithilfe dieser Gleichungen. Saari et al. untersuchten die Zusammendrückung von Flözen beim Auffahren eines Tunnels mit einer theoretischen Analyse und mit numerischen Berechnungen an nicht-dilatanten, kohäsiven und dilatanten Reibungsmaterialmodellen. Die Rechenresultate werden von den Autoren mit den Ergebnissen von Labormodellversuchen verglichen und diskutiert.

1.3 Entwurfsmethode

Als Entwurfsmethode für Tunnel hat die sog. Konvergenz - Begrenzungsmethode oder die Neue Österreichische Tunnelbauweise (NÖT) in den letzten Jahren große Aufmerksamkeit erregt. Einfachheitshalber soll diese Methode im folgenden als NÖT bezeichnet werden. Sie ist bemerkenswert wegen ihres umfangreichen felsmechanischen Hintergrundes und wegen des außerordentlich großen praktischen Nutzens, obwohl die herkömmlichen Tunnelbaumethoden, die auf langjähren Erfahrungen beruhen, ebenfalls weltweit ausgeführt werden. Das erstere Verfahren wird voraussichtlich auch für andere untertägige Hohlraumbauten angewendet werden. Da der Tunnelbau im Vergleich zu solchen Strukturen aus vielen wiederholten Schritten besteht, ist eine Standardisierung hierbei wahrscheinlicher.

Zwei Hauptmerkmale der NÖT können herausgestellt werden. Erstens, die Eigentragfähigkeit des umgebenden Gebirges wird so weit als möglich zur Stützung des Tunnelausbruches herangezogen, während der Ausbau eine eher nur ergänzende Rolle spielt. Es wird üblicherweise angenommen, daß der konventionelle Ausbau auf die von ihm zu tragenden Lasten auszulegen ist. Die NÖT läßt erwarten, daß der Ausbau nur noch ein Mindestmaß an Lasten aufzunehmen hat, was zu einem wirtschaftlichen Ausbausystem führt.

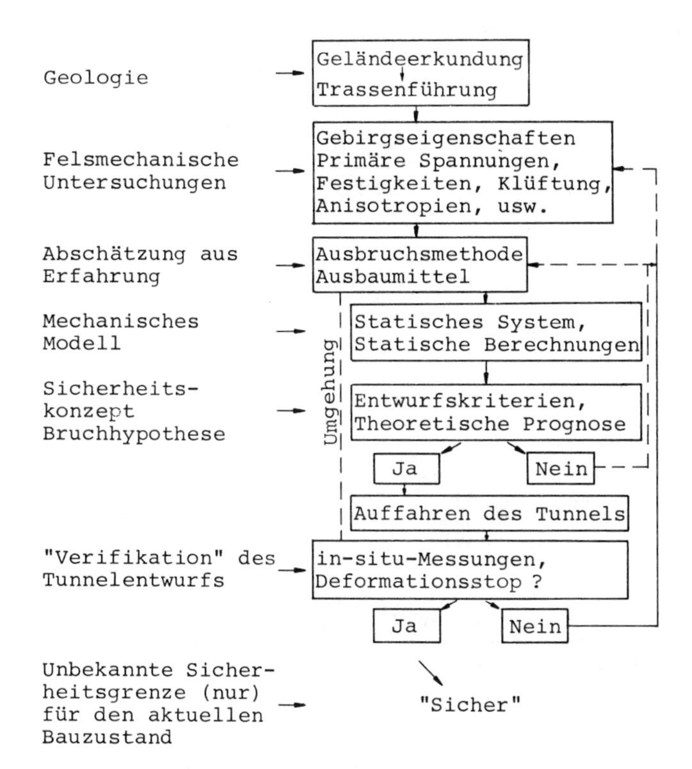

Fig. 1.2 : Entwurfsverfahren für Tunnel (Duddeck, 1980)

Es ist fast unmöglich, im Entwurfsstadium eines Tunnels das Verhalten des Gebirges vorherzusagen. Daher ist ein zweites Hauptmerkmal der NÖT, daß Messungen in situ durchgeführt werden, um aktuelle Daten über die Eigenschaften und das Verhalten des Gebirges zu gewinnen, die dann wiederum in die Berechnung einfließen und zu einem rationalen Entwurf des Tunnels beitragen. Fig. 1.2 zeigt das Beispiel eines Tunnelentwurfsverfahrens nach einem Vorschlag von Duddeck (1980).

Die Ausbautheorie der NÖT läßt sich überzeugend durch die beiden Kennlinien für den Ausbau und das Gebirge (Fig. 1.3) erklären (Fairhurst 1980). Die eine Kurve drückt die Abhängigkeit des erforderlichen radialen Ausbauwiderstandes von der Konvergenz des Tunnels aus. Die zweite Kurve ist die Kennlinie für den Ausbau. Sie gibt die Aktivierung des Ausbauwiderstandes in Abhängigkeit von der radialen Konvergenz des Ausbaues an. Das endgültige Gleichgewicht der Kräfte, aus dem sich die Ausbaubelastung bestimmen läßt, ergibt sich aus dem Schnittpunkt beider Kurven. Unter der Voraussetzung, daß beide Kurven gegeben sind, ist dieses Konzept für einen angemessenen Entwurf eines Tunnelausbaues sehr nützlich. Viele Forscher haben sich darum bemüht, die beiden Kennlinien rechnerisch oder experimentell zu ermitteln, doch ist dieses bis heute praktisch noch nicht gelungen (Kerisel, Lombardi, Egger, 1980). Es ist zu hoffen, daß das Problem mithilfe von in-situ-Messungen gelöst werden kann.

Es wird wichtig sein, die zeitabhängigen und die dreidimensionalen Faktoren für den Entwurf des Tunnelausbaues nach der NÖT-Theorie zu berücksichtigen. Fig. 1.4 zeigt ein Beispiel für

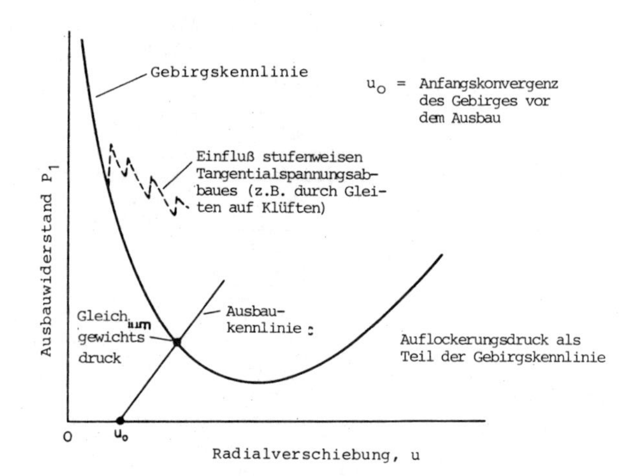

Fig. 1.3 : Idealisierte Gebirgs- und Ausbaukennlinien und ihre Wechselwirkung (Fairhurst, 1980)

die Veränderung der Kennlinien während der verschiedenen Ausbruchs- und Ausbaustufen eines Tunnels (Duddeck, 1980).

Zwei Kogreßbeiträge befassen sich mit dem NÖT-Verfahren. Gartung et al. berichten ihre Erfahrungen beim Bau von Zwillingstunneln in weichem, sedimentären Fels, die durch einen 36 cm schmalen Pfeiler voneinander getrennt waren. Der Erfolg bei dem wirtschaftlichen und sicheren Bau der Zwillingstunnel liegt wohl an den verfeinerten Techniken der NÖT. Maidl hat eine verbesserte Ausbaumethode sowie Ausbaumaterialien für Schächte untersucht, die aus den Erfahrungen mit den modernen Tunnelausbautechniken herrühren.

Eine Anzahl von Beiträgen behandelt das Problem des Tunnelausbaues. Romana et al. untersuchten das mechanische Verhalten von schwellenden Mergeln, um die entsprechenden Stoffparameter zu bestimmen; diese wurden als Grunddaten für die Berechnung der Kräfte, die auf Tunnelausbauten wirken, herangezogen. Altounyan et al. untersuchten den Druck auf den Ausbau eines Schachtes während des Abteufens durch tiefgelegene wasserführende Gebirgsschichten und gewannen mehrere instruktive Auskünfte. Kobayashi et al. studierten Ausbausysteme für Tunnel in extrem weichem Gebirge. Sie fanden heraus, daß für solche Fälle ein sehr starrer und stark bemessener Ausbau notwendig sei.

Holmgren berichtet von experimentellen Untersuchungen zur Festigkeit und Verformbarkeit von stahlfaserverstärktem Spritzbetonausbau, die ergaben, daß solch ein Ausbau mehr plastisch reagiert und daß durch die Kombination mit Felsankern das Ausbausystem sehr vereinfacht werden kann. Ortlepp untersuchte Ausbauentwürfe von tiefgelegenen Tunneln in hartem Fels aufgrund der Erfahrungen mit hohen Spannungen im Gebirge und mit Gebirgsschlagschäden. Er fand heraus, daß das Versagen von Ausbauelementen in enger Beziehung zu den Schäden aus den hohen Spannungen steht, und er schlägt das Konzept einer kritischen Einbindungslänge der Felsanker als Bestimmungsgröße für den Versagensfall vor.

Der untermeerische Seikan Tunnel wird in Japan gegenwärtig gebaut, um die Inseln Honshu und Hokkaido durch eine Eisenbahn zu verbinden. Die Konstruktion dieses Tunnels hat folgende Merkmale; (1) Es ist ein untermeerischer Tunnel; (2) er ist lang (Gesamtlänge 53 830 m, davon 23 300 m Länge von Küste zu Küste); (3) er ist tief (bis zu 240 m unter dem Meeresspiegel und 140 m unter dem Meeresboden); (4) der überwiegende Teil des Gebirges besteht aus lockerem sandigen Tonstein, und (5) der Querschnitt ist groß (88 m^2 lichter Querschnitt). Diese Eigenschaften sind vom Standpunkt der Felsmechanik her interessant, so daß die wichtigen Punkte hier vorgestellt werden sollen. Fig. 1.5 zeigt das geologische Profil (Nisugi, 1981).

Das Seikan-Tunnel-Projekt, dem viele Schwierigkeiten entgegenstanden, konnte diese durch die angesammelten Erfahrungen und den Stand der Technik überwinden. Am 27. Januar 1983 war der

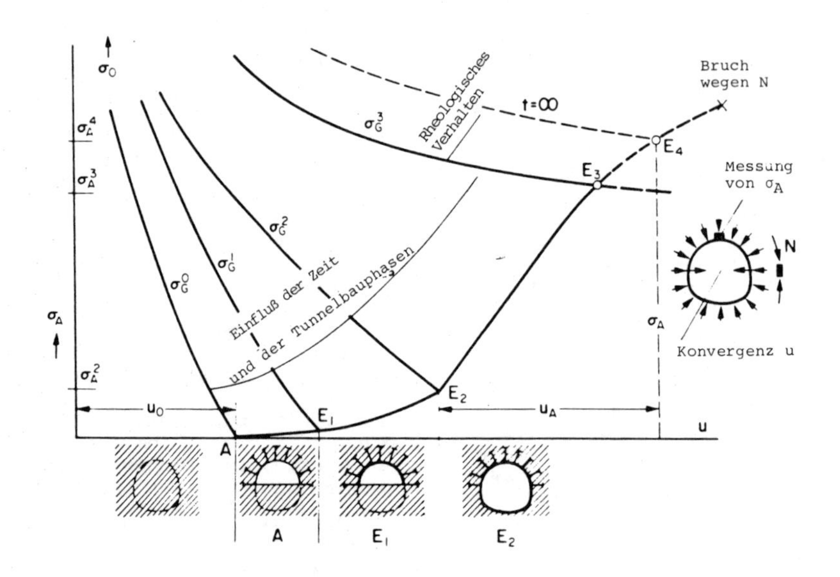

Fig. 1.4 : Entwicklung der Ausbaukennlinien (Duddeck, 1980)

Erkundungsstollen fertiggestellt, und die noch aufzufahrende Länge des Haupttunnels betrug zur Zeit der Berichterstattung nicht mehr als 2.7 km.

Die neuen Techniken beim Tunnelbau in untermeerischen weichen Gebirgsschichten wurden während der Durchführung des Seikan-Tunnelprojektes entwickelt. Sie lassen sich zusammenfassen als

(1) ein horizontales, langes Erkundungsbohrsystem,

(2) ein Injektionssystem zur Verhinderung von Wasserzufluß,

(3) eine Spritzbetonbauweise zur Sicherung der Ausbruchsoberflächen,

(4) ein Ausbausystem als Widerstand gegen den hohen Erddruck in den weichen Gebirgspartien.

1.4 Feldmessungen und Rückrechnung

Zahlreiche Verhaltensweisen des Gebirges und des Ausbaues werden während des Ausbruches eines Tunnels nicht notwendigerweise mit den Vorhersagen des Entwurfsstadiums übereinstimmen, und in den meisten Fällen sind sie in der Tat verschieden. Die Eingabedaten, die man durch in-situ-Messungen an einer bestimmten Stelle eines langen Tunnels ermittelt hat, lassen sich nicht auf die volle Länge des Tunnels übertragen, selbst wenn sie in angemessener Weise ausgewertet worden sind. Unter diesen Umständen ist es nahezu unmöglich, das Verhalten des Tunnels bereits in der Entwurfsphase richtig zu berechnen. Es ist sogar zu erstreben, den Originalentwurf nötigenfalls zu ändern, wenn man die Ergebnisse der Beobachtungen des Verhaltens des Tunnels während des Ausbruches betrachtet.

Möglicherweise kann man überprüfen, ob die für den Entwurf des Tunnels benutzten Eingangsdaten geeignet sind oder nicht, und untersuchen, ob die Tunnelkonstruktion nach dem ursprünglichen Plan ausgeführt werden kann oder nicht, indem man das Verhalten des Tunnels während des Bauens mit den ursprünglichen Vorhersagen vergleicht. Mit anderen Worten können wir den primären Spannungszustand und die mechanischen Parameter des Gebirges durch Verschiebungs-, Spannungs- und Verformungsmessungen einschätzen. Solch eine Analyse nennt man eine Rückrechnung und Probleme, die mit solch einer Rückrechnung angegangen werden, heißen entweder Identifikationsprobleme oder Charakterisierungsprobleme.

Viele Beiträge berichten über Rückrechnungen. Beispielsweise hat Kovari (1977 eine Methode zur Ermittlung des Erddruckes durch Rückrechnung aus den relativen Verschiebungen in axialer Richtung der Tunnelausbauten und aus den Veränderungen der Kurvatur entworfen. Er nennt sie Integrierte Meßtechnik. Der gleiche Gedanke jedoch war schon von Murayama (1976) vorgebracht worden, der die Dehnungen mit elektrischen Widerstands-Dehnungsmeßstreifen ermittelte, während Kovari mechanische Meßtechniken anwendete, um sie für den praktischen Einsatz im Tunnelbau einfacher handhaben zu können. Die Methode, den Erddruck auf ein unterirdisches Bauwerk durch Rückrechnung aus den Verschiebungen des Bauwerkes zu bestimmen, wurde gelegentlich auch in anderen Gebieten des Ingenieurwesens als im Tunnelbau vorgeschlagen, z.B. bei Pfählen oder bei Spundwänden (Gioda 1981). Darüber hinaus hat man versucht, den primären Spannungszustand und die Materialkennwerte des Gebirges aus Verschiebungs- und Verformungsmessungen des Gebirges beim Ausbruch des Hohlraumes durch Rückrechnung zu gewinnen. Gioda u.a. (198o, 1981) veröffentlichten viele Artikel über Rückrechnungsstudien auf dem Gebiet der Geomechanik. Ihre Forschungsarbeiten schließen die Rückrechnung auch nichtlinearer Problemstellungen ein. Es ist zu erwarten, daß die Rückrechnung der Ergebnisse von Feldmessungen eine zunehmend wichtige Rolle im geotechnischen Ingenieurwesen spielen wird.

Sakurai (1981, 1982) versuchte, die Verformungen des Bodens beim Ausbruch eines Tunnels aus Verschiebungsmessungen zu ermitteln. Er schlug eine Methode vor, die Tunnelkonstruktion durch Vergleich der Meßergebnisse mit der zulässigen Dehnung des Untergrundes zu handhaben, vorausgesetzt, daß die Anzahl der Meßpunkte sowohl an der Hohlraumoberfläche wie im Gebirge hinrei-

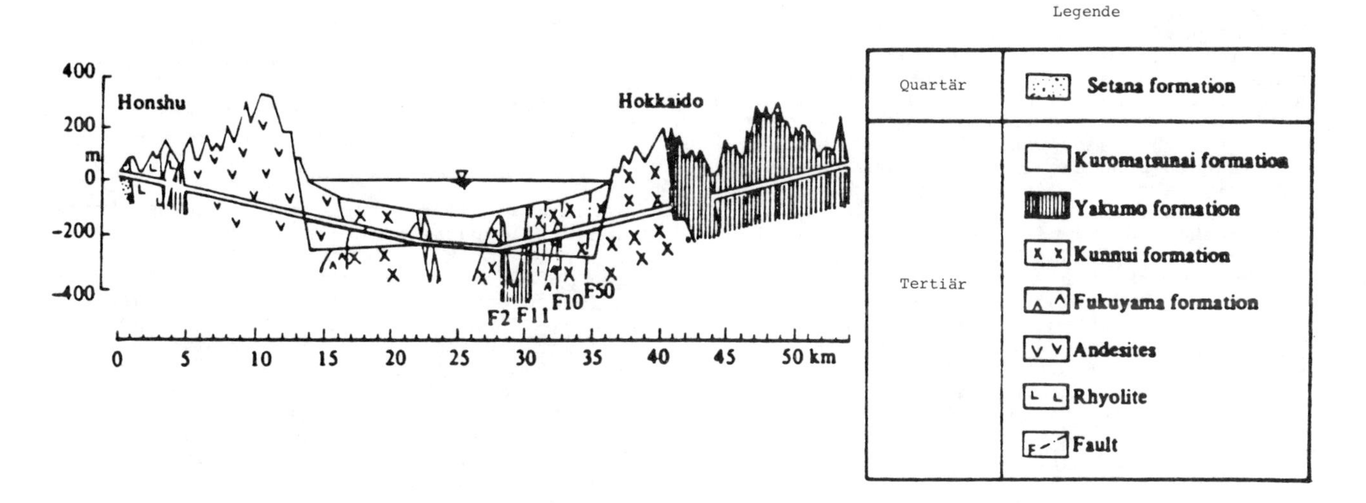

Fig. 1.5 : Geologisches Profil des untermeerischen Seikan Tunnels (Nisugi, 1981)

chend groß ist. Jedoch - selbst wenn die Anzahl der Meßpunkte nicht groß ist - kann es möglich sein, die Dehnungen über die Rückrechnung auszuwerten, um zu Informationen über den primären Spannungszustand und die mechanischen Kennwerte zu gelangen, mit deren Hilfe dann das Verformungsfeld um den Tunnel z.B. mit Methoden wie der FEM berechnet werden kann. Das Verfahren läßt sich sowohl bei der Entwurfsbearbeitung als auch bei der Bauausführung anwenden.

Nach dieser Methode kann die numerische Simulation selbst beim Auftreten von plastischen Zonen noch mit hinreichender Genauigkeit durchgeführt werden, wenn man den sog. äquivalenten Elastizitätsmodul und die primäre Gebirgsspannung eingibt. Da die Methode auf der Dehnung und nicht auf der Spannung beruht, hat sie den zusätzlichen Vorteil, daß sie nicht erst über eine Spannungsberechnung gehen muß.

Auf Grund der Studie von Kaiser et al. über die Konvergenzgeschwindigkeitsfunktion bei Modellversuchen unter hohen äußeren Spannungen wurde herausgefunden, daß sich das Tunnelverhalten in einer der drei Möglichkeiten ausdrückt:
(1) prefailure-Verhalten, (2) stabile Ausbreitung plastischer Zonen, (3) Instabilität oder Bruch. Mithilfe der Konvergenzgeschwindigkeitsfunktionen kann das Verhalten von Tunneln, die in vergleichbarem Gebirge aufgefahren werden, aus Konvergenzmessungen vorhergesagt werden.

Saito et al. diskutieren den Mechanismus von Gebirgsschlägen an der Ortsbrust tiefgelegener Tunnel, die tiefer als 1000 m unter der Erdoberfläche in hartem Eruptionsgestein aufgefahren werden. Aus der Zusammenstellung von Daten über den ursprünglichen Spannungszustand, über die Spannungsänderung durch das Vortreiben der Ortsbrust und über die Form der Scheibenbildung (Core discing) bei den Kernbohrungen längs der Tunneltrasse fanden sie, daß der Spannungszustand hochgradig konzentriert war und daß der Mechanismus der Gebirgsschläge ähnlich wie bei der Scheibenbildung an Bohrkernen sei.

2. Kohlebergbau

2.1 Kammer-Pfeilerbau

Einführung

Beim Kammer-Pfeilerbau läßt man Kohlepfeiler stehen, um die hangenden Schichten zu stützen. Da die Spannungsniveaus in der Regel niedrig sind, sind die örtlichen Ausbauanforderungen bescheiden und die Ausbaukosten entsprechend gering. Darüber hinaus sind Bergschäden an der Tagesoberfläche im Grunde genommen ausgeschlossen, wenn die Kammer- und Pfeilerabmessungen richtig ausgelegt sind. Der niedrige Kapitaleinsatz, die Anpassungsfähigkeit bei der Gewinnung und die Einfachheit der Durchführung machen den Kammerpfeilerbau zu einer bevorzugten Methode des Kohlenbergbaues. Da jedoch die Stärke der Pfeiler mit zunehmender Tiefe des Abbaues und mit größeren Mächtigkeiten der Flöze zunimmt, beschränkt sich das Verfahren auf geringe Teufen und mäßig dicke Kohleflöze. In den letzten Jahren wurden im wesentlichen zwei Abbaumethoden angewendet, nämlich der konventionelle Pfeilerbau und der Pfeilerbruchbau.

Auslegung der Kohlepfeiler

Die Pfeilergröße ist bei der Auslegung des Kammerpfeilerbaus grundlegend. Als erster Schritt ist die Spannungsverteilung im Kohlepfeiler zu bestimmen. Dieses wird man mithilfe der FEM- oder BEM-Spannungsanalysen vornehmen. Das numerische Finite-Differenzen-Verfahren für die Bewegungsgleichungen von elastisch gelagerten Platten, das Sheorey et al.(1982) für die Untersuchung von Pfeilerbelastungen in Indien herangezogen hat, steht ebenfalls zur Verfügung. Die einfachste Näherungsmethode zur Ermittlung der Pfeilerbelastung, oder richtiger der durchschnittlichen Pfeilerspannung, besteht jedoch in der Theorie der anteiligen Flächen, die eine obere Grenze für die durchschnittliche Pfeilerspannung liefert.

Der nächste Schritt der Entwurfsbearbeitung ist die Bestimmung des kritischen Spannungsniveaus oder der Festigkeit der Kohlepfeiler. Anstrengungen sind unternommen worden, die Festigkeit von Kohlepfeilern durch einachsige Druckversuche abzuschätzen (Salamon (1967), Bieniawski (1968) und viele andere). Hustrulid (1976) teilte die vorgeschlagenen Formeln in zwei Gruppen ein, nämlich in die Druckfestigkeits-/Größen-Beziehungen und in die Druckfestigkeits-/Gestalts-Beziehungen. Auf der Grundlage dieser Arbeiten und mit Hilfe der Theorie der anteiligen Flächen stellt Bieniawski zwei neue Näherungsverfahren für die Auslegung von Kohlepfeilern vor. Das Grenzfestigkeitskonzept behauptet, daß Pfeiler versagen, wenn die aufgebrachte Last die Druckfestigkeit der Pfeiler erreicht. Dabei wird angenommen, daß das Tragvermögen eines Pfeilers im gleichen Moment zu Null wird, wie die Spannung dessen Grenzfestigkeit erreicht. Das Konzept des progressiven Bruches hingegen betont die ungleichförmige Spannungsverteilung im Pfeiler. Der Bruch beginnt an der kritischsten Stelle und breitet sich schrittweise bis zum endgültigen Zusammenbruch aus. Die neue Pfeilerauslegungsmethode von Bieniawski nimmt als Versagenskriterium die Einleitung des Bruches anstelle des endgültigen Bruches an und wählt Sicherheitsfaktoren im Bereich zwischen 1,5 und 2,0.

Auslegung sicherer Hangendspannweiten

Die Standsicherheit der Firste ist ein grundlegendes Problem beim Kammer- und Pfeilerbau, das von einer Anzahl von Forschern untersucht worden ist. Der Kammerpfeilerbau mit konventionellen oder kontinuierlichen Gewinnungseinrichtungen ist ein zyklischer Vorgang, der aus folgenden grundsätzlichen Stufen besteht:

1) Kohlegewinnung und Verladung,
2) Setzen eines vorläufigen Ausbaues,
3) Einbringen des endgültigen Ausbaues.

Wade et al. (1982) berichtete, daß bei Kammer- und Pfeilerbauarbeiten in den Vereinigten Staaten von Amerika sich etwa 51 Prozent der schweren Unfälle durch Hangendbruch während der Zeit zwischen den o.g. Stufen 1 und 2 ereigneten, in der die Firste ungestützt war, daß sich etwa 19 Prozent der schweren Unfälle in der Zeit zwischen dem vorläufigen und dem endgültigen Ausbau zutrugen und daß etwa 30 Prozent der schweren Unfälle unter dem endgültigen Ausbau stattfanden. Diese Statistik zeigt, daß auch der Zeiteffekt bei der Einschätzung sicherer Firstspannweiten und bei der Auslegung des Ausbausystemes im Kammer- und Pfeilerbau zu berücksichtigen ist.

Als Mittel zur Bewertung der Firstenstabilität haben in letzter Zeit sog. Gebirgsklassifikationen zunehmende Aufmerksamkeit erregt. Gebirgsklassifikationen sind bekannte empirische Beiwerte zur Einschätzung der Standsicherheit untertägiger Felshohlräume. Sheorey et al. (1982) stellte ein Firstankersystem mit Drahtseilmaschen aus den indischen Kammerpfeilerbauarbeiten vor, das auf einer Gebirgsklassifikation beruht; er folgert, daß dieses Vorgehen sehr effektiv sei. In seiner Klassifikation werden sechs Parameter betrachtet, nämlich Gebirgsqualitätsbezeichnung, (RQD-Faktor), Schichtenabstand, Gebirgsfestigkeit, gefährliche strukturelle Besonderheiten, Verwitterungen und Wasserempfindlichkeit. Jeder Parameter wurde mit einer Wichtigkeitsbewertung versehen, die um so höher ausfüllt, je besser die Gebirgsbedingungen sind. Das vorgeschlagene Bewertungsschema für die Parameter ist eine Modifikation der Empfehlungen von Bieniawski (1976). Sie berücksichtigt die Einflüsse der Abbauteufe und die Spannungen in den Auflagern bei Arbeiten in deren Nähe.

Bieniawski schlägt eine Methode zur Einschätzung sicherer Firstweiten mittels eines Gebirgsklassifikationssystemes vor, und zwar speziell der geomechanischen Klassifikation, die für die Anwendung im Kohlebergbau abgeändert worden ist. Bieniawski und die Forscher an der Pennsylvania State Universität haben das Gebirgsklassifikationskonzept seit 1978 systematisch untersucht. Nach der Analysierung von 58 Fällen von Hangendbrüchen beim Kammer- und Pfeilerbau in den Vereinigten Staaten von Amerika stellten sie eine Beziehung zwischen der Standzeit und der Firstweite für verschiedene Gebirgsbewertungen vor. Diese geomechanische Klassifizierung soll die Richtlinien für die Einschätzung sicherer Firstweiten und die Auswahl geeigneter Firstenausbauten bereitstellen, wobei die geomechanischen Eigenschaften der individuellen Schichtung berücksichtigt werden.

2.2 Strebbau

Einführung

Im tiefen untertägigen Kohlebergbau ist der Strebbau dem traditionellen Kammer- und Pfeilerbau hinsichtlich der Leistung, der Rohstoffausbeute und der Sicherheit überlegen. Der moderne mechanisierte Langfront-Strebbau ist mit Walzenschrämladern oder Hobeln für die Gewinnung der Kohle sowie mit Schild- oder Bock-Schreitausbau, manchmal auch mit rahmenartigen Stützen ausgerüstet, um hohe Leistung und Sicherheit zu erreichen.

Nahezu alle Streben in Japan und in den Vereinigten Staaten von Amerika sind als Rückbau angelegt, während die Mehrzahl der Streben in Europa im Vorbau geführt werden. Die Hauptunterschiede zwischen diesen beiden Systemen mögen darin liegen, daß die Vorbaustreben sogleich mit der Gewinnung beginnen und die Abbaustrecken gleichzeitig mit dem vorwärtsschreitenden Abbau aufgefahren werden können. Umgekehrt müssen beim Rückbau Abbaustrecken im voraus aufgefahren werden, und der Abbau wird solange verzögert, bis die Vorrichtungsarbeiten abgeschlossen sind. Obwohl jede der Methoden ihre eigenen Vor- und Nachteile besitzt, hängt die Wahl zwischen den beiden Systemen üblicherweise von den behördlichen Vorschriften oder von den bergbaulichen und geologischen Bedingungen ab.

In der Bundesrepublik Deutschland dringt der Kohlebergbau mit einer jährlichen Rate von 14m stetig in größere Teufenbereiche vor, und die mittlere Abbauteufe liegt gegenwärtig bei etwa 900m. Einige der Bergwerke bauen Kohle bereits in 1400 m Teufe ab (Grotowsky, 1982). Japan und Frankreich haben eine vergleichbare Situation bei der Abbauteufe. Die Zunahme der Teufe und die intensive Konzentration der Arbeiten zwingen die Ingenieure zu scharfem Blick auf den Gebirgsdruck und seine Auswirkungen. Die Verhinderung von Gebirgeschlägen und Gasausbrüchen aus Kohle und Nebengestein ist zu einem der ernstesten Probleme in diesen Ländern geworden. Darüber hinaus haben verschiedene Tonschichten die Neigung zu zerfallen, wenn sie einer feuchten Umgebung ausgesetzt werden, und verursachen ausgeprägte Sohlhebungen (Singh et al.). Beim Strebbau in großen Teufen stellt man häufig Absetzen der Hangendschichten an der Abbaukante, Abplatzungen an den Stößen und allgemeine schlechte Beherrschung von Hangendem und Liegendem fest.

Probleme der Gebirgsbeherrschung erfährt man vor allem in den Abbaustrecken, insbesondere an den Übergängen von Strecken zum Streb. Auch gibt es im Strebraum verschiedene Schwierigkeiten der Gebirgsbeherrschung, z.B. mit Ausbrüchen aus dem Hangenden, die mit weichen Schichten und mit Störungen zusammenhängen. Je nach Heftigkeit dieser Probleme kann der Abbauvorgang ernsthaft behindert werden und einen erheblichen Produktionsrückgang zur Folge haben.

Planung der Streblänge und der Breite der Zwischenpfeiler

Die Streblänge ist ein grundlegender Faktor für die Gebirgsbeherrschung im Strebbau zur Erreichung eines einwandfreien Zubruchwerfens im Alten Mann des Hangenden und zur Minimierung der Anforderungen an den Strebausbau. Okamura et al. (1979) untersuchten die Standsicherheit der Dachschichten, die von den darüberliegenden Hangendschichten getrennt sind, in den Bruchhohlraum überhängen und sich durch ihr eigenes Gewicht absetzen können, durch druckdynamische Modellversuche. Sie stellten fest, daß der Hangendausbruch nicht einsetzte, wenn an den Auflagepunkten keine Drucknachgiebigkeit auftrat, und daß er sowohl von der Durchbiegung des Hangenden als auch von der Streblänge abhing. Wagner et al. (1982) schlugen ein Planungsverfahren aus Südafrika vor, wo Strebbau in Gebieten durchgeführt wird, die von massigen Doleritbänken überlagert sind. Diese Bänke können Hunderte von Metern überspannen, bevor sie zu Bruch gehen. Ist die Abbaufront kurz, kann ein Doleritgang die ganze Bauhöhe überspannen, ohne zu brechen, und das Gewicht des Dolerits sowie der darüberliegenden Schichten kann sich auf die Abbauränder einschließlich dem Abbaustoß absetzen. In südafrikanischen Gruben wurde die Streblänge durch die Notwendigkeit bestimmt, das Zubruchwerfen der massiven Doleritlager so früh als möglich nach dem Beginn des Strebbaubetriebes einzuleiten. Bestimmungsgleichungen für die Mindestgröße der Abbaufront wurden von Salamon et al. (1972) vorgeschlagen und von Galvin (1982) verfeinert.

Der Entwurf von Pfeilern zwischen den Bauhöhen hängt mit der Bestimmung der Streblänge zusammen. Zwei gegensätzliche Beurteilungsmerkmale werden angetroffen. Erstens soll der Pfeiler möglichst schmal sein, um Abbauverluste klein zu halten, zweitens sollte er breit genug sein, um die Abbaustrecken vor den hohen Auflagerdrücken zu schützen. Zur Verbesserung des Abbauausbringens sollte das Konzept der nachgiebigen Pfeiler geprüft werden, das die Kohlepfeiler in kontrollierter Weise zu Bruch gehen läßt. Die schlagartigen Brüche an Zwischenpfeilern in drei japanischen Bergwerken, von denen Fukuda et al. berichten, können als unkontrollierte Pfeilerbrüche betrachtet werden. Um ein kontrolliertes Brechen der Pfeiler sicherzustellen, muß die Steifigkeit des umgebenden Grubengebäudes größer sein als die Steigung der Spannungs-Dehnungs-Kennlinie der infolge Bruches nachgebenden Kohlepfeiler.

Für die Ermittlung der Standsicherheit von Tragpfeilern kann der "Integritätsfaktor", der mit dem Abbaufortschritt abnimmt, ein vernünftiger Kennwert sein. Barron (1978) hat den Pfeiler-Integritätsindex vorgeschlagen, der als das Verhältnis des Pfeilerkernquerschnittes zu der gesamten Pfeilerquerschnittsfläche definiert ist. Dieser Index kann ein nützliches quantitatives Mass für die Abschätzung der relativen Auflockerung des Pfeilers sein. Barron bestimmt die Kernfläche auf der Basis der Bruchindizes über den Querschnitt eines Pfeilers, die er aus Lufteinpreßversuchen ermittelte.

Lu schlägt einen neuen Weg für das gleiche Problem vor, indem er einen Integritätsfaktor als das Verhältnis des Restfestigkeitsgrades zu dem Belastungsintegral definiert. Er stellt eine Reihe von horizontalen und vertikalen Pfeilerdrücken vor, die mit hydraulischen Bohrloch-Druckmeßzellen gemessen worden sind. Die Restfestigkeit eines Pfeilers von voller Größe wird aus dem gemessenen horizontalen Druck in dem Pfeiler abgeschätzt. Dieses geschieht auf der Basis der aus Laborversuchen ermittelten dreiachsigen Druckfestigkeit, wobei der in situ gemessene horizontale Druck als Randbedingung angesehen wird, während die integrierte aktuelle Gesamtlast auf den Kohlepfeiler aus den in situ-Messungen abgeschätzt wird. Die Bestimmung einer durchschnittlichen Belastung nach der Theorie der anteiligen Fläche oder einer hypothetischen Belastung nach dem Prinzip des Spannungsgleichgewichtes, das von Wilson (1972, 1977 und 1981) empfohlen worden ist, wird dadurch umgangen. Obwohl der Sicherheitsfaktor, beispielsweise der von Bieniawski vorgeschlagene, für bergmännische Hohlräume großer Standdauer vernünftig sein mag, wird der Integritätsfaktor sich für vorübergehende Abbaustrukturen wie Zwischenpfeiler besser eignen.

Entwurf der Abbaustrecken und ihres Ausbaues

Die Einschätzung der Querschnittsverminderung ist für die Planung des Ausbausystemes von grundsätzlicher Bedeutung. Eine Anwendung der geomechanischen Klassifikation, von der Abad et al. berichten, hat gezeigt, daß die Konvergenz von Abbaustrecken mit der Kohleflözmächtigkeit, dem Streckentyp, der Abbauteufe, der Schichtenneigung, der Ausbaudichte (Stahlgewicht pro Kubikmeter Streckenhohlraum: kg/m^3) und dem RMR-Index von Bieniawski (1976) abhängt. Sie untersuchten die Konvergenzen, die an 75oo verschiedenen Punkten in 187 Strecken gemessen worden sind, und stellen eine Abschätzungsformel für die Endkonvergenz - ausgedrückt in Prozent der Anfangshöhe - vor. In dieser Formel hängt die Konvergenz mit der Ausbaudichte und einem neuen Index IGME 82 zusammen, der von der Klüftung, der Schichtneigung und der Kohleflözmächtigkeit abhängt. Die Ausbaudichte, wie sie von Abad et al. vorgeschlagen wird, ist ein Maß für die Stärke des Ausbaues, und diese Abschätzung ist nützlich für die Dimensionierung von Stahlausbauten, nämlich des Querschnittes, des Gewichtes und des Bauabstandes der Streckenbögen.

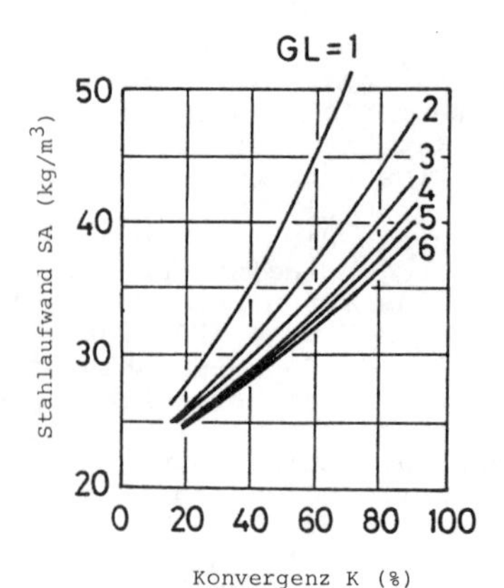

Fig. 2.1 : Stahlaufwand in Abhängigkeit von Konvergenz und der Liegendkennzahl GL (Irresberger, 1982, Grotowsky, 1982)

Die Berechnung mittels Ausbaudichte wurde vorher bereits von Irresberger (1982) und Grotowsky (1982) vorgestellt. Sie sprechen in ihren Artikels vom Stahlaufwand und schließen aus statistischen Aufnahmen, daß der geeignete Stahlaufwand von der Konvergenz und von der Liegendkennzahl abhängt, welche die geomechanischen Bedingungen der Streckensohle kennzeichnet. Fig. 2.1 zeigt die vorgeschlagene Beziehung zwischen diesen Größen. Die Autoren erörtern den Einfluß des Gebirgsdruckes auf die Konvergenz und die wichtige Bedeutung der geometrischen Beziehung zwischen der Abbaustrecke und dem Streb an Hand einer Reihe von Fallstudien in der Bundesrepublik Deutschland. Sie schlagen Abbauplanungen vor, die von der Forschungsstelle für Grubenausbau und Gebirgsmechanik der Bergbau-Forschung GmbH in Essen entwickelt worden sind. Diese Verfahren sind sehr systematisch und praktisch, und sie sind rückgeprüft durch statistische Aufnahmen und Verträglichkeitstests in einer Reihe von Bergwerken im Ruhrgebiet.

Die Abbauplanungsverfahren nach Irresberger (1982) verlaufen folgendermaßen: Zuerst wird die Abbaustrecke in Bereiche gleicher Belastungen aufgeteilt. Danach berechnet man die Konvergenz in jedem Bereich aus der sog. Grund-Konvergenz K_{EV}, die durch

$$K_{EV} = -78 + 0{,}066T + 4{,}3M \cdot SV + 24{,}3 \sqrt{GL} \pm 3\%$$

gegeben ist. Hierin bedeuten T die Abbauteufe, M die Mächtigkeit, SV die Art des Saumversatzes und GL die Liegendkennzahl. Die Grundkonvergenz K_{EV} ist diejenige Konvergenz, die sich in der Abbaustrecke einstellen würde, wenn sie vor der Abbaufront aufgefahren wäre, die Kohle auf einer Seite abgebaut würde und der Gebirgsdruck nicht durch irgendwelche anderen Abbaukanten beeinflußt sei. Wenn andere Arten von Abbaustrecken geplant sind, läßt sich die Konvergenz für jeden Fall aus K_{EV} nach der Methode aus dem Artikel von Irresberger abschätzen. Rückbaustrecken, die hinter dem Streb abgeworfen werden, zeigen nur die halbe Konvergenz. Sind Pfeiler vorhanden, muß die Konvergenz berichtigt werden durch den Druck, der sich nach der Methode von Everling et al. (1972) berechnen läßt. Ist die Konvergenz einmal eingeschätzt, kann man die passenden Ausbautypen und den Stahlaufwand ermitteln. Starre Bögen kann man für den Fall benutzen, daß die Konvergenz kleiner als 15% der ursprünglichen Streckenhöhe bleibt, und anderenfalls sollten Gleitbögen eingesetzt werden.

Hangendbeherrschung in Streben und Ausbauwirkungen

Bei den tiefgelegenen Streben führt der hohe Zusatzdruck zu starken Druckwirkungen in den Dachschichten und zum Abböschen des Kohlestosses. Letztere beeinträchtigen das Stützungsvermögen des Stoßes. Ferner vergrößern die Abschalungen und Abböschungen des Kohlestoßes die freie Stützweite der Firste und erleichtern lokale Ausbrüche. Das seitliche Auflager der gebrochenen Dachschichten am Bruchhaufwerk ist auch durch das Phänomen der stufenweisen diskontinuierlichen Absenkung begrenzt. Das Fehlen der seitlichen Einspannung in Verbindung mit hohen Drücken führt bevorzugt zum Entstehen von nahezu vertikalen Brüchen. Wegen des Fehlens von Reibungswiderständen können Gesteinskeile auf den steil geneigten Bruchflächen aus der Firste nahe am Stoß oder am Stoß selbst herausgleiten. Diese Ausbrüche können zu einem völligen Verlust der Hangendbeherrschung führen und einen größeren Strebbruch verursachen. Der gefährlichste Augenblick ist während oder kurz nach dem Schreiten des Ausbaues. Wagner et al. (1982) stellten fest, daß Ausbausysteme folgende Eigenschaften benötigen: die freie Stützweite soll so klein wie möglich sein, der Ausbau soll hohe Kräfte an den Kappenspitzen bei den meisten Betriebsbedingungen erzeugen können, und die gesamte ausgebaute Hangendfläche soll so klein sein wie praktisch möglich.

Die Auslegung eines geeigneten Ausbauwiderstandes ist der Hauptgegenstand beim Entwurf der Strebausrüstungen. In den letzten Jahren war die Mehrzahl der mechanisierten Streben mit hydraulischem Ausbau ausgerüstet. Der Hauptzweck dieses Ausbaues besteht in der Abstützung der Dachschichten im Strebraum. Dieses wird erstens durch den Ausbauwiderstand senkrecht zum Hangenden erreicht und zweitens durch die Möglichkeit, das Absenken des Hangenden zu behindern, wenn die freie Stützweite sich während des Abbaues vergrößert.

Deshalb wird der Ausbauwiderstand allgemein durch die mittlere Setzlastdichte dargestellt, die die Kraft bezeichnet, die der Ausbau pro Flächeneinheit des freigelegten Hangenden (vom rückwärtigen Ende der Kappe bis zum Kohlestoß) ausübt.

Um die Ausbauwirkungen zu erklären, wurden die Konvergenzen zwischen Hangendem und Liegendem sowie die Art der Rißbildung im Hangenden von einer Anzahl von Forschern untersucht (Gupta et al., und viele andere).

Christiaens (1982) faßte die Rißform im Hangenden nach Beobachtungen in belgischen Kohlegruben so zusammen wie in Fig. 2.2 gezeigt. In geringer Teufe findet man primär geneigte Sprödbrüche (R3), deren Neigung von den Reibungskräften abhängt, die von der Ausdehnung der Kohle in Richtung auf das Abbaufeld herrühren. Dieses primäre Muster kann von einem stärker geneigten - oft bis zu 45° steilem - sekundären Rißmuster überlagert sein, das von einer nachfolgenden Ausdehnung der Kohle verursacht worden ist. In tiefgelegenen Kohlelagern hingegen werden vertikale Sprödbrüche (R2) primär in den sandigen Schichten von Sandstein oder dichtem Tonstein gefunden. Diese vertikalen Bruchflächen können von geneigten sekundären Brüchen (R3) durchkreuzt sein, und die Querscherung trifft man in der gezeigten Form (R4) an, wie man sie hauptsächlich in Dachschichten von Schieferton vorfindet. Der Keil ABC wird durch einen nachfolgenden Sprödbruch (R3) und manchmal durch einen horizontalen Sprödbruch (R1) durchschnitten, die von der horizontalen Kompression des Keiles herrühren.

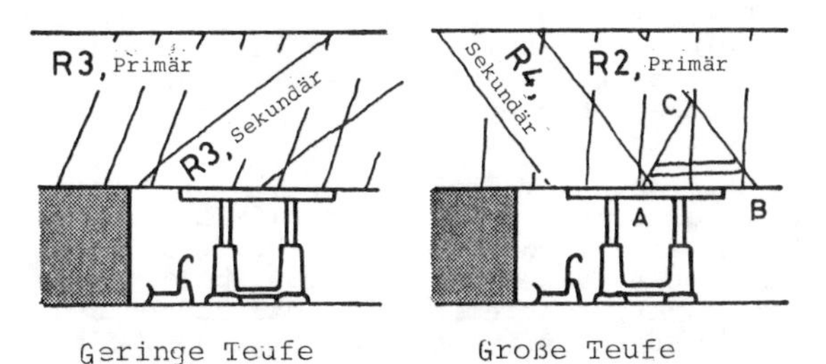

Fig. 2.2 : Zwei Typen von Bruchsystemen im Strebhangenden

Das Verhalten des Hangenden wird allgemein von zwei verschiedenen Gesichtspunkten aus betrachtet, nämlich für das Haupthangende und für die Dachschichten. Christiaens (1982) beschreibt, daß die Gewölbebildung einerseits von der Kohle und andererseits von dem ausgebrochenen Versatzmaterial getragen wird; die Dachschichten seien durch spröde Trennbrüche und manchmal durch Scherbrüche zerstört. Das Haupthangende ist dagegen entweder durch spröde Trennbrüche oder durch die Durchbiegung der festeren Schichten aufgespalten oder aber auch auf Grund älterer Abbauarbeiten. Die Standsicherheit der Schichtfolgen des Haupthangenden mag durch die dort aktiven horizontalen Druckkräfte sichergestellt sein. Die Hauptschubspannungen besitzen an den Auflagern ihre höchsten Intensitäten; Gleitungen können in den vertikalen Bruchflächen auftreten, insbesondere wenn der Strebausbau zu schwach ist.

Ein Näherungsverfahren zur Abschätzung des Strebausbaues unter Annahme vertikaler Dachschichtbrüche wurde von Christiaens (1982) vorgestellt. Er untersuchte die verschiedenen Fälle der Ausbaustellung und die Abstände der Hangendbrüche und schloß daraus folgendes:
Die Standsicherheit des Hangenden wird erleichtert, wenn man die Stempel näher an den Kohlestoß heransetzt und die Setzlastdichte so hoch wie möglich auslegt. Die besten praktischen Ergebnisse wurden immer mit den mechanisierten Ausbautypen erreicht, die den besten Sicherheitsfaktor (k) besitzen. Diesen kann man aus der Gleichung für das Kraftmoment des Ausbaues auf das Hangende abschätzen:

$$k = \{Fy + 0.15\ L\ (F - 100\ LM)\}\ /50\ ML^2$$

Hierin bedeuten F die vertikale Kraft des Schreitausbaues pro Einheitslänge des Strebes in [kN/m] , y die Hebelarmlänge von F in [m] bezogen auf das Kappenende, L die Breite der auszubauenden Dachschichten in [m] , M die Mächtigkeit in [m] , und die Höhe des zu stützenden Hangenden wird viermal so groß wie die Mächtigkeit angenommen.

Irresberger (1982) nennt Kriterien zur Bestimmung der Qualität der Hangendbeherrschung wie folgt: Das Hangende gilt als unter guter Kontrolle, wenn (1) während der Förderschichten die Grundfläche lokaler Ausbrüche in der unausgebauten Firstenfläche im Durchschnitt kleiner als 10% der Streblänge ist, wenn (2) der Flächenanteil des Hangenden vor dem Ausbau mit Ausbrüchen von über 30 cm Tiefe weniger als 5% der Streblänge beträgt, und wenn (3) der Flächenanteil mit Stufen von über 10 cm Höhe im Hangenden kleiner als 5% der Streblänge ausfällt. Umgekehrt ist das Hangende unter schlechter Kontrolle, wenn die entsprechenden Prozentanteile größer als 30%, 20% bzw. 10% sind. Die Ausbruchhäufigkeit ist vielleicht das wichtigste dieser Kriterien, da eine Zunahme der Hangendausbrüche notwenigerweise den Abbaufortschritt hemmen und damit die Gewinnung an dieser Stelle beeinträchtigen kann.

Der Gebirgsdruck am Stoß, der sich durch die Berechnungsmethode von Everling (1972) abschätzen läßt, wird von Irresberger (1982) als der kritische Faktor bei der Planung des Strebausbaues herausgestellt. Die Beziehungen zwischen der Standfestigkeit des Hangenden, dem Gebirgsdruck und den technischen Ausbauanforderungen werden von Irresberger (1982) auf der Basis von Erfahrungen in der Bundesrepublik Deutschland zusammengefaßt. Wo das Hangende feste Bänke von über 1o m Dicke enthält, besteht die Gefahr, daß die Dachschichten bei niedrigem Gebirgsdruck abtreppen. Um dieses zu verhindern, muß der Ausbauwiderstand relativ hoch sein. Ein minimaler Wert von 400 kN/m^2 wird vorgeschlagen. Jedoch kann der Spitzen-/Stoß-Abstand bei festen Dachschichten wegen der geringen Ausbruchempfindlichkeit bis zu 1 m betragen. In Fällen, bei denen der Gebirgsdruck hoch und eine Tonschieferbank von weniger als 2 m Mächtigkeit in den Dachschichten von festen Bänken überlagert ist, neigt die Firste zu häufigem Ausbruch. Der Ausbauwiderstand sollte deshalb größer als 300 kN/m^2 und der Spitze-/Stoß-Abstand kleiner als 30 cm sein. Ist die Tonschieferbank in der Dachschicht mächtiger als 2 m, sind die Anforderungen an den Strebausbau geringer. Je nach dem besonderen Gebirgsdruck sollten die Ausbauwiderstände zwischen 250 und 300 kN/m^2 betragen und die Spitze-/Stoß-Abstände sollten kleiner als 5o cm sein, um die Häufigkeit und die Ausdehnung der Ausbrüche aus dem Hangenden niedrig zu halten.

Eine Reihe von Beobachtungen an drei Rückbaufronten in England, die von Gupta et al. dargestellt sind, haben ebenfalls den Einfluß des Setzdruckes auf die Ausbauleistung klären helfen. Die ausführlichen Beobachtungen der Hangenddeformation haben gezeigt, daß die Zunahmen der Ausbausetzdichte die Verformung der Hangendschichten von einer ungleichförmigen Druckzone mit einer Zugzone vor den Streben zu einer gleichförmig verteilten Druckzone an der Abbaufläche veränderten. Die Zunahme der Setzlastdichte von 220 auf 420 kN/m^2 hat in diesem Zusammenhang auch die durchschnittliche Konvergenz über die ausgebaute Strebbreite ebenso wie das Hereinwandern des Kohlestoßes in den Streb und Stoßabplatzungen und Hangendausbrüche reduziert.

2.3 Überwachung des Gebirges

Überwachung ist ein nützliches Mittel, wichtiges Informationsmaterial für die Gebirgsbeherrschung in vielen Bergwerken zu gewinnen. Hangendsenkungen, Sohlhebungen und Flözdehnungen werden mit verschiedenen Geräten überwacht. Mehrfachextensometer werden in Horizontalbohrungen im Flöz und in Vertikalbohrungen in den Hangendschichten installiert. Hydraulische Druckmeßdosen und Schwingsaiten-Bohrlochspannungsmesser werden herangezogen, um Spannungsänderungen im Flöz und im Nebengestein zu überwachen. Manchmal wird der primäre Spannungszustand durch Spannungsentlastungsmethoden untersucht. Die Anwendungen dieser Standard-Feldüberwachungstechniken werden im folgenden erörtert und einige besondere Punkte eingeführt.

Die Änderung von Senkungsprofilen in den Abbaustrecken mit dem Vorschreiten des Strebabbaues ist von Kimura et al. (1982) und von Johnson et al. (1982) gezeigt worden. Johnson et al. untersuchten quantitativ den Einfluß von Versatz hinter der Abbaufront, indem sie die Beziehung zwischen Versatzdrücken und Konvergenzen analysierten, die in den britischen Kohlelagern gemessen worden waren. Eine ähnliche Messung in Japan stellten Sugawara et al. (1981) vor. Hebblewhite et al. (1982) beschrieben ein erfolgreiches Überwachungssystem, das aus einer Konvergenzüberwachung in den Abbaustrecken und zusätzlich einer Verformungs- und Spannungsänderungs-Überwachung im Flöz und in den Hangendschichten besteht. Sie benutzten manuelle teleskopische Stäbe zwischen Bolzen, die in Fels eingegossen waren, Mehrfach-Drahtextensometer mit Spreizhülsenankern und Schwingsaiten-Bohrlochspannungsmesser. Fisekci et al. (1982) zeigten ein Überwachungssystem für den hydraulischen Abbau von mächtigen und steilgelagerten Flözen. Sie erörterten die zeitabhängigen Eigenschaften der Stützlast und der Deformation der Strecke. Tinchon et al. (1982) berichteten über eine Anwendung einer Druckkissenmethode zur Messung der primären Spannungen in französischen Bergwerken. Sie zeigten, daß die Vertikalspannung mit der Teufe zunahm, aber hochgradig von den geologischen Verhältnissen abhing.

Beim gegenwärtigen Kongress werden verbesserte Feldüberwachungstechniken vorgestellt. Gupta et al. zeigen Messungen von Schichtenverschiebungen und von Dehnungen in den Deckschichten an der Abbaufront, wobei die Dehnungen aus den relativen Verschiebungen ermittelt wurden, die durch Ankerdrahtextensometer und magnetische Extensometer gemessen worden waren. Lu schlägt eine verbesserte Druckmessung mit hydraulischen Bohrloch-Druckzellen vor, um die Stabilität von Pfeilern zwischen den Bauhöhen einzuschätzen. Lu (1981) zeigte eine Technik zur Überwachung der biaxialen Gebirgsdrücke durch die Kombination einer zylindrischen und zweier flacher Druckzellen, die in einem einzigen Bohrloch installiert waren. Das Antwortverhältnis zwischen Zelldruck und Bodendruck wurde für Kohle etwa als eins-zu-eins bestätigt.

Von einer Überwachung des Stoßanbaues berichteten Hebblewhite et al. (1982). Sie installierten vier kontinuierliche Schreiber zur Registrierung der auf den Ausbau wirkenden Kräfte auf eine Auswahl der 89 Blockschildausbauten am Abbaustoß, und punktuelle Ablesungen wurden längs der gesamten Abbaufront in regelmäßigen Abständen vorgenommen. Der plötzliche Lastanstieg bekräftigte sich an den beiden front- und rückseitigen Stempeln, nachdem der Walzenschrämmlader die Stelle passierte, und es wurde klar, daß häufiges Nachgeben des Ausbaues besonders dann eintraf, wenn der Streb unter Oberflächenböschungen und tiefen Überdeckungen lag. Carrasco hat in seinem Kongreßbeitrag ein neues System zur Überwachung der dynamischen Spannungsänderungen von Lasten vorgestellt, die infolge von Versatzsprengungen durch Steinschlag auf den Ausbau wirken.

Datenfernübertragung wird bei der Gebirgsüberwachung im Bergbau im Zusammenhang mit der Entwicklung von Feldmeßsystemen für akustische Emissionen (AE) bevorzugt. Anwendung der AE-Methode sind von Watanabe (1981), Nakajama et al.(1981), Josien et al. (1982) und von vielen anderen aufgezeigt worden. Watanabe et al. hoben hervor, daß der bevorstehende Gasausbruch in einem Flözstreckenvortrieb mit hoher Wahrscheinlichkeit vorhergesagt werden kann, indem man den Grad der Zunahme der akustischen Emissionsereignisse nach dem Sprengen überwacht. Als Beitrag zum gegenwärtigen Kongreß behandelt Schuerman die Kostenwirksamkeit der Datenfernübertragung und stellt ein mannloses prozeßgesteuertes Überwachungssystem vor.

2.4 Mathematische Modellierung des Kohlenbergbaues

Mathematische Modellierungen mit finiten Elementen und Verschiebungsdiskontinuitäts-Methoden sind von einer Reihe von Forschern erprobt worden, um die Gebirgsdruckwirkungen in großen Teufen des untertägigen Kohlebergbaues zu untersuchen. Komplizierte geometrische Eigenarten der vollständigen Abbausysteme,z. B. nahe der Flözstrecke einer Langfront, können durch zweidimensionale Systeme nicht angemessen modelliert werden. Obwohl grundsätzlich dreidimensionale Modelle aus finiten Elementen oder allgemeinen Randintegral-Elementen für solche Geometrien herangezogen werden können, sind die hohen Kosten und Aufwendungen für solche Analysen zum gegenwärtigen Zeitpunkt selten gerechtfertigt. Andererseits wäre es unmöglich, eine geeignete mathematische Beschreibung des kontinuierlichen Brechens und der relativ ausgedehnten Bruchbewegungen zu erhalten. Aus diesem Grunde ist die Berechnung des Gebirgsdruckes weitgehend vereinfacht worden.

Everling et al. (1972) entwickelten eine einfache dreidimensionale Berechnungsmethode für den Gebirgsdruck, die auf homogener Elastizität beruht. Bei dieser Methode wird angenommen, daß die horizontalen Verschiebungen vernachlässigbar klein sind und daß die Verschiebung an jedem Punkt nur vertikal gerichtet ist. Aus der weiten Anwendung dieser Methode in der Bundesrepublik Deutschland wurde gefunden, daß die insitu Bruchbewegungen sich proportional zu der Verformung eines elastisch angenommenen Felskörpers einstellen (Grotowsky, 1982).

Yeates et al. wandten die Verschiebungsdiskontinuitäts-Methode an, um die Spannungszustände in Kettenpfeilern zu untersuchen. Sie benutzten die zweidimensionale, nichtlineare elasto-plastische Finite-Element-Methode, um die potentielle Standsicherheit von Abbaustrecken in Kohleflözen in verschiedenen Horizonten abzuschätzen. In ähnlicher Weise zogen Fukuda et al. die zweidimensionale Verschiebungsdiskontinuitäts-Methode (Crouch, 1976) zur Ermittlung der Spannungsänderung heran, die sich vor dem Stoß durch den Abbaufortschritt einstellten Diese Arbeiten haben die Nutzbarkeit der Verschiebungsdiskontinuitäts-Methode bestätigt. Die Methode beruht darauf, die aktuellen Abbaugeometrien durch dünne Schlitze der gleichen Flächenebene zu ersetzen. Dieses Näherungsverfahren ist vielversprechend und erlaubt dreidimensionale Spannungsberechnungen zu relativ niedrigen Kosten.

Verschiebungsdiskontinuitäts-Lösungen wurden zuerst von Berry (1964) benutzt, um die vom Bergbau verursachten Bergsenkungsprofile zu modellieren. Es wurde angenommen, daß der abgebaute Stoß durch eine einzige gleichförmige Verschiebungs-Diskontinuität angenähert werden kann.Die Erweiterung auf realistischere Randbedingungen und Abbaugeometrien wurde durch Aufteilung des Abbaugrundrisses in einzelne Elemente möglich, die jedes für sich eine gleichförmige Verschiebungsdiskontinuität darstellen (Plewman et al., 1969; Starfield and Crouch, 1973). Diering (1980) behandelte stückweise homogene isotrope Kontinuumseigenschaften durch Anwendung von Elementen längs der geologischen Trennflächen.

Wardle et al. erweiterten die Verschiebungsdiskontinuitäts-Methode auf ein geschichtetes anisotropes Gebirge. Dabei nahmen sie an, daß der Fels aus einer willkürlichen Anzahl von erdoberflächenparallelen Schichten besteht. Die elastischen Eigenschaften jeder Schicht sind transversal isotrop mit einer vertikalen Symmetrieachse. Die Trennflächen zwischen den Schichten werden als vollkommen kontinuierlich oder als völlig reibungsfrei angenommen. Das numerische Verfahren enthält eine Unterteilung der interessierenden Planfläche in rechtwinklige Elemente, und jedem Element wird ein Merkmal zugeordnet, das anzeigt, ob es sich um abgebaute Kohle, intakte Kohle oder um ausgebrochenes Versatzmaterial handelt. Für die Elemente, die Lasten zwischen Sohle und Firste übertragen können, werden eindimensionale Spannungs-Dehnungs-Beziehungen angesetzt. Das entstehende Gleichungssystem wird iterativ gelöst, wobei nichtlineare Kohle- und Versatzeigenschaften berücksichtigt werden können.

Die Zuverlässigkeit der Vorhersagen nach einem numerischen Modell ist nur mit der Verläßlichkeit der benutzten Eingabeparameter zu messen, so daß die Methode zur Bestimmung der relevanten Eingangsparameter besonders wichtig ist.Obwohl Wardle et al. eine Reihe von hoffnungsvollen experimentellen Vorgehensweisen vorgestellt haben, die die Messung der Verformbarkeit des ausgebrochenen Versatzmaterials durch Lastplattenversuche einschließen, bleiben viele Probleme noch ungelöst und zukünftige Arbeiten in dieser Richtung zu erwarten.

2.5 Flözschläge

In tiefgelegenen Kohlebergwerken werden Kohleschläge, d.h. spröde Brüche eines Teiles des Kohleflözes verbunden mit seismischer Aktivität, häufig festgestellt. Kohle- oder Gebirgsschläge sind die Ereignisse von gefährlichen dynamischen Brüchen.Sie passieren in verschiedenartigen Flözen unter unterschiedlichen Abbauformen.

Einige Symptome scheinen erkannt zu sein, doch da beträchtliche Unterschiede in den Sachlagen vorhanden sind, mag es nicht ratsam sein, die in einem bestimmten Gebiet erworbenen Erfahrungen zu verallgemeinern.

Fallstudien aus dem Ruhrkohlerevier zeigen, daß folgende geometrische Situationen das Auftreten von Flözschlägen begünstigen (Bräuner, 1979). (1) Abbaubeginn, wobei das Flöz unter zunehmende Last gerät, solange das Haupthangende noch nicht gebrochen ist. (2) Die Flözecke eines Strebbaufeldes in der Nachbarschaft eines früheren Abbaufeldes, wo der Druck im Strebvorfeld die Druckzone des früheren Abbaufeldes überkreuzt. (3) Ausbildung eines Pfeilers oder das Durchörtern der Auflagerdruckzone eines früheren Abbaufeldes unter Zusatzdrücken. (4) Abbaue unter oder über Restpfeilern anderer Flöze, die gewöhnlich beim Abbau der darunter oder darüber gelegenen Flöze hohe Druckkonzentrationen erfahren. (5) Das Auffahren zusätzlicher Strecken, das Spannungszunahmen im Flöz oder im Nebengestein verursacht. Bräuner (1979) schreibt, daß das wichtigste Merkmal von Flözschlägen stets ein dynamischer Längsschub unter großen Kräften im Kohleflöz sei. Jedoch sind das Ausmaß der Schubkräfte, der Bruchmasse und folglich der Zerstörungsgrad der Bergbauarbeiten unterschiedlich. Schwere Gebirgsschläge drücken das Flöz mehrere Meter heraus und haben z.B. eine Strecke über 100 m Länge vollständig verfüllt.

Sugawara et al. (1981) und Kimura et al. (1982) untersuchten die Flözschläge an Langfronten im Miike Kohlebergwerk. Schwerwiegende Kohleausbrüche traten bevorzugt nahe an den Flözstrekken auf, wie Fig. 2.3 zeigt. Die Ereignisse sind darin je nach der betroffenen Länge in vier Klassen unterteilt. Kimura et al. (1982) stellten fest, daß die Lage von Flözschlägen im Abbaustoß ziemlich regelmäßig mit dem Abbaufortschritt mitwanderte. Flözschlag trat anfangs nahe der Kopfstrecke auf, und danach wanderte der Ort des Auftretens in Richtung Fußstrecke und anschließend - wenn der Abbaustoß voranschreitet - in Richtung Kopfstrecke. Das Eintreten von Flözschlägen ist wahrscheinlich, wenn die Firste-/ Sohle-Konvergenz am Stoß nicht besonders groß ist und innerhalb eines bestimmten Bereiches liegt. Der Grund für die letztgenannte Tatsache ist wohl darin zu suchen, daß eine übermäßige Konvergenz durch die Auflockerung des Kohleflözes nahe am Stoß die Flözschlaggefahr ausschaltet.

Viele Forscher betonen, daß die wichtigste geologische Bedingung für Flözschläge das Vorhandensein einer mächtigen Bank im Hangenden sei, die üblicherweise aus Sandstein besteht. Bräuner (1979) beschrieb, daß solche Bänke mindestens 5 m dick sein und nicht weiter als 10 m über dem Flöz liegen dürfen. Fukuda et al. haben im gegenwärtigen Kongreß herausgestellt, daß die Flözschläge in Pfeilern eintreten, wenn ein Kohleflöz von einer dicken Sandsteinschicht überlagert ist, die mindestens viermal so mächtig wie das Kohleflöz ist. Josien et al. (1982) erläutern eine Methode zur Einschätzung der Flözschlaggefahr, die auf der Untersuchung der Felsverformung beruht. Sie schlagen einen W_{et}-Index vor, der durch das Verhältnis von gespeicherter elastischer Energie und dissipierter plastischer Verformungsenergie unmittelbar vor dem Bruch im einachsigen Druckversuch des Gesteines definiert ist. Die Zunahme des W_{et}-Indexes - insbesondere in mächtigen Bänken - kündigt die Flözschlaggefahr infolge dynamischer Energiefreisetzung in französischen Kohlegruben an.

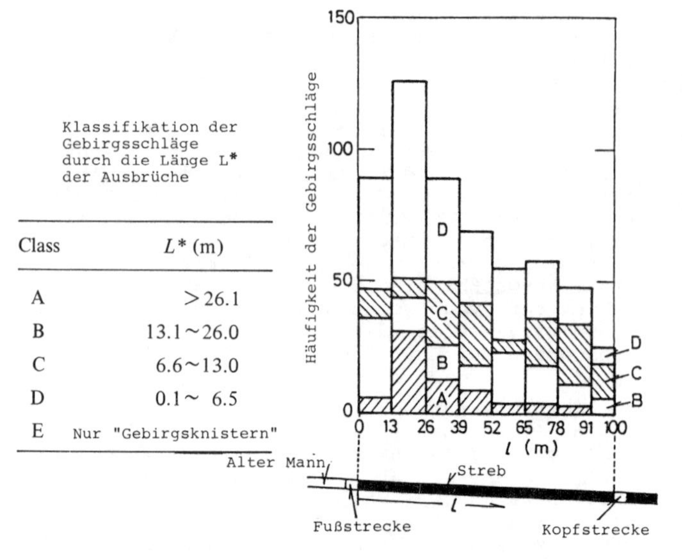
Klassifikation der Gebirgsschläge durch die Länge L* der Ausbrüche

Class	L^* (m)
A	>26.1
B	13.1~26.0
C	6.6~13.0
D	0.1~ 6.5
E	Nur "Gebirgsknistern"

Abb. 2.3: Klassifikation von Flözschlägen und Histogramm jeder Flözschlagklasse in Abhängigkeit von der Lokation des Ereignisses

In vielen Fällen vergeht die Flözschlaggefahr, wenn die Sohle zu brechen oder zu schwellen beginnt. Kimura et al. (1982) bestätigen, daß die Mächtigkeit der Sohlschichten in engem Zusammenhang mit dem Eintreten von Flözschlägen am Abbaustoß steht. Wie in Fig.2.4 gezeigt, traten die Sohlhebungen in einem begrenzten Bereich auf, wo die Sohlschichten dünner als 2 m sind, und zwar in jeder Kopfstrecke vor dem Durchfahren des unteren Streckenstoßes. Die Flözschläge ereigneten sich in dem Bereich, wo die Sandstein-Sohlschicht zwischen 2 m und 3 m mächtig ist. Andere geologische Faktoren, mit denen die Flözschläge zusammenhängen, sind Verwerfungen, Schwankungen in der Flözmächtigkeit und Änderungen im Einfallwinkel (Bräuner 1979, Kimura et al. 1982) sowie Unregelmäßigkeiten in der Festigkeit (Fukuda et al.).

Zur Überprüfung von hohen Spannungskonzentrationen im Flöz eignen sich Testbohrungen. Sugawara et al. (1981) schlugen ein Verfahren zur Ermittlung des Abbaufortschrittes A in Metern pro Schicht auf der Basis von Testbohrungen vor. Sei mit l^* die Breite der Zone bezeichnet, innerhalb derer Bohrlöcher schonend in das Flöz gebohrt werden können, und sei M die Mächtigkeit des Flözes, dann ist der zulässige Abbaufortschritt A gegeben durch

$$A = l^* - \alpha M,$$

worin der Faktor α aus Feldbeobachtungen abzuschätzen ist. In der Miike Kohlegrube wurde die Häufigkeit der Flözschläge in Abhängigkeit von dem Abbaufortschritt untersucht (Fig. 2.5), und man fand, daß mit der Wahl von $\alpha \geq 2$ die Flözschlaggefahr beseitigt werden kann.

Bräuner (1979) schlug Kontrollmessungen für den geomechanischen Zustand eines Kohleflözes unter

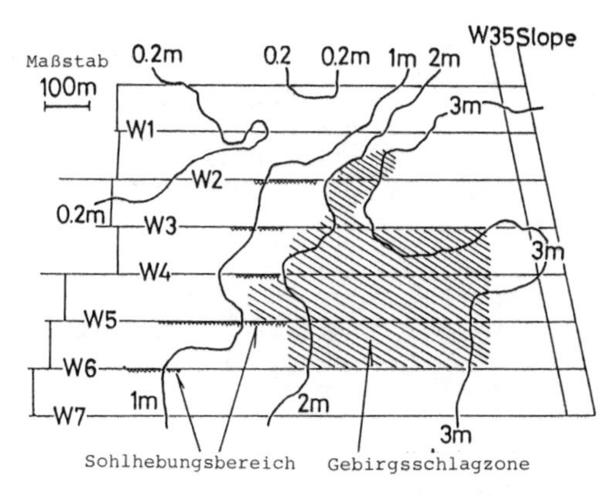

Fig. 2.4: Isopachenkarte der Sandstein-Sohlschicht mit der Flözschlagzone und den Sohlhebungsgrößen in den Abbaustrecken

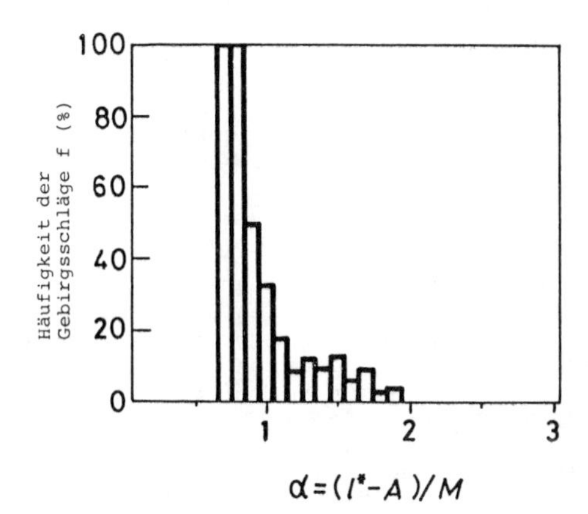

Fig. 2.5: Beziehung zwischen der Häufigkeit von Flözschlägen und dem Faktor α

Benutzung vorhandener Bohrlöcher wie folgt vor. Der Druck in den Bohrlöchern wird mit hydraulischen Druckzellen gemessen, und zur gleichen Zeit wird die longitudinale Ausdehnung in den Bohrlöchern mit Extensometern gemessen, die die volle Länge der Bohrlöcher abdecken. Fig. 2.6 zeigt einige der erhaltenen Resultate von einem Pfeiler zwischen einer Flözstrecke und einem benachbarten Abbaustoß. Die nahezu geraden Linien, wie die 1, 2 und 8 in der Fig. 2.6, deuten auf elastisches Verhalten des Flözes vor dem Bruch hin.

Bräuner (1975), Josien et al. (1982) und Sugawara et al. (1981) versuchten, den Zustand von Kohleflözen zu überwachen und die Flözschlaggefahr durch kontinuierliche Untersuchung der seismoakustischen Aktivität vorherzusagen, welche von Spröd- oder Scherbrüchen in den Auflagern im Strebvorfeld herrühren.Die Anzahl der Ereignisse und die gesammelte Energie dieser Ereignisse während der Dauer einer Stunde oder einer Schicht wurden als Parameter für die Intensitäten der seismo-akustischen Aktivitäten gewählt. Um die Dauer oder die Zone des Abbaues zu kennzeichnen, in der das Gebirge hohen Spannungen ausgesetzt ist, haben viele Forscher den gesamten Abstrahlungsumfang, die Änderungen in der seismo-akustischen Emission und die Verzögerung im Energieabbau untersucht.

Zur Verhinderung der Flözschläge sind systema-

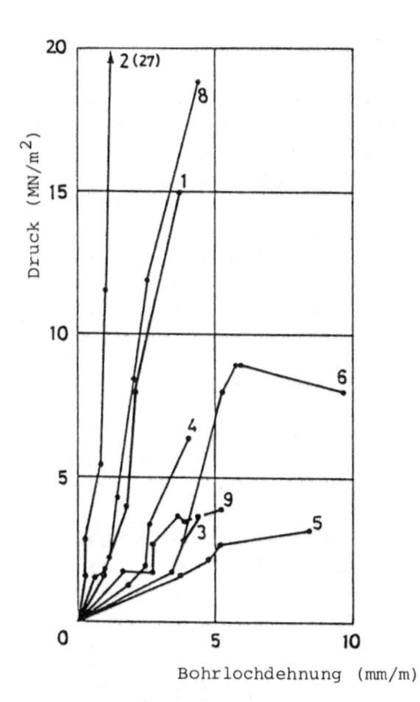

Fig. 2.6: Simultane Messungen des Druckes auf die Bohrloch-Druckzellen und der Ausdehnung derselben Bohrlöcher

tische Entspannungsbohrungen, Wassereinpressungen, Auflockerungs- oder Entspannungssprengungen und Kombinationen hiervon durchgeführt worden. Josien et al. (1982) berichtete, daß das Bohren von 24 m langen parallelen Entspannungsbohrlöchern von 10 cm Durchmesser und regelmäßigen Abständen von 2 m sich an der Kante eines Langfrontabbaues nahe eines alten Abbaues als sehr wirkungsvoll erwiesen haben. Die Wirksamkeit wurde durch Messungen mit einer Druckmeßzelle überwacht. Man fand heraus, daß während des Entspannungsvorganges ein plötzlicher Druckanstieg einsetzt, wenn ein beträchtliches Volumen von Feinkohle abgebaut wird, und daß im Bohrloch ein kleiner Flözschlag auftritt. Nakajima et al. (1981) untersuchten die Bruchaktivität im Flöz während der Entspannungsbohrung mit der Akustik-Emissions-Methode, um ihre vorbeugende Wirkung gegen Kohle- oder Gasausbrüche nachzuweisen. Andererseits erörterten Niu et al. (1982) die Wirksamkeit von Wassereinpressungen in die harten und mächtigen Sandstein- oder Konglomerat-Dachschichten. Diese Methode wird als effektiv für die Kontrolle plötzlicher ausgedehnter Hangendbrüche in den Alten Mann angesehen, die durch den Langfrontabbau entstehen, ebenso wie für die Entspannung des hohen Auflagerdruckes.

2.6 Gasausbrüche

Ein Gasausbruch kann als eine plötzliche Eruption von Gas, das sich im Gebirge mit zerstörtem Kohle- oder Nebengestein angesammelt hat, in den untertägigen Hohlraum definiert werden. Entweder wird die Menge des Berstgases oder das zerstörte Material verändert, sodaß es genauer ist, die Erscheinungen als Gasausstoß, Gas- und Kohleausbruch und Gas- und Felsausbruch je nach Menge und Art des zerstörten Materials zu bezeichnen. Im vorliegenden Bericht werden alle einfach Gasausbrüche genannt. Vier Kongreßbeiträge befassen sich mit diesem Thema.

Gasausbrüche sind häufig in einigen bestimmten Kohlegruben in vielen Ländern der Erde aufgetreten, die eine große Menge Kohle fördern. Ein schwerer Unfall eines Gasausbruches, bei dem 600.000 m^3 Methangas und 4000 t zerstörte Kohle ausgestoßen wurden, ereignete sich 1981 in einem japanischen Kohlebergwerk. Die Vorbeugung gegen Gasausbrüche und gegen die dadurch verursachten Katastrophen ist eines der wichtigsten Probleme in diesen Kohlebergwerken. Wissenschaftliche Forschungen werden hierzu seit den 1920er Jahren durchgeführt, und viele Hypothesen über die Ursachen für das Auftreten von Gasausbrüchen wurden seither aufgestellt.

Ursache für das Auftreten eines Gasausbruches

Die Hypothesen über die Ursache für das Auftreten von Gasausbrüchen lassen sich entsprechend den Hauptursachen in drei Kategorien einteilen, nämlich in die Gasdrucktheorie, die Gebirgsdrucktheorie und die Gas- und Gebirgsdrucktheorie. Obwohl diese Hypothesen nicht übereinstimmen, liegen die Unterschiede in der Ursache und in der Art der Kohlezerstörung, und sie scheinen sich in anderen Punkten vielfach zu gleichen. Durch Zusammenfassung dieser Hypothesen läßt sich die Ursache für das Auftreten von Gasausbrüchen wie in Fig. 2.7 dargestellt umreißen. Die Gassperre in diesem Diagramm ist eine imaginäre Zone im Kohleflöz, die durch Spannungskonzentration zusammengepreßt worden ist und eine Rolle bei der Abdämmung ausfließenden Gases aus dem inneren Teil des Kohleflözes spielt. Manche Forscher glauben nicht an die Existenz von Gasabsperrungen.

Hanes et al. berichten über die Merkmale und Ursachen von Gasausbrüchen in zwei australischen Kohlegruben. Sie meinen, daß - obwohl Spannung und Gas beide eine Rolle beim Auftreten von Gasausbrüchen spielen - eines von beiden je nach den lokalen Verhältnissen das andere überwiegt.

Um die Ursache für Gasausbrüche zu studieren, haben Hiramatsu et al. versucht, mehrere Aspekte zu den Gasausbrüchen zu erklären, die sich in bestimmten Kohlegruben in Japan ereignet haben. Zu diesem Zweck untersuchten sie die in den Flözen enthaltene Gasmenge, den Gasdruck in den Poren sowie die Freisetzungsrate von Gas längs verschiedener unterirdischer Wege. Sie stellten nach einer theoretischen Studie ein versuchsweises Kriterium für den Bruch von Kohle auf, die Hochdruckgas enthält.

Vorwarnung vor Gasausbrüchen

Es wird geglaubt, daß die Vorhersage des Auftretens von Gasausbrüchen im allgemeinen wegen der schwachen Voranzeichen sehr schwierig ist. Im folgenden werden einige Meßergebnisse zur Erklärung dieser Tatsache als Beispiel angeführt.

Oda et al. (1981) maßen die zunehmende Konvergenz während des Abbaufortschrittes in einer Abbaustrecke im Utashinai Kohlebergwerk auf Hokkaido, die in gestörtem Gebirge aufgefahren wurde. Sie benutzten für diese Messung 5 bis 7 Sätze von Konvergenzmeßgeräten, die in Abständen von etwa 2 m aufgestellt wurden. Das vorderste Meßgerät wurde immer wieder neu etwa 2 m hinter dem Abbaustoß aufgebaut, indem man das zurückbleibende hinterste Gerät nach vorne versetzte. Fig. 2.8 zeigt einen Teil der erhaltenen Ergebnisse. Während des Ausbruches traten Verwerfungen, Störzonen und Gasausbrüche im Zusammenhang mit den Sprengarbeiten auf. Jede gekrümmte Linie in dieser Abbildung zeigt den Verlauf der Konvergenz innerhalb eines schmalen Bereiches nahe dem Abbaustoß zu der Zeit, in der die vorderste Kon-

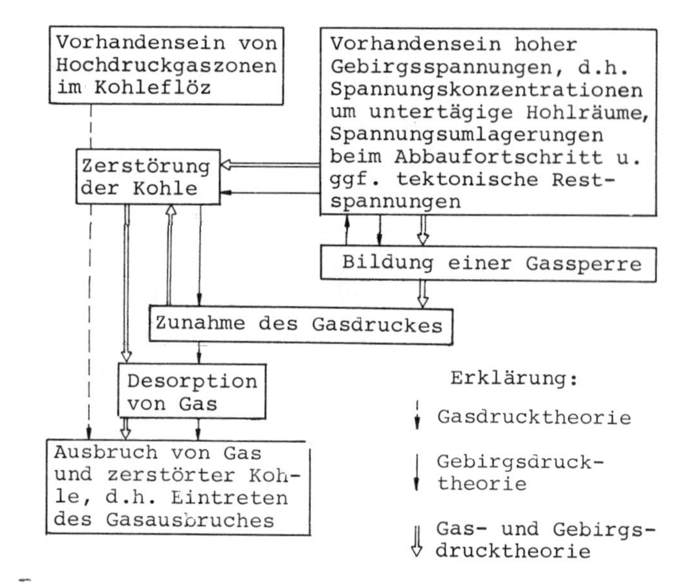

Fig. 2.7: Diagramm zu den Theorien über die Auftretensursache von Gasausbrüchen

vergenzmeßanordnung neu gesetzt wurde.

Aus dieser Abbildung läßt sich erkennen, daß die Konvergenz beim Eintreten des Gasausbruches gering ist, während große Konvergenzen entweder vor oder nach dem Gasausbruch gemessen werden. Ein sehr unregelmäßiger Konvergenzverlauf wird in der Verwerfungs- oder Störungszone festgestellt. In derselben Abbaustrecke wurde nach jeder Sprengung 5 Minuten lang die Menge des ausfließenden Gases gemessen, die bei drei vom Stoß in das Flöz gebohrten Bohrlöchern auf eine Länge von 2-3 m anfiel. Die durchschnittliche Rate des Gasaustrittes pro Bohrloch ist in Fig. 2.9 dargestellt. Ebenso sind dort die durchschnittlichen Konvergenzraten für die 2 m hinter dem Stoß liegenden Stationen aufgetragen, die unmittelbar vor der Sprengung beginnend 15 Minuten lang gemessen wurden. Auch ist die Rate der Gasabsaugung durch Ventilation nach jedem Sprengen in Fig. 2.9aufgetragen. Aus dieser Abbildung läßt sich keinerlei Anomalie bei den Meßwerten zum Zeitpunkt des Auftretens von Gasausbrüchen ablesen. Darüberhinaus sei angemerkt, daß die Messungen des Gasverhaltens und des Gebirgsdruckes Ergebnisse mit der gleichen Tendenz liefern. Sie lassen schließen, daß sowohl der Gas- wie der Gebirgsdruck das geomechanische Verhalten von schlaggefährdeten Gebirgsschichten beeinflussen.

Vorhersage von Gasausbrüchen

Zur Vorhersage von Gasausbrüchen sind viele Arten von Messungen ausgeführt worden. Jedoch sind die Erscheinungen am Abbaustoß sehr kompliziert. Zum Beispiel scheint es so, daß die Ausströmrate von Gas aus einem Kohleflöz oder die Gasdruckteilung in einem Kohleflöz durch den Gebirgsdruck beeinflußt und daß die Festigkeit und die mechanischen Eigenschaften eines Kohleflözes durch die Spannungsgeschichte bestimmt sind. Diese Erscheinungen hängen weitgehend von den geologischen, geomechanischen und bergbautechnologischen Bedingungen ab.

Darüberhinaus scheint das Auftreten von Gasausbrüchen mehr mit der Veränderung der gemesse-

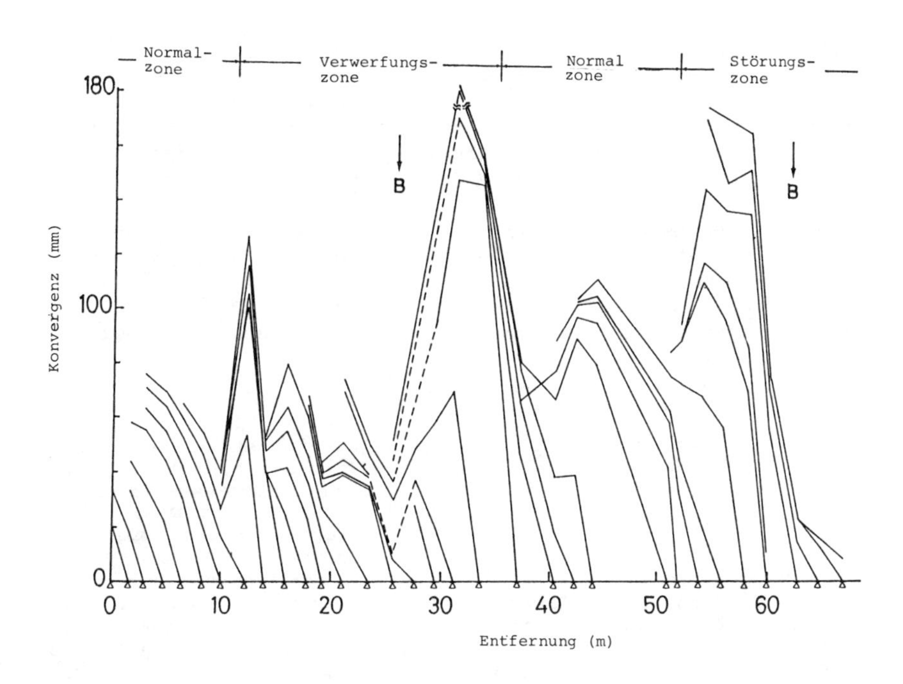

Fig. 2.8: Veränderung des Konvergenzverlaufes mit dem Auffahrfortschritt einer Abbaustrecke in gestörtem Gebirge. Kleine Dreiecke markieren die Position der Konvergenzmesser, und die Pfeile B deuten auf das Einsetzen von Gasausbrüchen.

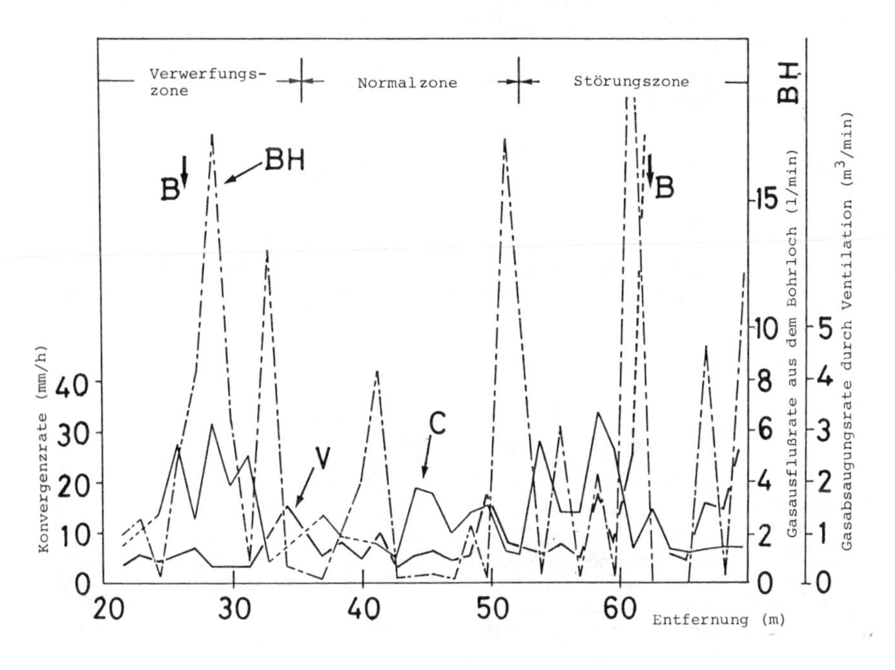

Fig. 2.9: Veränderungen der Konvergenzrate, der Gasausflußrate aus den Bohrlöchern und der Gasabsaugungsrate durch Ventilation in Abhängigkeit vom Abbaufortschritt.

nen Werte als mit den absoluten Werten zusammenzuhängen. Somit gibt es Raum für die Nachprüfung, ob der Ursprung eines Gasausbruches in einem bestimmten Gebiet liegt oder nicht.

Unter solchen Umständen sind Messungen, die genaues und zuverlässiges Material für die Vorhersage von Gasausbrüchen in allen Fällen erlauben, bisher nicht bekannt. Daher sind weitere Untersuchungen wünschenswert, bessere Meßmethoden und verbesserte Meßtechniken für verläßliche Vorhersagen von Gasausbrüchen zu finden.

Die hauptsächlichen bisherigen Meßmethoden sind mit den Kommentaren des Autors in Tab. 2.1 zusammengestellt.

Tab. 2.1 Messungen zur Vorhersage von Gasausbrüchen

Gegenstand	Meßobjekte	Meßmethoden	Bemerkungen	Literaturhinweise
Eigenschaften der Kohle	Mikrobruchdichte	Beobachtungen von Kohleschnitten unter einem Mikroskop	In gewissem Maße effizient	Vandeloise (1966)
	Δp-Index	Messung der Anfangsdesorptionsrate von Gas	In gewissem Maße nützlich, doch ist in japanischen Kohlegruben die Brauchbarkeit ungewiß	- " -
	Sprödheit	Anwendung des Trommeltests für Koks	Brauchbarkeit ungewiß	Oda (1970)
	Festigkeit	Prüfung der einachsigen Druckfestigkeit	- " -	Patching et al. (1966)
	Ausbruchsvolumen	Messung des Volumens von herausgebohrtem Gestein	In einigen Kohleflözen ist Vorsicht angeraten, wenn das Ausbruchsvolumen 3-4 oder noch mehr mal größer als das übliche Volumen ist	Gebirgsschlagrichtlinien (1976)
Gasinhalt der Kohleflöze und sein Ausfluß	Gaskonzentration in der Luft und Gasvolumen im Luftstromauslaß	Messung der Gaskonzentration und der Luftstromgeschwindigkeit sowie Berechnung des durch Wetterstrom abgesaugten Gasvolumens in 30 Minuten nach der Sprengung	In gewissem Maße erfolgreich. Die kritischen Werte scheinen 2% CH_4 in der Luft und 4 m^3 CH_4 im absaugenden Wetterstrom in 30 Minuten nach der Sprengung pro Tonne gewonnener Kohle zu sein.	Belin (1971)
	Gasvolumen in der Kohle	Messung des austretenden Gasvolumens beim Mahlen der Kohle in einer Kugelmühle	Effizienz wird untersucht	- " -
	Gasausflußrate aus einem Bohrloch und Gasdruckanstiegsrate in einem Bohrloch	Messung der Rate von freiem Gasausfluß aus einem Bohrloch und Messung der Anstiegsrate des Gasdruckes im Bohrloch nach Versiegelung des Bohrlochmundes	Effizienz unbestimmt	Cybulski et al. (1966)
	Temperaturabfall im Bohrloch	Messung der Temperaturabnahme im Bohrloch nach Beendigung der Bohrung	Bei Temperaturabnahmen von 1 - 2° C scheint Vorsicht geraten	Kuroiwa (1964)
Gebirgsspannung und Ausbruchsfolgeerscheinungen	Primäre Gebirgsspannung	Bestimmung der Gebirgsspannung durch eine geeignete Methode, z.B. durch Entspannungstechnik	Unbestimmte Effizienz	Takeuchi et al. (1976)
	Spannungsänderung beim Abbaufortschritt	Verschiedene Instrumente verfügbar	- " -	Novitshichin et al. (1975)
	Akustische Emission (AE) von Kohle und Nebengestein	Messung und Aufzeichnung akustischer Emissionen um das Abbaufeld herum mit geeigneter Instrumentierung	Effizienz wird untersucht	Watanabe et al. (1981)
	Konvergenz	Konvergenzmessungen in einer Abbaustrecke, insbesondere nahe dem Streb	Effizienz wahrscheinlich	Oda (1967)
	Schwellen des Kohlestoßes	Messung des Schwellvolumens	Effizienz unbestimmt	Szirtes (1966)
	Scheibenbildung	Beobachtung des Einsetzens von oder des Grades von Scheibenbildungen an Bohrkernen	Effizienz wird untersucht	Eckart et al. (1966)

Tab. 2.2 Vorbeugende Maßnahmen gegen Gasausbrüche

Vorbeugende Maßnahmen	Bemerkungen	Literaturhinweis
Vorheriger Abbau eines Kohleflözes, das nicht gasausbruchsgefährdet ist	Sehr wirkungsvoll	Belin (1971)
Begrenzung des Abbaufortschrittes	In gewissem Maß wirkungsvoll	- " -
Angemessene Abbauplanung zur Vermeidung von Spannungskonzentrationen	- " -	- " -
Erschütterungen zur Indikation von Gasausbrüchen	Geeignet zur Verhütung von Katastrophen in den meisten Fällen, jedoch Gefahr des Auftretens verzögerter Gasausbrüche	U.K. National Coal Board (1980)
Gebirgsentspannung durch Herstellung von Bohrlöchern mit großen Durchmessern	Wirkungsvoll in bestimmten Kohlegruben	Brouat (1966)
Entgasung der Kohleflöze vor dem Abbau	In gewissem Maß geeignet. Jedoch ist es sehr schwierig, das Gas aus unverritzten Lagerfeldern zu entziehen.	Garkavy et al. (1956)
Auswaschen der Kohle	Unbestimmte Wirksamkeit	Eckart et al. (1966)
Befeuchtung der Kohle	In gewissem Maß wirkungsvoll	Nedwiga (1966)

In bestimmten Fällen jedoch lassen sich zuverlässige Methoden erdenken. Hanes et al. berichten, daß in der West Cliff Kohlengrube, wo Gasausbrüche mit Querversetzungen zusammenhängen, eine geologische Kartierungstechnik für die Kohleklüfte entwickelt wurde, mit deren Hilfe Querversetzungen innerhalb von 45 m der dort festgelegten Ausbruchsschwelle vorhergesagt werden können. Die Autoren zeigen, daß auch die Gasdruckmeßtechnik unter verschiedenen anderen Methoden für die Vorhersage von Gasausbrüchen nützlich ist.

Vorbeugende Maßnahmen gegen Gasausbrüche

Die hauptsächlichen vorbeugenden Maßnahmen gegen Gasausbrüche, die praktiziert oder getestet worden sind, sind in Tab. 2.2 zusammengestellt.

Diese Maßnahmen können in unterschiedlichen Graden wirkungsvoll sein. Die Wirksamkeit von Entspannungsbohrungen kann je nach Natur des Gebirges schwierig sein, und die Entgasung gestörter Bereiche ist häufig sehr schwer. Erschütterungssprengungen zur Einleitung von Gasausbrüchen sind in den meisten Fällen tatsächlich geeignet, Katastrophen zu vermeiden, doch ist dieses Verfahren gegen die Gefahr verzögerter oder vorzeitiger Gasausbrüche empfindlich. Es muß betont werden, daß bisher keine zufriedenstellenden Maßnahmen gefunden worden sind.

Die wichtigsten hiernach zu unterscheidenden Themen sind die Entwicklung von Methoden zur Verhinderung von Gasausbrüchen in gestörten Gebieten und zur Verhinderung von verzögerten oder vorzeitigen Gasausbrüchen usw. Der vorherige Abbau eines sicheren Kohleflözes wäre eine vielversprechende Maßnahme. Falls kein sicheres Kohleflöz gefunden werden kann, kann diese Technologie der Untersuchung von ausgebrochenen Gebirgsschichten in der Nachbarschaft der infrage kommenden Kohleflöze dienen.

Khristianovich et al. überprüften ein weites Spektrum der gegenwärtigen Untersuchungen über Gasausbrüche des Institutes für Probleme der Mechanik an der Akademie der Wissenschaften der USSR. Die wichtigsten Punkte sind folgende: Eine Zone gerichteter Brüche mit einer bestimmten Breite kann sich nahe des Abbaustoßes durch aktive Spannungen und Gasdruck ausbilden, wenn ein gasausbruchsgefährdetes Flöz abgebaut wird. Diese Zone, die zwischen einer nahe am Stoß ausgebildeten plastischen Zone und einer elastischen Zone mit hohen Spannungskonzentrationen liegt, zeichnet sich durch viele stoßparallele Brüche aus. Die Brüche können freies Gas von hohem Druck enthalten, das Gasausbrüche verursachen kann, wenn die plastische Zone schmal ist. Vorheriges Abbauen eines sicheren Kohleflözes in der Nähe des gasausbruchsgefährdeten

Kohleflözes kann Gasausbrüche verhindern, da viele der flözparallelen Brüche im gasgefährdeten Kohleflöz bereits während der vorausgehenden Abbauarbeiten erzeugt werden. Das in den Bruchspalten enthaltene freie Gas kann beim Abbau dieses Flözes entweichen, so daß Gasausbrüche vermieden werden.

3. BERGSENKUNGEN

3.1 Einführung

Die Erscheinungen der Bodenbewegung infolge tiefer untertägiger Hohlraumbauten sind durch mehrere Bedingungen bestimmt wie z.B. durch die geologischen Gegebenheiten, die Geometrie der Hohlräume und die Technologie der Auffahrungen. Deshalb kann die Oberflächensenkung je nach den einzelnen Bedingungen unterschiedliche Formen annehmen. Tiefe untertägige Hohlraumbauten, welche Oberflächensenkungen verursachen, sind in den meisten Fällen ausgedehnte bergmännische Hohlräume zum Abbau von Erzlagern oder Kohleflözen.

Bergsenkungen verursachen Beschädigungen von übertägigen Einrichtungen wie etwa von Bauwerken und anderen Konstruktionen, Eisenbahnen, Autobahnen, Flüssen, Sportplätzen usw. Bergsenkungen lassen sich verhindern oder auf ein vernachlässigbares Maß einschränken, indem man untertägig hinreichend große Pfeiler stehenläßt. Dieses bedeutet allerdings einen Gewinnungsverlust an mineralischen Rohstoffen.

Ideal wäre, wenn man ein Maximum an mineralischen Rohstoffen mit einem Minumum an Schäden durch Bergsenkung erzielen könnte. In vielen Fällen jedoch sind wir gezwungen, kompromißweise die Menge des geförderten Erzes zu beschränken oder mögliche Maßnahmen zur Kontrolle der Bergsenkungen wie z.B. Bergeversatz zu treffen. Entsprechend gehören die Vorhersage, Kontrolle und Messung der Bergsenkung zu den wichtigsten Problemen bei der Planung eines rationalen Bergbaues.

Eine große Anzahl von Artikeln über Bergsenkungsprobleme ist veröffentlicht. Für den gegenwärtigen Kongreß wurden sieben Beiträge zu diesem Unterthema eingereicht. Davon befassen sich vier mit der Bergsenkung infolge Kohlebergbau, zwei mit der des Erzbergbaues, und der letzte Beitrag behandelt die Bergsenkung bei einem seichten Ausbruch.

3.2 Bergsenkung durch Kohleflözabbau

Um die allgemeinen Eigenarten von Bergsenkungen bei großräumigen untertägigen Grubenbauten zu umreißen, soll zuerst die Absenkung einer flachen Erdoberfläche beim Langfrontabbau eines söhligen Kohleflözes beschrieben werden. Nach dem Ausbruch eines Teiles des Kohleflözes stürzen die Dachschichten über dem Abbaubereich in den Alten Mann, woraufhin die darüberliegenden Schichten von unten nach oben nacheinander absacken, bis sich schließlich an der Erdoberfläche ein Senkungstrog einstellt.

Bereits in den frühen Jahren dieses Jahrhunderts war bekannt, daß die Fläche des Senkungstroges größer als das Abbaufeld ist und daß jeder Punkt dieses Troges mit Ausnahme des Randes und des Zentrums nicht nur Absenkungen sondern auch horizontale Versetzungen erfährt. Man nimmt deshalb an, daß die Gebirgsschichten zwischen dem Alten Mann und der Oberfläche sich in Richtung des Abbauraumes bewegen.

Entsprechend treten Neigungen und Dehnungen an jedem Punkt der Trogoberfläche auf. Die vertikalen bzw. horizontalen Verschiebungskomponenten seien mit s bzw. u_x und u_y bezeichnet, und x und y seien die horizontalen Koordinaten. Die Neigungskomponenten i_x und i_y und die Dehnungskomponenten ε_x, ε_y und γ_{xy} ergeben sich aus:

$$i_x = \partial s/\partial x, \quad i_y = \partial s/\partial y$$

$$\varepsilon_x = \partial u_x/\partial x, \quad \varepsilon_y = \partial u_y/\partial y, \quad \gamma_{xy} = \partial u_x/\partial y + \partial u_y/\partial x$$

Die wahre Neigung i und der Winkel ϕ zwischen der x-Achse und der Richtung von i folgt aus:

$$i = (i_x^2 + i_y^2)^{\frac{1}{2}},$$

$$\phi = \arctan\,(i_y/i_x).$$

Die Hauptdehnungen und ihre Richtungen lassen sich aus den Dehnungskomponenten durch herkömmliche Gleichungen bestimmen. Es sei angenommen, daß das obere Diagramm in Fig. 3.1 die Profile der endgültigen Absenkungen und der Horizontalverschiebungen längs einer Linie der Erdoberfläche darstellen, die senkrecht zu der breiteren Achse eines rechteckigen Langfront-Abbaufeldes verläuft. Die Neigungs- und Dehnungsverteilungen können mit Hilfe der obigen Formeln ermittelt werden wie im unteren Teil der Fig.3.1 gezeigt. In dem engen Gebiet an der Oberfläche, wo große Extensionen auftreten, ereignen sich häufig Zerrbrüche.

Substantielle Beschädigungen an untertägigen Gebäuden und anderen Konstruktionen durch Bergsenkungen rühren eher von überkritischen Neigungen oder Dehnungen her als von der Größe der vertikalen oder horizontalen Verschiebungen. Es scheint, daß der kritische Dehnungswert für Baustrukturen in der Größenordnung von 0,5 o/oo für Extensionen und 1 o/oo für Kompressionen liegt. Der kritische Neigungswert beträgt bei Maschinenfundamenten etwa 0,5 o/oo und bei dem Baugrund der neuen Superschnellbahnstrecke von Japan (Shinkansen) 0,25 o/oo.

Die Verteilungen von Oberflächensenkungen und Horizontalverschiebungen werden sich symmetrisch zum Zentrum des Troges einstellen, wenn man an den einfachen Fall denkt, daß die Erdoberfläche und das Kohleflöz flach und horizontal verlaufen, die geologische Struktur einfach und gleichförmig ist und das Abbaufeld rechteckig angelegt ist. Infolgedessen würden sich auch die Neigungs- und Dehnungsverteilungen symmetrisch ausbilden. In der Praxis sind jedoch diese Bedingungen nie erfüllt, so daß die Verteilungen der Bodensenkungen, der Horizontalverschiebungen, der Neigungen wie der Dehnungen in unterschiedlichen Maßen Abweichungen von der Symmetrie zeigen.

Das Flözeinfallen beeinflußt stark das Bergsenkungsprofil. Der Senkungstrog bewegt sich mit zunehmendem Fallwinkel des Flözes in Richtung des Einfallens, der Grenzwinkel nimmt am oberen Rand ab und am unteren Rand zu. Die Lage des Punktes mit der größten Oberflächenabsenkung bewegt sich vom Zentrum über dem Abbaufeld

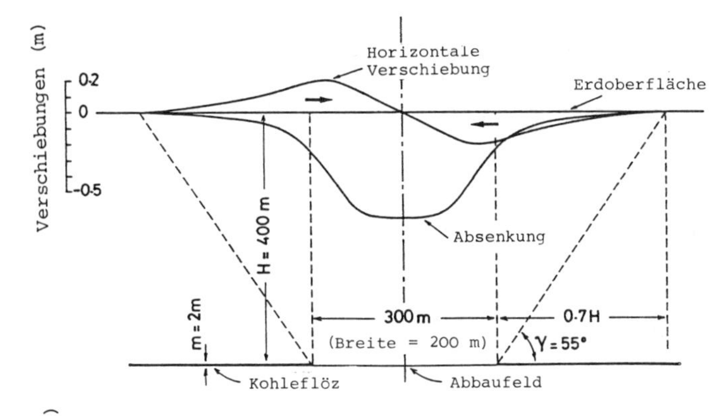

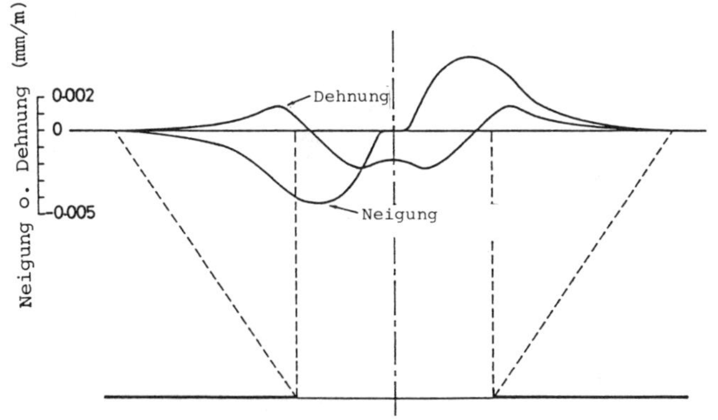

Fig. 3.1 Hypothetische Profile der Bergsenkungen und der horizontalen Verschiebungen (oberes Diagramm) sowie Verteilungen der entsprechenden Neigungen und Dehnungen (unteres Diagramm)

in Fallrichtung. Das U.K. National Coal Board (1975) stellte diese Beziehungen für den gesamten Neigungsbereich auf Grund von Messungen in britischen Kohlegruben auf. Hiramatsu et al. (1975) berichten über die Bergsenkung bei mäßig geneigten Flözen in japanischen Kohlegruben. Natürlich sind auch die horizontalen Verschiebungen von der Flözneigung beeinflußt.

Wäre das Gebirge ein kontinuierlicher Körper, dessen Verformung gewissen Gesetzen gehorcht, hingen die Absenkungen und die horizontalen Verschiebungen und entsprechend die Neigungen und Dehnungen miteinander zusammen. Das Gebirge ist beim Abbau eines Kohleflözes jedoch komplizierten Deformationen, Brüchen und Bewegungen unterworfen, wobei die Absenkungen und die horizontalen Verschiebungen unabhängig voneinander verlaufen können. Dennoch bemerkt man in der Praxis einige Ähnlichkeiten oder Beziehungen zwischen ihnen. Beispielsweise ist die Horizontalverschiebung etwa proportional zu der Neigung; die Berechnung der Absenkung nach der Elastizitätstheorie ergibt ein Bergsenkungsprofil, das dem tatsächlichen in gewissem Maße ähnelt. Deshalb kann man sagen, daß eine genaue Vorhersage für die Absenkung sehr schwierig ist, daß aber eine näherungsweise Vorhersage durch verschiedene Verfahren möglich ist.

3.3 Vorhersage der Bergsenkung beim Flözabbau

Vorhersage mit Hilfe der Einflußfaktoren

In den frühen Jahren dieses Jahrhunderts wurde bereits angenommen, daß die Absenkung s an einem bestimmten Punkt P der Oberfläche aus dem Kettenprodukt der Mächtigkeit m des abgebauten Flözes, dem Absenkungsfaktor a (= Verhältnis der maximalen Absenkung zu der Flözmächtigkeit), dem Zeitfaktor τ (bezogen auf die Zeit seit dem Abbau) und dem Einflußfaktor e (in Abhängigkeit von der Abbaufeldgröße und seiner relativen Lage zum Punkt P an der Oberfläche) berechnet werden kann. Die Faktoren a, τ und e reichen von O bis 1 und hängen von den geologischen Bedingungen ab. Außerdem hängt der Faktor a von dem Bergbauverfahren einschließlich der Bergeversatzmethode ab.

Eine andere Annahme, die einfach ist aber eine wichtige Bedeutung besitzt, ist das Superpositionsprinzip. Dieses besagt, daß die Absenkung eines bestimmten Oberflächenpunktes beim Abbau eines Teiles eines Kohleflözes sich durch die Addition aller hypothetischen Absenkungen ermitteln läßt, die sich beim Abbau eines gedanklich in kleine Bereiche geteilten Flözes einstellen würden.

Auf der Grundlage dieser Annahmen sind praktische Vorhersagemethoden von vielen Forschern untersucht worden, z.B. von Keinhorst (1925), Bals (1931), Flächenträger (1938), Perz (1948), Beyer (1945), Sann (1949), Hiramatsu et al. (1956, 1967) und Salamon (1963). Tatsächlich ist diese Art von Vorhersage von Bergsenkungen zweckmäßig und liefert recht annehmbare Ergebnisse. Allerdings ergibt sie wohl kein genaues Absenkungsprofil, so wie es wirklich gemessen wird. Zum Beispiel errechnet sich die Absenkung des Oberflächenpunktes genau über der Kante einer Bauhöhe, deren Breite gleich oder größer der halben kritischen Breite ist, in allen Fällen als halb so groß wie die maximale Absenkung, obwohl die wirkliche Absenkung geringer ist.

Empirische Methode

Auf der Grundlage von Beobachtungen der Bergsenkungen einer großen Anzahl britischer Kohlegruben ist vom U.K. National Coal Board (1975) eine graphische Methode zur Vorhersage des Absenkungsprofiles und der Dehnungsverteilung entwickelt worden.

Bei dieser Methode wird der Absenkungsfaktor a (= Verhältnis der maximalen Absenkung S zu der Flözmächtigkeit m) in Abhängigkeit von der Teufe h und der Frontbreite w graphisch dargestellt. Die Beziehung zwischen s/S und d/h ist für jeden Wert w/h tabelliert, wobei s die Absenkung eines Punktes P im Abstand d vom Zentrum des Troges bedeutet. Der Grenzwinkel wird zu 35° gesetzt, und die mögliche maximale Absenkung beträgt 0,9 m für den Bruchbau und 0,5 m für den Abbau mit Vollversatz in söhliger Schichtenlagerung. Die größtmögliche Neigung eines Absenkungstroges beim Abbau einer überkritischen Front wird zu 2.75 (S/h) abgeschätzt.

Andererseits ergeben sich die maximalen Extensionen und Kompressionen aus dem Produkt eines Faktors K_3, der sich graphisch aus den Verhältnissen w/h und S/h bestimmen läßt, und der Dehnungsverteilung, die für jeden Wert von w/h

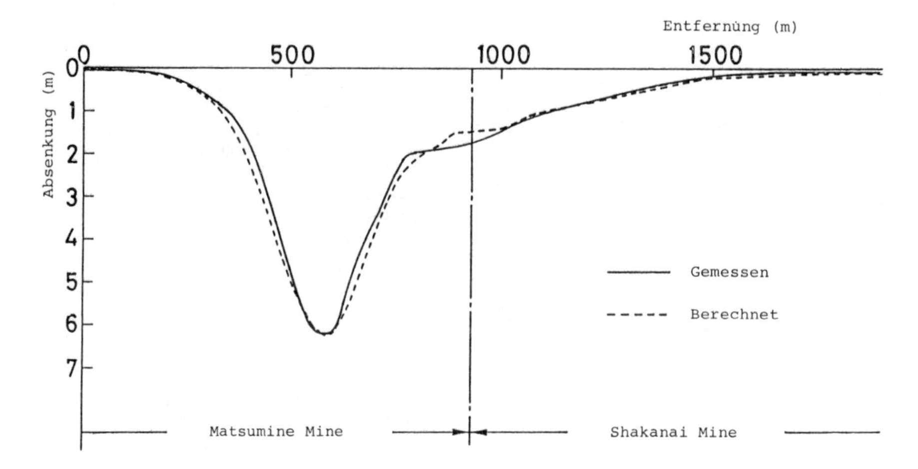

Fig. 3.2 Vergleich der vorhergesagten und gemessenen Bergsenkungsprofile längs der Linie No. 1 an der Erdoberfläche der Matsumine und Shakanai Gruben im Mai 1978

gegeben ist. Die maximalen Extensionen und Kompressionen über einer überkritischen Abbaufront werden zu 0,65 (S/h) bzw. -0,51 (S/h) gesetzt. Diese empirische Methode zur Vorhersage von Bergsenkungen und Verformungen wurde auch bei Kohlegruben in anderen Ländern angewandt, aber man fand nicht immer gute Resultate. Das mag an den unterschiedlichen geologischen Bedingungen und Bergbauverfahren liegen.

Mathematische Modellnäherungen

Die genannten Vorhersagemethoden hinterlassen Unsicherheiten wegen der fehlenden theoretischen Grundlagen. Eine empirische Vorhersagemethode für Bergsenkungen kann nur für diejenige Kohlegrube angewendet werden, aus der die Methode abgeleitet wurde. Deswegen sind schon seit über zwei Dekaden Anstrenungen unternommen worden, durch mathematische Modellierungen der Gebirgsbewegungen Daten zu gewinnen, die sich für die Vorhersage von Bergsenkungen eignen.

Berry (1964) stellte fest, daß es unter der Voraussetzung eines transversal isotropen elastischen Gebirgskörpers möglich ist, den endgültigen Deformationszustand des Gebirges qualitativ und in gewissem Maße auch quantitativ durch die Verschiebungs-Diskontinuitäts-Analyse vorherzusagen. Dahl und Choi (1972) untersuchten die Absenkung an einem elastoplastischen Modell mit gutem Ergebnis.

Jedoch sind die geologischen Strukturen sehr kompliziert, die mechanischen Gebirgseigenschaften sind unterschiedlich und die Randbedingungen verwickelt, so daß die Gebirgsbewegung beim Ausbruch von Hohlräumen keine einfache Erscheinung ist. Ein mathematisches Modell, das ein wirklichkeitstreues Bild der Absenkung liefert, wird sehr schwierig sein. Tatsächlich sind die Forscher gezwungen, geeignete Materialkonstanten durch Probieren so zu finden, daß die berechneten Absenkungen den gemessenen nahekommen.

In Anbetracht dessen, daß einfache mechanische Modelle ungeeignet sind, das komplizierte Gebirgsverhalten nachzubilden, während numerische Computermethoden die einzige praktische Alternative darstellen, untersuchten Mikula und Holt die Brauchbarkeit der Finite-Element-Methode zur Simulation der Bergsenkung. Als Materialeigenschaften des geotechnischen Modelles wurden realistische Werte aus Messungen und Datenverarbeitung herangezogen. Die Berechnung wurde mit einem Programm von finiten Elementen mit konstanten Verformungen durchgeführt, das auch die Kluftelemente von Goodman et al. enthält. Die Ergebnisse dieser Finite-Element-Analyse einer Bergsenkung waren leider nicht zufriedenstellend, doch hofft man bei weiterer Forschung auf Erfolg.

Salamon versuchte, die Oberflächensenkung beim Abbau eines Flözes durch elastische Modelle nachzubilden, deren Verformungen linearen Gesetzen gehorchen und Superpositionen erlauben. Die Gebirgsbewegung beim Abbau eines Kohleflözes wurde anhand dreier Modelle untersucht, nämlich einem homogenen isotropen Modell, einem homogenen transversal isotropen Modell und einem geschichteten elastischen Körper. Zur Beurteilung, ob die Modelle gute Simulationen ergaben oder nicht, wurden einige wenige kritische Messungen von Oberflächenbewegungen bei britischen Kohlegruben mit denen der Berechnung verglichen.

Um theoretische Oberflächenverschiebungen zu erhalten, die mit den kritischen Messungen verträglich sind, mußten außergewöhnliche Kennwerte für das mechanische Stoffverhalten der Modelle angesetzt werden. Daraufhin zog Salamon halb-empirische Einflußfunktionen heran, um annehmbare Vorhersagen der Bergsenkungen zu erzielen. Er meint, daß die Modelle die Gebirgsbewegungen qualitativ beschreiben können und daß es möglich sei, ein empirisches Modell aufzustellen, wenn die Abbauteufe nicht unwahrscheinlich gering ist.

Ein anderer Ansatz zur Vorhersage der Bergsenkung liegt in der Betrachtung des Gebirges als stochastischen Körper, der aus einer großen Anzahl von Felsblöcken unterschiedlicher Größen und Formen zusammengesetzt ist, die miteinander verbunden sind und in gewissem Maße zufällige Bewegungsfreiheitsgrade besitzen.

Litwiniszyn (1956) schlug eine Methode zur Berechnung der Felsverformung beim Abbau an dieser Stelle vor. In den mehr als 20 vergangenen Jahren wurde diese Methode durch festgesetzte theoretische und experimentelle Untersuchungen verbessert. Baoshen et al. (1979) berichten, daß die Ergebnisse der Bergsenkungsvorhersage nach dieser Methode gute Übereinstimmungen mit den gemessenen Werten zeigen, und daß sie eine Reihe von Problemen im Zusammenhang mit dem Bergbau unter geschützten Bauwerken, Eisenbahnen und Wassereinfassungen erfolgreich lösen konnten.

3.4 Bergsenkungen beim Erzkörperabbau

Erzkörper haben im Bergbau üblicherweise sehr unregelmäßige Formen. Die beim Abbau erzeugten Hohlräume sind in Form und Größe verschiedenartig und die geologischen Strukturen sind komplizierter und unregelmäßiger als bei den Kohlebergwerken. Deshalb haben die Bergsenkungen beim Abbau von Erzkörpern mannigfaltige Gestalt, und es gibt bisher keine systematische Untersuchung von Bergsenkungen dieser Art.

Goel beobachtete die Bergsenkung beim Abbau einer steilgelagerten tafeligen Kupferader, die sich etwa 5 km horizontal und bis etwa 1200 m Teufe vertikal erstreckte. Er stellte fest, daß der Grenzwinkel und der Bruchwinkel nicht konstant sind und daß die Lage des Punktes mit der höchsten Absenkung ebenso wie das Volumenverhältnis von Absenkung/Abbau sich sehr stark von denen beim Abbau söhliger Kohleflöze unterscheidet. Insbesondere sollte festgehalten werden, daß die Absenkungsrate auch bei nichtlinearer Abbaugeschwindigkeit jahrzehntelang konstant bleibt. Die Bergsenkungen beim Abbau von Erzkörpern unterscheiden sich in dieser Hinsicht von denen des Abbaues flacher Kohleflöze.

In den 1960-er Jahren wurden in der Präfektur Akita in Japan die sog. Schwarzerz-Lagerstätten entdeckt. Das Gebirge besteht im wesentlichen aus weichen Tuffen und Tonsteinen mit Intrusionen von Rhyoliten. An der Erdoberfläche über der Lagerstätte liegen Wohnhäuser, eine Eisenbahn, ein Fluß und Sportplätze. Da das Gebiet, über das sich die Lagerstätte erstreckt, von zwei benachbarten aber verschiedenen Bergwerksgesellschaften aufgeschlossen wurde, war eine genaue Senkungsvorausberechnung unbedingt erforderlich.

Dieses Problem wurde durch die Einrichtung einer Forschungskommission angegangen. Nach mehrjährigen Untersuchungen fand man heraus, daß eine genaue Vorhersage der Bergsenkung durch rein mathematische Betrachtungen zur Zeit unmöglich sei, daß die Bergsenkungsvorausberechnung in naher Zukunft jedoch mit Hilfe der Methode der Einflußfunktionen präzise durchgeführt werden könne, welche die Ergebnisse von Messungen der vergangenen zwei oder drei Jahre als Grundlagenmaterial für die Kompensation der lokalen Verhältnisse verwendet. Fig. 3.2 zeigt ein Vergleichsbeispiel für solch ein berechnetes und gemessenes Bergsenkungsprofil (Hiramatsu et al. 1979).

3.5 Kontrollierung der Bergsenkung

Es ist offenbar unmöglich, einen weiten Teil eines Kohleflözes ohne Bergsenkungen durch den Strebbau abzubauen. Um Schäden übertage zu vermeiden, ist es notwendig, die Bergsenkungen auf ein vertretbares Maß zu begrenzen. Der Kammerpfeilerbau oder das Verfahren mit angemessenen Pfeilern zwischen den Bauhöhen kann die Bergsenkungen und Bergschäden niedrig halten. Auch verringern Versatzberge die Oberflächenabsenkungen. Wenn besondere übertägige Bauwerke vor Beschädigungen zu schützen sind, muß man das Kohleflöz in einer solchen Weise abbauen, daß die dafür vorausberechneten Bergsenkungen entsprechend niedrig ausfallen.

Da die Oberflächenabsenkungen sich je nach dem Typ der Lagerstätte, nach den geologischen Bedingungen und dem Abbauverfahren unterschiedlich einstellen, und da die Ziele der Bergsenkungskontrolle unterschiedlich sind, fallen auch die Kontrollmethoden verschiedenartig aus. Deshalb sollen im folgenden nur die wichtigen Punkte aus den vorliegenden Kongreßbeiträgen zur Bergsenkungskontrolle vorgestellt werden.

Galvin behandelt ein einzigartiges Problem des kontrollierten Bruches von massigen (mehr als 30 m mächtigen) Doleritgängen in den Hangendschichten der südafrikanischen Kohlegruben. Um die Kohle vollständig abbauen zu können, ist es erforderlich, die Doleritgänge bald nach dem Beginn der Abbauarbeiten zu Bruch zu werfen. Galvin untersuchte die Voraussetzungen für solch eine Durchführung. Ausgehend von der Theorie einer dünnen elastischen Platte, modifizierte er sie unter Anwendung von Feldbeobachtungen und -messungen und erzielte eine einfache Gleichung zur Berechnung der für die Brucheinleitung im Dolerit erforderlichen Mindestbreite der Kohleabbaufront.

Arcamone und Poirot befassen sich mit Problemen des Kohlebergbaues in Frankreich, wo Tagebau und untertägiger Bergbau simultan nebeneinander betrieben werden.

Ein interessanter Beitrag von Stacey et al. beschreibt erfolgreiche Forschungsarbeiten zu der Bebauung über einem seicht liegenden Versatzfeld im Stadtgebiet von Johannesburg. Die Forschungen setzen sich aus Untersuchungen des unterhöhlten Baugrundes und nachfolgenden Berechnungen der Standsicherheits- und Auslegungsanforderungen für die Fundamente und die Hilfsmaßnahmen zusammen. Sie schließen die Stabilität des tiefergelegenen Versatzes und die Oberflächenverschiebungen ein. Zu diesem Zweck wurden hauptsächlich zweidimensionale Verschiebungs-Diskontinuitäts-Spannungsberechnungen durchgeführt, doch wurden auch dreidimensionale Bergbausimulationstechniken mit Verschiebungsdiskontinuitätselementen angewendet.

Danksagung

Der Autor ist den vielen Mitgliedern des Japanischen Kommittees der ISRM für ihre Unterstützung bei der Vorbereitung dieses Berichtes dankbar. Insbesondere dankt er Prof. S. Sakurai, Prof. S. Kobayashi, Dr. N. Oda, Dr. Y. Ohnishi, Dr. T. Saito, Dr. K. Sugawara und Dr. Y. Mizuta für ihre wertvollen Ratschläge und Mitarbeiten.

Danksagung des Übersetzers

Der Übersetzer dankt Herrn Dr.-Ing. G. Everling von der Forschungsstelle für Grubenausbau und Gebirgsmechanik der Bergbauforschung GmbH in Essen für die freundliche Prüfung der sachlichen Richtigkeit des übersetzten Textes insbesondere zum Kapitel Kohlebergbau.

G. Borm, Karlsruhe

DEEP UNDERGROUND EXCAVATIONS: TUNNELS, COAL MINING AND SUBSIDENCE

Excavations à grande profondeur: tunnels, l'exploitation minière des charbons et des affaissements du terrain

Tiefe unterirdische Hohlräume und Abbaue: Tunnel, Kohlenbergbau und Absenkungen

Yoshio Hiramatsu
Kyoto University, Kyoto, Japan

Thank you Mr. Chairman, ladies and gentlemen. I would like to give you the state of investigation on rock mechanics on these three topics and to introduce the papers submitted to this Congress on these topics. Concerning tunnels, these three may be the main topics. Theoretical research is highly important for the development of tunnel engineering. The next slide shows the constitution of research work on rock mechanics concerning tunnelling. Theoretical research by analysis and experiment. This investigation was made on continuous models and discontinuous models by mathematical methods and by numerical analysis. Numerical analysis will be applicable to the places where the boundary conditions and material properties are complex. Laboratory experiments act also important roles in studying the stability of the rock mass with discontinuities. Many active researches have been conducted by many investigators following these analysis and experimental researches, including the method of analysis itself. The recent research works in these fields of investigation are shortly described in the text. These research works should be highly estimated. Four papers have been submitted to this Congress for the present topics. The paper of Romana and others, Doucerain and Tang and Saari and others. Next I speak on design methods. The conventional methods of designing tunnel construction are based on the experiences over a long period of time, while the convergence-confinement method or the New Austrian Tunnelling Method, that is NATM, has recently been adopted widely because this method is based on the theory of rock mechanics and illustrates an idea of the interaction of rock and support. Several papers have been submitted to this Congress and most of them treat the tunnel lining, including NATM. They are the papers of Gartung, Minh, Romana, Altounan, Kobayashi, Holmgren, and Ortlepp. Lastly I would like to introduce the construction of the Seikan undersea tunnel in Japan joining Honshu and Hokkaido. The pilot tunnel was completed in January of this year. The feature of this tunnel is one undersea tunnel a length of 53 kilometres from shore to shore 23 kilometres, depth 240 metres from the sea level and 140 metres from sea bottom. The rock is an extremely weak, sandy mudstone. The tunnel has a large cross-sectional area, 88 square metres. The major technical development was of horizontal exploratory long boring system, lengths from 500 metres to 2 kilometres. 2. Grout system 3. shotcrete system 4. support system to resist high earth pressure.

Now I will proceed to the next theme - field measurements and inverse analysis. In the stage of designing, accurate prediction of the behaviour of the ground and supports will be difficult because the properties of ground and initial state of stress are generally unknown, therefore it would be advisable to alter the original design if necessary by considering the result of field measurements made during excavation. The analysis for this purpose is called the inverse analysis. Many studies have been reported so far on inverse analysis. From deformation or strain of support the earth's pressure is determined, c.f. the research of Kovari and Murayama. From displacement and strain of ground, the initial earth pressure is obtained by the D. Oda and others. From displacement, strain of ground is determined and this gives data used for administering the construction of the tunnel; e.g. the research of Sakurai. The papers classified in this category and submitted to this Congress are those of Kaiser and others and Saito and others.

The next topic is coal mining. I think the main problem in rock mechanics concerning coal mining will be illustrated as follows. For room and pillar mining, designing coal pillars and safe roof spans. For longwall working panel width and inter-panel, gate roads and supports, strata control in longwall faces, monitoring for strata control, mathematical modelling of coal mining, coal bursts, gas outbursts. Forty papers were submitted.

A great number of investigations have been carried out on these problems. Regarding the room and pillar mining Bieniawski has proposed a new pillar sizing method as well as a method for assessing the safe roof span. As to the stability of pillars in long wall workings the concept of yield pillars that fail in a stable manner, and integrity factor of a pillar have started to become considered. Lu has proposed a new technique to estimate the integrity factor of pillars. Concerning the gate road support it has been found that convergence is a most important factor for the steel factor, that is the support density. Abad and others have presented a new formula to assess the final convergence. Strata control at longwall faces and monitoring for it have been studied by many investigators. To these topics Gupta and others, Lu and Schuermann have submitted valuable papers to this Congress. On the other hand mathematical modelling by the finite element and displacement discontinuity methods have been tried by a number of investigators to analyse the rock pressure in deep coal mines; for example, by Yeates and others, Fukuda and others and Wardle and others.

Coal bursts are brittle fractures of coal seams accompanied by seismic activity, and happen to cause disasters. In some coal mines it is one of the most important problems to prevent occurrence of coal bursts.

I think these are the conditions for occurrence of coal bursts. Local stress concentration, thick strong bed in the roof, moderately thick bed in the floor, but floor heaves prevent bursts. Either geological conditions such as faults, changes in seam thickness, strength and dip angle. For assessing hazard of coal burst this method or idea is proposed. WET index measuring pressure and expansion in pillars. Test bore holes to find the depth of boreholes to be drilled smoothly. Acoustic emission.

For prevention stress relieving by boreholes or stress relieving by bursting or water infusion are tried, and effectiveness is monitored by several methods.

The term gas outbursts means in my report inclusively gas and coal outbursts and gas and rock outbursts. Gas outbursts have frequently occurred in some specified coal mines where prevention of occurrence of gas outbursts and disasters caused by them is one of the most important problems. The hypothesis concerning the cause of occurrence of gas outbursts may be classified into three categories, namely gas pressure theory, rock pressure theory, and gas and rock pressure theoryand are illustrated in the figure in my text. The differences among them are attributed to the cause and manner of crushing of coal. In the paper by Hanes and others and that of myself, the cause of gas outbursts is discussed. It is considered that prediction of occurrence of gas outbursts is generally difficult because of poor premonition of it. Measurements can be made of the convergence profiles in the vicinity of the face of a gate road which is being driven in disturbed strata and the position of gas outbursts indicated.

Convergence is measured in millimetres. The results of other measurements carried out at the same time and on the same gate road can be illustrated. The same gas outburst, and the rate of gas outburst from test boreholes previously drilled into coal seam, and the rate of gas exhaustion by ventilation after each blasting and the increase in convergence at the point 2 metres from the face for 15 minutes after each blasting can be demonstrated. We cannot find any anomaly at the time of the occurrence.

Many kinds of measurements have been conducted for the purpose of predicting gas outbursts and effectiveness of measurements is varied in my opinion. Again various major preventive measures for gas outbursts which have been practiced or are tested have varying degrees of effectiveness in my opinion. Khristianovich and others have submitted a paper in which they revealed the recent investigations on gas outbursts and suggested an interesting theory of oriented cracks.

The next topic is subsidence. Surface subsidence is inevitable when a large excavation is made underground so that when we have to limit surface subsidence to avoid or minimize damage caused by subsidence, design of mining should be based on the result of precise prediction of subsidence, therefore the technique of prediction of subsidence will be very important. Influence factor methods are much improved and convenient for using on computers, and empirical method developed in United Kingdom, and mathematical modelling approaches. These two methods are nice, but however, these two let us feel uncertainty because they have no theoretical basis therefore much effort has been made to analyse the ground movements by mathematical modelling.

Mathematical modelling, and method of control of subsidence have various values, depending on the geological and the mining condition and objects of subsidence control. Three papers have been submitted to this topic, namely by Galvin, Arcamone and Stacey and others.

DISCUSSIONS / DISCUSSIONS / DISKUSSIONEN

MINING EXCAVATIONS AND MINING METHODS INCLUDING CAVING

Excavation minière et méthodes minières, y compris l'ouvrage par éboulement

Bergbauliche Hohlräume und Bergbaumethoden, einschliesslich Bruchbau

J. L. Liebelt
Chairman

PAPER: Stress measurement and analysis for mine planning

AUTHORS: S. Bywater, R. Cowling and B.N. Black

PRESENTED BY: R. Cowling

DISCUSSION

Question: Mr Michael Wold, CSIRO, Australia.

My question is directed to the authors Bywater, Cowling and Black. Studies such as those reported by the authors which allow comparison of in situ behaviour with the predictions of now widely used numerical models are very valuable, and I thank them for their presentation. The results they have reported are encouraging, particularly in view of the assumption of linear elasticity. My question is: Have they gathered displacement data which provides similar encouragement? Now the combination of displacement data with measurements of absolute stress and stress changes would contribute to the fundamental problem of determining rock mass deformability, consequently providing more refined material properties for the application of these types of models.

R. Cowling:

I would like to call on my co-author, Mr. Steve Bywater, to answer that one please.

S. Bywater:

Yes, certainly in the future and in the past we have done a great deal of displacement monitoring. So far the results of these are particularly encouraging. Certainly at all of our sites in the future, of stress change monitoring, absolute stress measurement and displacement monitoring generally with extensometers is an essential part of our program.

Mr. Wold:

Well experience that I have is such that I believe with numerical modelling that displacements are sometimes rather difficult to get a match with, that if you can, you tend to get a rule of thumb of reduction of rock mass modulus compared to your assumed or measured properties from small specimens, and that is often interesting if we can find out what that reduction factor might be.

I suspect that would depend on the circumstances of each case.

Co-Chairman:

Thank you very much. I do not know whether Dr. Beer is around. I understand he has not put in a form, but please do put your questions to Dr. Brady if you would.

Question: Dr. H. Wagner, Chamber of Mines South Africa.

My first question is directed to the authors of the paper on stress measurements and analysis for mine planning, which was the first paper. In the paper the authors quote some values for the in situ strength of the rock mass and typically they indicated this was betwen 2 and 2 1/2 times lower than the strength determined in the laboratory. My question to the authors is: Are the stress values quoted by them field stress values, that is the stress value acting at the position of the excavation or do those stress values take into account the induced stresses around the excavation, due to the excavation effect itself?

R. Cowling:

They are the total stresses. This is the field stress plus the redistributed induced stress.

Dr. Wagner:

In South Africa we have found that in the hard quartzites the field stresses which are critical lead to tunnel damage. Typically they are also of the order of 2 1/2 times lower than the rock strength values. I have a second question to the authors of the paper and this is: It was also discussed to some extent the problem of predicting pillar strength and my question is: How does the confinement provided by the cemented fill influence the pillar strength?

R. Cowling:

We believe that the influence of the cemented fill is insignificant. All the results presented take no account of fill being there.

Question Dr. G. Beer, University of Queensland, Australia.

Thanks very much for allowing me to respond to the General Reporters' remarks on our paper. Unfortunately it was not included in the papers presented, but in the Poster Sessions. The work Dr. Brady has shown is very interesting and he has done very good work at the University of Minnesota using a combination of boundary element and blocky models. It is very interesting to note that we have had very similar results to what Dr. Brady showed for the two dimensional cases, using a different approach. In this case it was a combination of boundary elements and finite elements using special contact or joint elements which allowed a separation along weak bedding planes and also slip to take place. We found for example very similar results concerning the separation and the slip, with regard to location and also the development of tensile zones. However we have taken this approach a little bit further in so far as you will see in the poster presentation some results of some very recent studies where we used a 3-dimensional model. In this case instead of a beam we used plate elements and these plate elements are connected by contact elements which have a very similar basis for the 2-D case. So I want to point out that the model does consider rigid body movements. The only limitation is that these rigid body displacements are supposed to be small, and this ties in with our philosophy of hanging wall behaviour in so far as we think that the first displacements which produce the separation, the slip and eventually the formation of the hanging wall plates. The displacements are in the order of centimetres, may be up to 3 centimentres, but this is considered small in comparison with the overall size of the stope which is 30 x 200 metres. The second stage then should allow you to predict how stable these plates are and this is where blocky models come in as Dr. Brady outlined.

Thank you very much for giving me the time and opportunity to reply.

Dr. Brady:

I would like to respond to that. Yes I accept what Dr. Beer has said that there are alternative formulations. That the finite element formulation allowing slip and separation is possible and really the question is how important is it to incorporate these significant displacements. I would call two centimetres a reasonable sort of rigid body displacement, and that is the sort of question that arises, and that question can only be settled I would suggest by comprehensive physical modelling. Laboratory tests under controlled conditions under which one could demonstrate whether a necessity to accommodate these rigid body translations and rotations is necessary or whether the finite element formulation which is limited to small displacements, rotations, translations by rotation is acceptable, so there is something to be resolved there.

Question: **Renato Ribacchi, University of Rome, Italy.**

I would like to reply to some comments of the General Reporter upon the paper by Crea et.al on the Campiano Mine. The General Reporter expressed the opinion that the computed stresses within the fill were too high in comparison with the results found in North American mines. It is to be noted, however, that the fill which is to be applied in some parts of the Campiano mine is a cemented rock fill, not a tailing fill, and a cement content is about 150 kg/m^3. It has a strength and deformability which is comparable with a very weak concrete. It is to be noted also that in some other Italian mines where concrete rock fill was applied, high stresses within the rock fill were measured, up to 5 MPa.

Dr. Brady:

In reply to that I would not consider 5 MPa a high stress, but also my observation was that the modulus values that you used were significantly higher than modulus values that would have been determined for example in large scale field tests at Mount Isa Mines or Broken Hill Mines. That was the core of my argument. Clearly the higher modulus values you use, the higher are the stresses going to be and that was the comment I was making. That it might be useful for you to make a comparison between the modulus values you used and what were known to be generated in field tests.

Question: Dr. A.J. Hargraves, Australia.

This is directed to Professor Hiramatsu and I must say that he has done a mammoth job in compounding dozens of papers and reconciling so many diverse authors. I have a number of questions, but will ask one. Is migration of coal bursts across the longwall face as the face advances, as reported by Kimura compatible with the proposition of Brauner that location is related to the compounding of two rock pressure abutments?

The point Mr. Chairman is that one of those authors showed a graph and as the longwall face advanced the position at which the coal bursts occurred moved across the face, but the other author said that their location was at the position of the intersection of the two stress abutments.

In regard to instantaneous outbursts I see an omission in the list of predictions. Accepting that rate of advance has an influence on proneness the day to day monitoring for prediction is a compromise between what is possible and what prolonged delays will render mining uneconomic. The metre by metre semiquantitative assessment of gasiness of face coal has its place in prediction in Australia. Also the differences between shotfiring and machining advance are important.

Written answer to Dr. Hargraves:

Thank you Dr Hargraves for your valuable comments. Dr Brauner has pointed out that the hazard of coal burst exists at the tail gate side corner of a longwall face. Mr Kimura has experienced the same tendency, but he has also found that coal bursts can sometimes occur at the head gate side corner of a longwall face and that they can migrate across the face depending upon discontinuities in the geological conditions. Therefore I think that these two opinions are fundamentally compatible.

As for the prediction of instantaneous outbursts, I agree fundamentally with your opinion. I would like to explain, here, the position of Japanese coal mines prone to instantaneous outbursts. Various investigations have long since been carried out into this problem, and it has been found that the results of any measurement concerning desorption of gas from coal have little connection with occurrence of outbursts in Japanese coal mines, so that such measurements are nowadays not practiced. Measurements of the rate of gas outflow and the rate of gas pressure rise from/in a borehole made into a coal seam have been recommended for the prediction. However it has been found that the results of these measurements are often contradictory to occurrence of gas outbursts. Therefore, in addition to them, measurements of the volume of coal cuttings produced by boring and the rate of gas outflow from the coal face are recently suggested for the prediction.

Lastly I would like to mention that limiting the rate of advance is not a conventional way to prevent spontaneous outbursts. They are in a few cases practiced only on extremely outburst prone faces.

Question: Dr. Hargraves, Australia.

The following question was directed to Dr. Hiramatsu, General Reporter.

Talking about instantaneous outbursts and in regard to your reporting of Oda et.al. of convergences in the advancing longwall gate road at Utashinai, convergences showing no direct relation to the instantaneous outbursts occurring, I recall some experiments in room and pillar mining at Canmore, Canada in 1957 by the then Department of Mines and Technical Surveys. There was a strong suggestion of coincidence of sharp convergences and the instantaneous outbursts which occurred - both cases presumed with shotfiring advance. Would you comment please?

Answer: Dr. Hiramatsu

From Fig. 2.8 and 2.9 in the text, it may be thought that there is no direct relation between convergence and occurrence of spontaneous outbursts in Japanese coal mines. However it has been found that a large convergence can occur two or three shifts either before or after an outburst shift even when the convergence in the outburst shift is small.

I think that the cause and aspect of instantaneous outbursts are diverse according to the differences in geological and geomechanical conditions, and that the difference in the experiences about instantaneous outbursts between Canmore, Canada and Utashinai, Japan may be attributed to the differences in these conditions.

Lastly I would like to suggest that analysis of convergence may be helpful for prediction of spontaneous outbursts.

Question: Dr. J. Galvin, Australia directed to Ph. Weber. His paper on "An energy approach to room and pillar exploitations based on post-failure behaviour of pillars".

It is a fundamental principle of the rock mechanics of bord and pillar workings that the stability of the workings and the post-failure behaviour of pillars is a function of the ratio of the stiffness of the surrounding strata to the stiffness of the pillars. The author takes no account of the stiffness of the surrounding strata, employing only a simple model of deadweight loading on a beam. Could the author please comment?

Answer: Ph. Weber

I do of course agree with Dr. Galvin's statement concerning the role of the relative stiffness of roof and pillars in the equilibrium of the roof-pillars system; this system is not basically different from the classical one of a rod under compression with a deformable testing machine.

What I wish to underline in my paper is the importance of the "jump" when the equilibrium changes from stable to an unstable one. The proposed model, although rather "crude", takes into account a roof-stiffness, modelised by the elastic inertia EI of the beam: a judicious choice of the v (x) displacement function should moreover permit an accurate modelisation of the stiffness of any thick and massive roof. We are actually working on this point. Of course, a deadweight loading alone cannot take into account any post-peak behaviour of pillars (the same happens with an infinitely soft testing machine).

Question: By letter from Dr D. Nguyen Minh

I have participated to the last ISRM Symposium at Melbourne, and I read with special attention your general report on deep caverns, which I appreciated very much.

In your report, there is however a mistake in paragraph "1.2 Analysis". You mentioned that "visco-elasto-plastic body and strain softening characteristics have been reported (Tanimoto (1981), Brown (1982))". But neither of these two references, deal with any delayed behaviour of the rock mass; and to my own knowledge, I am not aware of such papers, and if there does exist any, I would like to have the references.

Answer: Ph. Weber

I appreciate your comment and general interest for my general report. As you have pointed out, the two references listed in the general report do not treat the time-dependent behaviour of the rock mass. As far as I know, the methods to analyse the visco-elasto-plastic problems have been reported by you, as well as by P. Fritz (Dr. Thesis, ETH Zurich, 1981) and T. Nakano (Int. J. Solids and Structures, Vol. 11, No. 10, 1981).

PERMANENT UNDERGROUND EXCAVATIONS: TUNNELS

Excavations souterraines permanentes: tunnels

Langlebige Grubenbaue: Tunnel

J. Müller
Chairman

PAPER: Excavation of the 21 km Walgautunnel by a full-face tunnelling machine

AUTHORS: G. Innerhofer and H. Loacker

PRESENTED BY: G. Innerhofer

DISCUSSION

Question: Professor A. Bello, Mexico

I want to address a question to Professor Innerhofer, about the flow of water shown in the pictures. I have the figure that was in the order of 30 - 40 litres per second in the place that you found water seepage through the tunnel. I would like to know if you have recorded variations in the quantities of water entering the face of the tunnel with time?

Answer: G. Innerhofer

At the face we had water inflows up to 500 litres per second. Most of them decreased within some days. The max. permanent inflow of one particular crack is more than 100 l/s in winter.

Question: Professor L.A. Endersbee, Australia

I was intrigued to notice that when you ran into difficulty with the Robbins machine with the limestone cavities that you went into the crown above the tunnelling machine and then went forward. Many years ago we used a Robbins machine for tunnelling through somewhat similar rocks, fortunately not karstic as yours were, but we had a door in the face of the Robbins machine which meant that when we did get into difficulties we could remove a portion of the face of the machine and get in front of the machine to solve the problems, and that also we used that for the purpose of periodically replacing the cutters. Now I would gather you would have a similar door in the face of your tunnelling machine and the question is: Why did you go up into the crown and not through the face of the machine?

Answer: G. Innerhofer

Four doors are found at the face of the Robbins TBM that was used to replace the cutters. However when difficulties were encountered in the crushed water-bearing limestones and dolomites, failures in the crown could not be avoided. The front of the face was more easily accessible in this case along the roof than through the doors in the cutter head. Furthermore, access along the roof provided safer rescue routes during this dangerous work.

Question: A. Neyland, Australia

For the predicted karst and dolomite zones, was consideration given to possible methods of predicting individual water filled cavities ahead of the machine - for example by probe drilling - and thus to pre-treat them?

It seems the method in fact used was to wait until a cavity was encountered and then to battle through it.

Answer: G. Innerhofer

Yes, you are right in general. We started to execute test drillings. However, the problem was that the advance rate of the boreholes was less than that of the fullface TBM. Furthermore the water-filled cavities were localized and were not necessarily found by isolated boreholes. The most dangerous cavities were close to the tunnel walls but outside the profile. In this case therefore the best advance strategy was to pay attention during fullface excavation.

PAPER: Swelling rocks and the stability of tunnels

AUTHOR: Professor Dr. Tan Tjong Kie

PRESENTED BY: Professor Tan

DISCUSSION

Question: Professor Ladanyi, Canada

Your finding that in swelling rocks the amount of swelling might increase if it is preceded by dilatancy, has undoubtedly very important practical implications. However, although both dilatancy and swelling tend to increase the overall volume of the rock mass, the dilatancy increases its porosity, which may accommodate a portion of swelling strain. This means in fact that, as far as the swelling pressure on the tunnel lining is concerned, the two effects might not be additive but compensating.

Answer: Professor Tan

When we have to deal with an underground cavity then due to dilatancy and swelling the rock will move inwards. By moving inwards some fissures will be closed again so there will be some hardening and in that respect you know from the NATM that a ring and arch will be formed. This is a completely non-linear problem, which I have not discussed in my presentation, but you are quite right.

Question: Professor Mueller

I also would like to ask a question to Professor Tan. It is an experience that in swelling rock, for instance in anhydrite, or in clay, the swelling phenomena mainly is heaving of the invert and we are always surprised that we have no swelling phenomena or little swelling phenomena around the crown. Have you any explanation for this effect?

Answer: Professor Tan

May I ask whether you mean that there are only little swelling phenomena in the walls and the roof and there are mainly swelling phenomena in the bottom. Is that your question?

Professor Mueller:

We have mainly swellinq phenomena in the invert and very seldom or very little in the root. Why?

Professor Tan:

I think it depends on many factors. First the stress condition. Second the presence of tectonic stresses. In my experience swelling is mainly in the floor and my opinion is this, that you have a cavity and the water will be infiltrated into the cavity, and most of the water will be absorbed in the bottom and not in the roof. That is my only explanation for the moment.

Question: Dr. M.J. Pender, New Zealand

I should like to return the discussion to Professor Tan's interesting paper on dilatancy. I too am interested in this problem and have done some rather simple analysis of the effect of pre-failure dilatancy - analysis which is much cruder that that done by Professor Tan. My analysis shows that, when compared with elastic behaviour, pre-failure dilatancy leads to an increase in the circumferential stress around the opening. I infer from Professor Tan's reply to Professor Ladanyi, that he observes this effect in his numerical results. My question is as follows: Is this dilatant increase in circumferential stress likely to initiate failure in a rather larger zone of rock than would occur from simply dealing with an elastic material?; or, if elastic analysis would predict failure around the opening, increase the volume of rock that is in a failure or post failure state?

Professor Tan:

If you have stresses which exceed upper yield failure, tunnelling dilatancy will occur. This dilatancy normally will increase with the time, but in the tunnel it is quite different. Then you have according to rheology a transfer of stress from the weaker part to the more rigid part. So the stresses in the periphery of the tunnel due to dilatancy will decrease and will be transferred to the most inner parts.

Question: Dr. H. Wagner, Austria

On one of the first tables there was shown among other factors, influencing swelling of rocks, the electro-chemical factor. As a special explanation regarding the E-C-factor was not given, it seems to be of special interest to mention the theoretical and practical works of Professor Veder of Austria, who gives a solution to this problem. According to his investigations in swelling tunnels, the high swelling pressures causing heaving deformations mainly in the invert are to be explained both with differences of the electric potentials between two different layers of clay soils, e.g. marls, and with the analogous differences between two different flat-shaped clay minerals. It has shown, that stabilization of the swelling is feasible with anchors.

Besides the mechanical effect which would not cover the swelling pressure economically, there is an electro-chemical effect, reducing the potential differences and stopping thus further deformations. Special measurements in advance to tunnel driving for an early heaving prognosis is possible assuming that corelations with additional parameters of swelling rocks could be done.

Professor Tan:

I had not known about this investigation. As you tell us it must be very important, therefore I am inclined to suggest to open a workshop on swelling rocks, may be in this conference, may be in the near future. It is a too complicated problem to be investigated. It is completely mechanically, mathematically and physically non-linear.

PAPER: Tunnel linings of steel fibre reinforced shotcrete

AUTHOR: B.J. Holmgren

PRESENTED BY: B.J. Holmgren

Question: Professor W. Wittke, West Germany

I have a question to the paper by Dr. Holmgren which was very interesting to me. I was a little concerned about the assumption with regards to the design of a wedge punching through the shotcrete membrane. We have for many cases found, also proven in practice that shotcrete tunnel linings are mainly subjected to normal forces, and only to a minor degree to bending and consequently also to shear. Did you perform tests on fibre reinforced shotcrete subjected to unconfined compression? If so, could you give some results, and compare them with the corresponding properties of shotcrete reinforced by two layers of square wire mesh. My second question is how were the stresses defined in your punching test in the membrane I mean, which stresses did you get in your membrane?

Answer: B.J. Holmgren

Of course it is correct that if you have a tunnel of a quite regular shape, circular for instance, the only possible final failure must be a compression shear failure. I have been thinking of making such tests and it was also suggested by the sponsor, the Swedish Rock Engineering Research Foundation that we should make such tests, but they have not been done yet. We consider the studied case important for our hard rock that we often have in Scandinavia. We never get a regular shape of the tunnel so that is why we are more interested in this case than you are on the European continent, where you have your soft rock.

I can say that the compressive strength of the shotcrete was rather high about 40 or 50 MPa and still more sometimes. The bending tensile strength was up to 12 MPa after 28 days curing.

Question: Professor Mueller

Normally one would think that most important effect of fibre reinforcement is to increase the tensile strength but I do not remember you have said something about the tensile strength.

Answer: B.J. Holmgren

Normally the tensile strength of concrete is not very much increased by the addition of fibres but the bending tensile strength may be increased up to two or three times, so that is the difference between these two loading cases.

Question: Professor Mueller

How have you investigated the effect of corrosion potential: are fibres protected against corrosion?

Answer: B.J. Holmgren

No I have not done that. Everybody asks about it and they say that there is no danger as long as the cracks are small. The fibres are protected in this alkaline media. Of course if you have that heavy cracking that I got in my tests the fibres cannot withstand corrosion for a long time. You must of course repair your lining if it looks like it did on my pictures.

Question: Dr. Benson, Canada

I have had some experience with this fibre reinforced shotcrete actually in situ in some tunnels and I find it performs not as satisfactorily in certain cases; for example when you put it in the arch and you test it afterwards it seems to be more drummy over the central portion, that is when you sound it, it is not adhered to the rock. This is in the case of shales essentially. I have also seen it where there was a significant amount of movement on the walls of the tunnel. Pieces of it actually rip out and fall on the floor. I have seen it strained and over-ridden by shear rupture failures and completely torn apart in tension essentially. This does not happen in mesh reinforced shotcrete. The question I would like to ask you is have you observed these kind of features in situ, never mind the testing?

Answer: B.J. Holmgren

I have not observed this phenomena in situ but I can say that there is really a difference between fibres and fibres and between fibre shotcreters and fibre shotcreters. You can use a fibre which gives nearly no reinforcement effect or it can give you a totally brittle failure and if you have a bad crew you can get of course no effect at all of these fibres. What I have used is the fibres that have been found to give the best results for the moment, and I also used a crew which really could do this kind of job, because the addition of fibre gives an extra dimension to this very difficult work that is called shotcreting. It is very difficult to judge from experience like this because there are so many difficult parameters involved.

Of course the addition of fibres cannot make the concrete unbreakable, of course if it is overloaded it must fail and it can fail in shear, it can fail in bending like if you have mesh and if you have a bad fibre which tears off instead of slip you will get that kind of brittle failure you were talking about.

Question: Professor Mueller

Do you know the model tests on shotcrete effect which Professor Sattler and Rabcewicz made in Graz. He made an excavation, then shotcreted in the model 3 by 3 metres and obtained only shear failure. Could you assume that you make a similar test with fibre shotcrete to compare the normal shotcrete action in soft rock or in medium rock, to compare the fibre shotcrete action with the normal shotcrete action.

Answer: B.J. Holmgren

I would love to because I made these small scale tests with cracked specimens which were sheared in order to go on with tests in the larger scale because I know from earlier discussions with you that there is this big interest for the shearing failures in shotcrete linings. As I said before those tests have not been done yet, but I hope to be able to do such tests too. There is a problem: you begin with something, discover several openings and you can not go through all of them at the same time. You have to take one at a time, but I hope to make such tests in the future, or perhaps someone else will do them. It is a very interesting material because you can never put in stirrups in a shotcrete lining and make it plastic in shear of course; it is impossible, but by using fibres you get both longitudinal and shear reinforcements and of course it could be tested more.

Question: Professor B. Ladanyi

I would like to ask you: (1) How much the steel fibres in shotcrete affect its tensile strength, and (2) Have you investigated the problem of corrosion of such fibres in a shotcrete tunnel lining?

Answer: B.J. Holmgren

(1) The steel fibres affect the pure tensile strength very little.
(2) No.

Question: Dr.R. Miller, Australia

Would the author please comment on the corrosion resistance of the fibre reinforced shotcrete. Due to the greater cross sectional area of the individual strands of a conventional steel mesh, would this type of reinforcement provide greater long term corrosion resistance for a partially failed (i.e. cracked) shotcrete lining?

Answer: B.J. Holmgren

I can't answer to that. My comment is that the steel fibre reinforced shotcrete normally has smaller cracks than the mesh reinforced shotcrete.

PERMANENT UNDERGROUND EXCAVATIONS: POWER STATIONS AND STORAGE CAVERNS

Excavations souterraines permanentes: centrales et les galeries de stockage

Langlebige Grubenbaue: Kraftwerke und unterirdische Speicherhohlräume

T. A. Lang
Chairman

PAPER: Behaviour of rocks around large caverns during excavation

AUTHORS: S. Hibino, M. Motojima and T. Kanagawa

PRESENTED BY: S. Hibino

DISCUSSION

Question: Sten G.A. Bergman, Sweden

Contribution: I am very impressed by the paper presented by Messrs. Hibino, Motojima and Kanagawa. They have been able to use a vast experience on the behaviour of the rock masses around large caverns based on both measurements and calculations. They have analyzed the material commendably and conveyed their conclusions in a form that is understandable to engineers in practice, and that is not very common in this brotherhood of rock mechanics. The conclusion that the stresses in the concrete linings in the ceilings depend on the cavern convergence may seem trivial but it is a valuable practical statement in the respect that it draws the attention to the often forgotten fact, that even heavy reinforced concrete support has only a marginal boundary-stabilizing effect in the large-scale inter-play of forces and deformations which is occurring in the surrounding rock mass when excavating a cavern. The conclusions offered about the wall displacements in a jointed rock correspond very well with a Swedish experience from the Brofjorden oil storage cavern, which is seen in preprint Section D, pp. 301-309, where the cross-sectional area was 30 x 20 m^2. In the Swedish caverns which were built in jointed hard granite, the displacements of the walls measured as convergence turned out to be three to ten times the calculated values, which is in accordance with the Japanese explanation. I suppose that there will be some who think that the authors could have gone deeper into their analysis of such a vast material. Certainly there is enough for writing one or two or three thick books on this material. However, I think it is as I said before highly commendable that you have been short and to the point and expressed your conclusions in the way that the production engineers can easily understand and I think that will make this paper very important for some time in the future.

Question: Mr. T. Lang, U.S.A.

I would be interested to know why a full concrete barrel arch lining was adopted for these structures together with the necessary re-entrant abutments for that arch as a basis design for these underground caverns.

Answer: S. Hibino

After we have excavated the arch part we have in practice to excavate the main parts. During excavation of the main part we have to have safety to work before excavation, so we had better protect the ceiling rock for the workmen to be able to excavate the main part. If it is necessary of course we can replace the concrete lining with shotcrete or another method to keep safety for workmen.

PAPER: Behaviour of rock around the Okuyoshino underground power house

PRESENTED BY: Mr. Yamashita

Question: Mr. Philip Pells, Australia

With reference to the calculation of the displacements: How are those calculations done and how was the rock mass modulus estimated?

Answer: Mr. Yamashita

The calculations of the body were by Dr. Hibino, who will answer your question.

Dr. Hibino:

The calculation was done by using the finite element method considering the known linear deformability of the rock masses. To obtain the deformability we usually carry out the plate loading test and in some cases we carried out triaxial compression test in situ. For example at Shintakase power station the size of the rock was 1 metre by 1 metre by 1 metre. A block was cut out from rock foundation and we applied triaxial compression stress to obtain the modulus of deformation.

Question: A. Bello, Mexico to authors of first paper: Hibino, Motojima and Kanagawa

I want to ask one question taking advantage of the answer by the authors of the first paper. Taking into account that the coincidence between the computed and the measured values is quite close at least in the first paper, I want to know if those calculations were made before or after the measured values were known.

Answer: S. Hibino

Of course, before.

PAPER: Exploration, design and excavation of the powerhouse cavern of Essangento Sallente

AUTHORS: Professor W. Wittke and Sorio

PRESENTED BY: Professor W. Wittke

Question: Dr. B. Chappell, Australia

Dr. Wittke, the diagrams you show were in the main for various horizontal and vertical stress ratios or load ratios on the cross section. We have found quite a few cases where you have the vertical load being the main one and the sides of the tunnel start to slab in. When I say slab in, they form vertical cracks. This is besides the blasting damage, so with the vertical cracking plus the blasting damage you start to develop block systems or you then start to slab out. Besides many other things, I enjoyed your paper immensely, and your presentation, but there are many other aspects which I feel are related to what you had there, the other ones being for instance where you spoke about the size relationships and you start getting jointed systems on your roof. We have found in time that you have mechanisms being developed with these joint systems and there are many cases where we have the invert coming in or rising up and the heading coming down on us. Again related to the stress relationships, the horizontal to vertical. I think it is very nice to have presentations which are clear, but there is still a lot more to know about them.

Horizontal and vertical loads are described in the presentation as setting up the requirement, of concentrating the supporting systems for the vertical and roofing respectively. If the vertical load causes the in situ stress to be at least 75% ultimate strength then vertical cracking and slabbing occurs. If the material stiffness decreases with depth (e.g. low to medium grade metamorphic rock) the invert rises and causes instability in the roof. These are just two mechanisms which besides the stress distribution which you discussed which are very important. There are many more examples of mechanistic deformation coupled with stress redistribution I could give as example. That is there are many mechanisms besides the classical stress distributions you stated.

Answer: Prof. Wittke

I may mention one condition that we usually (I think with only a few exceptions - I know only of one), do locate our underground powerhouses of large size in rock which is not subjected to stresses which exceed the intact rock strength and neither we try to do it in swelling rock. So if you have floor heave usually it is due to swelling or it is due to high horizontal stresses. If you have problems with your vertical walls it can be the result of vertical stresses and joints of course and it can be the result of horizontal stresses. I made this comment if you remember that you do have problems with your walls once your stresses exceed the strength of discontinuities for example, and probably you have to be more precise with your question so that I can answer it properly.

Mr. Lang:

Professor Wittke drew attention to something I found in my experience to be extremely important, that is the scaling-up factor where you go from a small exploratory tunnel where everything appears completely stable to a large excavation, and in my experience most of the problems that develop in construction have come from the lack of appreciation, both by engineers and contractors for the difference between small excavations for exploratory purposes and large construction openings.

C. Tanimoto, Japan:

Concerning modelling for interpretation of rock conditions: This is not a question to a specific person. We Japanese, have a lot of case histories in tunnelling works. Some have already been introduced and some others are going to be. From my personal point of view, as we have quite different rock conditions depending on site-locations, not only in Japan, sufficient discussion cannot be obtained without having the standard rock classification as a common base for consideration. So this is not the question but a request to the participants who are working in geological engineering. Through the representations I would like to have the opinions whether it is possible to establish the common rock classification which can prevent from giving misunderstandings on geologic conditions and specific features for characteristics of rocks.

COAL MINING INCLUDING GROUND CONTROL AND GAS OUTBURSTS

Les houillères, y compris le contrôle en surface et les coups de grisou

Kohlenbergbau, einschliesslich Gebirgsbeherrschung und Gasausbrüchen

T. C. Atchison
Chairman

PAPER: Several basic aspects of the forming of sudden outburst of coal (rock) and gas

AUTHOR: S. Khristianovich & R. Salganik

PRESENTED BY: R. Salganik

1. Question : R. Webster, Australia

Could you confirm my interpretation of your views of the effectiveness of longhole or shorthole in-seam drilling to reduce the possiblity of outbursting. As I understand in your paper you would not recommend that drilling be promoted as a means of solving the problem. Is this correct?

Answer: S. Khristianovich & R. Salganik

In outburst-prone situation characterized by the presence of zone with developed system of oriented-gas-filled cracks mainly parallel to stop face drilling of degassing holes is quite useful if the hole bottom is located approximately at the boundary between zone 1 and zone 2 (Fig.6). In such a case when a system of sufficiently developed oriented cracks occurs the system of the interconnected filtration canals is created through which the gas filling the cracks may escape from the fairly large area around a hole.

However, if the degassing hole extends far into zone 1 only slightly affected by destressing towards the stope, conditions similar to those presented in Fig.7 are produced around the hole and an outburst-prone situation may arise in respect to outbursts into the hole. Such an outburst if it occurs may develop later into a full-scale outburst. Such phenomena have been observed in practice. Therefore, drilling of advanced degassing holes from stope may serve as useful means for reducing the risk of outbursting, but it is only under the control of the hole depth with respect to location of the boundaries of the zone of oriented cracks, the prevention of penetration of the hole deep into non-damaged zone of the seam being provided. For such control step-by-step drilling can be used with determination of gas emission rate at each step. The hole advancing is feasible till gas emission rate increases, it should be stopped if the gas emission rate stops increasing (or starts decreasing). From the viewpoint of the scheme under consideration the latter indicates that the hole starts to sink into zone 1 (Fig.6).

2. Question : R. Webster, Australia

How many personnel are actively involved in outburst research in your country and how long has this level of research been proceeding.

Answer: S. Khristianovich & R. Salganik

In our country, a number of industrial and academic institutes deals with the problem of outbursts according to the state program including USSR Academy of Sciences Institute for Problems in Mechanics.

Statement by **A. Hargraves**, co-chairman

The **modelling** which **was** in the paper but not described is interesting that it **follows** the work of another USSR man.

Famin of USSR used large - 300 mm I think - intensively instrumented triaxially loaded coal briquettes charged with gas and instantaneoulsy unloaded one face and Botham of Canada used small coal cubes charged with gas like those of Khristianovich and Salganik and instantaneously unloaded them on one face. Both were about 25 years ago.

Famin with his test destroyed all windows in his laboratory and Botham filled his laboratory with coal dust.

Hargraves (1963) recorded some degradation of unstressed granular coal instantaneously released from equilibrium with carbon dioxide at 40 Atmospheres.

Now Khristianovich and Salganik have used a differentially triaxially stressed gas laden 50mm coal cube to model the principle of the protective seam which is a valuable addition.

The modelling which has been done and is described in the paper goes beyond what was done before, because it was able to confirm what was shown in one of the slides that the protective seam principle can be employed in models and show that laboratory outbursts can be suppressed too, and I congratulate the authors on reaching this stage.

Would the authors comment on any violence or destruction associated with the cavities produced in their tests.

3. Statement of A. Hargraves, Co-Chairman

Comments by authors.

Professor Hargraves is gratefully acknowledged by the authors for his useful notes. Indeed, in the experiments cited in our paper, the most important feature is the demonstration of the action of mechanism responsible for prevention of outbursting at sufficiently deep predestressing of the outburst-prone seam in the direction parallel to stope face. That explains the principle of action of both protecting seams and the majority of effective measures aimed at prevention of outbursts and associated with local destressing of the same type. As to the experiment themselves, they were carried out in the Donetsk Physical Technical Institute Ukranian Academy of Sciences, by Professor A.D. Alekseev and his colleagues, with our participation in elaboration of the program and discussion of the results (see relevant reference of their publication in the text of the report). No damage or collapse in the laboratory was observed while conducting these experiments (in these experiments only a small hatch in the centre of the relevant pressing opened instead of sudden opening of the whole sample face).

4. Question : Prof. Hargraves

Is 2.5% porosity a normal porosity for U.S.S.R. outbursting coals? In Australia a lower porosity - of the order of 1- is considered to be a characteristic of local outbursting coals.

Answer: Dr. Salganik

With the presence of free gas in the seam, it should occupy a certain volume filling pores and cracks. From this point of view, the porosity value of 2.5% used only for illustrative calculations is quite realistic. All basic results and conclusions are also valid in the case when free gas content in the seam corresponds to porosity value of 1%. However, of importance is to know how one determines the amount of free gas in undisturbed seam and its porosity.
Porosity determined with the help of samples extracted from the seam may give an absolutely different picture from this existing in reality, especially, in case of damaged coal of outburst-prone seams.

5. Question: Prof. Hargraves

In Australia the highest coal seam gas pressure known is about 40 atmospheres at 550m depth. For this reason the virgin seam gas pressure is considered to be about equal to hydrostatic head below the water table. What seam gas pressure measurements in U.S.S.R. support the notion of virgin gas pressure being equal to the pressure of superincumbent strata?

Answer: Dr. Salganik

Conventional methods of gas pressure measurement in coal seams are based on the assumption that seam has significant filtration capability. Stemming from the above, it is supposed that in the case when gas is prevented from escaping from the volume in seam where pressure is to be measured, the pressure measured reaches its true value in undisturbed seam in sufficient time. However, this should not take place in coal-seams at depths below a certain level (different for various ranks of coal). Owing to coal plasticity, coal-seams in their initial state should be fully impermeable at such depths, this is why free gas has been conserved there over long geological periods despite the fact that rocks covering these coal seams are often very permeable. Coal plasticity should lead (during geological time) to conditions where free gas pressure in seam would be equal to rock pressure at the given depth.
Under such conditions, resulting gas pressure in sealed cavity in outburst-prone seam is to be created by such gas which will emanate of restricted zone surrounding the hole (an important characteristic of the zone size is the amount of coal dust got out of the hole while drilling). Depending upon gas content in the seam and mechanical properties of the seam, the resulting pressure in the cavity may vary within a wide range, and in the majority of cases, it is much more lower than rock pressure for the given depth. There are many examples of the fact that at great depths, it was lower than hydrostatic pressure at the given depth (and even less then the limit sorption pressure). Therefore, it is impossible to know the initial value of free gas pressure in coal seam basing on the conventional gas pressure measurements in sealed cavity of coal seams in the conditions indicated.

6. Question : Prof. Hargraves

What is the nature of the analysis revealing that an overwhelming majority of outbursts could not have originated without free gas.

Answer: Dr. Salganik

The main argument lies in the fact that the time needed for outbursts to develop is too short to allow a significant share of sorbed gas to emanate and contribute to outburst formation.

7. Question : Dr. R.D. Lama

You have showed a slide with advance holes where you have indicated that a hole placed in the zone ahead of the abutment is liable to outbursts. I would have expected that if the stress and gas pressure theory is applied, then the hole placed in the abutment zone should be more susceptible to outbursts

Secondly, could you please comment on what percentage of outburts in U.S.S.R. occur on longwall compared to development headings?

Answer: Dr. Salganik

In our opinion, the location of maximum abutment pressure approximately coincides with the boundary between plastic zone and that of oriented cracks when the latter is created. This is the basis of our considerations. As for the second part of the question, it should be noted that the amount of outbursts in development workings is always much greater than at longwall faces, but percentage differs for various regions.

8. Question : John Shepherd, Australia

The authors associate outbursts with coal fracturing and the geometry of these indicates that they are mining or drilling induced. Have you any techniques for suppressing or additionally inducing these fractures (e.g. by cutting a different tunnel shape or by using a different mining method) and can you separate them from pre-existing geological fractures?

Answer: Dr. Salganik

Such methods are well known. They are based on preliminary relieving of outburst-prone seams of rock pressure in the direction parallel or almost parallel to stop face (perpendicular to seam roof and soil). Mechanism of protective action provided by these methods, both regional (associated with mining of protective seams) and local, consists from our standpoint, in creation of favourable conditions for formation of cracks perpendicular to stope face, thereby preventing dangerous splitting of the seam by cracks parallel to stope face. We consider that the main role in creation of outburst-prone situations belongs to systems of sufficiently developed dangerously oriented induced cracks. Formation of sufficiently developed oriented crack systems induced by mining leads to additional anisotropy of both mechanical and electric characteristics of the seam which can be used to detect such crack systems by geophysical methods.

PAPER: Linear models for predicting surface subsidence

AUTHOR: M.D.G. Salamon

PRESENTED BY: M.D.G. Salamon

Question: Dr. D. Maconochie, Australia.

Previous investigations have determined that in a transversely isotropic analysis, the Young's Modulus in the vertical direction had to be greater than the modulus in the horizontal direction in order to obtain better agreement with observed subsidence profiles.

As this arrangement is contrary to laboratory tests of coal measure rocks that are horizontally bedded the presence of vertical joints has been proposed to account for the softening in the horizontal direction. Could you comment on the selection of these elastic parameters and the relevance of laboratory values in modelling subsidence?

Answer: M.D.G. Salamon

First of all about the ratio or relative magnitude of the vertical and horizontal moduli. On page 312 in table 2 of the paper I actually tabulate some calculated moduli values which were obtained using a Monte Carlo calculation technique on the basis of highly stratified rock mass, and these calculations as you will see confirm what you said as regard to laboratory values, that horizontal modulus is larger than the vertical. Also as you will see that if you use these moduli values then you get a transversely isotropic model which doesn't discard the data we have from Britain. So one can do one of two things. Either one is a purist and says "look its obvious in this instance the problem is probably due to the fact that a large part of the rock mass is fractured, or jointed and the joints cause problems so the model is not strictly speaking valid". Now at that stage if you are a scientist you perhaps look for another model. If you are an engineer and you have to give an answer you say "well we go what I call the semi-empirical approach and accept those parameters but no longer claim that the transversely isotropic model actually describes the situation". We just say we do have a model which has been proven to be reasonably accurate. So I don't see any reason why one can't use the transversely isotropic model with parameters which cannot be substantiated in physical grounds, but then we can't claim that we are in fact using that model, we are using now an empirical influence function.

Comment: Dr. A. J. Hargraves.

It is clear from Dr. Salamon's paper and address that, after having had the most unplanned, most sudden and most disastrous subsidence in this century some 25 years ago, South Africa has passed into a situation where subsidence can be planned and where results are no longer orders of magnitude out from predictions.

PAPER: Investigations prior to the introduction of longwall mining

AUTHOR: R. Yeates, J. Enever and B. Hebblewhite

PRESENTED BY: Dr. Yeates.

Question: Prof. Walter Wittke, Germany

In your presentation you have shown a physical model of a longwall mining operation. Did you vary the mechanical properties such as strength of discontinuities e.g. in your tests? Could you please also comment on the costs of such large scale model testing. Also, I would be very interested to hear whether you also have performed numerical calculations and if so how they compared with the results of your model tests. Finally, I would like to ask you if you could give some information on the simulation of post failure behaviour in your calculation?

Answer: Dr. Yeates

The answer to the first question as to whether we modelled various properties in the physical model is that we selected one centrally located typical borehole and made the model represent that borehole. We did not vary the properties from that. We did however vary virgin stress levels and we varied the support setting loads. The second question, the cost of this sort of work; it is not cheap, this model was done in 1980, and I recall a figure of around $Aust50,000 as being the cost of that work. That was for the model test alone excluding the collection of data. The third question about numerical modelling: Yes, we did do some numerical modelling. In the paper there is a section where we in fact looked at the performance of the gate pillars, and this was done using the displacement discontinuity method. In fact there was quite good correlation between the results of this work and the physical model.

Question: Dr. Paul Lu, USA

Can you suggest some other applications of your results of the hydraulic pressure measurement for the face supports besides the support performance evaluation.

Answer: Dr. Yeates

There are two basic reasons for doing the monitoring. The first reason is purely a contractual reason, to monitor the performance of the supports to ensure the supports are behaving as per the supplier's specifications.

The other reason for doing the monitoring is to attempt to verify the results of these studies in terms of the support density requirements for the face, and by monitoring the leg pressures and taking other observations we should be able to better assess the support density requirements of the face.

Question: Dr. Lu

Do you have any measured results, or have you just started these measurements.

Answer: Dr. Yeates

The face is not operating yet. It will be starting shortly, so we have no results as yet.

Question: Mr. W. D. Ortlepp, South Africa to Dr. Yeates.

(1) On what criterion was the gate pillar dimension determined? (In the paper dimensions of 15 to 20 m were given while on the diagram it appeared to be 30 m - in either case it appears somewhat small for a seam thickness of 3m and a depth of 500m plus).

(2) Is it proposed to reinforce or support the gate road? And by what method?

Answer

1. The width of gate pillar selected was 24.5m. Computer modelling indicated a pillar width of 15 to 20m to be adequate and desirable (for later yielding). Current Australian practice is for the pillar width to be between 25 or 35m. The width chosen for Ellalong was a compromise between theory and practice.

2. The gateroads are relatively heavily supported as they are driven (5 to 7 roof bolts every 0.6m to 1.0m, depending on conditions). The tailgate, which is used only for ventilation and second egress will be centre legged. It is not intended to further reinforce the maingate, but this will be subject to review after face start up.

NUCLEAR WASTE DISPOSAL AND THERMAL BEHAVIOUR OF ROCKS

Traitement des déchets nucléaires et comportement thermique des roches

Beseitigung radioaktiven Abfalles und thermisches Verhalten von Felsen

T. A. Lang
Chairman

PAPER: Thermomechanical Room Region Analysis Of Four Potential Nuclear Waste Repository Sites In Salt

AUTHORS: R. A. Wagner, M.C. Loken and H.Y. Tammemagi

PRESENTED BY: R.A. Wagner

DISCUSSION:

Question: Prof. Natau, Germany

Mr. Wagner, you used laboratory tests for comparison of four sites and if I understood right, you used these results for numerical calculations. Our experience is that creeping results from small scale tests may be different from in-situ test results. Did you consider the problem and what is your experience in this field?

Answer: Dr. Wagner

I think that is a very pertinent question. What always comes up in this type of analysis is the validity of these calculations because that is simply all they are; calculations. Without some type of an effort to verify the numerical procedure they are subject to a lot of comment, but we do consider that type of an effort. In this analysis the emphasis of it was more or less a relative comparison of the behaviour of the sites. In the past within the program itself for nuclear waste disposal, particularly with the Company that I am involved with, we have done numerous studies where we have attempted to compare our numerical solutions with closed form analytical solutions, field tests and laboratory test data that becomes available. It seems to be a problem that you never get really rid of, because as you get involved with the problem you are continually updating for instance the constitutive relation for salt creep. The expression itself was updated about two years ago. Where in the past we have always considered a transient form now we have got into a transient steady state form that we feel is more physically correct. Consequently, the comparisons that we have made with the previous transient creep law must be updated again because in trying to modify your approach to better it, you have changed the constitutive relation form. So, in answer to your question, I guess there has been definite effort and there always will be. I think as the nuclear waste repository business gets closer and closer to an actual site selection there will be field experiments at the site and there will be correlations with that and there will be correlations going on between the numerical predictions and your in-situ and laboratory test data. I hope that touched on the point you brought up.

Question: Prof. G. Sonntag

Which temperature shouldn't be reached? What is the critical temperature of the salt?

Answer: Dr. Wagner

In the last six months there has been a big effort to come up with what we call performance constraints that you should try to strive for in numerical analyses. Those performance constraints deal with both the thermal and the thermomechanical behaviour. In the type of analyses we do the temperature constraints are more evident when we do the analyses around what we call the very near field or the cannister region itself, whereas this analysis deals solely with the disposal room. The feeling there is that the only rationale to have for a temperature constraint there is for ventilation purposes and they don't define that necessarily in the performance constraints, they emphasize more the temperature that should be reached in the waste package and within the salt for various brine migration and degradation of the waste package, so in that area as well as in the other phase of analysis, the far field phase of analysis, there are temperature constraints out in the outer regions of the salt and into the non-salt overlying regions that exist. So presently, there are performance constraints that are being established by a group of people familiar with this subject, on which the analyses can be based.

PAPER: Thermal Properties Of Stressed Rocks

AUTHORS: S. Ehara, M. Terada and T. Yanagidani

PRESENTED BY: M. Terada

DISCUSSION

Question: Dr. J. Franklin, Canada

Would the authors please confirm whether any problems were experienced in connection with the resistance strain gauge techniques used in their research. Did the strain gauges perform satisfactorily throughout the extreme range of temperature variations? What were the types of strain gauge and adhesive employed?

Answer: Dr. Terada

I didn't experience trouble but I think it might occur.

Question: Dr. J. Franklin

Could you please amplify a little on the type of strain gauge and the type of adhesive used in these experiments?

Answer: Dr. Terada

The type of strain gauge is cross type of foil gauge and amplifiers are made by ourselves in the laboratory.

PAPER: The Time-Dependent Behaviour Of Reservoir Rocks In Relation To Fluid Production

AUTHORS: T. W. Thompson, K.E. Gray and P.N. Jogi

PRESENTED BY: T. W. Thompson

DISCUSSION

Question: Prof. T. K. Tan, China

Your research is very interestinq, as it has an important economical value; further it is very valuable for basic research. Wehave to deal with a porous rheological body. From your experiments two type of creep:-

1. Volumetric creep with hardening.
2. Creep under deviatoric stresses.

So we have to deal with non-linear processes, which are governed by the current plastic volumetric strains. Have you studied these types of constitutive equations?

Answer: Dr. Thompson

That work is currently on-going. The constitutive relationship which I showed very quickly in the slides and is included in the paper is a strain hardening volumetric equation and we have done some work in trying to determine the parameters. Right now it doesn't look too successful. We have come up with another version. We still believe (and we have good evidence) that the material is strain hardening and that its strain rate depends upon strain but we haven't yet found the right formulation for the strain in the equation. So the short answer is yes, the work is continuing and I have some more information here if you want to talk about it at a later date.

TO ALL PARTICIPANTS IN THIS CONGRESS from Prof. M. Hoshi and Mr. M. Tsujita

We would like to propose that the next ISRM congress should employ a new session on earthquake engineering; since in seismically active zones, the safety of such structures as underground nuclear power stations, storage caverns for low-level nuclear waste disposal or oil storage caverns must be highly secured before their design and construction are to be carried out. Consequently many fundamental problems, either theoretical or experimental must be discussed.

We believe this topic must be attractive to many engineers.

GENERAL REPORTS / RAPPORTS GÉNÉRAUX / GENERALBERICHTE

ROCK DYNAMICS

Dynamique des roches

Felsdynamik

Per Anders Persson
Nitro Nobel AB, Sweden

Roger Holmberg
Swedish Detonic Research Foundation, Stockholm, Sweden

SYNOPSIS

This is a general report on the related subjects of rock breaking by blasting, mechanical cutting, and in situ fracturing. It presents briefly the recent general development within the subject area, and highlights specifically some recent developments. These include a new Swedish method for predicting rock damage due to blasting, both in the near and far region, and a Norwegian and an American technique for predicting rock drillability. The status of water jet assisted rock cutting is briefly reviewed, and a short survey of progress in in situ fracturing concludes the report.

Im Folgenden wird über verschiedene Methoden des Bergabbaus berichtet, wie Sprengung, mechanischen Ausbruch und Zerklüftung in situ. Auf Hintergrund einer kurzen Präsentation der letzten Fortschritte in diesem Bereich wird über einige der letzten Entwickelungen mehr eingehend berichtet. Somit handelt es sich teils um eine neue schwedische Methode Sprängbeschädigungen im Gebirge vorauszusehen, sowie in der unmittelbaren Nähe wie in einem weiteren Umkreis, teils um eine norwegische und amerikanische Technik für die Bestimmung von Bohrbarkeit. Die Verwendung von Wasser-jets beim Bergabbau wird flüchtig berührt und der Bericht schliesst mit einer kurzen Übersicht der Fortschritte in Zerklüftung in situ.

Le present est un rapport sur les différentes méthodes de creusment au rocher, comme tirs des mines, excavation méchanique et fracturation in situ. Sur le fond d'une courte présentation du développement général dans ce domaine quelques nouveautés récentes sont présentées plus en détail. Il s'agit donc d'une méthode suédoise de déterminer à l'avance l'endommagement apporté au roches par une charge explosive dans le voisinage comme à distance, et d'une technique norvégienne et américaine pour déterminer la pénétrabilité de forage. La question de l'emploi de jets d'eau dans les travaux d'excavation et touchée superficielment et le rapport s'achève par une courte présentation des progrès dans le domaine de la fracturation en situ.

ROCK BLASTING DYNAMICS

The use of explosives as a tool to remove rock requires controlled blasting to minimize damage to the remaining rock walls and neighbouring structures. In present-day open pit mining with shothole diameters in the range 250-500 mm (10 to 20") each shothole may contain one or two tons of explosive and a whole blast may involve the detonation of 200-500 tons of explosive. In underground mining, large shothole diameters in the range 150 to 200 mm are increasingly being used. In tunnelling, larger diameter (50-100 mm) and long (3-6 m) shotholes are also common. The development in all these areas towards larger blasts gives great savings in the cost of excavation, but also makes greater demands on methods to avoid damage.

In open pit mining, the stability of the pit slopes and the corresponding slope angles have a tremendous influence on the economy and safety of the operation. Two papers in this conference deal with methods to produce steeper slopes by controlled blasting that leaves the remaining rock strong enough for the increased stresses that result. Even steepening the slope by one degree means saving a considerable amount of waste rock removal in a deep and large open pit mine.

Similar savings can be realized by introducing controlled blasting in underground mining and tunnelling where damage to nearby tunnels and building structures representing large sums of money must be avoided. Controlling blasting damage requires an understanding of stress waves in rock.

Stress waves in rock

The detonation of an explosive charge in a borehole in rock gives rise to stress waves in the surrounding rock. For a drillhole fully charged with a strong explosive, the wave pressure exceeds the strength of rock and we get a very complicated shear deformation pattern, which ultimately leads to crushing of the rock around the borehole. Further out, the conditions are favourable for the formation of radial cracks.

As the wave moves radially out from the borehole, the amplitude (pressure) decreases and the wave becomes purely elastic. As a result of the interaction with the free surface, the different types of waves that we know from seismology develop, the p-wave, the s-wave, and the Rayleigh-wave. In this area, the structure of joints or fissures in the rock begin to influence both the wave propagation (wave velocity and stress amplitude) and the degree of fracturing more than in the region close to the drillhole.

When we are discussing wave strength in this far-field region, it becomes useful to use the peak particle velocity as measure. Figure 1 shows the approximate decrease of the peak particle velocity with distance away from a 15 m long charge in a 250 mm diameter borehole. At distances compared to which the charge dimension is small the peak particle velocity follows approximately the relation

$$u = K\frac{Q^{\alpha}}{R^{\beta}} \quad (1)$$

where Q is the charge weight, R the distance and K, α and β are constants (for hard bedrock K = 700, α = 0.7 and β = 1.5 if Q in kg, R in m and u in mm/sec). Figure 2 shows the considerable scatter due to rock structure-related differences in wave transmission of rock masses.

The stress waves move with different velocities, for hard bedrock typically

$$C_p \approx 5000 \text{ m/sec}$$
$$C_s \approx 3500 \text{ m/sec}$$
$$C_R \approx 2500 \text{ m/sec}$$

and somewhat lower for softer, more fissured rock masses.

Stress wave damage in rock

Depending upon the wave type, we can get an estimate of the stress σ or strain ε in the rock if we consider the motion as a simple harmonic oscillation (extension or bending)

$$\varepsilon = \frac{\sigma}{E} \approx \frac{u}{c} \quad (2)$$

Granite may be expected to fail in dynamic tension at a stress of perhaps 30 MPa, corresponding to a strain of ≈ 1 % , that is a particle velocity between 1000 and 2000 mm/sec depending on the wave type. But normal fissured rock will undoubtedly show tensile damage in the joints at lower stress levels(say around 700 mm/sec). For soft, sedimentary rocks, with relatively weak joints, damage may occur at a particle velocity of 400 mm/sec or less. Although the orientation of the fissures in relation to the wave propagation direction may also be important, the table below may be useful as a first guide to the range of the critical vibration velocity for rock damage. The typical

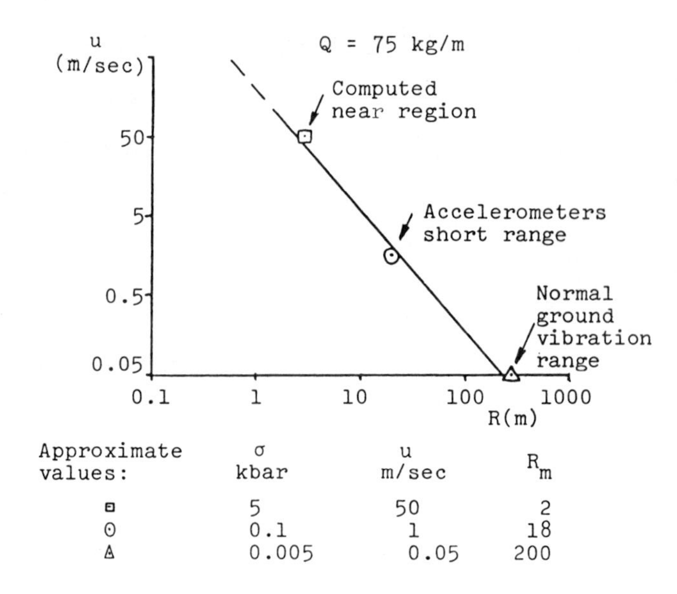

Approximate values:	σ kbar	u m/sec	R_m
⊡	5	50	2
⊙	0.1	1	18
Δ	0.005	0.05	200

Figure 1. Stress wave particle velocities from 250 mm charge 15 m long.

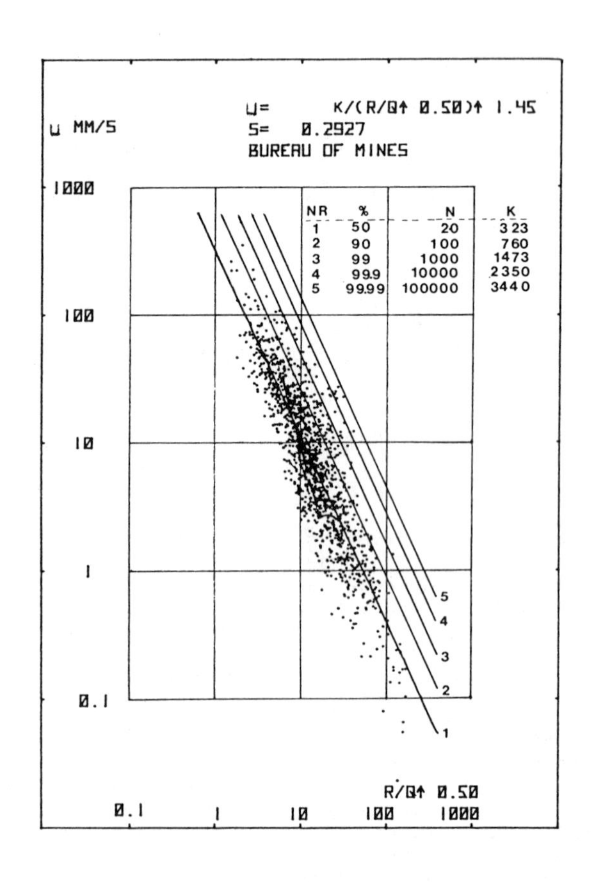

Figure 2. Peak particle velocity u as a function of $R/Q^{0.5}$ for U.S. Bureau of Mines measured data at different confidence levels and different number of rounds. Lundborg et al, 1979.

damage in fissured rock is an irreversible separation of the two fissure surfaces from each other, resulting in a decrease in shear strength of the fissure. This is accompanied by a slight swelling of the rock mass affected.

Table 1. Critical vibration velocity range for damage in different rock mass types.

rock/joint class	critical vibration velocity mm/sec
hard rock strong joints	$\geq$ 1000
medium hard rock no weak joints	800-700
soft rock weak joints	$\leq$ 400

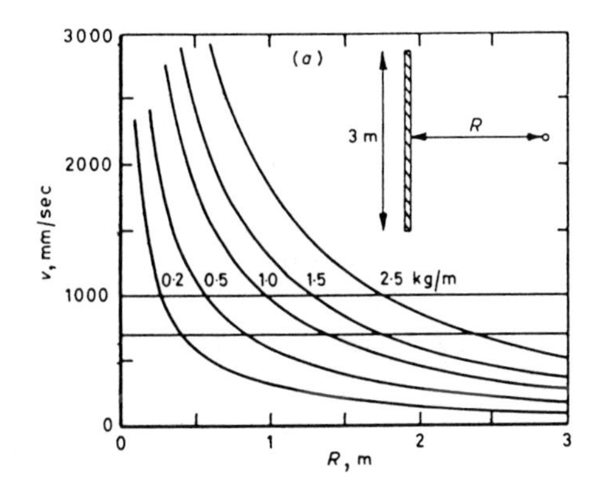

a)

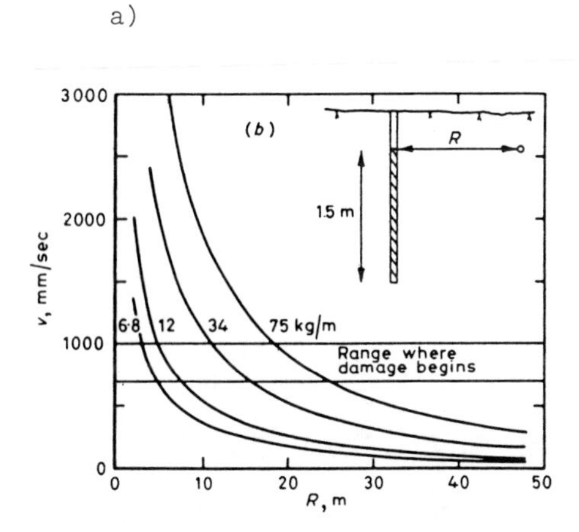

b)

Figure 3. Estimated vibration velocity in medium hard rock as a function of distance for different linear charge densities. Linear charge densities are given in kg ANFO per meter borehole; a) charge length 3 m typical range for tunnel blasting, b) charge length 15 m, typical range for open pit blasting.

Predicting vibration velocity near an extended charge

Based upon the relation (1) the authors have calculated the peak vibration velocity at different distances from explosive charges in medium hard rock at different charge lengths and diameters (Holmberg and Persson, 1978). Figures 3a and b show the results. The diagrams are very useful for blast pattern design when one wishes to keep damage to the remaining rock at a minimum. They allow, for example, the damage at a given point in the remaining rock close to the perimeter of a tunnel or near an open pit wall to be estimated, whether caused by the pre-split holes or by holes within the blast further away from the perimeter.

Design of perimeter blasts

The results given in figure 3 are in reasonable agreement with observed velocities over a large range of distances and hole diameters (Persson, Holmberg & Persson 1977). They can be used to design perimeter blasts to limit the extension of rock damage behind the contour row of holes. To do this it is not enough to keep the linear charge density in the contour holes low. The next few rows of holes away from the contour holes must also be given a reduced linear charge density. Otherwise, their damage zone will extend well past that of the contour holes. Details of these charge calculations have been published elsewhere (Holmberg and Persson 1979).

Vibration frequency spectra and damage criteria

Strictly, the use of a critical vibration velocity as a rock damage criterion is valid in an intermediate frequency range. The deciding factor is the vibration wave length in relation to the size of the vibrating rock mass. The low frequency range is where the vibrating rock mass in its entirety experiences a uniform acceleration because the vibration wave length is much larger than the size of the rock mass. There, the displacement is the important criterion. In the high frequency range, the acceleration may be the critical factor.

When in doubt, the ground vibration generated by a blast must be recorded at a site where rock or building damage may be suspected, and the peak values of vibration acceleration, velocity, and displacement must be determined together with the vibration frequency at which they occur. Then the size of the rock mass involved is determined. By comparing it with the wavelength of the vibration peak, the composite frequency-dependent critical vibration level envelope of figure 4 can be used to find the maximum allowed vibration level. This type of analysis is best left to an experienced consultant, and should also include an engineering geological survey of the direction, strength, and density of the major foliation, joints, and fissures occurring in the rock mass.

Presplitting and smooth blasting

To leave the surface of the remaining rock wall as free from damage as possible the normal methods are smooth blasting or presplitting. In both these techniques the linear charge density in the contour drillholes is made very low compared to that in ordinary drillholes. This gives a low initial drillhole pressure. In

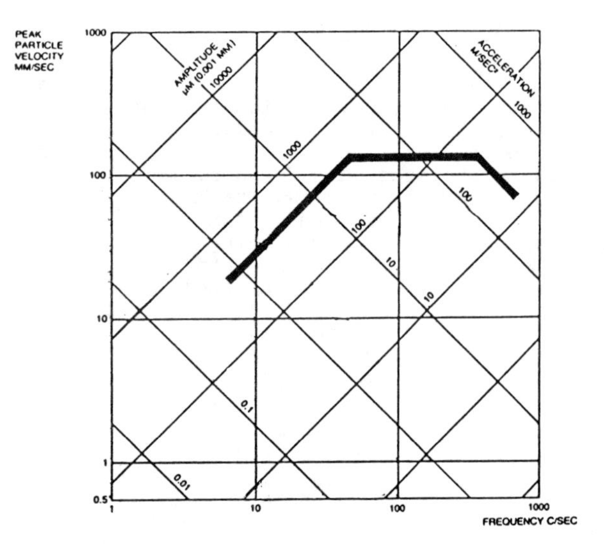

Figure 4. Example of composite frequency-dependent critical vibration velocity envelope for building damage.

this way the main result of the contour blast, if the charges in adjacent holes are allowed to cooperate, is a crack that runs from drillhole to drillhole. Then, the damage to the remaining rock is limited within a narrow zone close to the contour.

In smooth blasting, the contour charges are initiated last in the round. In presplitting, they are initiated before the rest of the charges, often in a separate round. Presplitting therefore requires a closer spacing of contour holes, about 50-75% of that for smooth blasting, and becomes more expensive than smooth blasting (figure 5). The minimum required linear charge density ℓ is the same for both techniques:

$$\ell = ad^2 \tag{3}$$

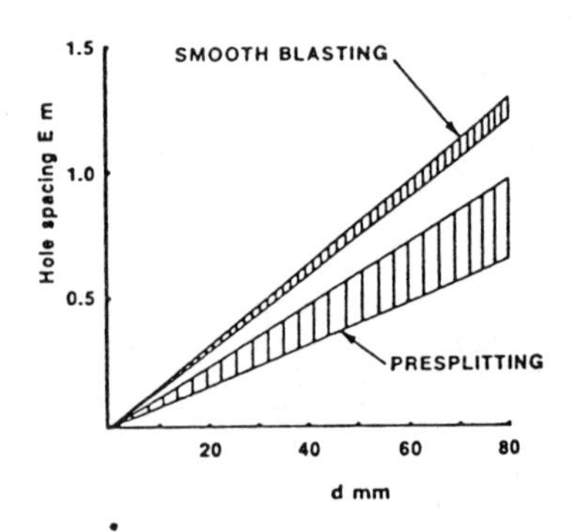

Figure 5. Recommended ranges of hole spacing as a function of hole diameter for smooth blasting and presplitting.
Smooth blasting: $E = k_s d$, where k_s = 15-16.
Presplitting: $E = k_p d$, where k_p = 8-12.

where a = 90 kg/m³ and ℓ will be in kg ANFO/m if d is expressed in m.

Equation (3) is equivalent to an initial borehole pressure of about 100 MPa (1000 bar). At this congress, Bauer and Calder present data indicating that for soft rock materials, the initial borehole pressure can be kept lower by still further reducing the linear charge density.

Crack guiding

An interesting development which is illustrated in one paper by Bjarnholt, Holmberg, and Ouchterlony 1982 is the technique of crack guiding in smooth blasting and presplitting using linear shaped charges. By using two linear metal jets produced by the shaped charge, two diametrically opposite grooves are formed in the rock along the drillhole wall. These serve as crack initiation guides, and by orienting the shaped charges in the drillholes, the cracks from adjacent boreholes can be directed towards each other.

Similar experiments were made previously (Thompson et al, 1979) with grooves cut by a tungsten carbide tool. Due to the high cost of grooving these experiments never reached large scale practical application.

At this conference, the paper by Pechalat and Lefin describes the use of water jets to form the grooves for crack guiding.

Although the economic feasibility of large scale use of crack guiding to improve the results of smooth blasting and presplitting, we may hope that the present developments will bring the price of grooving down to a level where crack guiding can be tried out in real life construction or mining blasting.

ROCK CUTTING

In this report we use the term rock cutting to describe processes in which rock is cut with mechanical means other than blasting. It thus includes drilling of blastholes, boring of raises and tunnels, and rock cutting with water jets. In situ fracturing will be dealt with under a separate heading.

Some years ago, great expectations were attached to research into unconventional methods of rock cutting, such as by projectile impact, laser beams, electron beams, and high pressure water jets. Of these, only water jet cutting has so far reached the stage of large scale practical application. In soft rock water jets are used to cut rock by themselves. In somewhat harder but still mostly sedimentary rock they are used mainly to assist the work of a steel or tungsten carbide cutting tool.

Great strides have been taken in recent years towards accurate methods to predict the rate of drilling or boring in different kinds of rock materials. We will review some work in this area that appears to have reached a stage of practical applicability.

Two papers in this session deal with water jet assisted rock cutting. They give in themselves a good idea of the state of the art in this field.

Mechanical drilling and boring

With the development in the 1950-s of tungsten carbide tipped steel drill rods came the dominance of mechanical percussive drilling over all other methods of making shotholes for rock blasting. Percussive drilling and blasting using conventional chemical explosives still dominates among today's rock excavation methods. The drilling machines have developed from the early simple hand-held pneumatic hammer machines drilling perhaps 0.2 m/min to the present-day self-propelled multi-boom drilling jumbo using hydraulic hammer machines with a hard-rock drilling rate of the order of 2 m/min each.

The bit may be either a chisel bit having 1, 3, or 4 cutting wedges or stud (button) bits having several tungsten carbide buttons.

Other means of mechanical rock cutting are, however, now gradually being introduced as alternatives to percussive drilling, and even to replace the conventional drilling and blasting method altogether. In relatively soft rocks, rotary spiral drilling with drill rods having fixed steel or tungsten carbide cutting edges like on a spiral drill is used more and more for small to medium diameter shotholes in the diameter range 30 to 150 mm. Larger shotholes of diameters from 150 mm up are produced by rotary crushing drilling using tricone roller bits studded with tungsten carbide buttons. These cut by crushing the rock into chips as they are pressed successively into the rock surface at the hole bottom by the force of the rotating drill rod. Similar studded roller bits are used to produce raise bore holes even in hard crystalline rock. The rollers are held by bearings on a rotating shield which is pulled and rotated by a drill rod which passes through a smaller pilot hole up from an underground tunnel. Such raise bore heads have diameters from 0.5 to 3 m diameter.

A major part of the world's long straight tunnels in sedimentary rock are excavated by full diameter boring using tunnel boring machines. For soft sedimentary rock materials these are usually of the ripping head type, having a rotating spiked ripping head, movable on a hydraulic boom, like a road header machine. For the intermediate and high hardness range, the disc cutter type machines dominate. These have a rotating shield of a diameter approaching that of the tunnel. On the shield are fastened heavy steel discs with bearings that allow the discs to roll heavily on the tunnel face as the rotating shield is hydraulically pressed against the tunnel face. The cutter discs are spaced over the shield so that the circular rolling grooves they leave on the rock face are evenly spaced 6-7 cm apart. With an average cutter load of 10-30 tons per cutter, hand-sized pieces of rock are chipped off on either side of each cutter, and the penetration per revolution may be of the order of 0.5-10 cm. For use in hard rock the cutter discs may have tungsten carbide tipped button bits around its cutting edge. Where the remaining rock is stable and dry, boring rates including necessary stops for retooling and repairs may be as high as 30 m per shift in limestone of compressive strength 200 MPa (2 kbar).

The fundamental mechanism of rock fragmentation under the action of a hard tool being pressed into a rock surface is similar, whether the tool is a blunt wedge, a hemispherical button or the edge of a cutting disc. Therefore, rock drillability and boreability can be described by essentially the same tests of rock material fragmentation and wear properties. However, for disc cutting tunnelboring, the natural fissures and joints in the rock mass have an additional influence on the boreability which must be taken into account.

Where nothing else is known about the rock to be drilled or bored, the unconfined compressive strength is a useful figure. In the first approximation, a high compressive strength means slow drilling or slow boring, high rates of tool wear, and expensive excavation. For tunnel boring there is a natural borderline between sedimentary rock which have generally low compressive strength and are easily boreable, and magmatic, crystalline rocks, which have high compressive strength heralding potentially high boring costs.

Drillability and boreability of rock materials

The terms drillability and boreability are used, often somewhat loosely, to describe the degree of ease and economy with which a rock mass lends itself to be drilled or bored by a given machine. The terms are thus functions mainly of the rock mass and its strength and structure, but they also to a certain degree depend on the economy and characteristics of the machine chosen, as well as the economy and characteristics of other available machines and methods. Drillability refers to shothole drilling in the diameter range 25 to 500 mm, boreability refers to raise and tunnelboring in the diameter range 1 to 12 m. In the following will be related the techniques for predicting drillability and boreability developed at the Norway Institute of Technology, Trondheim, Norway by Selmer-Olsen and Blindheim (1970) and Blindheim (1979), and also the techniques used by some manufacturers of tunnelboring machines.

Predicting drillability

Drillability is composed of the following factors:

- drilling rate (cm/min)
- bit wear (characterized by the length of borehole produced in a given rock between cutter grindings)
- bit life (the total length of hole drilled before the bit has to be scarapped)

According to Blindheim, the drilling rate can be estimated from a crushing test, the "Swedish Brittleness Test" in which an aggregate of the rock material to be tested is crushed in a cylinder by a falling weight and a wear test, the Sivers test, in which the penetration into a rock specimen of a small rotary tungsten carbide chisel drill is measured under standardized conditions. From these two simple tests one can determine the drilling rate index DRI which is a linear function of the S-value from

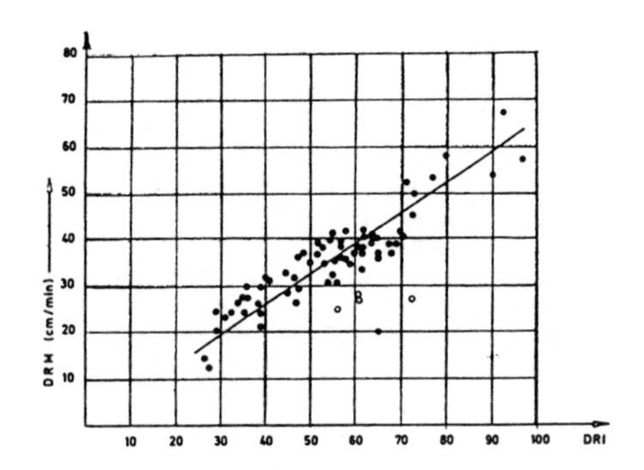

Figure 6. Correlation diagram for drilling rate index (DRI) and measured drilling rate in field tests (DRM) using light drilling equipment and Ø 33 mm chisel bits. Unfilled circles represent uncertain data.

the brittleness test, with the J-value from the wear test as a parameter. There is a good linear relationship between the DRI values for different rock materials and the measured drilling rate DRM using pneumatic percussion drills with 33 mm diameter tungsten carbide chisel bits. Figure 6 shows the scatter of the measured data around the linear relationship.

For larger borehole diameters than 33 mm, and heavier machines DRM is lower than shown in figure 6 at a given DRI. Reference must then be made to testing results in a rock with known DRI. The new generation of hydraulic percussion drilling machines give greatly increased drilling rates because they have a heavier bit load and a higher percussion frequency, by using stud bits instead of chisel bits. Figure 7 shows correlation curves between DRI and practical penetration rates for different drilling equipment. The diagrams are based on practical experience from a large number of construction sites in Norway.

The bit wear is the result of abrasion of the tungsten carbide bit by the hardest rock particles. The rate of bit wear increases with the content of quartzite or other equally hard minerals in the rock but also with the drilling rate. The abrasiveness of different rocks on a given type of tungsten carbide bit can be determined by an abrasion test, and it is found to depend not only on the quartz content but also on the other constituents of the rock.

Round quartz grains are often found to be less abrasive than sharp edged ones. Generally, sedimentary rocks are less abrasive than igneous and metamorphic rocks at a given quartz content.

The bit wear can be expressed in terms of the bit wear index BWI which is a linear function of the abrasion value with the drilling rate index DRI as a parameter. There is a good correlation between actual measured bit wear BWM and the drilling rate index DRI. The measured bit wear is expressed in the unit μm/m drilled length. For convenience bit wear is expressed as the length of borehole in m that can be

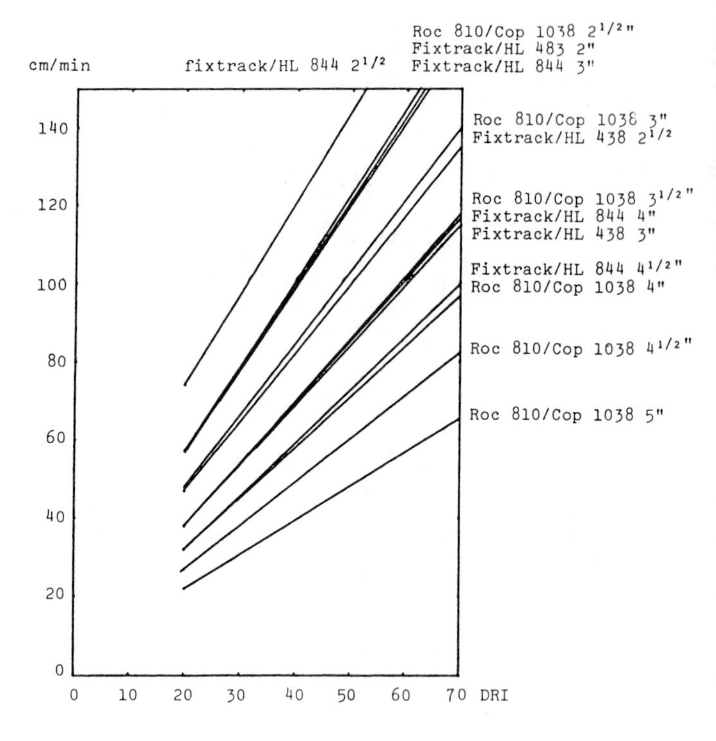

Figure 7. Estimated drilling rates for different drilling equipment as a function of rock drilling rate index (DRI) pneumatic drills at air pressure 60 MPa (6 bar) and drill rod length 3.05 m, hydraulic drills and drill rod length HL 844, HL 438: 3.05 m; COP 1038: 3.66 m. After Johannessen (1981).

produced in a given rock by a given equipment between bit grindings rather than the more scientific unit μm/m giving the thickness of bit worn off in drilling one meter of borehole.

The bit life is simply the total length of borehole that can be drilled by a given bit in a given machine configuration and at a given air or hydraulic pressure while drilling in a given rock mass. It is related to the bit wear in an obvious way mathematically, but the practical number of regrindings or the decision on what criterion to discard a worn-out bit also influence the actual bit life value.

Predicting boreability in tunnelboring

Because boring with disc cutter machines is the most frequently used technique the following section deals only with disc cutter boring.

Boreability is expressed in terms of the following factors:

- net penetration rate m/hour of true boring time
- cost of cutters, $/piece (due to bit and bearing wear)
- total excavation cost $/m tunnel length

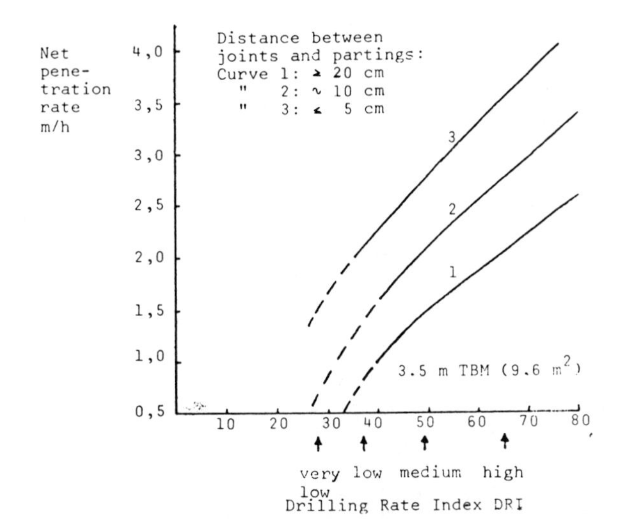

Figure 8. Net penetration as a function of the drilling rate index and the distance between joints and partings.

The net penetration rate* is the length of tunnel that can be bored in a given rock mass by a given tunnelboring machine per hour of actual boring time using normal cutter load and normal cutter head rpm.

$$P = 60\ N\frac{P_e}{100}$$

In tunnelboring much more than in shothole drilling the presence of weakness planes or joints in the rock mass influences the net penetration rate, especially when the distance between weakness planes (joint spacing) is of the same order as that between the cutter grooves. The crushability of the rock material itself can be accounted for in boring by the same technique as in drilling, using the drilling rate index DRI. Figure 8 shows the net penetration rate as a function of DRI with the distance between weakness planes as a parameter.

In the same way, the cutter cost, expressed in \$ per m³ of solid rock bored can be estimated by the bit wear index BWI derived as described above for drilling, again with joint spacing as a parameter.

With the great capital cost involved in aquiring a tunnelboring machine, the total excavation cost is greatly influenced by the ratio of productive to unproductive time during an average shift of working. The crucial figure is the number of new tunnelmeters produced in a shift, and we would like to recommend newcomers to the field to follow the very clear calculation of that figure given by an experienced contractor, Prader (1977).

*Note the distinct difference between the two terms penetration P_e which is expressed in mm/cutterhead revolution, and net penetration rate, P which is expressed in m/hour. They are related through N, the number of cutterhead revolutions per minute.

Rock strength testing for tunnel boreability predictions

It has already been mentioned that the first question to ask when faced with a new potential machine boring project is: What is the compressive strength of the different rock types to be encountered? The first answer comes from the early geological surveys. Additional information can be found from different kinds of button penetration tests or punch tests. (Handewith 1969 and Lindqvist 1982), or fracture toughness tests (Robbins 1977). Common for all these is that they can be made on core samples from exploration core drilling made during the site investigation period.

Handewith defines the penetration index δ_i as the ratio of force to permanent deformation as a button bit is pressed into a rock core sample mounted to that it does not crack apart. Figure 9 shows typical punch test curves, where repeated loading and unloading to gradually greater penetration values have been made. Table 2 shows a comparison of predicted and actual penetration rates for three different rock materials.

Table 2. Penetration rates from the Lawrence penetration index.

rock material	compressive strength PSI	penetration rate predicted FPH	actual FPH
dolomitic limestone	31,960	5.8	6.1
quartzite	36,170	2.1	2.5
argillite siliceous scale	19,200	6.1	6.4

Rock cutting with water jets

In open pit mining industry hydraulic monitoring has been used for a long time to wash away overburden alluvium deposits. Hydraulic mining operations have been applied to underground coal mines by Kaiser Resources in western Canada. The Kaiser Resources used continous miners in 1974, producing jets at pressures up to 13 MPa with a flow rate of 3800 l/m.

Discussions concerning advantages of underground hydraulic mining of oil sand has been carried out by Gates et al (1978). They outline the advantages and the problems likely to be encountered with hydraulic mining. Today however the economic development in oil business is low which is a disadvantage for this type of forthcoming projects.

Mathews (1980) report a borehole mining technique tested by U.S. Bureau of Mines. Access holes were drilled vertically from the surface into the coal strata and a horizontally rotating water jet nozzle was moved upwards as the coal was cut. The coal/water slurry was pumped up to the surface by a pump, see figure 10.

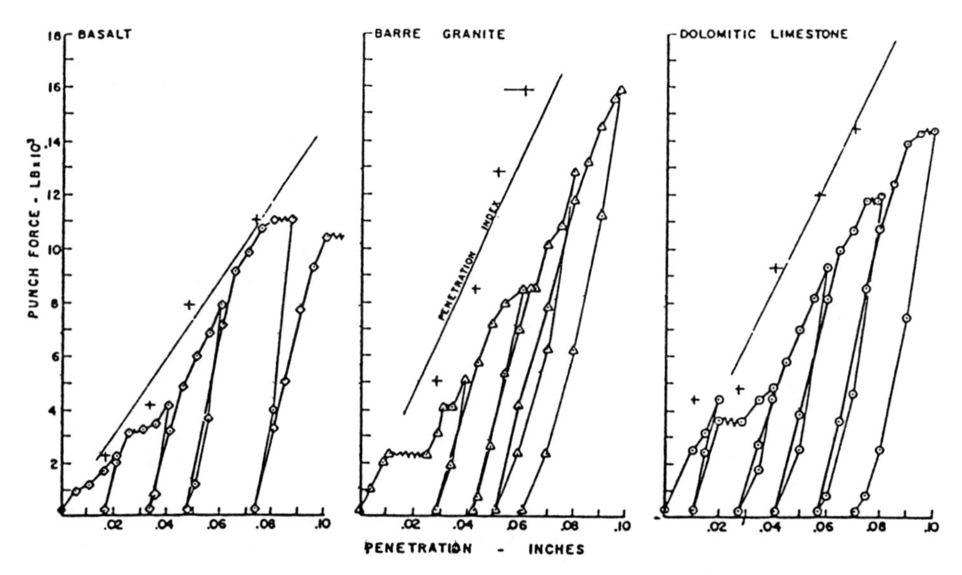

Figure 9. - Typical Punch Test Curves - The penetration index (δ_i) increases with the boring difficulty. The compressive strength (σ_c) and slope are only roughly related to the boring difficulty. The slope is calculated from the first loop to reach 5,000 pounds force without a noticeable spall. After Handewith (1969)

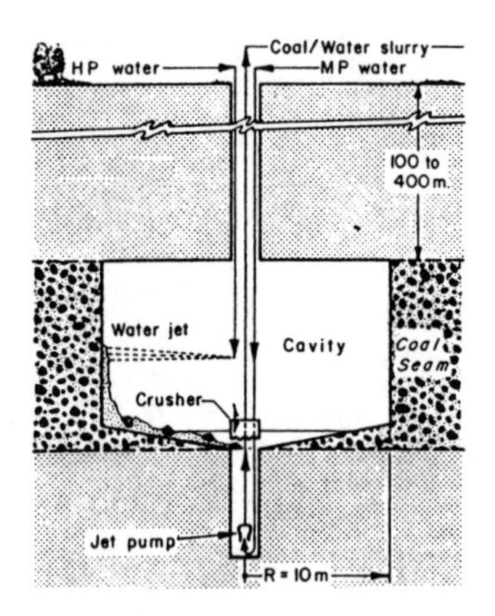

Figure 10. Cavity mining in coal strata. Mathews, 1980.

Much higher stagnation pressures are needed for cutting in hard rock. Rehbinder (1977) carried out experiments in high strength rocks with different grain sizes and permeabilities. Jet pressures in the range of 150-460 MPa were used. Rehbinder points out that the slot cutting rate is highly affected by the ratio of jet pressure and threshold pressure of the rock. If the stagnation pressure is greater than the threshold pressure of the rock, the grains are spalled at a rate which is equal to the mean rate at which the water passes a grain. Layer of grains will be cut as long as the stagnation pressure at the bottom of the slot is high enough. The conclusion is that erosion resistance of the rock is closely connected to its permeability.

Several authors have reported successful slotting in hard and abrasive rock but there exist problems which limit the use of water jet cutting alone for excavation.

a) In hard rock, two parallell slots can easy be cut close to each other but still the rock between the slots will remain undisturbed. This means that the water jet angle of attack must be changed continuously in order to fragment the rock.

b) The very high water jet pressures needed for hard rock requires automatic guidance which can be a problem due to the rough rock surface which is achieved in excavation in hard jointed rock.

c) The power requirement will be very large.

It definitely looks more promising to use a combination of high pressure water jets and mechanical tools for rock excavation. This approach has an obvious attraction as a supplement to mechanical excavation by drilling machines, tunnel boring machines, rippers and milling roadheaders. The water jets are here used to create additional free surfaces for the mechanical tools. Summers et al(1977) report water jet drilling in sedimentary and crystalline rock.Red granite was harder to penetrate due to its lower permeability compared with the Berea sandstone. At this congress Pechalat and Lefin present a paper about drilling equipment for mines tested by Cerchar. One interesting application concerning selective mining was mentioned. Notches along the boreholes for fracture control could be introduced by water jet. Notched holes together with a light linear charge can effectively separate the valuable ore from the remaining waste rock.

The utilization of water jet cutting heads instead of shearer drums on longwall mining machines for coal has been commented by Barker and Summers (1977).High pressure water jets are used for cutting slots at the bottom, back and top of the coal seam web. The coal plow can thereafter successfully advance into the slot and wedge coal cantilevers isolated by the water jets. Field trials with the developed Hydrominer were reported to be successful with no dust problem and an acceptable excavation rate.

It is interesting to see that experiments are carried out in order to break the present technical or rather economical limits for mechanical excavation by ripping and milling roadheaders. Today the technical limit is reached when the uniaxial compressive strength is in the interval of 120-150 MPa. The economical limit is around 70 MPa.

By using water jet together with the picks, advantages are reached for the excavation method. A high pressure water jet increases the penetration rate, lowers the tool cost by cooling the picks, contributes to better environment by reducing dust problems and lowers the ignition risk for methane/air mixtures. Fowell and Tecen report at this congress experiments carried out with water jet assisted drag tool cutting in sandstone and limestone with uniaxial compressive strengths from 40-150 MPa. Their experiments show how important it is to consider the water jet penetration in rocks with different strengths. If it is deeper than the cut taken by the pick the sides of the tool comes into contact with the rock (instead of the tip) which suppresses cracking ahead of the tool and lowers the cutting performance.

In 1976 Hood reported laboratory experiments where water jets were directed immediately ahead of the drag bit during the cutting operation. The bit force was reduced to 1/3 of what was needed without water jet assistance. As a result of the laboratory experiments some operating underground rock cutting machines were equipped with water jets to assist the cutting. Hood 1978, reported successful results indicating even better performance than the previous laboratory experiments.

Baumann et al (1980) mention a very promising method for efficient rock excavation. By utilizing two high pressure water jets, one on each side of a cutting disc (see figure 11) it is possible to lower the thrust per cutter. The water jets cut grooves and the rib left in between is removed by the shearing action caused by the disc.

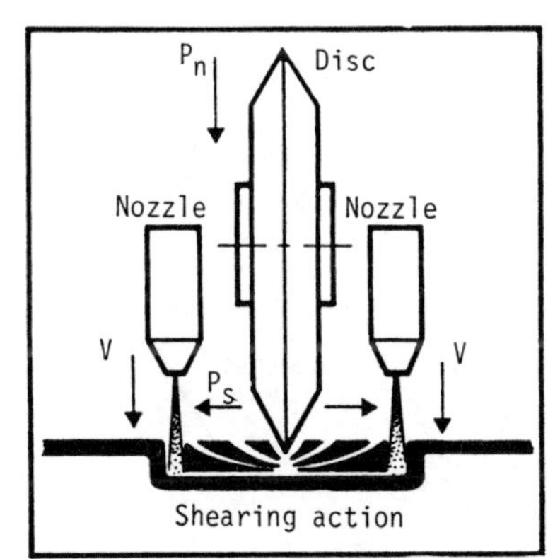

Figure 11. Dual high pressure water jets and a cutting disc. After Baumann et al, 1980.

The authors claims it was possible to reduce the required feed by more than 50 percent which, in turn, would entail a proportional decrease in machinery weight with beneficial effects on the tunnel machine flexibility.

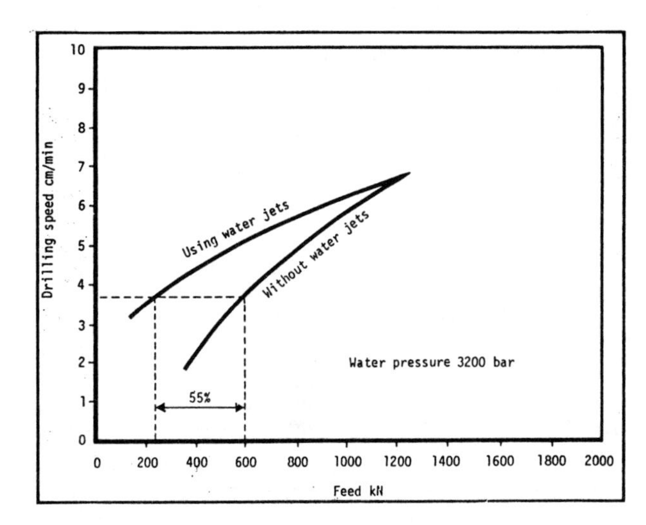

Figure 12. Drilling speed with and without assistance of water jets. Baumann et al, 1980.

The articles published about mechanical excavation by assistance of high pressure water jets clearly show that water jets are a versatile tool for improvement of the cutting performance. However the energy requirements per excavated volume of rock still seems to be much higher for the water jet production than for the mechanical excavation. Future improvement of energy utilization can probably be a reality through beneficial use of water additives for increased flow speed and better confinement of the water jet for the actual stand off distance.

IN SITU FRAGMENTATION

At the time of writing this report there is a temporary overproduction of oil due to the world recession. Over a longer time perspective, the demand for oil will undoubtedly be steadily increasing due to the growing energy demands. Oil shale is regarded as the major future source for extraction of oil and much work has been done during the last few years to develop methods for in situ recovery of oil shale.

In situ recovery methods are either true in situ or modified in situ retorting. In the true in situ method, no shale is mined before the retorting starts. The shale is fractured in situ to develop the required permeability using hydraulic or explosive fracturing techniques. After the necessary permeability is achieved the oil is extracted by pyrolysis and pumped up to the surface.

Hydraulic fracturing introduces single large fractures in the shale and these fractures either follow the bedding planes or are oriented parallel to the highest principal stress. No really successful experiments have been reported where enough permeability was achieved by use of only hydraulic fracturing.

In the modified in situ fragmentation techniques. about 20% of the oil shale is mined out by conventional mining methods and brought up to the surface to give room for the expansion of the 80% remaining rock which is rubblized in tall chimney stack retorts underground.

True in-situ fragmentation

The Bureau of Mines has run a number of laboratory experiments to determine the feasibility of using liquid explosives to fracture the oil shale. This is described in detail by Miller, et al, (1974). Experiments were done in an attempt to study if a dry porous sandstone would absorb sufficient amount of a NG (nitroglycerin) -EGDN (ethylene glycol dinitrate) mixture to yield a detonable charge and an high detonation velocity. Tests with a dry Berea sandstone having a density of 2.2 g/cm³ showed that this rock could absorb about 12% of its own volume of liquid. With a booster of 73 grams of NG-EGDN they succeded in initiating a sample (51x51x152 mm) that had absorbed 8.2% NG-EGDN. The detonation velocity was measured as 4,700 m/s. Fifteen percent gelled NG (ρ = 1.37 g/cm³) absorbed in sodium chloride with a diameter of 287 mm had a detonation velocity of 1,550 m/s. The Bureau of Mines also showed that NG-EGDN could be detonated with a detonation velocity of 7,500 m/s if it was confined in a 1.6 mm crack. NG-EGDN poured in a sand filled crack detonated with a detonation velocity of 2,100 m/s.

About 5.5 liters of desensitized NG was poured out into a presplit crack in limestone with an average width of 3 mm. The detonation extended the fracture about 40 m and the crack width was increased to about 70 mm when the limestone was displaced horizontally towards the vertical face in the limestone quarry.

Tests were run in oil shale (Rock Spring and Green River Sites) which indicated that NG will detonate and the explosion will propagate in water and sand filled natural and hydraulic fractures. The oil shale was fragmented to such an extent that retorting was indeed possible. However, the difficulties in controlling the NG flow pattern was such that its use is not recommended. Pelletized TNT was shot in wells and an extensive fracturing out to a radius of 15 m was disclosed by seismic methods. Air flow measurements between wells indicated the presence of fractures but the evaluation techniques did not indicate the extent of rock fragmentation.

Oil shale fracturing tests were investigated by the Bureau of Mines (Cambell et al, 1970) in the Green River Formation near Rock Springs, Wyoming. Five wells were drilled to a depth of 15-27 m and the wells were placed in a quadratic pattern with a hole distance of 7.6 m. One well was placed in the center of the quadrangle. At this test site the Bureau of Mines tested electrolinking, hydraulic fracturing without and with sand propping, and explosive fracturing. Liquid NG was used for the explosive fracturing.

Electrolinking and hydraulic fracturing without sand propping were relatively ineffective. Hydraulic fracturing with sand propping created horizontal fractures with desirable flow capacity. Hydraulic fracturing with sand propping took place in two wells at a depth of 22-24 m and 24-26 m, respectively. Almost 300 liters of a desensitized NG was poured into each well and was allowed to migrate into the hydrofractured rock. After the detonation the maximum surface elevation had increased about 50 mm directly above the wells. To evaluate the fracturing, airflows between selected wells were measured. The authors indicate that there was a significant increase in fracture permeability when an adequate NG shot was detonated (i.e. 300 liters). A 100 liter NG shot resulted in a lower permeability. The 300 liter shots increased the injection capacity up to eight fold. The report does not mention anything about the degree of vertical fracturing that occurred after the blasts. However, Burwell et al (1973) say that the detonation of the NG explosive undoubtedly created breakage of the shale on both sides of the fractures which resulted in a large surface area. The shale was apparently fractured to allow the in situ retorting process to proceed and the oxygen utilization and rate of burning were improving steadily at the time the test was terminated.

Burwell et al also report experience from another test, test site 7, where first only hydraulic fracturing took place before ignition. Two ignition attempts were made but they were terminated because the injection rates could not be maintained. When 160 kg of pelletized TNT was detonated in the wellbore the permeability increased and it became no problem to ignite the oil shale.

Coursen (1977)and McNamara et al (1979)have shown that true in situ blasting can increase the permeability and the explosion cavities if the depth is not too large and the tectonic stresses are not too high. However, a very large specific charge must be used (or several reshots) to establish a permeability large enough to allow a high recovery degree of oil.

It is hard to believe that true in-situ fragmentation methods will be economic for extraction of oil from kerogen in the near future. The major problem seems to be that insufficient porosity is generated in the shale (Parrish et al, 1981). Loading is costly and it is a problem to work with explosives having both a low enough sensitivity to be handled safety

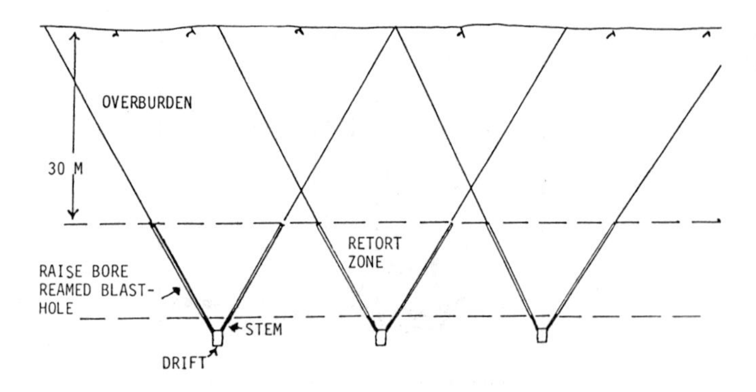

Figure 13. The LOFRECO-process. After Britton (1980).

and a small enough critical diameter that can be used for detonation in thin hydro fractures. Most likely the true in-situ explosive and/or hydraulic fragmentation will not be economic if the fracturing is intended to take place at large depth where the stress field is high. The two methods have their potentially most interesting application in shallow oil shale zones where the advantages of an upper free surface can be utilized, and where expensive development work can be avoided. One method is the LOFRECO-process, described by Lekas (1979) and Britton (1980). A heavy explosive load is used to heave the surface in order to create the necessary fractures and voids.

From their results, Geokinetics (Lekas, 1981) concluded that:

It is possible to drill a pattern of blastholes from the surface into the oil shale and fracture the shale with explosives to establish a zone of high permeability with a relatively impermeable zone between the fragmented shale and the surface.

It is possible to drill through the rubblized material and construct the various wells for the operation. A point ignition can be made in the rubblized shale and expanded into a burn front that covers the cross section of the retort.

The burn front can be made to move down the length of the retort as a cohesive temperature front with satisfactory sweep efficiency.

Produced oil can be recovered from a well drilled to the bottom of the rubblized zone.

Recovery of inplace oil of up to 50 percent can be achieved.

Modified in-situ fragmentation

Only a few years ago there were several large scale projects to produce oil by modified in-situ fragmentation and retorting. Today these activities have decreased and the modified in-situ methods (MIS) seem today to be hard to justify economically. Mignogna(1979), McCarthy and Cha (1976), Ridley (1978) and Ricketts (1980) have reported MIS methods with various results. The mined out void volumes have been in the order of 20% before "in-situ" fragmentation took place.

Due to an irregular fragmentation distribution and chimney effects preventing an effective (one level burn-out) the recoveries have often been too low. Modern underground mining methods, like room and pillar mining, combined with surface retorting and back fill of spent shale looks to be the most promising method for deeper oil shale formation - if the oil price goes up.

Numerical modelling of fragmentation

A number of codes for numerical modelling of rock fragmentation have been developed during the recent years. Some of them are very sophisticated and have all the material constitutive properties implemented as well as the fracture mechanics for the intact rock material.

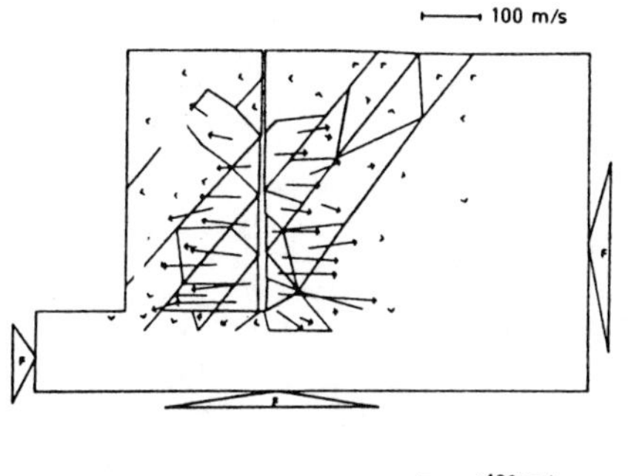

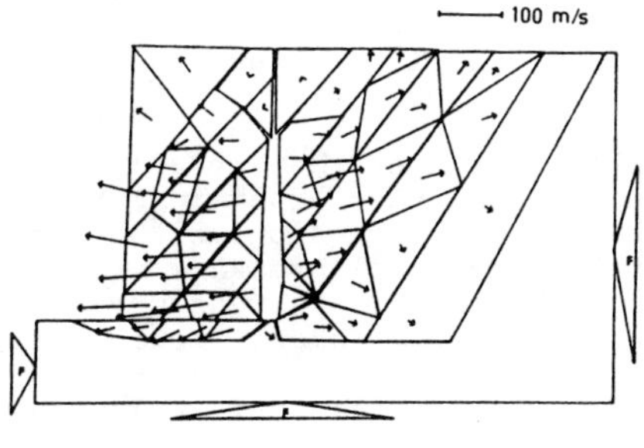

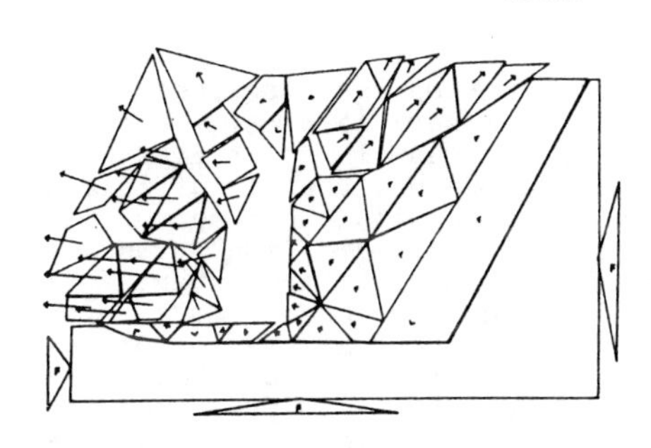

Figure 14. A bench blast simulation at: 1 ms, 5 ms and 50 ms.

We feel that future numerical models to a much higher degree must involve the inherent anisotropic effects. The rock mass is characterized by a large amount of joints, fractures and existing weakness planes. This geologic complexity plays the major roll in forming the post blast fragments whose size distribution is of vital interest. A well founded estimate will always help lowering the cost for investments in tailor made crushers and transportation systems for mining industry or help to consider the burn rate and degree of recovery in the oil shale business.

Testing of rock strength usually is carried out in small scale rock core samples selected from the intact core pieces available. The uniaxial tensile strength is often given with a standard deviation equal to fifty percent of the mean value. Predominant fracturing will occur in these non or low strength weakness planes when the rock mass is subjected to a dynamic load caused by a nearby detonation. All the weakness planes will have an individual strength and will break depending upon how far away from the nearest borehole it is situated. The geologic complexity will be more pronounced and will influence more upon the fragmentation more when larger hole diameters (i.e. larger spacings) are used for blasting.

It is desirable that future modelling codes can handle 3D- structure geology given by a geologic core mapping procedure (or a statistically simulated rock mass anisotropy). The strength of the weakness planes should preferably also be given as input.

The dynamic load can be modelled for various distances in the near field and this load together with the strength criteria for the weakness planes will describe major part of the blast induced fragment sizes.

Cooper (1981) has used the Cundall Block model program and implemented extensions to the model, such as joints with cohesion. With this approach it is possible to model breakage of existing weakness planes as well as mass movement of the broken rock. Figure 14 shows a simple 2D-bench blast with two free faces. Observe the fracturing that successively takes place and breaks existing weakness planes.

WELL STIMULATION

Hydraulic fracturing has long been used for stimulation of impermeable gas reservoirs and oil wells. For a successful result fracturing has to occur in such a way that a sufficiently large contact area with the reservoir is created. The method to use a fluid "mud" pressure in a drilled hole to initiate and extend a crack outwards into the surrounding rock formation has been used for several years. Well established theories exists where it is possible to predict necessary fluid pressure and crack growth direction. The in-situ stress field will however irrevokably steer the crack propagation into a plane perpendicular to the least principle stress and variations of this stress may cause unfavorable fracture growth into formations with no reservoir contact.

The hydraulic fracturing method has certain limitations because the method will only create one fracture plane with a surface contact area determined by the length of the driven fracture. Obviously it would be beneficial if we can find a method which initiates a multiple fracture pattern around the borehole or between several parallel hydraulic fractures. This would increase permeability and contact areas.

Randomly distributed radial cracks can be achieved by the use of an explosive charge but unfortunately the short time pulse loading will result in rather short fractures. The fracture lengths are also determined by the depth of overburden or rather the in-situ stress field.

A higher stress field result in less fracture propagation and the fractures outside the borehole will tend to be oriented towards the maximum principal stress in a non uniform stress field. High detonation velocity and pressure will (especially in low compressive and porous rock materials) damage the rock material in the near region and fines will be produced that effectively plug fractures. Decoupled charges where an air cushion is situated between the borehole and the explosive charge will effectively reduce the bore pressure.

The ratio between the borehole pressure for a decoupled charge (P_2) and a fully coupled charge (P_1) is approximately given by equation

$$\frac{P_2}{P_1} = \left(\frac{d_2}{d_1}\right)^{\gamma} \qquad (4)$$

where d_1 denotes the hole diameter, d_2 the charge diameter and γ can be approximated with 1.5 (γ is actually pressure dependent). The borehole pressure for a fully coupled charge is dependent upon detonation pressure and can be approximated with half of that pressure. By the decoupling the borehole pressure can be adjusted to the strength of the surrounding rock in order to avoid a crushing zone. Still however the short duration loading pulse will be to short to drive the fractures longer distances.

Alternatives to hydrofracturing and explosive fracturing of boreholes have been discussed by Schmidt et al (1980)and Swift and Kusubov (1981). By use of compressing fluid and a piston a repetitive loading condition can be achieved. Pulse durations can be shifted down to 1 ms and loading varied from 10^{-2} to 100 MPa/ms. Tests carried out in a sandstone with water as fluid indicate some interesting results. Multiple fractures will be formed without damage to the borehole if the sandstone is semi-dry or wet. Only two fractures are formed during dry conditions. Figure 15 shows the fracture patterns achieved for different loading conditions.

Warpinski et al (1979) report experiments conducted at the U.S. Nevada Test Site in a volcanic ashfall tuff formation. Three propellants with different burn times were used. The pressure loading rates and peak pressures are given in table 3 below. About 9 kg of each propellant was used.

The overburden was 420 m at the experimental site. Tensile strength for the ashfall tuff was reported to be 2.8 MPa and the fracture toughness K_{Ic} was 0.5 $MNm^{-3/2}$.

The length and the number of fractures were mapped after the test. Carbon black had been added to simplify identification of created fractures.

The slowest propellant GF # 1 resulted in a single fracture that was similar to fractures produced by hydro fracturing. The fracture formed extended to a length of 0.6 - 0.9 m from the borehole.

The intermediate propellant GF # 2 initiated twelve different fractures of which five only

Table 3. Propellants used by Warpinski et al, 1979.

Propellant	Burn time	Loading rate	Peak pressure
GF # 1	~ 900 ms	$6.21\cdot10^{-4}$ MPa/μs	43.1 MPa
GF # 2	~ 9 ms	$1.38\cdot10^{-1}$ MPa/μs	95.1 MPa
GF # 3	~ 1 ms	>10.3 MPa/μs	>138 MPa

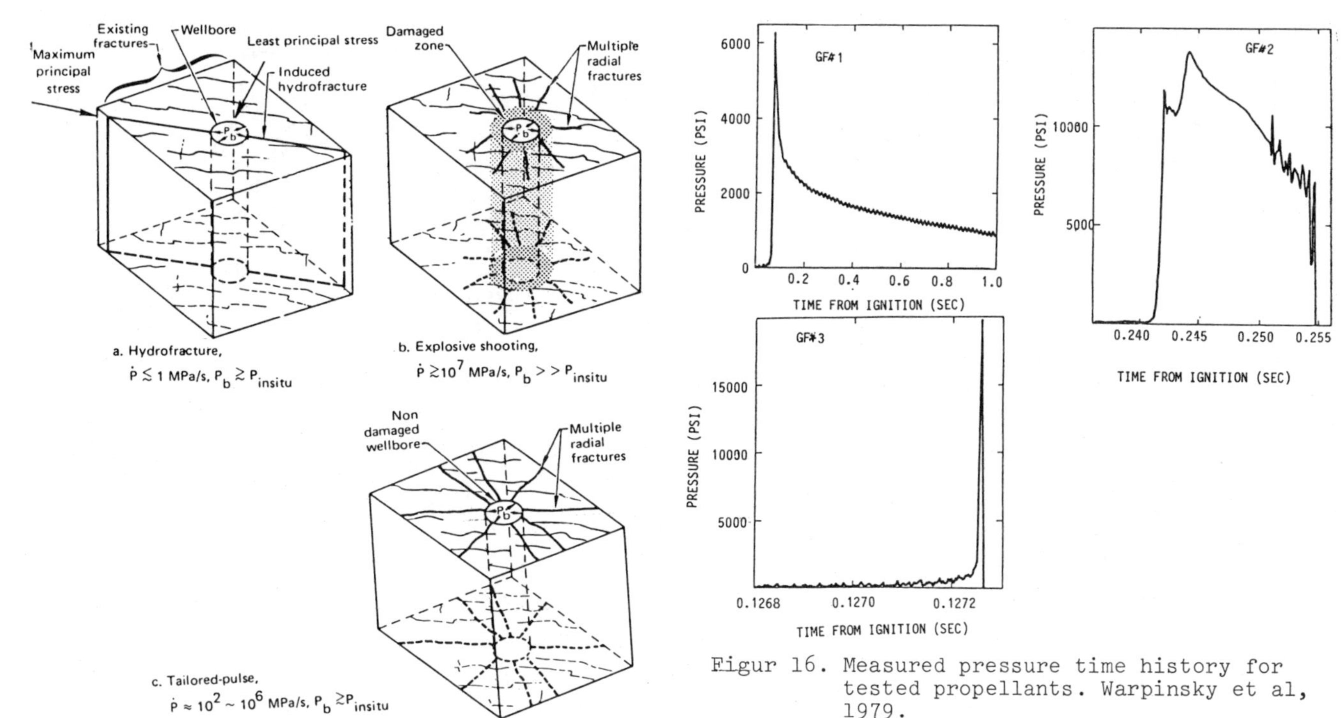

Figur 16. Measured pressure time history for tested propellants. Warpinsky et al, 1979.

Figure 15. Fracture patterns for different loading conditions, Swift and Kusubov, 1981.

extended less then 0.3 m. The other seven fractures had lengths running from 0.6 - 2.4 m.

Propellant GF # 3 initiated multiple fractures with lengths not exceeding 0.12 m and crushing was observed near the borehole. Only one fracture extended radially a longer distance(~1.2 m).

Fractures created were not found to be affected by the in-situ stress field.

Figure 16 shows the measured pressure time histories for the three propellants.

Considering the tests it is apparent that the fracture process is strongly dependent on loading rate. Pressure loads slightly higher than the in-situ stress will produce only one single fracture. This fracture will not extend very far if propellants are used because it is hard to maintain the pressure load for a longer period, with the limited amount of gas available from the propellant charge. Higher initial loading rates and pressure peaks helps to initiate multiple fractures. The number of fractures are dependent upon the number of flaws around the well or the grain size of the material. However in order to propagate the fractures longer distances it is necessary to keep the pressure level at a sufficiently high level for a longer period. The pressure should preferably be kept below the yield stress of the rock to prevent residual compressive stresses that help prevent crack growth during post peak pressure load. (Schmidt, R A et al, 1979).

A high loading rate and a high peak pressure evidently induce multiple fractures but unfortunately the residual compressive stress field and the fines produced effectively seal the initiated fractures and sometimes lower the pre-detonation permeability. If the in-situ stress field is low and the rock material is competent figure 3 can be used to estimate the maximum length of induced radial fractures from a coupled charge of different hights. Ouchterlony (1974) has shown that the pressure required to initiate cracks is independent of the number of cracks if the crack lengths are short. Figure 17 shows the relations between the normalized stress intensity ratio and the crack length parameter.

Using the equation valid for critical pressure P required to initiate cracks equal to a flaw of size a

$$P = K_{Ic}/2.24\sqrt{\pi a} \qquad (5)$$

it is obvious that very low pressures are required to initiate cracks. A suggested flaw size of less than 1 mm for the volcanic ashfall tuff (K_{Ic} = 0.5 $MNm^{-3/2}$) where Warpinskis gas fracture test was conducted indicates a required critical pressure of only a few MPa (static). A fully coupled charge which is detonated obviously is capable of producing borehole pressures several orders of magnitude larger than needed. Even decoupled charges used for smooth blasting (coupling ratio ≈ 9) give borehole pressures in the order of 70 to 150 MPa which still is higher than needed. Although this has been realized by people involved in well stimulation. Schmidt et al, 1979, have suggested that high explosives used for well stimulation should be positioned below the pay zone with a

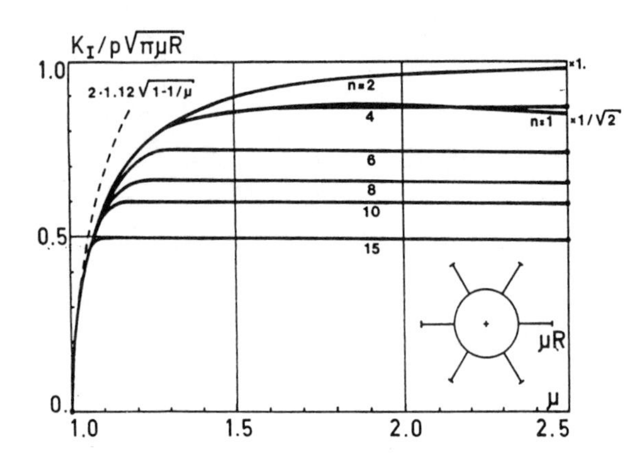

Figure 17. Stress intensity factor for a pressurized circular cavity as a function of crack length parameter.

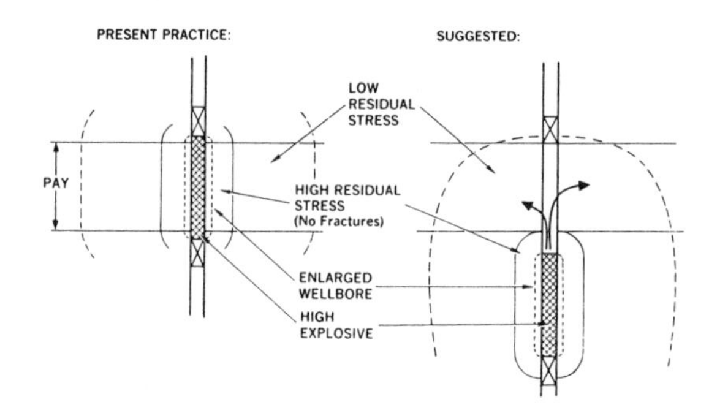

Figure 18. One alternative to conventional well shooting. Explosive is placed below pay zone. Schmidt et al, 1979.

stemming placed above the pay zone in order to utilize a lower peak pressure and longer pressure time history. See figure 18.

Fourney et al (1980) carried out high speed photography studies in a plexiglass model 51 mm high. At the bottom of the drillhole a small PETN charge was detonated. The hole was stemmed above the charge leaving an air cushion between the stemming and the charge equal to about 4 times the charge length. It was observed that fractures were initiated along the borehole but the fractures close to the stemming area grew at a faster rate than those in the charge area. Simultaneous pressure time records at different points along the borehole indicated a higher pressure close to the stemming and the high pressure was maintained for a longer time period here than at any other place along the borehole. This phenomenon favourably supported the crack propagation close to the stemming area and the cracks were not arrested as quickly as in the charge area where residual compressive stresses were developed, and the produced fines plugged the cracks.

Hydro fracturing alone in some strata can effectively increase permeability and thereby stimulate fluid production. It has been used for many years to stimulate production of petroleum

FRAME NO.	TIME (μs)	VELOCITIES BETWEEN FRAMES (ips) Charge Area	Stem Area
1	12.5	20,854 *	
2	34.0	13,760	
3	57.0	11,784	
4	80.5	12,493	
5	89.5	12,613	10,320 *
6	112.5	7,193	17,742
7	140.0	5,050	17,395
8	163.5		21,978
9	190.5		12,363
10	214.5		28,846
12	246.5		4,220
14	271.5		20,168
16	322.5		

* - Velocity probably low since fracture was not likely to initiate at the same instant the time count started.

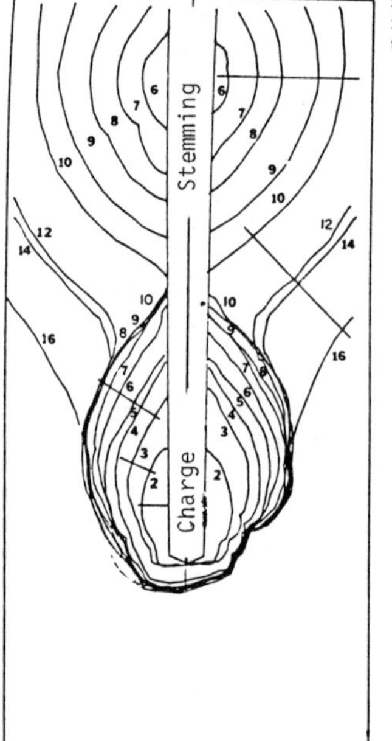

Figure 19. Crack propagation velocities in charge and stem area. After Fourney et al, 1980.

Other techniques have been used to control the fracture process. Bligh (1974) reports studies performed to control fracturing with a fuel-air mixture. Moore et al (1977) and Fitzgerald and Anderson (1978) describe methods where propellants are used to pressurize surrounding water or mud, push it into initiated cracks and propagate them. The application of metal-lined shaped charges to the chemical explosive stimulation of natural gas wells in Devonian shales has been investigated by Schott et al (1977).

and natural gas. Sand injected with fracturing fluid or mud is sometimes needed to prop the crack open. Unfortunately it is often difficult to control the hydrofracture so that a large contact area is achieved with the strata containing the heated water (geothermal and recirculation wells) or the petroleum reservoir. The in-situ stress field in the earth in deeper formations usually is such that the produced crack tends to orient itself vertically with its azimuthal orientation guided by the local tectonic stress field. This is a wellknown fact that is used for stress measurements. Further the hydraulic fracture will often be oriented parallel to already existing fractures in the rock mass.

The use of high explosive alone for fracture control seems be hard to justify sometimes. It looks more promising to control the fracture growth by a tailored pressure pulse from slow burning propellants. It is difficult, however, to keep the pressure at a sufficient level for a longer time span. To prevent fast gas-venting there is often a need to stem the well properly which makes it hard to shoot the well repetitively. The burning rate of propelants is greatly affected by surrounding pressure and temperature. When pressure increases, the burning rate will increase exponentielly which makes it impossible to use larger charges to increase the loading time. The pressure rise will increase the burning rate which increases the pressure and so on. This sometimes turns over the deflagration into a detonation.

Still we have to accept that numerous future applications can be found if we improve the understanding of effective in-situ fracturing and fragmentation by the use of explosive and hydraulic fracturing. There is a need to develop methods to control the fragmentation mechanism for deep in-situ recovery methods for geothermal energy and fuel recovery.

There is a tremendous wealth of oil and gas to be won from already existing wells if even marginal improvements in present-day well stimulation techniques can be made.

REFERENCES

Barker, C.R. and Summers, D.A. (1977). Consideration in the Development of a Water Jet Cutting Head - Energy Resources and Excavation Technology: Proc. 18th U.S. Symp. on Rock Mechanics, editors Fun - Den Wang and George B. Clark, Colorado School of Mines Press, Golden, CO, USA.

Baumann, L. and Heneke, J. (1980). High Pressure Water Jets Aid TBM's: 5th International Symposium on Jet Cutting Technology, June -80, Hannover, West Germany.

Bjarnholt, G., Holmberg, R., Ouchterlony, F. (1982). A System for Contour Blasting with Directed Crack Initiation: Swedish Detonic Research Foundation, SveDeFo Report DS 1982:3 (in Swedish), Stockholm, Sweden (1982).

Bligh, T.P. (1974). Principles of Breaking Rock Using High Pressure Gases, Advances in Rock Mechanics: Proc. 3rd Congress of the International Society for Rock Mechanics, Denver, CO, 1974, Vol. II, Part B, pp. 1421.

Blindheim, O.T. (1977). Preinvestigations, Resistance to Blasting and Drillability Predictions in Hard Rock Tunnelling. In "Mechanical Boring or Drill and Blast Tunnelling?: 1st US-Swedish Underground Workshop, Stockholm, December 5-10, 1976. Document D3:1977 - Statens Råd för Byggnadsforskning (Swedish Council for Building Research), Stockholm 1977, pp. 81-97.

Blindheim, O.T. (1979). Drillability Predictions in Hard Rock Tunnelling . Preprint, Paper 15, 2nd Int. Symposium Tunnelling '79, Institution of Mining and Metallurgy, London, 1979.

Britton, K., (1980). Principles of Blast Design Developed for In Situ Retorts of the Geokinetics Surface Uplift Type: Proc. 13th Oil Shale Symposium, Golden Colorado, April 16-18, 1980, pp. 169-180, James H. Gary ed., Colorado School of Mines Press, Golden, CO, USA.

Burwell, E.L., Sterner, T.E. and Carpenter, H.C. (973). In Situ Retorting of Oil Shale, Results of Two Field Experiments. Bureau of Mines: Report of Investigation 7783, U.S. Dept. of the Int. Library, 1973.

Cambell, G.G., Scott, W.G. and Miller, J.S. (1970). Evaluation of Oil Shale Fracturing Tests Near Rock Springs, Wyo., Bureau of Mines, Rep. of Inv. 7397, U.S. Dept. of Int. Library, June, 1970.

Cooper, T. (1981). An Implementation of the Cundall Block Program on the HP 2100 with Extensions to the Physical Model: SveDeFo Report DS 1981:7, Stockholm, Sweden.

Coursen, D.L. (1977). Cavities and Gas Penetrations from Blasts in Stressed Rock with Flooded Joints: 6th Int. Colloquium on Gasdynamics of Explosives and Reactive Systems, Stockholm, Sweden, 22-26 August, 1977.

Fitzgerald, R. and Anderson, R. (1978). Kine-Frac: A New Approach to Well Stimulation: ASME Paper 78-PET-25, ASME Energy Technology Conference and Exhibition, Houston, TX, November 5-19, 1978.

Fourney, W.L., Holloway, D.C. and Barker, D.B. (1980). Pressure Decay in Propagating Cracks, University of Maryland, USA.

Gates, E.M. and Gilpen R.R. (1978). Jet Cutting in Oil Sands: AOSTRA Seminar on Underground Excavation of Oil Sands, Edmonton, Canada.

Handewith, H.J. (1969). Predicting the Economic Success of Continous Tunneling in Hard Rock: 71'st Annual General Meeting of the CIM, Montreal, Canada.

Holmberg, R. and Persson, P.A. (1978). The Swedish Approach to Contour Blasting: Proc. 4th Conf. on Explosives and Blasting Technique arr. by the Soc. of Explosives Engineers, New Orleans, pp. 113-127.

Holmberg, R. and Persson. P.A. (1979). Design of Tunnel Perimeter Blasthole Patterns to Prevent Rock Damage: Proc. Tunnelling '79, editor Jones M.J., Inst. of Mining and Metallurgy, London, March 12-16.

Hood, M. (1976). Cutting Strong Rock with a Drag Bit Assisted by High Pressure Water Jets: Journal of South African Institute of Mining and Metallurgy. Vol. 77, No. 4, Nov. pp. 79-90.

Hood, M. (1978). A Study of Methods to Improve the Performance of Drag Bits Used to Cut Hard Rock, Ph.D. Thesis, University of Witwatersrand.

Johannessen, O. (1981). Drillability, Drilling Rate Index Catalogue: Project Report 8-79, The University of Trondheim, Trondheim, Norway.

Lekas, M.A. (1979). Progress Report on the Geokinetics Horizontal In-Situ Retorting Process: Proc. 12th Oil Shale Symposium, James H. Gary, editor, Colorado School of Mines Press, Golden, Colorado, pp. 228-236.

Lekas, M.A. (1981). The Geokinetics Horizontal In-Situ Retorting Process: Proc. 14th Oil Shale Symposium, James H. Gary, editor, Colorado School of Mines Press, Golden, Colorado, pp. 146-153.

Lindqvist, P.A. (1982). Stress Fields and Subsurface Crack Propagation of Single and Multiple Rock Indentation and Disc Cutting. Rock Fragmentation by Indentation and Disc Cutting. University of Luleå. Ph.D. Thesis 1982:20D. Luleå, Sweden.

Lundborg. N., Holmberg, R. and Persson, P.A. (1978). The Distance-Charge Size Dependence of Ground Vibrations. Report R11:1978, Statens Råd för Byggnadsforskning (Swedish Council for Building Research), Stockholm (In Swedish).

Mathews, K.E. (1980). Potential for the Underground Mining of Oil Sands in Canada: Proceedings of the International Symposium Rockstore '80, Pergamon Press, pp. 673-681. Editor Bergman, M., Stockholm, Sweden.

McCarthy, H.E. and Cha. C.Y. (1976). Oxy Modified In-Situ Process Development and Update. Quarterly of the Colorado School of Mines: Proc. 9th Oil Shale Symposium. Volume 71, no. 4, pp. 85-101, 1976.

McNamara, P.H., Peil, C.A. and Washington, L.J. (1979). Characterization, Fracturing and True In Situ Retorting in the Antrim Shale of Michigan: 12th Oil Shale SymposiumProceedings, Golden, Colorado, April 18-20, 1979, pp. 353 and 365, editor James H. Gary, Colorado School of Mines Press, Golden, CO, USA.

Mignogna, R.P. (1979). Conceptual Design of a Horizontal Modified In Situ Oil Shale Retort: Dept. of Mining Engineering, Colorado School of Mines, Golden, CO, USA, March 30, 1979.

Miller, J.S., Walker, C.J. and Eakin, J.L. (1974). Fracturing Oil Shale for In Situ Oil Recovery. Bureau of Mines Report of Investigation 7874, U.S. Dept. of the Int. Library, 1974.

Moore, E.T., Mumma, D.M. and Seifert, K.D. (1977). Dynafrac - Application of a Novel Rock Fracturing Method to Oil and Gas Recovery. Physics International Final Report 827.

Ouchterlony, F. (1974). Fracture Mechanics Applied to Rock Blasting: Proc. 3rd International Congress of the ISRM. Volume 2, part B, pp. 1377-1383, Denver, CO, USA.

Parrish, R.L. Stevens, A.L. and Turner. T.F. (1981). A True In Situ Fracturing Experiment - Final Results: Journal of Petroleum Technology. July 1981, pp. 1297-1304.

Persson, P.A., Holmberg, R. and Persson, G. (1977). Careful Blasting of Slopes in Open Pit Mines. Swedish Detonic Research Foundation, Stockholm, SveDeFo Report 1977:4 (in Swedish).

Prader, D. (1977). Boring of Tunnels: Remark on Borability. In Mechanical Boring or Drill and Blast Tunnelling?: 1st US-Swedish Underground Workshop Stockholm, December 5-10, 1976. Document D3:1977 - Statens Råd för Byggnadsforskning (Swedish Council for Building Research), Stockholm 1977, pp. 129-131.

Rehbinder, G. (1977). Slot Cutting in Rock with a High Speed Water Jet: Research Report TULEA 1977:03, Luleå University of Technology, Luleå, Sweden.

Ricketts, T. (1980). Occidental's Retort 6 Rubblizing and Rock Fragmentation Program. Oil Shale Symposium, Golden, CO, USA, April 16-18, 1980, pp. 46-61, editor James H. Gary, Colorado School of Mines Press, Golden, CO, USA.

Ridley, R.D. (1978). Progress in Occidentals' Shale Oil Symposium: Proceedings, Golden, CO, April 12-14, 1978, pp. 169-175, editor James H. Gary, Colorado School of Mines Press, Golden, CO, USA.

Robbins, R.J. (1977). Development Trends in Tunnel Boring Machines for Hard Rock Application. In Mechanical Boring or Drill and Blast Tunnelling: 1st US - Swedish Underground Workshop Stockholm, December 5-10, 1976. Document D3:1977 - Statens Råd för Byggnadsforskning (Swedish Council for Building Research), Stockholm, 1977, 27-39.

Schmidt, R.A., Boade, R.R. and Bass, R.C. (1979). A New Perspective on Well Shooting - The Behaviour of Contained Explosives and Deflagrations: Society of Petroleum Engineers of AIME, SPE 8346.

Schmidt, R.A., Warpinski, N.R. and Cooper, P.W. (1980). In Situ Evaluation of Several Tailored - Pulse Well Shooting Concepts: SPE Paper 8934, Soc. of Petroleum Eng., Dallas, TX, USA.

Schott, G.L., Carter, W.J. and Vanderborgh, N.E. (1977). Stimulation and Characterization of Eastern Gas Shales: Los Alamos Report LA-7320-PR, Los Alamos, NM, USA.

Selmer-Olsen, R. and Blindheim, O.T. (1970). On the Drillability of Rock by Percussive Drilling: Proc. 2nd Congress Int. Soc. Rock Mechanics, Belgrade, 1970.

Summers, D.A. and Lehnhoff, T.F. (1977). Water Jet Drilling in Sandstone and Granite: Proc. 18th U.S. Symposium on Rock Mechanics, editor Wang and Clark, Colorado School of Mines Press, Golden, CO, USA.

Swift, R.P. and Kusubov, A.S. (1981). Tailored Pulse Loading Conditions for Multiple Fracturing of Boreholes: 22'nd U'S. Symposium on Rock Mechanics, Massachusetts Institute of Technology, USA.

Thompson et al (1979). Field Evaluation of Fracture Control in Tunnel Blasting: Report No. UMTA-MA-06-0100-79-14, U.S. Depmartment of Transportation, Administration.

Warpinski, N.R. et al (1978). High Energy Gas Frac.: Report SAND 78-2342. Sandia Laboratories, Albuquerque, NM, USA.

Warpinski, N.R. et al (1979). High Energy Gas Frac. Multiple Fracturing in a Wellbore: Proc. 20th U.S. Symposium on Rock Mechanics, June 4-6, 1979, Austin, TX, USA.

FELSDYNAMIK

Rock dynamics

Dynamique des roches

Per Anders Persson
Nitro Nobel AB, Schweden

Roger Holmberg
Swedish Detonic Research Foundation, Stockholm, Schweden

ZUSAMMENFASSUNG

Im Folgenden wird über verschiedene Methoden des Felsabbaus berichtet, wie Sprengung, mechanischen Ausbruch und des Zerbrechens in situ. Auf dem Hintergrund einer kurzen Darstellung der jüngsten Fortschritte in diesem Bereich wird auf einige der modernsten Entwicklungen näher eingegangen. Es handelt sich sowohl um eine neue schwedische Methode, Sprengeinwirkungen im Gebirge vorauszusehen, in der unmittelbaren Nähe und im weiteren Umkreis, als auch um eine norwegische und amerikanische Technik zur Bestimmung der Bohrbarkeit. Die Verwendung von Wasser-jets beim Bergbau wird kurz angesprochen. Der Bericht schließt mit einer knappen Übersicht über Fortschritte in Gesteinszerbrechen in situ.

DYNAMIK DES FELSSPRENGENS

Um beim Gebrauch von Sprengstoffen zum Lösen von Fels die Beanspruchung der stehenbleibenden Felswände und benachbarten Gebäude gering zu halten, muß gebirgsschonend gesprengt werden. In heutigen Tagebauen enthält bei Verwendung von Schussbohrlöchern mit Durchmessern zwischen 250-500 mm jedes Schussbohrloch ein oder zwei Tonnen Sprengstoff. Für die gesamte Sprengung können 200-500 Tonnen Sprengstoff benötigt werden. Im Untertagebau werden dagegen in zunehmendem Maße Schussbohrlochdurchmesser zwischen 150-200 mm gewählt. Auch im Tunnelbau sind größere Durchmesser (50-100mm) und lange Schussbohrlöcher (3-6m) gebräuchlich. Die Entwicklung auf allen genannten Gebieten hin zu größeren Sprengungen führt zu Zeit- und Kostenersparnissen, erfordert aber größere Anstrengungen auf Methoden, die Schäden verhindern.

Im Tagebau haben Standsicherheit der Böschungen und entsprechende Böschungswinkel einen großen Einfluß auf Wirtschaftlichkeit und Sicherheit des Abbaues. Zwei Beiträge dieser Konferenz behandeln Methoden, steilere Böschungen durch gebirgsschonendes Sprengen herzustellen, wobei der stehenbleibende Fels genügend Festigkeit besitzt, um zunehmende Spannungen aufzunehmen. Selbst das Versteilen der Böschungen um ein Grad bedeutet eine deutliche Einsparung an Abraummaterial im tiefen und großen Tagebau.

Im Untertagebau und Tunnelbau können durch die Einführung des gebirgsschonenden Sprengens ähnlich große Einsparungen erzielt werden, weil Schäden an nahegelegenen Tunnel und Gebäuden große Geldsummen verschlingen würden. Gebirgsschonendes Sprengen erfordert Kenntnis über Art und Verlauf der Druckwellen im Fels.

Druckwellen im Fels

Die Detonation einer Ladung in einem Bohrloch im Fels führt zu Druckwellen im umgebenden Gebirge. Bei Besetzung eines Bohrloches mit einem starken Sprengstoff überschreitet der Explosionsdruck die Festigkeit des Gebirges und es entstehen komplizierte Scherdeformationen, die letztlich zum Zerstören des Gebirges in der Umgebung des Bohrloches führen. Weiter entfernt davon entstehen radiale Risse.

Mit zunehmender Entfernung vom Bohrloch nimmt die Amplitude der Druckwelle ab und die Welle wird rein elastisch. Durch das Zusammenwirken mit der freien Oberfläche entwickeln sich die in der Seismik bekannten p-Wellen, s-Wellen und Rayleigh-Wellen. Weiter vom Bohrloch entfernt beeinflussen die vorhandenen Klüfte oder Risse im Fels sowohl die Wellenausbreitung (Wellengeschwindigkeit und Druckhöhe) als auch den Grad der Gebirgszerlegung stärker, als in Bohrlochnähe.

Will man die Stärke der Wellen in weiter entfernten Bereichen betrachten, sollte man die maximale Teilchengeschwindigkeit als Maß gebrauchen. Abb. 1 zeigt die ungefähre Abnahme der maximalen Teilchengeschwindigkeit mit zunehmender Entfernung von einem 250 mm Ø Bohrloch, das mit einer 15 m Ladung besetzt ist. Bei Entfernungen, für die die Besetzungslänge klein ist im Verhältnis zu eben dieser Entfernung, gilt angenähert für die maximale Teilchengeschwindigkeit die folgende Beziehung:

$$u = K \frac{Q^{\alpha}}{R^{\beta}} \qquad (1)$$

mit Q = Gewicht der Ladung, R = Abstand vom Bohrloch und den Konstanten K, α und β (für harten Fels K = 700,

α = 0,7 und β = 1,5). Abb. 2 zeigt eine beträchtliche Streuung, die auf variierende Felsverhältnisse und damit verbundener unterschiedlicher Wellenausbreitung zuzurückzuführen ist.

Die Druckwellen pflanzen sich mit unterschiedlichen Geschwindigkeiten fort. Für einen harten Fels können typische Kennwerte mit

$$c_p = 5000 \text{ m/s}$$

$$c_s = 3500 \text{ m/s}$$

$$c_R = 2500 \text{ m/s}$$

angegeben werden. Bei einem weichen, mehr geklüfteten Fels liegen die Werte etwas niedriger.

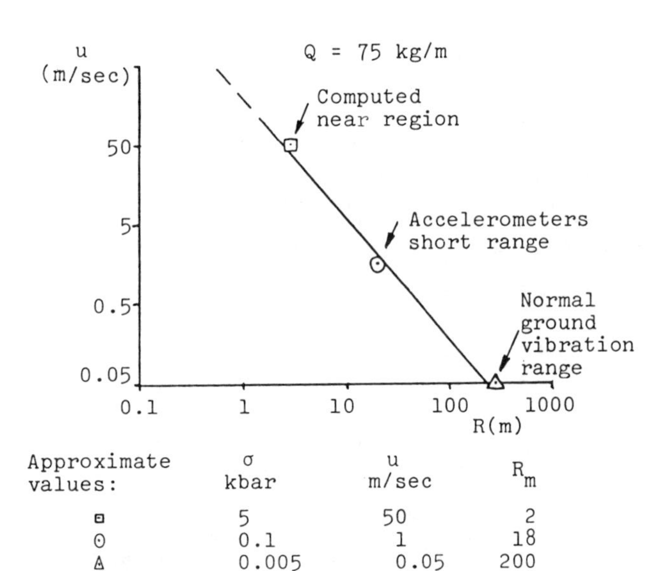

Abb. 1 Druckwellengeschwindigkeit einer 250 mm Ladung, Länge 15 m

Zerbrechen des Gebirges durch Druckwellen

Betrachtet man den Gang als eine einfache harmonische Schwingung, so kann man in Abhängigkeit vom Wellentyp die Spannung ρ oder die Verformung ε im Fels bestimmen.

$$\varepsilon = \frac{\sigma}{E} \approx \frac{u}{c} \tag{2}$$

Ein Granit wird voraussichtlich bei dynamischer Beanspruchung und einem Druck von vielleicht 30 MPa, das einer Verformung von ca. 1‰ entspricht, zu Bruch gehen. Abhängig von dem jeweiligen Wellentyp bedeutet das eine Teilchengeschwindigkeit zwischen 1000 und 2000 mm/s. Aber bei normal geklüftetem Fels zeigt sich eindeutig eine Überschreitung der Zugfestigkeit innerhalb der Klüfte bei schon niedrigen Spannungsniveaus (etwa bei 700 mm/s). Bei weichen Sedimentgesteinen mit relativ schwacher Klüftung tritt ein Zerreissen bereits bei Teilchengeschwindigkeiten von 400 mm/s oder weniger auf.

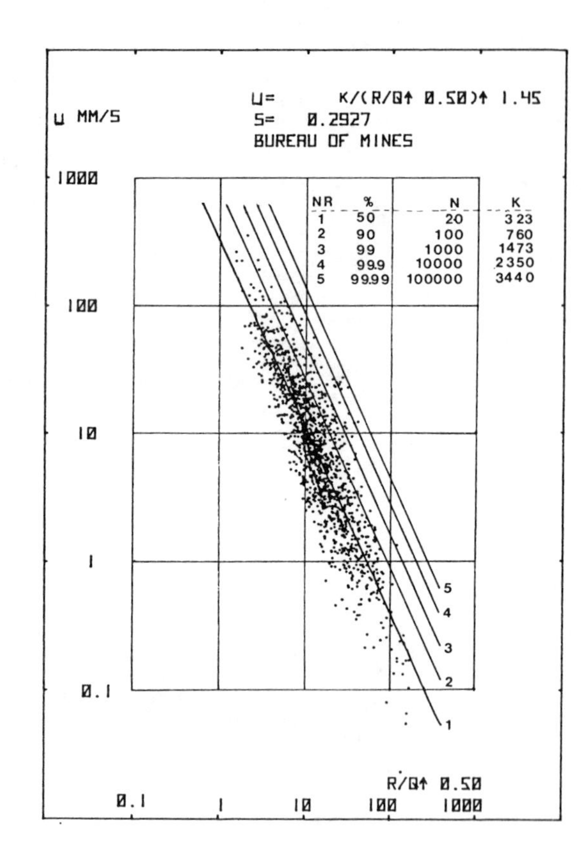

Abb. 2 Maximale Teilchengeschwindigkeit $\overline{U}$ als Funktion von R/Q bei verschiedenen Vertrauensbereichen und verschiedenen Durchgängen

Obgleich die Raumstellung der Klüfte im Verhältnis zur Ausbreitungsrichtung der Wellen eine wichtige Rolle spielt, kann untenstehende Tabelle benutzt werden, um die Spannweite der kritischen Schwingungsgeschwindigkeiten zum Zerbrechen von Fels abzuschätzen. Der typische Bruchvorgang in einem geklüfteten Fels besteht aus einem irreversiblen Aufreissen zweier Kluftflächen, wodurch eine Abnahme der Scherfestigkeit in dieser Kluft eintritt. Dieser Vorgang wird von einem leichten Schwellen des betroffenen Gebirges begleitet.

Tab. 1 Bereich der kritischen Schwingungsgeschwindigkeit zur Bruchinitiierung in verschiedenen Gebirgstypen.

Felsklasse/Kluftzustand	kritische Schwingungsgeschwindigkeit (mm/s)
harter Fels, feste Klüfte	≥ 1000
mittelharter Fels, keine schwachen Klüfte	800 - 700
weicher Fels, schwache Klüfte	≤ 400

Vorhersage der Schwingungsgeschwindigkeit in der Nähe einer gestreckten Ladung

Auf der Basis der Beziehung 1 haben die Autoren die maximale Schwingungsgeschwindigkeit unter verschiedenen Entfernungen von den Sprengladungen in mittelhartem Fels und bei verschiedenen Ladungslängen und Durchmessern berechnet (Holmberg und Persson, 1978). Abb. 3a und b zeigen die Ergebnisse. Die Diagramme dienen dem Entwurf von Besetzungsschemata, wenn Auflockerung des stehenbleibenden Felsens minimal sein soll. Damit kann man zum Beispiel den Auflockerungsgrad an einem gegebenen Punkt hinter dem Tunnelprofil oder nahe einer Tagebauwand bestimmen, wobei die Auflockerung sowohl von Vorspaltlöchern als auch von Sprenglöchern, die weiter vom Profil entfernt liegen, herrühren kann.

Entwurf der Profil-Sprengung

Die in Abb. 3 dargestellten Ergebnisse stimmen gut mit

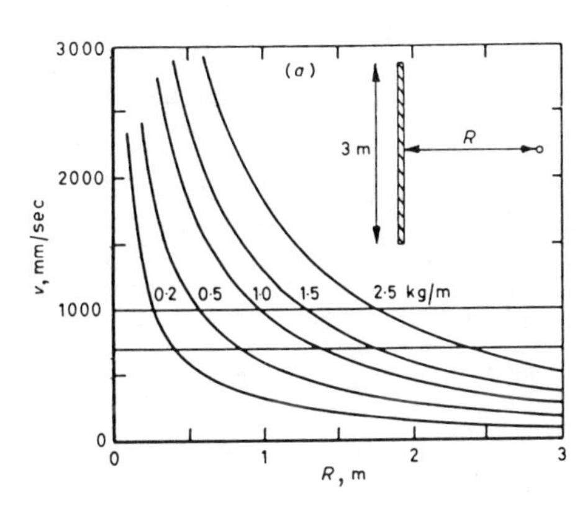

a)

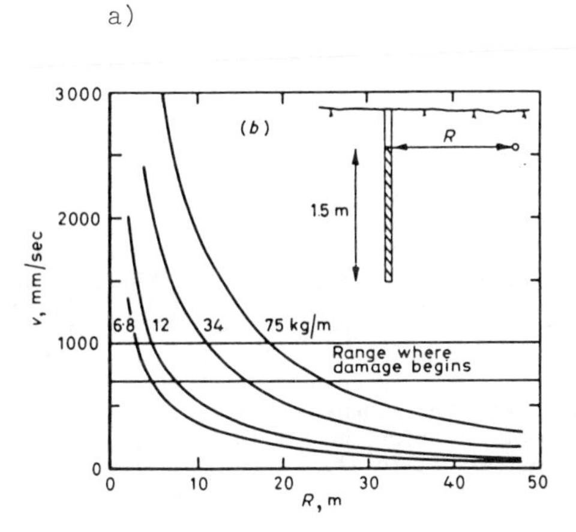

b)

Abb. 3 Darstellung der abgeschätzten Schwingungsgeschwindigkeit in mittelhartem Fels als Funktion der Entfernung für unterschiedliche Sprengladungen (kg/m Bohrloch).
a) Ladungslänge 3 m, typisch für Schussbohrlöcher im Tunnelbau,
b) Ladungslänge 15 m, typisch für den Tagebau.

beobachteten Geschwindigkeiten eines großen Bereichs von Entfernungen und Bohrlochdurchmessern überein (Persson, Holmberg & Persson, 1977). Sie können zur Festlegung der Sprengtechnik für Profil-Sprengungen herangezogen werden, um das Gebirge außerhalb des zu lösenden Bereiches möglichst zu schonen. Es reicht dabei nicht aus, die lineare Ladungsdichte in den Kranzbohrlöchern niedrig zu halten. Auch die nächsten, von den Kranzbohrlöchern entfernt liegenden Lochreihen müssen ebenfalls mit einer niedrigen Ladungsdichte versehen werden. Andernfalls dehnt sich die von ihnen ausgehende Sprengwirkung über die Kranzbohrlöcher aus. An anderer Stelle sind Einzelheiten hierzu veröffentlicht worden (Holmberg und Persson, 1979).

Das Schwingungsfrequenzspektrum und Kriterien für den Bruch

Genaugenommen kann die kritische Schwinggeschwindigkeit als Sprengkriterium nur in einem mittleren Frequenzbereich herangezogen werden. Der entscheidende Faktor ist die Wellenlänge der Schwingung im Verhältnis zur Größe des in Schwingung versetzten Gebirgsteiles. Der niedrige Frequenzbereich herrscht dort, wo der gesamte Fels einheitlich beschleunigt wird, da die Wellenlänge der Vibration viel größer ist als die Größe der Felsmasse. Deshalb ist hier die Verstellung das wichtige Kriterium. In hohen Frequenzbereichen dagegen kann die Beschleunigung der kritische Faktor sein.

Im Zweifelsfall müssen dort, wo Schäden am Fels oder an Gebäuden erwartet werden, die Bodenschwingungen aufgezeichnet werden. Zusätzlich sind die Spitzenwerte der Schwingungsbeschleunigung, die Geschwindigkeit und die Verstellung, zusammen mit der Schwingungsfrequenz, bei welcher diese Werte auftreten, zu bestimmen. Ebenfalls wird die Größe des in Schwingung versetzten Felsbereiches bestimmt. Durch Vergleich mit der Wellenlänge des Schwingungsmaximums kann die zusammengesetzte Umhüllende in Abb. 4 benutzt werden, um die maximal zulässige Schwingungsstärke zu bestimmen. Die Umhüllende grenzt den frequenzabhängigen kritischen Schwingungsbereich ein. Diese Art der Analyse sollte jedoch von einem erfahrenen Fachmann durchgeführt werden. Zudem sollte sie eine ingenieurgeologische Erkundung über Richtung, Größe und Häufigkeit der Haupttrennflächen, Störungen und Risse im Gebirge mit einschließen.

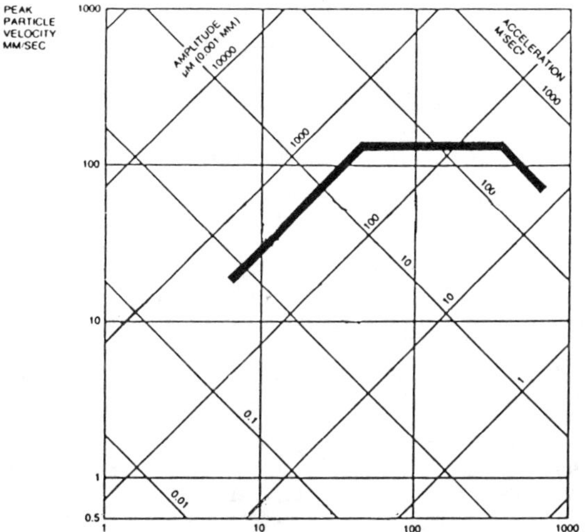

Abb. 4 Beispiel für die Bestimmung des kritischen Schwingungsbereiches für Gebäude

Vorspalten und Abspalten

Die Vorspalt- und Abspaltverfahren werden normalerweise angewandt, um den verbleibenden Fels beim Sprengvorgang möglichst zu schonen. Bei beiden Techniken ist die Besetzungsdichte in den Profilbohrlöchern viel niedriger als in normalen Sprengbohrlöchern. Das Ergebnis ist ein von Bohrloch zu Bohrloch verlaufender Riß, ohne daß der stehenbleibende Fels nennenswert gestört wird.

Beim Abspalten werden die Ladungen in den Profilbohrungen zum Schluß gezündet; beim Vorspaltverfahren dagegen zu Beginn des Sprengvorganges, oftmals in einem getrennten Gang. Das Vorspaltverfahren erfordert daher einen etwa um 50 bis 75 % engeren Profilbohrlochabstand, als das Abspaltverfahren und ist deswegen teuerer (Abb. 5). Die minimal benötigte lineare Ladungsdichte ℓ ist für beide Techniken gleich:

$$\ell = ad^2 \qquad (3)$$

mit $a = 90\ kg/m^3$ und d = Abstand der Bohrlöcher in m.

Die Ladungsdichte gemäß Gleichung (3) bewirkt einen Anfangsdruck im Bohrloch von etwa 100 MPa (1000 bar).

Auf diesem Kongress legen Bauer und Calder Ergebnisse vor, wonach der Bohrlochdruck bei weniger festem Fels durch weiteres Herabsetzen der linearen Besetzungsdichte noch niedriger gehalten werden kann.

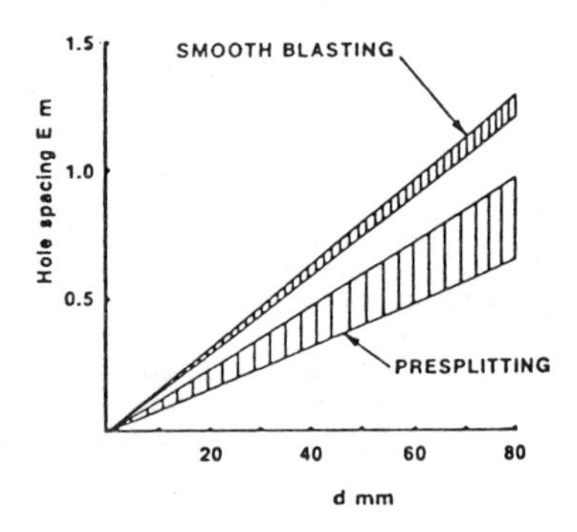

Abb. 5 Empfohlene Bereiche für Bohrlochabstände als Funktion des Bohrlochdurchmessers für die verschiedenen Vorspaltverfahren

Vorspaltverfahren $E = k_p d$,

mit $k_p = 8 - 12$

Abspaltverfahren $E = K_s d$,

mit $k_s = 15 - 16$.

Herstellen orientierter Risse

Eine interessante Entwicklung wird in einem Beitrag von Bjarnholt, Holmberg und Ouchterlony (1982) aufgezeigt. Durch linear geformte Ladungen werden in beiden Techniken richtungsorientierte Risse erzeugt. Durch Einkerben der Bohrlochwand und entsprechender Orientierung der Ladungen kann die Richtung der entstehenden Risse so festgelegt werden, daß Risse zweier benachbarter Bohrungen aufeinander zu laufen. Ähnliche Versuche wurden bereits früher von Thompson et al (1979) durchgeführt. Die orientierten Nuten im Bereich der Bohrlochwand wurden dabei mit einem Hartmetall-Gerät eingeschnitten. Wegen der hohen Fräskosten fanden diese Versuche nie eine großmaßstäbliche praktische Anwendung.

Pechalat und Lefin beschreiben in vorliegenden Beiträgen den Einsatz von Wasserdüsen zur Herstellung der Nuten.

Wir hoffen, daß die gegenwärtigen Entwicklungen zu einer kostengünstigeren und damit wirtschaftlichen, für praktische Zwecke geeigneten Anwendung der beschrieben Methoden führen.

LÖSEN DES FELS DURCH SCHNEIDEN

In diesem Bericht wird mit dem Ausdruck „rock cutting" das mechanische Lösen von Fels ohne Sprenganwendung verstanden. Somit schließt es das Bohren für Sprengbohrlöcher, das Bohren von (aufwärts getriebenen) Schächten, Tunnels, sowie das Schneiden von Fels mittels Wasserdüsen (sog. Jet-Cutting-Verfahren) ein. Die In situ Zertrümmerung (Zerkleinerung) wird in einem getrennten Abschnitt behandelt.

Vor einigen Jahren wurden große Hoffnungen an die Entwicklung unkonventioneller Schneideverfahren geknüpft. Dazu zählten unter anderem Versuche mit Geschossen, Laser- und Elektronenstrahlen, sowie mit Hochdruck-Wasserdüsen (Hydromechanik). Von diesen Verfahren hat sich für die praktische Anwendung allerdings nur das letztere durchgesetzt. In weicherem Gestein wird es allein durch den Wasserstrahl eingeschnitten. In härteren, jedoch meist sedimentären Formationen wird das Verfahren der Hydromechanik meist als Hilfsmittel bei der Lösearbeit mittels Stahl- oder Hartmetallwerkzeugen benutzt.

Große Anstrengungen wurden in den vergangenen Jahren im Hinblick auf die genaue Vorhersage des Bohrfortschrittes in verschiedenen Felsarten unternommen. Es will uns scheinen, daß auf diesem Gebiet praktische Anwendungsreife erreicht sei.

Zwei Arbeiten handeln diesmal von „water-jet"-Verfahren (Hydromechanik) als Hilfsmittel der Gesteinslösung und geben einen guten Einblick in den erreichten Stand auf diesem Gebiet.

Mechanisches Bohren

Mit der Entwicklung von mit Hartmetall besetzten Stahlkronen in den 50er Jahren wurde das schlagende Bohrverfahren vor allen anderen Verfahren immer häufiger zur Herstellung von Schussbohrlöchern im Fels eingesetzt. Noch heute dominiert das schlagende Bohren zusammen mit dem Einsatz konventioneller chemischer Sprengstoffe im Felsbau. Die Bohrmaschinen wurden von den ersten einfachen handgehaltenen pneumatischen Hammermaschinen mit einem Bohrfortschritt in der Größenordnung von 0,2 m/min zu den heutigen automatischen Bohrwagen mit hydraulischen Hämmern und einem Bohrfortschritt in der Größenordnung von 2 m/min weiter entwickelt. Das eigentliche Bohrwerkzeug kann dabei aus einem ein- bis vierschneidigem Meißel oder einer hartmetallbesetzten Krone bestehen.

Zögernd werden jetzt auch andere mechanische Felsschneideverfahren als Alternativen zum schlagenden Bohren und sogar als Alternativen zum konventionellen Bohren und Sprengen eingeführt. In relativ weichem Fels werden zunehmend Spiralbohrer mit Hartmetallschneiden für kleine bis mittlere Durchmesser für Schussbohrlöcher (30-150 mm) verwendet. Für größere Bohrlöcher von 150 mm aufwärts werden mit Hartmetall besetzte Rollenmeißel herangezogen. Durch den entsprechenden Andruck wird der Fels in der Bohrlochsohle zerkleinert und damit der Bohrfortschritt erreicht. Ähnliche Rollenmeißel finden bei der Herstellung von untertägigen, nach oben gerichteten Bohrlöchern selbst in hartem kristallinen Fels Anwendung. Die Rollen des Meißels sind dabei durch Lager auf einem rotierenden Schild befestigt. Dieses Schild wiederum ist mit einem Bohrgestänge verbunden, das durch ein kleineres Pilotloch aufwärts zu einem Tunnel führt. Solche Bohrköpfe haben Durchmesser von 0,5 - 3,0 m.

Der größte Teil der in der Welt aufgefahrenen langen und geraden Tunnel im Sedimentgestein werden durch Vollschnittmaschinen hergestellt. In weichem Sedimentgestein kommen gewöhnlich Maschinentypen infrage, deren rotierende Köpfe mit Reißzähnen besetzt sind, die hydraulisch vor und zurück bewegt werden können (Teilschnittmaschinen). Im mittleren bis hohen Festigkeitsbereich des Gebirges werden Vollschnittmaschinen eingesetzt. Diese besitzen ein rotierendes, scheibenartiges Schneideschild von annähernd der Größe des Tunnelquerschnittes. Auf diesem Schild sind schwere Stahlscheiben beweglich gelagert. Beim Bohrvorgang lösen diese Rollen im Bereich der Ortsbrust den Fels. Der rotierende Schild wird dabei hydraulisch gegen die Ortsbrust gedrückt. Die Schneidescheiben sind so auf dem Schild angeordnet, daß die durch sie hervorgerufenen kreisförmigen Vertiefungen auf der Felsoberfläche etwa 6-7 cm auseinanderliegen. Bei einem Andruck von 10-30t/Schneide werden handgroße Felsstücke an jeder Seite der Schneiden gelöst. Der Vortrieb pro Umdrehung liegt dabei etwa in der Größenordnung von 0,5-10 cm. Für die Anwendung in hartem Fels werden die Schneidscheiben mit Hartmetallkronen an ihrem Schneiderand besetzt sein. In einem nichtwasserführenden, standfesten Kalkstein (Druckfestigkeit 200MPa) kann dann z.B. ein Bohrfortschritt von ca. 30 m/Schicht einschließlich Arbeits- und Reparaturpausen erreicht werden.

Der grundlegende Ablauf der Felszerkleinerung unter der Wirkung eines harten Werkzeuges, das gegen die Felsoberfläche gepresst wird, bleibt gleich, ob das Werkzeug nun ein stumpfer Keil, ein halbkugelförmiger Dorn oder die Kante einer Schneidscheibe ist. Aus diesen Gründen kann die Bohrbarkeit durch exakt die gleichen Druck- und Verschleißversuche an Felsproben beschrieben werden. Dabei ist jedoch zu beachten, daß beim Einsatz von Tunnelbohrmaschinen mit Schneidscheiben das natürliche Trennflächengefüge des Felsens einen zusätzlichen Einfluß auf die Bohrbarkeit ausübt.

Falls keine anderen Kennwerte über den zu bohrenden Fels vorliegen, kann die einaxiale Druckfestigkeit zur Beurteilung herangezogen werden. In erster Näherung bedeutet eine hohe Druckfestigkeit ein langsames Bohren, hohen Verschleiß der Geräte und damit kostenintensive Herstellung des Hohlraumes. Bei der Herstellung eines Tunnels durch Bohren gibt es eine natürliche Grenze zwischen Sedimentgesteinen, die im Prinzip eine niedrige Druckfestigkeit besitzen und die leicht bohrbar sind sowie zwischen magmatischen, kristallinen Felsgesteinen mit einer hohen Druckfestigkeit und deswegen hohe Bohrkosten erfordern.

Bohrbarkeit des Felsmaterials

Der Begriff Bohrbarkeit wird meist etwas locker verwandt, um den Grad der Leichtigkeit und Wirtschaftlichkeit, mit der ein Fels durch eine bestimmte Mschine gebohrt werden kann, zu beschreiben. Er ist hauptsächlich eine Funktion des Gebirges, seiner Festigkeit und seines Trennflächengefüges. Er hängt aber auch zu einem gewissen Grad von der Wirtschaftlichkeit und den Eigenschaften der ausgewählten Maschine ab. Generell wird unterschieden zwischen Schussbohrlöchern im Durchmesserbereich von 25 - 500 mm (Drillability) und Schacht- und Tunnelbohrungen im Durchmesser zwischen 1 - 12 m (Boreability). Im folgenden werden die Techniken zur Vorhersage der Bohrbarkeit zusammengestellt, die z.T. von Selmer-Olsen und Blindheim (1970) und Blindheim (1979) am Norwegischen Institut für Technologie, Trondheim, und ebenso von einigen Herstellern von Tunnelbohrmaschinen entwickelt wurden.

Die Bohrbarkeit in Bezug auf kleinere Durchmesser (25 - 500 mm) wird beeinflußt von folgenden Faktoren:

Bohrfortschritt	(cm/min)
Meißelverschleiß	(charakterisiert durch die Länge eines Bohrloches in einem gegebenen Fels zwischen zwei Schneideschliffen)
Lebensdauer des Meißels	(Die Gesamtlänge des gebohrten Loches bis zum Aussondern der Krone)

Der Bohrfortschritt kann nach Blindheim durch den „Swedish-Brittleness-Test" bestimmt werden, bei dem eine Felsprobe durch ein fallendes Gewicht in einem Zylinder getestet wird. Außerdem durch einen Verschleißtest, den sogenannten Sivers-Test, bei dem die Eindringrate eines kleinen Hartmetallbohrers in die Felsprobe unter genormten Bedingungen gemessen wird. Diese beiden einfachen Versuche ermöglichen es, den sogen. Bohrfortschrittsindex DRI zu bestimmen, der eine lineare Funktion des S-Wertes des Brittleness-Tests ist, mit dem J-Wert des Verschleißtests als Parameter. Es besteht eine gute lineare Beziehung zwischen den DRI-Werten für verschiedene Felsarten und der gemessenen Bohrfortschrittsrate DRM unter Verwendung eines pneumatischen Schlagbohrgerätes mit einer Hartmetallkrone von 33 mm Durchmesser. Abb. 6 zeigt die Streuung der gemessenen Daten gegenüber der linearen Beziehung.

Für größere Bohrlochdurchmesser als 33 mm und schwerere Maschinen sind die DRM-Werte bei einem bestimmten DRI-Wert niedriger als in Abb. 6. In diesem Falle müssen Versuchsergebnisse von Gesteinen mit bekanntem DRI-Wert herangezogen werden. Mit der neuen Generation hydraulischer Schlagbohrmaschinen werden höhere Bohrfortschrittsraten erreicht, da sie einen größeren Meißelandruck besitzen und eine höhere Schlagfrequenz aufweisen. Abb. 7 zeigt die Korrelationskurven zwischen DRI und praktischen Bohrfortschrittsraten für verschiedene Bohrausrüstungen. Diese Diagramme resultieren aus der praktischen Erfahrung bei zahlreichen Baustellen in Norwegen.

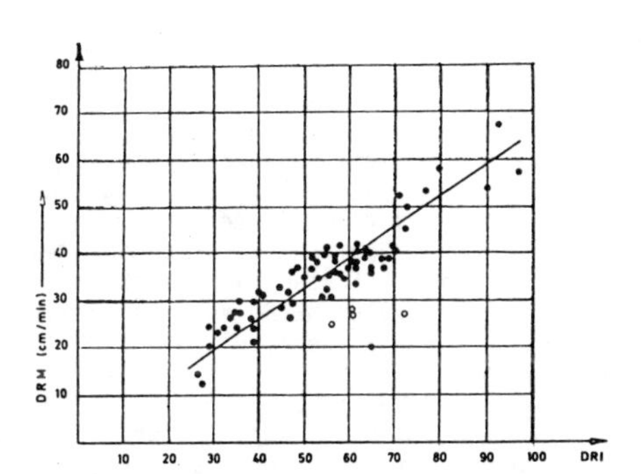

Abb. 6 Korrelationsdiagramm für den Bohrfortschrittsindex (DRI) und dem gemessenen Bohrfortschritt (DRM) in Feldversuchen. Die Bohrungen wurden mit einer leichten Bohrausrüstung und einem 33mm-Meißel durchgeführt.

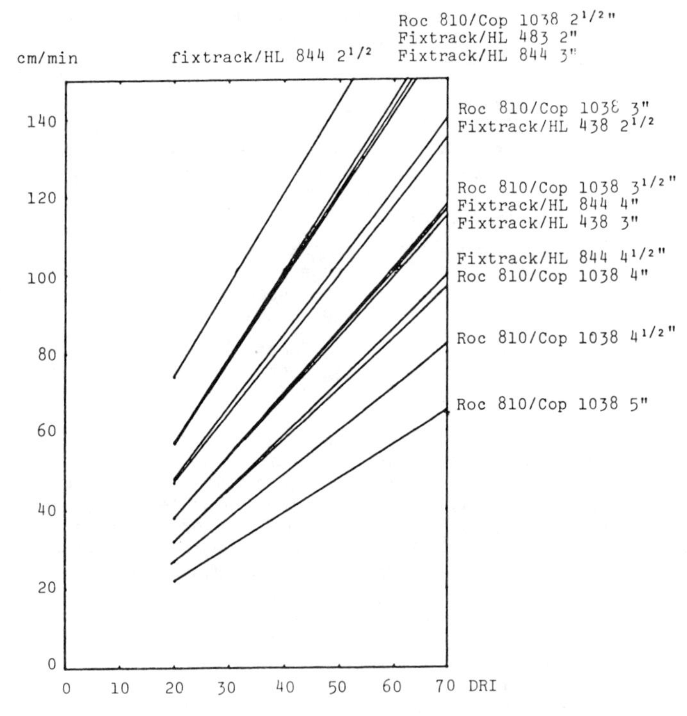

Abb. 7 Geschätzter Bohrfortschritt für verschiedene Bohrausrüstungen als Funktion des Bohrfortschrittsindex (DRI). Pneumatische Bohrungen: Luftdruck 60MPa, Marschlänge 3,05 m; hydraulische Bohrungen: Marschlänge bei HL 844 und HL 438 3,05 m, bei COP 1038: 3,66 m (nach Johannessen, 1981).

Der Meißelverschleiß ist das Ergebnis der Abrasion des Hartmetallmeißels durch die härtesten Gesteinspartikel. Der Verschleiß steigt sowohl mit dem Gehalt an Quarzit oder anderen, gleich harten Mineralien im Fels, als auch mit dem Bohrfortschritt. Die Abrasivität verschiedener Felsarten bei Bearbeitung durch einen bestimmten Hartmetallmeißel kann durch einen Abrasionstest bestimmt werden. Dabei läßt sich erkennen, daß die Abrasion nicht nur vom Quarzgehalt, sondern auch von anderen Bestandteilen des Gebirges abhängt.

Gerundete Quarzkörper sind oft weniger abrasiv als scharfkantige. Im allgemeinen gilt, daß Sedimentgesteine weniger abrasiv sind als Erguss- oder metamorphe Gesteine bei gegebenem Quarzgehalt.

Der Meißelverschleiß kann als Meißelverschleißindex BWI, der eine lineare Funktion des Abrasionswertes ist, ausgedrückt werden, wobei der Bohrfortschrittsindex DRI einen Kennwert darstellt. Es besteht eine gute Korrelation zwischen dem tatsächlich gemessenen Meißelverschleiß BWM und dem Bohrfortschrittsindex DRI. Der gemessene bzw. festgestellte Meißelverschleiß wird in der Einheit μm/m Bohrlänge ausgedrückt. Günstig ist ebenfalls die Festlegung des Meißelverschleiß als die Länge des Bohrloches in m, das in einem gegebenen Fels bei einer gegebenen Ausrüstung jeweils zwischen 2 Meißelneuschärfungen hergestellt werden kann. Diese Festlegung ist empfehlenswerter als die mehr wissenschaftliche Einheit μm/m Bohrlänge, die die Dicke des Kronenabriebes während eines gebohrten Meters angibt.

Unter der Lebenszeit eines Meißels versteht man die Gesamtlänge von Bohrlöchern, die von einem bestimmten Meißel mit einem bestimmten Bohrgerät unter bestimmten Luft- oder hydraulischen Druckbedingungen während des Bohrens in einem bestimmten Gebirge gebohrt werden können. Naturgemäß besteht eine Beziehung zwischen der Lebensdauer eines Meißels und dem Meißelverschleiß, die formelmäßig ausgedrückt werden kann. Allerdings beeinflußt die in der Praxis tatsächlich vorgenommene Häufigkeit des Meißelschärfens oder die Entscheidung, wann ein Meißel unbrauchbar geworden ist, die Ergebnisse.

Vorhersage der Bohrbarkeit bei Tunnelbohrungen

Beim mechanischen Vortrieb wird im Tunnelbau am häufigsten die Vollschnittmaschine eingesetzt. Der folgende Abschnitt befaßt sich daher nur mit dieser Vortriebsart.

Die Bohrbarkeit läßt sich durch folgende Faktoren beschreiben:

Effektiver Bohrfortschritt in m/h

Kosten des Schneidewerkzeuges in $/Stck (Abhängig vom Verschleiß)

Gesamte Ausbruchkosten in $/m Tunnellänge.

Die effektive Vortriebsgeschwindigkeit ist durch die Länge eines Tunnels gegeben, der in einem bestimmten Gebirge unter Verwendung einer bestimmten Tunnelbohrmaschine pro Stunde hergestellt werden kann. Dabei wird nur die tatsächliche Bohrzeit gerechnet. Dies gilt für normalen Andruck und normalen Schneidkopf.

Die formelmäßige Beziehung für den effektiven Bohrfortschritt P lautet wie folgt:

$$P = 60 \cdot N \frac{P_e}{100}$$

Die effektive Vortriebsgeschwindigkeit wird bei Tunnelbohrungen mehr noch als bei Bohrungen für Schussbohrlöcher von geologischen Schwächezonen bzw. Klüften beeinflußt, besonders, wenn der Abstand der Klüfte in der Größenordnung des Abstandes zwischen den einzelnen Schneidezähnen liegt. Die Lösbarkeit des Felsmaterials dagegen kann sowohl beim Tunnelbohren als auch für Bohrungen mit kleinem Durchmesser unter Anwendung des Bohrfotschrittsindex DRI bestimmt werden. Abb. 8 zeigt den Zusammenhang zwischen der effektiven Bohrgeschwindigkeit und dem DRI-Index in Abhängigkeit vom Kluftabstand.

Vergleichsweise kann auch der Kostenaufwand für das Schneidematerial - ausgedrückt als $/m³ gebohrten Fels - unter Verwendung des oben beschriebenen Meißelverschleißindex BWI und des Kluftabstandes als zusätzlicher Kennwert bestimmt werden.

Wegen der großen Vorhaltekosten für eine Tunnelbohrmaschine werden die Gesamtausbruchkosten vom Verhältnis der Arbeitszeiten zu den Stillstandszeiten während einer durchschnittlichen Schicht beeinflußt. Ein schwieriger Punkt ist die tatsächliche Fortschrittsleistung während einer Schicht und wir möchten Neulingen auf diesem Gebiet hierzu empfehlen, sich an die entsprechenden Ausführungen eines erfahrenen Unternehmers, Duri Prader (1977) zu halten.

Gesteinsfestigkeitsversuche zur Vorhersage der Tunnelbohrbarkeit

Wir haben bereits eingangs erwähnt, daß beim Einsatz einer Tunnelbohrmaschine die Frage nach der einaxialen Druckfestigkeit der verschiedenen Gesteinstypen im Vordergrund steht. Erste Angaben hierzu stammen aus der geologischen

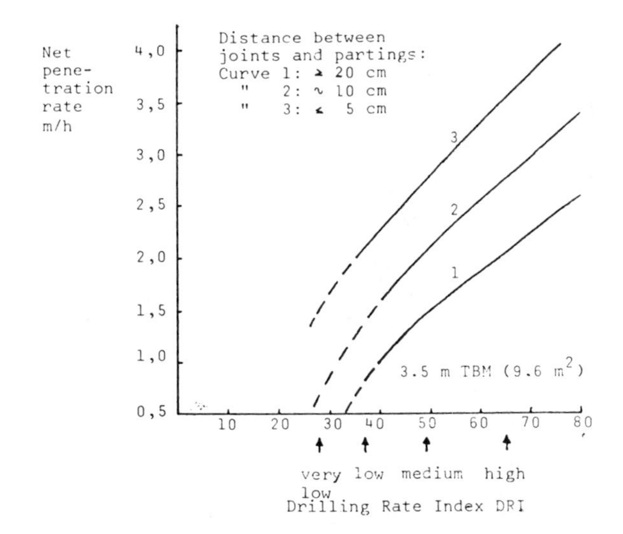

Abb. 8 Effektive Bohrgeschwindigkeit als Funktion des Bohrfortschrittsindex und des Kluftabstandes

Vorerkundung. Zudem können unterschiedliche Gesteinsversuche an Bohrkernen aus der Erkundungsphase durchgeführt werden (s. hierzu Handewith 1969, Lindqvist 1982 und Robbins 1977). Handewith definiert für Kernproben einen Penetrationsindex δ_i als das Verhältnis der aufgebrachten Kraft zur bleibenden Verformung. Beim Versuch wird ein Meißel in eine Felskernprobe getrieben, wobei die Probe so befestigt wird, daß sie nicht auseinanderbrechen kann. Abb. 9 zeigt typische Versuchskurven mit zunehmenden Be- und Entlastungszyklen. Tab. 2 enthält einen Vergleich von vorhergesagten und tatsächlichen Vortriebsgeschwindigkeiten für 3 verschiedene Gesteinsarten.

Tab. 2 Vorhergesagter und tatsächlicher Bohrfortschritt unter Verwendung des Lawrence-Fortschrittsindex

Felsmaterial	Druckfestigkeit in PSI	Bohrfortschrittsrate vorhergesagt FPH	tatsächlich FPH
Kalkstein, dolomitisch	31,960	5,8	6,1
Quarzit	36,170	2,1	2,5
Kieselschiefer	19,200	6,1	6,4

Felsschneiden unter Verwendung des Jet-Cutting-Verfahrens (Hydromechanik)

In der Tagebauindustrie werden hydraulische Methoden schon seit langer Zeit angewandt um Überlagerungsmaterial wegzuspülen. Diese Methoden sind dann auch im Untertagekohlebau von 'Kaiser Resourses'in Westkanada angewandt worden.

Gates et al (1978) diskutierten die Vorteile des Einsatzes hydraulischer Methoden für den untertägigen Abbau von Ölsand. Sie stellten dabei die hiermit verknüpften Vor- und Nachteile heraus. Wegen der heutigen langsamen Entwicklung im Ölgeschäft kommen diese Methoden allerdings nicht richtig zum Zuge.

Mathews (1980) berichtet über eine Bohrlochabbautechnik, die vom US-Büro of Mines getestet wurde. Vertikale Verbindungsbohrlöcher wurden hierzu von der Erdoberfläche bis in die Kohlenflöze abgeteuft, ein rotierendes Düsensystem schnitt dann im Bereich des Flözes bei hohem Wasserdruck die Kohle heraus. Das Kohlen-Wassergemisch wurde dann zur Tagesoberfläche hochgepumpt (s. Abb. 10).

Um harten Fels zu schneiden, werden allerdings viel höhere Wasserdrücke benötigt. Rehbinder (1977) führte Versuche in sehr festem Felsmaterial mit unterschiedlicher Korngröße und Durchlässigkeiten durch. Dabei kamen Wasserdrücke im Bereich von 150 bis 460 MPa zur Anwendung. Rehbinder betont, daß die Geschwindigkeit für das Schneiden eines Schlitzes sehr stark vom Verhältnis des Wasserdruckes zum Schwelldruck des Felsens abhängt. Falls der Wasserdruck größer ist alsder Schwelldruck des Felsens, werden Gesteinspartikel mit einer Geschwindigkeit gelöst, die mit der Wassergeschwindigkeit vergleichbar ist. Gesteinspartikel werden nur solange gelöst, wie der Wasserdruck am Ende des Schlitzes groß genug ist. Die Erosionsresistenz vom Fels ist also eng verbunden mit seiner Durchlässigkeit.

Einige Autoren berichteten zwar über eine erfolgreiche Anwendung der o.g. Schneideverfahren in hartem und abrasivem Fels, aber es sind einige Probleme damit verbunden, die den Anwendungsbereich des Jet-Cutting-Verfahrens als einzige Möglichkeit zum Ausbruch oder zum Lösen des Felsens einschränken.

a) in hartem Fels können zwar zwei dicht nebeneinanderliegende parallele Schlitze leicht angelegt werden, aber der Fels zwischen den Schlitzen bleibt dabei im wesentlichen ungestört. Das aber heißt, daß der Einwirkwinkel des Wasserstrahls ständig geändert werden muß, um den Fels zu zerkleinern.

b) Die sehr hohen Wasserdrücke, die bei hartem Fels gebraucht werden, erfordern eine automatische Führungsvorrichtung. Wegen der oft rauhen (gezackten) Felsoberfläche, die bei Ausbrüchen in hartem, geklüftetem Fels erzielt wird, kann dies zu Problemen führen.

c) Die aufzuwendende Kraft oder Energie muß sehr groß sein.

Vielversprechender erscheint es daher, das Jet-Cutting-Verfahren zusammen mit anderen mechanischen Möglichkeiten für den Felsausbruch einzusetzen. Dies könnte eine gute Alternative zum einseitigen Anwenden der üblichen Schnitt- oder Bohrmaschinen werden. Mit Hilfe des Jet-Cutting-Verfahrens werden zusätzlich freie Oberflächen geschaffen, die den Lösevorgang für mechanische Werkzeuge erleichtern. Summers et Al (1977) berichten über ein Wasserdruckbohrverfahren in Sedimentgesteinen und kristallinem Fels. Wegen seiner niedrigen Durchlässigkeit war der rote Granit schwerer bohrbar als der Berea-Sandstein. Pechelat und Lefin stellen auf diesem Kongress einen Beitrag über Bohrausrüstungen für den Untertagebau vor. Eine interessante Anwendung, die den selektiven Abbau betrifft, wird erwähnt. Dabei werden in Bohrlöchern, die für orientierte Risse eingesetzt werden, Kerben mit Hilfe des Jet-Cutting-Verfahrens eingebracht. Zusammen mit einer leichten Sprengladung kann hiermit ziemlich exakt das Erz vom tauben Gestein getrennt werden.

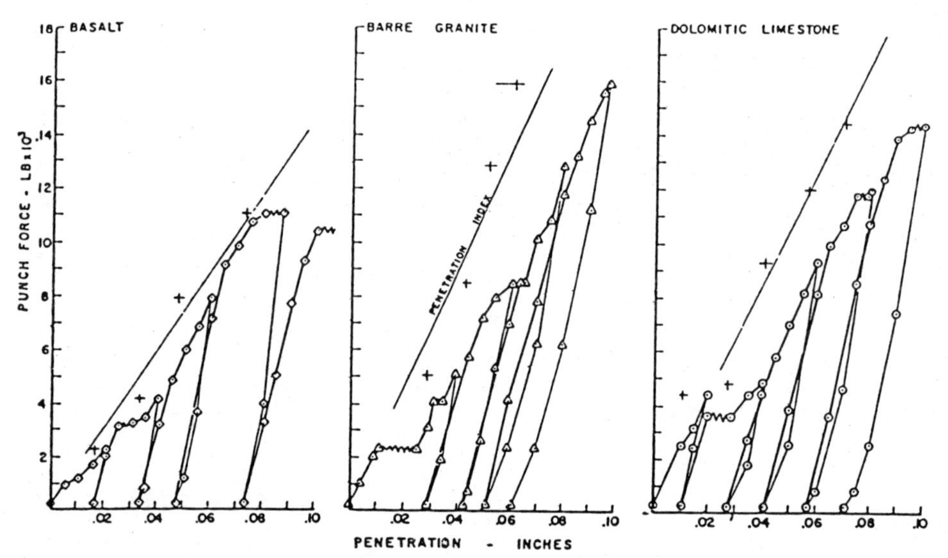

Abb. 9 Ergebnisse von Bohrkernstoßversuchen. Der Penetrationsindex (δ_i) steigt mit der Bohrschwierigkeit. Nähere Einzelheiten s. englische Version (nach Handewith, 1969).

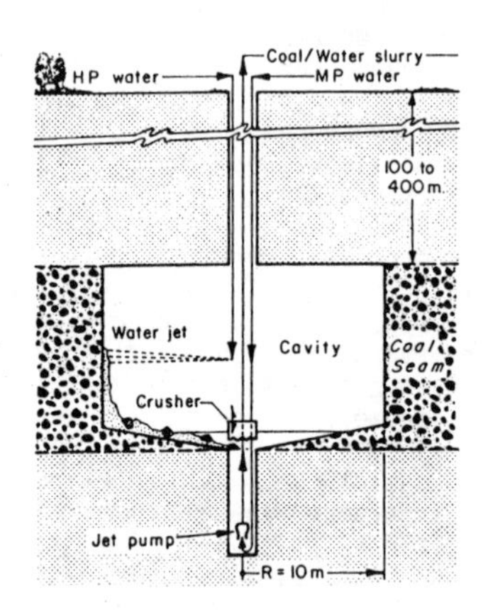

Abb. 10 Abbau eines Flözes durch Bohrungen

Über den Einsatz von Schneidköpfen, die mit dem Hochdruckwasserverfahren an Stelle der Schrämköpfe bei Strebabbaumaschinen für Kohle arbeiten, wird von Barker und Summer (1977) berichtet. Das Hochdruckwasserverfahren wird dabei eingesetzt, um verschieden orientierte Schlitze im Flöz zu treiben. Der Kohlehobel kann danach leichter in dem bereits vorzerlegten Kohlekörper arbeiten. Die Feldversuche mit dem entwickelten Hydrominer verliefen erfolgreich ohne Staubprobleme und mit einer akzeptablen Vortriebsrate.

Interessanterweise werden auch Versuche durchgeführt, um den technischen und wirtschaftlichen Anwendungsbereich für mechanischen Ausbruch durch Reiß- und Fräsmaschinen auszuweiten. Zur Zeit ist die technische Grenze dann erreicht, wenn die einaxiale Druckfestigkeit im Bereich von 120 - 150 MPa liegt. Die wirtschaftliche Grenze liegt etwa bei 70 MPa. Günstig wirkt sich ebenfalls das Zusammenspiel von der Jet-Cutting-Methode und sogenannten Picks aus. Ein hoher Wasserdruck erhöht die Eindringgeschwindigkeit, erniedrigt die Werkzeugkosten durch Kühlung der Picks, trägt ferner zu besseren Arbeitsbedingungen durch Reduzierung des Staubproblems bei und erniedrigt das Explosionsrisiko für Methan-Luftgemische. Fowell und Tecen berichten auf diesem Kongress über Experimente, die mit Wasserdruck unterstützten Werkzeuge im Sandstein und Kalkstein mit einaxialen Druckfestigkeiten von 40 - 150 MPa durchgeführt wurden. Die Ergebnisse zeigen, wie wichtig es ist, die Eindringtiefe des Druckwassers in Fels mit verschiedenen Festigkeiten zu berücksichtigen. Wenn diese Eindringtiefe größer ist als der Einschnitt, der durch das Bohrwerkzeug hervorgerufen wird, kommen dessen Seitenteile anstelle der Spitze in Kontakt mit dem Felsen. Dies führt zum Zerbrechen des Werkzeuges und verlangsamt damit den gesamten Schneidevorgang.

Hood (1976) berichtet über Laboruntersuchungen, bei denen Wasserdüsen direkt auf den Bereich vor einem Meißel während der Schneideoperation gerichtet waren. Die Meißelkraft wurde dadurch auf etwa 1/3 gegenüber fehlender Unterstützung der Wasserdüsen reduziert. Als Ergebnis dieser Laboruntersuchungen wurden zur Unterstützung des Schneidevorganges einige Untertageschneidemaschinen mit Wasserdüsen ausgerüstet. 1978 veröffentlicht Hood Resultate, die sogar eine noch bessere Arbeitsleistung ergaben, als die vorhergegangenen Laborversuche. Baumann et al (1980) beschreiben eine vielversprechende Methode für einen Felsausbruch mit hohem Wirkungsgrad. Durch den Einsatz zweier Hochdruckwasserdüsen, an jeder Seite einer Schneidscheibe (s. Abb. 11), läßt sich der Andruck des Schneidwerkzeuges reduzieren. Durch den Wasserstrahl werden Löcher im Fels erzeugt. Das zwischen den Löchern stehenbleibende Material wird durch die Scherwirkung der Schneidscheibe entfernt. Nach Ansicht der Autoren kann der erforderliche Andruck um mehr als 50 % vermindert werden. Das bedeutet eine entsprechende Verringerung des Maschinengewichtes und eine größere Beweglichkeit der Tunnelmaschine.

Die veröffentlichten Beiträge über mechanischen Ausbruch mit Hochdruckwasserdüsen zeigen sehr deutlich, daß die Wasserdüsen ein vielseitiges Instrument zur Verbesserung des Schneidevorganges darstellen. Allerdings scheint der spezifische Energiebedarf je Volumen gelösten Fels viel höher zu sein, als bei mechanischem Aushub. In Zukunft

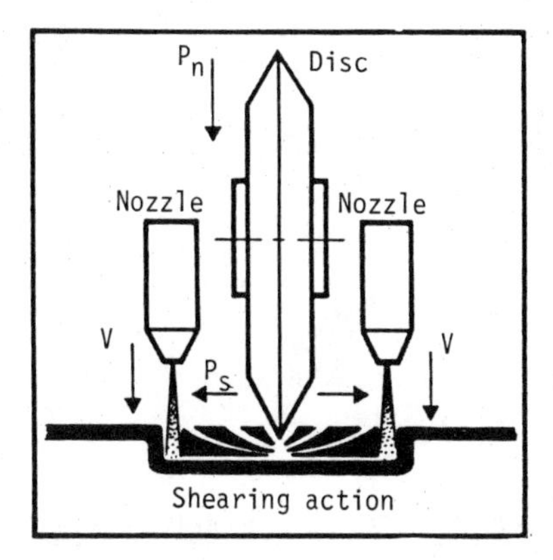

Abb. 11 Kombination des Jet-Cutting-Verfahrens mit einer Schneidscheibe. Nach Baumann et al, 1980

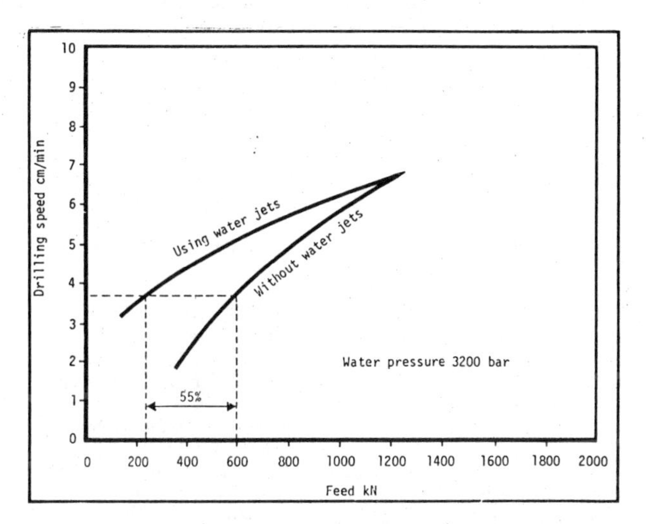

Abb. 12 Bohrfortschritt mit und ohne Unterstützung durch Druckwasser. Baumann et al, 1980

kann der Energiebedarf durch geeignete Wasserzusätze, welche die Fließgeschwindigkeit erhöhen, gesenkt werden.

FELSZERKLEINERUNG VOR ORT

Zum Zeitpunkt der Berichtsabfassung führt die weltweite Rezession zu einer zeitlichen Überproduktion von Erdöl. Langfristig gesehen wird jedoch die Nachfrage nach Erdöl durch den wachsenden Energiebedarf zweifellos ständig steigen. Da in Zukunft der Ölschiefer die Hauptquelle der Gewinnung von Erdöl sein wird, sind in den letzten Jahren neue Gewinnungsmethoden entwickelt worden.

Bei den Gewinnungsmethoden unterscheidet man zwischen einer eigentlichen in situ Methode und einer abgewandelten in situ - retorting. Bei der eigentlichen in situ Methode wird der Schiefer nicht gewonnen, sondern über Bohrungen an Ort und Stelle durch Wasser- oder Explosionsdruck aufgebrochen, um die gewünschte Durchlässigkeit zu erreichen. Anschließend wird das Öl durch Pyrolyse extrahiert und an die Oberfläche gepumpt.

Bei der Wasserdruckmethode reißen im Schiefer einzelne große Risse auf, die entweder den Schichtflächen folgen oder parallel zur größten Hauptnormalspannung ausgerichtet sind. Es fehlen jedoch Versuchsergebnisse darüber, daß durch diese Methode allein eine genügend große Durchlässigkeit erreicht wurde.

Bei der abgewandelten Gewinnungsmethode werden zunächst etwa 20 % des Ölschiefers mit konventionellen Bergbaumethoden gewonnen und zur Oberfläche gefördert. Dadurch entstehlt so viel Raum, daß sich die 80 % restlichen Gesteins ausdehnen können und in kaminartigen Schachtretorten im Untergrund zerkleinert werden.

Eigentliche in-situ Zerkleinerung

Um die Einsatzmöglichkeiten von flüssigen Sprengstoffen zum Aufbrechen der Ölschiefer zu untersuchen, sind vom 'Bureau of Mines' eine Reihe von Laborversuchen durchgeführt worden. Darüber wird detailliert von Miller et al (1974) berichtet. Man versuchte festzustellen, ob ein trockener, poröser Sandstein soviel einer Mischung von NG (Nitroglycerin) EGDN (Äthylenglycol Dinitrat) absorbieren könne, um eine explosive Ladung mit einer hohen Detonationsgeschwindigkeit zu erzeugen. Versuche mit einem trockenen Berea-Sandstein mit einer Dichte von 2,2 g/cm^3 haben gezeigt, daß dieses Gestein etwa 12 % seines Volumens an Flüssigkeit absorbieren konnte. Mit einem Zusatz von 73 g NG-EGDN gelang es ihnen, eine Probe (51x51x152mm), die 8,2 % NG-EGDN absorbiert hatte, zu zünden. Die Detonationsgeschwindigkeit betrug 4700 m/s. 15 % gelartiges NG (ρ = 1,37 g/cm^3), absorbiert in Steinsalz mit einem Durchmesser von 287 mm, führte zu einer Detonationsgeschwindigkeit von 1550 m/s. Das 'Bureau of Mines' konnte ebenfalls zeigen, daß NG-EGDN eine Detonationsgeschwindigkeit von 7500 m/s ergibt, wenn es in einem 1,6mm breiten Riss eingeschlossen war. Versuche in einer sandgefüllten Spalte führten zu Detonationsgeschwindigkeiten von 2100 m/s.

Etwa 5,5 l unempfindlich gemachtes NG wurde in eine vorgegebene Spalte eines Kalksteins mit einer mittleren Öffnungsweite von 3 mm geschüttet. Durch die Detonation wurde der Riss auf etwa 40 m verlängert und die Spaltenweite vergrößerte sich auf 70 mm. Der Kalkstein wurde horizontal in Richtung auf die senkrechte Wand des Kalksteinbruches verschoben. Die in Ölschiefer (Rock Spring und Green River) ausgeführten Versuche zeigten, daß NG detonationsfähig ist und daß sich die Explosion sowohl in mit Wasser und mit Sand gefüllten, natürlichen und durch 'Hydraulic-Fracturing' entstandenen Klüfte ausbreitet. Der Ölschiefer wurde dabei derart zerkleinert, daß eine Retoring-Verarbeitung tatsächlich möglich wurde. Durch die Schwierigkeiten bei der Kontrolle des Fließverhaltens des NG kann jedoch sein Gebrauch nicht empfohlen werden. In Bohrungen gezündetes, gekörntes TNT führte in einem Radius von bis zu 15 m zu einem intensiven Zerbrechen des Gebirges. Der Einwirkbereich konnte mit seismischen Methoden festgestellt werden. Luftstrommessungen zwischen den Bohrungen zeigten zwar das Vorhandensein von Rissen an, allerdings konnte der Grad des Verbruches nicht festgestellt werden.

Aufsprengversuche in Ölschiefer wurden vom 'Bureau of Mines' (Cambell et al, 1970)in der Green-River-Formation nahe Rock Springs, Wyoming, durchgeführt. Fünf Bohrungen wurden dabei bis zu einer Tiefe von 15 - 27 m abgeteuft. Vier wurden quadratisch mit einem Abstand von 7,6 m angeordnet. Die fünfte wurde im Mittelpunkt des Quadrates angeordnet. Bei diesem Versuch wurde 'Elektrolinking', Hydraulic-Fracturing mit und ohne Sandverschluß und Sprengzerkleinerung getestet.

Elektrolinking und Hydraulic-Fracturing ohne Sandverdämmung führte nur zu geringem Erfolg. Mit Sandverdämmung wurden horizontale Risse erzeugt, die brauchbare Fließkapazität aufwiesen. Versuche von Hydraulic-Fracturing mit Sandverdämmung wurden in zwei Bohrungen bei Tiefen zwischen 22-24 m und 24-26 m durchgeführt. Anschließend wurden fast 300 Liter unempfindlich gemachten NG in jede

Bohrung gegegeben, wobei die Flüssigkeit in die neu-angelegten Trennflächen eindrang. Direkt im Bereich der Bohrungen konnte nach der Detonation eine Bodenerhebung von etwa 50 mm festgestellt werden. Um den Grad der Zerkleinerung festzustellen, wurden Luftstromversuche zwischen ausgewählten Bohrungen ausgeführt. Es zeigte sich, daß die Durchlässigkeit signifikant zunahm, wenn die Ladungsmenge an NG etwa 300 Liter betrug. Eine 100 Liter Ladung führte zu einer niedrigeren Durchlässigkeit. Die 300 Liter Schüsse erhöhten die Aufnahmefähigkeit bis zum Achtfachen. Im Bericht wird allerdings nichts über das Ausmaß der vertikalen Zerreissung nach der Sprengung gesagt. Nach Burwell et al (1973) soll jedoch bei der Detonation von NG der Schiefer zu beiden Seiten der Brüche zerstückelt worden sein. Der Schiefer war derart zerbrochen, daß eine Retorting-Verarbeitung an Ort und Stelle möglich war. Im Verlauf des Versuches nahm die Sauerstoffausnutzung und die Brenngeschwindigkeit bis zum Testende ständig zu.

Burwell et al berichten auch über Erfahrungen eines anderen Versuches, (Versuch an der Stelle 7) wo vor dem Zünden der Schiefer nur mit der Hydraulic-Fracturing-Methode behandelt wurde. Zwei Zündversuche wurden durchgeführt, aber sie wurden abgebrochen, weil die Injektionsraten nicht aufrechterhalten werden konnten. Bei Sprengungen von 160 kg gekörntem TNT konnte die Durchlässigkeit in den Bohrungen gesteigert werden. Der Ölschiefer konnte problemlos weiter behandelt werden.

Coursen (1977) und McNamara et al (1979) haben gezeigt, daß durch echtes Bohrlochsprengen die Durchlässigkeit und die durch Explosion geschaffenen Hohlräume vergrößert werden können, falls die Versuchstiefen nicht zu groß und die tektonischen Spannungen nicht zu hoch sind. Es wird jedoch eine große spezielle Ladung benötigt (oder mehrere Schüsse), um eine genügend hohe Durchlässigkeit zu erzielen, damit eine hohe Ölgewinnungsrate entsteht.

Es ist wenig wahrscheinlich, daß die Bohrlochsprengmethoden die Extraktion von Öl aus Kerogen in naher Zukunft wirtschaftlich machen. Dabei scheint das Hauptproblem das zu sein, daß der Schiefer nur eine ungenügend große, natürliche Porosität aufweist (Parrish et al, 1981). Das Sprengen ist kostenintensiv und es ist schwierig, die Sprengstoffe so in den Griff zu bekommen, daß sie sowohl zur sicheren Handhabung eine genügend niedrige Empfindlichkeit aufweisen und zugleich in sehr engen Spaltendurchmessern eingesetzt werden können. Sehr wahrscheinlich sind die Bohrlochspreng- und/oder Hydraulic-Fraturing-Methoden dann nicht wirtschaftlich, wenn sie in größeren Tiefen bei entsprechend hohen Spannungen eingesetzt werden. Der Anwendungsbereich beider Methoden liegt damit zweifelsohne in oberflächennahen Ölschieferzonen, wo die Vorteile der freien Oberfläche genutzt werden können und wo eine aufwendige Entwicklungsarbeit vermieden werden kann. Eine besondere Methode stellt das sogenannte LOFRECO-Verfahren (Abb. 13) dar, das von Lekas (1979) und Britton (1980) beschrieben wird. Es wird dazu eine schwere Sprengladung eingesetzt, wobei die erforderlichen Risse und Hohlräume durch Hebung der Oberfläche entstehen.

Die Geokinetiker schlossen aus ihren Ergebnissen folgendes (s. Lekas, 1981):

Es ist möglich, Bohrungen oder Sprengbohrungen derart von der Oberfläche aus in den Ölschiefer anzusetzen, daß dort durch den Sprengvorgang eine Zone mit hoher Durchlässigentsteht und darüber bis zur Erdoberfläche eine rel. v undurchlässige Zone bleibt.

Es möglich ist, das zersprengte Gebirge zu durchbohren

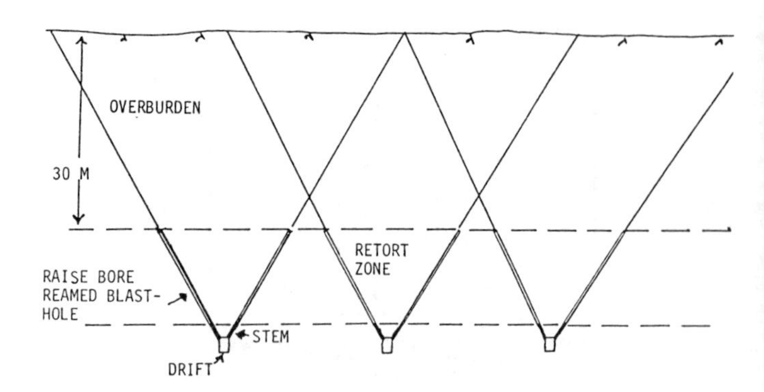

Abb. 13 Das LOFRECO-Verfahren nach Britton (1980)

und Bohrungen für die Ölgewinnung auszurüsten. Zudem kann der Schiefer in der gesprengten Zone punktuell entzündet werden, wobei sich anschließend eine über den gesamten Querschnitt der zersprengten Zone reichende Brennfront ausbreitet. Die Brennfront kann dann mit genügender Wirksamkeit über der Gesamtlänge des Abschnittes Ausbreitung finden.

Das entstandene Öl kann durch eine bis zum Boden der zersprengten Zone geführten Bohrung gewonnen werden.

Abgewandelte Gewinnungsverfahren (Modifiziertes in-situ-Verfahren)

Noch vor wenigen Jahren liefen mehrere großmaßstäbliche Versuche,um Erdöl durch eine modifizierte in-situ-Zerkleinerung und Verarbeitung zu gewinnen. Heute haben die Aktivitäten auf diesem Gebiet nachgelassen, da die Verfahren nicht sehr wirtschaftlich sind. Über unterschiedliche Ergebnisse dieses Verfahren berichteten Mignogna (1979), McCarthy und Cha (1976)Ridley (1978) und Ricketts (1980). Dabei hat man zunächst einen untertägigen Raum in der Größenordnung von 20 % des Gesamtvolumens geschaffen, ehe das übrige Gestein an Ort und Stelle zerkleinert wurde. Wegen des unregelmäßigen Zerklüftungsgrades und sogenannter Schornsteineffekte, die ein wirksames und auf eine Ebene beschränktes Ausbrennen verhinderten, war die Ausbeute oft zu niedrig. Moderne Untertagebaumethoden, wie etwa der Kammerpfeilerbau kombiniert mit einer Verarbeitung an der Oberfläche und dem Wiedereinfüllen des Bergematerials, scheint eine zukunftsträchtige Entwicklung für tiefergelegene Ölschiefervorkommen zu sein - wenn der Ölpreis wieder steigt.

Numerische Modelle für den Zerkleinerungsprozess

Vielfältige numerische Modellvorstellungen für das Zerbrechen von Fels sind in den letzten Jahren entwickelt worden. Einige berücksichtigen sogar alle Materialeigenschaften und die Bruchmechanismen des intakten Felsmaterials.

Wir glauben, daß in zukünftigen numerischen Modellen viel stärker als bisher die Anisotropieeffekte berücksichtigt werden müssen. Art, Ausbildung und Orientierung geologischer Trennflächen und Schwächezonen beeinflussen sehr stark die durch das Sprengen hervorgerufene äusserst wichtige Größenverteilung der einzelnen Felsbruchstücke. Eine gut fundierte Abschätzung der Verhältnisse hilft immer, die Kosten von Investitionen in maßgeschneiderte Brechanlagen und Transportsysteme zu senken oder sichert die Vorhersage über Brenngeschwindigkeit und Ausbeutungsgrad im Ölschiefergeschäft ab.

Die Festigkeit des Felsmaterials wird in der Regel an kleinen, intakten Kernstücken überprüft. Die Zugfestigkeit wird dabei oft mit einer Standardabweichung angegeben, die etwa 50 % der Größe des Mittelwertes entspricht. Ein Aufreißen findet vornehmlich in Schichten mit niedriger oder fehlender Festigkeit statt, wenn diese nahe am Detonationsort liegen. Alle Schwächezonen besitzen eine individuelle Festigkeit und reißen in Abhängigkeit von der Entfernung zum nächsten Bohrloch auf. Sofern größere Bohrlöcher zum Sprengen benutzt werden, wird der Grad der Zerstückelung noch stärker von den komplexen geologischen Verhältnissen beeinflußt.

Es ist anzustreben, in Zukunft auch die durch eine Bohrkernaufnahme gewonnenen räumlichen geologischen Daten zu berücksichtigen (oder ein statistisches Modell der Gebirgsanisotropie zu entwickeln). Außerdem sollte der Grad der Festigkeit von Schwächezonen in den Modellen berücksichtigt werden.

Die dynamische Belastung durch den Sprengvorgang kann für verschiedene Entfernungen im Umfeld der Sprengbohrungen nachvollzogen werden. Die Festigkeitskriterien der geologischen Schwächezonen zusammen mit der erwähnten dynamischen Belastung sind für die durch das Sprengen entstandenen Blockgrößen maßgeblich. Das Cundall-Block-Programm wurde von Cooper (1981) um Klüfte mit Kohäsion erweitert, so daß im Modell das Aufbrechen von Schwächeflächen und die Bewegungsabläufe in zerbrochenem Fels dargestellt werden kann. Abb. 14 zeigt einen einfachen Sprengvorgang im Bereich einer zweidimensional dargestellten Berme mit zwei freien Oberflächen. Man beachte das fortschreitende Zerbrechen des Modellkörpers, das zusätzlich das Aufbrechen geologischer Schwächezonen mit einschließt.

VERBESSERUNG VON BOHRLÖCHERN

Schon seit längerer Zeit werden in Bohrungen künstliche Risse erzeugt, um die Ausbeute undurchlässiger Gas- oder Ölfelder zu erhöhen. Um befriedigende Ergebnisse zu erhalten, müssen die künstlich erzeugten Risse eine genügend große Kontaktfläche zum Lager herstellen. Man hatte dazu eine Flüssigkeit in einer Bohrung einem so hohen Druck unterworfen, daß sich im Bohrloch Risse ausbilden, die sich bis in das umgebende Gebirge fortsetzen. Aufgrund abgesicherter Theorien kann man den erforderlichen Flüssigkeitsdruck und die Richtung der Rissausbildung vorhersagen. Da sich jedoch wegen des natürlichen Spannungsfeldes Risse in einer Ebene senkrecht zur kleinsten Hauptnormalspannung ausbreiten, kommt es zu einem unerwünschten Risswachstum ohne Kontakt zum Lager.

Einschränkungen bei der Anwendung der Hydraulic-Fracturing-Methode sind dadurch gegeben, daß sich nur ein Bruchsystem ausbildet, dessen Verbindungsbereich mit der Oberfläche von der Länge des Bruches bestimmt wird. Es wäre daher besser, eine Methode zu finden, die vielfältige Bruchstrukturen um das Bohrloch herum oder zwischen mehreren parallelen Brüchen erzeugt. Dies würde sowohl die Durchlässigkeit als auch die Kontaktzonen vergrößern.

Durch Anwendung der Sprengtechnik können zwar unregelmäßig verteilte radiale Risse erzeugt werden, die aber wegen des extrem kurzzeitigen Impulses nur eine begrenzte Ausdehnung erlangen. Die Länge der Risse wird durch die Überlagerungshöhe und mehr noch durch das natürliche Spannungsfeld bestimmt.

Ein höheres Spannungsfeld beeinträchtigt die Bruchausbreitung. Gleichzeitig richten sich die Risse bei einem nicht gleichförmigen Spannungsfeld außerhalb des Bohrloches nach der Hauptnormalspannung aus. Besonders in

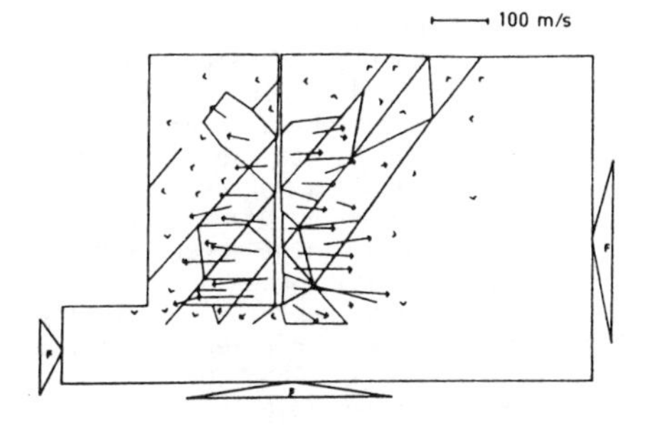

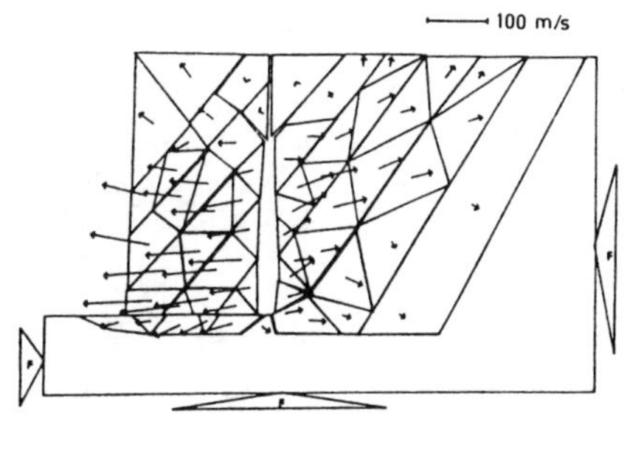

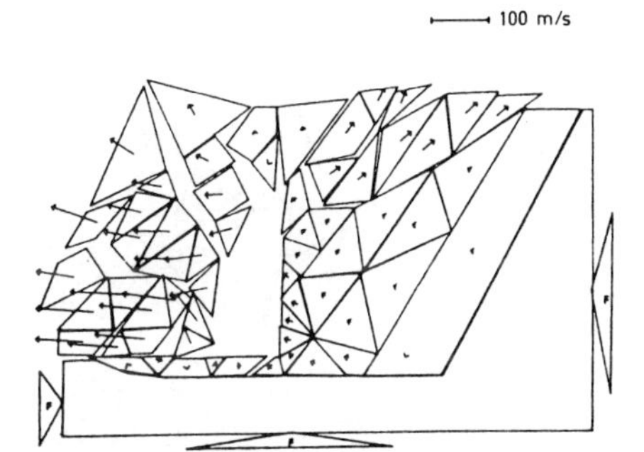

Abb. 14 Simulation einer Sprengung im Bereich einer Berme nach 1 ms, 5 ms und 50 ms.

weichem und porösem Gebirge entwickelt sich durch hohen Deformationsdruck und große Detonationsgeschwindigkeit feinkörniges Material, das vorhandene Risse verschließen kann. Der Druck im Bohrloch kann wirksam vermindert werden, wenn sich zwischen der Ladung und der Bohrlochwandung ein Luftpolster befindet.

Das Verhältnis zwischen dem Druck im Bohrloch bei einer Ladung mit Luftpolster (P2) und einer eng an der Bohrlochwandung anliegenden Ladung (P1) wird annähernd durch die folgende Gleichung gegeben

$$\frac{P_2}{P_1} = \left(\frac{d_2}{d_1}\right)^{\gamma} \qquad (4)$$

wobei mit d_1 der Bohrlochdurchmesser, mit d_2 der Ladungsdurchmesser bezeichnet wird. γ kann annähernd mit 1,5 festgelegt werden (γ ist in Wirklichkeit aber druckabhängig). Der Bohrlochdruck hängt bei einer voll anliegenden Ladung vom Detonationsdruck ab und kann näherungsweise mit der Hälfte dieses Druckes angegeben werden. Durch den Einsatz einer nicht anliegenden Ladung kann der Bohrlochdruck an die Festigkeit des umgebenden Felsens angepasst werden. Hierdurch können stark zerstörte Zonen vermieden werden. Aber immer noch ist der Detonationsimpuls zu kurz, um längere Risse zu erzwingen.

Alternativen zum flüssigkeits- oder sprengstoffunterstützten Aufbrechen von Bohrlöchern sind von Schmidt et al (1968) sowie Swift und Kusobov (1981) diskutiert worden. Mit einer Flüssigkeit in einem Druckkolben können wiederholte Belastungszyklen erreicht werden. Impulslängen können bis auf 1 ms verkürzt werden. Die Belastung variiert von 10^{-2} - 100 Mpa/ms. Versuche in Sandstein mit Wasser ergaben einige interessante Ergebnisse. Wenn der Sandstein halbtrocken oder feucht ist, ergeben sich vielfache Bruchstrukturen ohne Zerstörung des Bohrloches. Unter wasserfreien Bedingungen bilden sich nur zwei Risse aus. Abb. 15 zeigt die Rissausbildung unter verschiedenen Belastungsbedingungen.

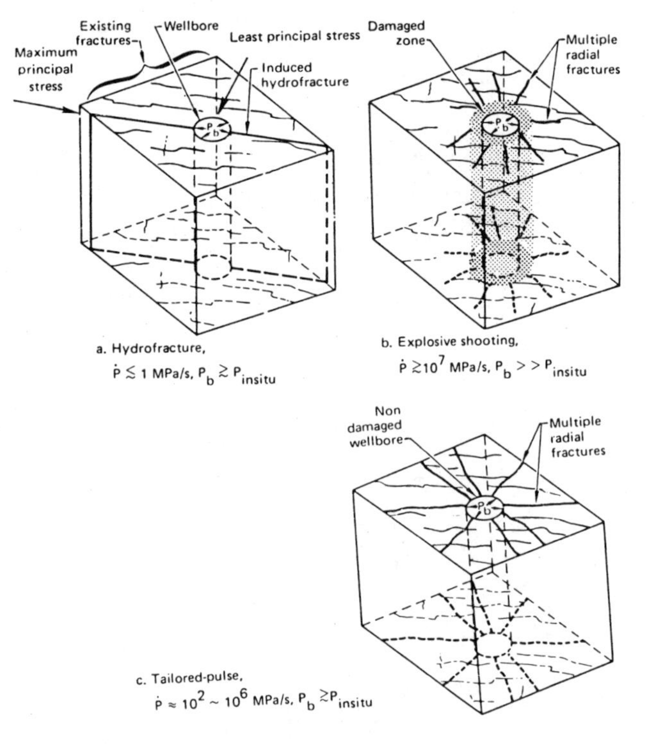

Abb. 15 Bruchmuster für verschiedene Belastungsbedingungen, Swift und Kusubov, 1981.

Warpinski et al (1979) berichten über Experimente, die im US-Nevada-Testgelände in einer Gesteinsfolge aus vulkanischem Tuff durchgeführt wurden. Drei Treibstoffe mit unterschiedlichen Brennzeiten kamen zum Einsatz. Die Druckbelastungsraten und die Spitzendrücke sind in Tab. 3 wiedergegeben. Von jeder Treibstoffart wurden etwa 9 kg gebraucht.

Tab. 3 Treibstoffe, die Warpinski et al 1979 benutzten

Treibstoff	Brennzeit	Ladungsgröße	Spitzendruck
GF # 1	~ 900 ms	$6.21 \cdot 10^{-4}$ MPa/μs	43.1 MPa
GF # 2	~ 9 ms	$1.38 \cdot 10^{-1}$ MPa/μs	95.1 MPa
GF # 3	~ 1 ms	< 10.3 MPa/μs	> 138 MPa

Im Untersuchungsgebiet war die Überdeckung 420 m mächtig. Die Zugfestigkeit des Tuffs betrug 2,8 MPa und die Festigkeit der Trennflächen wurde mit K_{ic} = 0,5 $MNm^{-3/2}$ angegeben.

Die Länge und die Anzahl von Trennflächen wurden nach dem Test auskartiert. Mittels eines beigegebenen Farbstoffes wurden auf einfache Weise neugebildete Risse erkundet.

Der langsamste Treibstoff GF#1 konnte lediglich eine Spalte aufreißen. Die Wirkung war also ähnlich dem Hydraulic-Fracturing-Verfahren. Der Riss vom Bohrloch aus hatte eine Länge von etwa 0,6 - 0,9 m.

Der mittlere Treibstoff GF#2 erzeugte 12 verschiedene Risse, von denen nur 5 eine Länge von < 0,3 m aufwiesen. Die anderen 7 Risse erreichten Längen von 0,6 bis 2,4 m.

Treibstoff GF#3 schuf viele kleine Risse von Längen < 0,12 m. In Bohrlochnähe war das Material stark zerbrochen. Lediglich ein Riss erstreckte sich radial bis zu einer Entfernung von ca. 1,2 m.

Die entstandenen Risse schienen nicht vom natürlichen Spannungsfeld beeinflußt worden zu sein.

Abb. 16 zeigt den gemessenen zeitlichen Druckverlauf für die drei Treibstoffe.

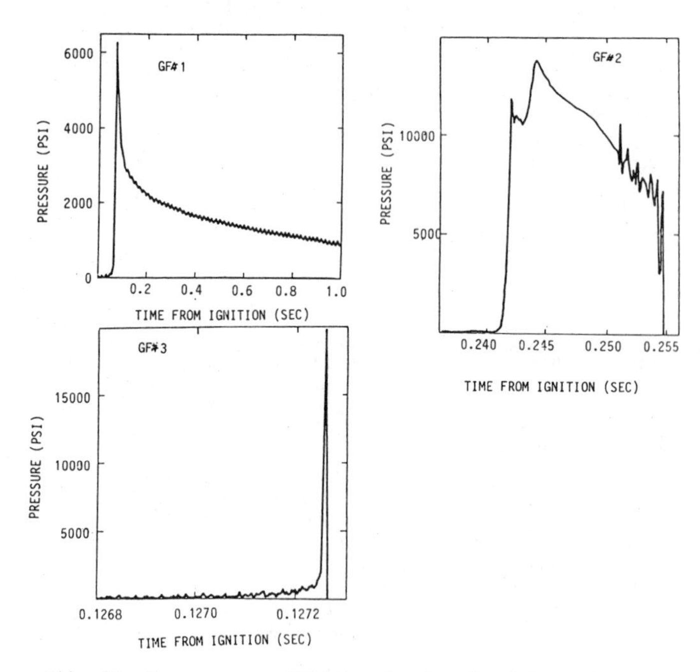

Abb. 16 Gemessener zeitlicher Druckverlauf der untersuchten Treibstoffe

Die Versuchsergebnisse machen deutlich, daß der Rissbildungsprozess stark von der Ladungsstärke abhängt. Liegt die Druckbelastung etwas höher als die natürliche Spannung, wird lediglich ein einziger Riss erzeugt. Wird Treibstoff benutzt, so dehnt sich dieser Riss nicht sehr weit aus, weil es unmöglich ist, die Druckbelastung wegen der begrenzten Gasmenge aus der Treibstoffladung für einen längeren Zeitraum aufrecht zu erhalten. Höhere Anfangsbelastungen und Druckspitzen führen zur Ausbildung vielfältiger Brüche. Die Anzahl der Risse hängt dabei von der Anzahl bereits vorhandener Risse oder Defekte im Bereich der Bohrlochwand, oder von der Korngröße des Materials ab. Damit sich die Risse verlängern, muß die Druckhöhe während einer längeren Zeit genügend hochgehalten werden. Der Druck sollte nach Möglichkeit unterhalb der Fließgrenze des Felsens gehalten werden, um

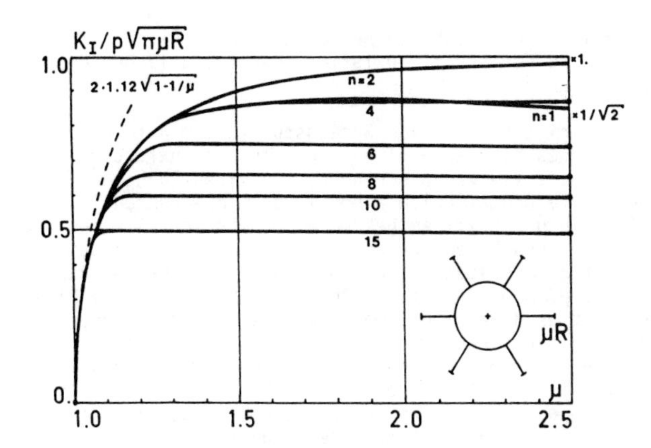

Abb. 17 Spannungsintensitätsfaktor für einen unter Druck gesetzten kreisförmigen Hohlraum, als Funktion der Risslängenparameter.

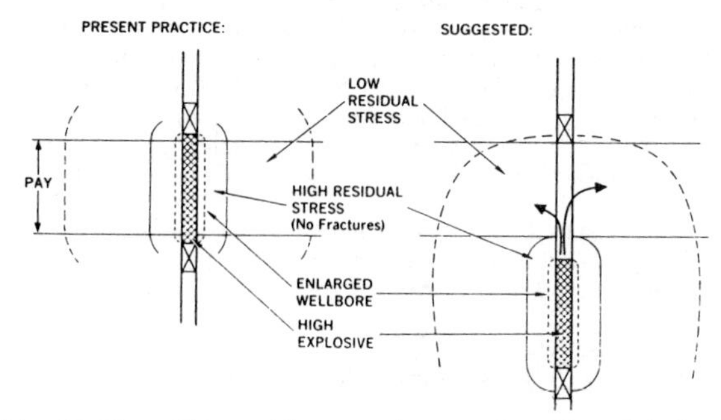

Abb. 18 Eine Alternative zum konventionellen Bohrlochschießen. Der Sprengstoff wird unterhalb der Abbauzone plaziert.

FRAME NO.	TIME (μs)	VELOCITIES BETWEEN FRAMES (ips) Charge Area	Stem Area
1	12.5	20,854 *	
2	34.0	13,760	
3	57.0	11,784	
4	80.5	12,493	
5	89.5	12,613	10,320 *
6	112.5	7,193	17,742
7	140.0	5,050	17,395
8	163.5		21,978
9	190.5		12,363
10	214.5		28,846
12	246.5		4,220
14	271.5		20,168
16	322.5		

* - Velocity probably low since fracture was not likely to initiate at the same instant the time count started.

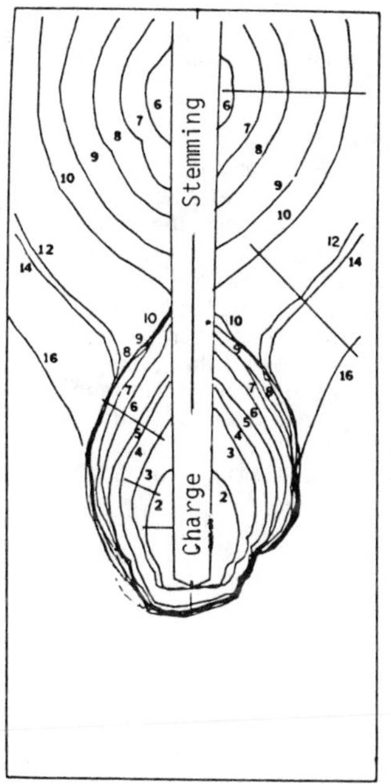

Abb. 19 Rissausbreitungsgeschwindigkeiten in der Ladungs- und Dämmzone. Nach Fourney et al, 1980.

Restspannungen zu vermeiden, durch die Rissbildung während der Phase nach der Spitzendruckbelastung verhindert würde (Schmidt et al, 1979).

Eine hohe Ladungsstärke und ein hoher Spitzendruck führt offensichtlich zu vielfältigen Bruchstrukturen. Leider aber bewirkt das Restspannungsfeld und das entstandene feinkörnige Material eine Verheilung der erzwungenen Klüfte. In einigen Fällen ist die Durchlässigkeit nach der Detonation sogar geringer als vorher. Wenn das natürliche Spannungsfeld niedrig ist und kompetentes Gestein vorliegt, kann man nach Abb. 3 die maximale Länge der hervorgerufenen radialen Risse bei voll anliegenden Ladungen unterschiedlicher Stärken abschätzen. Ouchterlony (1974) konnte zeigen, daß der zur Rissbildung erforderliche Druck unabhängig von der Anzahl der Risse ist, wenn es sich um kurze Risse handelt. Abb. 17 zeigt die Zusammenhänge zwischen Druckintensitätsverhältnis und der Risslängenparameter.

Gleichung 5 zeigt den Zusammenhang zwischen dem kritischen Druck P, um einen Bruch der Größe A zu erzeugen.

$$P = K_{Ic}/2{,}24\sqrt{\pi a} \qquad (5)$$

Dabei wird deutlich, daß bereits sehr niedrige Drücke zu Rissbildungen führen. Im Falle des o.g. vulkanischen Tuffs ($K_{Ic} = 0{,}5\ MNm^{-3/2}$) werden zur Erzeugung eines Defektes von weniger als 1 mm nur wenige MPa (statisch) als kritischer Druck benötigt. Eine vollanliegende Ladung kann Bohrlochdrücke entwickeln, die um ein Vielfaches größer sind als nötig. Selbst nichtanliegende Ladungen, die beim schonenden Sprengen Verwendung finden, führen zu viel zu hohen Bohrlochdrücken in der Größenordnung von 70 - 150 MPa. Schmidt et al (1979) haben vorgeschlagen, Sprengstoffe in Bohrlöchern unterhalb der Abbauzone einzubringen. Oberhalb der Abbauzone wird das Bohrloch abgedämmt, um niedrigen Spitzendruck und längere Druckzeit auszunutzen (s. Abb. 18).

Fourney et al (1980) haben in einem 51 mm hohen Plexiglasmodell hochgeschwindigkeitsphotographische Studien durchgeführt. An der Sohle eines Bohrloches wurde eine kleine PETN-Ladung gezündet. Das Bohrloch wurde oberhalb der Ladung abgedichtet, so daß ein Luftpolster zwischen der Ladung und der Abdichtung entstand. Der Abstand zwischen der Ladung und der Abdichtung betrug etwa das Vierfache der Ladungslänge. Entlang des Bohrloches konnte dann die Bildung von Rissen beobachtet werden. Dabei ergab sich, daß sich die Risse in der Nähe des abgedichteten Bereiches schneller entwickelten als im Bereich der Ladung. Gleichzeitig durchgeführte zeitabhängige Druckmessungen an verschiedenen Punkten entlang des

Bohrloches zeigten, daß nahe beim Abdichtungsbereich höhere Drücke auftraten und daß diese Drücke über einen längeren Zeitraum anhielten. Ein vergleichbarer Druckzeitverlauf konnte an keiner anderen Stelle des Bohrloches beobachtet werden. Dieses Phänomen unterstützte wirksam die Rissentwicklung nahe dem abgedichteten Bereich. Dabei wurden die Risse nicht so schnell wieder geschlossen wie im Bereich der Ladung, wo sich Restspannungen entwickelten und feinkörniges Material die Risse verheilte.

Auch andere Verfahrensweisen sind angewandt worden, um eine kontrollierte Bruch- oder Rissbildung zu erzeugen. Bligh (1974) berichtet über Versuche mit einem Brennstoff-Luft-Gemisch. Moor et al (1977) und Fitzgerald und Andersson (1978) beschreiben Methoden, bei denen Treibstoffe benutzt wurden, um Wasser oder Spülung so unter Druck zu setzen, daß sie sich in die bildenden Risse pressen, um sie zu erweitern. Schott et al (1977) beschreiben Versuche mit metallumhüllten, geformten Ladungen. Die Versuche mit chemischen Sprengstoffen wurden in Bohrlöchern zur Gewinnung von natürlichem Gas in devonischen Schiefern durchgeführt.

Allein die Anwendung der Hydraulic-Fracturing-Methode führte in einigen Schichten zur Erhöhung der Durchlässigkeit und der Flüssigkeitsproduktion. Diese Methode ist bereits seit vielen Jahren zur Steigerung der Produktion von Petroleum oder Gas eingesetzt worden. Leider ist es oft schwierig, den Prozess des Hydraulic-Fracturing-Verfahrens zu kontrollieren. Deswegen muß eine große Kontaktzone zu dem Schichtenkomplex, der z.B. Thermalwasser enthält oder zum Erdölreservoir führt, angelegt werden. Insbesondere in tiefer gelegenen Formationen bewirkt das natürliche Spannungsfeld, daß sich die erzwungenen Risse etwa senkrecht einstellen, wobei ihre Streichrichtung durch das örtliche tektonische Spannungsfeld bestimmt wird.

Weiterhin ordnen sich die neu entstandenen Risse oft parallel zu den bereits existierenden Trennflächen des Gebirges an.

Der alleinige Einsatz von Sprengstoffen zum Aufbrechen des Gebirges ist manchmal nur schwer zu rechtfertigen. Es erscheint vielversprechender, die Rissbildung durch abgestimmte Druckimpulse eines langsam brennenden Treibstoffes in den Griff zu bekommen. Jedoch ist es schwierig, einen genügend hohen Druckpegel über eine längere Zeit aufrecht zu erhalten. Um einen schnellen Gasverlust zu verhindern, muß man oft das Bohrloch gut abdichten. Dies erschwert es allerdings, mehrere Sprengungen hintereinander durchzuführen. Die Brennrate von Treibstoffen wird erheblich vom Umgebungsdruck und der Temperatur beeinflußt. Bei steigendem Druck erhöht sich die Brennrate exponentiell. Deswegen ist es unmöglich, zur Erhöhung der Belastungszeit größere Ladungen einzusetzen. Der Druckanstieg bewirkt einen Anstieg der Brennrate, die wiederum den Druck erhöht usw. Daher geht manchmal ein schnell ablaufender Verbrennungsprozess in eine Detonation über.

Es ist davon auszugehen, daß in der Zukunft zahlreiche Anwendungsmöglichkeiten entwickelt werden können, wenn wir erst unser Verständnis über gesteinsaufbrechende Vorgänge durch den Einsatz von Sprengstoffen und der Hydraulic-Fracturing-Methode erweitern, und wir müssen auch für tiefliegende Lagerstätten Methoden entwickeln, die ein kontrolliertes Aufbrechen im Gebirge ermöglichen.

Auch aus bereits bestehenden Bohrungen kann noch eine enorme Menge Öl und Gas gewonnen werden. Hierzu bedarf es nur kleinerer Verbesserungen der heute bestehenden Techniken.

Literaturverzeichnis s. englische Fassung

DISCUSSIONS / DISCUSSIONS / DISKUSSIONEN

ROCK DYNAMICS

Dynamique des roches

Felsdynamik

E. T. Brown
Chairman

GENERAL REPORT: Presented by Dr. Per Anders Persson and Roger Holmberg.

PAPER: Analytical Contribution To Tunnel Behaviour Caused By Blasting

AUTHORS: M. Hisatake, S. Sakurai, T. Ito and Y. Kobayashi

PRESENTED BY: M. Hisatake

DISCUSSION

Question: Dr. W. E. Bamford, Australia

Was it contemplated that you should measure dynamic strain as well on the inside of the tunnel lining? You measured particle velocity and simulated that in your computer program; you showed results for stresses which I presume were computed, not measured, and it seems to me that you had an excellent opportunity very seldom available to strain gauge the inner walls of the tunnel lining and measure dynamic strains. Could you comment on that please?

Answer: Prof. S. Sakurai, Japan.

I am a co-author and very pleased to answer your question.

In our paper we tried to demonstrate the mathematical computed result for stress distribution, calculated by finite element analysis. We did field measurement by using velocity gauge so that we measured only the particle velocity, and we tried to compare the particle velocity with strength. Stress is only a calculated value and particle velocity is computed and measured and the purpose of the paper is to demonstrate what a difference. Some people are quite interested to measure only the particle velocity, but particle velocity is not a good index for failure of material. Failure of material is due to the strain so the place were the maximum particle velocity occurs is different from the place where the maximum strain occurs. That's why we tried to demonstrate the position of where the large stress concentration occurs and which is different from the place where the large particle velocity occurs.

Question: Dr. Bamford

Well I don't think you really answered my comment that if the purpose of the investigation was to validate a computer code which would enable you to design blasts so as not to cause tunnel damage, or damage to existing tunnel linings, and if it's conceded that damage is caused by strain, why did you not think of measuring the strain to get even more value out of your investigation?

Answer: Professor Sakurai

Well I think that is true but we have quite a lot of experience at measuring particle velocity and strain in tunnel linings. But any time we measure the strain we have quite big problems. As you know in a tunnel there is very high humidity and such a very sophisticated measurement technique is not good in a tunnel.

That's why we like to measure only the particle velocity. This result is not included in our paper, but we finally found out the maximum allowable particle velocity for the concrete lining may be about 30 or 35cm per second, so I think this is a very good allowable value if we are involved in monitoring a project.

Question: Dr. C. Tanimoto, Japan

I think it is possible to determine the magnitude of allowable stress caused in lining in terms of measuring particle velocity, by referring the dynamic stress wave equation of $\sigma = \rho.C.V.$, where σ : dynamic stress, ρ: density of material, C: specific longitudinal wave velocity, V: particle velocity. I think this is a different opinion from Prof. Sakurai. I have several experiences like that so as Dr Bamford pointed out it's quite possible to measure the vibration in the lining also on the lining it gives the critical situation for the failure of the lining.

Answer: Prof. Sakurai

The equation you mentioned is that obtained for an infinite elastic ground without a tunnel when a plane wave propagates. Therefore, by using the equation, it is not possible to get accurate lining stresses produced by the blasting waves.

PAPER: Studies In Water Jet Assisted Drag Tool Rock Excavation

AUTHORS: R. J. Fowell And O.Tecen

PRESENTED BY: R. J. Fowell

DISCUSSION

Question: Mr. B. P. Knoop, Australia

I would like to ask Dr. Fowell about the amount of water to be used in his cutting arrangement in rock. He only showed us his experiments which were in the laboratory scale but he did show one large coal cutter in one of his slides. I'd like to know what is the amount of water he uses say per cubic metre of coal excavated.

Answer: Dr. R.J. Fowell

The results in the laboratory are just with one jet. The slide that I showed was of a Dosco Mark 2A road header adapted for high pressure water jet cutting. This was work done by the National Coal Board in Britain in a limestone mine. The quantity was 9 gallons per minute from 9 jets at 10,000 psi. I accept the inference that you are making that if there is a lot of water coming out of the head it is perhaps possibly going to damage the floor. The indications are that we are talking about quantities of water which are of the same order of magnitude for dust suppression by water and that is not considered to be a draw back of the system. If that covers your point, I think one of the points Dr. Holmberg mentioned against this particular method was that it uses a lot of energy and certainly it's a couple of orders of magnitude greater than you would normally get with normal mechanical excavation tools. The way I have approached it in terms of a philosphy is that we are going for an effective excavation technique which can be used where other techniques can't be used, so high specific energies, taking the energy of the jet and the energy put into the tool, are a minor consideration if there is no alternative.

The obvious way forward is where the alternative is possibly to use a full face drilling machine which is expensive, cuts a circular tunnel, takes a lot of time and is restricted in diameter and the radius which it will negotiate. So in this sense, effectiveness is more important than the efficiency.

PAPER: Limit Blast Design Evaluation

AUTHORS: C.K.McKenzie, P.D.Forbes, G.E.LeJuge, I.H. Lewis, P.A. Lilly and J.D. Lilly

PRESENTED BY: C.K. McKenzie

DISCUSSION

Question: Dr. T. L. By, Norway.

I have several questions to Cameron McKenzie's presentation but I will discuss them with him later. I have however, some comments to Fig. 8 in this paper. McKenzie denies the separation in successive vibration packets. To me it looks like there is a fairly good separation. The dominant frequency fits well to the delay time, so this frequency seems to me is nothing but the signal detonation of rows. There is however an increase in the amplitudes. I am not sure that this increase is due to the inadequate separation as the authors claim. The signal displayed in Fig. 8 shows sign of resonance problems somewhere in the equipment-chain. It could be interesting to hear what the authors mean about this.

The second question: there is a clear difference between the signal forms from the pre and post blast signals in Fig. 9. What experience do you have utilising the frequency spectra for rock quality evaluation, for example by using a transmitter with a known frequency?

Answer: Mr. C.K. McKenzie:

Thank you Dr. By. Your question is in fact many questions and I will attempt to answer them one by one. The Fig. to which you refer in the paper does in fact show the best observed data. There is typical vibration separation observed at Mount Isa and that bears little resemblance to fig. 8 in the paper. Enhancement is certainly not responsible for the high amplitude region in the centre of these records. This is the result of monitoring a large blast from a single centrally located point with the nearest holes producing the maximum amplitude. We would not expect enhancement effects in this blast to account for more than a 10% increase in vibration level. The enhancement effect is in fact better shown in this slide where there is clearly no separation whatsoever and we can see that at the point of detonation of row 3 for example the ambient conditions already indicate a significant vibration level. The possibility of resonance is ruled out but was certainly very closely considered but ruled out for the following reasons. Firstly, the geophones are heavily shunted to remove resonance.

Secondly, they have been compared and in fact overlaid with integrated signals from accelerometers, thirdly the modelling confirms the basic shape of the wave forms, which have been observed, fourthly, the vibration period which we measure is not that of the resonant frequency of the geophones and certainly this slide is not typical at all of a resonating geophone. The quoted frequency in this figure does in fact as you point out relate to the delay interval. The deconvolution of a vibration record to obtain the spectral characteristics of a single row or hole is most difficult, but Fig. 8 indicates that at this dominant frequency, the individual packets would be approximately 50 Herz.

GENERAL REPORT / RAPPORT GÉNÉRAL / GENERALBERICHT

SPECIAL TOPICS IN ROCK MECHANICS

Aspects particuliers de la mécanique des roches
Sonderthemen der Felsmechanik

F. H. Cornet
Institut de Physique du Globe, University of Pierre and Marie Curie, Paris, France

ABSTRACT

The stress field which exists in rock masses can be apprehended either thru direct measurements or seismotectonics observations. The most popular techniques for direct measurements are overcoring and hydraulic fracturing. Results indicate a continuous increase of the deviatoric component down to depth at least equal to 5 km. Lateral variations of the stress field suggest that intraplate tectonics, and therefore seismic activity, does not depend only on the stress field at the plate contacts ; its variation with time remains badly understood. A second important datum is the geometrical characterization of discontinuities. Although statistical methods provide means for local investigations, no satisfactory solution has yet been found for large scale domains (larger than a few km^3). Whilst large discontinuities are treated as such in the mechanical modelling of rock masses, small ones are integrated in the definition of equivalent continua. These equivalent continua are sometimes supposed to behave elastically ; however for large stress deviatoric components, elasto-plasticity must be considered in order to take into account microcracking and dislocation motion. Unfortunately the time dependency of microcracking, including stress corrosion effects, is not well understood yet. The modelling of large scale discontinuities involves always dilatancy with shear deformation, a mechanism for which the effective stress law is not that usually accepted. Once the three above steps have been overcome, solutions to rock mechanics problems may be tackled by physical modelling provided proper scaling factors are considered. However the advent of high speed computers has priviledged the development of numerical solutions : finite elements, boundary elements and jointed block models.

1. INTRODUCTION

Solving a continuum mechanics problem proceeds classicaly in three steps :
1. Definition of the geometry of the problem and identification of the boundary conditions ;
2. Choice of appropriate constitutive equations describing the rheology of the various materials under consideration.
3. Solution of the set of partial differential equations corresponding to the law of motion, the compatibility conditions (expressed in terms of displacements or stresses) and the constitutive equations so as to satisfy the boundary conditions.

The difficulty encountered in rock mechanics problems is that rocks are discontinuous, heterogeneous, polyphased materials which are not directly tractable by the classical continuum mechanics approach. Simplifying assumption must be made at the expense of a loss of accuracy in the solution.

However, thanks essentially to the coming of the age of electronics and computers more and more sophisticated numerical models are being developed and more and more reliable measurements are being conducted in-situ. The validity of the simplifying assumptions made in solving a particular problem can now often be established, or proved wrong, by comparison between expected and observed values of displacements or stresses. Although in many respects still and art, rock mechanics is developing more and more as a science. The purpose of this report is to present some recent developments which may help accelerate this evolution.

First the problem of stress determination is discussed. A few measuring techniques are presented along with some results relative to spatial and temporal stress variations. Then the problem of the geometrical characterization of discontinuities in rock masses is approached. Finally various models proposed for the representation of rock mass behavior are examined. They include finite element models, boundary element methods as well as various physical and numerical models for large deformation investigations.

2. STRESSES IN ROCK MECHANICS

In many rock mechanics problems, part of the boundary conditions are specified by stresses at " infinity " , that is far away from the domain under investigation. This has prompted the engineer to develop means for ascertaining the local stress field before undertaking large excavations. This knowledge of the stress field at infinity may involve only the determination of principal stress orientations ; it often requires also an estimation of their magnitude.

More recently the design of large hydraulic fractures in oil reservoirs has created the need to obtain a sound appraisal of stress variations with depth : injected fluid pressures must be adjusted so as to prevent undesired vertical fracture expansion and to maintain propagation within a chosen horizon (see e.g. Cleary, 1980).
Stress measurements have also been undertaken for a better comprehension of local seismicity and its corollary, induced seismicity caused by injection of fluid under pressure (see e.g. Handin et al., 1972). Further, measurements are conducted now in conjunctions with geodynamics for a better understanding of plate tectonics (e.g. Zoback and Zoback, 1980, Mc Garr 1982).
Considering these various problems, one may wonder whether they all deal with the same stress concept. In fact, in rock mechanics, the scale at which a stress is defined must always be specified. The global mechanical behavior of rock masses is analysed by considering equivalent continua which are supposed to represent the real rock mass only at a given scale. These equivalent continua loose their significance for volumes smaller than a critical value which may be considered as the physical equivalent of the point in the mathematical analysis. Within this critical domain the stress, in the equivalent continuum, is supposed to be uniform whilst in reality it is not. The problem of stress determination is to determine the value which exists in the equivalent continuum. It is therefore essential to insure that the scale of the stress measurement is the same as that of the actual problem .

By defining volumes large enough to overcome the problem of local heterogeneity, one is led to ignore local stress concentrations ; however these may influence drastically the behavior of the rock mass. When only global effects are considered, the influence of these local stress concentrations is taken into account by adjusting the constitutive equation. However when stability is concerned, the scale of the problem is dictated by that of local stress concentrations and this may complicate drastically the analysis. In fact, this choice of the proper scale is probably one of the main difficulties in rock mechanics problems as will be shown in this report.

2.1. Various stress determination procedures

Leeman (1964) conducted a detailed review of the various stress measurement techniques then available (borehole deformation cells, borehole inclusion stress meters, door-stopper, hydraulic jack methods, the photo stress method, resistivity and sonic methods). Fairhurst (1968) updated the review and added a detailed discussions of the hydraulic fracturing technique. In 1969 Leeman proposed another review of the then recent developments on measurement of stress in rocks (more than 350 references). More techniques have been presented in Kovari's " field measurements in rock mechanics " (1977). Accordingly rather than conducting a detailed review of all available techniques, only the general trends of present stress determination methods will be presented here, with special emphasis on the hypotheses implied by the various techniques along with the means available to ascertain the accuracy and the precision of the stress determination. However this last point is only seldom discussed in the literature and it is hoped that this report will encourage more work on this topic.

Stress relief methods

The classical stress relief methods are based on measurements of displacements caused by total stress relief. Relationships between displacements and stresses, determined from the linear theory of elasticity, are then used to compute some components of the " far field " stress tensor.
In the usual procedure, an initial borehole is drilled and a displacement measuring device installed. This hole is then overcored so as to release completely the stresses around the rock where the measuring device is installed. Various types of apparatus have been designed for this purpose ; all are supposed to be compliant enough to let the rock deform freely upon stress relief.
With borehole deformation cells, different combinations of diametral and longitudinal displacements are measured : three diametral measurements with the U.S. Bureau of Mines cell (Hooker and Bickel, 1974) as well as with the CERCHAR cell (Hellal and Dejean, 1981), four diametral measurements (Crouch and Fairhurst, 1967) ; four diametral and eight longitudinal measurements (Bonnechère, 1971).
With soft inclusion cells, instrumented either with strain gauges or inductive gauges, combinations of circumferential, longitudinal and inclined displacements measurements are conducted (Leeman, 1968, Rocha and Silverio, 1969, Blackwood, 1977, Pahl, 1977).
Whilst cells providing only diametral measurements imply that boreholes in at least three different directions be drilled for a complete stress determination, cells which combine radial and longitudinal or circumferential, inclined and longitudinal, displacements measurements require theoretically only one overcoring operation for the complete determination. All these methods are based on the solution of the infinite cylindrical hole in an isotropic elastic medium submitted to a triaxial field at infinity. The wellknown solution of this problem (see e.g. Kirsch, 1898, Hiramatsu and Oka, 1968) is given by equations (1) for stresses and (2) for displacements.
If σ^{∞}_{ij} , i,j, = 1,2,3 are the stress components at infinity (expressed in the geographical frame of reference) ;
ρ , θ , z are cylindrical coordinates (z is the axis of the borehole oriented positively toward the free end of the borehole) ;
E, ν are Young's modulus and Poisson's ratio of the rock ;
r is the borehole radius, then :

$$
1\begin{cases}
\sigma_{\rho\rho} = (1-\frac{r^2}{\rho^2})\frac{\sigma_{11}^{\infty}+\sigma_{22}^{\infty}}{2}+(1-4\frac{r^2}{\rho^2}+3\frac{r^4}{\rho^4})(\frac{\sigma_{11}^{\infty}-\sigma_{22}^{\infty}}{2}\cos 2\theta+\sigma_{12}^{\infty}\sin 2\theta) \\
\sigma_{\theta\theta} = (1+\frac{r^2}{\rho^2})\frac{\sigma_{11}^{\infty}+\sigma_{22}^{\infty}}{2}-(1+3\frac{r^4}{\rho^4})(\frac{\sigma_{11}^{\infty}-\sigma_{22}^{\infty}}{2}\cos 2\theta+\sigma_{12}^{\infty}\sin 2\theta) \\
\sigma_{zz} = \sigma_{33}^{\infty}-4\nu\frac{r^2}{\rho^2}(\frac{\sigma_{11}^{\infty}-\sigma_{22}^{\infty}}{2}\cos 2\theta+\sigma_{12}^{\infty}\sin 2\theta) \\
\sigma_{\theta z} = (1+\frac{r^2}{\rho^2})(\sigma_{23}^{\infty}\cos\theta-\sigma_{31}^{\infty}\sin\theta) \\
\sigma_{z\rho} = (1-\frac{r^2}{\rho^2})(\sigma_{31}^{\infty}\cos\theta+\sigma_{23}^{\infty}\sin\theta) \\
\sigma_{\theta\rho} = (1+2\frac{r^2}{\rho^2}-3\frac{r^4}{\rho^4})(\frac{\sigma_{22}^{\infty}-\sigma_{11}^{\infty}}{2}\sin 2\theta+\sigma_{12}^{\infty}\cos 2\theta)
\end{cases}
$$

$$
2\begin{cases}
u_{\rho} = \frac{1+\nu}{E}(\frac{1-\nu}{1+\nu}\rho+\frac{r^2}{\rho})\frac{\sigma_{11}^{\infty}+\sigma_{22}^{\infty}}{2}-\frac{\nu}{E}\rho\sigma_{33}^{\infty}+ \\
\quad +\frac{1+\nu}{E}(\rho+4(1-\nu)\frac{r^2}{\rho}-\frac{r^4}{\rho^3})(\frac{\sigma_{11}^{\infty}-\sigma_{22}^{\infty}}{2}\cos 2\theta+\sigma_{12}^{\infty}\sin 2\theta) \\
u_{\theta} = \frac{1+\nu}{E}(\rho+2(1-2\nu)\frac{r^2}{\rho}+\frac{r^4}{\rho^3})(\frac{\sigma_{11}^{\infty}-\sigma_{22}^{\infty}}{2}\sin 2\theta+\sigma_{12}^{\infty}\cos 2\theta) \\
u_z = \frac{2(1+\nu)}{E}(\rho+\frac{r^2}{\rho})(\sigma_{31}^{\infty}\cos\theta+\sigma_{23}^{\infty}\sin\theta)-\frac{2\nu}{E}z\frac{\sigma_{11}^{\infty}+\sigma_{22}^{\infty}}{2}+\frac{z}{E}\sigma_{33}^{\infty}
\end{cases}
$$

This solution has been extended to the case of planar isotropy (5 elastic constants) by Berry (1968) and later by Hiroshima and Koga (1977). With " Doorstoppers " a rosette of strain gauges is glued to the flattened end of a borehole. When coring is resumed, displacements caused by the stress relief are measured. The complete stress tensor determination requires at least three different borehole orientations. The elastic solution is based on empirical relationships derived for isotropic materials either by experimental techniques (Galle and Wilhoit, 1962 ; Leeman 1964 ; Bonnechère and Fairhurst, 1967) or finite element computations (De la Cruz, 1969). The solution for a transverselly isotropic medium (planar isotropy) is presented at this congress (Rahn, 1983) for the case of a borehole lying parallel or perpendicular to the planes of isotropy of the rock ; coefficients have been determined by finite element computations.
The main theoretical drawback of the stress relief methods is that the rock is supposed to be linearly elastic (and isotropic in most cases) and that elastic constants must be known. Another limitation comes from the fact that these measurements are very time consuming and therefore fairly costly.
In addition, in most instances stress determinations conducted with these methods have been restricted to fairly shallow depth boreholes (usually shorter than 50 m) although results presented by Martna et al. (1983) refer to measurements as deep as 500 m. Finally for rock submitted to high stress regimes, core discing occurs which does not allow sufficient lengths of overcoring ; in such instances only the doorstopper can be used.
The main advantage of these methods is that the precision of the determination can be estimated, although this is not done in most papers published on this topic.
With all these instruments, displacements are either measured at a point (borehole deformation cells) or along very short basis (strain gauges are not longer than one or two centimeters). Thus results are very much influenced by local heterogeneities and a large scattering is often observed. Blackwood (1978) has proposed to perform continuous displacement recording during the overcoring operation so that, by comparison of the displacement-overcoring advancement curves with theoretical ones obtained by finite element computations, only those measurements which are theoretically admissible be retained for the stress determination.
Cornet et al. (1979) have proposed to use all the theoretically admissible displacement measurements simultaneously to determine the stress field, along with the error of the determination. Indeed if {E} is a vector the components of which are all the admissible measurements, {X} is a vector the components of which are the components of stress at infinity expressed in the geographical frame of reference, [A] is a matrix the coefficients of which are those of the stress-displacement relationship relevant to the measuring technique, [B] is a matrix which rotates the stress tensor from the geographical frame of reference to the cell frame of reference for each of the measurements and [C] is a matrix the coefficients of which are the stress concentration factors associated to the proximity of the cavity from which measurements are made, then all the admissible results can be expressed as :

$$[A] \quad [B] \quad [C] \quad \{X\} \quad = \quad \{E\} \qquad (3)$$

The linear system of equation (3) is overdetermined if more than six measurements have been obtained ; accordingly the components of {X} can be determined by a least squares techniques and the error of the determination estimated. In practice [A] is seldom known and only those measurements which are far away from the cavity can be delt with simultaneously. For closer measurements, only those which are at the same distance from the cavity are treated simultaneously.
Quite clearly, with such an inversion technique, the displacement measurements, which apply at a point only, are averaged to obtain an estimation of the mean stress tensor at the scale of the volume in which measurements were made. Influence of large discontinuities (i.e. with respect to the scale of the volume in which measurements have been made) may not be negligible. Then those measurements which are not coherent with the overall stress determination can be eliminated. The remaining collection can then be used for a more refined stress determination. An example of such a back analysis is illustrated on figure 1 . The direction of the long axis of the deformed borehole cross sections, where measurements were made, is compared with that calculated from the global stress determination. Quite clearly measurements made at 1.80 m from the cavity wall (a 3 m diameter horizontal shaft) are not in agreement with the overall stress determination. These measurements, influenced by the proximity of the cavity, should be removed from the collection of " admissible " displacements.

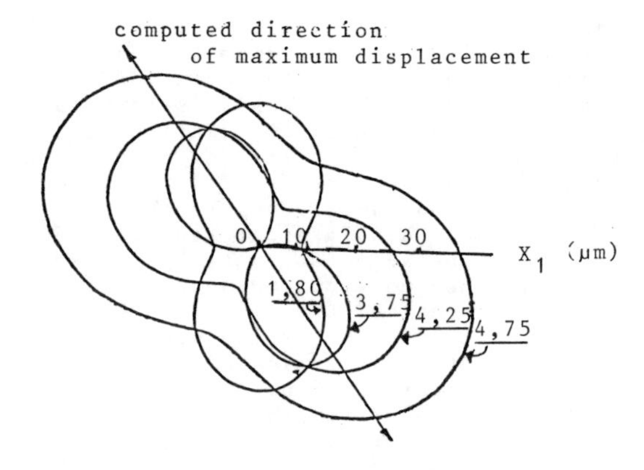

Fig. 1 Comparison between computed and measured deformation of a borehole after overcoring. The curves are obtained from actual diametral displacements measurements ; the computed axis is established from the stress determination. The reference axis (X_1) is the horizontal axis in the cross section of the borehole.

Stiff inclusions

Stiff inclusions, also referred to as " stress meters " , are built to measure directly one stress component. They must be very rigid as compared to the host rock (more than 4 to 5 times the rock modulus according to Leeman, 1964) so that stress variations in the host rock are directly transmitted to the inclusion. Deformation measuring devices (vibrating wires for Hawkes and Hooker 1974 and Pariseau and Eitani 1977, strain gauges for Potts 1957 and Filak and Cyrul 1977, photo elastic cell for Roberts et al. 1964) set up inside the inclusion are then used to determine the change of stresses supported by the inclusion. The main advantage of stress meters over soft inclusions is that they do not require a detailed knowledge of the rock moduli ; they even do not require the rock to be linearly elastic provided the inclusion is much stiffer than the rock.
They are well suited to investigate stress variations in time but have not been used as extensively as soft inclusions or borehole deformation cells, for absolute stress determination because of contact problems (see e.g. Fairhurst 1968). They are presently restricted to fairly shallow boreholes. Further, problems on long term stability may be encountered when dealing with stress variation with time (Fairhurst, 1968). No means to evaluate their precision has been described.

Strain, stress, compensation methods

The oldest stress compensation technique is that developed by Mayer et al. (1951) and Tincelin (1951). First two plugs are inserted on a free surface and the distance between the plug carefully measured. After relieving the stress field by a saw cut set-up perpendicularly to the plug base, the distance between the plugs is measured again. Then a flat jack is inserted in the saw cut and pressurized to such a value that the distance between the plug is brought back to its original value. The pressure required to cancel the displacements caused by the saw cut provides a measurement of the normal stress which was exerted onto that plane before the cut. Hoskins (1966) has discussed the best plug distribution which minimizes the effect of shear stress relief.
The main drawback of the technique is that it is limited to short distances from the free surface. In this domain, non elastic deformations often occur so that difficulties arise to relate measured stresses to the far field stress state ; in addition viscosity effects must be negligible for the method to be accurate.
Balthazar and Wenz (1983) propose adaptation of this method to deep boreholes. Thus not only can measurements be conducted some distance away from the free surface but also the complete stress tensor can be determined.
Kuznetzov et al. (1983) have developed a new strain compensation technique for nonlinearly elastic rocks. A pressure is applied onto the flattened end of the borehole in order to prevent its axial deformation when coring proceeds, that is when the stress is being relieved. Determination of the pressure required to preclude any axial deformation can be related to the diagonal components of the stress tensor at this point when three orthogonal boreholes are used. The value of these diagonal components, defined in the frame of reference associated with the three axis of the boreholes, can be determined. Additional boreholes drilled in directions different from that of the three first ones are required for the complete stress tensor determination. The main advantage of the method is that Young's modulus of the rock does not need to be known ; however Poisson's ratio must be measured. The main draw back is the large amount of drilling needed for a complete determination of the stress tensor.
The method of differential strain curve analysis proposed by Ren and Roegiers (1983) assumes that the complete relief of stresses on a rock sample generates micro-cracks, the density of which, in a given direction, is proportional to stress changes perpendicularly to this direction ; thus it is expected that the induced crack spectrum will reflect the stress history. Submitting the rock sample to a hydrostatic pressure should induce a non hydrostatic strain field the principal axis of which are identical to those of the stress field which was exerted onto the specimen before it was extracted from the ground. Further, the relative values of the strain along the principal axes should provide information on the relative values of the principal stresses. Results presented by Ren and Roegiers are a good demonstration of the potentials and limitations of the technique. In particular the influence of paleostresses remains difficult to be dealt with, as well as that of dilatancy. However, when it works, this method provides a very inexpensive way of determining the stress field at depth.

Fracturing procedures.

For a long time now, hydraulic fracturing has been applied to stress determination at depth (see e.g. Hubbert and Willis 1957 ; Scheidegger 1962 ; Fairhurst 1964 ; Kehle 1964 ; Haimson and Fairhurst 1969).
In the original version of this technique, a portion of the borehole is sealed off with a straddle packer and then pressurized up to

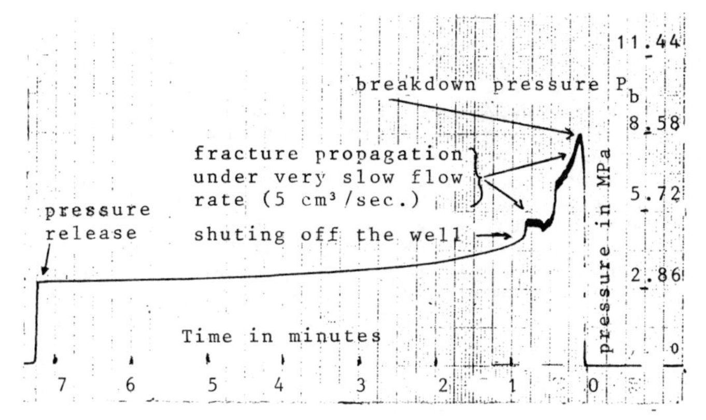

Fig. 2 Typical pressure-time record during a hydraulic fracture test in granite (depth is 200m)

fracture at the so-called breakdown pressure (P_b). When injection stops, a shut-in pressure (P_s) can be observed if the wellbore remains sealed (see figure 2).
If the borehole is parallel to one of the principal stress directions (which is often assumed to be vertical) and if the rock is linearly elastic and isotropic with respect to both its elastic behavior and its " strength ", the breakdown pressure and the shut-in pressure can be used to determine some of the components of the local stress tensor. Because most straddle packers are inflattable, only vertical fractures are initiated from a vertical boreholes (Haimson and Fairhurst, 1969) even in zones very close to the ground surface. From equation 1 it can be shown that, for dry impervious rocks :

$$\sigma^t - \sigma_1 + 3\sigma_2 = P_b \qquad (4\ a)$$

$$\sigma_2 = P_s \qquad (4\ b)$$

where σ_1 and σ_2 are the principal stress components perpendicular to the borehole axis ($\sigma_1 > \sigma_2$) and σ^t is the so-called " tensile strength " of the rock. Determination of P_b, P_s and σ^t, either in the laboratory or in the field, provide means to estimate σ_1 and σ_2. From Griffith's theory of fracture (Griffith, 1921) it can be shown that, for isotropic rocks, the fracture extends perpendicularly to σ_2 so that determination of the fracture orientation at the wellbore provides the orientation of σ_1.
Equation (4 a) must be modified for porous rocks in order to take into account the influence of pore pressure variation in the volume surrounding the borehole (see e.g.Haïmson, 1968, Rice and Cleary, 1976).
If the rock mass is saturated by a fluid under pressure P_p but impervious enough to prevent significant leakage into the formation, equation (4 a) must be replaced by :

$$P_b = -\sigma_1 + 3\sigma_2 + \sigma^T - P_p \qquad (5)$$

Indeed, as shown on figure 3, the in-situ stress field can be decomposed into two components :

Component A which is hydrostatic ;
Component B which involves only effective stresses at infinity :

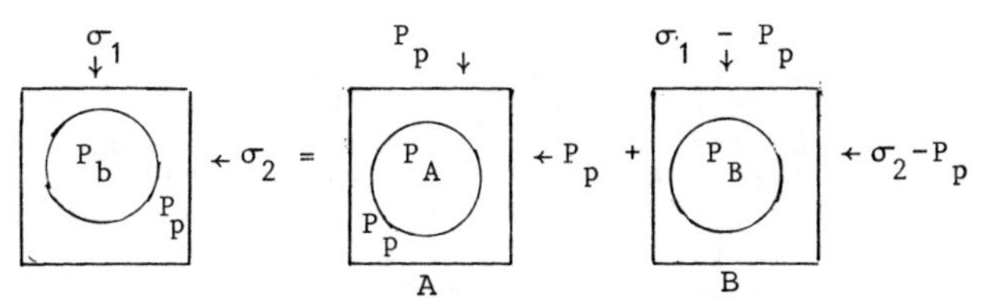

Fig.3 Influence of pore-pressure on the breakdown pressure with impervious rocks.

The borehole pressure P_B required to create a fracture for component B is :

$$P_B = -(\sigma_1 - P_p) + 3(\sigma_2 - P_p) + \sigma^T \qquad (6)$$

whilst the borehole-pressure P_A required to maintain a uniform hydrostatic pressure P_p is $P_A = P_p$ so that for combined components A and B, equation (5) is obtained.

For the shut-in pressure, no such problem arises. If in figure 3, the borehole is replaced by a fracture perpendicular to the σ_2 direction, it is found that the pressure required to barely open the fracture under component B is $\sigma_2 - P_p$ so that under combined components A and B the shut-in pressure for which the fracture stabilizes is σ_2.
In this formulation, the hydraulic fracturing technique suffers a few limitations :
1. The rock mass must be linearly elastic and isotropic with respect to both its elastic properties and its " strength " ;
2. the borehole must be parallel to one of the principal stress directions ;
3. the " strength " of the rock must be known ;
4. the peak of the pressure-time record must correspond to the pressure value at which fracture initiates ;
5. accuracy of the determination relies only on the consistency of the results.

Bredehoeft et al. (1976) pointed out that the problem of strength determination can be eliminated by dealing only with reopening pressure (Pr). Indeed, after the fracture has been initiated and the pore pressure has returned to its original value in the total rock volume, since the tensile strength is now zero along the fracture plane, equation (5) becomes :

$$P_r = 3\sigma_2 - \sigma_1 - P_p \qquad (7)$$

For equation (7) to be valid no fluid percolation along the fracture plane prior to reopening must occur (Cornet, 1976) ; thus high pressurization rates, or highly viscous fluids, must be used. But for such conditions, fracture reopening does not coincide with the peak of the pressure-time record.
In order to obtain an accurate measurement of the reopening pressure, Hickman and Zoback (1982) propose to pump at the same rate as that used for fracturing the rock. Then the reopening pressure is defined as the point at which the pressure-time record deviates from that obtained initialy. They advise to use the second reopening cycle in order to avoid problems with incomplete fracture development during the first pressurization test.

Often no sharp shut-in pressure is identified on pressure-time records and various means have been proposed to improve the reading (see e.g. Mc Lennan and Roegiers 1982 ; Gronseth and Kry 1982 ; Aamodt and Kuriyagawa 1982). They all refer to instantaneous shut-in pressure, i.e. the pressure drop observed right after shutting off the well, and propose various semi-log plotting techniques. An alternative procedure is to increase the borehole pressure steps by steps after the fracture has been well developed ; the initial step is set at a pressure value much lower than the expected shut-in pressure. Then, by plotting the flow rate required to maintain a constant pressure into the well versus the value of the pressure, the pressure value at which the fracture starts to open is determined with good accuracy : at that pressure level, flow along the fracture occurs so that the flow rate required to maintain the pressure constant drastically increases. This pressure value is exactly equal to the normal stress exerted onto the fracture plane if the steps are long enough to avoid pressure gradients into the fracture. This assumes of course that the fracture remains planar as it extends away from the borehole ; in particular it is required that no preexisting fracture deviates the fracture from its original path.
In order to eliminate limitations 1, 2 and 5 Cornet (1982) has proposed to take advantage of the weakness planes that always exist in a rock mass : If the straddle packer is set-up in front of a weakly recemented joint, it is possible to reopen these joints by using slow flow rates. Then equating the shut-in pressure (P_s), or the step by step reopening pressure, to the normal stress supported by the fracture plane, the following equation is obtained :

$$\underset{\sim}{\sigma}\underline{n}\cdot\underline{n} = P_s \qquad (8)$$

where $\underset{\sim}{\sigma}$ is the local stress, $\underline{n}$ the normal to the fracture plane and P_s the shut-in pressure. In the most general case six different fracture orientations at the same depth are required to determine the six independent components of $\underset{\sim}{\sigma}$. However if it is assumed that one of the principal stresses is vertical (but not necessarily parallel to the borehole axis) then only three different planes are needed. When more than three fractures are available a standard least squares technique is used to solve the linear system of equations.
This new method is well adapted to stress measurements from underground cavities where many boreholes can be drilled. When only one borehole is available, the stress variation with depth is assumed to be linear :

$$(\underset{\sim}{\sigma}) = \begin{pmatrix} \sigma_1 + \alpha_1 z & 0 & 0 \\ 0 & \sigma_2 + \alpha_2 z & 0 \\ 0 & 0 & \rho g z \end{pmatrix} \qquad (9)$$

where ρg is the vertical gradient of the vertical stress. If it is assumed, as a first approximation, that $\alpha_1 = \alpha_2$ (measurements are conducted very close from one another so that only gravity effects are considered), equation (8) can be linearized by a suitable change of variable so that a standard least squares technique can still be used. When fracture planes are some distance apart equation (8) is no longer linear and more elaborate inversion methods must be used (Valette and Cornet, 1983).
When most fractures are parallel to the borehole direction and exhibit the same orientation with respect to the geographical frame of reference, the classical interpretation method is applied. Tests on inclined fractures provide only a mean to validitating the determination. Thus questions relative to the possible influence of rock mass strength anisotropy caused by preexisting weakness planes on hydraulic fracture propagation can readily be answered. Further it provides means for evaluating the determination accuracy.

It is worthwhile to note that the scale at which stresses are being determined with the hydraulic fracturing method is quite flexible. Depending on the distance between the packers, the borehole diameter and the quantity of liquid injected. The stress measurement can concern volumes as small as a few cubic centimeter (laboratory conditions) or as large as a few cubic meters. For larger volumes, the influence of stress gradients caused by gravity becomes significant ; further the problem of the interaction between preexisting weakness planes and the artificial hydraulic fracture becomes more difficult to resolve.

Other rock fracture procedures for stress measurements have been proposed (see e.g.Jaeger and Cook, (1963), De la Cruz, (1976). In this respect the use of pressuremeters seems to constitute an interesting new development (see e.g. Huergo, 1983, Charlez, 1983) for it combines in-situ measurements of rock deformability with that of stresses.
Stephansson (1983) discusses the applicability of the Colorado School of Mines pressuremeter (Hustrulid and Hustrulid, 1975) to in-situ stress determination. In particular he proposes to use the pressure required to reopen the fracture developed during a preliminary loading cycle as a measurement of the far field least principal stress perpendicular to the borehole axis (the borehole is assumed to be parallel to one of the principal stresses).
However it seems that, in hydraulic fracturing, this pressure value depends on both principal stresses σ_1 and σ_2 (see equation 7) along with the formation pore pressure. It is hoped that this point will be discussed at the congress.
It has also been proposed to use the natural fracturing processes caused by tectonic stresses to determine them. These are now standard techniques in seismology and in microtectonics. They will be described as a separate general stress determination technique for they refer to very large rock volumes (from a few cubic decameters to few cubic kilometers).

Seismo-tectonic procedures

For a long time now, seismologists have been able to define so-called "fault plane solution" for earthquakes, also called "focal mechanism". They refer to the determination of the orientation of the shearing fault which gave rise

to the earthquake as well as to the direction of motion along that plane (see e.g. Stauder, 1962). Although at first empirical, the mehod has been refined and is now of common use in seismology. These solutions have been used in turn to infer the principal directions of the stress field around the earthquake source. Before discussing this last point let us first briefly recall the principle of fault plane solutions.
It had been known for a long time (Byerly, 1926) that the first motion caused by P waves (compressional waves polarized in the direction of motion of the radiation) generated by an earthquake was compressional in some areas and dilational in others. This radiation pattern can be explained if the source is equated to some point forces : Nakano (1923) showed, for example that a couple of point forces sends alternate compression and dilatations into quarter spaces separated by two orthogonal planes (known as the nodal planes ; see fig.4).

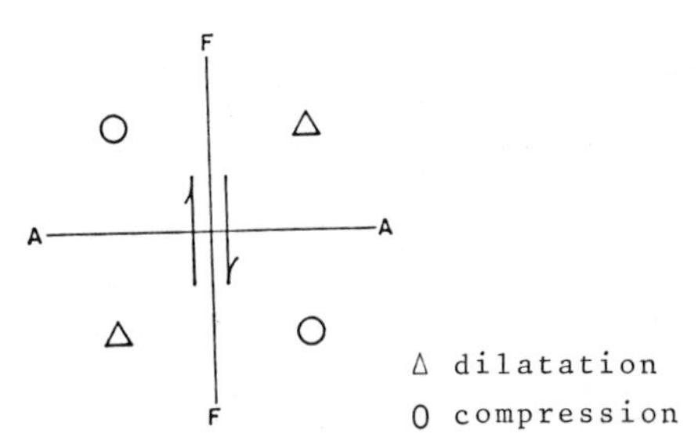

Fig.4 Distribution of compressions and dilatations resulting from movement along a fault FF (after Stauder 1962)

Later, various authors (e.g. Knopoff and Gilbert 1960) pointed out that the problem of characterization of the seismic radiations caused by an earthquake could be connected to that of a dislocation, i.e. a discontinuity in the tangential displacement field at the fault plane. Burridge and Knopoff (1964) showed that the radiation caused by such a dicontinuity in the displacement field is equivalent to that of a double couple (point forces with a moment) applied at the focus of the earthquake ; one of the nodal planes of the radiation so created coincides with the fracture plane. This is known in seismology as the representation theorem. For such a representation to be valid, the wave lengths considered in the radiation must be larger than that of the source so that disturbances caused by end effects of the displacement discontinuity as well as by local heterogeneities in the medium do not alter the first motion of the observed radiation. Thus the theoretical basis for the observed first motions direction was established.
Currently determination of the "focal mecanism" of an earthquake proceeds as follow :

1. Locate the earthquake focus and consider a sphere of unit radius centered on the focus.

2. Define the direction of first arrival on as many stations as possible ; only wave length that are long with respect to source dimension should be used. The source dimension can be established from the corner frequency of the Fourier spectrum of the radiation (see e.g. Madariaga, 1979).

3. Perform ray tracing for each station and determine the point where the ray intersects the sphere. Ray tracing means finding the path followed by the seismic ray from its source up to the observational point. This path depends strongly on the wave velocity structure of the medium and may be difficult to determine in very heterogeneous media.

4. Using an equal area stereographic projection procedure, plot on the equatorial plane of the sphere the projection of the points of intersection of the rays with the lower half sphere. When stations close to the epicenter (the vertical projection of the focus on ground surface) are being used, the rays going to these stations intersect the upper half sphere. These points must be transfered to the lower half-sphere by taking their symmetrical counterpart with respect to the focus before plotting them on the stereographic projection.

5. Define on stereographic projection the two orthogonal planes which best separate quadrants where first motion is identical (compressional or dilational). These orthogonal planes are the nodal planes ; one of them is the fault plane. It is impossible to determine from radiation analysis which of the nodal planes is that of the fault ; additional information is needed to make the proper choice (trace of the fault on ground surface or location of aftershocks).

Although tedious to perform by hand, fault plane solutions can be determined on a routine basis using adequate computer codes now available. An example of such a determination is shown on figure 5.

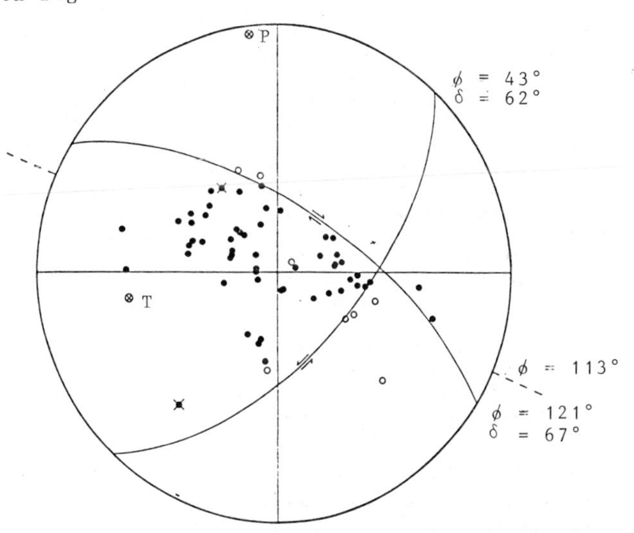

Fig.5 Fault plane solution for the shock of August 19, 1966 in Eastern Turkey. Diagram is an equal area projection of the lower hemisphere of the radiation field. Solid circles represent compressions, open circles dilatations, crosses indicate station is near a nodal plane. ϕ and δ are the strike and dip of the nodal planes, and arrows indicate the sense of motion. The dotted line shows the strike of the major right lateral surface break which accompanied the earthquake (after Mc Kenzie 1969).

More informations on the rupture that caused the earthquake can be obtained from spectral analysis of the seismic waves. For example the size of the source can be estimated as well as the stress drop which occured during failure (see e.g. Madariaga 1979). It may be worth pointing out that these stress drops are found to vary from 1 MPa to 10 MPa irrespective of the earthquake magnitude (e.g. Hanks 1972). Once the nodal planes are known, three axis are defined : the P axis is oriented in the direction where compressive motion is a maximum perpendicular to the nodal plane intersection ; the T axis, also perpendicular to the nodal planes intersection, is oriented in the direction where dilatations are maximum ; the null axis lies along the intersection of the two nodal planes. It has been proposed to equate these three axes with the principal stress directions (see e.g. Scheidegger, 1964). This proposition is based on the hypothesis that the fault plane corresponds to the surface where the shear stress is a maximum and that displacement along the fault is parallel to the shear component supported by the fracture. In fact this simplistic approximation is usually not valid for brittle rocks (see e.g. Mc Kenzie 1969, Raleigh et al. 1972) since, first of all, their fracture in compression does not obey a Tresca type fracture criterion but rather a Mohr-Coulomb type (the macroscopic fracture surface is inclined by an angle which varies from 20 to 45° with respect to the maximum principal stress direction depending on the value of σ_3) and secondly rock masses are usually anisotropic with respect to their strength because of the existence of preexisting weakness surfaces. Mc Kenzie pointed out that sliding along preexisting discontinuities governed the shear behavior of the rock mass so that the error in the principal stress direction determination could be as high as 90°. Raleigh et al. argued that, even when preexisting faults exist, new fracture surfaces may form if the shear stress condition which must be satisfied for sliding to accur along preexisting fractures implies larger principal stresses than those necessary for the development of a new fracture. They considered that the maximum error on the principal stress direction determination is of the order of 20 to 30°.

The microtectonics procedure (see e.g. Angelier 1979 , Armijo et al. 1982) for identification of the principal stress directions as well as the "shape factor" (also called " stress ratio ") $R = (\sigma_3 - \sigma_1) / (\sigma_2 - \sigma_1)$, where σ_1 , σ_2 and σ_3 are the eigen values of the stress tensor, is derived from principles similar to those of the fault plane solutions procedure. Identification of slickensides direction on faults provides a direct measurement of the direction of slip motion along the fault. If it is assumed that this motion occured in the same direction as that of the shear force supported by the fault, and that friction characteristic along the faults are independent of fault orientation and direction of slip, then the Euler angles, which define the stress tensor principal directions, along with the shape factor R can be determined.

if $\quad \underline{f}_i = \underline{\tau}_i \,/\, |\underline{\tau}_i| \qquad (10\ a)$

where $\quad \underline{\tau}_i = \underset{\sim}{\sigma}\underline{n}_i - (\underset{\sim}{\sigma}\underline{n}_i.\underline{n}_i)\underline{n}_i \qquad (10\ b)$

then $\quad \underline{f}_i = \underline{s}_i^j \quad ; \quad i = 1,N \quad j = 1,M \qquad (10\ c)$

$\underline{f}_i$ is a unit vector parallel to the resolved shear stress in the ith plane ;

$\underset{\sim}{\sigma}$ is the " average " stress tensor exerted in the volume under consideration ;

$\underline{n}_i$ is the unit normal to the ith fault plane;

$\underline{s}_i^j$ is a unit vector parallel to the jth slickenside in the ith plane

N is the total number of fault planes

M is the total number of slickenside in the ith fault plane.

Equation (10 b) reduces to

$$|\tau_i| = (\sigma_2 - \sigma_1)A_\ell \qquad (11)$$

with $\ell = 1,2,3$; $A_1 = -kn_1$, $A_2 = (1-k)n_2$, $A_3 = (R-k)n_3$

$$k = n_2^2 + Rn_1^2$$

Thus the main hypothesis of this stress determination technique is equation (10 c). Further it is assumed that all fracture plane motions are independent of each other.

When N is large enough (between 100 and 200 slickenside measurements) equation (10c) constitutes a system of non linear equation which can be inverted after linearization of equation (10 c) (see e.g. Armijo et al. 1982). In this procedure, the stress tensor $\underset{\sim}{\sigma}$ is assumed to be uniform within a volume of the order of a few cubic hectometers, or even a few cubic kilometers (slickenside measurements are usually conducted along faults observed in quarries or large opening). Difficulties have been encoutered when the faults have been activated by more than one tectonic stress field ; however inversion techniques have been developed to identify these various stress tensors (see e.g. Etchecopar et al. 1981, Armijo et al. 1982).

Since fault plane solutions provide informations on the direction of slip along faults, they can be compared to slickenside measurements: where many slickensides are necessary to define with some certainty the direction of slip, a fault plane solution provides information on the overall motion at once. It is not affected by small scale heterogeneities and provide informations at the same scale as that for which the stress tensor is being defined. Accordingly, analysis of fault plane solutions with a similar technique as that proposed in microtectonics should provide informations on the principal stress directions at depth, as well as on the " aspect ratio " of the stress tensor. This determination should be more accurate than that which is presently obtained with the Tresca type criterion of shear fracture. However some difficulty may arise for small earthquakes because of the difficulty encountered in defining which of the nodal planes is the actual fault plane.

2.2. Some results

Comparison of results obtained with various techniques

In a few instances now, stresses have been determined at the same place with various means.

Haimson (1982) reports on results obtained by hydraulic fracturing and overcoring of borehole deformation gages. At the Nevada test site, for example, measurements were conducted from two tunnels in some 450 m thick series of tuff and tuffaceous sandstone overlying massive paleozoic rocks. Results compare very favorably with each other (see table 1) as well as with local fault plane solutions.

Table 1. Results from the Nevada test site (after Haimson, 1982).

	σ_H max. MPa direction	σ_h min. MPa direction	σ_v MPa direction dip	depth in m
Hydraulic fracturing	9.0/N 35°E	3.5/N 55°W	7.	380
overcoring	8.0/N 45°E	2.5/N 45°W	5.7 N 42°E 20°	
hydraulic fracturing	7.5/N 40°E	3.0/N 50°W	7.3	426
overcoring	8.5/N 22°E	2.6 N 68°W dip 83°	6.8 N 83°W dip 7	
focal mechanism	N 45°E	N 45°W		?

Results obtained at five other sites with both hydraulic fracturing and overcoring techniques are fairly consistent. Scatter on the least horizontal principal stress magnitude is of the order of 30 % whilst that for the maximum horizontal stress may be as high as 50 %. This large error for the determination of σ_H may be attributed to the difficulty encountered with the classical hydraulic fracturing technique to determine with accuracy either the tensile strength of the rock or the reopening pressure. Mean direction for the maximum principal horizontal stress as obtained with both techniques are typically within 10° but the scatter may reach 120° for the overcoring technique and 50° for hydraulic fracturing.
Comparison of results obtained at the Stripa mine in Sweden by hydraulic fracturing and overcoring of the Leeman's triaxial cell, the C.S.I.R.O. soft inclusion cell and the University of Lulea triaxial cell is presented by Doe (1982) ; detailed description of the scatter of the results is provided. Whilst some cementing problems were encountered with the C.S.I.R.O. cell, other techniques provided very consistent results when comparing mean values even though some fairly significant scatter was observed among each individual results.
Li Fang Quan et al. (1982) also report on a comparison of stress measurements near Yi Xio in the Hebei province (China) by overcoring and hydraulic fracturing ; results are fairly consistent

$\sigma_{H\ max} = 3.4 + 0.03\ z$ MPa, $\sigma_{h\ min} = 2.5 + 0.031\ z$ MPa,

direction of $\sigma_{H\ max}$ = N 64°W ± 20° for overcoring ,

$\sigma_{H\ max} = 4.9 + 0.028\ z$ MPa , $\sigma_{h\ min} = 2.6 + 0.021\ z$ and

direction of $\sigma_{H\ max}$ = N 80°W ± 10° for hydraulic fracturing ; depth z ranges from 0 to 90 m from surface.
Results presented by Kuznetsov et al. (1983) provide a comparison between determination obtained by a strain compensation method and microtectonic analysis. Both techniques provide consistent results.

Thus, results obtained by various researchers in various different geological backgrounds seem to indicate that reasonably uniform stress fields exist in rock masses and that present techniques provide fairly satisfactory means for their determination, although some efforts are still needed to improve the precision of the determination.

Stress variation with depth

Knowledge of the magnitude of principal stresses at depth have improved drastically in the last few years thanks essentially to deep stress measurements conducted either from deep mines or in deep wells (see e.g. Mc Garr and Gay, 1978 ; Brown and Hoek, 1978 ; Jamison and Cook, 1980).
McGarr (1980), using worlwide data, concluded that, on the average, the maximum shear stress increases linearly with depth to at least 5.1 km in soft rocks, such as shale or sandstone, and to 3.7 km in hard rock such as granite or quartzite (these values correspond to the deepest measurement). He observed that a regression analysis of these data suggests a vertical gradient of the maximum shear equal to 3.8 MPa/km in soft rock and 6.6 MPa/km in hard rock.
Assuming elastic behavior for the upper crust, he showed that equilibrium conditions combined with Beltrami-Mitchell compatibility equations for the stresses in an homogeneous isotropic elastic body can be used to provide functional constraints on stress variation with depth. Thus, for domains where no lateral stress variation is observed (see next section) these conditions imply a linear increase with depth for the principal stresses magnitude but allow a monotonic rotation of principal directions. This would suggest that, even for such simple conditions, surface stress determination may not provide information on principal stress directions at depth. For domains where the stress field is assumed to be independent of only one of the horizontal coordinate (say x_2 , the other horizontal axis is x_1), these conditions imply that the diagonal components of the stress tensor increase linearly with depth but that the shear components σ_{12} and σ_{23} are not linear. This clearly suggests that, even at the scale considered in this analysis, inference of stresses at depth from surface measurements is not a trivial matter and that the vertical stress component at great depth may not be a principal stress.
Obviously for heterogeneous materials only direct measurements can presently be used to obtain meaningfull data. For example, using the slow rate reopening pressure technique of hydraulic fracturing for the determination of the least principal stress magnitude in evaporite deposits, Cornet (1982) observed variations equal to 2.5 MPa within less than 10 meters (the average value was 13 MPa and the precision of each individual measurement was better than 2 MPa ; depth was around 700 m).
Within these soft deposits (alternance of salt beds 1 to 10 meters thick and shale layers 0.01 to 1 m thick), millimetre to centimetre thick stiff anhydrite beds exist ; their amount per unit length is highly variable. Clearly those stiff anhydrite beds modify significantly the mechanical behavior of the deposits ; the measured maximum shear of the order of 5 MPa is not compatible with the plastic type behavior of

this material. This local stiffening effect is considered to be the cause of the fast variation in horizontal principal stress magnitude : the larger values were observed close to zones of high density in anhydrite beds.
Martna et al. (1983) report much more drastic stress variation measured by overcoring in a homogeneous granitogneiss. A uniform stress field with maximum principal horizontal stress of the order of 15 to 20 MPa was observed in a deep borehole down to 320 m. Then, after a fault zone had been encountered, the value jumped to 65 MPa and remained stable down to 500 m ; no rotation of the principal direction is reported. These examples illustrate the fact that, independently of the scale for which stresses are defined, stresses at depth cannot be estimated by simple extrapolation from surface measurements. However, very grossly speaking, a monotonic variation for the maximum shear is to be expected down to depth at least equal to 5 km.

Lateral stress variation

Beside the wellknown superficial lateral stress variations caused by topography (see e.g. Scheidegger 1977, Haimson 1978, Buyle-Bodin 1980) horizontal variations of the stress tensor at a given depth can be anticipated from simple mechanical consideration. For example heterogeneities in rock properties may cause an otherwise uniform stress field to deviate locally ; discontinuities like faults may generate stress variations both in space and time. In this respect, results obtained by Li Fang-Quan et al. (1982) indicate that the maximum principal stress measured near an earthquake of magnitude 7.9 was, after the earthquake, of the order of 2.5 MPa and 6.0 MPa some distance away.
Yet despite all these reasons for non uniformity the large amount of stress measurements conducted up to now suggest that, in fact, broad regions exist where stresses are fairly uniform.
Zoback and Zoback (1980) compiled a host of stress measurements conducted in the conterminous Unitated States and were led to define stress provinces, with linear dimension ranging from 100 to 2000 km, where directions and relative magnitude of principal stresses were fairly uniform (see fig. 6). Froidevaux et al. (1980) and Rummel et al. (1982) were driven to similar conclusions for western Europe although most of their data were obtained at shallow depth (less than 300 m).
These results are clear indications that, superposed on broad stress fields generated by the stress acting on lateral plate boundaries (see e.g. Solomon et al. 1980) some other sources of stress exist. Mc Garr (1982) proposes to associate Zoback and Zoback's stress provinces to various fields of shear tractions acting on the base of the elastic-brittle crust whilst Fleytout and Froideveaux (1982) have proposed to consider the influence of spacial density variations in the lithosphere as well

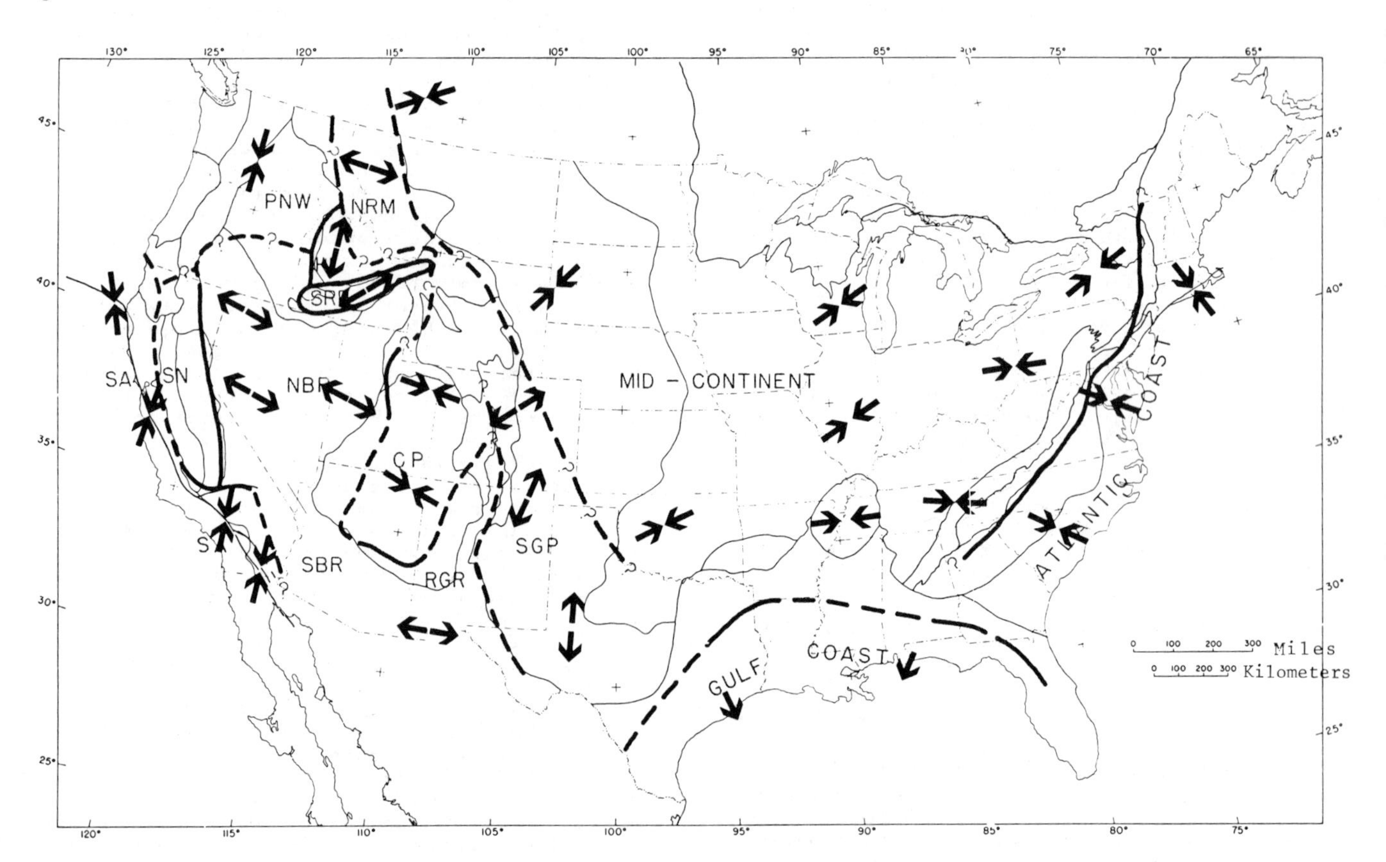

Fig. 6 Generalized stress map of the conterminous United States. Arrows represent direction of either least (outward directed) or greatest (inward directed) principal horizontal compression after Zoback and Zoback (1980).

as in the upper mantle to explain these intraplate stress domain.
Quite clearly stress measurements remain, at the moment, necessary data to constrain geodynamic models and it is hoped that more and more measurements will be conducted to improve these models It is at present too early to inverse the process and to use geodynamic models to infer local stress fields. However, when only orders of magnitude and general direction of principal stresses are sought, stresses at a given location can be inferred from neighbouring measurements when continuity and homogeneity is assumed. In fact, multiplication of stress measurements could allow draw stress maps at various scales ; these in turn could be used to obtain coarse estimation of stresses at any point within the map and thus spare the cost of a regional stress determination. It is hoped that such maps will be developed in the near futur for large parts of the continents.

Stress variations with time

Stresses do vary with time otherwise earthquake would not occur twice at the same place and some understanding of these time variations is needed. Yet, although a few instruments (stiff inclusion for example) have been developed to measure stress variations with time, few results have been published in this regard (see e.g. Clark 1982).
An intriguing phenomenon associated with earthquakes may soon shed some new light to this question : In a few instances significant magnetic field variations have been observed in areas affected by large earthquakes (e.g. Moore 1964, Smith and Johnston 1976, Rikitake 1979). An example of such a variation is presented on figure 7 ; it was associated with a magnitude 5 earthquake which occured in the area of Dushambe in Tadjikistan (U.S.S.R.).

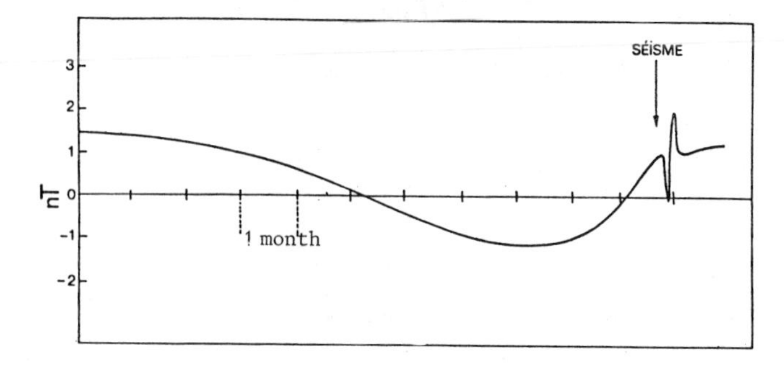

Fig.7 Magnetic anomaly observed in the region of Dushambe (Tadjikistan U.S.S.R.) before a magnitude 5 earthquake (after Skovorotkin personnal communication).

A possible explanation for such magnetic variations may be found in the piezomagnetism exhibited by magnetic rocks : when submitted to a stress variation, both the magnetic susceptibility of the rock and its remanent magnetization are modified (see e.g. Martin et al.1978, Zlotnicki et al. 1981). For example Zlotnicki et al. observed that the change of induced magnetization caused by stress variations can be represented by a tensor law :

$$\Delta J_i = P_{ijkn} \Delta \sigma_{kn} J_{oj} \qquad (13)$$

where J_{oj} is the jth component of the induced magnetization caused by a magnetic field $\underline{H}$ ($\underline{J}_o = \underset{\sim}{S}_o \underline{H}$, where $\underset{\sim}{S}_o$ is the magnetic susceptibility tensor of the rock) ;

$\Delta \sigma_{kn}$ is the variation of stress component σ_{kn}

P_{ijkn} are the components of a fourth order tensor with 36 constants in the most general case.

When the rock is isotropic, P_{ijkn} reduce to two constants so that equation (13) can be written

$$\Delta J_i = (p_1 \Delta \sigma_{Kk} \delta_{ij} + 2p_2 \Delta \sigma_{ij}) J_{oj} \qquad (14)$$

Determination of p_1 and p_2 in the laboratory provides a means of obtaining some information on stress variations.
Although no simple relation of this type has been derived for remanent magnetization, parameterization of laboratory curves could help to obtain some quantitative figures on stress variations with time in broad regions. Indeed numerical computations conducted by Jin-Qi Hao et al. (1982) show that changes of the magnetic field observed after earthquakes are of the same order of magnitude as those estimated from dislocation models.
Accordingly detailed survey of the magnetic field in earthquake prone fault areas combined with detailed laboratory investigation of the piezomagnetic properties of the local rocks could provide informations on stress variations with time. Similar procedures could be applied to stress variations caused by human activity.

3. THE PROBLEM OF GEOMETRICAL CHARACTERIZATION OF DISCONTINUITIES

Discontinuities affect the mechanical behavior of rock masses in various ways :
- by providing a preferential path to fluid flow they affect the pore pressure distribution and thereby the effective stresses that exist in the mass ;
- because they exhibit much lower " strength " characteristics than does the rock matrix, they are of paramount importance in stability analysis ;
- their dilatant behavior combined with their very soft elastic characteristics influences drastically the deformation of the whole rock mass.

Much attention has been devoted to improve the understanding of the mechanical properties of discontinuities, and although much more work in this respect is needed, many advances have been made. However the characterization of their geometry has proved to be a much more intractable difficulty. Yet it is only when this problem is solved that accurate modeling of the mechanical behavior of rock masses (as well as that of a host of other physical characteristics) will be possible.

The geometrical characterization of rock discontinuities may be undertaken in two different contexts :

- In the first , the face of a rock mass is amenable to observation and an attempt is made to obtain some characterization of the volume behind the face.This will be referred to as local characterization.
- In the second, only small surfaces with respect to the volume under consideration can be studied and this spars information is used to infer the discontinuity structure in the whole volume. Most of the time, in addition to observations retrievable from exposed surfaces, some information can be obtained from boreholes. This will be referred to as global characterization.

A few attempts have been made to clarify the first problem ; the second one seems to be nearly untouched.

3.1. Local characterization

In order to obtain some appreciation of the density of fractures in a given volume Deere (1964) defined the Rock Quality Designation (R.Q.D.) as the proportion of borehole core of intact length of 0.1 m, or longer, per core length :

$$RQD = 100 \sum_{i=1}^{n} x_i / L \qquad (15)$$

where x_i is the length of the ith piece of core with length > 0.1 m ;

n is the total number of pieces the length of which is larger than 0.1 m ;

L is the toal core length.

Further, he suggested that a scanline (measuring tape) used at the face of a rock mass may be regarded as directly analogous to a borehole core since the RQD can be found in this case too. A fairly detailed report has been issued by the International Society of Rock Mechanics on suggested methods for the quantitative description of discontinuities in rock masses (Barton 1978). It refers essentially to direct measurement techniques concerning spacing, persistence, roughness, wall strength, aperture, filling, seepage, number of sets and block size.

Hagan (1980) proposed some photogrammetric means to map rock fracture orientations. He pointed out that this mapping technique may cut the underground, or outside, mapping time by 90 % and that the area coverage obtained with this technique remains more or less constant so that biased sampling of fractures is prevented.

With this method stereographic pictures are taken from the face, with two normal cameras. Then a computer is used for the calculation of coordinates of any number of points lying on the overlapping area of the photograph so that average linear trend can be calculated using a least squares technique. The amount and direction of dip of this average plane is then determined. This data collection technique is directly applicable to scanline or area sampling of discontinuities.

Thomas (1981) define the directional density of fracture by :

$$\delta_v = \frac{\Sigma s_{\theta_s}}{v} \quad ; \quad \delta_s = \frac{\Sigma s_{\theta_\ell}}{A}$$

Σs_{θ_s} = sum of fracture areas for surfaces in solid angle θ_s

Σs_{θ_ℓ} = sum of fracture length in direction θ_ℓ

v = volume of investigation

A = area of investigation

where θ_s and θ_ℓ are respectively solid and plane angles of the measurement. He proposed a new method for obtaining the directional density of a discontinuity network as observed on the face of a rock mass.

The idea is to find a statistical estimate of directional density from a computation of the number of intersections between the discontinuity network, as observed on a photograph, and a chosen grid. For a network involving only one family of parallel discontinuities, it can be shown that the density of intersection points, $d_i(\theta)$, between the discontinuities and a line inclined at θ with respect to the discontinuity direction is an estimate of $\delta_s \sin\theta$; the longer the line the more precise the determination. Thus instead of using one scanline he proposed to use a series of parallel lines. More generally, for a network of q families (each family is reffered to as F_j) of discontinuities with density δ_j, making an angle α_j with a reference axis (see fig.8)

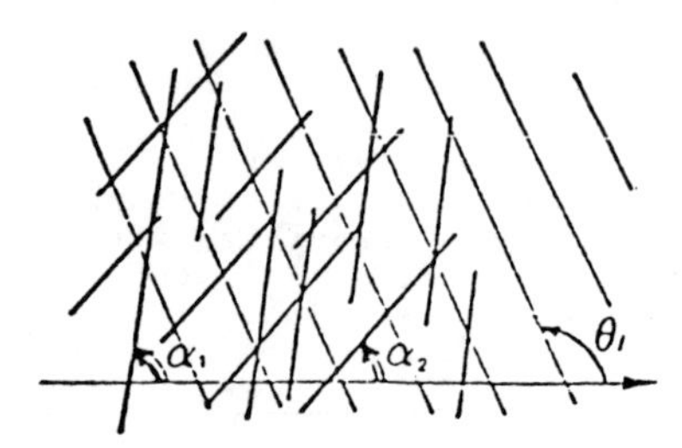

Fig.8 Discontinuity density analysis with a grid of parallel lines α_1 and α_2 are relative orientations of two fractures sets with respect to an arbitrary reference axis, θ is the inclination of the grid with respect to the reference axis (after Thomas 1981).

the density of intersection points observed with a grid of parallel lines making an angle θ_i with respect to the reference axis is given by :

$$d(\theta_i) = \sum_{j=1}^{q} \delta_j \left|\sin(\theta_i - \alpha_j)\right| \qquad (16)$$

the curvature of the curve $D(\theta)$ versus θ is a continuous function of θ except for the values which coincide with angle α_j (see fig.9).

Thomas has proposed a method based on discrete Fourier's series for determining the values α_j. This method which is directly applicable to an automatic texture analyser, has been applied in a coal mine.Pictures were taken on vertical planes the strikes of which were chosen so as to provide about 20° azimuthal rotation increments.Some appraisal of the directional volumetric density could thus be obtained.

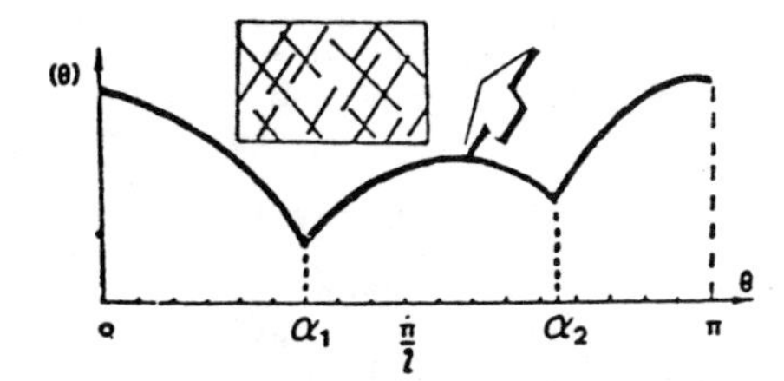

Fig.9 Theoretical shape of a density curve D(θ) for two sets of discontinuities oriented α_1 and α_2 with respect to an arbitrary reference axis (after Thomas 1981)

Warburton (1980 a,b) proposed a statistical model to derive analytical prediction of the pattern of discontinuity traces as observed in a geological survey. The discontinuities are supposed to be quite flat, very thin, circular or rectangular shaped and parallel to a given direction. Their spatial distribution characterized by that of their centroid, is supposed to be random and described by a Poisson distribution for the volume density of the centroids. The Poisson distribution is supported by experimental observation that trace spacings along a sampling line are often distributed exponentially and this is precisely what would be expected from results for Poisson flats in geometrical probability theory. The sizes of joints have a statistical distribution g(x) that is related to the distribution of trace length. This relationship can then be used to determine the joint size distribution from the measured distribution of trace length as observed on a face.

In practice the sample distribution of trace length is usually both censored and truncated. Truncation is caused by the fact that only traces larger than a certain length are mapped. Censoring comes from the fact that traces often extend outside the zone of investigation. Solutions for both area sampling and line sampling are proposed, including allowance for truncation, although this should be kept as low as possible because of its influence on the determination of g(x).

For area sampling for example when no truncation occurs the solution for circular joints is given by :

$$h(y) = \frac{y}{m} \int_y^{\infty} \frac{g(x)dx}{\sqrt{x^2-y^2}} \qquad (17)$$

where h(y) is the probability density for trace length y ;

m is the mean trace length

x is the diameter of a circular joint ;

g(x) is the probability density that a joint exhibits diameter x.

Methods for solving equation (17) for g(x), when h(y) is known, are discussed by Warburton. For convenience, he suggests assuming an analytical expression for g(x) and proposes the log normal function.

Priest and Hudson (1976) and Hudson and Priest (1979) investigated the possibility of using scanline surveys to derive a probability density distribution of block length, block area and block volume. They showed that spacing between discontinuities, as observed along a scanline, can be represented, in most cases, by a negative exponential distribution. The validity of this proposition was verified for exposures of eight sedimentary rocks and two crystalline-rocks. Wallis and King (1980) confirmed this result for granite. If f(x) is the probability density, λ is the mean discontinuity frequency and x a spacing value, then the cumulative length proportion of discontinuity spacings larger than a certain length ℓ is :

$$L(\ell) = \int_0^{\ell} \lambda x f(x) dx \qquad (18)$$

Accordingly, since the Rock Quality Designation is defined as the percentage of scanline consisting of intact length over 0.1 m, it can be shown that :

$$RQD = 100^{-0.1\lambda}(0.1\lambda + 1) \qquad (19)$$

if the probability density is represented by a negative exponential function ($f(x) = \lambda e^{-\lambda x}$). Considering two sets of parallel, continuous, joints, orthogonal to each other Hudson and Priest showed that the RQD, for such geometry, depends on the scanline orientation with respect to the discontinuity sets. These results were extended later (Priest and Hudson, 1981) to take into account the possibility of biasing caused by truncation and censoring .These authors also presented means to estimate both the accuracy and the precision of RQD determination as well as that of mean discontinuity trace length.

Once a statistical representation of joints has been obtained, the problem of their connectivity which is directly correlated to that of rock mass permeability may be approached. Hudson and Lapointe (1980) have proposed to use electrical printed circuits for such an investigation when the problem is two dimensional ; Priest and Sameniego (1983) propose, at this congress,a numerical technique. However applicability of these two-dimensional models to real life three dimensional conditions is not discussed. It is hoped that this point will be raised at the congress.

3.2. The global characterization

Bertrand et al. (1982) have tried to determine the permeability and its variation within granite masses from an investigation of the fracture pattern combined with in-situ stress and permeability measurements. Although this problem has not been solved yet, some interesting results with respect to fracture pattern analysis are reported.

The fracture pattern is described either by a scalar (area or volume density) or a vector (fracture density in a given direction). The purpose of the study is to identify these characteristics at a few locations and then to use geostatistics (statistics of regionalized variables) to infer their value at other locations.

Two types of data were considered : first, area fracture densities as well as angular fracture densities as determined from aerial photographs, each photograph covering one square kilometer and secondly volume fracture densities as determined from granite outcrops and underground exposures (at a 100 to 1000 m^2 scale). Results have shown that the fracture pattern depends on the scale at which it is studied, hence that observed at one scale cannot help deduce simply that to be expected at a different scale.

4. MODELING THE MECHANICAL BEHAVIOR OF ROCK MASSES

The need for modeling the mechanical behavior of rock masses arises in very different problems. Sometimes the model is supposed to be quantitative and implies an evaluation of both displacements and stresses in the whole domain under investigation. These data may be used for stability analysis, for the design of large structures or that of artificial fracturing schemes, or for the understanding of natural processes such as earthquake source mechanisms or intraplate deformation. Sometimes the model is qualitative and helps only to identify the relevant parameters affecting the deformation process. This report does not intend to cover all the facets of this very broad theme ; it rather focusses attention on a few particular topics concerning quasistatic deformations.
An important step in this modeling process is the choice of appropriate constitutive equations to represent both the equivalent continuum and the large scale discontinuities. Ideally this choice should be governed only by laboratory and in-situ tests as well as by a sound understanding of the rock mass discontinuity pattern. Indeed, only those discontinuities which are large compared to the size of the volume under investigation are treated as such ; the others are ignored but the constitutive equations for the equivalent continuums are defined so that the global effects of small scale discontinuities such as dilatancy, low shear strength in a given direction, etc.. are taken into account. Practically, until recently, this choice was dictated mostly by the mathematical technique available for solving the partial differential equations : the necessity to rely on analytical closed form solutions or on linear elastic two dimensional models imposed unrealistic oversimplifications. Nowadays the development of high speed computers has opened the way to more refined numerical solutions so that non linear constitutive equations and more accurate geometrical shapes can be handled. However many of these sophisticated models are applicable only to two dimensional situations. Even when the two dimensional models are justified, it always should be kept in mind that the accuracy of the solution depends strongly on that of the input data. Thus it may be unrealistic to adopt too sophisticated a model if the parameters on which it is based are not accessible to direct observations or to accurate evaluation. For engineering problems, ideally, simple large scale in-situ tests should be used to validate a numerical model before it is applied. Unfortunately such tests are often too costly to be carried out ; in-situ measurements conducted during the early development stages of a project are then used to adjust the model to the real life situation, or at least, to provide means for evaluating its " degree " of accuracy.
Further, long term instrumentation is always desirable so as to provide elements for comparison between observed and computed results ; this helps to validate, or modify, the model for later use.
For geological problems,means to validate a solution are more subjective. Usually the model is considered realistic when its results compare favorably with surface observations. However more and more data on plate motion, deformation rate or stress magnitude are available so that some quantitative validation is possible, in a few instances.

4.1. Constitutive equations

In most engineering situations temperature remains below 200°C and depth is much less than 5000m although higher temperatures may be encountered in some geothermal fields (e.g. 350° at Los Alamos hot dry rock site in the U.S.A.) and greater depth may be reached in the oil industry (e.g. 9000 m in Southern U.S.A.). Yet this domain appears very limited when one deals with deformations in the earth crust where depth may reach 50 to 60 km and temperature may get as high as 1000°C.Further, the time scale of quasi-static deformation encoutered in engineering practice may vary from half a day to a hundred years (e.g. the Simplon Tunnel in Switzerland is more than 100 years old) whilst in geological situations millions of years must be considered. Interestingly enough, relatively few models have been proposed to analyse the mechanical behavior of the many rock types encountered in the earth crust, for the very broad spectrum of temperature, stress and strain rate conditions described here above. Some of them are presented now.

Elastic models

The linear isotropic, or planar isotropic, elastic model is of course the oldest and still remains the favorite model. Indeed, in many problems most of the volume under consideration is submitted to sufficiently low temperatures and stress levels that the linear elastic hypothesis be verified. Further, because of its simplicity, this model provides a means for readily identifying the main difficulties to be anticipated such as zones of high stress concentrations or regions where tensile stresses may appear. These difficulties are then analyzed in a second step with a more refined model.
Elasticity fails to take into account the influence of large scale discontinuities ; when these are likely to have a significant influence on the deformation process, the elastic model is coupled with a specific model for these discontinuities, as is discussed later.
Because rock masses are also very often affected by numerous small scale discontinuities with virtually no tensile strength, Zienckiewicz et al. (1968) have proposed to consider that the equivalent continuum (which includes in this case the discontinuities) does not sustain tensile stres ; it is considered as a bilinear elastic model (zero stiffness when submitted to tensile stress , standard elastic behavior when submitted to compressive stress). In this case,because joints exhibit much lower

stiffness than does the rock matrix, moduli for the continuum equivalent to the rock mass must take into account joint set orientation, joint spacing within each set and joint stiffness. For example Heuzé et al. (1982) propose to represent a granite with three orthogonal sets of joints with roughly equal spacing by an equivalent isotropic medium with equivalent Young's modulus given by :

$$E_g^{-1} = E_1^{-1} + (Sk_n)^{-1} \quad (21)$$

where E_g is the Young's modulus of the rock mass

E_1 is the Young's modulus of intact granite

s is the joint spacing

k_n is the normal stiffness of the joints.

Gerrard (1982) conducted a detailed analysis of the influence of one, two, or three orthogonal sets of joints. The joints are supposed to be continuous with orthorhombic elastic characteristics (9 elastic constants) and roughly equally spaced. Further, his analysis assumes that the normal to the three planes of symmetry of the material, in each layer, are respectively parallel to the three axis x_1, x_2, x_3 of the frame of reference (the joints are perpendicular to x_3). The equivalent material is then found to be orthorhombic.

Brace et al. (1968) were the first group to describe the dilatancy effect (i.e. the increase of bulk volume) observed after the linear elastic limit of rock has been reached but before the peak strength is attained. Brace and Martin (1968) discussed the influence of this dilatancy on the mechanical behavior of fluid saturated rocks and found a stabilizing effect. Stuart and Dietrich (1972) proposed to consider a non linear isotropic elastic equivalent material to take into account this volume increase :

$$\underset{\sim}{\varepsilon} = \phi_o \underset{\sim}{1} + \phi_1 \underset{\sim}{\sigma} + \phi_2 \underset{\sim}{\sigma}^2 \quad (22)$$

where ϕ_o, ϕ_1 and ϕ_2 are polynomial functions of the first invariants of the stress tensor $\underset{\sim}{\sigma}$ ($\underset{\sim}{\varepsilon}$ is the linearized deformation tensor).
However, to this writer's knowledge, no practical application of this model has been proposed yet.

Elasto-plastic models

The development of stiff, or servo-controlled, laboratory testing systems (e.g. Cook and Hojem 1966, Wawersik 1968, Rummel and Fairhurst 1971, Hudson et al. 1972) has opened the way for detailed experimentation on the post-peak strength mechanical behaviour of rock samples. Wawersik and Fairhurst (1970), Crouch (1970), Cornet and Fairhurst (1972), Rummel (1974) and many others have thus investigated the variation of dilatancy, and more generally the stress-strain relationship, in the post-peak domain, for various rock types with various confining pressure conditions. Typical results are shown on figures 10 to 13.
As can be seen in figures 10 and 12 an important characteristic of rocks in their ability to maintain significant loads after their peak strength has been reached : they exhibit, for some stress conditions, a strain softening elasto-plastic behavior associated with dilatancy.

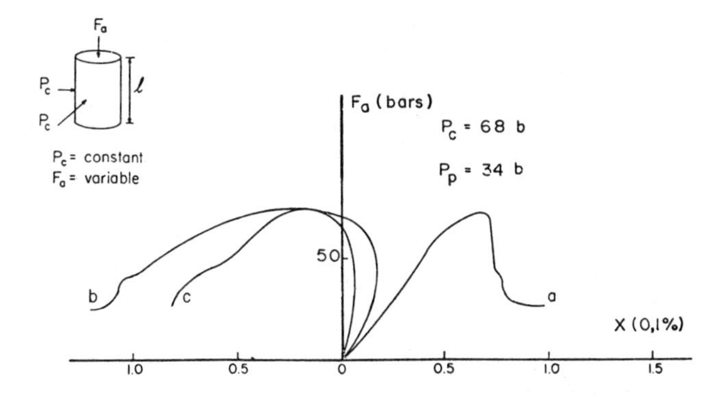

Fig.10 Results from triaxial test on fluid saturated Berea Sansdone; curve a = X abcissa is axial deformation whilst for curve b it is volumetric deformation and pore volume for curve c ; effective confining pressure is 3.4 MPa.

For larger compressive stresses dilatancy disappears (see fig. 13).
However the behavior exhibited on figure 11 does not comply with this rule : Once a certain amount of deformation has been reached after peak strength, the rock sample failed in an unstable manner, also called class II behavior by Wawersik (1968): the specimen had to be unloaded progressively so as to maintain failure quasistatic. This fracturing process was observed with Berea sandstone for confining pressures ranging from about 7 MPa to 25 MPa, i.e. for values typical of many engineering situations (they correspond to depth ranging from 500 m to 2000 m). Similar results have been observed with limestone, coal or granite for different confining pressure domains.
In none of the reviewed papers dealing with the modeling of rock post-peak behavior has this point been raised and it is hoped that it will be discussed at the congress.
Consequences of this dilatancy on the post-peak behavior of fluid saturated rocks have been discussed by this reporter (Cornet 1977).

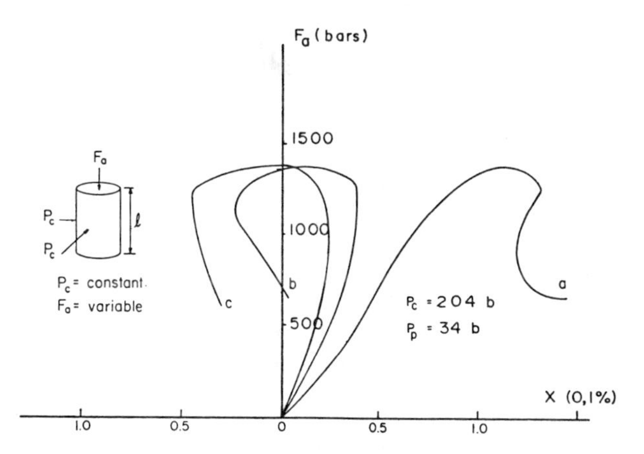

Fig. 11 Same as fig. 10 but effective confining pressure is 17.0 MPa.

It was shown for example, that for such materials the confining pressure domain where a class II fracturing process is observed is larger than that for dry rocks when drainage is only partial: Unstable fracture propagation induces a decrease in pore volume and therefore an increase in pore pressure so that the effective confining pressure is reduced.

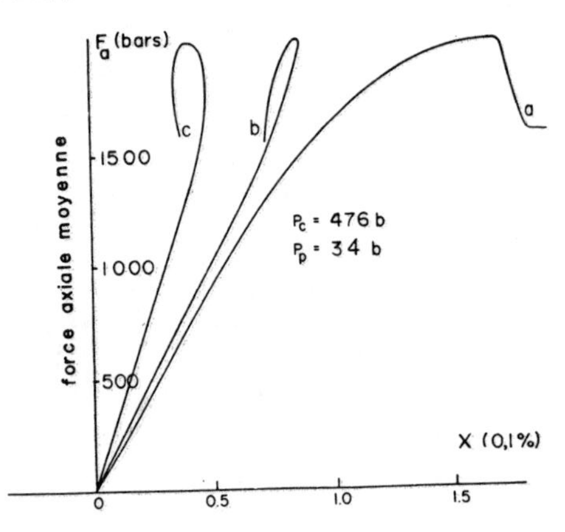

Fig.12 Same as fig.10 but effective confiming pressure is 44.2 MPa.

Jia-Shou Zhuc and Yin-Tang Wang (1983, this congress) propose three types of failure criteria to characterize the peak strength :

(23)

Pure tensile fracture : $\sigma_1-\sigma_t = 0$ when $3\,\sigma_1+\sigma_3 \leqslant 0$

Tensile tortional failure : $(\sigma_1-\sigma_3)^2+8\,\sigma_t(\sigma_1+\sigma_3)=0$

when $3\sigma_1+\sigma_3 > 0$; $\sigma_1+(3-2\sqrt{2})\sigma_3 \leqslant 0$

Compressional torsional failure :

$$(\sqrt{1+f^2}+f)\;\sigma_1-(\sqrt{1+f^2}-f)\sigma_3-4\,\sigma_t = 0$$

when $\sigma_1+(\sqrt{1+f^2}-f)^2\,\sigma_3 \geqslant 0$

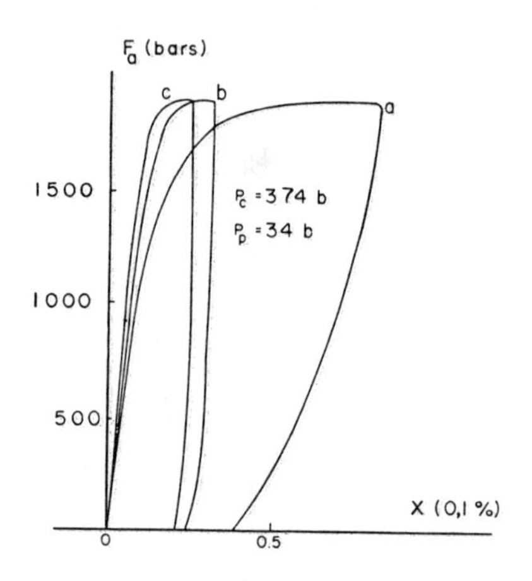

Fig.13 Same as figure 10 but rock is Indiana limestone and effective confining pressure is 34 MPa.

where compressive stresses are positive, f is the friction coefficient of the rock and σ_t its tensile strength.
Applying the incremental theory of plasticity they obtain :

$$\{d\sigma\} = ([D] - (1-r)[D_p])\{d\varepsilon\} \quad (24)$$

where r = 1 in the elastic domain or when unloading occurs ;

r = 0 in the plastic region ;

0 < r < 1 in the transient region ;

[D] is the elastic matrix ;

$$[D_p] = [D]\left\{\frac{\partial F}{\partial\{\sigma\}}\right\}(A + \left\{\frac{\partial F}{\partial\{\sigma\}}\right\}^T [D]\left\{\frac{\partial F}{\partial\{\sigma\}}\right\})^{-1}\left\{\frac{\partial F}{\partial\{\sigma\}}\right\}^T [D] \quad (25)$$

F is the yield function ;

A corresponds to the strain hardening effect.

Pierce and Ryder (1983), this congress) prefer to use a deformation formulation which is obtained by a parameterization of laboratory results. These idealized stress-strain curves are assumed to be independent of loading history and loading path. They assume that pure elastic behavior persists up to peak strength which is supposed to be described by the linear law $\hat{\sigma}_1 = \sigma_c + k\sigma_3$ (σ_c is the maximum stress drop supposed to be independent of σ_3).

The post-failure behavior follows a law with constant negative modulus ; when the shear reaches the residual strength, deformation occurs with no dilatation.
Whilst in both these models the material is supposed to be isotropic with respect to its strength, Guenot and Panet (1983, this congress) propose to consider two different Mohr-Coulomb failure criteria for rocks which exhibit a strong strength planar anisotropy. The first one ($|\tau| = c + \sigma_n \operatorname{tg}\varphi$) is defined for all fracture surfaces except for those which are parallel to the weakness planes ; for these, a second criterion is used $|\tau| = \bar{c} + \sigma_n \operatorname{tg}\varphi$ where $\bar{c} \ll C$.

The main drawback of all these models is their failure to recognize the very significant influence of strain rate on the peak strength as well as on the post-peak behavior. Further, it is not clear whether the yield criteria chosen by these various authors is really corroborated by field or laboratory evidence .

Time dependent models for low temperatures and low minimum principal stress conditions.

Time dependent effects observed at low temperature (say smaller than 300° C) and low minimum principal stresses (smaller than 500 MPa) may have two different physical origins : motion of dislocations within the crystals that constitute the rock and stable microcracking. The first mechanism is encountered in rock salt for example, even under uniaxial compression conditions and room temperature, or calcite and serpentine when the minimum principal stress is larger than a few tens of megapascals.

This deformation process is often investigated in the laboratory with creep tests, in which the load is maintained constant, or constant loading rate tests (see e.g. Langer 1979, Wawersik 1980, Vouille et al. 1981). For given loading conditions, the creep velocity may decrease because of strain hardening, may remain constant or may increase and lead to some sort of instability. These various stages are called primary, secondary, and tertiary creep.
Andrade (1910) proposed a general equation for creep behavior :

$$\varepsilon = A + B.E(t) + Ct \qquad (26)$$

where A , B and C are constants whilst E(t) is an empirical function. Morlier (1966) and Parson and Hedley (1966) found that E(t) should be logarithmic as well as Langer (1979) for primary creep. For secondary creep Langer proposes that

$$\varepsilon = A' \exp(-u/kT).f(\sigma) \qquad (27)$$

where T is temperature, σ is the applied uniaxial stress, u is the activation energy, k Boltzman's constant and A a material characteristic constant.
Vouille et al. (1981) found that after the elastic limit has been reached, the viscoplastic strain observed in the case of creep may be expressed as

$$\varepsilon_{vp} = \frac{(\sigma_1 - \sigma_3 - 2c)^{\beta}}{k} t^{\alpha} \qquad (28)$$

where ε_{vp} is the visco-plastic strain,

σ_1 is the maximum principal stress,

σ_3 is the minimum principal stress,

c is the cohesion,

k, β, α, are constant which are determined from the tests,

t is the time.

They propose as rheological law for this rock salt :

$$< \sigma_1 - \sigma_3 - 2c > = \frac{K}{(\alpha)^{\alpha/B}} (\varepsilon_{vp})^{(1-\alpha)/\beta} (\overset{1}{\varepsilon}_{vp})^{\alpha/\beta} \qquad (29)$$

where $< x > = \frac{1}{2}(x + |x|)$

Further they observe that the internal friction angle is zero for this rock whilst the apparent cohesion is less than 1.5 MPa ;
this last value decreases as temperature and confining pressure increase.
During tertiary creep some microcracking appears which leads ultimately to complete failure (see e.g. Scholz 1968, Kranz and Scholz 1977).
Influence of strain rate on this stable microfracturing process is of primary importance for all rocks once they reach the post peak domain. Bieniavski (1970), Peng and Podmieks (1972), Hudson and Brown (1973), Houpert (1974) investigated experimentally in the laboratory this strain rate dependency. Results (see fig.14) show that decreasing the strain rate produces a decrease in peak strength and a " softening " of the post-peak behavior ; the more heterogeneous the rock the larger the softening affect.
Kaiser and Morgenstern (1981) proposed a simple rheological model to explain phenomenologically this effect. They consider an elementary model (see fig. 15) which consists of an elastic, brittle A unit, which exhibits instantaneous

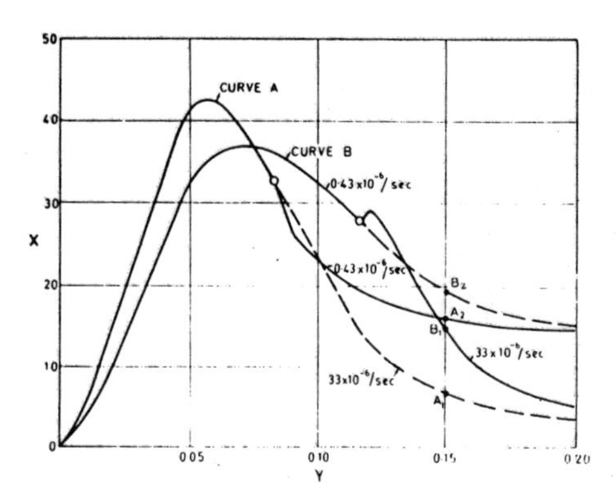

Fig.14 effect of changing strain rate on the behavior of fractured rock (after Bieniawski 1970)

strength loss of magnitude C_A at a strain equal to E_A and a constant ultimate strength F_A , and, in parallel, a time-dependent B unit.

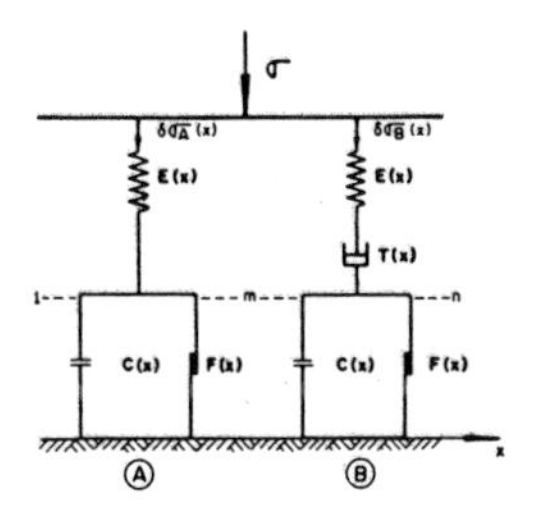

Fig.15 Elementary model for tim dependency of post-peak failure after Kaiser and Morgenstern (1981)

The later exhibits an instantaneous strength loss C_B and a constant ultimate strength F_B. Unit B may fail before or after unit A.
They propose to model the time dependency of failure development in rock masses by considering several of these elementary models to act in parallel.
Tan and Kang (1983, this congress) consider that the total strain exhibited by a dilatant material can be viewed as the superposition of three components :

$$\varepsilon(t)_{ij} = \varepsilon^{E}_{ij} + \varepsilon^{D}_{ij} + \varepsilon^{D}(t)_{ij} \qquad (30)$$

$$i,j, = 1,2,3$$

where ε^{E}_{ij} is the elastic strain

ε^{D}_{ij} is the instantaneous dilatational strain

$\varepsilon^{D}(t)_{ij}$ is the time dependent dilatational strain

Onset of dilatancy is assumed to occur when $\sigma_{oct} > f_3$, where f_3 is a material constant.
They apply this formulation to the analysis of regional upheaval observed before some large earthquakes in China .
Dragon and Mroz (1979) propose to relate the creep behavior observed in laboratory uniaxial

compression tests to a time dependent crack density (β) :

$$\dot{\beta} = k_1 \dot{\varepsilon}_{kk}^{(i)} + k_2(\beta) g(\sigma_{k\ell}, \beta) \qquad (31)$$

where $\varepsilon_{kk}^{(i)}$ denotes the inelastic volumetric strain rate ;

k_1 is a material constant ;

$k_2(\beta)$ and $g(\sigma_{k\ell}, \beta)$ are functions describing time progressive fracturing ($k, \ell = 1,2,3$).

The total strain rate is assumed to be the sum of an elastic and an inelastic strain rate. Inelastic strain rate is assumed to occur when the following yield condition is satisfied :

$$f(\sigma_{ij}, k, \beta) = S_{ij} S_{ij} - 2p(k,\beta)(I_1^\circ - \sigma_{kk}) = 0 \qquad (32)$$

where $p(k,\beta) = \bar{p}_o + k - \beta^2$

$\dot{k} = \sigma_{mn} \dot{\varepsilon}_{mn}$ = rate of strain hardeming,

S_{ij} is the stress deviator component,

I_1° is a positive constant,

P_o determines the configuration of an initial yield surface for $k = \beta = 0$.

Martin (1972), Henry et al. (1977), Atkinson (1979) among others, pointed out the influence of stress corrosion on tertiary creep ; that is the influence of chemically active species on micro-crack growth (the main corroding agent being water). This concept of stress corrosion has been applied by Das and Scholz (1981) in explaining a few phenomena associated with earthquakes, like multiple events, delayed multiple events, foreshocks, aftershocks, postseismic rupture extension or " slow " earthquakes. Their reasoning assumes that earthquakes correspond to combined mode II and mode III fractures, i.e. pure shear, and that the fracture, source of the earthquake, propagates in its own plane. This assumption is supported by field evidence which indicates that usually earthquakes occur along preexisting faults.

According to classical fracture mechanics (see Irwin 1957) a fracture extends when the stress intensity factors (k), characterizing the stress field at the crack edge, reaches a critical value K_c^* which is a material property. For mode I fractures (tensile) it is often observed that fracture propagation occurs in a stable manner for stress intensity factors much lower than K_c^*. This is attributed to the development of a corrosion effect which lowers the value of the critical stress intensity factor to $K_o \leqslant K_c^*$ (crack propagates when $k \geqslant K_o$). This fracture propagation is quasistatic, with an extension velocity of the form (Atkinson 1979) :

$$\dot{X} \ \alpha \ k^n \qquad (33)$$

where n is a constant called the stress corrosion index.

Das and Scholz assume that similar laws exist for modes II and mode III. Since k is proportionnal to $\sqrt{X}$ it is readily seen that this stress corrosion effect leads ultimately to unstable fracture propagation. They argue that unstable crack growth occurs when k reaches a critical K_c which is close to the theoretical K_c^* value since, under dynamic condition, no corrosion has any time to develop. The stress intensity factor k for a circular, plane, shear crack of radius X in an infinite, homogeneous, linear, elastic medium is given by, for both modes II and III :

$$k = C \Delta\tau \sqrt{X} \qquad (34)$$

where c is a geometrical constant and $\Delta\tau$ the static stress drop. Since, for a circular crack, both modes II and III exist simultaneously at points along the crack edge, k is taken as the mean square root of these stress intensity factors so that $C = \frac{20\sqrt{2}}{7\pi}$ (Sih, 1973).

From experimental results (Atkinson 1979), onset of stable propagation occurs when :

$$k = K_o \left(\frac{\dot{X}}{v_o} \right)^{1/n} \qquad (35)$$

where K_o and v_o are the value of k and $\dot{X}$ at the initiation of quasi-static growth. If $\Delta\tau$ is independent of time then, since $K_o = C \Delta\tau \sqrt{X_o}$,

$$\dot{X} = v_o \left[\left(\frac{X}{X_o} \right)^{1/2} \right]^n \qquad (36)$$

so that

$$X = \left[X_o \frac{2-n}{2} - \frac{n-2}{2} \frac{V_o t}{X_o^{n/2}} \right]^{2/(2-n)} \qquad (37)$$

thus the duration of stable crack growth is given by

$$t_f = \frac{X_o}{V_o} \frac{2}{n-2} \qquad (38)$$

Equation (36) is obtained by setting the expression in brackets in equation (35) equal to zero ; indeed this condition corresponds to infinite value of $\dot{X}$ since $n > 2$. The derivation of stable crack growth, t_f, depends on the initial conditions as well as on n, the stress corrosion index. Das and Scholz have applied this result to a few typical conditions and found that stable fracture growth preceeding earthquakes could vary from 48 hours, for an initial fracture length equal to 1 km and a stress corrosion index equal to 12.5, to 278 days for an initial crack length equal to 0.5 cm and the same corrosion index ; $\Delta\tau$ was equal to 10 MPa in both cases. The crack length before unstable propagation was 1.4 km in the first case and 14.2 cm in the second one. Their results appear to be in agreement with dilatometer records obtained before a magnitude 3.8 earthquake.

Modeling the mechanical behavior of large natural discontinuities

In many rock mechanics problems the behavior of rock masses is strongly influenced by preexisting large discontinuities like joints. Joints cannot support any tensile stress ; their shear strength is highly variable and their shear deformation is usually associated with volume variations. As mentioned earlier in this report, only those discontinuities which are large compared to the size of the volume under investigation are treated as such ; the smaller ones are integrated in the equivalent continuum.

Large joints influence both the elastic response

of a rock mass and its resistance characteristics. For its influence on the elastic response, the joint is usually represented by a special stiffness matrix (see e.g. Goodman et al. 1968, Heuze et al. 1971, Gerrard 1982, Walsh and Grosenbaugh 1979) which relates normal and shear tractions exerted onto the joint surface to normal and shear displacements at this surface. Desai et al.(1983, this congress) propose to model the joint by thin layer elements with appropriate stiffness matrix. They propose further to retain only the diagonal terms of this matrix because of the difficulty encountered for the practical determination of non diagonal terms.
Barton (1976) pointed out that the mechanical behavior of joints depends strongly on the value of the normal stress acting on them. For low effective normal stress magnitudes, rock joints exhibit a wide spectrum of shear strength because of the strong influence of surface roughness and variable rock strength (see fig.16).

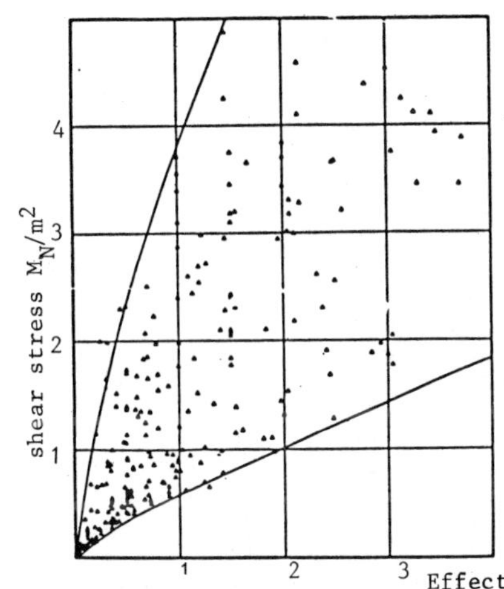

Fig.16 Peak shear strength of unfilled rock joints from the published results of direct shear tests in the laboratory and in-situ (after Barton 1976).

For high effective normal stress levels, the spectrum of shear strength of joints or artificial faults is narrow (see fig.17), despite the wide variation in shear strength of intact rocks, and well represented by the classical Coulomb friction criterion (Byerlee 1975) :

$$\tau = C + \mu(\sigma_n - P) \qquad (39)$$

where τ is the shear strength, σ_n the normal stress, P the interstitial pore pressure, μ the coefficient of friction, C is the cohesion which is usually negligible for large values of σ_n.
For even higher effective normal stresses, natural joints do not influence any longer the shear behavior of rock masses.
Efforts have been made to quantify this variation of shear strength with effective normal stress as well as the change of volume observed during shear deformation. Barton (1973, 1976) performed direct shear tests on a variety of artificial tension fractures generated in realistic brittle model materials. He observed that if μ is the total friction angle of the material ($\mu = \mathrm{arctg}\,\tau|\sigma_n$ where τ is the peak shear strength and σ_n the effective normal stress) and d_n is the peak dilation angle which is defined as the instantaneous inclination of the shearing path, at peak strength, relative to the mean joint plane, then

$$\mu = 2\, d_n + 30 \qquad (40\ a)$$

$$d_n = 10 \log_{10}\left(\frac{\sigma_c}{\sigma_n}\right) \qquad (40\ b)$$

where σ_c is the uniaxial compression strength of the intact material ; the value 30 in equation (40 a) corresponds to the basic friction angle of the material.
For the 0.1 to 10 MPa range of normal stress magnitude investigated in these experiments, the peak shear strength envelope is found to be :

$$\tau = \sigma_n \mathrm{tg}\left[20 \log_{10}\left(\frac{\sigma_c}{\sigma_n}\right) + 30\right] \qquad (41)$$

Barton proposed a generalized form of equation (41) for natural joints submitted to low normal stresses :

$$\tau = \sigma_n \mathrm{tg}\left[\mathrm{JRC} \log_{10}\left(\frac{\mathrm{JCS}}{\sigma_n}\right) + \phi_b\right] \qquad (42)$$

where JRC is the joint roughness coefficient defined on an empirical scale ranging from 20, for

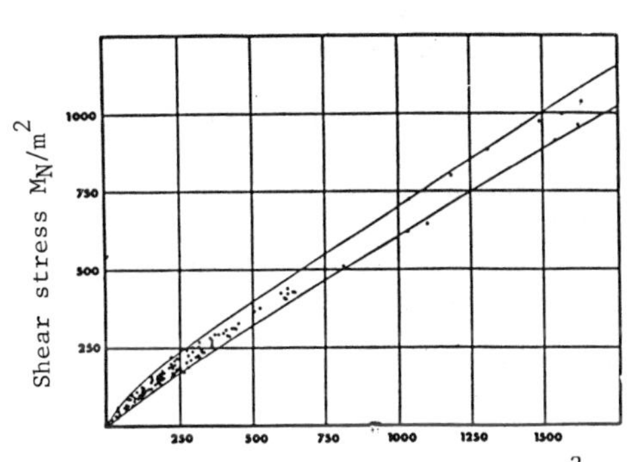

Fig.17 The shear strength of faults and tension fractures in rock under high effective normal stress (after Byerlee 1975, 1967, 1968) ; Data from Westerly granite, Solenhofen, limestone, oak Hale limestone, Mahant gabbro, Spruce pine dunite, cabramurra serpentinite, weber sandstones. (figure is from Barton 1976).

very rough joints, to 0 for smooth surfaces (see Barton and Choubey 1977 , Barton and Bandis 1980), for a detailed discussion of this empirical scale) ;
JCS is the joint wall uniaxial compression strength ;
ϕ_b is the basic friction angle defined as the value of arctg ($\tau|\sigma_n$) obtained from residual shear tests on flat unweathered rock.

The value of JCS may vary from σ_c, the uniaxial compression strength of the material, to $\sigma_c/4$ for weathered surfaces.

For larger normal stresses, Barton proposes to replace the uniaxial compression strength by the triaxial compression strength so that equation (42) becomes :

$$\tau = \sigma_n \mathrm{tg}\left[\mathrm{JRC} \log_{10}\left(\frac{\sigma_{TC} - \sigma_3}{\sigma_n}\right) + \phi_b\right] \qquad (43)$$

where σ_{TC} is the peak strength in triaxial compression tests with confining pressure equal to σ_3.

Results obtained with equation (43) for shear along fractures developed in laboratory triaxial tests with sandstone, limestone, gabbro dunite, serpentinite and granite specimens were in very satisfactory agreement with experimental results for confining pressures up to 500 MPa and for JRC values set equal to 20.
Whilst equation (43) corresponds to a generalization of equation (41) that for equation (37 b) (the prediction of the dilatation angle) is, for $\sigma_n > \sigma_c$:

$$d_n = 10 \log_{10} \frac{\sigma_{TC} - \sigma_3}{\sigma_n} \qquad (44)$$

Results obtained from the laboratory tests mentionned here above yield dilation angles of the order of 2° for faults sheared at the brittle ductile transition (ductile characterizes, here, shear deformation with no stress drop).
Pande and Gerrard (1983, this congress) propose to model the joint shear strength by the Coulomb criterion but with a variable cohesion, C_e , which depends on joint opening :

$$C_e = C_o \frac{\langle \varepsilon_c - \varepsilon_t \rangle}{c} \qquad (45)$$

where ε_t is the visco-plastic tensile strain normal to the joint ;

C_o is the cohesion of the joint when it is closed ;

ε_c is a critical tensile strain which depends on the joint roughness ;

$$\langle \varepsilon_c - \varepsilon_t \rangle = \begin{cases} \varepsilon_c - \varepsilon_t & \text{if } \varepsilon_c - \varepsilon_t > 0 \\ 0 & \text{if } \varepsilon_c - \varepsilon_t \leqslant 0 \end{cases}$$

Although this report is not concerned with problems dealing with fluid flow in rock masses, an interesting result obtained by Walsh (1981) concerning the effective stress concept as applied to the permeability of joints, or fractures, should be mentionned. In his analysis, Walsh shows that the change in pore volume dv_p within a joint, caused by variations in both hydrostatic confining pressure and pore pressure, is :

$$dv_p = \frac{\partial v_p}{\partial P}(dP_c - dP_p) + v_p \beta_S dP_p \qquad (46)$$

where $\frac{\partial v_p}{\partial p}$ is the rate of change of pore volume with applied hydrostatic pressure for a joint with no pore fluid ;

dP_c is the change of hydrostatic confining pressure exerted on host rock ;

dP_p is the change of pore pressure ;

v_p is the original pore volume in the joints ;

β_s is the bulk compressibility of the rock matrix

From (46) it can be shown that the effective stress, Pe , on which depends the joint permeability is :

$$P_e = P_c - sP_p \qquad (47)$$

where $s = 1 - v_p \beta_S / (\partial V_p / \partial P)$

s can be as low as 0.5 and as high as 5. Since the permeability is somewhat related to joint opening it may be anticipated that this joint opening depends on effective stresses given by equation similar to (47) rather than the classical $(\underset{\sim}{\sigma} - \underset{\sim}{P1})$ expression which is usually assumed.

4.2. Some examples of solutions

Because the geometry is usually quite complex and because constitutive equations are non-linear in many instances, the solution of most rock mechanics problems cannot be obtained by analytical closed form solutions. When the material exhibits a linear isotropic elastic behavior (i.e. when the complexity of the problem comes only from its geometry), the photoelastic method can be used, for two and three dimensional problems, for identifying in a fairly inexpensive manner zones of high stress concentration. But the advent of high speed computers provides, now, means to obtain, in nearly as inexpensive a manner, more accurate numerical solutions, so that numerical models are nearly always preffered to photo elastic models. However for the analysis of large deformations physical models have been shown to be very useful for investigating the gross aspects of deformation processes. For example, Tapponnier et al.(1982) have developed a plasticine model to study the punching of the Asian continent by the Indian sub-continent. With this model they have been able to obtain a global understanding of the kinematics of this plate deformation process like the opening of the South China sea before late Miocene time, or the rotation and extrusion of Indochina to the south east along the then-lateral Red River fault during the first 20 to 30 millions years of the collision, as well as the opening of the Mergui basin and Andaman sea up to the present time. Their experiment suggests that a major left lateral strike-slip fault and rift system will propagate across the Tien-Shan, Mongolia and Baikal up to the sea of Okhotk. On a smaller scale Deramont et al. (1983, this congress) have investigated with a two material model the effect of a deep shear on a sedimentary layer. The lower layer is made of plasticine because of its cohesion, the upper one being a cohesionless granular material. Their model appears to predict with some accuracy the features observed during sliding-spreading of nappes. However the main difficulty encoutered with these physical models is the proper scaling of the various parameters involved : a proper model must satisfy rules of similarity.

This similarity principle is probably more adequately satisfied by base friction models (Bray and Goodman, 1981) in which the effect of gravity acting in the plane of a two dimensional physical model is modelled by a drag force acting along its base. Various materials can be used to scale properly the rheological characteristics of interest. Further a uniform gas pressure applied over the entire model surface allows the simulation of confinement (Egger and Gindroz 1979). This model is well adapted for problems involving slow motion for which inertia forces are small compared with gravity; it is not applicable to general dynamic conditions. Lessi and Sarda (1980) have discussed the problem of scaling the fracture toughness of rocks (K_{IC}) for model simulation of hydraulic fracture propagation; they show that if λ is the geometrical reduction factor, then fracture toughness for mode I must be scaled by :

$$K_{IC\ model} = \sqrt{\lambda}\ K_{IC\ real}$$

and they developed model materials, based on a mixture of cement and bentonite, which satisfy this condition for a reduction factor λ equal to about 2×10^{-3}.

The difficulty for properly satisfying scaling rules is the main limitation of physical models, a drawback which does not affect numerical models. Further, numerical models are directly amenable to investigating the influence of such or such variable by simply changing its value so as to conduct parametric studies. This approach is particularly useful when an input parameter is not defined with good accuracy. In addition the advent of efficient graphic display terminals provides very efficient means for directly visualizing the solutions. This is probably why most problems are now tackled by numerical modeling.

Finite element solutions

" Finite elements " must be understood here in the usual engineering connotation i.e. the equivalent continuum and the major discontinuities are discretized into a finite number of parts the behavior of which is specified by a finite number of parameters. This definition is reconciled with that given in applied mathematics by stating that the solution of the complete system, as an assembly of its elements, follows precisely the same rules as those applicable to a standard mathematical discrete problem. A detailed presentation of the application of this method to mechanical problems is given by Zienckiewikz (1977) and no attempt will be made here to discuss this very powerful numerical procedure. Let us only recall that, as pointed out by Jaeger and Cook (1979), advantages of finite elements are :

i) The elements can be chosen in any way and of any size so that irregular boundaries can be fitted without trouble, and a greater density of nodal points can be chosen in regions of greater stress concentration.

ii) variable surface and body forces can be easily handled.

iii) Each element has its own stiffness matrix so that the material need not be homogeneous or isotropic. Also non linear and bilinear materials can easily be studied.

iv) The method is not restricted to elastic behavior, and other rheological models can be introduced.

v) Friction and dilatancy on joint surfaces can be allowed for.

vi) The final output can be processed in the computer in any desired fashion.

Papers presented at this congress are good examples of the potentials of the method. For example Pande and Gerrard present an elasto-visco-plastic two-dimensional model for rock masses which also includes specific modelling of sets of parallel continuous evenly spaced joints; these are supposed to exhibit a given stiffness in the elastic domain, to have no tensile strength and to slide in compression, according to a Coulomb friction criterion with possible dilatancy. Reinforcement elements can be introduced; they are supposed to exhibit a given elastic response and then yield according to a strain hardening law. Carol and Alonso propose a new joint element for two dimensional analysis of fractured rock which requires no less than eleven parameters to describe its complete mechanical behavior.

Practical applications of finite element analysis are also presented. For example Casteleiro et al. propose a three dimensional analysis of no tension materials as applied to the design of a dam. Sousa discusses the advantages of combining three dimensional elastic solutions with two dimensional non-linear ones to optimizing the sequences of construction of a large underground power station as well as investigating its long term stability. Quiblier and Ngokwey use a three-dimensional, visco-elastic, multiple layered, orthorhombic model to investigate, under large deformation, zones of high probability of intense fracturing in deformed sedimentary formations for the forecasting of potential oil reservoirs.

Boundary element methods

Boundary element methods refer to numerical procedures in which only the boundary of the continuum is discretized; from a mathematical point of view they correspond to a class of finite element methods. They use the superposition of known influence functions along boundary contours to produce a system of simultaneous integral equations relating boundary tractions and displacements. These equations are solved numerically, or analytically, by assuming that the tractions, or the displacements, or both, are constant over small areas of the boundary. Displacements and stresses at points in the continuum are then computed from the boundary data.

In addition to the simplicity of the discretization process, the main advantage of boundary elements methods over finite elements rests essentially in the fact that conditions at infinity are implicitely taken into account with boundary elements solutions whilst, for finite element solutions, first the mesh must be extended far away from the domain of stress perturbation and secondly some hypothesis must be made with respect to displacement conditions. Further, boundary elements techniques are particularly well adapted for solving problems with evolutionary boundaries like fracture propagation investigations or problems involving specific joint behaviors.

A basic treatment of integral equation methods as applied to elastostatic is given for example by Jawson and Symm (1977); detailed presentations of boundary element methods has been proposed recently by Banerjee and Butterfield (1981) and Crouch and Starfield (1983). Only a brief overview of some applications of these methods as proposed in the rock mechanics literature is presented here.

The displacement discontinuity method (Starfield and Crouch 1973, Crouch 1976, 1978) consists of placing N displacement discontinuities of unknown magnitude along a boundary, then setting up and solving a system of linear algebraic equations to find the discontinuity values that produce the prescribed boundary conditions. When the geometry involves closed contours, the interior problem and the exterior problem are solved simultaneously. Closed form solution for the problem of a uniform normal, or tangential displacement discontinuity along a finite plane boundary element are used to obtain influence coefficients. For example, for two dimensional problems in which the boundary conditions are specified by tractions, the following linear system is obtained :

$$\begin{cases} \sum_{j=1}^{M} A_{NN}(i,j)DN_j + \sum_{j=1}^{M} A_{NT}(i,j)DT_j = TN_i \\ \sum_{j=1}^{M} A_{TN}(i,j)\,DN_j + \sum_{j=1}^{M} A_{TT}(i,j)DT_j = TT_i \end{cases} \qquad (48)$$

where $A_{NN}(i,j)$ is the normal component of the traction generated at the center of the jth segment of the boundary by a unit normal displacement discontinuity at the ith segment whilst A_{NT} refers to the tangential component of the same traction. Coefficients A_{TN} and A_{TT} refer to the normal and tangential components of tractions generated by a unit tangential displacement discontinuity. DN_j and DT_j are the unknown normal and tangential displacement discontinuities at the jth segment whilst TN_i and TT_i are the components of the known boundary traction at the center of the ith segment. The system (48) of 2N linear equations can be inverted so that, once DN_j and DT_J are known, they can be used to determine the stress field, and therefore the displacement field, throughout the domain by direct solution of a system of linear equations similar to (48).
Similar equations are obtained if the boundary conditions are expressed in terms of displacements so that mixed boundary value problems can be solved.
This method has been adapted to three dimensional cases by Sinha (1979) and to problems involving two different elastic materials by Crouch (1982) and Bouchez (1981). Crawford and Curran (1982) have considered higher order functional variation displacement discontinuity elements so that tractions can be specified at two or three points per segment. These higher order elements provide better accuracy for the same amount of elements as compared to the constant displacement discontinuity method.
The displacement discontinuity technique is particularly well adapted for modelling extraction schemes in planar or non-planar seams (Crouch 1978) or for the investigation of quasistatic fracture propagation problems (Cornet 1979). In these problems, provision is made for friction along the discontinuities when slip occurs ; adaptation to dilating joints is also fairly straightforward provided the dilation law is known.
A different approach is followed by Brady and Bray (1978 a and b) and Brady (1979). These authors observe that frequently the main axis direction of underground cavities does not coincides with a principal stress direction and propose a numerical solution to what they call complete plane strain conditions, i.e. situations in which displacements depend only on coordinates in the plane perpendicular to this main axis direction. If S^* is the surface of the openings, these problems can be considered as the superposition of two separate sub-problems :

i) an infinite medium subjected to a homogeneous stress field equivalent to that which existed prior to excavation (i.e. the undisturbed pre-existing stress field);

ii) the same infinite medium with a given distribution of loads $\underline{t}$ and displacement $\underline{u}$ applied along a surface S identical to S^*. $\underline{t}$ and $\underline{u}$ represent tractions and displacements which must be induced on the surfaces in the othewise unloaded continuum to simulate excavation of the opening.

Solution of the second problem involves the determination of

1) unknown boundary values, i.e. excavation induced displacement where final boundary tractions are imposed and, conversely ;

2) total stresses around excavation boundaries ;

3) total stresses and displacements at selected internal points in the continuum.

Let Y be the main cavity axis and X and Z the two other reference frame axes. Total stresses at internal points and around excavation boundaries are obtained by superposing the undisturbed field stresses with those resulting from loading of the surface S. As shown by Brady and Bray (1978) problem ii) consists of two uncoupled subsidiary problems :

a) a plane problem involving t_x, t_z, u_x, u_z ;

b) an antiplane problem involving t_y, u_y.

Solutions to these problems is sought thru a discretization of the surface S. They can be formulated in two different ways. In the first one, referred to as the indirect formulation, distributions of load singularities are applied over the surface and the intensities of these distributions are set up to satisfy simultaneously the boundary conditions on all the elements of the surface. The most suitable form of singularity is chosen by trial and error (Brady and Bray 1978 a, b). In the direct formulation, solutions are obtained from equations which relate tractions and displacements on the elements of the surface S (Brady 1979) as described here after.
The objective is to set up and solve boundary constraint equations for the subsidiary problems a) and b). Let us first consider problem a) within the frame of reference X, Z ; tractions at all points of the boundary have components t_x and t_z and give rise to displacements, at all points, with components u_x, u_z. Let us now consider local axes X', Z' at point i of the boundary S (X' is tangent to S whilst Z' is normal to it) ; tractions and displacements at all points j of S become t'_{xj}, t'_{zj}, u'_{xj}, u'_{zj} which is referred to as load case 1. Next, consider a line load, of unit intensity per unit length in the Y direction, exerted at i in the X' direction. This load induces at j, on S, tractions and displacements $T'^{x_i}_{xj}$, $T'^{x_i}_{zj}$, $U'^{x_i}_{xj}$, $U'^{x_i}_{zj}$ (expressed in the X', Z' frame of reference). The surface can be

considered to be loaded by tractions $T'^{x_i}_{xj}$, $T'^{x_i}_{zj}$ giving rise to displacements $U'^{x_i}_{xj}$, $U'^{x_i}_{zj}$ (this is called case 2). Now, by application of the reciprocal work theorem, one may write :

$$\int_S (T'^{x_i}_{xj}\, u'_{xj} + T'^{x_i}_{zj}\, u'_{zj})\,ds = \int_S (t'_{xj}\, U'^{x_i}_{xj} + t'_{zj}\, U'^{x_i}_{zj})\,ds \tag{49}$$

or, if traction and displacements of load case 1 are considered constant on small boundary elements

$$\sum_{j=1}^{n} \int_{s_j} (T'^{x_i}_{xj}\, u'_{xj} + T'^{x_i}_{zj}\, u_{zj})\,ds_j = \sum_{j=1}^{n} \int_{s_j} (t'_{xj}\, U'^{x_i}_{xj} + t'_{zj}\, U'^{x_i}_{zj})\,ds_j \tag{50}$$

where S_j is the length of the jth boundary element.

Now letting :

$$F'^{x_i}_{xj} = \int_{s_j} T'^{x_i}_{xj}\,ds_j\ ;\ V'^{x_i}_{xj} = \int_{s_j} U'^{x_i}_{xj}\,ds_j\ ;\ \text{etc...,}$$

equation (50) becomes :

$$\sum_{j=1}^{n} \begin{bmatrix} F'^{x_i}_{xj} & F'^{x_i}_{zj} \end{bmatrix} \begin{bmatrix} u'_{xj} \\ u'_{zj} \end{bmatrix} = \sum_{j=1}^{n} \begin{bmatrix} V'^{x_i}_{xj} & V'^{x_i}_{zj} \end{bmatrix} \begin{bmatrix} t'_{xj} \\ t'_{zj} \end{bmatrix} \tag{51}$$

The same process can be repeated for a line load exerted at i but in the Z'direction, and repeating this process for all the centers of the n elements of S and expressing the results in the original X-Z plane, we get :

$$\begin{bmatrix} F \end{bmatrix} \{u\} = \begin{bmatrix} V \end{bmatrix} \{t\} \tag{52}$$

Equation (52) expresses the required boundary constraint equation in the X-Z plane for problem a). A similar procedure can be followed for the antiplane problem b) :

$$\begin{bmatrix} F_y \end{bmatrix} \{u_y\} = \begin{bmatrix} V_y \end{bmatrix} \{t_y\} \tag{53}$$

Solution for unknown components for $\underline{u}$ or $\underline{t}$ at all points of the boundary is obtained by solving the linear system of equations (52)+(53).

The quantities $\int_{s_j} T'^{x_i}_{xj}\,ds_j$, $\int_{s_j} U'^{x_i}_{xj}\,ds_j$, etc..

can either be found by direct analytical integration or numerically. Brady proposes a numerical solution using Gaussian quadrature. Once all the tractions and displacements are known at the surface S, components of boundary stresses as well as displacements and stresses at all internal points can be obtained fairly simply.

Situations involving non linear constitutive equations in some parts of the domain under investigation can be solved by coupling boundary elements and finite elements : finite elements are used to solve the non-linear constitutive equations whilst boundary elements are introduced to deal with the infinite boundary conditions (see e.g. Brady and Wassyng 1981, Lovan et al. 1983, this congress). Lin DezHang and Liu Baoshen (1983, this congress) propose to extend the boundary element formulation to linear viscoelastic materials by using the Laplace transform to get rid of the time dependency. Pierce and Ryder (1983 this congress) extend the two dimensional boundary element formulation to piecewise linear constitutive equations by introducing " initial strain " or " initial stress " conditions in some surface elements. These special surface elements are supposed to accomodate for strain softening plastic type deformations.

Jointed blocks models

With the finite element or boundary element models, small discontinuities are included within the equivalent continuum and only the large ones are treated as such. Such an approach is well suited for problems in which deformation affects the joint system as well as the rock matrix, that is when the stress field is large enough.

For low stress fields, two phenomena occurs usually simultaneously :

- first the density of joints, at various scales, increases drastically so that the rock mass becomes an agregate of blocks of various sizes and shapes ;
- second the deformation process affects the joints primarily such that the elastic deformation of the blocks becomes negligible compared to that of the discontinuities.

Clearly in such conditions, modeling the rock mass by an equivalent continuum is not appropriate. This is why jointed blocks models have been developed.

These models, originally developed by Cundall (1971, 1974) assume that all the displacements occur along joints or discontinuities whilst the blocks remain rigid (although elastic deformation has been introduced in the latest models proposed by Cundall 1980). They are based on two laws :

- the force-displacement law ;
- the law of motion.

The force displacement law relates the forces between contacts to the incremental displacements across the contacts. These displacements are derived from the incremental displacements of the blocks. The law of motion is simply Newton's second law : forces and moments acting on each block cause acceleration that are integrated twice over a time step Δt to give new incremental displacements. The equivalent forces and moments acting on each block are the summation of the contact forces acting around the periphery of the blocks. The system of equations is solved by finite differences with respect to time. These models handle unlimited rotations and translations for arbitrarily shaped two-dimensional blocks. It has been applied, in addition to slope stability problems, to fluid flow in heavily fractured medium (Cundall 1982). Warburton (1983 this congress) proposes a program for constructing three dimensional block geometries and for analyzing the stabilities of individual polyhedral blocks

Now the time may not be far when, by combination of the statistical characterization of joints geometry, along lines similar to those exposed in section 3 of this report, and three dimensional blocks models, the mechanical behavior of heavily fractured rock masses will be amenable to analysis.

5. CONCLUSION

The natural stress field existing in undisturbed rock masses is an important datum for engineering problems as well as for a better understanding of present day tectonics. It can be appraised by direct in-situ measurements or be inferred from seismotectonic observations. The most widely used in-situ determination techniques are overcoring and hydraulic fracturing methods ; the former ones are essentially applied from underground openings and are limited to fairly shallow boreholes (less than 100 m in most instances) whilst the second one can be carried out at any depth. Because of its cost effectiveness and its depth independency, hydraulic fracturing tend to be the most attractive technique in most instances. Its precision is of the order of 15 % but some improvement may still be expected especially with respect to the accuracy of the shut-in pressure determination.
All results obtained, up to now indicate an increase of the deviatoric stress component with depth, at least down to 5 km, the depth limit of present direct measurements. However an important problem remains to be solved, namely the survey of stress variations with time, especially in earthquake prone regions.
A second important datum for the understanding of the mechanical behavior of rock masses is the geometrical characterization of natural discontinuities. Although some progress has been made with respect to the local characterization of this geometry (i.e. the local spatial extrapolation of surface exposure observations), thanks to appropriate statistical investigation, an accurate method for the appraisal of large scale fractures geometry remains to be developed.
The third important question raised by rock mechanics problems is the definition of appropriate constitutive equations for the equivalent continuums as well as for the large scale discontinuities.
The most favorite model is linear elasticity although some models are now being developed for elasto-visco-plasticity with dilatancy.
The modelling of joints assumes dilatancy with shear deformation. Appropriate representation of time effects for both the rock matrix and the joints behavior, when microfracturing is involved, remains a difficulty, especially when stress corrosion is involved. Some data have been obtained in the laboratory but their application to in-situ conditions remains problematical and some in-situ testing is desirable.
In this respect, application of the latest developments in seismology, as far as source mechanism modelling is concerned, may be very fruitful for interpreting the seismo-acoustic activity observed during rock mass deformation. This could at least provide a better understanding of quasi-stable fracturing processes which precede large scale instabilities.
These observations may be coupled to advantage with in-situ deformation measurements and, eventually, temporal stress variations.

Once these three questions have been answered (i.e. determination of local stress field, geometrical characterization of discontinuities, definition of appropriate constitutive equations), rock mechanics problems are usually solved with numerical techniques although some physical models are sometimes efficient problem solvers, especially when proper scaling coefficients have been adopted. When linear elasticity is assumed, boundary element formulation is a powerful technique especially when dealing with infinite, or semi-infinite rock masses or for evalutionary boundary geometries. When non-linear constitutive equations must be considered, the most favorite method involves finite elements models. However a new approach is now being developed for heavily fractured masses submitted to low stresses, namely the jointed blocks model in which deformation occurs only along the discontinuities. Although these models remain presently two dimensional, the advant of three dimensional geometrical models for rock masses, coupled with a statistical definition of joints pattern, may soon provide a powerful tool for investigating the mechanical behavior of such rock masses.
The role of fluid pressures has not been discussed in this report but results obtained with respect to pressure dependency of permeability along fractures suggest that, for this problem, effective stresses may not obey the classical formulation proposed by Terzaghi (Terzaghi 1945). Consequences of these results to stability analysis, and more generally deformation problems, remain to be investigated.

Aknowledgements

I wish to thank C. Fairhurst, G. Jobert and R. Madariaga for suggesting corrections for some parts of this report.
This is I.P.G. contribution n°641.

BIBLIOGRAPHY

AAMODT A.a. and M. KURIYAGAWA (1982). Measurement of instantaneous shut-in pressure in crystalline rock ; Workshop on hydraulic fracturing stress measurements ; Monterey U.S.G.S Report nb 82-1075.

ANDRADE E.N.(1910). On the viscous flow of metals and allied phenomena ; Proc. Roy. Soc. London, A-84, pp.1-12.

ANGELIER J. (1979). Determination of the mean principal directions of stresses for a given fault population ; tectonophysics, 56 p.T 17-T 26.

ARMIJO R., E. CAREY and A. CISTERNAS (1982). The inverse problem in microtectonics and the separation of tectonic phases ; tectonophysics, 82, p.145-160.

ATKINSON B.K. (1979). A fracture mechanics study of subcritical tensile cracking of quartz in wet environments ; Pageoph. vol.117, nb 5, 1011-1024.

BALTHASAR K. and E. WENZ (1983). The determination of the complete state of stress with the compensation method as a contribution for investigations of tectonically influenced stresses ; This congress.

BANERJEE P.K. and BUTTERFIELD R.(1981). Boundary elements methods in engineering Science ; Mc Graw Hill, London.

BARTON N. (1973). Review of a new shear strength criterion for rock joints ; Eng. Geology vol.7, p.287-332.

BARTON N. (1976). The shear strength of rock and rock joints ; Int. Jour. Rock Mech.Min.Sc. vol.13,pp.255-279.

BARTON N.(coordinator)(1978). Suggested methods for the quantitative description of discontinuities in rock masses ; Int. Jour. Rock Mech. Min. Sc. vol.15, nb.6, p.319.

BARTON N. and V. CHOUBEY (1977). The shear strength of rock joints in theory and practice ; Rock Mechanic ; vol.10,p.1-54.

BARTON N. and S. BANDIS (1980). Some effects of scale on the shear strength of joints ; Int. Jou. Rock Mech. Min. Sc. ; vol.17,nb.1,p.69.

BERRY D.S. (1968). Calculation of the stress in the surface of a circular hole in an infinite transverselly isotropic elastic medium subjected to general stresses at infinity and hydrostatic pressure at the hole,Appendix 1. in Methods of determining in-situ rock stress at great depth, by Fairhurst C. (see this author's reference).

BERTRAND L., H. BEUCHER, D. CREUTON, B.FEUGA, J. CANDRY D. THIERY (1982). Essai de détermination de la distribution régionale du tenseur de perméabilité du milieu poreux equivalent ; colloque-jubilée Castany ; BRGM-Orléans.

BIENIAWSKI Z.T.(1970). Time dependent behavior of fractured rock ; Rock Mechanics ; vol.2, p. 123.

BLACKWCOD R.L. (1977). An instrument to measure the complete stress field in soft rock or coal in a single operation ; Field measurements in Rock Mechanics, Int. Symp. Zurich, Kovari editor. Balkema Rotherdam

BLACKWOOD R.L. (1978). Diagnostic stress relief curves in stress measurements by overcoring; Int. Jou. Rock Mech. Min. Sc. vol.15,nb.4, p.205.

BONNECHERE F.J.(1971). Contribution à l'étude de la détermination de l'état de contrainte des massifs rocheux, thèse de doctorat Faculté des Sciences Appliquées ; Liege.

BONNECHERE F.J. and C. FAIRHURST (1968). Determination of the regional stress field from " Doorstopper " measurements ; Jou. South Afr. Inst. Min. and Met ; july.

BOUCHEZ J. (1981). Résolution des problèmes d'élasticité plane par la méthode des discontinuités de déplacement en milieu non homogène ; Inst. Phys. du Globe, Paris ; int. Report.

BRACE W.F., B.W. PAULDING Jr, and C. SCHOLTZ (1966). Jou. Geophys. Res. vol.71,nb 16 p.3939.

BRACE W.F. and R.J. MARTIN III (1968). A test of effective stress law for crystalline rocks of low porosity ; Int. Jou. Rock Mech. Min. Sc., vol. 5 p.415.

BRAY J.W. and R.F. GOODMAN (1981). The theory of base frictions models ; Int. Jou. Rock Mech. Min. Sc. ; vol.18 nb 6 ; p.453.

BRADY B.H.G. (1979). A direct formulation of the boundary element method of stress analysis for complete plane strain ; Int. Jou. Rock Mech. Min. Sc. vol.16, nb4, p.235.

BRADY B.H.G. and J.W. BRAY (1978).
a) The boundary element method for determining stresses and displacements around long openings in a triaxial stress field.
b) The boundary element method for elastic analysis of tabular orebody extraction, assuming complete plane strain.
Int. Jou. Rock Mech. Min. Sc. vol.15 nb.1 p.21-p.29.

BRADY B.H.G. and A. WASSYNG (1981). A coupled finite element boundary element method of stress analysis Int. Jou. Rock. Mech. Min. Sc., vol. 18 nb 6 ; p.475.

BROWN E.T. and E. HOEK (1978) Trends in relationship between measured in-situ stresses and depth ; Int. Jou. Rock Mech. Min. Sc. vol.15, nb.4, p.211.

BREDEHOEFT J.D., R.G. WOLFF, W.S. KEYS and E. SCHUTER (1976). Hydraulic fracturing to determine the regional in-situ stress field. Piceance Basin, Colorado ; Geol. Soc. Am. Bull. vol.87, p.250-258.

BURRIDGE R. and L. KNOPOFF (1964). Body force equivalents for seismic dislocations ; Bull. Seism. Soc. Am.,vol.54, nb.6,p.1875.

BUYLE-BODIN F., F.J. BONNECHERE and F.H. CORNET (1983). Mesures de contrainte dans la vallée d'Aspe, Revue Fr. de Geotechnique n°20, p.31.

BYERLY P. (1926). The Montana earthquake of june 28, 1925 ; Bull. Seism. Soc. An., vol.16, p.209.

BYERLEE J.D. (1967). Frictional characteristics of granite under high confining pressure , Jou. Geophy. Res., vol.72, p.3639.

BYERLEE J.D. (1968). Brittle ductile transition in rocks, Jou. Geophys. Res. vol.73,p.4741.

BYERLEE J.D. (1975). The fracture strength and frictional strength of Weber sandstone Int. Jou. Rock. Mech.Min. Sc.,vol.12, nb1, p.1.

CAROL I. and E.E. ALONSO (1983). A new joint Element for the analysis of fractured rock ; this congress.

CASTELEIRO M., E. ONATE, A. HUERTA, J.ROIG,and E. ALONSO (1983). Three dimensional analysis of no tension materials ; This congress.

CHARLEZ Ph. (1983). Une méthode de mesures in-situ des contraintes dans les massifs rocheux ; Thèse Ing. Doct. Univ. PARIS 6.

CLARK B.C. (1982). Monitoring changes of stress along active faults in southern California. Jou. Geophys. Res. ; vol.87 nb.B 6,p.4645.

CLEARY M.P. (1980). Analysis of mechanisms and procedure for producing favourable shape of hydraulic fractures ; 55 th Annual fall technical Conf. Soc. Petr. Eng. AIME paper SPE 9260.

COOK M.G.W. and J.P.M. HOJEM (1966). A rigid 50-ton compression and tension testing machine, South. Afr. Mech. Eng. vol.14, p.89-92.

CORNET F.H. (1976) Crack propagation in rock normal to the maximum compression stress ; N.S.F. report for contact nb. G.K. 41220 (Bligh and Fairhurst ppl. investigators).

CORNET F.H. (1977). Etude du comportement elastique et fragile des roches saturées par un liquide; Revue française de Géotechnique n°2.

CORNET F.H. (1979). Comparative analysis by the displacement discontinuity method of two energy criteria of fracture ; Jou. App. Mech. vol.46, nb.2,p.349.

CORNET F.H. (1982). Analysis of injection tests for in-situ stress determination ; proceedings of the workshop on hydraulic fracturing stress measurements U.S.G.S. open-file rep.82-1075. Haimson and Zoback editors.

CORNET F.H. and C. FAIRHURST (1972). Variation of pore volume in disintegrating rock ; Percolation thru fissured rock. Symp. of the I.S.R.M. Stuttgart Wittke editor,p.T2-A.

CORNET F.H., F.J. BONNECHERE and F.BUYLE-BODIN (1979). Discussion sur la détermination de l'état de contrainte dans un massif rocheux ; 4 th Cong. of the ISRM, Montreux vol.3, p.137.

CRAWFORD A.M. and J.H. CURRAN (1982) Higher order functional variation displacement discontinuity elements ; Int. Jou. Rock Mech. Min. Sc. ; vol.19 nb3 , p.143.

CROUCH S.L.(1970). Experimental determination of volumetric strain in failed rock ; Int.Jou. Rock. Mech. Min. Sc. vol.7, p.589.

CROUCH S.L. (1976). Solution of plane elasticity problems by the displacement discontinuity method ; Int. Jou. Num. Meth. Eng. , vol.10 p.301.

CROUCH S.L. (1976). Analysis of stresses and displacements around underground excavations: an application of the displacement discontinuity method ; Univ. of Minn.Geomechanics report ; NSF-RANN program(grant nb GI-37923).

CROUCH S.L. and C. FAIRHURST (1967). A four component deformation gauge for the determination of in-situ stress in rock masses ; Int. Jou.Rock Mech. Min. Sc. vol.4,p.209-217.

CROUCH S.L. and A. STARFIELD (1983). Boundary element methods in solid mechanics; George Allen and Unkin publisher ; London.

CUNDALL P.A. (1971). A computer model for simulating progressive large scale movement in blocky rock systems ; Int. Soc. Rock.Mech. Symp. Nancy.

CUNDALL P.A. (1974). Rational design of tunnel supports : A computer model for rock mass behavior using interactive graphics for the input and output of geometrical data. U.S. Army corp of Engineers, Technical Report MRD-2-74.

CUNDALL P.A. (1980). UDEC, a generalized distinct element program for modelling jointed rock. Final technical report ; European Research office ; U.S. Army London. Contract nb DAJA 37-79-C-0548.

CUNDALL P.A. (1982). The FRIP users mannal. Camborne School of Mines, Geothermal project.

DAS.S and C.H.SCHOLZ (1981).Theory of time dependent rupture in the earth ; Jou.Geophys.Res. vol.86,nb. B 7,p.6039.

DEERE D.U. (1964). Technical description of rock cores for engineering purposes ; Rock Mech. Eng. Geol.vol.6 , p.1.

DE LA CRUZ R.V. (1977). Jack fracturing technique of stress measurement ; Rock Mechanics vol.9, p.27.

DE LA CRUZ R.V. and R.E. GOODMAN (1969). The borehole deepening method of stress measurement ; Int. Symp. on the determination of stress in rock masses. Lisboa.

DERAMOND J., P. SINEYS and J.C. SOULA (1983), Mechanismes de déformation de l'écorce terrestre : structure et anisotropie induites ; This congress.

DESAI C.S., M.M. EITANI and C. HAYCOCKS (1983). An application of finite element procedure for underground structures with non linear materials and joints ; This congress.

DOE T.W. (1982) Determination of the state of stress of the Stripa Mine, Sweden ; Proc. Workshop on hydraulic fracturing stress measurements;U.S.G.S.open file report 82-1075.

DRAGON A. and Z. MROZ (1979). A model for plastic creep of rock like materials accounting for the kinetics of fracture ; Int. Jou. Rock Mech. Min. Sc. vol.16, nb.4,p.253.

EGGER R. and C. GINDROZ (1979). Tunnels ancrés à faible profondeur, étude comparative sur models physiques et mathématiques ; Proc. 4 th Int. Soc. Rock Mech. Congress vol.2, p.121.

ETCHECOPAR A., G. VASSEUR, M. DAIGNIERES (1981) An inverse problem in microtectonics for the determination of stress tensors from fault striation analysis ; Jou. Structural Geol. vol.3 nb1, p.51.

FAIRHURST C. (1964). Measurement of in-situ rock stresses with particular reference to hydraulic fracturing ; Rock Mech. Eng. Geol., vol.11, p.129.

FAIRHURST C. (1968). Methods of determining in-situ rock stress at great depth ; Tech. Rep.1-68.; U.S. Corps of Engineers ; Omaha, Nebraska.

FILCEK H. and T. CYRUL (1977). Rigid inclusion with high sensitivity, Field Measurements in Rock Mechanics. Int. Symp. Zürich Kovari Editor, vol.1,p.219 ; Balkema Rotterdam.

FLEITOUT L. and C. FROIDEVAUX (1982) Tectonics and topography for a lithosphere containing density heterogeneities ; Tectonics, vol.1, nb.1,p.21.

FROIDEVAUX C., C. PAQUIN and M. SOURIAU Tectonic stresses in France : in situ measurements with flat jack. Jou. Geoph. Res. vol.85 nb B 11,p.6342.

GALLE E.M. and J.C. WILHOIT Jr. (1961). Stress around a wellbore due to internal pressure and unequal principal geostatic stresses ; 36 th Annual meeting of Soc. Pet. Eng. Dallas Texas.

GERRARD C.M. (1982). Elastic models of rock masses having one, two and three sets of joints ; Int. Jour. Rock Mech. Min. Sci. vol.19 nb 1, p.15.

GOODMAN R.E., R.L. TAYLOR and T.L. BREKKE (1968) A model for the mechanics of jointed rock ; Jou. Soil. Mech.and Found.Div. A.S.C.E, vol.94, n°SM 3, p.637.

GRIFFITH A.A. (1921). The phenomenon of rupture and flow in solids ; Phil. Trans. Roy. Soc. London A 221.

GRONSETH J.M. and P.R. KRY (1982).Instantaneous shut-in pressure and its relationship to the minimum in-situ stress ; Proc.workshop hydraulic fracturing stress measurements; U.S.G.S. open file report 82-1075 ; Menlo Park Calif.

GUENOT A. et M. PANET (1983). Etude numérique d'ouvrage en massif rocheux a structure planétaire ; This congress.

HAGAN T.O. (1980). A case of terrestrial photogrammetry in deep-mine rock structure studies ; Int. Jou. Rock Mech. Min. Sci. vol.17, nb 4, p.191.

HAIMSON B.C. (1968). Hydraulic fracturing in porous and non porous rock and its potential for determining in-situ stress at great depth ; U.S.corps of Engineers,Tech. Rep. 4-68, Omaha, Nebraska.

HAIMSON B.C. (1978). The hydrofracturing stress measuring method and recent field results, Int.Jou.Rock Mech.Min.Sc.vol.15,nb.4,p.167.

HAIMSON B.C. (1982). A comparative study of deep hydrofracturing and overcoring stress measurements at six locations with particular interest to the Nevada test site;Proc. workshop on hydraulic fracturing stress measurements ; U.S.G.S. openfile rep.82-1075.

HAIMSON B.C. and C. FAIRHURST (1969). In-situ stress determination at great depth by means of hydraulic fracturing ; 11 th Symp. on Rock Mech.Berkeley p.559 AIME.

HANDIN J. and C.B. RALEIGH (1972). Manmade earthquakes and earthquake control ; Percolation thru fissured rock. Int. Soc. Rock. Mech. Symp. Stuttgart, Wittke Editor.

HANKS T.C.(1977). Earthquake stress drops, Ambient tectonic stresses and stresses that drive plate motions ; Pageoph.vol.115 p.441.

HELAL H. et M. DEJEAN (1981). Mise au point d'une technique de mesure des contraintes in-situ par surcarottage ; Comité Français de Mécanique des Roches.

HENRY J.P., J.PAQUET and J.P.TANCREZ (1977). Experimental study of crack propagation in calcite rocks ; Int. Jou. Rock Mech.Min.Sci. vol.14 nb.2, p.85.

HEUZE F.E. (1971). Sources of errors in rock mechanics field measurements and related solutions ; Int. Jou. Rock Mech. Min. Sci. vol. 8, p.297.

HEUZE F.E., W.C. PATRIK, T.R. BUTKOWICH, J.C. PETERSON, R.V. DE LA CRUZ and C.F. VASS (1982). Rock Mechanics studies of mining in the climax granite; Int.Jou. Rock Mech.Min. Sc. vol.19 nb.4, p.167.

HICKMAN S.H. and M.D. ZOBACK (1982).The interpretation of hydraulic fracturing pressure time data for in-situ stress determination. Proc.workshop on hydraulic fracturing stress measurements ; U.S.G.S. open file report n°82-1075. Menlo Park Calif.

HIRASHIMA K. and A. KOGA (1977). Determination of stresses in anisotropic elastic medium unaffected by boreholes from measured strains or deformations, in field measurements in Rock Mechanics. Int.Symp. Kovari Editor ; Balkema Rotterdam.

HIRAMATSU Y. and Y.OKA (1968). Determination of the stress in rock unaffected by boreholes or drifts, from measured strains or deformations ; Int.Jou. Rock Mech.Min.Sc. vol.5, july.

HOOKER V.E. and D.L. BICKEL (1974). Overcoring equipment and techniques used in Rock Stress determination ; information circular 8618, Denver Mining res. center ; U.S.B.M.

HOSKINS E.R.(1966). An investigation of the flat jack method of measuring rock stress; Int. Jou.Rock Mech.Min.Sci.vol.3, p.249.

HOUPERT R. (1974). Le rôle du temps dans le comportement à la rupture des roches ; Proc. 3d congress of the Int. Soc.Rock Mech. Denver ; vol.2 part A p.325.

HAWKES I. and V.E. HOOKER (1974). The vibrating wire stress meter ; Proc. 3d Int. Soc. Rock Mech. Congress. Denver vol.II-A,p.439.

HUBBERT M.K. and D.G. WILLIS (1957). Mechanics of hydraulic fracturing ; Petrol. Trans. of Am. Inst.Min. Eng.vol.210, p.153.

HUDSON J.A., E.T. BROWN and C. FAIRHURST (1971). Optimizing the control of rock failure in servo-controlled laboratory test ; Rock Mechanics vol.3 p.217.

HUDSON J.A., E.T. BROWN (1973). Studying time dependent effects in failed rock; 14 th Symp. on Rock Mech. Pensylvania State Univ; Amer. Soc.Civil.Eng., p.25.

HUDSON J.A. and S.D. PRIEST (1979).Discontinuities and rock mass geometry ; Int.Jou.Rock Mech. Min.Sc. vol.16 nb 6, p.339.

HUDSON J.A., P.R. LA POINTE (1980). Printed circuits for studying rock mass permeability; Int. Jou.Rock Mech.Min.Sci.vol.17,nb 5,p.297.

HUERGO P.J.(1983). Evaluation des contraintes naturelles dans les couches supérieures d'un massif schisteux ; This congress.

HUSTRULID W. and A. HUSTRULID (1975). The CSM cell a borehole device for determining the modulus of rigidity of rock ; 15 th Symp. on Rock Mech. Ann. Soc. Civ.Eng.p.181.

IRWIN G.R.(1957). Analysis of stresses and strains near the end of a crack traversing a plate ; Jou. Appl. Mech. vol.24 nb 3,p.361.

JAEGER J.C. and N.G.W. COOK (1963). Pinching-off and discing of rocks ; Jou. Geophys. Res. vol.68, p.1759.

JAEGER J.C. and N.G.W. COOK (1979). Fundamentals of Rock Mechanics, 3nd Ed. Chapman and Hall, London.

JAMISON D.B. and N.G.W. COOK (1980). Note on measured values for the state of stress in the Earth's crust. Jou. Geophys. Res.vol.85. nb B.4 p.1833.

JASWON M.A. and G.T. SYMM (1977). Integral equation methods in potential theory and elasto-statics ; Academic Press, London.

JIA-SHOU ZHUC and YIN TONG WANG (1983). Non-linear analysis of rock foundation with soft interfaces. This congress.

JIM-QI HAO , L.M. HASTIE and F.D. STACEY (1982). Theory of seismomagnetic effect : a reassessment. Physics. earth.Plan.Int. vol.28, p.129.

KAISER P.K. and N.R. MORGENSTERN (1981). Time dependent deformation of small Tunnels part II, Int.Jou.Rock Mech.Min.Sc. vol.18 nb2, p.153.

KEHLE R.D. (1964). Determination of tectonic stresses through analysis of hydraulic well fracturing;Jou. Geophys. Res. vol.69 p.259.

KIRSH G. (1898). Die theorie der elastizität und die bedürfnisse der Festigkeitslehre Z.V.D.I., vol.42,nb.29,p.113.

KNOPOFF L. and F. CULBERT (1960). First motion from seismic sources. Bull.Seis.Soc.Am . vol.50, p.117.

KOVARI K. (1977). Field measurements in rock mechanics ; Proc. Int. Symp. Zurich, Balkena, Rotterdam.

KRANZ R.L. and C.M. SCHOLZ (1977). Critical dilatant volume of rocks at the onset of tertiary creep ; Jou. Geophys. Res. vol.82 nb.30, p.4893.

KUZNETSOV S.V., D.M. BRONNIKOV, I.A. PARABUCHEV V.D. PARPHENOV, I.T. AITMATOV, G.A. MARKOV (1983). The state of stress in rock and methods of its determination. This congress.

LANGER M. (1979). Rheological behavior of rock masses, General Report Theme I, 4 th Int. Soc. Rock Mech. Congress ; Montreux ; Balkema , Rotterdam.

LEEMAN E.R (1964). The measurement of stress in Rock, Part I, II, III ; Jou. South. African Inst. Min. Met.,p.45,p.82,p.254.

LEEMAN E.R. (1968). The determination of the complete state of stress in rock in a single borehole. Laboratory and underground measurements, Int. Jou. Rock. Mech. Min. Sc. vol.5, p.31.

LEEMAN E.R. (1969). The measurement of stress in rock. A review of recent development (and a bibliography). Int. Symp. on the determination of stresses in Rock Masses; Lisboa.

LESSI J. and J.P. SARDA (1980). Scale model studies of well linking by hydraulic fracturing and fatigue microfracturing P.957, Proc. 2ème Int.Sem.Results of EC Geothermal Energy Res. ; Reidel Pub. Dordrecht.

LI FANG-QUAN, LI YAN-MEI, WANG EN-FU, ZHAI QING-SHAN, BI SHANG-SU, ZHANG-JUN, LIU-PENG WI QING-YON and ZHAO SHI-GUANG (1982). Experiments in the in-situ stress measurements using the stress relieving and hydraulic fracturing techniques ; Proc. Workshop on hydraulic fracturing stress measurements, p.332, U.S.G.S., open file report 82-1075.

LIN DESHANG and L. BAOSHEN (1983). Boundary element method for linear viscoelastic stress analysis in rock mass and its application in rock engineering ; This congress.

LOVAN P., J. OKSANEN and K. AIKAS (1983). Element methods in planing of mine openings in highly stressed precambrian bedrock ; This congress.

MADARIAGA R. (1979). Seismic radiation from earthquake models based on fracture mechanics, Am. Math. Soc. vol.12 p.59.

MARTIN R.J.III (1972). Time dependent crack growth in quartz and its application to the creep of rocks ; Jou. Geophys. Res. vol.77, p.1406.

MARTIN R.J.III, R.E. HABERMANN and M. WYSS (1978). The effect of stress cycling and inelastic volumetric strain on remanent magnetization ; Jou. Geophys.Res. vol.83, nb B7, p.3485.

MARTNA J., R. HILSCHER and K. INGEVALD (1983). Triaxial rock stresses in two deep boreholes at Forsmark ; This congress.

MAYER A., P. HABIB et R. MARCHAND (1951). Conférence Int. sur la pression de terrains et le soutenemènt dans les chantiers d'exploration, Liège, p.217.

Mc GARR A. (1980). Some constraints on levels of shear stress in the crust from observations and theory;Jou.Geophys.Res.vol.85,nb.B11,p.6231.

(1982). Analysis of state of stress between provinces of constant stress; Jou. Geophys. Res.,vol 87, nb B11, p.9279.

Mc KENZIE D.P. (1969). The relation between fault plane solutions for earthquakes and the directions of the principal stresses,Bull. Seism. Soc. An. vol.59 n°2, p.591.

Mc LENNAN J.D. and J.C. ROEGIERS (1982). Do instantaneous shut-in pressure accurately represent the minimum principal stress. Proc. workshop on hydraulic fracturing stress measurements ; p.181 ; U.S.G.S. open file Rep.82-1075,Menlo Park Calif.

MOORE G.W. (1964). Magnetic disturbances preceding the 1964 Alaska earthquake ; Nature vol.203, p.508

MORLIER P. (1966). Le fluage des roches ; Ann.de Inst. Tech. Bat. Tr. Publ.vol.19 p.89.

NAKANO M. (1923). Notes on the nature of the forces which give rise to the earthquake motions; Centr. Meteor. Observ. Japan Seism.Bull. vol.1.,p.92.

PAHL A.(1977). In situ stress measurements by overcoring inductive gages ; p.161, Field measurements in rock mechanics. Int. Symp. Zurich,Kovari Editor;Balkema , Rotterdam.

PANDE G.M. and C.M. GERRARD (1983). The behavior of reinforced jointed rock masses under various simple loading states,This congress.

PARISEAU W.G. and I.M. EITANI (1972). Post Elastic vibrating wire stress measurements in coal; p.255, Field measurements in rock mechanics Int.Symp.Zurich, Kovari Editor Balkema ,Rotterdam.

PARSONS R.C. and D.G.F. HEDLEY (1966). The analysis of viscous property of rocks for classification. Int.Jou.Rock Mech.Min.Sc. vol3,p.325.

PENG S. and E.R. PODNIEKS (1972). Relaxation and the behavior of failed rock ; Int.Jou.Rock Mech.Min.Sc.,vol.9,p.699.

PIERCE A.P. and J.A. RYDER (1983). Extended boundary element methods in the modelling of brittle rock behavior. This congress.

POTTS E.L.J. (1957). Underground instrumentation quart.color.,school of mines vol.52,p.135.

PRIEST S.D. and J.A. HUDSON (1976). Discontinuity spacing in rock ; Int.Jou.Rock Mech. Min.Sci.,vol.13,p.135.

PRIEST S.D. and J.A. HUDSON (1981). Estimation of discontinuity spacing and trace length using scanline surveys ; Int.Soc.Rock Mech. Min.Sc. vol.18,nb.3,p.183.

PRIEST S.D. and A. SAMEMIEGO (1983). A model for the analysis of discontinuity characteristics in two dimensions ; This congress.

QUIBLIER J. and K. MGOKWEY (1983). Modèles mathématiques et étude de la fracturation naturelle ; this congress.

RAHN W. (1983). Analysis of potential errors of interpretation of in-situ stress measurements in anisotropic rocks ; This congress.

RALEIGH C.B.,J.H.HEALY and J.D.BREDEHOEFT(1972). Faulting and crustal stress at Rangely, Colorado ; in Flow and Fracture of Rocks. Geophysical Monogr. series,p.275,Am.Geoph. Union, Washington D.C.

REN N.K. and J.C. ROEGIERS (1983). Differential strain curve analysis : A new method for determining the preexisting in-situ stress state from rock core measurements;This congress.

RICE J.R. and M.P. CLEARY (1976). Some basic stress diffusion solutions for fluid saturated elastic porous media with compressible constituents, reviews of Geophys. and space phys.,vol.14,nb.2,p.227.

RIKITAKE T. (1979). Changes in the direction of magnetic vector of short period ; geomagnetic variations before the 1972 Sitka Alaska earthquake,Jou.Geomagn.Geoelectr. vol.31,p.441.

ROBERTS A.,I.HAWKES,F.T.WILLIAMS and R.K.DHIR (1964). A laboratory study of the photoelastic stress-meter;Int.Jou.Rock Mech.Min. Sc.vol.1,nb.3,p.441.

ROCHA M. and A.SILVERIO (1969). A new method for the complete determination of the state of stress in rock masses ; Int.symp. on the determination of stress in rock masses. Lisboa.

RUMMEL F. (1974). Changes in the P.wave velocity with increasing inelastic deformation in rock specimens under compression ; p.517 3d congress of the Int. Soc. Rock Mech. Denver,Nat.Acad.Sciences; Washington.

RUMMEL F. and C. FAIRHURST (1970). Determination of the post-failure behavior of brittle rock using a servo-controlled testing machine ; rock mechanics,vol.2,p.189.

RUMMEL F.,J. BAUMGARTNER and H.J. ALHEID (1982). Hydraulic fracturing stress measurements along the eastern boundary of the SW-German block ; Proc.workshop on hydraulic fracturing stress measurements ; p.1 ; U.S.G.S open file report 82-1075.

SCHEIDEGGER A.E. (1962). Stress in Earth crust as determined from hydraulic fracturing data ; Geologie und Bauwesen, vol.27,p.45.

SCHEIDEGGER A.E. (1964). The tectonic stress and tectonic motion direction in Europe and western Asia as calculated from earthquake fault plane solutions ; Bull. Seism.Soc.Ann.vol.54 p.1519.

SCHEIDEGGER A.E. (1977). Geotectonic stress determinations in Austria ; Field Measurements in rock mecanics,p.197 ; Int.Symp. Zurich,Kovari editor,Balkenna Rotterdam.

SCHOLZ C.H. (1968). Microfracturing and the inelastic deformation of rock in compression Jou.Geophys.Res.vol.73,nb.3,p.1417.

SIH G.C. (1973). Handbook of stress intensity factors.Institute of fracture and solid Mechanics,Lehigh Univ. Bethlehem,Pa.

SINHA K.(1979). Ph.D. Thesis Univ. of Minnesota unpublished.

SMITH B.E.and M.J.S. JOHNSTON (1976). A tectonomagnetic effect observed before a magnitude 5.2 earthquake near Hallister, California, Jou. Geophys.Res.vol.81,p.3556.

SOLOMON S.C., R.M. RICHARDSON and E.A.BERGMAN (1980). Tectonic stress : models and magnitude, Jou.Geophys.Res.vol.85,nb.B4 p.6086.

SOUSA L.R. (1983). Three dimensional analysis of large underground power stations ; This congress.

STARFIELD A.M. and S.L. CROUCH (1973). Elastic analysis of single seam extraction: 14 th Rock Mechanics Symp. Hardy and Stefanko editors. p.421; Am. Soc. Civ. Eng.

STAUDER S.J. (1962). The focal mechanism of earthquakes. Advances in Geophysics,vol.9 p.1, Am. Geophys. Union.

STEPHANSSON O.(1983). Rock stress measurement by sleeve fracturing ; This congress.

STUART W.D. and J. DIETRICH (1974). Continuum theory of rock dilatancy ; 3d congress of the Int.Soc.Rock Mech.,Denver Theme 2 Nat.Acad.Sc. Washington.

TAN TJONG KIE and KANG SEN FA (1983). Time dependent dilatancy prior to rock failure and earthquakes ; This congress.

TAPPONNIER P., G. PELTZER, A.Y. LE DAIN and R. ARMIJO (1982). Propagating extrusion tectonics in Asia ; New insights from simple experiments with plasticine ; Geology, vol.10 p.611.

TERZAGHI K. (1945). Stress conditions for the failure of saturated concrete and rock ; Proc.Am.Soc.Test Mat. ,vol.45,p.727.

THOMAS A., P. THERME and P. RICHARD (1981). A new method for quick measuring and computing of rock jointing ; 22 nd U.S. Rock Mech. Symposium ; M.I.T.; Einstein Editor.

TINCELIN M.E. (1952). Mesure de pressions de terrain dans les mines de fer de l'est ; Ann. Inst. Tech. Bat. Tra. Publ., vol.58 p.972.

VALETTE B. and F.H. CORNET (1983). The inverse problem of stress determination from hydraulic injection tests data ; submitted to Jou. Geophys.Res.

VOUILLE G., S.M. TIJANI and F. de GRENIER (1981). Experimental determination of the rheological behavior of Tersanne rock salt ; Proc. 1st conf. on the mechanical behavior of salt ; Pensylvanie State Univ. Nov.1981.

WALLIS P.F. and M.S. KING (1980). Discontinuity pacing in a crystalline rock ; Int. Jou. Rock Mech. Min.Sc.,vol.17, nb.1,p.63.

WALSH J.B. (1981). Effect of pore pressure and confiming pressure on fracture permeability ; Int.Jou.Rock Mech.Min.Sci.,vol.18 nb.5,p.429.

WALSH J.B. and M.A. GROSENBOUGH (1979). A new model for analyzing the effect of fractures on compressibility ; Jou. Geophys. Res. vol.84,nb.B.7,p.3535.

WARBURTON P.M. (1980).A stereological interpretation of joint trace data ; Int.Jou.Rock Mech.Min.Sc.,vol.17,nb.4 ,p.181.

WARBURTON P.M. (1980) A Stereological interpretation of joint trace data : influence of joint shape and implications for geological surveys, Int.Jou.Rock Mech.Min. Sc.,vol.17, nb.6, p.305.

WARBURTON P.M. (1983). Applications of a new computer model for reconstructing blocky rock geometry, analyzing single block stability and identifying keystones ; This congress.

WAWERSIK W.R. (1968). Detailed analysis of rock failure in laboratory compression tests ; Ph.D. Thesis Univ. of Minnesota, unpublished.

WAWERSIK W.R. and C. FAIRHURST (1970). A study of brittle rock fracture in laboratory compression experiments ;Int.Jou.Rock Mech. Min.Sci.,vol.7,nb.5,p.561.

WAWERSIK W.R. and D.W. HANNUM (1980).Mechanical behavior of New Mexico rock salt in triaxial compression up to 200°C ; Jou.Geophys.Res. vol.85,nb.B2,p.891.

ZIENKIEWICZ O.C. (1977). The finite element method, 3d edition,787 p.Mc Graw Hill London.

ZIENKIEWICZ O.C.,S.VALLIAPEN and I.P. KING(1968) Stress analysis of rock as a no tension material ; Geotechnique,vol18,p.56.

ZLOTNICKI J.,J.P. POZZI and F.H. CORNET (1981). Investigation of induced magnetization variations caused by triaxial stresses ; Jou.Geophys.Res.,vol.8 ,nb.B 12,p.11899.

ZOBACK M.L. and M. ZOBACK (1980). State of stress in the conterminous United States Jou. Geophys. Res.,vol.85, nb.B 11,p.6113.

ASPECTS PARTICULIERS DE LA MÉCANIQUE DES ROCHES

Special topics in rock mechanics

Sonderthemen der Felsmechanik

F. H. Cornet
Institut de Physique du Globe, Université de Pierre et Marie Curie, Paris, France

RESUME. L'état de contrainte qui existe dans un massif rocheux peut être déterminé soit au moyen de mesures directes soit par l'interprétation d'observations sismotectonique. Les techniques le plus généralement utilisées pour les mesures directes sont les méthodes par surcarottage et la fracturation hydraulique. Les résultats obtenus à ce jour montrent une augmentation continue de la composante déviatorique avec la profondeur au moins jusqu'à 5.000 m. Les variations latérales de contrainte indiquent que les contraintes intraplaques, du point de vue tectonique, et par conséquent l'activité sismique, ne dépendent pas seulement des conditions de chargement aux frontières de plaques. Leurs variations temporelles restent mal comprises.
Une deuxième donnée importante, dans un problème de mécanique des roches, est fournie par la caractérisation géométrique des discontinuités du massif. Bien que certaines méthodes statistiques fournissent un moyen de décrire localement ces caractéristiques, leur extrapolation à de grands volumes, où l'observation ne peut se faire qu'en certains points particuliers, reste problématique.
Du point de vue de la modélisation mécanique des massifs rocheux, alors que les grandes discontinuités sont traitées en tant que telles, les petites sont intégrées dans la définition de milieux continus équivalents. Ces milieux équivalents sont souvent supposés élastiques; toutefois pour les massifs où la composante déviatorique du tenseur des contraintes est grande, des modèles élasto-visco-plastiques sont considérés de façon à prendre en compte les déformations permanentes liées à la microfissuration ainsi qu'au mouvement des dislocations. Cependant la dépendance en fonction du temps du développement de la microfissuration, y compris l'effet de la corrosion sous tension, reste mal comprise. La modélisation des larges discontinuités comporte, elle, toujours un effet de dilatance associé aux déformations en cisaillement ; les contraintes effectives qui doivent être considérées dans ces modèles ne sont pas celles qui sont généralement retenues pour les milieux continus équivalents.
Une fois précisées les conditions aux frontières, la géométrie des discontinuités et les lois de comportement des milieux continus équivalents et des grandes discontinuités, la solution des problèmes mécaniques est abordée par une modélisation soit physique soit numérique. Dans le premier cas il est essentiel de tenir compte des facteurs de similarité ; dans le second, les méthodes employées sont soit les éléments finis, soit les équations intégrales, soit les différences finies.

I. INTRODUCTION

La résolution d'un problème de mécanique des milieux continus comporte classiquement trois étapes :

- définition de la géométrie du problème et identification des conditions aux frontières ;
- choix des lois de comportement appropriées pour les différents matériaux mis en jeu ;
- résolution du système d'équations aux dérivées partielles constitué par les conditions d'équilibre (ou plus généralement du mouvement), les conditions de compatibilité (en déplacement ou en contrainte) et les lois de comportement, de façon à satisfaire les conditions aux frontières.

La difficulté rencontrée en mécanique des roches provient du fait que les massifs rocheux sont discontinus, hétérogènes et polyphasés de sorte que les méthodes de mécanique des milieux continus ne sont pas directement applicables. Des hypothèses simplificatrices doivent être formulées qui engendrent une perte de précision pour la solution finale retenue.
Cependant, grâce essentiellement à l'avènement de l'électronique et des ordinateurs, des modèles numériques de plus en plus sophistiqués sont maintenant développés et les mesures in-situ et de laboratoire deviennent de plus en plus fiables. La validité des hypothèses simplificatrices nécessaires à la résolution d'un problème peut maintenant être établie, ou rejetée, grâce à la comparaison des résultats obtenus par le modèle et ceux recueillis sur le terrain qu'il s'agisse de déplacements ou de contraintes. Bien que par bien des aspects la mécanique des roches soit encore un art, elle tend de plus en plus à se transformer en une science rigoureuse. L'objet de ce rapport est de présenter un certain nombre de développements récents qui favorisent cette évolution.

Tout d'abord le problème de la détermination in-situ des contraintes est discuté.

Quelques méthodes de mesure sont présentées ainsi que certains résultats relatifs aux variations spatiales et temporelles de l'état de contrainte. Puis le problème de la caractérisation géométrique des discontinuités affectant les massifs rocheux est abordé. Enfin quelques modèles proposés pour l'analyse du comportement mécanique des massifs rocheux sont décrits ; certaines modélisations des lois de comportement sont présentées. Les modèles physiques et numériques mentionnés dans la littérature pour résoudre certains problèmes de mécanique des roches sont esquissés.

2. LES CONTRAINTES EN MECANIQUE DES ROCHES.

Pour de nombreux problèmes, une partie des conditions aux frontières sont fournies par les contraintes " à l'infini " c'est-à-dire loin du domaine d'investigation. De ce fait un certain nombre de méthodes de mesures in-situ ont été développées pour déterminer, avant tous travaux, cet état de contrainte. Cette détermination peut ne concerner que l'orientation des directions principales; le plus souvent elle implique aussi l'estimation de l'amplitude des contraintes principales.
Récemment, la réalisation de grandes fractures hydrauliques développées dans l'industrie pétrolière a créé le besoin de connaître précisément la variation de contrainte avec la profondeur : les pressions d'injection de fluide doivent être ajustées pour limiter l'extension verticale de ces fractures et maintenir leur propagation dans l'horizon producteur choisi (Cleary 1980).
Des mesures de contrainte ont également été entreprises pour obtenir une meilleure compréhension de la sismicité et, son corollaire, la sismicité induite liée à l'injection de fluides sous pression (Handin et al. 1972). De plus des mesures sont effectuées maintenant dans le cadre d'études géodynamiques pour obtenir une meilleure compréhension de la tectonique des plaques (Zoback et Zoback 1980, Mc Garr 1982).
Considérant ces nombreux problèmes, on peut se demander s'ils sont tous concernés par le même concept de contrainte. En fait, en mécanique des roches, l'échelle pour laquelle un état de contrainte est défini doit toujours être spécifié. Le comportement mécanique des masses rocheuses est analysé en considérant des milieux continus équivalents qui ne sont supposés représenter la réalité que pour une échelle donnée. Ces milieux continus équivalents perdent leur représentativité pour des volumes plus petits qu'un volume critique qui peut être considéré comme l'expression physique du point mathématique de l'analyse. A l'intérieur de ce volume critique, la contrainte définie pour le milieu équivalent est supposée uniforme alors que dans la réalité elle ne l'est pas. Le problème de la détermination de l'état de contrainte est d'évaluer la contrainte qui existe dans le milieu équivalent. Il est donc indispensable de s'assurer que l'échelle de la mesure est compatible avec celle du problème traité.
En considérant des volumes suffisamment grands pour s'affranchir de la difficulté que pose l'hétérogénéité du matériau, on est amené à ignorer les concentrations de contraintes ponctuelles. Cependant celles-ci peuvent influencer de façon très significative le comportement du massif. Lorsque seules les déformations globales sont considérées, l'influence de ces concentrations locales est prise en compte en ajustant les lois de comportement du matériau équivalent. Lorsqu'il s'agit de définir les conditions de stabilité d'un massif, l'échelle du problème est dictée par celle de ces concentrations locales et ceci peut compliquer terriblement l'analyse. En fait le choix de l'échelle appropriée à laquelle doit se faire l'analyse est probablement l'une des difficultés majeures rencontrées en mécanique des roches ainsi que cela sera discuté dans ce rapport.

2.1. Procédures de détermination in-situ de l'état de contrainte.

Leeman (1964) a effectué une revue détaillée des différentes techniques de mesure des contraintes alors disponibles (cellules de mesure de déformation de forages, inclusions rigides en forages, "door-stopper" , vérins plats, photo-élasticité, mesures de résistivité et de célérité des ondes de fréquences supérieures à 1000 Hz). Fairhurst (1968) a proposé une nouvelle étude comparée de ces méthodes en y incluant une discussion détaillée de la fracturation hydraulique. En 1969, Leeman a mis à jour le travail qu'il avait entrepris en 1964 (plus de 350 références). D'autres techniques sont présentées dans les comptes-rendus de congrès sur les mesures in-situ en mécanique des roches (Kovari, 1977).
Aussi, plutôt que de proposer une nouvelle revue détaillée des différentes techniques actuellement disponibles, nous nous attacherons à ne présenter que celles qui sont le plus couramment utilisées en insistant sur les hypothèses sur lesquelles elles reposent ainsi que sur les méthodes qui permettent d'estimer la précision de ces déterminations. Ce dernier point n'est d'ailleurs que très rarement abordé dans la littérature et nous espérons que ce rapport incitera des études plus approfondies sur ce sujet.

Méthodes par surcarottage

Les méthodes de mesure de contrainte par surcarottage sont basées sur la mesure des déplacements à la paroi d'un forage engendrés par le relâchement complet des contraintes autour de celui-ci. Les relations contraintes/déplacements, établies d'après la théorie de l'élasticité linéaire, permettent de calculer l'état de contrainte "à l'infini" lorsque les caractéristiques élastiques du matériau sont connues. Dans la procédure habituelle, un premier forage est réalisé et un appareil de mesure des déplacements est mis en place. Le forage est ensuite surcarotté de façon à relâcher complètement les contraintes dans le volume de roche qui entoure l'appareil de mesure.
Différents types d'appareils ont été construits à cette fin, tous sont supposés être suffisamment souples pour laisser la roche se déformer librement lors du relâchement des contraintes. Avec les cellules de mesure de déplacement, différentes combinaisons de mesures sont possibles : trois mesures de déplacements diamétraux avec la cellule de l'U.S. Bureau of Mines (Hooker et Bickel 1974) ainsi qu'avec la cellule du CERCHAR (Hellal et Dejean 1981), quatre mesures de déplacements diamétraux avec la cellule de Crouch et Fairhurst (1967), la mesure de quatre déplacements diamétraux et huit déplacements longitudinaux avec la cellule de Bonnechère (1971). Avec les inclusions souples équipées soit de jauges de

déformation soit de jauges inductives, diverses combinaisons de mesures de déformation circonférentielles, longitudinales, ou inclinées par rapport à l'axe du forage, sont obtenues (voir par exemple Leeman 1968, Rocha et Silverio 1969, Blackwood 1977, Pahl 1977).
Tandis que les cellules qui ne fournissent que des mesures de déplacements diamétraux supposent que celles-ci soient effectuées dans au moins trois forages d'inclinaison différentes pour la détermination complète du tenseur des contraintes, les cellules qui combinent des mesures diamétrales et longitudinales, ou circonférentielles longitudinales et inclinées, ne requièrent, elles, qu'une seule opération de surcarottage.
Toutes ces méthodes sont construites sur la solution du cylindre circulaire creux infini percé dans un milieu linéaire élastique et isotrope. La solution bien connue de ce problème (voir par exemple Kirsch 1898, Hiramatsu et Oka 1968) est fournie par les équations (1) pour les contraintes et (2) pour les déplacements.
Si σ_{ij} , i,j = 1,2,3 sont les composantes du tenseur des contraintes à l'infini (exprimées dans le repère géographique) ; ρ, θ, z, sont les coordonnées cylindriques (où z est confondu avec l'axe du forage et est orienté positivement vers l'extrémité libre de celui-ci) ; E, ν sont le module de Young et le coefficient de Poisson du matériau ; r est le rayon du forage ; alors :

$$
(1)\quad
\begin{cases}
\sigma_{\rho\rho} = (1-\frac{r^2}{\rho^2})\frac{\sigma_{11}+\sigma_{22}}{2} + (1-4\frac{r^2}{\rho^2}+3\frac{r^4}{\rho^4})(\frac{\sigma^\infty_{11}-\sigma^\infty_{22}}{2}\cos 2\theta + \sigma^\infty_{12}\sin 2\theta) \\
\sigma_{\theta\theta} = (1+\frac{r^2}{\rho^2})\frac{\sigma^\infty_{11}+\sigma^\infty_{22}}{2} - (1+3\frac{r^4}{\rho^4})(\frac{\sigma^\infty_{11}-\sigma^\infty_{22}}{2}\cos 2\theta + \sigma^\infty_{12}\sin 2\theta) \\
\sigma_{zz} = \sigma^\infty_{33} - 4\nu\frac{r^2}{\rho^2}(\frac{\sigma^\infty_{11}-\sigma^\infty_{22}}{2}\cos 2\theta + \sigma^\infty_{12}\sin 2\theta) \\
\sigma_{\theta z} = (1+\frac{r^2}{\rho^2})(\sigma^\infty_{23}\cos\theta - \sigma^\infty_{31}\sin\theta) \\
\sigma_{z\rho} = (1-\frac{r^2}{\rho^2})(\sigma^\infty_{31}\cos\theta + \sigma^\infty_{23}\sin\theta) \\
\sigma_{\theta\rho} = (1+2\frac{r^2}{\rho^2}-3\frac{r^4}{\rho^4})(\frac{\sigma^\infty_{22}-\sigma^\infty_{11}}{2}\sin 2\theta + \sigma^\infty_{12}\cos 2\theta)
\end{cases}
$$

$$
(2)\quad
\begin{cases}
u_\rho = \frac{1+\nu}{E}(\frac{1-\nu}{1+\nu}\rho + \frac{r^2}{\rho})\frac{\sigma^\infty_{11}+\sigma^\infty_{22}}{2} - \frac{\nu}{E}\rho\sigma^\infty_{33} + \\
\quad + \frac{1+\nu}{E}(\rho + 4(1-\nu)\frac{r^2}{\rho} - \frac{r^4}{\rho^3})(\frac{\sigma^\infty_{11}-\sigma^\infty_{22}}{2}\cos 2\theta + \sigma^\infty_{12}\sin 2\theta) \\
u_\theta = \frac{1+\nu}{E}(\rho + 2(1-2\nu)\frac{r^2}{\rho} + \frac{r^4}{\rho^3})(\frac{\sigma^\infty_{11}-\sigma^\infty_{22}}{2}\sin 2\theta + \sigma^\infty_{12}\cos 2\theta) \\
u_z = \frac{2(1+\nu)}{E}(\rho + \frac{r^2}{\rho})(\sigma^\infty_{31}\cos\theta + \sigma^\infty_{23}\sin\theta) - \frac{2\nu}{E}z\frac{\sigma^\infty_{11}+\sigma^\infty_{22}}{2} + \frac{z}{E}\sigma^\infty_{33}
\end{cases}
$$

La solution de ce problème a été étendue au cas des roches à isotropie planaire (5 constantes d'élasticité) par Berry (1968) et plus tard par Hirashima et Koga (1977).
Avec la méthode dite du " Doorstopper " une rosette de jauges de déformations est collée au fond du forage après qu'il ait été rendu parfaitement plan. Lorsque le carottage du forage est poursuivi les déplacements causés par le relâchement des contraintes sont mesurés. La détermination complète du tenseur des contraintes nécessite au moins trois forages d'orientations différentes. La solution élastique de ce problème a été obtenue de façon empirique pour des matériaux isotropes soit par des méthodes expérimentales (Galle et Wilhoit 1962, Leeman 1964 , Bonnechère et Fairhurst 1967), soit par la méthode des éléments finis (De la Cruz 1969). La solution pour les matériaux à isotropie planaire est présentée à ce congrès (Rahn 1983) pour le cas d'un forage perpendiculaire ou parallèle au plan d'isotropie du matériau ; les coefficients ont été obtenus par la méthode des éléments finis.
La principale limitation théorique de ces méthodes est que le matériau est supposé se comporter de façon linéairement élastique et que les constantes d'élasticité sont supposées connues. On notera en outre que ces mesures demandent beaucoup de temps ce qui les rend très coûteuses. De plus, dans la plupart des cas, les mesures effectuées par ces méthodes sont limitées à des forages peu profonds (inférieurs à 100 m le plus généralement) ; notons toutefois que Martna et al. (1983, ce congrès) présentent des résultats obtenus dans des forages de 500 m de profondeur.Mentionnons enfin que dans le cas de fortes contraintes perpendiculairement à l'axe du forage, le phénomène de discage pose de graves difficutés pour l'obtention de surcarottes de longueur suffisante. Dans ce dernier cas, seule la méthode du " Doorstopper " est applicable.
L'avantage principal de ces méthodes est que la précision de la détermination peut être établie, point qui n'est généralement pas discuté dans les articles publiés sur ce sujet.
Avec tous ces appareils les déplacements sont mesurés soit en un point soit le long de très courtes bases (1 à 2 cm). De ce fait les mesures sont très influencées par les hétérogénéités locales ce qui entraîne une grande dispersion des résultats. Blackwood (1978) a proposé d'enregistrer de façon continue les déplacements lors de l'opération de surcarottage en sorte que, par comparaison des courbes déplacement mesuré-avancement du surcarottage avec les courbes théoriques calculées par la méthode des éléments finis, seules sont retenues pour le calcul des contraintes les mesures théoriquement admissibles.
Cornet et al. (1979) ont proposé d'utiliser l'ensemble de ces mesures théoriquement admissibles simultanément pour déterminer l'état de contrainte ainsi que l'incertitude associée à cette détermination. En effet si {E} est un vecteur dont les composantes sont l'ensemble des mesures admissibles (exprimées dans le repère de la cellule de mesure), {X} estun vecteur dont les composantes sont celles du tenseur des contraintes à l'infini (exprimées dans le repère géographique), [A] est une matrice dont les coefficients sont ceux de la relation déplacement-contrainte explicités dans l'équation (2) (ou les relations empiriques dans le cas du doorstopper), [B] est une matrice de rotation qui permet d'obtenir les composantes du vecteur [X] dans le repère de la cellule et [C] est une matrice dont les coefficients sont les facteurs de concentration de contrainte associés à l'effet de la cavité d'où sont effectués les forages sur l'état de contrainte

local, alors l'ensemble des résultats admissibles peut s'exprimer par :

$$[A] \quad [B] \quad [C] \quad \{X\} = \{E\} \qquad (3)$$

Le système d'équation linéaire (3) est surdéterminé dès que plus de six mesures admissibles sont obtenues dans six directions différentes. Ainsi, en résolvant ce système par une méthode de moindres carrés, le tenseur des contraintes et l'incertitude associée à sa détermination peuvent être évalués . Dans la pratique, la matrice [A] est rarement connue et seules les mesures non affectées par la proximité de la cavité sont traitées simultanément. Pour les mesures plus proches de la paroi libre, seules celles qui en sont équidistantes sont traitées simultanément. Il apparaît ainsi qu'avec cette méthode d'inversion les mesures de déplacement, bien qu'effectuées seulement en quelques points, permettent d'obtenir une estimation de la valeur moyenne des composantes du tenseur des contraintes pour l'ensemble du volume où des mesures ont été effectuées. L'influence de larges discontinuités à l'intérieur de ce volume peut ne pas être négligeable, aussi les mesures qui ne sont pas cohérentes avec la détermination moyenne sont elles rejetées et une nouvelle détermination est effectuée à partir de la collection réduite. La figure (1) fournit un exemple pour lequel il a été possible de rejeter effectivement une mesure incohérente avec la détermination moyenne.Les directions du grand axe de la déformée des sections transverses du forage, du fait du surcarottage, sont comparées à la direction calculée d'après la détermination du tenseur des contraintes obtenue avec l'ensemble des mesures admissibles. Il apparaît clairement que les mesures obtenues lors du surcarottage à 1 m 80 de la paroi de la galerie sont influencées par cette dernière. Elles ont été éliminées de la collection de mesures admissibles et une nouvelle détermination a été effectuée à partir de la collection réduite.

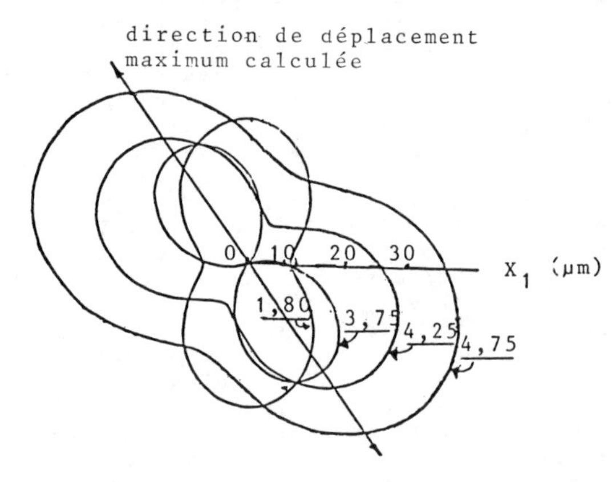

Fig.1 Comparaison entre valeurs calculées et valeurs mesurées pour la déformation d'un forage liée à une opération de surcarottage.Les courbes sont obtenues à partir des mesures de déplacements radiaux.L'orientation de la direction diamétrale pour laquelle le déplacement est maximum a été calculée à partir de la détermination de la contrainte moyenne.L'axe de référence est la direction horizontale dans le plan perpendiculaire à l'axe du forage

Inclusions rigides

Les inclusions rigides sont construites pour mesurer directement une ou plusieurs composantes de la contrainte. Elles doivent être beaucoup plus rigides que la roche (au moins 4 à 5 fois d'après Leeman 1964) pour que les variations de contrainte affectant celle-ci soient transmises directement à l'inclusion. Des capteurs de mesure de déformation ou de déplacement sont placés à l'intérieur de l'inclusion pour permettre de déterminer directement ces variations de contrainte (cordes vibrantes pour Hawks et Hooker 1974 et Pariseau et Eitani 1977, jauges de déformation pour Potts 1957 et Filak et Cyrul 1977, cellule photo élastique pour Roberts et al. 1964).Lorsque ces variations de contraintes sont causées par un surcarottage, les mesures ainsi obtenues peuvent permettre de déterminer l'état de contrainte local. L'avantage principal des inclusions rigides par rapport aux inclusions souples réside dans le fait qu'aucune connaissance de la loi rhéologique de la roche n'est nécessaire. Elles sont bien adaptées aux études de variation de contrainte dans le temps mais n'ont pas été aussi utilisées que les inclusions souples pour la détermination de l'état de contrainte, du fait de problèmes de contact (voir par exemple Fairhurst (1968) à ce sujet). Elles sont de plus limitées aux forages peu profonds et peuvent poser des problèmes de stabilité à long terme (Fairhurst 1968). Aucune méthode pour établir la précision de ces mesures n'est décrite dans la littérature.

Méthodes de mesure par compensation

La plus ancienne méthode par compensation est la méthode du vérin plat développé par Mayer et al. (1951) et Tincellin (1951). Tout d'abord deux plots sont mis en place sur la surface libre d'une galerie, ou d'un pilier de mine, et la distance entre ces plots est mesurée soigneusement. Cette mesure est ensuite reprise après que la contrainte s'exerçant sur un plan perpendiculaire à la droite joignant les deux plots ait été libérée au moyen d'un trait de scie. En insérant un vérin plat dans ce trait de scie puis en exerçant une pression normalement au plan de la saignée, la distance séparant les deux plots est ramenée à sa valeur originelle. La valeur de la pression requise pour rétablir cette distance originelle fournit une mesure directe de la contrainte normale qui existait sur le plan du trait de scie avant que ce dernier ne soit effectué. Hoskins (1966) a discuté la configuration de plots optimal qui permet de supprimer l'effet du relâchement de la composante de cisaillement dans le plan de la saignée.
L'inconvénient majeur de cette technique est qu'elle n'est applicable qu'à de faibles distances de surfaces libres. Dans ce domaine, des déformations non élastiques se produisent généralement de sorte que l'interprétation des résultats en termes de contrainte à l'infini est très délicate. De plus les effets de viscosité du matériau peuvent perturber de façon significative la mesure.
Kuznetzov et al. (1983, ce congrès) ont développé une nouvelle technique de mesure par compensation. Une pression est appliquée sur le fond aplani d'un forage de façon à prévenir toute déformation axiale lorsque le

forage est prolongé, c'est-à-dire lorsque les contraintes sur la portion de roche qui devient la carotte sont relâchées. La détermination de la pression requise pour supprimer les déformations axiales peut être mise en relation avec les composantes diagonales du tenseur des contraintes. Lorsque trois forages orthogonaux sont utilisés, la valeur de ces composantes diagonales, définies dans le repère associé aux trois axes des forages, peuvent être déterminées. Des forages complémentaires réalisés dans d'autres directions que les trois premières permettent d'obtenir une détermination complète des composantes de la contrainte. L'avantage principal de la méthode est que le module d'Young de la roche n'a pas besoin d'être déterminé ; par contre le coefficient de Poisson doit être connu. Le principal inconvénient de la méthode provient du nombre élevé de forages nécessaires à la détermination.

La méthode d'analyse de déformations différentielles proposée par Ren et Roegiers (1983, ce congrès) est également apparentée aux méthodes par compensation. Elle suppose qu'un relâchement complet des contraintes engendre dans un échantillon de roche des microfissures dont la densité dans un plan donné est proportionnelle à la variation de contrainte dans la direction perpendiculaire à ce plan. Ainsi il est supposé que le spectre de la microfissuration induite reflète l'état de contrainte auquel était soumis l'échantillon. Si cet échantillon est soumis à une pression hydrostatique, les axes principaux de la déformation résultant de ce chargement sont identiques à ceux de l'état de contrainte originel. De plus les valeurs relatives des déformations dans la direction des axes principaux fournissent des indications sur la valeur relative des composantes principales du tenseur des contraintes.

Les résultats présentés par Ren et Roeggiers sont une bonne démonstration des potentialités et des limitations de la méthode. En particulier l'influence des paléocontraintes et de la dilatance qui a pu en résulter semble difficile à éliminer. Cependant, quand elle est applicable, cette méthode fournit un moyen très bon marché pour déterminer l'état de contrainte en profondeur.

Méthodes par fracturation

Depuis longtemps maintenant, la fracturation hydraulique a été appliquée à la détermination de l'état de contrainte en profondeur (voir par exemple Hubbert et Willis 1957, Scheidegger 1962, Fairhurst 1964, Kehle 1964, Haimson et Fairhurst 1969).

Dans la version originale de cette technique, une portion de forage est isolée par un double obturateur puis est pressurisée jusqu'à ce qu'apparaisse une fracture à la pression dite de rupture (P_b). Lorsque l'injection de fluide est arrêtée, la pression se stabilise à la pression dite de fermeture (P_s) (voir fig.2). Si le forage est parallèle à une des directions principales de la contrainte (la direction verticale dans la plupart des cas) et si la roche est linéairement élastique et isotrope vis à vis tant de ses propriétés élastiques que de son comportement à la rupture, la mesure des pressions de rupture et de fermeture permet de déterminer certaines composantes de la contrainte. Du fait que les obturateurs sont

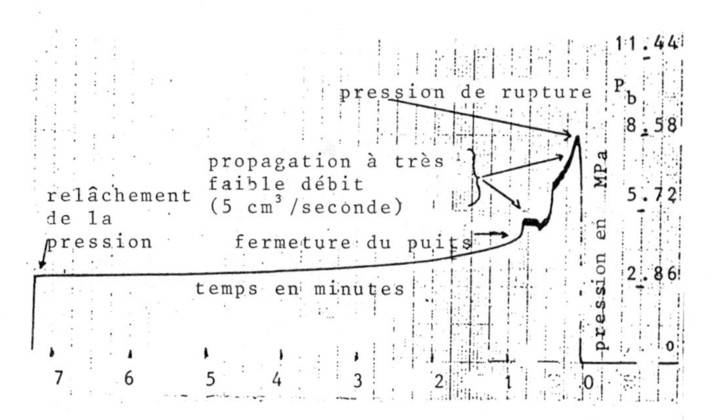

Fig.2 Enregistrement pression-temps caractéristique d'un essai de fracturation hydraulique effectué dans un granite (profondeur de 200 m).

généralement gonflables, seules des fractures verticales sont initialisées à partir d'un forage vertical (Haimson et Fairhurst 1961) même dans les zones très proches de la surface du sol. On montre alors, en utilisant les équations (1), que :

$$\sigma^t - \sigma_1 + 3\,\sigma_2 = P_b \qquad (4\ a)$$

$$\sigma_2 = P_s \qquad (4\ b)$$

où σ_1 et σ_2 sont les contraintes principales perpendiculaires à l'axe du forage ($\sigma_1 > \sigma_2$) et σ^t est la résistance à la traction de la roche. La mesure de P_b et P_s et la détermination, soit au laboratoire, soit in-situ, de σ^t fournissent les éléments nécessaires au calcul de σ_1 et σ_2. D'après la théorie de Griffith sur la rupture (Griffith 1921), il est possible de montrer que, pour les roches isotropes, la fracture se propage perpendiculairement à la direction de σ_2 de sorte que la détermination de l'orientation de la fracture au niveau du forage fournit l'orientation de σ_1.

L'équation (4 a) doit être modifiée pour les roches poreuses pour tenir compte des variations de pression interstitielle au voisinage du forage (voir par exemple Haimson 1968, Rice et Cleary 1976).

Si la roche est saturée par un fluide sous pression P_p mais est suffisamment imperméable pour prévenir toute percolation dans la formation durant l'opération de mise sous pression, l'équation (4 a) doit être remplacée par :

$$P_b = -\,\sigma_1 + 3\,\sigma_2 + \sigma^T - P_p \qquad (5)$$

En effet, ainsi que cela est indiqué sur la figure 3, l'état de contrainte peut être décomposé en deux composantes :

- la composante hydrostatique A ,
- la composante B qui n'implique que des contraintes effectives à l'infini.

La pression P_B requise pour créer une fracture

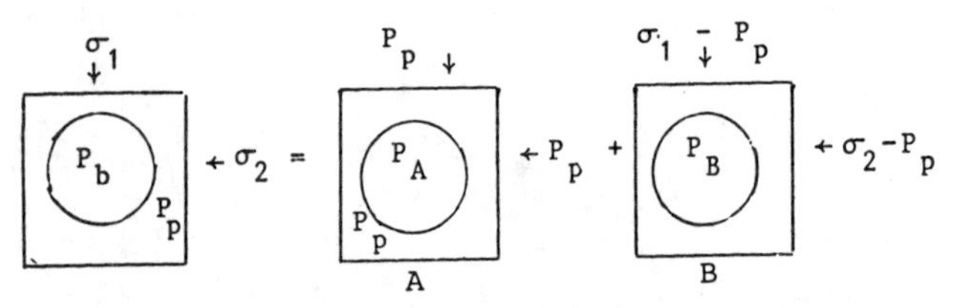

Fig. 3 Influence de la pression de pore sur la pression de rupture dans le cas de roches imperméables

dans le cas du chargement B est :

$$P_B = -(\sigma_1 - P_p) + 3(\sigma_2 - P_p) + \sigma^t \qquad (6)$$

tandis que la pression P_A requise pour maintenir hydrostatique la contrainte dans le chargement A est P_p . Ainsi par superposition des chargements A et B, on obtient l'équation (5). Pour la pression de fermeture l'équation (4 b) reste valable.
Dans cette formulation, la méthode de mesure souffre d'un certain nombre de limitations :

1. Le massif rocheux doit être linéairement élastique et isotrope tant vis à vis de ses propriétés élastiques que de son comportement à la rupture ;
2. le forage doit être parallèle à l'une des directions principales de la contrainte ;
3. la résistance à la traction du matériau doit être connue ;
4. le pic de l'enregistrement pression-temps doit coincider avec la valeur de la pression pour laquelle la fracture démarre ;
5. la précision de la détermination n'est établie qu'à partir de la cohérence des résultats.

Bredehoeft et al. (1976) ont montré que le problème de la détermination de la résistance à la traction de la roche peut être éliminée en considérant la pression dite de réouverture (P_r) plutôt que la pression de rupture. En effet, après que la fracture a été formée, la résistance à la traction de la roche à cet endroit est nulle ; aussi, après que la pression de pore est retournée à sa valeur originelle, la pression requise pour réouvrir la fracture est :

$$P_r = 3\,\sigma_2 - \sigma_1 - P_p \qquad (7)$$

Pour que l'équation (7) soit valable aucune percolation de fluide le long du plan de fracture ne doit intervenir avant sa réouverture (Cornet 1976). Cette condition peut être satisfaite en utilisant de forts débits d'injection ou des fluides très visqueux. Pour de telles conditions, la pression de réouverture ne coincide pas avec le pic de l'enregistrement pression-temps. La pression de réouverture peut néanmoins être mesurée précisément par comparaison des courbes pression-temps obtenues lors de la première et de la seconde montée en pression (Hickman et Zoback 1982). En effet, si ces deux montées en pression ont été effectuées au même débit, le point où la seconde se sépare de la première correspond au moment où la fracture se réouvre. Ces auteurs suggèrent d'ailleurs d'utiliser la troisième montée en pression plutôt que la deuxième de façon à limiter les difficultés liées à une extension asymétrique de la fracture.

Fréquemment aucune pression de fermeture n'est identifiable sur les enregistrements. Aussi plusieurs méthodes ont été proposées pour améliorer cette détermination (voir par exemple Mc Lennan et Roegiers 1982, Gronseth et Kry 1982, Aamodt et Kuriyagawa 1982). Toutes font référence à la pression de fermeture instantanée, c'est-à-dire la variation de pression observée à la fin de l'injection. Différentes techniques semi-logarithmiques sont proposées pour identifier précisément cette pression de fermeture instantanée. Une autre méthode de mesure est d'utiliser la pression de réouverture quasi-statique : la pression est augmentée par échelons de pression constante successifs, chaque niveau de pression étant maintenu durant quelques minutes pour limiter les effets de gradient de pression liés à l'écoulement dans la fissure. Le premier échelon doit correspondre à une pression nettement inférieure à la pression de fermeture. En reportant sur un diagramme pression-débit les résultats obtenus lors d'une telle montée en pression, il est possible de mettre en évidence très précisément la réouverture de la fracture : cette réouverture se manifeste par un coude marqué de la courbe pression-débit. Cette procédure suppose évidemment que la fracture reste plane loin du forage ; plus précisément elle n'est pas supposée être déviée de sa direction originelle par des fractures préexistantes loin du forage.
Afin d'éliminer les limitations 1, 2 et 5 de la méthode de mesure originale, Cornet (1982) a proposé d'exploiter les plans de faiblesse préexistants qui affectent toujours une masse rocheuse : si l'obturateur double est mis en place au niveau d'un tel plan, il peut être possible de le réouvrir à condition d'utiliser de très faibles débits. Pour de tels plans, la pression de fermeture ou encore la pression de réouverture quasistatique, est exactement égale à la contrainte normale supportée par le plan de fracture ; aussi obtient-on l'équation suivante :

$$\underset{\sim}{\sigma}\underline{n} \cdot \underline{n} = P_s \qquad (8)$$

où $\underset{\sim}{\sigma}$ est l'état de contrainte local, $\underline{n}$ est la normale au plan de fracture et P_s la pression de fermeture.
Dans le cas le plus général, six plans de fracture d'orientations différentes doivent être utilisés pour déterminer le tenseur complet des contraintes. Cependant si l'une des directions principales est supposée verticale (mais pas nécessairement parallèle à l'axe du forage) alors trois plans seulement sont nécessaires pour la détermination .
Lorsque plus de trois fractures sont disponibles, une méthode classique d'inversion par moindres carrés permet de déterminer l'état de contrainte.
Cette nouvelle méthode est particulièrement bien adaptée pour les mesures effectuées à partir de cavités souterraines dans lesquelles des forages peuvent être effectués dans plusieurs directions de façon à recouper, à la même profondeur, des fractures préexistantes d'orientations différentes. Lorsqu'un seul forage est disponible, l'inversion des données suppose que les variations de contrainte avec la profondeur sont linéaires :

$$(\underset{\sim}{\sigma}) = \begin{pmatrix} \sigma_1 + \alpha_1 z & 0 & 0 \\ 0 & \sigma_2 + \alpha_2 z & 0 \\ 0 & 0 & \rho g z \end{pmatrix} \quad (9)$$

où ρg est le gradient vertical de la contrainte verticale. Si l'on suppose, en première approximation, que $\alpha_1 = \alpha_2$ (les mesures sont très proches l'une de l'autre en sorte que seuls les effets de gravité sont pris en compte) l'équation (8) peut être linéarisée par un simple changement de variable de sorte que la méthode classique de moindres carrés est encore applicable. Lorsque les plans de fracture sont éloignés les uns des autres l'équation (8) n'est plus linéaire et des méthodes d'inversion plus élaborées doivent être utilisées (Valette et Cornet, 1983).

Lorsque toutes les fractures sont parallèles à la direction du forage et présentent la même orientation par rapport au repère géographique, les résultats sont interprétés selon la méthode classique. Les essais le long d'un plan de fracture préexistant permettent alors d'obtenir une vérification de cette détermination et d'écarter ainsi les questions possibles relatives à l'influence éventuelle de l'anisotropie de résistance du massif rocheux sur l'orientation des fractures artificielles.

Notons de plus que l'échelle pour laquelle les contraintes sont déterminées avec la méthode de fracturation hydraulique est assez flexible ; elle dépend de la distance entre les obturateurs, du diamètre du forage et de la quantité de liquide injecté. Ces mesures de contrainte peuvent ne concerner que des volumes de quelques centimètres cubes (au laboratoire par exemple) mais toucher des volumes de quelques mètres cubes en conditions de terrain. Il n'est pas recommandé de toucher des volumes plus grands du fait de l'influence du gradient de contrainte lié aux effets de la gravité. En outre, pour les grands volumes, l'influence des plans de faiblesse de la roche sur la propagation des fractures loin du forage peut poser des difficultés d'interprétation.

D'autres méthodes de mesure des contraintes basées sur des études de fracturation ont été proposées (voir par exemple Jaeger et Cook 1963, De la Cruz 1976). Dans ce domaine, l'utilisation de pressiomètres semble assez prometteuse (voir par exemple Huergo 1983, ce congrès ; Charlez 1983) car elle combine les mesures de déformations et les mesures de contraintes. Stephanson (1983, ce congrès) présente une discussion sur l'applicabilité du pressiomètre de l'Ecole des Mines du Colorado (Hustrulid et Hustrulid 1976) pour la mesure des contraintes. Plus précisément il propose d'interpréter la pression requise pour réouvrir une fracture développée durant un cycle de chargement préliminaire comme une mesure de la contrainte principale minimum perpendiculaire à l'axe du forage (le forage est supposé parallèle à l'une des directions principales de la contrainte). Or, nous avons vu dans la discussion ci-dessus (voir équation 7) que cette valeur dépend de deux contraintes principales et de la pression de pore.

Nous espérons que ce point pourra être discuté lors du congrès.

Méthodes sismo-tectoniques

Les sismologues associent à chaque séisme un mécanisme au foyer c'est-à-dire qu'ils déterminent l'orientation du plan de faille ayant donné lieu au séisme ainsi que la direction du mouvement dans ce plan (voir par exemple Stauder 1962). Bien qu'initialement empirique, cette méthode a été progressivement affinée ; elle est maintenant de pratique courante en sismologie. Ces mécanismes sont utilisés fréquemment pour évaluer l'orientation des directions principales de la contrainte dans le domaine où le séisme a eu lieu. Avant de discuter ce dernier point, il convient de rappeler le principe sur lequel repose la détermination des mécanismes au foyer.

Il est connu depuis longtemps (Byerly 1926) que le premier mouvement causé par les ondes P (ondes de compression polarisées longitudinalement) engendrées par un séisme est en compression pour certaines régions et en dilatation pour d'autres. Cette distribution peut être expliquée si la source est assimilée à des forces ponctuelles. En effet Nakano (1923) a montré qu'un couple de forces ponctuelles engendre soit des compressions soit des dilatations selon le quart d'espace considéré. Les sous-espaces sont séparés par deux plans orthogonaux connus sous le nom de plans nodaux (voir fig. 4).

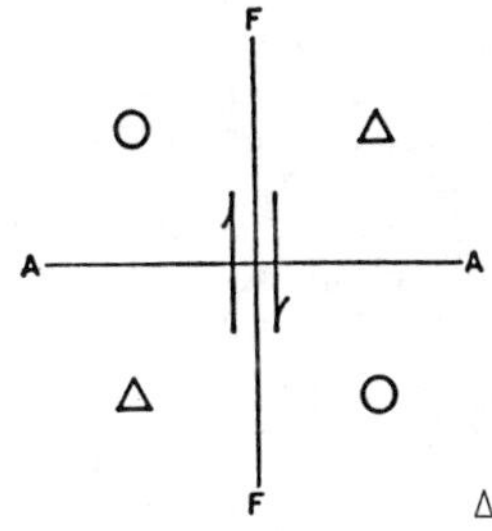

Δ dilatation

O compression

Fig. 4 Distribution des compressions et dilatations résultant d'un mouvement de cisaillement le long de la faille FF (d'après Stauder 1962).

Il a été montré (Knopoff et Gilbert 1960) que le problème de la caractérisation de ces schémas de radiation peut être rapproché de ceux liés à une dislocation, c'est-à-dire une discontinuité tangentielle du champ des déplacements le long d'une surface d'extension finie.

Burridge et Knopoff (1964) ont ainsi pu montrer que le champ des déplacements causés par une dislocation est équivalent à celui engendré par un double couple (forces ponctuelles ayant un moment) appliqué au foyer du séisme ; un des plans nodaux du champ de radiation ainsi créé coincide avec le plan de fracture. Ce résultat est connu en sismologie sous le nom de théorème de la représentation. Pour qu'une telle représentation soit possible, les longueurs d'ondes considérées doivent être plus grandes que la source de façon à limiter les perturbations liées aux effets de bord ainsi qu'aux hétérogénéités du terrain.

La détermination des mécanismes au foyer procède de la façon suivante :

1. localiser le foyer du séisme et considérer une sphère de rayon unité centrée sur le foyer.

2. définir le sens du premier mouvement pour le plus grand nombre de stations possibles ; seules les longueurs d'ondes supérieures à la taille de la source doivent être considérées. Les dimensions de la source peuvent être estimées d'après la "fréquence coin " du spectre de Fourier du déplacement (voir par exemple Madariaga 1979).

3. Déterminer le chemin de propagation des ondes entre la source et le point d'observation et déterminer le point d'intersection de ce rai avec la sphère. Le rai dépend essentiellement des vitesses de propagation dans le milieu et la structure de ce dernier doit donc être connue.

4. Utilisant la méthode de projection stéréographique qui conserve les aires, reporter sur le plan équatorial de la sphère la projection des points d'intersection des rais avec la demi-sphère inférieure. Pour les stations proches de l'épicentre, les rais intersectent la demi-sphère supérieure ; les symétriques de ces points sur la demi-sphère inférieure sont reportés en projection stéréographique.

5. Définir sur la projection stéréographique les deux plans orthogonaux qui séparent au mieux les quadrants pour lesquels des sens de premier mouvement identiques sont observés. Ces directions orthogonales sont les plans nodaux. Il est impossible de déterminer à partir de la seule étude du champ de déplacement lequel des deux est la faille qui a donné lieu au séisme. Aussi a-t-on recours à des observations complémentaires : trace de la faille en surface, distribution des répliques.

Bien qu'assez fastidieuse à effectuer à la main, la détermination du mécanisme au foyer peut être menée de façon routinière sur ordinateur. Un exemple de mécanisme au foyer est fourni sur la figure 5.

Des informations complémentaires sur la rupture à l'origine du séisme peuvent être obtenues par l'analyse spectrale des ondes enregistrées. Par exemple il est possible de déterminer les dimensions de la source ainsi que la chute de contrainte le long de la surface de glissement (Madariaga 1979). Il est d'ailleurs intéressant de souligner que ces chutes de contrainte varient de 1 MPa à 10 MPa, indépendamment de la magnitude du séisme (voir Hanks 1972), ce qui pose un problème de mécanique des roches intéressant.

Une fois connu les plans nodaux, trois axes sont définis : l'axe P, orienté perpendiculairement à la droite d'intersection des plans nodaux dans la direction où les mouvements de compression sont maximum ; l'axe T , également perpendiculaire à la droite d'intersection des plans nodaux, est orienté dans la direction où les mouvements de dilatation sont maximum ; l'axe "nul" coincide avec la droite d'intersection des plans nodaux. Il a été proposé d'assimiler ces trois axes aux directions principales de contrainte (Scheidegger 1964). Cette proposition est basée sur l'hypothèse que le plan de faille correspond à la surface pour laquelle le cisaillement est maximum et que les déplacements le long de la faille sont parallèles à la composante de cisaillement supportée par la faille. En fait cette approximation n'est généralement pas valable (voir par exemple Mc Kenzie 1969, Raleigh et al. 1972) car la fracturation des roches en compression n'obéit

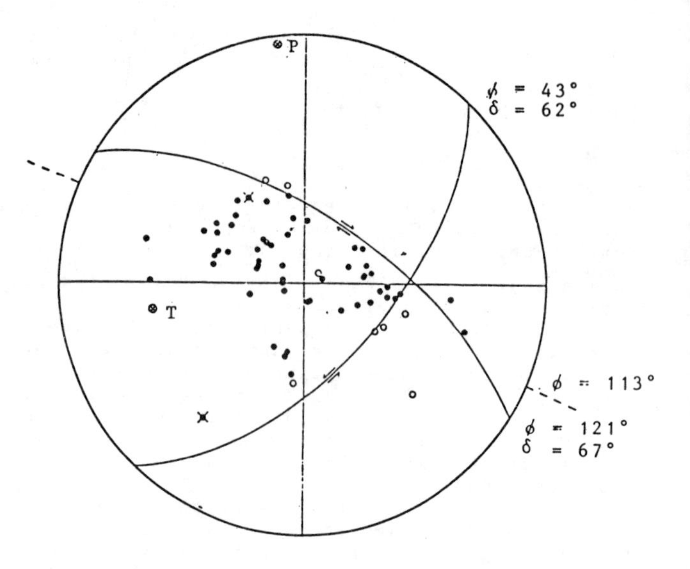

Fig.5 Mécanisme au foyer pour le séisme du 19/8/66 de Turquie-orientale. Ce diagramme est une projection stéréographique de l'hémisphère inférieur. Les cercles pleins représentent les compressions tandis que les cercles ouverts correspondent aux dilatations. ϕ et δ sont la direction et le pendage des plans nodaux et les flèches indiquent les sens du mouvement. La ligne pointillée indique la direction de la faille observée en surface à la suite du séisme. (d'après Mc Kenzie 1969).

généralement pas à un critère de rupture du type Tresca mais plutôt à un critère de type Mohr-Coulomb (la surface macroscopique de rupture est inclinée d'un angle variant de 20 à 45° par rapport à la direction de la contrainte principale maximum). En outre les masses rocheuses sont généralement anisotropes du point de vue de leur résistance du fait de l'existence de fractures préexistantes plus ou moins bien recimentées ce qui influence leur comportement en cisaillement. Mc Kenzie considère qu'il existe toujours des plans de rupture privilégiés et que ce sont les cisaillements le long de ces plans qui contrôlent l'équilibre du massif. Il est ainsi amené à conclure que l'erreur faite sur la détermination des directions principales de la contrainte locale, à partir des mécanismes au foyer, peut atteindre 90°. Raleigh et al. considèrent eux, que même s'il existe des plans de fractures préexistants de nouvelles surfaces de rupture apparaissent si les conditions de chargement requises pour leur apparition sont satisfaites avant celles nécessaires pour induire un glissement le long des plans préexistants.

Ils considèrent de ce fait que l'erreur maximum qui résulte de l'interprétation d'un mécanisme au foyer en termes de directions principales reste inférieur à une vingtaine de degrés.

Les méthodes microtectoniques (voir par exemple Angelier 1979, Armijo et al. 1982) pour identifier les directions principales ainsi que le facteur de forme $R = (\sigma_3 - \sigma_1) / (\sigma_2 - \sigma_1)$

(où σ_1 , σ_2 , σ_3 sont les valeurs principales du tenseur de contraintes) sont dérivées de principes un peu similaires à ceux de la méthode précédente : La mesure des directions et sens des stries qui sont observées sur une surface de faille permet de déterminer la direction et le sens des mouvements qui ont affecté cette faille. Si l'on suppose que le mouvement s'est produit dans une direction colinéaire à la direction du cisaillement supporté par ce plan, c'est-à-dire que les caractéristiques de la résistance au cisaillement le long des plans sont indépendants de leur orientation et de la direction du mouvement dans ces plans, alors les directions principales de la contrainte moyenne et le facteur R peuvent être déterminés.

Si

$$\underline{f}_i = \underline{\tau}_i \;/\; |\tau_i| \qquad (10\ a)$$

$$\underline{\tau}_i = \underset{\sim}{\sigma}\, \underline{n}_i - (\underset{\sim}{\sigma}\, \underline{n}_i \cdot \underline{n}_i)\, \underline{n}_i \qquad (10\ b)$$

alors

$$\underline{f}_i = \underline{s}_i^j \;\; ; \; i = 1,N \quad j = 1,M_i \qquad (10\ c)$$

où

- $\underline{f}_i$ est un vecteur unitaire parallèle à la direction de la composante de cisaillement sur la ième faille ;
- $\underset{\sim}{\sigma}$ est le tenseur moyen des contraintes exercées dans le volume considéré ;
- $\underline{n}_i$ est la normale au plan de faille i;
- $\underline{s}_i^j$ est un vecteur unitaire parallèle à la $j^{\text{ème}}$ strie dans le $i^{\text{ème}}$ plan;
- N est le nombre total de plans de failles ;
- M_i est le nombre total de stries dans le ième plan de faille.

L'équation (10 b) peut se ramener à :

$$|\tau_\ell| = (\sigma_2 - \sigma_1) A_\ell \qquad (11)$$

où $\ell = 1,2,3$; $A_1 = -kn_1$; $A_2 = (1-k)n_2$; $A_3 = (R-k)n_3$

$$k = n_2^2 + Rn_1^2 \ .$$

Si N est grand (de l'ordre de 100 à 200) le système d'équations (10 c) qui est non linéaire, peut être inversé après linéarisation (voir par exemple Armijo et al. 1982).

L'hypothèse principale faite pour ce calcul est résumée par l'équation (10 c). De plus il est supposé que les mouvements le long des plans de faille sont indépendants les uns des autres. Cette méthode suppose en outre que le tenseur $\underset{\sim}{\sigma}$ est uniforme à l'intérieur de volumes de l'ordre de quelques hectomètres cubes, voir quelques kilomètres cubes. Un problème d'interprétation peut se poser lorsque les fractures ont été soumises à des champs de contraintes tectoniques qui ont varié au cours des âges géologiques. Ce problème est maintenant résolu en adaptant la méthode d'inversion des données ; diverses techniques à cet effet sont présentées dans la littérature (Etchecopar et al. 1981, Armijo et al. 1982).

Etant donné que les mécanismes au foyer fournissent des informations sur la direction et le sens du mouvement le long des failles, ils peuvent être comparés aux mesures de stries : là où de nombreuses stries définissent la direction du mouvement, un seul mécanisme au foyer est nécessaire. Cette information est moins dépendante des petites hétérogénéités locales et, de plus, l'information concerne des grandeurs à la même échelle que celle à laquelle la contrainte moyenne est définie. Aussi l'interprétation d'une collection de mécanismes au foyer menée de façon similaire à celle effectuée en microtectonique devrait fournir des informations sur les directions principales de la contrainte moyenne plus exactes que celles recueillies avec un seul mécanismes au foyer. Cependant pour les petits séismes la difficulté de sélectionner le plan de faille à partir des deux plans nodaux reste encore mal maîtrisée.

2.2 Quelques résultats

Comparaison des résultats obtenus avec diverses techniques

Dans un certain nombre de cas les contraintes ont été déterminées au même endroit par différentes techniques. Haimson (1982) présente des résultats obtenus par fracturation hydraulique et par surcarottage. Au Nevada test site par exemple des mesures ont été effectuées à partir de deux tunnels creusés sous 450 m de séries de tufs et de grès. Les résultats obtenus avec ces deux méthodes sont assez comparables (voir table 1) et concordent avec des déterminations de mécanisme au foyer de séismes locaux.

Table 1. Résultats obtenus sur le site d'essais du Nevada (d'après Haïmson, 1982).

fracturation hydraulique	σ_H max. MPa direction	σ_h min. MPa direction	σ_v MPa direction pendage	Profondeur en m.
fracturation hydraulique	9.0/N 35°E	3.5/N 55°W	7.	380
surcarottage	8.0/N 45°E	2.5/N 45°W	5.7 N 42°E 20°	380
fracturation hydraulique	7.5/N 40°E	3.0/N 50°W	7.3	426
surcarottage	8.5/N 22°E	2.6 N 68°W pendage 33°	6.8 N 83°W pendage 7°	426
mécanisme au foyer	N 45°E	N 45°W		?

Les résultats obtenus sur cinq autres sites sont également très cohérents entre eux. La dispersion sur l'amplitude de la contrainte principale minimum horizontale est de l'ordre de 30% tandis que pour la contrainte principale maximum la dispersion atteint 50%. Cette dernière valeur peut s'expliquer par la difficulté rencontrée, avec la méthode classique de fracturation hydraulique, à mesurer avec précision soit la résistance à la traction de la roche soit la pression de réouverture. L'écart entre les directions moyennes pour la contrainte principale maximum obtenue avec chaque technique reste inférieur à 10° mais la dispersion pour les déterminations par surcarottage peut atteindre 120° alors qu'elle reste inférieure à 50° pour la fracturation hydraulique.

Doe (1982) présente une comparaison détaillée,

entre les résultats obtenus, à la mine de Stripa en Suède, par fracturation hydraulique et par surcarottage de la cellule triaxiale de Leeman, de l'inclusion souple du C.S.I.R.O. et de la cellule triaxiale de l'Université de Lulea. Alors que des problèmes de cimentation ont été rencontrés avec la cellule du C.S.I.R.O., les autres techniques ont fourni des résultats très homogènes entre eux quant aux valeurs moyennes ; une dispersion importante étant observée néanmoins parmi les résultats individuels.
Li Fang Quan et al. (1982) présente également une comparaison entre des mesures effectuées par fracturation hydraulique et par surcarottage dans la province de Hebei, près de Yi Sio.
Les résultats sont assez cohérents. Pour le surcarottage :

$$\sigma_{H\,max} = (3.4 + 0.03\,z)\,MPa \; ; \; \sigma_{h\,min} = (2.5 + 0.031\,z)\,MPa$$

direction de $\sigma_{H\,max}$ = N 64° ± 20°

pour la fracturation hydraulique :

$$\sigma_{H\,max} = (4.9 + 0.028z)\,MPa, \sigma_{h\,min} = (2.6 + 0.021\,z)\,MPa ,$$

direction de $\sigma_{H\,max}$ = N 80°W ± 10°

Les profondeurs varient de 0 à 90 m de la surface.
Les résultats présentés par Kuznetzov et al. (1983, ce congrès) fournissent une comparaison entre les déterminations obtenues par leur méthode de compensation et par la méthode microtectonique. Les résultats sont là encore, cohérents entre eux.
Ainsi, les résutats obtenus par différentes équipes dans différents environnements géologiques semblent indiquer que des états de contrainte relativement homogènes peuvent exister dans différents types de masses rocheuses. Les méthodes actuellement disponibles semblent fournir les moyens adéquats pour les déterminer même si certains progrès sont encore à réaliser pour améliorer la précision de ces déterminations.

Variations des contraintes avec la profondeur.

Notre connaissance de l'amplitude des contraintes principales en profondeur s'est améliorée considérablement au cours de ces dernières années grâce essentiellement aux mesures directes profondes réalisées dans les mines et surtout dans les forages (voir Mc Garr et Gay 1978, Brown et Hoek 1978, Jamison et Cook 1980).
Mc Garr (1980), considérant l'ensemble des données mondiales, a été amené à conclure que, en moyenne, le cisaillemnt maximum augmente linéairement avec la profondeur au moins jusqu'à 5.1 km pour les roches tendres telles que les calcaires marneux et les grès et 3.7 km pour les roches dures telles que le granite ou les quartzites (ces valeurs correspondent aux mesures les plus profondes). Il observe qu'une régression linéaire effectuée sur ces données suggère un gradient vertical du cisaillement maximum égal à 3.8 MPa/km dans les roches tendres et 6.6 MPa/km pour les roches dures.
Supposant un comportement élastique pour la croûte supérieure, il a montré que les conditions d'équilibre combinées aux conditions de compatibilité de Beltrami-Mitchell pour les contraintes permettent de définir des relations fonctionnelles sur les variations de contrainte avec la profondeur.
Ainsi par exemple, pour des domaines où aucune variation latérale de contrainte n'est observée (voir la section suivante à ce sujet) les relations fonctionnelles impliquent une augmentation linéaire des composantes principales avec la profondeur mais permettent une rotation monotone des directions principales. Ce résultat suggère donc que, même pour des conditions aussi simples, la détermination des contraintes en surface ne permet pas de prévoir à coup sûr les directions principales en profondeur.
Pour des domaines où le champ de contrainte reste indépendant de seulement une des coordonnées horizontales (disons la composante x_2, l'autre coordonnée dans le plan horizontal étant x_1), ces conditions impliquent que seules les composantes diagonales du tenseur de contrainte croissent linéairement avec la la profondeur ; les variations des composantes σ_{12} et σ_{23} étant elles non linéaires.
Ce résultat suggère donc que même à l'échelle considérée dans cette analyse, l'inférence des contraintes en profondeur à partir des seules mesures de surface est hasardeuse et que, de plus, en profondeur la composante verticale peut ne pas être principale.
Bien évidemment, pour des matériaux hétérogènes seules des mesures directes peuvent permettre d'obtenir des résultats significatifs. Par exemple, en utilisant la méthode de la pression de réouverture quasistatique pour déterminer la contrainte principale minimum horizontale existant dans des évaporites, Cornet (1982) a observé des variations atteignant 2.5 MPa en moins de 10 m de profondeur (la valeur moyenne était de 13 MPa et la précision de chaque détermination meilleure que 2 MPa ; la profondeur était de l'ordre de 700 m). A l'intérieur de ces dépôts (alternance de bancs de sel de 1 à 10 m d'épaisseur et de bancs de marne de 0.01 à 1 m d'épaisseur) existent des lits d'anhydrite (matériau très raide) dont l'épaisseur reste de l'ordre de quelques millimètres à quelques centimètres ; leur densité par unité de longueur est très variable. Ces bancs rigides modifient à l'évidence de façon très significative le comportement mécanique de l'ensemble des terrains. Le cisaillement maximum mesuré, de l'ordre de 5 MPa, n'est pas compatible avec le comportement de type visco-plastique du sel et des marnes. Un effet de rigidification imposé par l'anhydrite est considéré être la cause des variations rapides des contraintes avec la profondeur, les valeurs les plus élevées de la contrainte étant observées dans les zones proches des domaines à forte densité en anhydrite.
Martna et al. présentent à ce congrès des résultats de mesure de surcarottage dans un granito-gneiss homogène. Un état de contrainte uniforme pour lequel la contrainte principale maximum horizontale atteint 15 à 20 MPa est observé jusqu'à 320 m de la surface. Puis, après qu'une zone fortement faillée a été rencontrée, cette valeur passe à 65 MPa pour rester constante jusque vers 500 m, profondeur maximum des mesures. Aucune rotation des directions principales n'est mentionnée.
Ces résultats illustrent ainsi le fait que, indépendemment de l'échelle considérée, les

contraintes en profondeur ne peuvent pas être estimées par une simple extrapolation de mesures de surface. Cependant très grossièrement parlant, on peut considérer que le cisaillement maximum augmente de façon monotone avec la profondeur jusqu'à au moins 5 km de profondeur.

Variations latérales de contrainte

Hormis les variations de contrainte liées à la topographie (Scheidegger 1977, Haimson 1978, Buyle-Bodin 1980) des variations latérales de contrainte, à une profondeur donnée, peuvent être anticipées pour de simples raisons mécaniques. Par exemple l'hétérogénéité des propriétés mécaniques des roches engendre des rotations locales des directions principales ; des discontinuités telles que les failles peuvent créer des variations de contrainte aussi bien dans l'espace que dans le temps. A ce sujet les résultats obtenus par Li Fang Quan et al. (1982) ont montré que la contrainte principale maximum mesurée dans une région proche d'une faille ayant donné lieu à un séisme de magnitude 7.9 était de l'ordre de 2.5 MPa alors qu'elle atteignait 6.0 MPa dans une région plus éloignée du séisme.
Néanmoins, malgré toutes ces raisons, l'ensemble des mesures effectuées à ce jour suggère qu'il existe en fait de vastes régions où les contraintes sont relativement uniformes. Zoback et Zoback (1980) ont rassemblé un grand nombre de résultats de mesure obtenu sur l'ensemble du territoire continental des Etats-Unis et ont été amenés à définir des provinces de contrainte dont les dimensions linéaires varient de 100 à 2000 km, provinces pour lesquelles les directions principales et l'amplitude relative des contraintes principales restent relativement uniformes(voir fig.6). Froidevaux et al. (1980) et Rummel et al. (1982) ont été conduits à des conclusions similaires pour l'Europe de l'ouest même si l'ensemble de leur données sont relativement superficielles (moins de 300 m).
Ces résultats sont une claire indication que, superposées à un champ de contrainte lié aux efforts qui agissent sur les frontières de plaques (voir Solomon et al. 1980), d'autres sources de contrainte existent.
Mc Garr (1982) propose d'associer les provinces de contrainte définies par Zoback et Zoback à des champs de cisaillement agissant à la base de la croûte élastique fragile alors que Fleytout et Froidevaux ont proposé de considérer plutôt l'influence des variations spatiales de densité dans la lithosphère ainsi que dans le manteau supérieur pour expliquer ces provinces intraplaques.
A l'évidence, des mesures directes de contrainte constituent encore pour le moment des données nécessaires pour contraindre les modèles géodynamiques. Nous espérons que de telles mesures seront effectuées en nombre toujours plus grand pour améliorer ces modèles. Il est trop tôt actuellement pour inverser le processus et utiliser les modèles géodynamiques pour estimer les directions principales de contrainte en un point donné. Toutefois, losque seuls des ordres de grandeur et des directions générales de contrainte principales sont désirés, l'extrapolation à partir de mesures avoisinantes

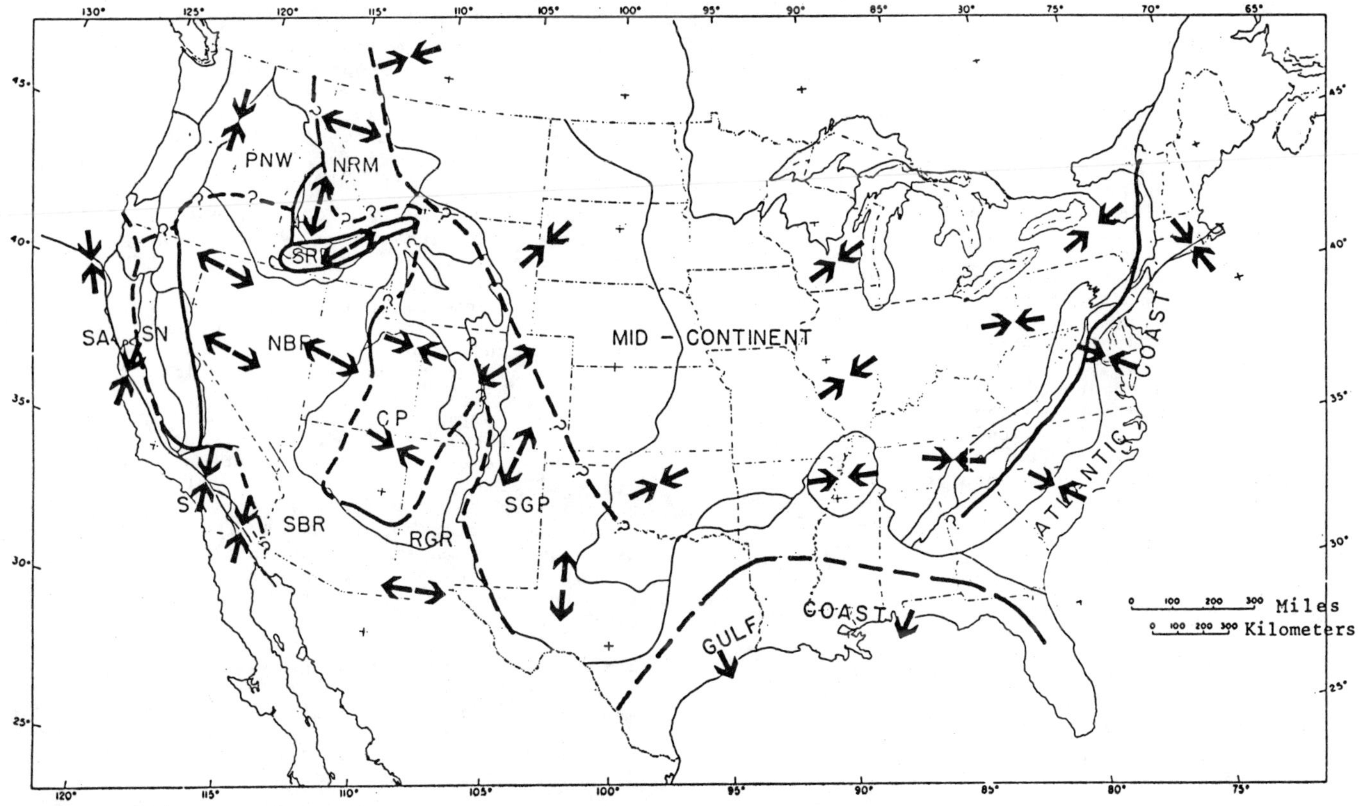

Fig.6 Carte de contraintes pour les Etats-Unis. Les flèches représentent les directions de la composante principale minimum (flèche vers l'extérieur) ou miminum (flèche vers l'intérieur) d'après Zoback et Zoback (1980).

peut fournir des résultats acceptables si le milieu est homogène. En multipliant les mesures, on peut espérer réaliser des cartes de contraintes qui, une fois établies, pourraient permettre d'estimer, sans avoir recours à la mesure directe, l'état de contrainte en n'importe quel point des continents.

Variations temporelles des contraintes

Les contraintes varient dans le temps, et une compréhension de ces variations temporelles est nécessaire. Cependant, bien que quelques instruments aient été développés pour les étudier (les inclusions rigides par exemple), peu de résultats à ce sujet ont encore été publiés (voir néanmoins à ce sujet Clark 1982). Un phénomène intéressant observé en association avec les séismes va peut être permettre prochainement d'améliorer notre connaissance à ce sujet : des variations du champ magnétique local ont été observées en association avec certains séismes (Moore 1964, Smith et Johnston 1976, Rikitake 1979). Un exemple est présenté sur la figure 7 ; l'anomalie observée était associée à un séisme de magnitude 5 qui a secoué la région de Dushambé dans le Tadjikistan (U.R.S.S.).

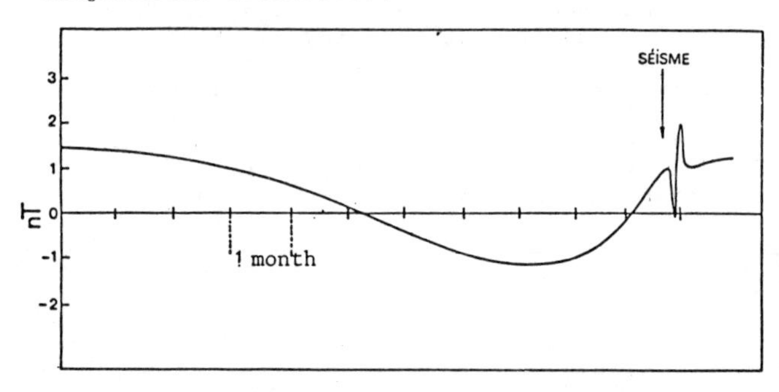

Fig.7 Anomalie magnétique observée dans la région de Dushambé (Tadjikistan, U.R.S.S.) avant un séisme de magnitude 5 ; d'après Skovorotkin (communication personnelle).

Une explication possible pour de telles variations du champ magnétique peut être fournie par le piezomagnétisme observé pour les roches magnétiques. En effet, lorsqu'un échantillon de roche est soumis à une variation de contrainte sa suceptibilité magnétique ainsi que son aimantation rémanente sont modifiées (Martin et al. 1978, Zlotnicki et al. 1981). Zlotnicki et al., par exemple, ont observé que les variations d'aimantation induite causées par des variations de contrainte peuvent être représentées par une loi tensorielle :

$$\Delta J_i = P_{ijkn}\, \Delta\sigma_{kn} J_{oj} \qquad (13)$$

où J_{oj} est la $j^{ème}$ composante de l'aimantation induite du fait du champ magnétique $\underline{H}$ ($\underline{J}_o = \underset{\sim}{S}_o \underline{H}$ où $\underset{\sim}{S}_o$ est le tenseur de susceptibilité magnétique de la roche) ;

$\Delta\sigma_{kn}$ est la variation de la composante kn de la contrainte ;

P_{ijkn} sont les composantes d'un tenseur du 4 ème ordre (36 composantes dans le cas général).

Lorsque la roche est isotrope, P_{ijkn} se ramène à deux coefficients indépendants de sorte que l'équation (13) devient :

$$\Delta J_i = (P_1 \Delta\sigma_{kk} \delta_{ij} + 2P_2 \Delta\sigma_{ij}) J_{oj} \qquad (14)$$

La détermination de P_1 et P_2 au laboratoire fournit le moyen d'obtenir certaines informations sur les variations de contrainte. Bien qu'aucune relation de ce type n'ait été mise en évidence pour l'aimantation rémanente, une paramétrisation des courbes expérimentales obtenues au laboratoire pourrait permettre d'obtenir des informations quantitatives sur les variations temporelles de contrainte dans de vastes régions. En effet les calculs effectués par Jim-Qi Hao et al. (1982) montrent que les variations du champ magnétique à attendre de ces effets sont du même ordre de grandeur que ceux effectivement observés. Ainsi il apparaît que des mesures précises de variations du champ magnétique dans les zones susceptibles d'être le siège de séismes, combinées à des études détaillées de laboratoire pour préciser le comportement piezomagnétique des roches de la région considérée, pourraient fournir des informations sur les variations temporelles de contraintes. L'application de telles méthodes pourrait également être envisagée dans le cas de variations de contrainte liées à l'activité humaine.

3. CARACTERISATION DE LA GEOMETRIE DES DISCONTINUITES

Les discontinuités qui existent dans les massifs rocheux affectent le comportement mécanique de ces derniers de plusieurs manières :

- en constituant des chenaux d'écoulement privilégié elles affectent la distribution de pression interstitielles et donc des contraintes effectives qui existent dans le massif ;
- de par leur comportement à la rupture, elles jouent un rôle de première importance pour la stabilité des massifs ;
- la souplesse de leur comportement élastique et leurs propriétés de dilatance influencent très profondément le processus de déformation de l'ensemble du massif.

Un travail considérable a déjà été effectué pour améliorer notre compréhension du comportement mécanique des discontinuités et bien que beaucoup de travail reste à faire, celui-ci commence à être assez bien modélisable. Cependant le problème posé par la caractérisation de la géométrie de ces discontinuités s'est révélé constituer une difficulté beaucoup plus délicate à surmonter. Ce n'est cependant que lorsque ce problème aura été résolu qu'une modélisation précise du comportement mécanique des masses rocheuses sera possible. La caractérisation de la géométrie des discontinuités peut être entreprise dans deux contextes différents :

- Dans le premier, la masse rocheuse se prête à l'observation sur une certaine surface et le problème consiste à extrapoler ces observations de surface dans le volume sous-jacent. C'est le problème de la caractérisation locale tel qu'il peut se poser dans certains problèmes de stabilité de pentes par exemple.
- Dans le second, l'observation ne peut se faire que sur de petites surfaces, comparées à l'échelle du volume considéré, et cette information discontinue doit être extrapolée

à l'ensemble du volume. Dans la plupart des cas, en plus d'observations de surface, il est possible de disposer d'informations recueillies le long des lignes que constituent les forages. C'est le problème de la caractérisation globale des discontinuités affectant un massif.

Si quelques résultats ont été obtenus dans le cadre du premier problème, le second reste quasiment vierge.

3.1. La caractérisation locale

Deere (1964) a défini un indice de qualité (R.Q.D.) destiné à fournir une caractérisation qualitative de la densité de fracturation affectant un certain volume : le R.Q.D. (Rock Quality Designation) est défini comme la proportion de longueurs de carottes intactes dépassant 10 cm par longueur totale carottée :

$$RQD = 100 \sum_{i=1}^{n} x_i \, / \, L \qquad (15)$$

où x_i est la longueur du ième morceau de carotte de longueur supérieure ou égale à 10 cm ;
n est le nombre total de morceaux de longueur supérieure ou égale à 10 cm ;
L est la longueur totale carottée.

Il suggère d'élargir cette définition aux mesures réalisables le long de lignes droites tracées sur un affleurement.
La Société Internationale de Mécanique des Roches a fourni un rapport détaillé sur les méthodes qui permettent d'obtenir une description quantitative des discontinuités affectant une masse rocheuse (Barton 1978). Ce rapport mentionne essentiellement les techniques de mesure directe d'un certain nombre de paramètres tels que l'espacement, la continuité, la rugosité la résistance des épontes, l'ouverture, le remplissage, les conditions d'écoulement, le nombre de directions privilégiées, la taille moyenne des blocs.
Hagan (1980) propose une méthode photogrammétrique pour cartographier l'orientation des fractures. Il souligne que cette méthode , qui présente l'avantage de diminuer considéblement le temps d'observation sur le terrain, fournit une méthode d'échantillonnage qui recouvre des aires plus au moins constantes ce qui limite le biais souvent obtenu lors de la récolte de l'observation sur le terrain. Des photos stéréographiques sont prises avec deux caméras normales ; un ordinateur est ensuite utilisé pour calculer les coordonnées d'un certain nombre de points situés sur la zone de recouvrement des deux photos de sorte que les directions et pendages des linéaments observés peuvent être obtenus par moindres carrés. Ces données peuvent être échantillonnées le long soit de segments de droite soit de surfaces d'aire donnée.
Thomas (1981) définit la densité directionnelle de fracture par :

$$\delta_v = \frac{\Sigma S_{\theta_s}}{V} \; ; \; \delta_s = \frac{\Sigma S_{\theta_\ell}}{A}$$

où ΣS_{θ_s} est la somme des aires des fractures comprise dans l'angle solide θ_s ;
ΣS_{θ_ℓ} est la somme des longueurs de fracture comprises dans l'angle θ_ℓ . V et A sont respectivement le volume et l'aire échantillonnés. θ et θ_ℓ sont les angles solides et plans de l'échantillonnage.
Pour obtenir une estimation statistique de la densité directionnelle d'un réseau de discontinuités, Thomas propose d'utiliser le nombre d'intersections observées entre un maillage donné et le réseau de discontinuités observable sur la photographie d'un affleurement. Pour un réseau impliquant une seule famille de discontinuités parallèles, il peut être montré que la densité des points d'intersection , $d_i(\theta)$, entre les discontinuités et une ligne droite inclinée de θ par rapport à la direction des discontinuités est une estimation de $\delta_s \sin \theta$.

Cette détermination est d'autant plus précise que la longueur d'échantillonnage est longue. Aussi, plutôt que d'utiliser un seul segment de droite, il propose de considérer une série de segments parallèles.
Plus généralement, pour un réseau de q familles de discontinuités F_j , de densité δ_j et faisant un angle α_j avec un axe de référence (voir fig.8), la densité des points d'intersection obtenue avec un réseau de lignes parallèles faisant un angle θ_i par rapport à l'axe de référence est donné par :

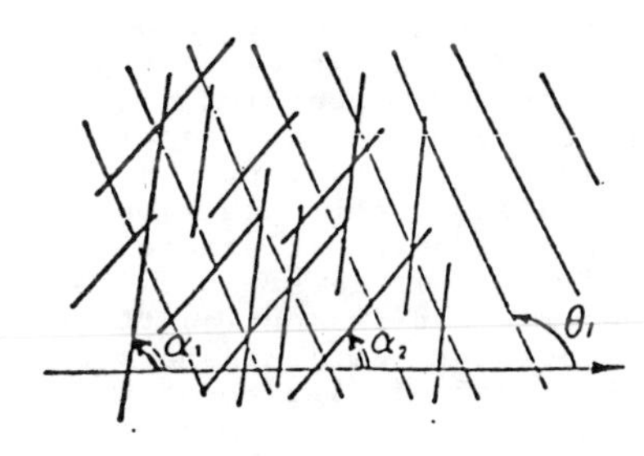

Fig. 8 Analyse de la densité de discontinuités au moyen d'un réseau de lignes parallèles. α_1 et α_2 sont les orientations relatives de deux réseaux de discontinuité par rapport à un axe de référence arbitraire ; θ est l'inclinaison du réseau de lignes parallèles par rapport à l'axe de référence (d'après Thomas 1981).

$$D(\theta_i) = \sum_{j=1}^{q} \delta_j \; |\sin(\theta_i - \alpha_j)| \qquad (16)$$

Le rayon de courbure de la courbe représentative de la fonction $D(\theta)$ est une fonction continue de θ sauf pour les valeurs de θ qui coincident avec les angles α_j (voir fig.9).
Thomas propose une méthode basée sur les séries de Fourier pour déterminer les valeurs α_j .
Cette méthode, qui est directement applicable à un analyseur automatique de texture , a été utilisée dans le cadre d'une étude de

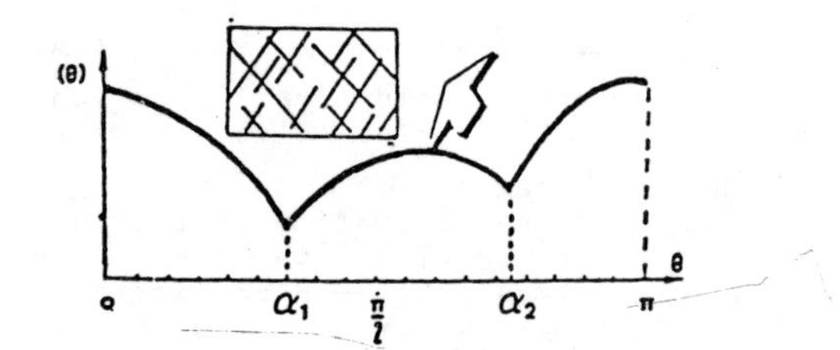

Fig. 9 Forme théorique d'une courbe de densité de fracturation D(θ) pour deux réseaux de discontinuités orientées à α_1 et α_2 par rapport à un axe de référence arbitraire (d'après Thomas 1981).

stabilité entreprise dans une mine de charbon. Des photos de plans subverticaux, dont les azimuts étaient choisis pour effectuer un échantillonnage de la masse rocheuse selon un pas angulaire de 20° environ, ont fourni les informations nécessaires à l'estimation de la densité directionnelle volumique du massif.
Warburton (1980 a,b) propose un modèle probabiliste pour représenter les réseaux de fracturation et plus précisément la distribution des traces des discontinuités telles qu'elles sont observables sur un effleurement. Les discontinuités sont supposées être plates, très fines, de forme circulaire ou rectangulaire et parallèles à une direction donnée.
Leur distribution spatiale caractérisée par celles de leur centre, est supposée obéir à une loi de Poisson. Cette hypothèse est justifiée par l'observation expérimentale.
En effet, les espacements des traces des discontinuités tels qu'ils sont observables le long de lignes d'échantillonnage sont souvent distribués exponentiellement et ceci est précisément ce que l'on doit observer si la distribution des centres des discontinuités obéit à une loi de Poisson. Les dimensions des discontinuités obéissent à une loi de distribution g(x) qui est reliée à la distribution des longueurs de traces des discontinuités observées à l'affleurement. Si elle est connue, cette relation peut être utilisée pour déterminer la distribution de la taille des discontinuités à partir de la mesure de la distribution des longueurs de trace observées.
Dans la pratique, l'échantillonnage des longueurs de trace est à la fois tronqué et censuré. La troncature provient du fait que seules les traces de longueur supérieure à une valeur critique sont échantillonnées. La censure provient de ce que les traces s'étendent souvent en dehors du champ d'observation. Warburton propose des solutions pour le cas d'un échantillonnage linéaire ou surfacique ; la troncature de l'information peut être prise en compte dans le modèle, elle doit cependant rester la plus faible possible du fait de son influence sur la détermination de g(x). Dans le cas d'un échantillonnage surfacique, par exemple, pour lequel aucune troncature n'est effectuée, la solution pour le cas de discontinuités circulaires est donnée par :

$$h(y) = \frac{y}{m} \int_y^\infty \frac{g(x)\,dx}{\sqrt{x^2 - y^2}} \qquad (17)$$

où h(y) est la densité de probabilité qu'une longueur de trace soit égale à y ;
m est la longueur de trace moyenne ;
x est le diamètre des joints circulaires ;
g(x) est la densité de probabilité qu'un joint possède un diamètre égal à x.

Différentes méthodes pour déterminer la fonction g(x) à partir de l'équation (17), lorsque h(y) est connu, sont discutées par Warburton. Il suggère, dans un but pratique, de ne retenir que des solutions analytiques pour g(x) et propose plus précisément une fonction log normale. Priest et Hudson (1976) et Hudson et Priest (1979) ont étudié la possibilité d'utiliser l'échantillonnage par ligne pour obtenir la densité de probabilité de la distribution des longueurs, des aires et des volumes de blocs. Ils ont montré que l'espacement entre les discontinuités tel qu'il est observé le long d'un segment de droite peut être représenté dans la plupart des cas par une distribution exponentielle négative. Ce résultat a pu être établi pour des affleurements de huit types de formation sédimentaires et deux types de roches cristallines. Wallis et King (1980) ont confirmé ce résultat pour le granite. Si f(x) est la densité de probabilité, λ est la fréquence moyenne des discontinuités et x la valeur d'un espacement, alors la proportion de longueur cumulée des espacements (entre les discontinuités) de longueur plus grande qu'une valeur ℓ donnée s'exprime par :

$$L(\ell) = \int_0^\ell \lambda x \, f(x)\,dx \qquad (18)$$

Ainsi, étant donné que le R.Q.D. est défini comme le pourcentage de ligne d'échantillonnage consistant en longueur intacte supérieure ou égale à 0,1 m, il peut être montré que :

$$RQD = 100^{-0.1\lambda}(0.1\lambda + 1) \qquad (19)$$

si la densité de probabilité est représentée par une exponentielle négative ($f(x) = \lambda e^{-\lambda x}$).
Considérant deux ensembles de joints parallèles et continus orthogonaux entre eux, Hudson et Priest ont montré que le RQD, pour une telle géométrie, dépend de l'orientation de la droite d'échantillonnage par rapport à l'orientation des réseaux de joints. Ces résultats ont été généralisés par la suite (Priest et Hudson 1981) aux conditions d'échantillonnage pour lesquels il y a troncature et censure.
Une fois obtenue une représentation statistique des discontinuités d'un massif, se pose le problème de la connectivité de ces discontinuités; élément dont dépend directement la perméabilité du massif. Hudson et Lapointe (1980) proposent d'utiliser une modélisation par circuit électrique imprimé pour une approche bidimensionnelles du problème, tandis que Priest et Sameniego proposent à ce congrès une solution numérique. Toutefois l'applicabilité de ces modèles bidimensionnels aux conditions tridimensionnelles effectivement rencontrées dans la réalité n'est pas discutée par ces auteurs.

3.2. La caractérisation globale

Bertrand et al. (1982) ont essayé de déterminer la perméabilité, et ses variations à l'intérieur d'un massif rocheux, à partir d'études des réseaux de fracturation, de déterminations de l'état de contrainte et de mesures ponctuelles de perméabilité. Bien que ce problème n'ait pas encore été résolu, certains résultats intéressants

vis àvis de l'analyse des réseaux de fracturation naturelle sont présentés.
Dans cette analyse un réseau de fractures est représenté soit par un scalaire (densité d'aire ou de volume), soit par un vecteur (densité de fracturation dans une direction donnée). Le propos de l'étude était d'identifier ces caractéristiques pour un certain nombre limité de sites puis d'utiliser les méthodes géostatiques pour déterminer ces grandeurs dans les domaines non touchés par l'observation directe. Deux types de données ont été considérées :

- les densités surfaciques, ou angulaires, de fracturation déterminées à partir de photos aériennes couvrant chacune environ 1 km^2 ;

- les densités volumiques déterminées à partir d'affleurement de surface et de sites d'observation souterrains, définies à une échelle de 100 à 1000 m^2.

Les résultats ont montré que la géométrie du réseau de fracturation dépend de l'échelle à laquelle elle est étudiée et que l'observation effectuée à une échelle n'est pas simplement transposable à d'autres échelles.

4 - MODELISATION DU COMPORTEMENT MECANIQUE DES MASSIFS ROCHEUX

Le besoin de modéliser le comportement mécanique d'un massif rocheux peut apparaître dans des contextes très variés. Le plus souvent la modélisation doit être quantitative ; elle implique une évaluation tant des contraintes que des déplacements dans l'ensemble du volume considéré, tout au moins en un certain nombre de points privilégiés.
Ces données peuvent être utilisées soit pour une étude de stabilité, soit pour le calcul de larges structures souterraines ou encore pour la mise au point d'un plan de fracturation artificielle ; elles peuvent aussi être recherchées pour permettre une meilleure compréhension de certains phénomènes naturels tels que les mécanismes à la source des séismes ou encore les champs de déformation intraplaque. Parfois la modélisation ne poursuit qu'un but qualitatif pour aider à mettre en évidence les éléments significatifs qui gouvernent un type de déformation. Ce rapport n'a pas pour but de discuter tous les aspects de ce très large thème mais plutôt de focaliser l'attention sur un certain nombre de points particuliers touchant aux déformations quasistatiques des massifs rocheux.
Une étape importante dans ce processus de modélisation est le choix des lois de comportement appropriées pour représenter correctement les caractéristiques des milieux continus équivalents impliqués ainsi que celles des larges discontinuités. Théoriquement, ce choix devrait être dicté seulement par les résultats d'essais de laboratoire et de terrain ainsi que par la connaissance détaillée du réseau de discontinuités. En effet, seules les discontinuités qui sont grandes comparées au volume considéré devraient être traitées en tant que telles, les plus petites étant prises en compte par la définition de lois de comportement ad hoc pour le milieu continu équivalent. Comme illustration de ce principe on peut citer par exemple la modélisation de la dilatance ou de l'anisotropie de résistance des massifs. Dans la pratique ce choix était dicté jusqu'à un passé récent par les techniques mathématiques disponibles pour résoudre les équations aux dérivées partielles définies dans l'introduction de ce rapport. La nécessité de recourir à des solutions analytiques ou à des modèles d'élasticité linéaire bidimensionnels, amenait à effectuer des simplifications excessives rendant finalement suspecte la solution retenue. Actuellement le développement de l'informatique permet la mise au point de solutions numériques pour résoudre les problèmes impliquant des lois de comportement non-linéaire ou des géométries tridimensionnelles. Il est indispensable de garder à l'esprit néanmoins que la précision de la solution obtenue avec de tels modèles dépend pour beaucoup de celle des données qui sont utilisées. Il peut ainsi être tout à fait irréaliste de faire appel à des modèles numériques très sophistiqués si les paramètres utilisés comme données de bases ne sont pas accessibles à l'observation directe ou du moins à une évaluation précise. En ce qui concerne les problèmes d'ingéniérie, des essais simples, menés in-situ à l'échelle 1 devraient être utilisées normalement pour calibrer le modèle.
Dans la pratique de tels essais sont souvent trop coûteux ; des mesures in-situ effectuées durant les premières phases des travaux permettent alors d'ajuster le modèle retenu à la réalité du terrain. Notons que l'instrumentation à long terme est toujours recommandée pour que les prévisions effectuées avec le modèle puissent être comparées a posteriori avec le résultat réel ; de telles comparaisons permettent toujours d'améliorer le modèle.
Pour la modélisation de phénomènes géologiques, les éléments utilisés pour vérifier la validité d'un modèle sont souvent plus subjectifs. Généralement le modèle est considéré comme satisfaisant lorsque ses résultats sont comparables aux observations géologiques de surface. Cependant des données quantitatives de plus en plus nombreuses sur le mouvement des plaques, leur vitesse de déformation, leur champ de contraintes, sont maintenant disponibles en sorte que des validations quantitatives deviennent progressivement possibles.

4.1. Lois de comportement

Dans la plupart des problèmes de l'ingénieur, la température reste inférieure à 200°C et la profondeur est généralement plus petite que 5000 m même si des températures plus élevées peuvent être rencontrées dans les champs géothermiques(350° sur le site expérimental de Fenton Hill dans le Nouveau Mexique) et des profondeurs plus grandes sont atteintes dans les champs pétroliers (9000 m dans le Texas). Ces valeurs sont évidemment largement dépassées lorsqu'il s'agit d'étudier les déformations de la croûte terrestre, problème où la profondeur peut atteindre 50 à 60 km et la température devenir supérieure à 1000° C. Ajoutons de plus, que si l'échelle de temps rencontrée dans les problèmes quasi-statiques de l'ingénieur peuvent varier de la demi-journée à la centaine d'années (le tunnel du Simplon, en Suisse, est plus que centenaire) celle des problèmes géologiques se définit généralement en millions d'années. Néanmoins un nombre relativement limité de modèles permet d'analyser le comportement mécanique des roches pour le vaste spectre de température, contrainte et vitesse de déformation décrit ci-dessus. Nous en présentons quelques uns maintenant.

Les modèles élastiques

L'élasticité linéaire des milieux isotropes, ou à isotropie planaire, constitue un des modèles les plus utilisés. En effet dans de nombreux problèmes, la plus grande part du volume considéré est soumise à un état de contrainte suffisamment faible pour que les hypothèses de l'élasticité soient vérifiées. De plus, de par sa simplicité, ce modèle fournit le moyen d'identifier rapidement les zones où des difficultés sont à attendre : zones de forte concentration de contrainte ou zones en traction. Ces difficultés peuvent être analysées dans une deuxième étape avec des modèles plus raffinés. Les modèles d'élasticité homogène ne permettent pas de prendre en compte les grandes discontinuités ; celles-ci doivent être modélisées indépendemment ainsi que cela sera discuté dans une section spéciale de ce rapport. Par contre les petites discontinuités sont généralement assimilées au milieu continu équivalent. Elles ne supportent généralement aucune résistance à la rupture en traction ; aussi Zinckewicz et al. (1968) proposent-ils de considérer que le matériau continu équivalent ne peut pas supporter de contraintes de traction. Ce dernier est alors traité comme un matériau bilinéaire : raideur nulle, ou infiniment petite, dans le domaine des tractions, raideur normale dans les milieux en compression. Dans les champs de compression, du fait que les joints sont toujours beaucoup moins raides que la matrice rocheuse, les modules du milieu équivalent sont définis en fonction de l'orientation des réseaux de discontinuités, de l'espacement des joints à l'intérieur d'un réseau et de la raideur des joints. Heuzé et al. (1982) par exemple proposent de représenter un granite dans lequel on observe trois réseaux de joints orthogonaux ayant un espacement entre les joints à peu près uniforme et identique pour les trois réseaux, par un milieu isotrope équivalent de module d'Young

$$E_g^{-1} = E_1^{-1} + (Sk_n)^{-1} \qquad (21)$$

où E_g est le module d'Young de la masse rocheuse,

E_1 est le module d'Young du granite intact,

S est l'espacement entre les joints,

k_n est la raideur normale des joints

Gerrard (1982) propose une étude détaillée de l'influence de un, deux, ou trois réseaux de joints orthogonaux. Les joints sont supposés continus, d'espacement régulier et de caractéristiques élastiques du type orthorhombique (9 constantes élastiques). De plus l'analyse suppose que la normale aux trois plans de symétrie du matériau sont respectivement parallèles aux trois axes du système de référence.
Le matériel équivalent est alors trouvé être également orthorombique. Brace et al. (1968) ont été les premiers à décrire le phénomène de dilatance c'est-à-dire l'augmentation de volume observée lorsque la limite élastique de la roche, en conditions de compression, est dépassée mais avant que le maximum de résistance ne soit atteint. Brace et Martin (1968) discutent de l'influence de cette dilatance sur le comportement mécanique des roches saturées par un fluide et montre l'effet stabilisateur de la dilatance. Stuart et Dietrich (1972) ont proposé de modéliser cette dilatance pour un comportement élastique isotrope non linéaire de façon à rendre compte des variations de volume associées à un cisaillement :

$$\underset{\sim}{\varepsilon} = \phi_o \underset{\sim}{1} + \phi_1 \underset{\sim}{\sigma} + \phi_2 \underset{\sim}{\sigma}^2 \qquad (22)$$

où ϕ_o, ϕ_1 et ϕ_2 sont des fonctions polynomiales des premiers invariants de la contrainte $\underset{\sim}{\sigma}$ ($\underset{\sim}{\varepsilon}$ est le tenseur de déformation linéaire). Cependant à notre connaissance, aucune application pratique de ce modèle n'a encore été proposée.

Modèles élasto-plastiques

Le développement de presses très rigides, ou servo-asservies (Cook et Hojem 1966, Wawersik 1968, Rummel et Fairhurst 1971, Hudson et al. 1972) a permis de mener des études de laboratoire détaillées sur le comportement à la rupture d'échantillons rocheux. Wawersik et Fairhurst (1970), Crouch (1970), Cornet et Fairhurst (1972), Rummel (1974) et beaucoup d'autres ont ainsi pu étudier les variations de dilatance et plus généralement les relations efforts-déformations, dans le domaine des déformations atteintes après le pic de la courbe effort-déformation et ce pour divers types de roche et pour différentes conditions de pression de confinement et de pression de pore. Des résultats typiques sont indiqués sur les figures 10 à 13.

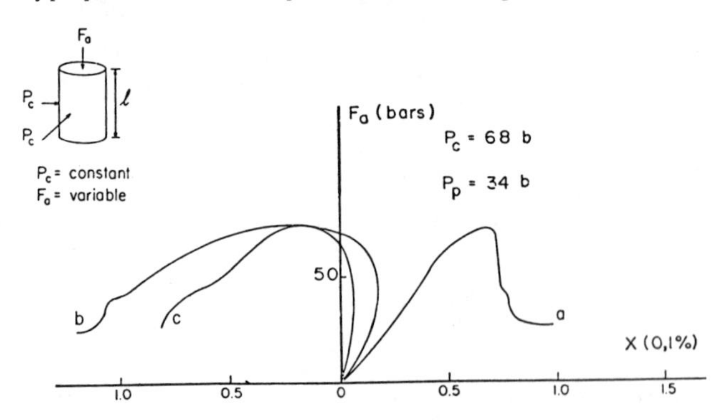

Fig.10 Résultats d'essais triaxiaux sur les échantillons de grès de Berea ; pour la courbe a, l'axe x représente la déformation axiale tandis qu'il représente la déformation volumique pour la courbe b et la variation de volume de pore pour la courbe c : la pression de confinement effective était de 3.4 MPa.

Ainsi que l'indiquent les figures 10 et 12, une caractéristique importante des roches est le fait qu'elles peuvent supporter des charges de compression après que le pic de la relation effort-déformation a été atteint. Pour certaines conditions de contrainte, elles présentent un comportement de type élasto-plastique avec adoucissement et dilatance. Pour des pressions de confinement élevées la dilatance disparait (voir fig. 13). Cependant le comportement illustré sur la figure 11 est moins connu. En effet, après qu'un certain niveau de déformation dans le domaine "post pic" a été atteint , l'échantillon se rompt de façon instable, appelée de classe II par Wawersik (1968) : l'échantillon doit être déchargé progressivement pour que le développement de la rupture reste quasistatique.

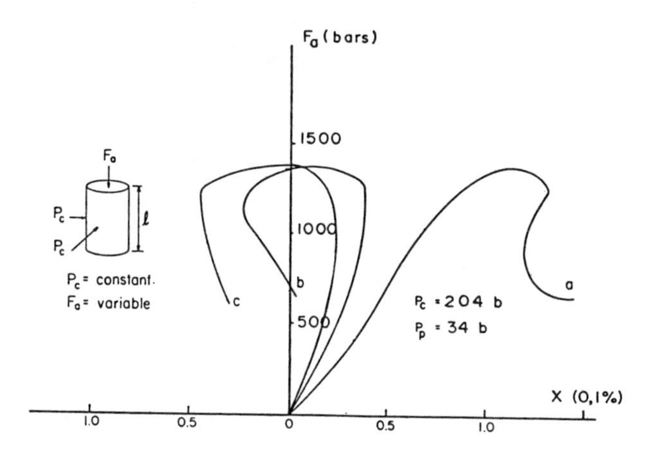

Fig.11 idem que pour la figure 10 mais la pression de confinement effective est de 170 MPa.

Ce comportement correspond au cas où l'énergie de déformation élastique, en supposant les plateaux de la presse parfaitement fixes (raideur infinie), est supérieure à la quantité d'énergie de surface impliquée par la propagation de la rupture. Cette situation donne naissance à une seule surface de cisaillement c'est-à-dire à une localisation du développement de la rupture. Ce comportement a été observé pour le grès de Berea pour des pressions de confinement comprises entre 7 MPa et 24 MPa c'est-à-dire pour des valeurs assez caractéristiques des conditions rencontrées par l'ingénieur (ces valeurs correspondent à des profondeurs variant de 500 à 2000 m). Des résultats similaires ont été observés avec des calcaires, du charbon ou du granite pour différents domaines de pression de confinement. Dans aucun des articles présentés à ce congrès ce point n'a été remarqué .

Les conséquences de ces variations de dilatance sur le comportement " post-pic " d'échantillons saturés par un liquide sous pression sont discutées par Cornet (1977). Il a montré par exemple que le domaine de pression de confinement pour lequel un comportement de classe II est observé est plus grand dans le cas de roches saturées que dans le cas de roches sèches ; ceci du fait des conditions de drainage partiel imposées par la vitesse de déformation.

Jia-Shou Zhuc and Yin-Tang Wang (1983, ce congrès) proposent trois types de critères de rupture pour caractériser la résistance maximum d'un échantillon pour une pression de confinement donnée :

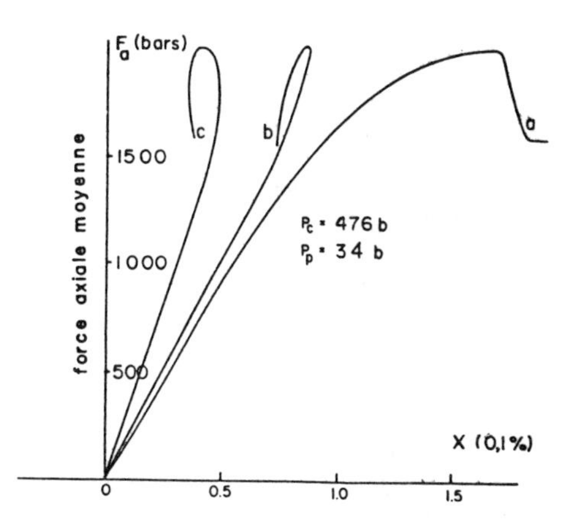

Fig. 12 idem que pour la figure 10 mais la pression de confinement effective est de 44.2 MPa.

$$
(23)\begin{cases}
\text{fracture de traction pure } \sigma_1-\sigma^t=0 \text{ quand } 3\,\sigma_1+\sigma_3<0 \\
\text{fracture de traction en torsion } (\sigma_1-\sigma_3)^2+8\,\sigma^t(\sigma_1+\sigma_3)=0 \\
\qquad \text{quand } 3\,\sigma_1+\sigma_3>0\ ;\ (\sigma_1+3-2\sqrt{2})\sigma_3<0 \\
\text{fractures de compression en torsion :} \\
\qquad (\sqrt{1+f^2}+f)\ \sigma_1-(\sqrt{1+f^2}-f)\sigma_3-4\,\sigma_t=0 \\
\text{quand } \quad \sigma_1+(\sqrt{1+f^2}-f)^2\,\sigma_3 \geqslant 0
\end{cases}
$$

les contraintes de compression sont supposées positives, f est le coefficient de frottement de la roche et σ^t sa résistance à la traction. Appliquant alors le théorème de la plasticité incrémentale, ces auteurs obtiennent :

$$\{d\sigma\} = ([D]-(1-r)[D_p])\ \{d\varepsilon\} \qquad (24)$$

où r = 1 dans le domaine élastique ou lors du déchargement ;
r = 0 dans le domaine plastique ;
0 < r < 1 dans le régime transitoire ;
D est la matrice de raideur élastique :

$$[D_p] = [D]\left\{\frac{\partial F}{\partial\{\sigma\}}\right\}\left(A + \left\{\frac{\partial F}{\partial\{\sigma\}}\right\}^T [D]\left\{\frac{\partial F}{\partial\{\sigma\}}\right\}^{-1}\right)\left\{\frac{\partial F}{\partial\{\sigma\}}\right\}^T [D] \qquad (25)$$

F est la fonction d'écoulement ; A est une constante qui représente l'effet d'écrouissage. Pierce et Ryder (1983, ce congrès) préfèrent une paramétrisation directe des résultats de laboratoire pour modéliser le comportement mécanique décrit ci-dessus. Ils idéalisent les courbes efforts-déformations en supposant que celles-ci sont indépendantes du chemin de chargement et du temps. Les courbes correspondent à un comportement élastique non dilatant jusqu'au pic lequel est supposé être caractérisé par une loi linéaire :

$\hat{\sigma}_1 = \sigma_c + k\sigma_3$ où $\hat{\sigma}_1$ est le pic de la contrainte, σ_c est la chute de contrainte maximum observée après le pic ; elle est supposée indépendante de σ_3 . Le comportement " post-pic " est supposé

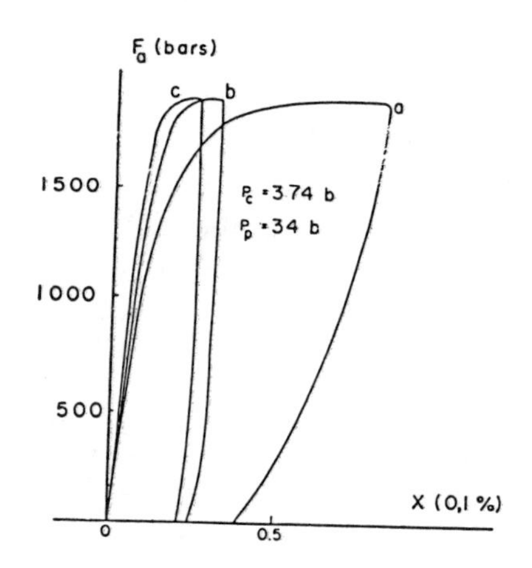

fig.13 idem que pour la figure 10 mais la roche est un calcaire de l'Indiana et la pression de confinement effective est égale à 34 MPa.

suivre une loi linéaire à module négatif constant puis, lorsque le cisaillement atteint sa valeur résiduelle, la déformation se produit sans variation de volume.
Alors que pour ces deux modèles le matériau est supposé isotrope vis à vis de son comportement à la rupture, Guenot et Panet (1983, ce congrès) proposent de considérer deux critères de Mohr-Coulomb différents pour les roches qui possèdent une forte anisotropie planaire. Le premier ($|\tau| = c + \sigma_n$ tg) est défini pour toutes les surfaces de ruptures à l'exception de celles qui sont parallèles aux plans de faiblesse de la roche (plans d'anisotropie de résistance) ; pour ces derniers un second critère est utilisé : $|\tau| = \bar{c} + \sigma_n \operatorname{tg} \bar{\varphi}$ où $\bar{c} \ll c$.
La limitation principale de tous ces modèles provient du fait qu'ils ne tiennent pas compte du rôle du temps dans la déformation "post-pic". En outre, il n'est pas évident que les critères de rupture retenus soient effectivement vérifiés par l'expérience.

<u>Modèles dépendant du temps pour les basses températures et les faibles pressions de confinement.</u>

L'effet de la vitesse de déformation sur les courbes efforts-déformations à basse température (disons plus petite que 300°) et faibles pressions de confinement (inférieures à 500 MPa) peut avoir deux origines : des mouvements de dislocation à l'intérieur des minéraux qui constituent la matrice rocheuse et une microfissuration stable. Le premier mécanisme se rencontre dans le sel gemme par exemple, même sous contrainte uniaxiale et à la température ambiante, ou encore dans la calcite ou bien la serpentine dès que la contrainte principale minimum atteint quelques dizaines de mégapascals. Ce processus de déformation est souvent étudié au laboratoire au moyen d'essais de fluage, c'est-à-dire d'essais où la charge est maintenue constante dans le temps, ou encore d'essais à vitesse de chargement constante (Langer 1979, Wawersik 1980, Vouille et al. 1981). Selon les conditions de chargement, la vitesse de fluage peut diminuer du fait de l'écrouissage du matériau, peut rester constante ou peut augmenter jusqu'à l'apparition d'une instabilité entraînant la ruine de l'éprouvette. Ces différents types de fluages sont appelés fluage primaire, secondaire, ou tertiaire. Andrade (1910) a proposé une équation générale pour représenter le fluage :

$$\varepsilon = A + B.E(t) + Ct \tag{26}$$

où A, B et C sont des constantes tandis que E(t) est une fonction empirique. Morlier (1966) et Parson et Hedley (1966) ont montré que E(t) devrait être logarithmique de même que Langer (1979) pour le fluage primaire. Pour le fluage secondaire, Langer suggère que :

$$\varepsilon = A \exp(- u/kT).f(\sigma) \tag{27}$$

où T est la température, σ est la contrainte uniaxiale appliquée, u est l'énergie d'activation, k est la constante de Boltzman et A est une constante caractéristique du matériau. Vouille et al. (1981) ont trouvé qu'après que la limite élastique a été atteinte, la déformation visco-plastique observée dans le cas du sel de Tersanne, lors d'un essai de fluage, est donnée par :

$$\varepsilon_{vp} = \frac{(\sigma_1 - \sigma_3 - 2\ C)^{\beta}}{k}\ t^{\alpha} \tag{28}$$

où ε_{vp} est la déformation visco-plastique,
σ_1 est la contrainte principale maximum
σ_3 est la contrainte principale minimum,
c est la cohésion,
k, β, α, sont des constantes déterminées expérimentalement,
t est le temps.

Ils proposent comme loi de comportement pour le matériau :

$$< \sigma_1 - \sigma_3 - 2c > = \frac{K}{(\alpha)^{\alpha/B}} (\varepsilon_{vp})^{(1-\alpha)/\beta} (\varepsilon_{vp})^{\alpha/\beta} \tag{29}$$

où $< x > = \frac{1}{2} (x + |x|)$

De plus ils observent que l'angle de frottement interne est nul pour cette roche tandis que la cohésion apparente est inférieure à 1.5 MPa ; cette dernière valeur décroît lorsque la température et la pression de confinement augmentent.
Quel que soit le matériau considéré, durant le fluage tertiaire des microfissures apparaissent progressivement et induisent finalement la rupture complète de l'éprouvette (Scholz 1968, Kranz et Scholz 1977). L'influence de la vitesse de déformation sur cette microfissuration stable devient très importante lorsque le domaine " post-pic " est atteint.
Bieniawski (1970), Peng et Podmieks (1972), Hudson et Brown (1973), Houpert (1974) ont étudié expérimentalement au laboratoire cette influence de la vitesse de déformation. Les résultats (fig. 14) montrent qu'une diminution de cette dernière induit une diminution du pic et un adoucissement du comportement en cours de rupture ; ce phénomène est d'autant plus marqué que l'échantillon est hétérogène. Kaiser et Morgenstern (1981) ont proposé un modèle rhéologique simple pour expliquer phénoménologiquement cet effet. Ils considèrent un modèle élémentaire (voir fig. 15) qui consiste en une composante élastique A pour laquelle, à la rupture qui intervient pour une déformation

E_A ,la perte de résistance C_A est instantanée et la résistance résiduelle est égale à F_A, et, en parallèle, une composante B dont le comportement dépend du temps.

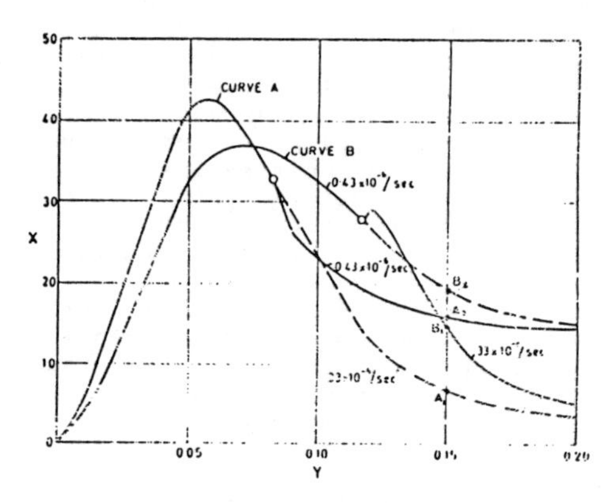

Fig. 14 Influence d'un changement de la vitesse de déformation sur le comportement à la rupture d'un échantillon (d'après Bieniawski 1970)

Cette composante B présente, à la rupture une perte de résistance instantanée C_B et une résistance résiduelle F_B. La composante B peut se rompre avant ou après la composante A. Kaiser et Morgenstern proposent de modéliser la dépendance de la rupture en fonction du temps d'un massif en mettant de nombreux modèles élémentaires de ce type en parallèle.

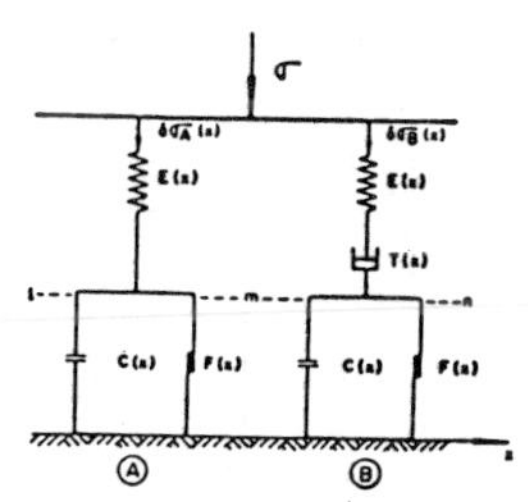

Fig.15 Modèle élémentaire pour représenter l'influence de la vitesse de déformation sur le comportement "post-pic" d'une roche d'après Kaiser et Morgenstern (1981).

Tan et Kang (1983, ce congrès) considèrent que la déformation totale d'un matériau dilatant peut être vue comme la superposition de trois composantes :

$$\varepsilon(t)_{ij} = \varepsilon^E_{ij} + \varepsilon^D_{ij} + \varepsilon^D(t)_{ij} \quad (30)$$

où ε^E_{ij} est la déformation élastique, ε^D_{ij} est la déformation dilatante instantanée et $\varepsilon^D(t)_{ij}$ la déformation dilatante qui dépend du temps. L'amorce de la dilatance est supposée apparaître lorsque $\sigma_{oct} > f_3$ où f_3 est une constante dépendant du matériau. Cette formulation est appliquée à l'analyse des soulèvements régionaux observés avant certains gros séismes de Chine. Dragon et Mroz (1979) proposent de relier le comportement en fluage observé au laboratoire durant des essais de compression simple à une densité de microfissuration (β) qui dépend du temps :

$$\dot{\beta} = k_1 \dot{\varepsilon}^{(i)}_{kk} + k_2(\beta)\, g(\sigma_{k\ell},\beta) \quad (31)$$

où $\dot{\varepsilon}^{(i)}_{kk}$ représente la vitesse de déformation volumique inélastique ;

k_1 est une constante caractéristique du matériau ;

$k_2(\beta)$ et $g(\sigma_{k\ell},\beta)$, $k,\ell = 1,2,3$ sont des fonctions décrivant le développement de la fracturation dans le temps.

La vitesse totale est supposée résulter de la somme d'une vitesse de déformation élastique et d'une autre de déformation inélastique (ceci dans le cas où la charge varie dans le temps). La vitesse de déformation inélastique est supposée apparaître lorsque :

$$f(\sigma_{ij},k,\beta) = S_{ij}S_{ij} - 2\,p(k,\beta)(I^\circ_1 - \sigma_{kk}) = 0 \quad (32)$$

où $p(k,\beta) = p_o + k - \beta^2$,

$\dot{k} = \sigma_{mm}\,\dot{\varepsilon}_{nn}$ = vitesse d'écrouissage ,

S_{ij} composante ij du déviateur de l'état de contrainte,

I°_1 = constante positive,

p_o détermine la configuration des conditions de rupture pour $k = \beta = 0$.

Martin (1972), Henry et al. (1977), Atkinson (1979) et d'autres chercheurs ont souligné l'influence de la corrosion sous tension sur le fluage tertiaire ; c'est-à-dire l'influence d'éléments chimiquement actifs sur la croissance des microfissures (le plus souvent il s'agit d'eau). L'effet de la corrosion sous tension a été invoqué par Das et Scholz (1981) pour expliquer certains phénomènes associés aux séismes tels que évènements multiples, évènements multiples retardés, précurseurs, répliques ou ruptures stables post-sismiques. Leur raisonnement suppose que les séismes correspondent à des ruptures de modes II et III combinés, c'est-à-dire des ruptures en cisaillement pur, et que la fracture reste plane lors de sa propagation. Cette supposition est étayée par les observations de terrain qui montrent que généralement les séismes affectent des failles préexistantes. Selon le critère d'Irwin (Irwin 1957) une fracture se propage lorsque le facteur d'intensité de contrainte k qui caractérise le champ de contrainte au voisinage du front de la fracture atteint la valeur critique k^*_c considérée comme une grandeur caractéristique du matériau. Pour le mode I (rupture en traction) il est souvent observé que la propagation de la fracture intervient de façon stable pour des facteurs d'intensité de contrainte bien inférieurs à K^*_c . Ce mécanisme est attribué à un effet de corrosion qui abaisse à une valeur k_o la valeur du facteur d'intensité de contrainte critique pour laquelle il y a propagation ($k_o \leqslant k^*_c$). Tant que la propagation de la fracture est associée à cet effet de corrosion elle reste quasistatique c'est-à-dire sans libération d'énergie cinétique, la vitesse de propagation étant proportionnelle à k^n (Atkinson 1979) :

$$\dot{X} \propto k^n \quad (33)$$

où X est la longueur de la fracture et n une constante appelée indice de corrosion sous tension.

Das et Scholz supposent que des lois similaires existent pour les modes II et III. Ainsi, du fait que k, le facteur d'intensité de contrainte, est proportionnel à la racine carrée de la longueur de la fissure, il apparaît que l'effet de la corrosion sous tension amène progressivement la fracture à des conditions de propagation instable. Ils considèrent que l'amorce de cette instabilité apparaît lorsque le facteur d'intensité de contrainte atteint une valeur critique k_c proche ou égale à la valeur de k_c^* car, en conditions dynamiques, la corrosion n'a plus le temps de se développer.
Le facteur d'intensité de contrainte k pour une fracture plane circulaire de rayon X, soumise à un cisaillement pur et affectant un milieu infini homogène, linéairement élastique, est donné pour les modes II et III par :

$$k = C \, \Delta\tau \sqrt{X} \qquad (34)$$

où C est une constante qui ne dépend que de la géométrie et $\Delta\tau$ est la chute quasistatique de contrainte de cisaillement à travers une fissure. Etant donné que pour une fracture circulaire les modes II et III existent simultanément le long de son contour, k est pris égal à la moyenne quadratique des deux facteurs d'intensité de contrainte ; $C = \frac{20\sqrt{2}}{7\pi}$ d'après Sih (1973) pour cette géométrie.
Les résultats expérimentaux d'Atkinson (1979) suggèrent que la propagation stable intervient lorsque :

$$k = k_o \left(\frac{\dot{X}}{v_o} \right)^{1/n} \qquad (35)$$

où k_o et v_o sont les valeurs de k et $\dot{X}$ lors de l'amorce quasistatique de la croissance. Si l'on suppose de plus que $\Delta\tau$ est indépendant du temps, étant donné que $k_o = C \, \Delta\tau \sqrt{X_o}$, alors :

$$\dot{X} = v_o \left| \left(\frac{X}{X_o}\right)^{1/2} \right|^n \qquad (36)$$

si bien que

$$X = \left| X_o^{\frac{2-n}{2}} - \frac{n-2}{2} \, \frac{V_o t}{X_o^{n/2}} \right|^{2/(2-n)} \qquad (37)$$

de sorte que la durée de la propagation stable est donnée par :

$$t_f = \frac{X_o}{V_o} \, \frac{2}{n-2} \qquad (38)$$

on obtient l'équation (36) en faisant $\frac{\dot{X}}{\dot{V}_o} = 0$, condition qui revient à supposer X infini étant donné que $n > 2$. La valeur de t_f dépend donc des conditions initiales et de l'indice de corrosion sous tension. Das et Scholz ont appliqué ce résultat pour évaluer quelques ordres de grandeur. Ils trouvent par exemple que la période de rupture stable avant séisme peut varier de 48 heures , pour une longueur de fracture initiale de 1 km et un indice de corrosion sous tension égal à 12.5 , à 278 jours pour une longueur de fracture initiales de 0.5 cm et le même indice de corrosion sous tension ; $\Delta\tau$ était supposé égal à 10 MPa dans les deux cas. La longueur de fracture atteinte avant l'amorce de l'instabilité était de 1.4 km dans le premier cas et 14.2 cm dans le second. Ces ordres de grandeur semblent être en accord avec des enregistrements d'extensométrie observés avant un séisme de magnitude 3.8.

Modélisation du comportement mécanique des grandes discontinuités.

Le comportement mécanique des masses rocheuses est généralement affecté de façon significative par celui des grandes discontinuités : Elles ne supportent aucune résistance à la traction, leur résistance au cisaillement est très variable et leur déformation en cisaillement est généralement associéeà des variations de volume. Ainsi que cela a été déjà dit seules les discontinuités qui sont grandes par rapport aux dimensions du volume considéré sont traitées en tant que telles. Les petites discontinuités sont intégrées dans le milieu continu équivalent. L'influence des grandes discontinuités sur le comportement élastique du milieu est généralement modélisé au moyen de matrices de raideur spéciales (voir par exemple Goodman et al. 1968, Heuzé et al. 1971, Gerrard 1982, Walsh et Grosenbaugh, 1979) qui mettent en relation les efforts normaux et de cisaillements supportés par ces plans et les déplacements normaux et tangentiels induits le long de ces plans. Desai et al. (1983, ce congrès) proposent eux de modéliser un joint par une couche très fine de matrice de raideur appropriée. Rencontrant certaines difficultés pour mesurer les termes non-diagonaux de cette matrice de raideur, ces auteurs préfèrent attribuer des valeurs nulles à ces coefficients.
Barton (1976) a souligné le fait que le comportement mécanique des joints dépend fortement des efforts normaux auxquels ils sont soumis. Pour les faibles contraintes normales, la résistance au cisaillement est très variable du fait de l'influence de la rugosité des surfaces. (voir fig.16)

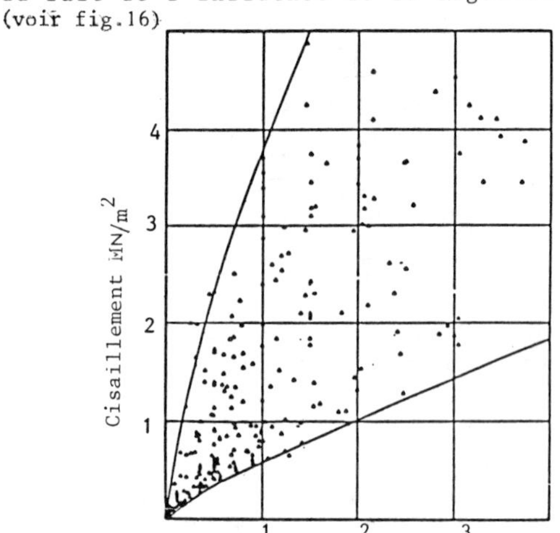

Fig.16 Pic de la résistance au cisaillement de joints rocheux non remplis (d'après les résultats d'essais de cisaillement direct mesurés au laboratoire et in-situ ; Barton 1976).

Pour des contraintes effectives élevées, le spectre des résistances au cisaillement des discontinuités est beaucoup plus étroit (voir fig. 17) malgré les grands écarts de résistance au cisaillement de la roche saine ; il est bien représenté par le critère de Coulomb (Byerlee 1975) :

$$\tau = C + \mu(\sigma_n - P) \qquad (39)$$

où τ est la résistance au cisaillement, σ_n la

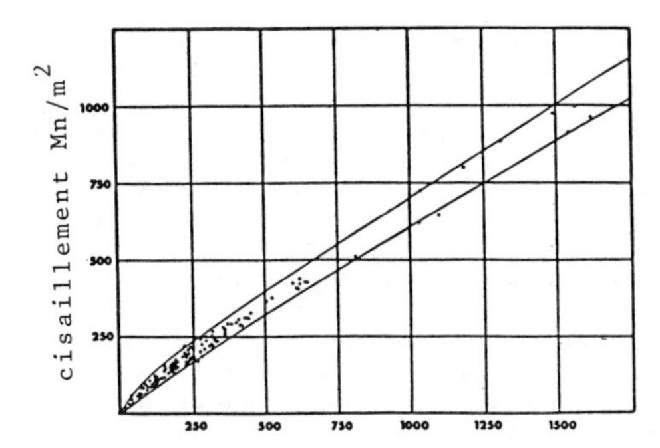

Fig.17 Résistance au cisaillement de failles et de fractures de traction pour des roches soumises à des contraintes normales effectives élevées (d'après Byerlee 1967, 1968, 1975). Les résultats correspondent à des essais sur le granite de Westerly, les calcaires de Solenhofen et de Oak Hale, le gabbro de Mahant, la dunite de Spruce pine, la serpentinite de Cabramurra et les grès de Weber (figure d'après Barton 1976).

contrainte normale, P la pression interstitielle, μ le coefficient de frottement et C la cohésion; cette dernière valeur est généralement négligeable pour les valeurs élevées de σ_n. Pour des contraintes effectives normales très élevées, les joints naturels n'influencent plus le comportement au cisaillement des masses rocheuses.
Des travaux ont été effectués pour essayer d'obtenir une représentation quantitative des variations de résistance au cisaillement en fonction de la contrainte effective normale ainsi que des variations de volume observées durant la déformation en cisaillement. Barton (1973, 1976) a réalisé des essais de cisaillement direct sur un grand nombre de fractures de traction créées artificiellement dans un matériau modèle de comportement élastique et fragile. Il a observé que si μ est l'angle de frottement total du matériau (μ = arctg τ/σ_n où τ est le pic de la résistance au cisaillement et σ_n la contrainte effective normale) et d_n l'angle de dilatance au pic du cisaillement (défini comme l'inclinaison instantanée du déplacement en cisaillement par rapport au plan moyen du joint), alors :

$$\mu = 2\, d_n + 30 \qquad (40\ a)$$

$$d_n = 10 \log_{10}\left(\frac{\sigma_c}{\sigma_n}\right) \qquad (40\ b)$$

où σ_c est la résistance en compression uniaxiale du matériau intacte. La valeur 30 dans l'équation (40 a) correspond à l'angle de frottement interne du matériau. Pour le domaine 0.1 - 10 MPa de contrainte normale considéré dans cette étude expérimentale, l'enveloppe des maxima de résistance au cisaillement est décrite par la relation :

$$\tau = \sigma_n \mathrm{tg} \left| 20 \log_{10}\left(\frac{\sigma_c}{\sigma_n}\right) + 30 \right| \qquad (41)$$

Barton a proposé de généraliser ce résultat aux joints naturels en écrivant :

$$\tau = \sigma_n \mathrm{tg}\left[\mathrm{JRC} \log_{10}\left(\frac{\mathrm{JCS}}{\sigma_n}\right) + \phi_b\right] \qquad (42)$$

où JRC est le coefficient de rugosité du joint défini sur une échelle empirique variant de 20 pour les surfaces très rugueuses à 0 pour les surfaces lisses (voir Barton et Choubey 1977, Barton et Bandis 1980 pour une discussion détaillée de cette échelle empirique) ;

JCS est la résistance en compression simple des épontes du joint ;
ϕ_b est l'angle de frottement interne défini comme la valeur arctg(τ/σ_n) obtenu lorsque la résistance au cisaillement du joint atteint sa valeur résiduelle.
La valeur de JCS peut varier de σ_c (la résistance en compression simple du matériau) à $\sigma_c/4$ pour les surfaces altérées .
Pour les contraintes normales élevées, Barton propose de remplacer la valeur de la résistance en compression simple par la résistance en compression triaxiale de sorte que l'équation (42) devient :

$$\tau = \sigma_n \mathrm{tg}\left[\mathrm{JRC} \log_{10}\left(\frac{\sigma_{TC}-\sigma_3}{\sigma_n}\right) + \phi_b\right] \qquad (43)$$

où σ_{TC} est la résistance à la compression en conditions triaxiales lorsque la pression de confinement est égale à σ_3 .

Les résultats obtenus avec l'équation (43) pour des essais de cisaillement le long de fractures artificielles développées au laboratoire au cours d'essais triaxiaux sur des grès, un calcaire, un gabbro, une dunite , une serpentinite et un granite sont tout à fait en accord avec les résultats expérimentaux pour des pressions de confinement pouvant atteindre 500 MPa et pour une valeur de JRC prise égale à 20.
Tandis que la généralisation de l'équation (41) a mené à l'équation (43) celle de l'équation (37 b) , la prédiction de l'angle de dilatance, est donnée par

$$d_n = 10 \log_{10} \frac{\sigma_{TC}-\sigma_3}{\sigma_n} \qquad (44)$$

pour $\sigma_n > \sigma_c$. Pour les résultats expérimentaux décrits ci-dessus des angles de dilatance de 2° sont observés dans le cas de cisaillement effectués pour des conditions de pression de confinement équivalant à la transition fragile-ductile (on entend par comportement ductile des déformations de cisaillement sans perte de résistance).
Pande et Gerrard (1983, ce congrès) proposent de représenter la résistance au cisaillement des joints par un critère de Coulomb mais avec une cohésion variable, qui dépend de l'ouverture du joint :

$$C_e = C_o \frac{\langle \varepsilon_c - \varepsilon_t \rangle}{c} \qquad (45)$$

où ε_t est la déformation visco-plastique normale au plan du joint ,

C_o est la cohésion du joint quand celui-ci est fermé ;

ε_c est une déformation critique qui dépend de la rugosité du joint :

$$\langle \varepsilon_c - \varepsilon_t \rangle \begin{cases} \varepsilon_c - \varepsilon_t & \text{si } \varepsilon_c - \varepsilon_t > 0 \\ 0 & \text{si } \varepsilon_c - \varepsilon_t \leqslant 0 \end{cases}$$

Bien que ce rapport ne soit pas concerné par les problèmes d'écoulement de liquides, un résultat intéressant obtenu par Walsh (1981) au sujet du concept de contrainte effective tel qu'il doit être appliqué à la perméabilité des joints ou des fractures mérite d'être mentionné. Dans son analyse, Walsh montre que la variation du volume des pores (dv_p) à l'intérieur d'un joint, causée par des variations de contrainte hydrostatique et des variations de pression interstitielle, est donnée par :

$$dv_p = \frac{\partial v_p}{\partial p} (dP_c - dP_p) + V_p \beta_s \, dP_p \qquad (46)$$

$\frac{\partial v_p}{\partial p}$ est le taux de variation de volume de pore causé par une variation de contrainte hydrostatique lorsqu'il n'y a pas de fluide interstitiel ;

dP_c est le changement de contrainte hydrostatique exercée sur la roche ;

dP_p est la variation de pression de pore ;

v_p est le volume originel des pores ;

β_s est la compressibilité de la roche

L'équation (46) permet de montrer que la contrainte effective, Pe , de laquelle dépend la perméabilité du joint, est donnée par :

$$P_e = P_c - sP_p \qquad (47)$$

où $s = 1 - v_p \, \beta_S \, / \, (\partial V_p / \partial P)$

s peut varier de 0.5 à 5 . Or du fait que la perméabilité est reliée à l'ouverture des joints on doit s'attendre à ce que cette ouverture des joints dépende de contraintes effectives similaires à celles définies par l'équation (47) plutôt qu'à celles classiquement utilisées en mécanique des roches ($\underset{\sim}{\sigma} - \underset{\sim}{P}1$).

4.2. Quelques exemples de solution

Dans la plupart des problèmes de mécanique des roches, du fait que la géométrie est assez complexe et du fait que les lois de comportement ne sont pas linéaires au moins en ce qui concerne les grandes discontinuités, la solution ne peut pas être obtenue de façon exacte. Lorsque le matériau peut être considéré comme linéairement élastique et isotrope c'est-à-dire lorsque la complexité du problème provient essentiellement de sa géométrie, la méthode photo-élastique peut être utilisée pour mettre en évidence à faible coût les zones à fortes concentrations de contrainte , que le problème soit bi ou tridimensionnel. Toutefois le développement d'ordinateurs de grande puissance permet maintenant d'obtenir des solutions numériques plus précises et ce, de façon presqu'aussi bon marché que pour les méthodes photo-élastiques de sorte que ces dernières tendent maintenant à être supplantées par les solutions numériques. Cependant pour l'analyse des problèmes impliquant de grandes déformations les modèles physiques restent très utiles ne serait-ce que pour mettre en évidence les traits principaux de ces déformations.
Par exemple Tapponnier et al. (1982) ont mis au point un modèle à base de plasticine pour étudier le poinçonnement du continent asiatique par le sous-continent indien. Ce modèle leur a permis de mettre en évidence certaines caractéristiques cinématiques du processus de déformation tels que, par exemples, l'ouverture de la mer de Chine du sud avant la période tardi-miocène ou bien encore la rotation et l'extrusion de la péninsule indochinoise vers le sud-est le long de la faille de la Rivière Rouge durant les 20 ou 30 premiers millions d'années de la collision. Leurs expériences suggèrent qu'une faille majeure en décrochement senestre accompagnée d'un système de rifts va se propager dans la région du Tien-Shan puis en Mongolie et jusqu'à la mer d'Okhotk. A une plus petite échelle Deramont et al. (1983, ce congrès) ont étudié à l'aide d'un modèle à deux matériaux l'influence d'un cisaillement horizontal profond sur le processus de déformation d'une couverture sédimentaire. La couche inférieure de leur modèle est constituée de plasticine, produit ayant une certaine cohésion, la couche supérieure étant modélisée par un matériau pulvérulant. Ce modèle semble bien mettre en évidence les phénomènes d'écoulement-glissement des nappes.
La difficulté majeure rencontrée avec ces modèles physiques est celle du facteur d'échelle : un modèle physique pour être représentatif d'un problème donné, doit satisfaire aux règles de similitude .
Ce principe de similitude a été étudié dans le détail pour les modèles à frottement basal (Bray et Goodman 1981) pour lesquels l'effet de la gravité agissant dans un plan vertical est modélisé par une force de frottement agissant à la base du modèle. Divers types de matériaux peuvent être utilisés pour modéliser correctement les caractéristiques rhéologiques du massif rocheux. De plus une pression de gaz uniforme appliquée sur la surface du modèle permet de modéliser les effets de confinement (Egger et Gindoz 1979). Ce modèle semble bien adapté à la modélisation des problèmes pour lesquels les effets d'inertie sont négligeables devant les effets de la gravité.
Lessi et Sarda (1980) se sont intéressés à la modélisation de la ténacité des roches (K_{IC}) pour étudier sur modèle physique certains problèmes de fracturation hydraulique. Ils ont montré que si λ est le facteur de similitude de la géométrie, le facteur de similitude de la ténacité est $\sqrt{\lambda}$ ($K_{IC\ \text{modèle}} = \sqrt{\lambda} . K_{IC\ \text{roche}}$) et ils ont développé des matériaux modèles, à base d'un mélange de ciment et de bentonite, qui permettent d'obtenir des facteurs de réduction géométrique de l'ordre de $2 \; 10^{-3}$.
Ces problèmes de similitude n'existent évidemment pas avec les modèles numériques ; de plus ceux-ci se prêtent particulièrement facilement aux études paramétriques qui doivent être menées lorsque les grandeurs considérées ne sont pas directement accessibles à des mesures précises.

Solutions par éléments finis

Les termes " éléments finis " doivent être

compris ici au sens de l'ingénieur : le milieu continu équivalent et les discontinuités majeures sont discrétisées en un nombre fini d'éléments dont le comportement est représenté par un nombre fini de paramètres. Cette définition peut être conciliée avec celle qu'en donnent les mathématiciens en mentionnant que la solution du système complet d'équations obtenues obéit aux mêmes lois que celles qui régissent les problèmes de discrétisations mathématiques classiques. Une présentation détaillée de l'application de cette méthode aux problèmes de mécanique est fournie par Zienckiewicz (1977) ; aucune discussion de cette puissante méthode de résolution ne sera entreprise ici. Nous nous contenterons de rappeler ses avantages principaux ainsi qu'ils ont été énoncés par Jaeger et Cook (1979) :

1. Les éléments peuvent être choisis de n'importe quelle taille et de géométrie variée (bien que l'on ne considère généralement que des éléments triangulaires ou quadrangulaires pour les problèmes bidimensionnels et tétrahédraux ou octahédraux pour les problèmes tridimensionnels) de sorte que n'importe quelle forme de frontière peut être discrétisée , une plus grande densité d'éléments étant généralement requise dans les zones à fort gradient de contrainte.

2. N'importe quel système de forces de surface ou de volume peut être introduit.

3. Chaque élément possède sa propre matrice de raideur de sorte que le milieu peut être hétérogène et anisotrope. Des comportements non linéaires ou bi-linéaires peuvent être considérés.

4. N'importe quelles propriétés de frottement et de dilatance peuvent être introduites pour les discontinuités mécaniques.

5. Les résultats peuvent être présentés, grâce au traitement informatique, sous diverses formes pratiques.

Les présentations faites à ce congrès sur ce sujet sont un bon exemple de la puissance de cette méthode. Pandé et Gerrard par exemple présentent un modèle élasto-visco-plastique bidimensionnel qui inclut une modélisation spécifique de réseaux de joints distribués régulièrement. Ces derniers sont supposés posséder une certaine raideur dans le domaine élastique, ne supporter aucune résistance à la traction et obéir à un critère de Coulomb lorsqu'ils sont soumis à un cisaillement suffisamment élevé, une certaine dilatance étant possible lors de la éformation en cisaillement. Des éléments de renforcement peuvent être simulés ; ces derniers se comportent de façon élastique tout d'abord puis se déforment selon une loi avec écrouissage. Carol et Alonso proposent quant à eux un nouvel élément pour modéliser le comportement mécanique des joints, élément qui ne requiert pas moins de onze paramètres pour modéliser son comportement mécanique complet.
Des applications pratiques sont également présentées. Casteleiro et al. par exemple proposent une analyse tridimensionnelle appliquée à l'étude d'un barrage dans laquelle le matériau n'est supposé supporter aucune résistance à la traction . Susa montre l'avantage qu'il y a à combiner une solution élastique tridimensionnelle et une étude bidimensionnelle non linéaire pour optimiser les séquences de construction d'une grande centrale électrique souterraine de même que pour en étudier sa stabilité à long terme . Quiblier et Ngokwey utilisent quant à eux un modèle visco-élastique tridimensionnel incluant plusieurs matériaux stratifiés pour identifier, dans le domaine des grandes déformations, les zones à haute probabilité de fracturation qui peuvent exister dans un bassin sédimentaire déformé ; ces zones pouvant être les hôtes potentiels de réservoirs d'huile.

Les méthodes par équations intégrales

Les méthodes par équations intégrales se réfèrent aux méthodes numériques pour lesquelles seules la frontière du milieu est discrétisée ; d'un point de vue mathématique ces méthodes forment une classe de méthodes par éléments finis. Elles sont basées sur la superposition de fonctions d'influence connues le long des frontières de façon à produire un système d'équations intégrales qui, résolues simultanément, permettent de relier les tensions et les déplacements sur ces frontières. Ces équations sont résolues en supposant que les tensions, ou les déplacements, ou les deux, sont constants sur de petits éléments de la frontière.

Les déplacements et les contraintes en tout point du milieu peuvent alors être calculés à partir de ces données.
En sus de la simplicité du processus de discrétisation l'intérêt principal des méthodes par équations intégrales par rapport à celles par éléments finis provient essentiellement du fait que les conditions à l'infini sont implicitement prises en considération. Avec la méthode par éléments finis le maillage doit être poursuivi loin de la zone d'intérêt lorsque le problème implique des milieux infinis ; en outre , certaines hypothèses sur les déplacements à l'infini doivent souvent être introduites. Notons de plus que les méthodes par équations intégrales sont particulièrement bien adaptées aux problèmes ayant des frontières évolutives tels que les problèmes de propagation de fracture, ou les problèmes faisant intervenir des comportements spécifiques pour un petit nombre de joints ou autres discontinuités.
Une description détaillée des méthodes par équations intégrales appliquées aux problèmes de l'élasto-statique est fournie par exemple par Jawson et Symm (1977). Une présentation plus détaillée des applications dans le cadre de la mécanique des roches a été proposée récemment par Banerjee et Butterfield (1981) et Crouch et Starfield (1983).
La méthode des discontinuités de déplacement (Starfield et Crouch 1973, Crouch 1976) consiste à placer N discontinuités de déplacement, d'amplitude inconnue, le long de la frontière puis à construire et à résoudre un système d'équations algébriques linéaires de façon à déterminer les valeurs des discontinuités qui produisent les conditions aux frontières prescrites dans le problème. Lorsque la géométrie implique un contour fermé, le problème intérieur et le problème extérieur sont résolus simultanément.

La solution analytique du problème posé par une discontinuité normale, ou tangentielle, du champ des déplacements sur un élément de surface plan est utilisée pour obtenir des coefficients d'influence. Par exemple, pour un problème bidimensionnel dans lequel les conditions aux frontières sont exprimées par les composantes des tensions, le système linéaire ci-après est obtenu :

$$\begin{cases} \sum_{j=1}^{M} A_{NN}(i,j)DN_j + \sum_{j=1}^{M} A_{NT}(i,j)DT_j = TN_i \\ \\ \sum_{j=1}^{M} A_{TN}(i,j)\,DN_j + \sum_{j=1}^{M} A_{TT}(i,j)DT_j = TT_i \end{cases} \qquad (48)$$

où $A_{NN}(i,j)$ est la composante normale de la tension induite au centre du jème segment de la frontière par une discontinuité normale unitaire du champ des déplacements sur le ième segment tandis que A_{NT} réfere à la composante tangentielle de la même tension. Les coefficients A_{TN} et A_{TT} représentent les composantes normales et tangentielles des tensions induites par une discontinuité tangentielle unitaire du champ des déplacements. DN_j et DT_j sont les discontinuités normale et tangentielles inconnues du déplacement sur le jème segment tandis que TN_i et TT_i sont les composantes de la tension connue appliquée au centre du ième segment. Le système (48) de 2N équations linéaires peut être inversé de sorte que, une fois connues les valeurs de DN_j et DT_j celles-ci peuvent être utilisées pour déterminer les contraintes et déplacements en tout point du milieu en résolvant un système d'équations linéaires similaire à (48) : on cherche maintenant les valeurs TN_i et TT_i au point i.
Des équations similaires sont obtenues lorsque les conditions aux frontières sont exprimées en termes de déplacement de sorte que les problèmes ayant des conditions aux frontières mixtes peuvent être résolus de la même manière. Cette méthode a été adaptée aux situations tridimensionnelles (Sinha 1979) ainsi qu'aux problèmes faisant intervenir deux matériaux différents (Crouch 1982, Bouchez 1981).
Crawford et Curran (1982) ont introduit des éléments pour lesquels la discontinuité du champ des déplacements n'est pas constante mais est une fonction du second ordre des coordonnées locales. Cette formulation permet de préciser les tensions en deux ou trois points du segment et non plus seulement en son centre comme dans le modèle de Crouch. Ces éléments d'ordre supérieur fournissent une meilleure précision pour le même nombre d'éléments que la solution à discontinuités constantes ; de plus cette méthode est mieux adaptée aux problèmes faisant intervenir des angles aigus le long de la frontière.
Ces techniques de discontinuités de déplacement sont très utilisées pour étudier les plans d'exploitation de gisements en couches fines que celles-ci soient planes ou non (Crouch 1976) ou pour l'étude de la propagation quasistatique des fractures (Cornet 1979).
Pour les problèmes où il est tenu compte du frottement lorsque des discontinuités tangentielles se produisent sous champ de compression, l'introduction d'un comportement dilatant le long de ces frontières pourrait se traiter de façon tout aussi aisée.
Une approche différente est proposée par Brady et Bray (1978 a et b) et Brady (1979). Ces auteurs observent que généralement l'axe principal des structures étudiées ne coincide pas avec une direction principale de contrainte de sorte que les hypothèses de déformation plane ne sont généralement pas satisfaites. Ils définissent donc des conditions de déformations planes généralisées pour lesquelles les déplacements ne sont fonction que de deux des trois coordonnées spatiales (les coordonnées dans le plan perpendiculaire au grand axe de la structure considérée). Si S^* est la surface de la cavité étudiée, ces problèmes peuvent être considérés comme la superposition de deux sous-problèmes séparés :

1. un milieu infini soumis à un champ de contrainte homogène équivalent à celui qui existerait en l'absence de toute cavité (c'est-à-dire le champ de contraintes naturelles).
2. Le même milieu infini avec une distribution de charge $\underline{t}$ et déplacements $\underline{u}$ appliquée le long d'une surface S, identique à S^* . $\underline{t}$ et $\underline{u}$ représentent les tensions et déplacements qui doivent être appliqués le long de S pour simuler l'excavation à étudier.

La solution du second problème implique la détermination de

a) la valeur des déplacements à la frontière si les tensions sont imposées et inversement, les valeurs des tensions si ce sont les déplacements qui sont imposés ;
b) l'état de contrainte complet sur l'ensemble du pourtour de la cavité ;
c) les déplacements et les contraintes en tout point du milieu continu équivalent.

Soit Y le grand axe de la cavité , X et Z les deux autres axes de référence. Les contraintes en tout point du milieu, y compris à la frontière, sont obtenus en superposant le champ de contrainte naturel à celui imposé par le chargement de la surface S . Ainsi que l'ont montré Brady et Bray (1978) le problème 2 peut se résoudre en considérant deux sous-problèmes découplés :

i) un problème plan ne faisant intervenir que t_x , t_z , u_x , u_z ;
ii) un problème antiplan ne faisant intervenir que t_y et u_y .

La solution de ces problèmes est obtenue en discrétisant la frontière S ; elle peut être obtenue selon deux méthodes différentes.
Avec la première, appelée formulation indirecte, des distributions de singularité de chargement sont appliquées le long de la frontière et l'intensité de ces singularités est choisie de façon à satisfaire l'ensemble des conditions aux frontières. La meilleure forme de singularité est choisie en en testant un certain nombre et en ne retenant que la meilleure (Brady et Bray 1978 a,b). La seconde méthode de résolution, appelée formulation directe, consiste à résoudre des équations reliant les tensions et les déplacements le long des éléments discrets de la frontière S (Brady 1979). L'objectif est de construire et résoudre les équations obtenues pour les deux sous-problèmes i) et ii). Considérons tout d'abord le problème i) défini dans le repère X, Z décrit ci-dessus. Les tensions en tout point de la frontière ont les composantes t_x et tz ; elles donnent lieu en tout point de la frontière aux déplacements u_x

et u_z . Considérons maintenant les axes locaux X' et Z' au point i de la frontière S (Z' est la composante selon la direction normale) ; les tensions et déplacements en tout point j de S deviennent t'_{xj} , t'_{zj} , u'_{xj} , u'_{zj} que l'on appellera conditions de chargement 1. Considérons maintenant un chargement d'intensité unité dans la direction X' appliqué de façon uniforme le long d'une ligne parallèle à la direction de Y et passant par le point i. Cette charge induit en j, sur S, les tensions et déplacements $T'^{x_i}_{xj}$, $T'^{x_i}_{zj}$, $U'^{x_i}_{xj}$, $U'^{x_i}_{zj}$ exprimés dans le repère local X' , Z'. La surface S peut donc être considérée comme étant chargée par des tensions $T'^{x_i}_{xj}$, $T'^{x_i}_{zj}$ qui donnent naissance à des déplacements $U'^{x_i}_{xj}$, $U'^{x_i}_{zj}$; nous appelerons ces conditions de chargement les conditions de chargement 2 . Appliquons maintenant le thèorème des travaux réciproques : on peut écrire :

$$\int_s (T'^{x_i}_{xj}\, u'_{xj} + T'^{x_i}_{zj}\, u'_{zj})ds = \int_s (t'_{xj}\, U'^{x_i}_{xj} + t'_{zj}\, U'^{x_i}_{zj})ds \qquad (49)$$

qui peut encore s'exprimer, si les tractions et les déplacements des conditions de chargement 1 sont considérées constantes le long des éléments de discrétisation de la frontière :

$$\sum_{j=1}^{n} \int_{s_j} (T'^{x_i}_{xj}\, u'_{xj} + T'^{x_i}_{zj}\, u_{zj})ds_j = \sum_{j=1}^{n} \int_{s_j} (t'_{xj}\, U'^{x_i}_{xj} + t'_{zj}\, U'^{x_i}_{zj})ds_j \qquad (50)$$

où s_j est la longueur du jème segment de discrétisation de la frontière.
Si l'on définit les termes suivants :

$$F'^{x_i}_{xj} = \int_{s_j} T'^{x_i}_{xj} ds_j \; ; \; V'^{x_i}_{xj} = \int_{s_j} U'^{x_i}_{xj}\, ds_j \; ; \; \text{etc..}$$

l'équation (50) peut s'écrire :

$$\sum_{j=1}^{n} \begin{bmatrix} F'^{x_i}_{xj} & F'^{x_i}_{zj} \end{bmatrix} \begin{bmatrix} u'_{xj} \\ u'_{zj} \end{bmatrix} = \sum_{j=1}^{n} \begin{bmatrix} V'^{x_i}_{xj} & V'^{x_i}_{zj} \end{bmatrix} \begin{bmatrix} t'_{xj} \\ t'_{zj} \end{bmatrix} \qquad (51)$$

Le même processus peut maintenant être repeté pour un chargement unitaire dans la direction z' et appliqué en i le long d'une ligne parallèle à la direction y . Répétant ce raisonnement pour l'ensemble des n centres de segments de discrétisation de s et superposant les résultats on obtient dans le repère X - Z :

$$[F] \; \{u\} = [V] \; \{t\} \qquad (52)$$

L'équation (52) exprime les relations recherchées entre les conditions aux frontières, exprimées dans le repère X - Z pour le problème i).
Une procédure similaire peut être suivie pour le problème antiplan ii) :

$$[F_y] \; \{u_y\} = [V_y] \; \{t_y\} \qquad (53)$$

La détermination des composantes inconnues de <u>u</u> ou de <u>t</u> le long de la frontière est obtenue en résolvant le système d'équations linéaires (52) + (53).
Les quantités :

$$\int_{s_j} T'^{x_i}_{xj}\, ds_j \; , \; \int_{s_j} U'^{x_i}_{xj}\, ds_j \; , \text{ etc ...}$$

peuvent être obtenues soit par intégration directe soit par intégration numérique ; Brady propose une solution numérique . Lorsque toutes les tensions et tous les déplacements sont connus sur la surface S, l'état de contrainte et le champ des déplacements peut être calculé en tout point du milieu.
Dans leur essence même, ces méthodes par équations intégrales utilisent des superpositions de solutions élémentaires ce qui suppose un comportement linéaire pour le matériau. Dans le cas de problèmes impliquant des comportements non-linéaires après que certains seuils de contrainte ont été atteints, Brady et Wassyng (1981), Lovan et al. (1983, ce congrès) proposent de coupler une méthode par éléments finis et une méthode par équations intégrales : la première permet de résoudre le problème non linéaire tandis que la seconde permet de bien prendre en compte des conditions à l'infini. Lin Dezhang et Liu Baoshen (1983, ce congrès) proposent d'élargir le champ d'application des méthodes par équations intégrales aux matériaux à visco-élasticité linéaire en utilisant les transformations de Laplace appropriées. Pierce et Ryder (1983, ce congrès) proposent une extension de ces méthodes au cas de lois de comportement multilinéaires c'est-à-dire ayant des matrices de raideurs différentes selon le domaine de déformation considéré. Ils introduisent pour ce faire des éléments spéciaux pour lesquels sont définies des conditions de contrainte initiales, ou de déformations initiales,sur lesquelles viennent se superposer les contraintes et déplacements du problème. Ces éléments "spéciaux" sont supposés pouvoir traiter le cas des déformations plastiques.

Modèles par blocs jointifs

Pour les méthodes par éléments finis ou par équations intégrales les petites discontinuités sont incluses dans le matériau continu équivalent et seules les grandes discontinuités sont traitées individuellement. De telles modélisations sont bien adaptées pour les problèmes où la déformation affecte aussi bien les joints que le milieu continu, c'est-à-dire lorsque les contraintes sont suffisamment élevées. Pour les faibles contraintes, deux phénomènes interviennent simultanément : d'une part la densité des joints augmente considérablement si bien que la masse rocheuse peut être considérée comme un agrégat de blocs de tailles et de formes variées ; d'autre part le processus de déformation affecte essentiellement les joints , la déformation élastique des blocs devenant négligeable. Pour de telles conditions la notion de milieu continu équivalent n'est pas valable, c'est la raison pour laquelle la modélisation par blocs jointifs a été développée. Ces modèles initialement mis au point par Cundall (1971, 1974) supposent que la totalité de la déformation est supportée exclusivement par les joints et autres discontinuités. Ces modèles sont construits sur deux lois :

- les relations force-déplacement ;
- la loi du mouvement .

Les relations force-déplacement lient les forces supportées par les points de contact des blocs aux déplacements qui interviennent le long de ces contacts. Ces déplacements sont calculés à partir de la connaissance du mouvement des blocs. La loi du mouvement est simplement la deuxième loi de Newton : les forces et les moments qui agissent sur chaque bloc engendrent une accélération qui, intégrée deux fois sur l'intervalle de temps Δt, donne le déplacement incrémental utilisé pour le calcul des déplacements aux contacts des blocs. Les forces et moments agissant sur chaque bloc sont la somme des forces et des moments agissant au contact des blocs. Le système d'équation ainsi obtenu est résolu par différences finies sur le temps.

Ces modèles peuvent traiter des rotations et des translations illimitées pour des blocs bidimensionnels de forme quelconque. Ils ont été appliqués à des problèmes aussi variés que les études de stabilité de pente ou l'analyse de l'écoulement des liquides en milieu très fracturé (Cundall 1982). Warburton (1983) propose à ce congrès, un programme permettant de construire des géométries tridimensionnelles et d'analyser les conditions de stabilité de chaque bloc polyhédral.

Il est permis de penser que, par combinaison des modèles probabilistes pour la caractérisation géométrique des discontinuités affectant un massif et des modèles de blocs jointifs tridimensionnels, le comportement mécanique des masses rocheuses intensément fracturées pourra bientôt être modélisé avec précision.

5. CONCLUSION

Le champ de contrainte naturel qui existe dans un massif rocheux est une donnée importante pour de nombreux problèmes d'ingénierie de même que pour la compréhension de la tectonique actuelle. Il peut être déterminé soit au moyen de mesures directes soit par l'interprétation des observations sismotectoniques. Les techniques de mesures in-situ les plus employées sont le surcarottage et la fracturation hydraulique. Les méthodes par surcarottage sont essentiellement utilisées pour les mesures à partir de cavités souterraines ; elles sont généralement limitées à des forages peu profonds (inférieurs à 100 m dans la plupart des cas). Les méthodes par fracturation hydraulique s'appliquent en forages superficiels et en forages profonds (profondeur maximum actuelle = 4.200 m). Du fait de son relativement faible coût de réalisation et de son indépendance vis à vis de la profondeur, la méthode par fracturation hydraulique est dans bien des cas la plus attractive. Sa précision atteint 15 % mais des améliorations sont encore à attendre, spécialement en ce qui concerne la précision des mesures de pression de fermeture en milieu très fracturé.

L'ensemble des résultats de mesure obtenus à ce jour indique que le cisaillement maximum augmente avec la profondeur au moins jusqu'à 5 km, profondeur limite des mesures directes actuellement réalisées. Toutefois l'étude des variations de contrainte dans le temps notamment dans les zones à fort risque sismique, rencontre encore des difficultés d'observation.

Une deuxième étape importante dans l'analyse du comportement mécanique des masses rocheuses est la caractérisation de la géométrie des discontinuités naturelles. Bien que certains progrès aient été réalisés vis à vis de la caractérisation locale de ces discontinuités (c'est-à-dire l'extrapolation locale à trois dimensions des observations effectuées sur des surfaces d'affleurement) grâce essentiellement à l'analyse statistique et à la modélisation probabiliste, aucune méthode précise pour caractériser à grande échelle la fracturation naturelle d'un massif n'a encore été développée ; les méthodes géostatistiques considérées actuellement sont gênées par le fait que l'observation effectuée à une échelle donnée n'est pas directement transposable à d'autres échelles.

La troisième étape d'un problème de mécanique des roches est la définition de lois de comportement appropriées pour représenter le comportement mécanique de milieux continus équivalents ainsi que celui des grandes discontinuités qui affectent le massif. Le modèle le plus fréquemment utilisé est bien sûr l'élasticité linéaire mais des modèles élasto-visco-plastique sont de plus en plus fréquemment considérés. La modélisation des discontinuités suppose généralement une loi de Coulomb pour le cisaillement maximum sous forte contrainte bien que des relations plus détaillées faisant intervenir la rugosité et la résistance en compression des parois soient maintenant envisagées. Les mouvements de cisaillement le long de ces discontinuités mobilisent généralement une certaine dilatance dont la représentation reste actuellement empirique.

Une représentation satisfaisante de l'influence de la vitesse de déformation, tant sur le comportement de la matrice rocheuse que sur celui des discontinuités rencontre encore des difficultés lorsque la déformation relève du développement de microfissures sujettes en particulier à l'effet de corrosion sous tension.

Des résultats de laboratoire ont été obtenus à ce sujet mais leur transposition aux conditions de terrain reste problématique et des études in-situ seraient souhaitables. Pour ce faire l'application des derniers développements effectués en sismologie sur la modélisation des sources sismiques (détermination de l'orientation du plan de faille, du sens du mouvement, de la chute de contrainte, des dimensions de la rupture) devrait permettre une meilleure compréhension de l'activité sismo-acoustique associée à la déformation des massifs donc des mécanismes mis en jeu. De telles études pourraient en particulier aider à mieux comprendre les déformations quasistatiques qui précèdent les instabilités catastrophiques et par conséquent fournir un moyen de les prévoir.

Ces observations devraient être couplées à des mesures in-situ de déformation ainsi qu'à un contrôle des variations de contrainte dans le temps.

Une fois franchies ces trois étapes, les problèmes de mécanique des roches sont généralement résolus au moyen de méthodes numériques bien que certains modèles physiques soient utilisés. Dans ce dernier cas les facteurs de similitude appropriés doivent être utilisés.

Si les matériaux considérés sont linéairement élastiques la formulation par équations intégrales constitue une méthode puissante de résolution et ceci plus particulièrement dans le cas de milieux infinis ou semi-infinis ou lorsque la géométrie de la frontière est évolutive.
Dans le cas de milieux à comportement non linéaire, la méthode la plus efficace est celle des éléments finis. Toutefois dans le cas de massifs fortement fracturés et peu contraints pour lesquels l'essentiel de la déformation n'affecte que les discontinuités, une modélisation par blocs jointifs peut se révèler très efficace. Bien que ces derniers modèles ne soient encore que bidimensionnels, le développement d'une méthode tridimensionnelle couplée à une caractérisation probabiliste de la distribution des discontinuités devrait fournir très prochainement un outil puissant d'investigation.
Le rôle des pressions de fluide n'a pas été discuté dans ce rapport, cependant l'étude de la dépendance de la perméabilité des fractures en fonction de la pression interstitielle suggère que la théorie des contraintes effectives à prendre en compte dans l'étude du comportement mécanique des massifs rocheux fracturés n'obéit probablement pas à la loi proposée par Terzaghi (1945). Les conséquences de ces résultats sur l'analyse de la stabilité des massifs rocheux reste à être étudiées.

REMERCIEMENTS

Je tiens à remercier ici G. Jobert qui a bien voulu m'aider à corriger ce texte.

BIBLIOGRAPHIE

AAMODT A.a. and M. KURIYAGAWA (1982). Measurement of instantaneous shut-in pressure in crystalline rock ; Workshop on hydraulic fracturing stress measurements ; Monterey U.S.G.S Report nb 82-1075.

ANDRADE E.N.(1910). On the viscous flow of metals and allied phenomena ; Proc. Roy. Soc. London, A-84, pp.1-12.

ANGELIER J. (1979). Determination of the mean principal directions of stresses for a given fault population ; tectonophysics, 56 p.T 17-T 26.

ARMIJO R., E. CAREY and A. CISTERNAS (1982). The inverse problem in microtectonics and the separation of tectonic phases ; tectonophysics, 82, p.145-160.

ATKINSON B.K. (1979). A fracture mechanics study of subcritical tensile cracking of quartz in wet environments ; Pageop vol.117, nb 5, 1011-1024.

BALTHASAR K. and E. WENZ (1983). The determination of the complete state of stress with the compensation method as a contribution for investigations of tectonically influenced stresses ; This congress.

BANERJEE P.K. and BUTTERFIELD R.(1981). Boundary elements methods in engineering Science ; Mc Graw Hill, London.

BARTON N. (1973). Review of a new shear strength criterion for rock joints ; Eng. Geology vol.7, p.287-332.

BARTON N. (1976). The shear strength of rock and rock joints ; Int. Jour. Rock Mech.Min.Sc. vol.13, pp.255-279.

BARTON N.(coordinator)(1978). Suggested methods for the quantitative description of discontinuities in rock masses ; Int. Jour. Rock Mech. Min. Sc. vol.15, nb.6, p.319.

BARTON N. and V. CHOUBEY (1977). The shear strength of rock joints in theory and practice ; Rock Mechanic ; vol.10, p.1-54.

BARTON N. and S. BANDIS (1980). Some effects of scale on the shear strength of joints ; Int. Jou. Rock Mech. Min. Sc. ; vol.17, nb.1, p.69.

BERRY D.S. (1968). Calculation of the stress in the surface of a circular hole in an infinite transverselly isotropic elastic medium subjected to general stresses at infinity and hydrostatic pressure at the hole, Appendix 1. in Methods of determining in-situ rock stress at great depth, by Fairhurst C. (see this author's reference).

BERTRAND L., H. BEUCHER, D. CREUTON, B.FEUGA, J. CANDRY D. THIERY (1982). Essai de détermination de la distribution régionale du tenseur de perméabilité du milieu poreux equivalent ; colloque-jubilée Castany ; BRGM-Orléans.

BIENIAWSKI Z.T.(1970). Time dependent behavior of fractured rock ; Rock Mechanics ; vol.2, p. 123.

BLACKWOOD R.L. (1977). An instrument to measure the complete stress field in soft rock or coal in a single operation ; Field measurements in Rock Mechanics, Int. Symp. Zurich, Kovari editor. Balkema Rotherdam

BLACKWOOD R.L. (1978). Diagnostic stress relief curves in stress measurements by overcoring; Int. Jou. Rock Mech. Min. Sc. vol.15, nb.4, p.205.

BONNECHERE F.J.(1971). Contribution à l'étude de la détermination de l'état de contrainte des massifs rocheux, thèse de doctorat Faculté des Sciences Appliquées - Liège.

BONNECHERE F.J. and C. FAIRHURST (1968). Determination of the regional stress field from " Doorstopper " measurements ; Jou. South Afr. Inst. Min. and Met ; july.

BOUCHEZ J. (1981). Résolution des problèmes, d'élasticité plane par la méthode des discontinuités de déplacement en milieu non homogène ; Inst. Phys. du Globe, Paris ; int. Report.

BRACE W.F., B.W. PAULDING Jr, and C. SCHOLTZ (1966). Jou. Geophys. Res. vol.71, nb 16 p.3939.

BRACE W.F. and R.J. MARTIN III (1968). A test of effective stress law for crystalline rocks of low porosity ; Int. Jou. Rock Mech. Min. Sc., vol. 5 p.415.

BRAY J.W. and R.F. GOODMAN (1981). The theory of base frictions models ; Int. Jou. Rock Mech. Min. Sc. ; vol.18 nb 6 ; p.453.

BRADY B.H.G. (1979). A direct formulation of the boundary element method of stress analysis for complete plane strain ; Int. Jou. Rock Mech. Min. Sc. vol.16, nb4, p.235.

BRADY B.H.G. and J.W. BRAY (1978).
a) The boundary element method for determining stresses and displacements around long openings in a triaxial stress field.
b) The boundary element method for elastic analysis of tabular orebody extraction, assuming complete plane strain.
Int. Jou. Rock Mech. Min. Sc. vol.15 nb.1 p.21-p.29.

BRADY B.H.G. and A. WASSYNG (1981). A coupled finite element boundary element method of stress analysis Int. Jou. Rock. Mech. Min. Sc., vol. 18 nb 6 ; p.475.

BROWN E.T. and E. HOEK (1978) Trends in relationship between measured in-situ stresses and depth ; Int. Jou. Rock Mech. Min. Sc. vol.15, nb.4, p.211.

BREDEHOEFT J.D., R.G. WOLFF, W.S. KEYS and SCHUTER (1976). Hydraulic fracturing to determine the regional in-situ stress field. Piceance Basin, Colorado ; Geol. Soc. Am. Bull. vol.87, p.250-258.

BURRIDGE R. and L. KNOPOFF (1964). Body force equivalents for seismic dislocations ; Bull. Seism. Soc. Am.,vol.54, nb.6,p.1875.

BUYLE-BODIN F., F.J. BONNECHERE and F.H. CORNET (1983). Mesures de contrainte dans la vallée d'Aspe, Revue Fr. de Geotechnique n°20, p.31.

BYERLY P. (1926). The Montana earthquake of june 28, 1925 ; Bull. Seism. Soc. An., vol.16, p.209.

BYERLEE J.D. (1967). Frictional characteristics of granite under high confining pressure , Jou. Geophy. Res., vol.72, p.3639.

BYERLEE J.D. (1968). Brittle ductile transition in rocks, Jou. Geophys. Res. vol.73,p.4741.

BYERLEE J.D. (1975). The fracture strength and frictional strength of Weber sandstone Int. Jou. Rock. Mech.Min. Sc.,vol.12, nb1, p.1.

CAROL I. and E.E. ALONSO (1983). A new joint Element for the analysis of fractured rock ; this congress.

CASTELEIRO M., E. ONATE, A. HUERTA, J.ROIG, and E. ALONSO (1983). Three dimensional analysis of no tension materials ; This congress.

CHARLEZ Ph. (1983). Une méthode de mesures in-situ des contraintes dans les massifs rocheux ; Thèse Ing. Doct. Univ. PARIS 6.

CLARK B.C. (1982). Monitoring changes of stress along active faults in southern California. Jou. Geophys. Res. ; vol.87 nb.B 6,p.4645.

CLEARY M.P. (1980). Analysis of mechanisms and procedure for producing favourable shape of hydraulic fractures ; 55 th Annual fall technical Conf. Soc. Petr. Eng. AIME paper SPE 9260.

COOK M.G.W. and J.P.M. HOJEM (1966). A rigid 50-ton compression and tension testing machine, South. Afr. Mech. Eng. vol.14, p.89-92.

CORNET F.H. (1976) Crack propagation in rock normal to the maximum compression stress ; N.S.F. report for contact nb. G.K. 41220 (Bligh and Fairhurst ppl. investigators).

CORNET F.H. (1977). Etude du comportement elastique et fragile des roches saturées par un liquide; Revue française de Géotechnique n°2.

CORNET F.H. (1979). Comparative analysis by the displacement discontinuity method of two energy criteria of fracture ; Jou. App. Mech. vol.46, nb.2,p.349.

CORNET F.H. (1982). Analysis of injection tests for in-situ stress determination ; proceedings of the workshop on hydraulic fracturing stress measurements U.S.G.S. open-file rep.82-1075. Haimson and Zoback editors.

CORNET F.H. and C. FAIRHURST (1972). Variation of pore volume in disintegrating rock ; Percolation thru fissured rock. Symp. of the I.S.R.M. Stuttgart Wittke editor,p.T2-A.

CORNET F.H., F.J. BONNECHERE and F.BUYLE-BODIN (1979). Discussion sur la détermination de l'état de contrainte dans un massif rocheux ; 4 th Cong. of the ISRM, Montreux vol.3, p.137.

CRAWFORD A.M. and J.H. CURRAN (1982) Higher order functional variation displacement discontinuity elements ; Int. Jou. Rock Mech. Min. Sc. ; vol.19 nb3 , p.143.

CROUCH S.L.(1970). Experimental determination of volumetric strain in failed rock ; Int.Jou. Rock. Mech. Min. Sc. vol.7, p.589.

CROUCH S.L. (1976). Solution of plane elasticity problems by the displacement discontinuity method ; Int. Jou. Num. Meth. Eng. , vol.10 p.301.

CROUCH S.L. (1976). Analysis of stresses and displacements around underground excavations: an application of the displacement discontinuity method ; Univ. of Minn.Geomechanics report ; NSF-RANN program(grant nb GI-37923).

CROUCH S.L. and C. FAIRHURST (1967). A four component deformation gauge for the determination of in-situ stress in rock masses ; Int. Jou.Rock Mech. Min. Sc. vol.4,p.209-217.

CROUCH S.L. and A. STARFIELD (1983). Boundary element methods in solid mechanics; George Allen and Unkin publisher ; London.

CUNDALL P.A. (1971). A computer model for simulating progressive large scale movement in blocky rock systems ; Int. Soc. Rock.Mech. Symp. Nancy.

CUNDALL P.A. (1974). Rational design of tunnel supports : A computer model for rock mass behavior using interactive graphics for the input and output of geometrical data. U.S. Army corp of Engineers, Technical Report MRD-2-74.

CUNDALL P.A. (1980). UDEC, a generalized distinct element program for modelling jointed rock. Final technical report ; European Research office ; U.S. Army London. Contract nb DAJA 37-79-C-0548.

CUNDALL P.A. (1982). The FRIP users mannal. Camborne School of Mines, Geothermal project.

DAS.S and C.H.SCHOLZ (1981).Theory of time dependent rupture in the earth ; Jou.Geophys.Res. vol.86,nb. B 7,p.6039.

DEERE D.U. (1964). Technical description of rock cores for engineering purposes ; Rock Mech. Eng. Geol.vol.6 , p.1.

DE LA CRUZ R.V. (1977). Jack fracturing technique of stress measurement ; Rock Mechanics vol.9, p.27.

DE LA CRUZ R.V. and R.E. GOODMAN (1969). The borehole deepening method of stress measurement ; Int. Symp. on the determination of stress in rock masses. Lisboa.

DERAMOND J., P. SINEYS and J.C. SOULA (1983), Mechanismes de déformation de l'écorce terrestre : structure et anisotropie induites ; This congress.

DESAI C.S., M.M. EITANI and C. HAYCOCKS (1983). An application of finite element procedure for underground structures with non linear materials and joints ; This congress.

DOE T.W. (1982) Determination of the state of stress of the Stripa Mine, Sweden ; Proc. Workshop on hydraulic fracturing stress measurements;U.S.G.S.open file report 82-1075.

DRAGON A. and Z. MROZ (1979). A model for plastic creep of rock like materials accounting for the kinetics of fracture ; Int. Jou. Rock Mech. Min. Sc. vol.16, nb.4,p.253.

EGGER R. and C. GINDROZ (1979). Tunnels ancrés à faible profondeur, étude comparative sur models physiques et mathématiques ; Proc. 4 th Int. Soc. Rock Mech. Congress vol.2, p.121.

ETCHECOPAR A., G. VASSEUR, M. DAIGNIERES (1981) An inverse problem in microtectonics for the determination of stress tensors from fault striation analysis ; Jou. Structural Geol. vol.3 nb1, p.51.

FAIRHURST C. (1964). Measurement of in-situ rock stresses with particular reference to hydraulic fracturing ; Rock Mech. Eng. Geol., vol.11, p.129.

FAIRHURST C. (1968). Methods of determining in-situ rock stress at great depth Tech. Rep.1-68.; U.S. Corps of Engineers Omaha, Nebraska.

FILCEK H. and T. CYRUL (1977). Rigid inclusion with high sensitivity, Field Measurements in Rock Mechanics. Int. Symp. Zürich Kovari Editor, vol.1,p.219 ; Balkema Rotterdam.

FLEITOUT L. and C. FROIDEVAUX (1982) Tectonics and topography for a lithosphere containing density heterogeneities ; Tectonics, vol.1, nb.1,p.21.

FROIDEVAUX C., C. PAQUIN and M. SOURIAU Tectonic stresses in France : in situ measurements with flat jack. Jou. Geoph. Res. vol.85 nb B 11,p.6342.

GALLE E.M. and J.C. WILHOIT Jr. (1961). Stress around a wellbore due to internal pressure and unequal principal geostatic stresses ; 36 th Annual meeting of Soc. Pet. Eng. Dallas Texas.

GERRARD C.M. (1982). Elastic models of rock masses having one, two and three sets of joints ; Int. Jour. Rock Mech. Min. Sci. vol.19 nb 1, p.15.

GOODMAN R.E., R.L. TAYLOR and T.L. BREKKE (1968) A model for the mechanics of jointed rock ; Jou. Soil. Mech.and Found.Div. A.S.C.E, vol.94, n°SM 3, p.637.

GRIFFITH A.A. (1921). The phenomenon of rupture and flow in solids ; Phil. Trans. Roy. Soc. London A 221.

GRONSETH J.M. and P.R. KRY (1982).Instantaneous shut-in pressure and its relationship to of the minimum in-situ stress ; Proc.workshop hydraulic fracturing stress measurements; U.S.G.S. open file report 82-1075 ; Menlo Park Calif.

GUENOT A. et M. PANET (1983). Etude numérique d'ouvrage en massif rocheux a structure planétaire ; This congress.

HAGAN T.O. (1980). A case of terrestrial photogrammetry in deep-mine rock structure studies ; Int. Jou. Rock Mech. Min. Sci. vol.17, nb 4, p.191.

HAIMSON B.C. (1968). Hydraulic fracturing in porous and non porous rock and its potential for determining in-situ stress at great depth ; U.S.corps of Engineers,Tech. Rep. 4-68, Omaha, Nebraska.

HAIMSON B.C. (1978). The hydrofracturing stress measuring method and recent field results, Int.Jou.Rock Mech.Min.Sc.vol.15,nb.4,p.167.

HAIMSON B.C. (1982). A comparative study of deep hydrofracturing and overcoring stress measurements at six locations with particular interest to the Nevada test site;Proc. workshop on hydraulic fracturing stress measurements ; U.S.G.S. openfile rep.82-1075.

HAIMSON B.C. and C. FAIRHURST (1969). In-situ stress determination at great depth by means of hydraulic fracturing ; 11 th Symp. on Rock Mech. Berkeley p.559 AIME.

HANDIN J. and C.B. RALEIGH (1972). Manmade earthquakes and earthquake control ; Percolation thru fissured rock. Int. Soc. Rock. Mech. Symp. Stuttgart, Wittke Editor.

HANKS T.C.(1977). Earthquake stress drops, Ambient tectonic stresses and stresses that drive plate motions ; Pageoph.vol.115 p.441.

HELAL H. et M. DEJEAN (1981). Mise au point d'une technique de mesure des contraintes in-situ par surcarottage ; Comité Français de Mécanique des Roches.

HENRY J.P., J.PAQUET and J.P.TANCREZ (1977). Experimental study of crack propagation in calcite rocks ; Int. Jou. Rock Mech.Min.Sci. vol.14 nb.2, p.85.

HEUZE F.E. (1971). Sources of errors in rock mechanics field measurements and related solutions ; Int. Jou. Rock Mech. Min. Sci. vol. 8, p.297.

HEUZE F.E., W.C. PATRIK, T.R. BUTKOWICH, J.C. PETERSON, R.V. DE LA CRUZ and C.F. VASS (1982). Rock Mechanics studies of mining in the climax granite; Int.Jou. Rock Mech.Min. Sc. vol.19 nb.4, p.167.

HICKMAN S.H. and M.D. ZOBACK (1982).The interpretation of hydraulic fracturing pressure time data for in-situ stress determination. Proc.workshop on hydraulic fracturing stress measurements ; U.S.G.S. open file report n°82-1075. Menlo Park Calif.

HIRASHIMA K. and A. KOGA (1977). Determination of stresses in anisotropic elastic medium unaffected by boreholes from measured strains or deformations,in field measurements in Rock Mechanics. Int.Symp. Kovari Editor ; Balkema Rotterdam.

HIRAMATSU Y. and Y.OKA (1968). Determination of the stress in rock unaffected by boreholes or drifts, from measured strains or deformations;Int.Jou. Rock Mech.Min.Sc. vol.5,july.

HOOKER V.E. and D.L. BICKEL (1974). Overcoring equipment and techniques used in Rock Stress determination ; information circular 8618, Denver Mining res. center ; U.S.B.M.

HOSKINS E.R.(1966). An investigation of the flat jack method of measuring rock stress; Int. Jou.Rock Mech.Min.Sci.vol.3, p.249.

HOUPERT R. (1974). Le rôle du temps dans le comportement à la rupture des roches ; Proc. 3d congress of the Int. Soc.Rock Mech. Denver ; vol.2 part A p.325.

HAWKES I. and V.E. HOOKER (1974). The vibrating wire stress meter ; Proc. 3d Int. Soc. Rock Mech. Congress. Denver vol.II-A,p.439.

HUBBERT M.K. and D.G. WILLIS (1957). Mechanics of hydraulic fracturing ; Petrol. Trans. of Am. Inst.Min. Eng.vol.210, p.153.

HUDSON J.A., E.T. BROWN and C. FAIRHURST (1971). Optimizing the control of rock failure in servo-controlled laboratory test ; Rock Mechanics vol.3 p.217.

HUDSON J.A., E.T. BROWN (1973). Studying time dependent effects in failed rock; 14 th Symp. on Rock Mech. Pensylvania State Univ; Amer. Soc.Civil.Eng., p.25.

HUDSON J.A. and S.D. PRIEST (1979).Discontinuities and rock mass geometry ; Int.Jou.Rock Mech. Min.Sc. vol.16 nb 6, p.339.

HUDSON J.A., P.R. LA POINTE (1980). Printed circuits for studying rock mass permeability; Int. Jou.Rock Mech.Min.Sci.vol.17,nb 5,p.297.

HUERGO P.J.(1983). Evaluation des contraintes naturelles dans les couches supérieures d'un massif schisteux ; This congress.

HUSTRULID W. and A. HUSTRULID (1975). The CSM cell a borehole device for determining the modulus of rigidity of rock ; 15 th Symp. on Rock Mech. Ann. Soc. Civ.Eng.p.181.

IRWIN G.R.(1957). Analysis of stresses and strains near the end of a crack traversing a plate ; Jou. Appl. Mech. vol.24 nb 3,p.361

JAEGER J.C. and N.G.W. COOK (1963). Pinching-off and discing of rocks ; Jou. Geophys. Res. vol.68, p.1759.

JAEGER J.C. and N.G.W. COOK (1979). Fundamentals of Rock Mechanics, 3nd Ed. Chapman and Hall, London.

JAMISON D.B. and N.G.W. COOK (1980). Note on measured values for the state of stress in the Earth's crust. Jou. Geophys. Res.vol.85. nb B.4 p.1833.

JASWON M.A. and G.T. SYMM (1977). Integral equation methods in potential theory and elasto-statics ; Academic Press, London.

JIA-SHOU ZHUC and YIN TONG WANG (1983). Non-linear analysis of rock foundation with soft interfaces. This congress.

JIM-QI HAO , L.M. HASTIE and F.D. STACEY (1982). Theory of seismomagnetic effect : a reassessment. Physics. earth.Plan.Int. vol.28, p.129.

KAISER P.K. and N.R. MORGENSTERN (1981). Time dependent deformation of small Tunnels part II, Int.Jou.Rock Mech.Min.Sc. vol.18 nb2, p.153.

KEHLE R.D. (1964). Determination of tectonic stresses through analysis of hydraulic well fracturing;Jou. Geophys. Res. vol.69 p 259.

KIRSH G. (1898). Die theorie der elastizität und die bedürfnisse der Festigkeitslehre Z.V.D.I., vol.42,nb.29,p.113.

KNOPOFF L. and F. CULBERT (1960). First motion from seismic sources. Bull.Seis.Soc.Am . vol.50, p.117.

KOVARI K. (1977). Field measurements in rock mechanics ; Proc. Int. Symp. Zurich, Balkena, Rotterdam.

KRANZ R.L. and C.M. SCHOLZ (1977). Critical dilatant volume of rocks at the onset of tertiary creep ; Jou. Geophys. Res. vol.82 nb.30, p.4893.

KUZNETSOV S.V., D.M. BRONNIKOV, I.A. PARABUCHEV V.D. PARPHENOV,I.T. AITMATOV, G.A. MARKOV (1983). The state of stress in rock and methods of its determination. This congress.

LANGER M. (1979). Rheological behavior of rock masses, General Report Theme I, 4 th Int. Soc. Rock Mech. Congress ; Montreux ; Balkema , Rotterdam.

LEEMAN E.R (1964). The measurement of stress in Rock, Part I, II, III ; Jou. South. African Inst. Min. Met.,p.45,p.82,p.254.

LEEMAN E.R. (1968). The determination of the complete state of stress in rock in a single borehole. Laboratory and underground measurements, Int. Jou. Rock. Mech. Min. Sc. vol.5, p.31.

LEEMAN E.R. (1969). The measurement of stress in rock. A review of recent development (and a bibliography). Int. Symp. on the determination of stresses in Rock Masses; Lisboa.

LESSI J. and J.P. SARDA (1980). Scale model studies of well linking by hydraulic fracturing and fatigue microfracturing P.957, Proc. 2ème Int.Sem.Results of EC Geothermal Energy Res. ; Reidel Pub. Dordrecht.

LI FANG-QUAN, LI YAN-MEI, WANG EN-FU, ZHAI QING-SHAN, BI SHANG-SU, ZHANG-JUN, LIU-PENG WI QING-YON and ZHAO SHI-GUANG (1982). Experiments in the in-situ stress measurements using the stress relieving and hydraulic fracturing techniques ; Proc. Workshop on hydraulic fracturing stress measurements, p.332, U.S.G.S., open file report 82-1075.

LIN DESHANG and L. BAOSHEN (1983). Boundary element method for linear viscoelastic stress analysis in rock mass and its application in rock engineering ; This congress.

LOVAN P., J. OKSANEN and K. AIKAS (1983). Element methods in planing of mine opening in highly stressed precambrian bedrock ; This congress.

MADARIAGA R. (1979). Seismic radiation from earthquake models based on fracture mechanics, Am. Math. Soc. vol.12 p.59.

MARTIN R.J.III (1972). Time dependent crack growth in quartz and its application to the creep of rocks ; Jou. Geophys. Res. vol.77, p.1406.

MARTIN R.J.III, R.E. HABERMANN and M. WYSS (1978). The effect of stress cycling and inelastic volumetric strain on remanent magnetization ; Jou. Geophys.Res. vol.83, nb B7, p.3485.

MARTNA J., R. HILSCHER and K. INGEVALD (1983). Triaxial rock stresses in two deep boreholes at Forsmark ; This congress.

MAYER A., P. HABIB et R. MARCHAND (1951). Conférence Int. sur la pression de terrains et le soutenement dans les chantiers d'exploration, Liège, p.217.

Mc GARR A. (1980). Some constraints on levels of shear stress in the crust from observations and theory; Jou.Geophys.Res.vol.85,nb.B11,p.6231.

(1982). Analysis of state of stress between provinces of constant stress; Jou. Geophys. Res.,vol 87, nb B11, p.9279.

Mc KENZIE D.P. (1969). The relation between fault plane solutions for earthquakes and the directions of the principal stresses,Bull. Seism. Soc. An. vol.59 n°2, p.591.

Mc LENNAN J.D. and J.C. ROEGIERS (1982). Do instantaneous shut-in pressure accurately represent the minimum principal stress. Proc. workshop on hydraulic fracturing stress measurements ; p.181 ; U.S.G.S. open file Rep.82-1075,Menlo Park Calif.

MOORE G.W. (1964). Magnetic disturbances preceding the 1964 Alaska earthquake ; Nature vol.203, p.508

MORLIER P. (1966). Le fluage des roches ; Ann.de Inst. Tech. Bat. Tr. Publ.vol.19 p.89.

NAKANO M. (1923). Notes on the nature of the forces which give rise to the earthquake motions; Centr. Meteor. Observ. Japan Seism.Bull. vol.1.,p.92.

PAHL A.(1977). In situ stress measurements by overcoring inductive gages ; p.161, Field measurements in rock mechanics. Int. Symp Zurich,Kovari Editor,Balkema , Rotterdam.

PANDE G.M. and C.M. GERRARD (1983). The behavior of reinforced jointed rock masses under various simple loading states,This congress.

PARISEAU W.G. and I.M. EITANI (1972). Post Elastic vibrating wire stress measurements in coal; p.255, Field measurements in rock mechanics Int.Symp.Zurich, Kovari Editor Balkema ,Rotterdam.

PARSONS R.C. and D.G.F. HEDLEY (1966). The analysis of viscous property of rocks for classification. Int.Jou.Rock Mech.Min.Sc. vol3,p.325.

PENG S. and E.R. PODNIEKS (1972). Relaxation and the behavior of failed rock ; Int.Jou.Rock Mech.Min.Sc.,vol.9,p.699.

PIERCE A.P. and J.A. RYDER (1983). Extended boundary element methods in the modelling of brittle rock behavior. This congress.

POTTS E.L.J. (1957). Underground instrumentation quart.color.,school of mines vol.52,p.135.

PRIEST S.D. and J.A. HUDSON (1976). Discontinuity spacing in rock ; Int.Jou.Rock Mech. Min.Sci.,vol.13,p.135.

PRIEST S.D. and J.A. HUDSON (1981). Estimation of discontinuity spacing and trace length using scanline surveys ; Int.Soc.Rock Mech. Min.Sc. vol.18,nb.3,p.183.

PRIEST S.D. and A. SAMEMIEGO (1983). A model for the analysis of discontinuity characteristics in two dimensions ; This congress.

QUIBLIER J. and K. MGOKWEY (1983). Modèles mathématiques et étude de la fracturation naturelle ; this congress.

RAHN W. (1983). Analysis of potential errors of interpretation of in-situ stress measurements in anisotropic rocks ; This congress.

RALEIGH C.B., J.H.HEALY and J.D.BREDEHOEFT(1972). Faulting and crustal stress at Rangely, Colorado ; in Flow and Fracture of Rocks. Geophysical Monogr. series,p.275,Am.Geoph. Union, Washington D.C.

REN N.K. and J.C. ROEGIERS (1983). Differential strain curve analysis : A new method for determining the preexisting in-situ stress state from rock core measurements;This congress.

RICE J.R. and M.P. CLEARY (1976). Some basic stress diffusion solutions for fluid saturated elastic porous media with compressible constituents, reviews of Geophys. and space phys.,vol.14,nb.2,p.227.

RIKITAKE T. (1979). Changes in the direction of magnetic vector of short period ; geomagnetic variations before the 1972 Sitka Alaska earthquake,Jou.Geomagn.Geoelectr. vol.31,p.441.

ROBERTS A.,I.HAWKES,F.T.WILLIAMS and R.K.DHIR (1964). A laboratory study of the photoelastic stress-meter;Int.Jou.Rock Mech.Min. Sc.vol.1,nb.3,p.441.

ROCHA M. and A.SILVERIO (1969). A new method for the complete determination of the state of stress in rock masses ; Int.symp. on the determination of stress in rock masses. Lisboa.

RUMMEL F. (1974). Changes in the P.wave velocity with increasing inelastic deformation in rock specimens under compression ; p.517 3d congress of the Int. Soc. Rock Mech. Denver,Nat.Acad.Sciences; Washington.

RUMMEL F. and C. FAIRHURST (1970). Determination of the post-failure behavior of brittle rock using a servo-controlled testing machine ; rock mechanics,vol.2,p.189.

RUMMEL F.,J. BAUMGARTNER and H.J. ALHEID (1982). Hydraulic fracturing stress measurements along the eastern boundary of the SW-German block ; Proc.workshop on hydraulic fracturing stress measurements ; p.1 ; U.S.G.S open file report 82-1075.

SCHEIDEGGER A.E. (1962). Stress in Earth crust as determined from hydraulic fracturing data ; Geologie und Bauwesen, vol.27,p.45.

SCHEIDEGGER A.E. (1964). The tectonic stress and tectonic motion direction in Europe and western Asia as calculated from earthquake fault plane solutions ; Bull. Seism.Soc.Ann.vol.54 p.1519.

SCHEIDEGGER A.E. (1977). Geotectonic stress determinations in Austria ; Field Measurements in rock mecanics,p.197 ; Int.Symp. Zurich,Kovari editor,Balkenna Rotterdam.

SCHOLZ C.H. (1968). Microfracturing and the inelastic deformation of rock in compression Jou.Geophys.Res.vol.73,nb.3,p.1417.

SIH G.C. (1973). Handbook of stress intensity factors.Institute of fracture and solid Mechanics,Lehigh Univ. Bethlehem,Pa.

SINHA K.(1979). Ph.D. Thesis Univ. of Minnesota unpublished.

SMITH B.E.and M.J.S. JOHNSTON (1976).A tectonomagnetic effect observed before a magnitude 5.2 earthquake near Hallister, California, Jou. Geophys.Res.vol.81,p.3556.

SOLOMON S.C., R.M. RICHARDSON and E.A.BERGMAN (1980). Tectonic stress : models and magnitude, Jou.Geophys.Res.vol.85,nb.B4 p.6086.

SOUSA L.R. (1983). Three dimensional analysis of large underground power stations ; This congress.

STARFIELD A.M. and S.L. CROUCH (1973). Elastic analysis of single seam extraction: 14 th Rock Mechanics Symp. Hardy and Stefanko editors. p.421; Am. Soc. Civ. Eng.

STAUDER S.J. (1962). The focal mechanism of earthquakes. Advances in Geophysics,vol.9 p.1, Am. Geophys. Union.

STEPHANSSON O.(1983). Rock stress measurement by sleeve fracturing ; This congress.

STUART W.D. and J. DIETRICH (1974). Continuum theory of rock dilatancy ; 3^{d} congress of the Int.Soc.Rock Mech.,Denver Theme 2 Nat.Acad.Sc. Washington.

TAN TJONG KIE and KANG SEN FA (1983). Time dependent dilatancy prior to rock failure and earthquakes ; This congress.

TAPPONNIER P., G. PELTZER, A.Y. LE DAIN and R. ARMIJO (1982). Propagating extrusion tectonics in Asia ; New insights from simple experiments with plasticine ; Geology, vol.10 p.611.

TERZAGHI K. (1945). Stress conditions for the failure of saturated concrete and rock ; Proc.Am.Soc.Test Mat. ,vol.45,p.727.

THOMAS A., P. THERME and P. RICHARD (1981). A new method for quick measuring and computing of rock jointing ; 22 nd U.S. Rock Mech. Symposium ; M.I.T.; Einstein Editor.

TINCELIN M.E. (1952). Mesure de pressions de terrain dans les mines de fer de l'est ; Ann. Inst. Tech. Bat. Tra. Publ., vol.58 p.972.

VALETTE B. and F.H. CORNET (1983). The inverse problem of stress determination from hydraulic injection tests data ; submitted to Jou. Geophys.Res.

VOUILLE G., S.M. TIJANI and F. de GRENIER (1981). Experimental determination of the rheological behavior of Tersanne rock salt ; Proc. 1st conf. on the mechanical behavior of salt ; Pensylvanie State Univ. Nov.1981.

WALLIS P.F. and M.S. KING (1980). Discontinuity pacing in a crystalline rock ; Int. Jou. Rock Mech. Min.Sc.,vol.17, nb.1,p.63.

WALSH J.B. (1981). Effect of pore pressure and confiming pressure on fracture permeability ; Int.Jou.Rock Mech.Min.Sci.,vol.18 nb.5,p.429.

WALSH J.B. and M.A. GROSENBOUGH (1979). A new model for analyzing the effect of fractures on compressibility ; Jou. Geophys. Res. vol.84,nb.B.7,p.3535.

WARBURTON P.M. (1980).A stereological interpretation of joint trace data ; Int.Jou.Rock Mech Min.Sc.,vol.17,nb.4 ,p.181.

WARBURTON P.M. (1980) A Stereological interpretation of joint trace data : influence of joint shape and implications for geological surveys, Int.Jou.Rock Mech.Min. Sc.,vol.17, nb.6, p.305.

WARBURTON P.M. (1983). Applications of a new computer model for reconstructing blocky rock geometry, analyzing single block stability and identifying keystones ; This congress.

WAWERSIK W.R. (1968). Detailed analysis of rock failure in laboratory compression tests ; Ph.D. Thesis Univ. of Minnesota,unpublished.

WAWERSIK W.R. and C. FAIRHURST (1970). A study of brittle rock fracture in laboratory compression experiments ;Int.Jou.Rock Mech. Min.Sci.,vol.7,nb.5,p.561.

WAWERSIK W.R. and D.W. HANNUM (1980).Mechanical behavior of New Mexico rock salt in triaxial compression up to 200°C ; Jou.Geophys.Res. vol.85,nb.B2,p.891.

ZIENKIEWICZ O.C. (1977). The finite element method, 3d edition,787 p.Mc Graw Hill London.

ZIENKIEWICZ O.C.,S.VALLIAPEN and I.P. KING(1968) Stress analysis of rock as a no tension material ; Geotechnique,vol18,p.56.

ZLOTNICKI J.,J.P. POZZI and F.H. CORNET (1981). Investigation of induced magnetization variations caused by triaxial stresses ; Jou.Geophys.Res.,vol.8 ,nb.B 12,p.11899.

ZOBACK M.L. and M. ZOBACK (1980). State of stress in the conterminous United States Jou. Geophys. Res.,vol.85, nb.B 11,p.6113.

SONDERTHEMEN DER FELSMECHANIK

Special topics in rock mechanics

Aspects particuliers de la mécanique des roches

F. H. Cornet
Institut de Physique du Globe, Universität Pierre und Marie Curie, Paris, Frankreich

I. EINLEITUNG

Die klassische Lösung eines Problems der Kontinuumsmechanik erfolgt in drei Stufen :

1. Bestimmung der Geometrie des Problems sowie Untersuchung der Grenzbedingungen;
2. Wahl der entsprechenden zugrundeliegenden Gleichungen, mit der die Rheologie der in Frage stehenden Stoffe ausgedrückt wird;
3. Lösung der Teildifferenzialgleichungen, die dem Gesetz der Bewegung, den Bedingungen der gegenseitigen Verträglichkeit (ausgedrückt in Werten von Verschiebungen oder Spannungen) sowie der zugrunde liegenden Gleichungen, um den Grenzbedingungen zu genügen.

Die bei Felsmechnik begegneten Probleme basieren auf diskontinuierlichen, heterogenen, vielphasigen Stoffen, die nicht mit dem Verfahren Kontinuummechanik unmittelbar zu erfassen sind. So bleibt keine andere Wahl, als den Weg der vereinfachenden Annahme auf Kosten der Genauigkeit zu wählen. Andererseits können Dank des elektronischen Zeitalters und der Vervollkommnung des Hochrechners zunehmend anspruchsvolle digital gesteuerte Verfahren eingesetzt werden, die zuverlässige Messungen vor Ort ermöglichen, sodaß die Gültigkeit der vereinfachenden Annahmen oft unter Beweis gestellt, jedoch auch durch Vergleich der erwarteten und beobachteten Werte für Verschiebungen oder Spannungen als unrichtig erkannt werden.
Obschon sie eine Kunst mit vielen Gesichtern ist hat die Felsmechanik sich zunehmend den Rang einer Wissenschaft erworben, und der Zweck dieser Ausführungen ist es, einige kürzliche Entwicklungen vorzustellen und sie möglicherweise zu fördern.
An erster Stelle wird das Problem der Spannungsbestimmung behandelt. Einige der Meßtechniken werden gleichzeitig mit Ergebnissen der räumlichen und zeitlichen Spannungsänderungen vorgestellt. Dann wird das Problem der geometrischen Kennzeichnung von Diskontinuitäten im Gebirge ausgeleuchtet. Schließlich folgen verschiedene zur Darstellung des Gebirgsverhaltens vorgeschlagene Muster. Diese schließen gleichzeitig finite Elementmodelle, Grenzelementverfahren und ebenfalls verschiedene physikalische und digitale Muster für die Untersuchung großer Verformungen ein.

2. Spannungen in der Felsmechanik.

Bei vielen Felsmechanischen Problemen werden einige Grenzbedingungen für Spannungen "im Unendlichen" bestimmt; dies liegt jedoch außerhalb des Untersuchungsbereichs. Der Ingenieur hat aus diesem Grunde, Wege entwickelt, um vor der Inangriffnahme größerer Abbauarbeiten das örtliche Spannungsfeld festzustellen. Diese Kenntnis des Spannungsfelds "im Unendlichen" kann lediglich eine Bestimmung der Hauptspannungsrichtungen beinhalten, während die Bestimmung seiner Größe häufig genug ebenfalls erforderlich ist.
In letzter Zeit hat die Vorsehung größerer hydraulischer Risse in Erdöllagerstätten das Erfordernis einer richtigen Einschätzung von Spannungsänderungen mit zunehmender Tiefe ergeben : Injektionssdrücke müssen zur Verhinderung unerwünschter senkrechter Rißausbildung bezw. zur Aufrechterhaltung der Rißfortpfalnzung in einem ausgewählten Horizont ausgeglichen berechnet werden (s. auch Cleary, 1980). Es sind ebenfalls Spannungsmessungen zur besseren Erfassung örtlicher Seismizität und deren Begleiterscheinungen, sowie der durch Injektionsdruck hervorgerufenen seismischen Aktivitäten, unternommen worden (s. auch Handin u.A., 1972). Weitere Messungen werden bei gleichzeitigem Einsatz von Geodynamik durchgeführt, um so die Plattentektonik besser zu erfassen (s. Zoback und Zoback, 1980; McGarr, 1982). Angesichts der Vielfalt dieser Probleme, wird man sich fragen, ob überall das gleiche Spannungskonzept angewendet wird. Grundsätzlich muß die Größenbereich in dem die Spannung bei Felsmechanik bestimmt immer vorgegeben sein.
Das allgemeine mechanische Verhalten der Gebirge wird untersucht, indem man entsprechende Kontinua betrachtet, die das tatsächliche Gebirge jedoch nur für einen bestimmten Größenbereich darstellen sollen. Die äquivalenten Kontinua verlieren ihre Bedeutung bei Volumen, die einen kritischen Wert unterschreiten, der als das physikalische Äquivalent des Punktes in der mathematischen Analyse angesehen werden darf. Innerhalb dieses kritischen Bereichs wird die Spannung im äquivalenten Kontinuum als einheitlich angesehen, während dies den Tatsachen effektiv nicht entspricht. Das Problem der Spannungsbestimmung liegt in der genauen Feststellung des Werts, der im äquivalenten Kontinuum vorhanden ist. Es ist daher von wesentlicher Bedeutung sicherzustellen, daß der bei der Spannungs-

messung angewendete Maßstab der gleiche ist, wie für das behandelte Problem.
Bei der Bestimmung von Volumen, die groß genug sind, um dem Problem der örtlichen Verschiedenartigkeit zu begegnen, neigt man dazu örtliche Spannungskonzentrationen zu übersehen, während diese doch das Verhalten der Felsmasse wesentlich beeinflussen können. Werden nur globale Wirkungen in Betracht gezogen, berücksichtigt man die Auswirkung dieser Spannungskonzentration durch Berichtigung der zugrundeliegenden Gleichung. Soweit jedoch Stabilität betroffen ist, wird die Größe des Problems durch die örtliche Spannungskonzentration vorgegeben, und die Analyse kann dadurch erheblich erschwert werden. Tatsächlich ist, wie in diesem Bericht weiter ausgeführt werden soll, die Wahl der richtigen Größe wahrscheinlich eine der Hauptschwierigkeiten bei den Problemen der Felsmechanik.

2.1 Verschiedene Verfahren zur Bestimmung von Spannungen.

(1964) stellte Leeman eine ausführliche Zusammenstellung der verschiedenen derzeitigen Spannungsmeßverfahren vor. (Bohrlochverformungszellen, Bohrlocheinschlußspannungsmeßgeräte, Doorstopper, hydraulische Hebevorrichtungen, Photo-Spannungsmeßverfahren, Widerstands- und Schallverfahren). (1964) brachte Fairhill die Zusammenfassung dann auf den neuesten Stand und berichtete gleichzeitig über die hydraulische Aufbrechmethode. (1969) schlug Leeman wieder einen weiteren Beitrag über die letzten Entwicklungen auf dem Gebiet der Spannungsmessung im Fels (mit mehr als 350 Quellennachweisen) vor. Weitere Verfahren gehen aus Kovaris Bericht über Feldmessungen der Felsmechanik (1977) hervor.
Es sollen daher unter Verzicht auf eine vollständige Zusammenstellung aller verfügbaren Verfahren, hier nur die allgemeinen Trends bei den heutigen Methoden zur Spannungsbestimmung vorgestellt werden, wobei außer auf die Bestimmung der Genauigkeit der Spannungsermittlung durch die verschiedenen Techniken noch besonders auf die jeweils zugrundeliegende Hypothese eingegangen wird. Letzteres wird jedoch nur selten in der Literatur erwähnt, und es steht zu erwarten, daß durch diesen Beitrag eine weitere Behandlung dieses Themas ermutigt werden wird.

Methoden der Spannungsentlastung.

Die klassischen Verfahren der Spannungsentlastung beruhen auf der Messung der durch die vollständige Spannungsentlastung hervorgerufenen Verschiebungen. Die durch die lineare Theorie der Elastizität ermittelten Beziehungen zwischen den Verschiebungen und Spannungen werden dann eingesetzt, um einige Komponenten des "Fernen Feld" -Spannungstensors hochzurechnen.

Beim üblichen Verfahren wird ein erstes Bohrloch abgeteuft und ein Verschiebungsmeßgerät eingebaut. Dieses erste Loch wird dann zur vollständigen Entspannung in dem Felsbereich, wo das Meßgerät eingesetzt ist, überbohrt. Zu diesem Zweck sind verschiedene Geräte entwikkelt worden, von denen alle als weich genug ten, um zu gewährleisten, daß der Fels sich nach erfolgter Spannungsentlastung frei verformen kann.

Bei Bohrloch-Deformationszellen werden verschiedene Kombinationen von Quer- und Längsverformungen gemessen : drei diametrale Messungen mit Meßzelle US-amerik. Bureau of Mines (Hooker and Bickel, 1974) sowie mit der Cherchar-Zelle (Hellat und Dejean, 1981), vier diametrale Messungen (Crouch and Fairhurst, 1967), vier diametrale Messungen und acht Längenmessungen (Bonnechère, 1971).

Bei weichen Einschlußzellen, die entweder mit Verformungsmessern oder induktiven Meßgeräten besetzt werden, werden kombinierte Umfangs-, Längs- und Schrägverschiebungsmessungen durchgeführt (Leeman, 1968; Rocha und Silverio, 1969; Blackwood, 1977; Pahl, 1977).

Während Zellen, die nur Durchmessermessungen zur vollständigen Bestimmung der Spannung zulassen, Bohrlöcher in mindestens drei verschiedenen Richtungen erfordern, ist bei Zellen, die Radial- und Längs- bezw. Umfangsverschiebungsmessungen kombinieren, lediglich ein Überbohren für die vollständige Bestimmung erforderlich. Alle diese Verfahren begründen sich auf das infinite zylindrische Loch ineinem elastischen isotropen Stoff, der einem dreiachsigen Feld "ad infinitum" unterliegt. Die bekannte Lösung dieses Problems (s. Kirsch, 1968; Hiramatsu und Oka, 1968) ist hinsichtlich Spannungen durch Gleichung (1) und hinsichtlich Verschiebungen durch Gleichung (2) gegeben.

Wenn σ^{∞} ij, i,j, = 1,2,3 die unbeeinflußten -Spannungskomponenten (ausgedrückt im geographischen Bezugsrahmen) sind,

dann sind
ρ, θ, z zylindrische Koordinaten (z ist die Achse des Bohrlochs, die positiv zum freien Ende des Bohrlochs weist),

E, ν, sind der Elastizitätsmodul und das Querdehnungsverhältnis für den Fels.

r ist der Bohrlochradius, dann :

$$
1\begin{cases}
\sigma_{\rho\rho} = (1-\frac{r^2}{\rho^2})\frac{\sigma_{11}+\sigma_{22}}{2} + (1-4\frac{r^2}{\rho^2}+3\frac{r^4}{\rho^4})(\frac{\sigma^{\infty}_{11}-\sigma^{\infty}_{22}}{2}\cos 2\theta + \sigma^{\infty}_{12}\sin 2\theta) \\
\sigma_{\theta\theta} = (1+\frac{r^2}{\rho^2})\frac{\sigma^{\infty}_{11}+\sigma^{\infty}_{22}}{2} - (1+3\frac{r^4}{\rho^4})(\frac{\sigma^{\infty}_{11}-\sigma^{\infty}_{22}}{2}\cos 2\theta + \sigma^{\infty}_{12}\sin 2\theta) \\
\sigma_{zz} = \sigma^{\infty}_{33} - 4\nu\frac{r^2}{\rho^2}(\frac{\sigma^{\infty}_{11}-\sigma^{\infty}_{22}}{2}\cos 2\theta + \sigma^{\infty}_{12}\sin 2\theta) \\
\sigma_{\theta z} = (1+\frac{r^2}{\rho^2})(\sigma^{\infty}_{23}\cos\theta - \sigma^{\infty}_{31}\sin\theta) \\
\sigma_{z\rho} = (1-\frac{r^2}{\rho^2})(\sigma^{\infty}_{31}\cos\theta + \sigma^{\infty}_{23}\sin\theta) \\
\sigma_{\theta\rho} = (1+2\frac{r^2}{\rho^2}-3\frac{r^4}{\rho^4})(\frac{\sigma^{\infty}_{22}-\sigma^{\infty}_{11}}{2}\sin 2\theta + \sigma^{\infty}_{12}\cos 2\theta)
\end{cases}
$$

$$
2\begin{cases}
u_{\rho} = \frac{1+\nu}{E}(\frac{1-\nu}{1+\nu}\rho + \frac{r^2}{\rho})\frac{\sigma^{\infty}_{11}+\sigma^{\infty}_{22}}{2} - \frac{\nu}{E}\rho\sigma^{\infty}_{33} + \\
\quad + \frac{1+\nu}{E}(\rho + 4(1-\nu)\frac{r^2}{\rho} - \frac{r^4}{\rho^3})(\frac{\sigma^{\infty}_{11}-\sigma^{\infty}_{22}}{2}\cos 2\theta + \sigma^{\infty}_{12}\sin 2\theta) \\
u_{\theta} = \frac{1+\nu}{E}(\rho + 2(1-2\nu)\frac{r^2}{\rho} + \frac{r^4}{\rho^3})(\frac{\sigma^{\infty}_{11}-\sigma^{\infty}_{22}}{2}\sin 2\theta + \sigma^{\infty}_{12}\cos 2\theta) \\
u_{z} = \frac{2(1+\nu)}{E}(\rho + \frac{r^2}{\rho})(\sigma^{\infty}_{31}\cos\theta + \sigma^{\infty}_{23}\sin\theta) - \frac{2\nu}{E}z\frac{\sigma^{\infty}_{11}+\sigma^{\infty}_{22}}{2} + \frac{z}{E}\sigma^{\infty}_{33}
\end{cases}
$$

Die Lösung ist auf den Fall der ebenen Isotropie (5 elastische Konstanten) ausgedehnt worden (Berry, 1968) und später durch Hiroshima und Koga (1977). Beim "Doorstopper-Verfahren" wird eine Dehnungsmeßstreifenrosette an das abgeflachte Ende des Bohrlochs geklebt. Sobald wieder mit dem Kernbohren fortgefahren wird, mißt man die Verschiebungen, die durch die Spannungsentlastung entstehen. Die vollständige Spannungstensorbestimmung erfordert mindestens drei verschiedene Bohrrichtungen. Die elastische Lösung basiert auf den Erfahrungswerten für isotrope Stoffe, die entweder experimentell ermittelt worden sind (Galle und Wilhoit, 1962); Leeman, 1964; Bonnechère und Fairhurst, 1967) oder durch finite Elementberechnungen (De la Cruz, 1969). Die Lösung für querisotropische Medien (planare Isotropie) wurde während dieser Konferenz für den Fall des parallel oder lotrecht zur isotropen Ebene des Fels liegenden Bohrlochs vorgestellt (Rahn, 1983). Beiwerte sind mittels finiter Elementberechnungen bestimmt worden. Der wesentlichste Nachteil der Spannungsentlastungsmethode liegt darin, daß der Fels als linear elastisch angenommen wird (sowie ebenfalls in denmeisten Fällen isotrop), und daß elastische Konstanten bekannt sein müssen. Eine weitere Einschränkung beruht auf der Tatsache, daß die Messungen außerordentlich zeitraubend sind und daher entsprechend kostspielig.

Darüber hinaus sind die nach diesem Verfahren durchgeführten Spannungsmessungen vorwiegend auf Bohrlöcher einer geringen Tiefe beschränkt geblieben (meist kürzer als 50 m), obschon Ergebnisse, die Martna u.A. (1983) erzielt haben, von Tiefenmessungen bis 500 m berichten. Schließlich tritt aber auch bei Fels, der unter hohen Drücken steht, ein Abblättern des Kerns auf, der das Erreichen ausreichend langer überbohrter Kerne verhindert. In diesen muß dann auf das "Doorstopper"-Verfahren zurückgegriffen werden.
Der wesentliche Vorteil dieses Verfahrens ist, daß die Genauigkeit der Bestimmung geschätzt werden kann, obschon diese Tatsache in den meisten Berichten über dieses Thema unerwähnt bleibt.
Bei Verwendung der genannten Instrumente werden Verschiebungen entweder an einem Punkt gemessen (Bohrlochverformungszellen) oder entlang einer sehr kurzen Basis (Druckmeßdosen sind nicht länger als 1 oder 2 cm). Die Ergebnisse werden von örtlichen Inhomogenitäten stark beeinflußt, und es werden daher häufig starke Streuungen beobachtet. Blackwood (1978) hat vorgeschlagen, während des Überbohrvorgangs Aufzeichnungen über Verschiebungen zu führen und durch Vergleich der während des fortschreitenden Überbohrens ermittelten Kurven mit den theoretischen Kurven, die durch finite Elementberechnung erzielt werden, nur die theoretisch zulässigen Messungen zur Spannungsermittlung heranzuziehen.
Cornet u.A. (1979) schlugen vor, alle theoretisch zulässigen Verschiebungsmessungen gleichzeitig einzusetzen, um das Spannungsfeld und ebenfalls eventuelle Bestimmungsfehler zu ermitteln. Tatsächlich darf gesagt werden, daß wenn (E) ein Vektor ist, der sich aus allen zulässigen Messungen zusammensetzt, (X) ein Vektor ist, dessen Komponenten die der Spannung außerhalb des betrachtungsrahmens sind, (A) eine Matrix ist, deren Beiwerte die des Spannungs-Verschiebungsverhältnisses bezogen auf die Meßtechnik sind, (B) eine Matrix ist, die für jede Messung den Spannungstensor vom geographischen Bezugsrahmen zum Zell-Bezugsrahmen dreht, und (C) ein Matrix ist, deren Beiwerte die Faktoren der Spannungskonzentration in Abhängigkeit von der Entfernung zum Meßloch darstellen, dann dürfen alle zulässigen Ergebnisse wie folgt ausgedrückt werden : -

$$[A] \quad [B] \quad [C] \quad \{X\} = \{E\} \qquad (3)$$

Das lineare Gleichungssystem wird überbestimmt, wenn mehr als sechs Messungen erzielt worden sind; demzufolge können die Komponenten von (X) auf dem Wege der kleinsten Fehlerquadrate ermittelt und damit der Fehler der Bestimmung abgeschätzt wird. In der Praxis ist (A) selten bekannt, und es können nur die Messungen, die in weiter Entfernung vom Hohlraum vorgenommen wurden gleichzeitig berücksichtigt werden. Bei Messungen aus näherer Entfernung können nur die gleichzeitig verwendet werden, die in gleicher Entfernung vom Hohlraum vorgenommen worden sind. Es geht ganz klar hervor, daß mit dieser Inversionstechnik, die Verschiebungsmessungen, die punktuell sind, zum Erzielen eines Schätzwerts für den mittleren Spannungstensor im Maßstab des Volumens, bei dem die Messungen vorgenommen wurden, gemittelt werden. Der Einfluß großer Diskontinuitäten (d.h. bezogen auf den Maßstab des Volumens in dem die Messungen vorgenommen wurden) darf nicht übersehen werden. Dann können die Messungen, die in keinem Zusammenhang mit der Spannungsermittlung stehen, entfallen. Die verbleibende Sammlung kann nunmehr für eine feinere Spannungsermittlung eingesetzt werden. Abb. 1 stellt ein Beispiel für eine derartige Rückanalyse dar. Die Richtung der Längsachse des des verformten Bohrlochquerschnitts, an dem Messungen vorgenommen werden wird mit der aus der globalen Spannungsberechnung resultirenden Richtung verglichen.
Es ergibt sich ganz klar, daß die bei 1.80 Abstand von der Hohlraumwand (es handelt sich hierbei um einen horizonzalen Schacht von 3 m Ø) vorgenommenen Messungen nicht mit der Bestimmung der Gesamtspannung übereinstimmen. Diese durch die Nähe des Hohlraums beeinflußten Messungen dürfen nicht in die Sammlung der zulässigen Verschiebungen eingehen.

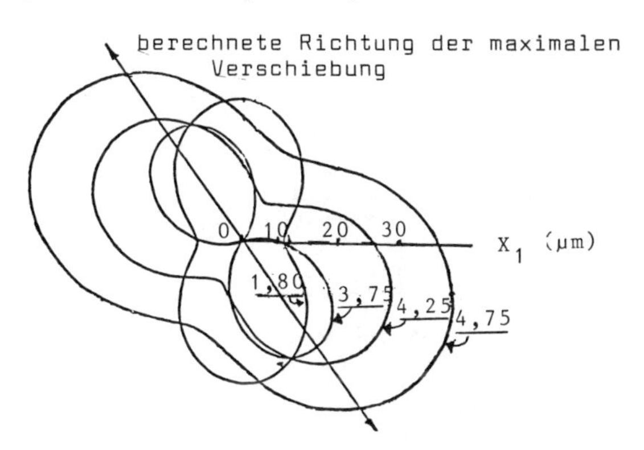

Abb. 1 Vergleich zwischen der berechneten und gemessenen Verformung eines Bohrlochs nach dem Überbohren. Die flachen Kurven werden durch Messungen der effektiven Querverschiebungen und die errechnete Achse aus der Spannungsbestimmung ermittelt. Die Bezugsachse (X_1) ist die horizontale Achse des Bohrlochquerschnitts.

Steife Einschlüsse.

Starre Einschlüsse, die auch als "Spannungsmesser" bezeichnet werden, sind so konstruiert, daß sie jeweils eine Komponente messen. Sie müssen gemessen am Mutterfels sehr steif sein (nach Leeman, 1964) wenigstens 4 bis 5 mal so hoch wie der Felsmodul, damit Spannungsänderungen im Mutterfels unmittelbar auf den Einschluß übertragen werden. Verformungsmeßgeräte (nach Hawkes und Hooker, 1974 sowie Pariseau und Eitani, 1977 - Schwingungsdrähte; nach Potts 1957 und Filak und Cyrul 1977 - Meßzellen; nach Roberts u.A., 1964 - Photo-Elastische Zellen), die direkt in den Einschluß eingesetzt sind, werden zur Ermittlung der vom Einschluß getragenen Spannungsänderungen angewendet. Der Hauptvorteil der Spannungsmesser gegenüber weichen Einschlüssen, besteht darin, daß sie keine genaue Kenntnis des Felsmodul voraussetzen. Für sie ist es nocht nicht einmal erforderlich, daß der Fels linear elastisch ist, solange der Einschluß erheblich steifer ist als der Fels.
Sie sind gut zur Ermittlung zeitabhängiger Spannungsänderung zu verwenden, jedoch wegen der bestehenden Kontaktprobleme noch keineswegs auf so breiter Ebene für die sichere Spannungssbestimmung eingesetzt worden, wie die weichen Einschlüsse oder Bohrloch-Verformungszellen (s. auch Fairhurst, 1968). Zur Zeit werden sie nur für verhältnismäßig flache Bohrlöcher verwendet. Darüber hinaus können Probleme mit der Langzeitstandfestigkeit auftreten, wenn sie für die Bestimmung von zeitabhängigen Spannungsänderungen eingesetzt werden (Fairhurst, 1968). Ein Verfahren zur Eichung ihrer Genauigkeit ist noch nicht beschrieben worden.

Verfahren zur Spannungs- und Verformungskompensation.

Das älteste Spannungsausgleichverfahren ist das von Mayer u.A. (1951) und Tincelin (1951) entwickelte. Es werden erst auf einer freien Fläche zwei Pflöcke eingesetzt, und die Entfernung zwischen den Pflöcken wird genau gemessen. Nach Entlastung des Spannungsfeldes durch einen Sägeschnitt senkrecht zur Fläche, wird der Abstand zwischen den Pflöcken nochmals gemessen. Dann wird ein hydraulisches Hebegerät in den Sägeschnitt eingeführt und so mit Druck beaufschlagt, daß der Abstand zwischen den Pflöcken wieder der gleiche ist wie zuvor. Der für die Wiederherstellung des vorherigen Abstands erforderliche Druck spiegelt die Normalspannung wider, die vor der Ausführung des Schnitts auf diese Fläche wirkte. Hoskins (1966) erläutert die wohl beste Verteilung der Pflöcke, die die Wirkung der Scherspannungsentlastung auf ein Mindestmaß herabsetzt.

Ein Nachteil des Verfahrens, besteht in der Begrenzung auf kurze Entfernung von der freien Fläche. In diesem Bereich treten häufig nichtelastische Verformungen auf, und es ergeben sich daraus Schwierigkeiten, die gemessenen Spannungen zu dem Spannungszustand des "Fernen Felds" in Beziehung zu setzen. Darüber hinaus müssen auch die Viskositätseinflüsse sehr gering sein, damit das Verfahren genau ist.

Balthasar und Wenz (1983) schlagen eine Anwendung dieser Methode auf tiefe Bohrlöcher vor. So könnten dann Messungen nicht nur in einiger Entfernung von der freien Fläche durchgeführt werden, es wäre ebenfalls möglich, den ganzen Spannungstensor zu erfassen.
Kutznetzov u.A. (1983) haben eine neue Spannungsausgleichstechnik für nicht linearen elastischen Fels entwickelt. Das abgeflachte Ende des Bohrlochs wird mit Druck beaufschlagt, um beim Weiterbohren, d.h. während der Spannungsentlastung, eine axiale Verformung desselben zu verhindern. Die Bestimmung des zur Verhinderung jeder Art von Verformung erforderlichen Drucks kann bei drei orthogonalen Bohrlöchern auf den Spannungstensor an diesem Punkt bezogen werden. Die Werte dieser diagonalen Komponenten, die in dem den drei Achsen der Bohrlöcher zugeordneten Bezugsrahmen bestimmt sind, lassen sich ermitteln. Zur vollständigen Bestimmung (* Bohrlöcher in andere Richtungen, als die zuvor genannten gebohrt werden. Ein Vorteil ist hier, daß die Elastizitätszahl des Fels nicht bekannt sein muß. Andererseits muß jedoch das Querdehnungsverhältnis gemessen werden. Der wesentliche Nachteil sind die vielen Bohrungen, die für eine vollständige Bestimmung des Spannungstensors erforderlich sind.
Die Untersuchungsmethode unter Verwendung der differentialen Verformungskurve die Ren und Roegiers (1983) vorschlagen geht von der Annahme aus, daß vollständige Spannungsentlastung an einer Gesteinsprobe Haarrisse hervorruft, deren Dichte in einer gegebenen Richtung proportional den Spannungsänderungen in der Normalen zu dieser Richtung ist. Demzufolge wird angenommen, daß das induzierte Rißspektrum die Spannungsgeschichte widerspiegelt. Das Beaufschlagen der Gesteinsprobe mit einem hydrostatischen Druck sollte ein nicht hydrostatisches Verformungsfeld hervorrufen, dessen Hauptachsen denen des ehemaligen Spannungsfelds identisch sind, dem die Gesteinsprobe ausgesetzt war, ehe sie entnommen wurde. Darüber hinaus sollten die relativen Werte der Verformungen entlang der Hauptachse Aufschluß über die relativen Werte der Hauptspannungen geben. Die von Ren und Roegiers vorgelegten Ergebnisse sind eine aufschlußreiche Darstellung der Möglichkeiten und Grenzen des Verfahrens. Vor allem bleibt der Einfluß tektonischer Restspannungen schwer zu handhaben, und dies trifft gleichermaßen auf den Einfluß der Dilatanz zu. Nichtsdestotrotz bietet dieses Verfahren, wenn es zum Einsatz kommt, einen preiswerten Weg zur umfassenden Ermittlung des Spannungsfelds an.

Methode der hydraulischen Rißbildung.

Hydraulische Rißbildung wird bereits seit längerer Zeit zur Tiefenspannungsermittlung eingesetzt (s. auch Hubbert und Willis, 1957; Scheidegger, 1962; Fairhurst, 1964; Kehle, 1964; Haimson und Fairhurst, 1969).
Bei diesem Verfahren in seiner ursprünglichen Form wurde ein Teil des Bohrlochs mit einem Spreizpacker verschlossen und dann mit Druck bis zum Aufreißen beim sog. Aufreißdruck (P_b) beaufschlagt. Sobald die Druckzufuhr beendet war, konnte bei noch verschlossenem Bohrloch ein Einschlußdruck (P_s) beobachtet werden (s. Abb. 2).
Liegt das Bohrloch parallel zu einer der Hauptspannungsrichtungen (sie wird oft als vertikal angenommen), und wenn der Fels linear elastisch und bezogen sowohl auf sein elastisches Verhalten und seine "Festigkeit" isotrop ist, können Aufreiß- und Einschlußdruck eingesetzt

---

(*des Spannungstensor müssen weitere...

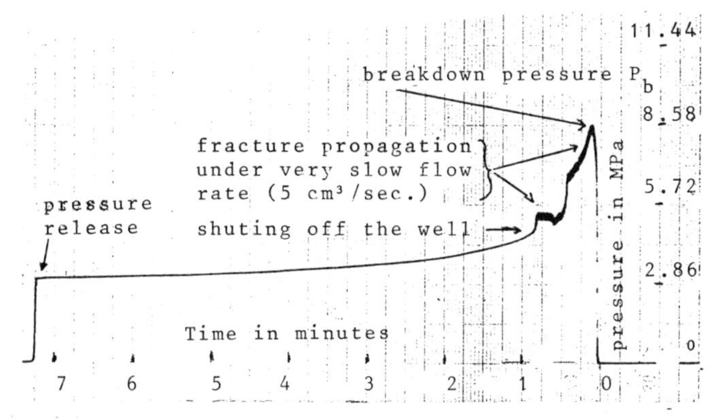

Abb. 2 Eine typische Druck:Zeit Aufzeichnung während einer hydraulischen Rißbildung in Granit bei 200 m Tiefe.

werden, um einige Komponenten des örtlichen Spannungstensors zu bestimmen. Da meistens pneumatische Packer eingesetzt werden, entstehen vertikale Risse bei vertikalen Bohrlöchern, sogar im oberflächennahen Bereich (Haimson und Fairhurst, 1969). Aus Gleichung 1 läßt sich für trockenen, undurchlässigen Fels beweisen, daß

$$\sigma^t - \sigma_1 + 3\sigma_2 = Pb \qquad (4a)$$

$$\sigma_2 = P_s \qquad (4b)$$

worin σ_1 und σ_2 die Hauptspannungskomponenten lotrecht zur Bohrlochachse ($\sigma_1 > \sigma_2$) sind und σ^t die sogenannte Zugfestigkeit des Fels darstellt. Bestimmung von P_b, P_s und σ^t im Labor oder vor Ort ermöglichen eine Schätzung von σ_1 und σ_2. Nach Griffiths Theorie der Bruchbildung (Griffith, 1921) läßt sich beweisen, daß bei isotropem Fels der Riß sich lotrecht zu σ_2 erstreckt, und daß dementsprechend die Richtung des Risses am Bohrloch, die Richtung von σ_1 bestimmt. Die Gleichung (4 a) muß bei durchlässigem Gestein zur Berücksichtigung der Wirkung von Porenwasserdruckänderung im Bereich um das Bohrloch berichtigt werden (s. auch Haimson, 1968; Rise und Cleary, 1976). Ist die Felsmasse mit einer unter Druck P_p stehenden Flüssigkeit gesättigt, andererseits jedoch undurchlässig genug, um eine Versickerung in die Gesteinsformation zu verhindern, dann ist Gleichung (4 a) wie folgt zu ersetzen :

$$P_b = -\sigma_1 + 3\sigma_2 + \sigma^T - P_p \qquad (5)$$

Tatsächlich kann, wie Abb. 3 zeigt, das statische Spannungsfeld in zwei Komponenten zerlegt werden : -
Komponente A, die hydrostatisch ist,
Komponente B, die nur effektive Spannungen des ungestörten Feldes enthält.

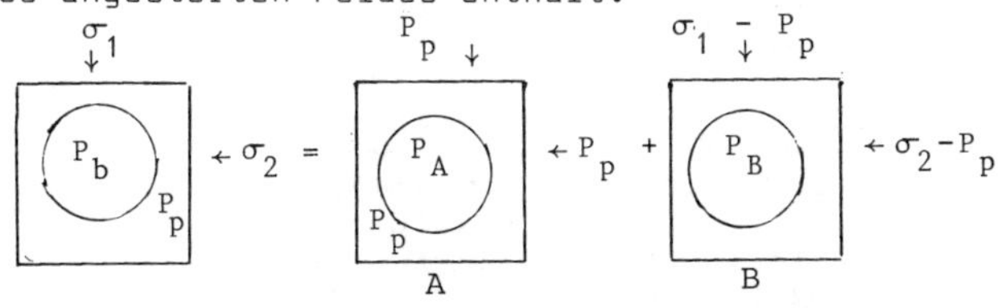

Abb. 3 Der Einfluß von Porendruck auf den Aufreißdruck im Falle von undurchlässigem Fels.

Der zur Rißbildung für Komponente B erforderliche Bohrlochdruck P_B lautet : -

$$P_B = -(\sigma_1 - P_p) + 3(\sigma_2 - P_p) + \sigma^T \qquad (6)$$

während der zur Aufrechterhaltung eines gleichmäßigen hydrostatischen Drucks P_B erforderliche Bohrlochdruck P_A lautet : $P_A = P_B$, und es ergibt sich für die Komponenten A und B die Gleichung (5).
Was den Einschlußdruck anbelangt, entsteht ein derartiges Problem nicht. Wird Abb. 3, das Bohrloch, durch einen Riss in der Normalen zur σ_2 -Richtung ersetzt, stellt man fest, daß der Druck, der erforderlich ist, um den Riß unter Komponente B gerade noch zu öffnen gleich $\sigma_2 - P_p$ ist, sodaß der Einschlußdruck unter den kombinierten Komponenten A und B, bei dem der Riß ausgeglichen ist, = σ_2 ist.
In dieser Formulierung werden dem hydraulischen Rißbildungsverfahren gewisse Grenzen gesetzt :
1. die Felsmasse muß bezogen sowohl auf ihre elastischen Eigenschaften und ihre Festigkeit, linear elastisch und isotrop sein;
2. das Bohrloch muß parallel zu einer der Hauptspannungsrichtungen liegen;
3. die Festigkeit des Fels muß bekannt sein;
4. der Spitzenwert des Druck:Zeit Diagramms muß dem Druckwert entsprechen, bei dem die Rißbildung entsteht,
5. die Genauigkeit der Bestimmung hängt ab von der Wiederspruchsfreiheit der Ergebnisse.

Brodehoeft u.A. (1976) wies darauf hin, daß das Problem der Festigkeitsbestimmung durch Behandlung ausschließlich des Wiederöffnungsdrucks (P_r) gelöst werden kann, indem man nur den Wiederöffnungsdruck heranzieht. Tatsächlich wird Gleichung (5) zu :

$$P_r = 3\sigma_2 - \sigma_1 - P_p \qquad (7)$$

nachdem der Riß entstanden ist, und der Porendruck wieder seinen ursprünglichen Wert im Gesamtfelsvolumen erreicht hat, da die Zugfestigkeit entlang der Rißfläche jetzt 0 ist.

Es darf, damit Gleichung (7) gültig ist, entlang der Rißebene vor dem Wiederöffnen kein Fließen auftreten (Cornet, 1976), und es muß daher eine hohe Druckbeaufschlagung bezw. eine hoch viskose Flüssigkeit eingesetzt werden. Unter diesen Bedingungen tritt die Wiederöffnung des Risses nicht gleichzeitig mit Erreichen des Spitzenwerts der Druck:Zeit Aufzeichnung ein.

Um eine genaue Messung des zur Wiederöffnung erforderlichen Drucks zu erzielen, schlagen Hickman und Zoback (1982) vor, mit dem gleichen Druck zu fahren, der für die Rißbildung im Fels eingesetzt wurde. Dann wird der Wiedereröffnungsdruck als der Punkt festgestellt, bei dem die Druck:Zeit Aufzeichnung von der ursprünglichen abweicht. Sie raten ferner, den zweiten Wiedereröffnungstakt zu verwenden, um so das Problem einer unvollständigen Rißentwicklung während des ersten Druckversuchs zu vermeiden.

Häufig ist den Druck:Zeit Aufzeichnungen keine schlüssige Auskunft über den Einschlußdruck zu entnehmen, und es sind verschiedene Wege zur Verbesserung der Ergebnisse vorgeschlagen worden (s. auch McLennan und Roegiers, 1982; Gronseth und Kry, 1982; Aamodth und Kuriyawagawa, 1982). Alle beziehen sie sich auf einen unmittelbaren Einschlußdruck, dem Druckabfall, der direkt nach dem Verschließen des Bohrlochs festzustellen ist. Hierfür werden verschiedene Formen der Darstellung auf Millimeterpapier vorgeschlagen. Eine weitere Möglichkeit besteht darin, den Druck im Bohrloch schrittweise zu steigern, nachdem sich der Riß gut entwickelt hat. Der erste Schritt erfolgt bei einem Druckwert, der weit unter dem erwarteten Einschlußdruck liegt. Dann kann durch Auftragung der Durchflußrate, die erforderlich ist, um einen Druck gegen den Druckwert aufrecht zu erhalten, derjenige Druck möglichst genau gefunden werden, bei dem die Rißbildung einsetzt. Bei diesem Druckpegel setzt entlang des Risses Fließen ein, und die Fließrate, die zur Aufrechterhaltung eines konstanten Drucks erforderlich ist, wird erheblich erhöht. Dieser Druckwert ist, sofern die Stufen lang genug sind, um Druckgradienten im Riß zu verhindern, genau gleich der Normalspannung, die auf die Trennfläche wirkt. Hierbei wird natürlich unterstellt, daß der Riß während er sich vom Bohrloch ausgehend fortpflanzt, eben bleibt, und es ist hierbei besonders wichtig, daß kein bereits zuvor vorhandener Riß den neuen Riß aus seiner ursprünglichen Richtung ablenkt.

Um die Einschränkungen 1, 2 und 5 aufzuheben, hat Cornet (1982) vorgeschlagen, die immer vorhandenen Schwächeflächen eines Gebirges zu nutzen, indem ein Spreitzpacker vor einer leicht zementierten Fuge aufgesetzt wird. Diese Fuge läßt sich durch den Einsatz langsamer Fließraten erneut wieder öffnen. Durch eine Angleichung des Einschlußdrucks (P_s) oder des schrittweisen Wiederöffnungsdrucks auf die Größe der Normalspannung, der die Trennfläche ausgesetzt ist, wird die folgende Gleichung erzielt :

$$\underset{\sim}{\sigma} \cdot \underline{n} = P_s$$

in der $\underset{\sim}{\sigma}$ die örtliche Spannung, $\underline{n}$ die Normale zur Trennfläche und P_s der Einschlußdruck sind. In dem zumeist üblichen Fall werden sechs verschiedene Rißrichtungen in gleicher Tiefe benötigt, um die sechs unabhängigen Komponenten von $\underset{\sim}{\sigma}$ zu bestimmen.

Wird jedoch angenommen, daß eine der Hauptspannungen vertikal ist (jedoch nicht unbedingt parallel zur Bohrlochachse verläuft), dann sind nur drei verschiedene Fläche erforderlich. Stehen mehr als drei Risse zur Verfügung wird eine normale Technik kleinster Fehlerquadrate angewendet, um das lineare Gleichungssystem zu lösen. Dieses neue Verfahren läßt sich leicht für die Spannungsmessungen in untertägigen Hohlräumen einsetzen, wo viele Bohrlöcher abgeteuft werden können. Ist jedoch nur eines vorhanden, wird die Spannungsänderung mit der Tiefe als linear angenommen :

$$(\underset{\sim}{\sigma}) = \begin{pmatrix} \sigma_1 + \alpha_1 z & 0 & 0 \\ 0 & \sigma_2 + \alpha_2 z & 0 \\ 0 & 0 & \rho g z \end{pmatrix} \qquad (9)$$

in der ρg der vertikale Gradient der vertikalen Spannung ist. Wird als erste Annäherung angenommen, daß $a_1 = a_2$ (Messungen werden sehr nahe bei einander durchgeführt, sodaß nur der Einfluß von Schwerkraft berücksicht wird). Gleichung (8) kann durch eine angemessene Änderung der Variablen linear gemacht werden, und es kann somit nach wie vor eine normale Technik des kleinsten Fehlerquadrats angewendet werden. Sobald Rißflächen etwas von einander entfernt sind, ist Gleichung (8) nicht mehr linear, und es müssen aufwendige Umwandlungsmethoden eingesetzt werden (Valette und Cornet, 1983).

Sobald die meisten Risse parallel zur Bohrlochrichtung sind und bezogen auf den geographischen Bezugsrahmen die gleiche Orientierung haben, wird die klassische Auslegung der Methode angewendet. Die Prüfung geneigter Trennflächen liefert nur ein Mittel für die Stichhaltigkeiteit der Feststellung. So lassen sich Fragen nach der möglichen Auswirkung der Gebirgsanisotropie auf hydraulische Rißfortpflanzung, die durch natürliche Schwächeflächen hervorgerufen werden, leicht beantworten. Darüber hinaus ergibt sich hieraus ein Mittel zur Bewertung der Genauigkeit der Bestimmung.

Es dürfte sich lohnen, darauf hinzuweisen, daß der Größenbereich nach dem die Spannung mit der Methode hydraulischer Rißbildung gemessen wird, recht flexibel ist. In Abhängigkeit von Abstand zwischen den Packern, vom Bohrlochdurchmesser und der Aufnahmemenge an abgepreßter Flüssigkeit, kann es sich bei der Spannungsmessung um kleine Volumen von nicht über einigen Kubikzentimetern (Laborbedingungen) und um größere Volumen von einigen Kubikmetern handeln. Bei größeren Volumen kommt dem Einfluß des durch Schwerkraft erzeugten Spannungsgradienten eine größere Bedeutung zu. Darüber hinaus wird es wieder schwieriger dem Problem der Wechselwirkung zwischen natürlichen Schwächeflächen und der künstlichen hydraulischen Rißbildung zu begegnen.

Es sind noch weitere Verfahren zur Erzeugung von Rißbildung für die Spannungsmessung vorgeschlagen worden (s. auch Jaeger und Cook, 1936; De La Cruz, 1976). In dieser Hinsicht scheint die Anwendung von "Pressiometern" neue Aspekte zu bieten (s. auch Huergo, 1983; Charlez, 1983), da dies Verfahren gleichzeitig mit den stationären Messungen der Felsverformung auch die Spannung mißt.

Stephansson (1983) bespricht den Einsatz des Pressiometers der Colorado School of Mining (Hustrulid und Hustrulid, 1975) bei Spannungsmessungen vor Ort. Er schlägt hier insbesondere vor, den zum Wiederöffnen des Risses erforderlichen Druck, der während eines ersten Lasttaktes erzeugt worden ist, als Messung der kleinsten Hauptnormalspannung des natürlichen Spannungsfeldes lotrecht zur Bohrachse zu verwenden (das Bohrloch wird dann als parallel zu einer der Hauptspannungen verlaufend angenommen). Andererseits hat es den Anschein, daß dieser Druckwert bei der hydraulischen Rißbildung in Abhängigkeit zu beiden Hauptspannungen $\sigma 1$ und $\sigma 2$ sowie der Bildung von Porendruck (s. Gleichung 7) steht. Man erhofft sich von der Konferenz eine Diskussion dieses Punkts.

Weiter ist vorgeschlagen worden, die natürlichen durch tektonische Spannungen hervorgerufenen Vorgänge der Rißbildung zur Bestimmung der Spannung einzusetzen. Hierbei handelt es sich um in der Seismologie und Mikrotektonik geläufige Verfahren. Sie werden als getrennte allgemeine Spannungsermittlung beschrieben, da sie sich auf größere Gebirge (von einigen Kubikdekametern bis zu einigen Kubikkilometern) beziehen.

Seismisch-tektonische Verfahren.

Seit langem schon sind Seismologen in der Lage, die sog. Störflächenlösung für Erdbeben, die auch als "Herdmechanik" bezeichnet werden kann, einzusetzen. Sie beziehen sich hierbei auf die Lagebestimmung der Scherfläche, die das Erdbeben verursachte, sowie ebenfalls auf die Bewegungsrichtung entlang der Fläche (s. auch Stauder, 1962). Obschon diese Methode ursprünglich empirisch angewendet wurde, ist sie inzwischen verfeinert worden und stellt jetzt in der Seismologie ein gebräuchliches Verfahren dar. Diese Lösungen sind dann zur Bestimmung der Hauptrichtungen des Spannungsfelds um das Zentrum des Bebens angewendet worden. Ehe wir zu diesem letzten Punkt übergehen, möchten wir zuvor auf das Prinzip der Störungsflächenlösungen zurückkommen. Es war bereits lange bekannt (Byerly, 1926), daß bei einem Erdbeben die erste durch P-Wellen (Druckwellen), die in der Ausbreitungssrichtung der Emission polarisiert sind) hervorgerufenen Erschütterungen, in einigen Bereichen verdichtend wirken und in anderen ausdehnend. Dieses Emissionsmodell läßt sich damit erklären, daß die Quelle auf einige punktuelle Kräfte zurückzuführen ist. Nakano (1923) bewies z.B., daß ein Paar punktueller Kräfte abwechselnd Dilatation und Kompression in einem durch zwei orthogonale Ebenen geviertelten Raum bewirken (s. Abb. 4).

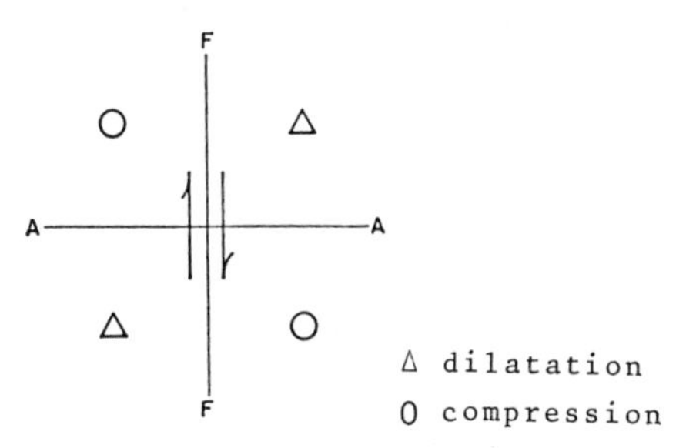

Abb. 4 Verteilung der aus Bewegungen entlang einer Störungsfläche herrührenden Kompressionen und Dilatationen (Stauder, 1962).

Später haben dann einige Autoren (Knopoff und Gilbert, 1960; Balkina, Shirokowa und Vvedenskaja, 1960) darauf hingewiesen, daß das Problem der Beschreibung seismischer Emissionen durch Erdbeben mit einer Verschiebung erklärt werden kann, d.h. einer Diskontinuität im tagentialen Versetzungsfeld der Störungsfläche. Burridge und Knopoff (1964) haben bewiesen, daß die aus einer derartigen Unstetigkeit im Versetzungsfeld herrührende Emission gleichwertig der einer Doppelkopplung (punktuelle Kräfte mit einem Moment) ist, die für den Brennpunkt des Bebens angesetzt wird. Eine der Knotenflächen der so gebildeten Emission stimmt mit der Rißfläche überein. In der Seismologie spricht man hier von der Darstellungstheorie. Damit eine solche Darstellung als gültig anerkannt werden kann, müssen die für die Emission eingesetzten Wellenlängen größer sein, als die Länge des Herdes, sodaß die durch die Endeffekte der Versetzungsunstetigkeit sowie durch die örtlichen Heterogenitäten im Medium hervorgerufenen Turbulenzen die erste Bewegung der festgestellten Emission nicht verändern können. Auf diese Weise wurde die theoretische Grundlage für die beobachtete Verlaufsrichtung der ersten Bewegungen festgestellt.
Derzeit geschieht die Bestimmung des "Brennpunktmechanismus" eines Bebens wie folgt : -
1. Feststellung des Brennpunkts, der in den Mittelpunkt einer Lagenkugel gelegt wird;
2. Bestimmung der Richtung des ersten Eintreffens an so vielen Stationen, wie möglich. Es dürfen nur lange Wellenlängen bezogen auf die Größe des Ausgangspunktes verwendet werden. Die größe des Ausgangspunkts wird aus der Eckfrequenz des Fourier'schen Emissionsspektrums ermittelt (s. Madariaga, 1979).
3. Die Richtung der Wellen an jeder Stelle aufzeichnen und als Polpunkt in die Lagenkugel eintragen. Mit Richtungsbestimmung ist das Aufspüren des Wegs gemeint den die seismische Emission von ihrem Ausgangspunkt bis zum Beobachtungspunkt zurückgelegt hat. Dieser Weg hängt weitgehend von den durch das Gefüge des Mediums vorgegebenen Wellenausbreitungsgeschwindigkeit ab und kann bei sehr heterogenen Medien u.U. schwer feststellbar sein.
4. Durch den Einsatz einer flächentreuen Projektion, wird die Projektion der Verschneidungspunkte der Wellen mit der unteren Halbkugel auf der äquatorialen Fläche der Kugel aufgetragen. Werden Stationen eingesetzt, die nahe dem Epizentrum des Bebens liegen (vertikale Projektion des Brennpunkts auf die Oberfläche), dann durchstoßen die an diesen Stationen ankommenden Wellen die obere Halbkugel, und die Punkte müssen entsprechend auf die untere Halbkugel übertragen werden. Dies geschieht durch Übertragung auf ihren symmetrischen Gegenpunkt bezogen auf den Brennpunkt und Auftragen in der stereographischen Projektion.
5. Auf der stereographischen Projektion die beiden orthogonalen Flächen bestimmen, die am besten geeignet sind, die Quadranten zu trennen, bei denen die erste Bewegung identisch ist (verdichtend oder ausdehnend). Diese orthogonalen Flächen sind die Knotenflächen, eine von ihnen ist die Störungsfläche. Die Emissionsanalyse kann keinen Aufschluß darüber geben, welche der beiden Flächen die Störungsfläche ist. Es ist also mehr Information erforderlich, um hier die richtigen Feststellungen zu treffen. (Ausbiß der Störung auf der Bodenfläche bezw. örtliche Bestimmung der Folgestöße.
Obschon dies von Hand zeitraubend ist, kann immerhin die Störungsfläche dank der enstprechenden Hochrechnereingaben routinemäßig ermittelt werden. In Abb. 5 ist ein Beispiel für diese Bestimmung aufgezeigt .
Mehr Aufschluß über den Riß, der das Erdbeben hervorrief, ist der Spektralanalyse der Erschütterungswellen zu entnehmen. Es können z.B. sowohl die Größe des Ausgangspunkts als auch der Spannungsabfall, der während des Ereignisses auftrat, geschätzt werden (s. auch Madariaga, 1979). Es muß noch erwähnt werden, daß diese in größen von zwischen 1 MPa und 10 MPa schwanken können und zwar unabhängig von der Stärke des Bebens (s. Hanks, 1972). Sobald erst die Knotenflächen bekannt sind, werden 3 Achsen bestimmt : die P-Achse weist in die Richtung wo die Kompression am lotrechtesten zur Knotenflächenverschneidung ist; die T-Achse, die ebenfalls lotrecht zur Knotenflächenverschneidung verläuft, weist in die Richtung maximaler Dilatation; die O-Achse verläuft entlang der Verschneidung zweier Knotenflächen. Ein Vorschlag ging dahin, diese drei Achsen in der Hauptspannungsrichtung verlaufen zu lassen (s. auch Scheidegger, 1964). Dieser Vorschlag beruht auf der Hypothese, daß die Störungsfläche der Fläche mit der größten Scherspannung ent spricht, und daß die Verschiebung entlang der Störung parallel zu der Scherkomponente verläuft, die vom Bruch gefördert wird. Tatsächlich gilt diese verinfachte Annahme im Allge-

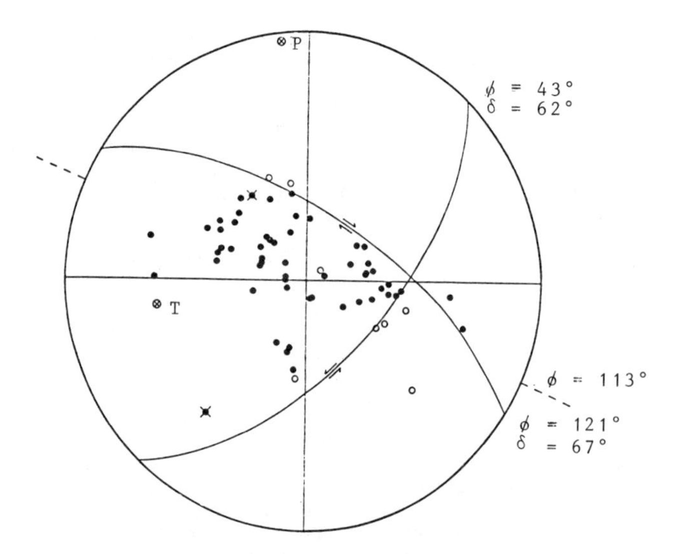

Abb. 5 Störungsflächenbestimmung des Erdbebens in der Türkei im August 1966. Das Diagramm stellt eine flächentreue Projektion der unteren Halbkugel des Emissionsfeld dar. Die Punkte sind die Kompressionen, die Kreise die Dilatationen, Kreuze bedeuten, daß die Stelle nahe einer Knotenfläche lag, ø und δ stellen den Einfallen bezw. das Streichen der Knotenfläche dar, und die Pfeile deuten die Bewegungsrichtung an. Die gestrichelte Linie ist das Streichen des rechten Hauptflächenaufbruchs, der das Beben begleitete (nach McKenzie, 1969).

meinen nicht für sprödes Gestein (s. auch McKenzie, 1969); Raleigh u.A., 1972), zumal deren Risse bei Verdichtung nicht den Kriterien des "Tresca-Bruchs", sondern eher dem Mohr-Coulomb'schen Bruch entspricht (die makroskopische Bruchfläche ist geneigt zwischen einem Winkel von 20° bis 45° zur Hauptspannungsrichtung, während zudem Gebirge meist bezogen auf ihre Festigkeit infolge natürlicher Schwächeflächen anisotropisch sind. McKenzie weist auf das Gleiten entlang natürlicher Trennflächen hin, das bestimmend auf das Scherverhalten des Gebirges wirkt, wodurch die Fehler bei der Bestimmung der Hauptspannungsrichtung bis zu 90° betragen können. Raleigh u.A. halten dem entgegen, daß sich sogar im Falle bestehender natürlicher Störungen, neue Trennflächen bilden können, sobald die Scherspannung vorliegt, die die Voraussetzung für ein Gleiten entlang natürlicher Klüfte ist. Dies bedingt jedoch wesentlich höhere Spannungen als die, die für die Entwicklung eines neuen Risses erforderlich sind. Sie wiederum siedeln die erheblichsten Fehler in der Ermittlung des Spannungsverlaufs bei 20° bis 30° an.

Das Mikrotektonik-Verfahren (s. auch Angelier, 1979; Armijo u.A., 1982) zur Feststellung der Hauptspannungsrichtungen sowie des "Formfaktors" (auch Spannungsfaktor genannt) $R = (\sigma_3 - \sigma_1) / (\sigma_2 - \sigma_1)$, wobei σ_1, σ_2 und σ_3 die Eigenwerte des Spannungstensors darstellen, wird von Lehrsätzen abgeleitet, die denen des Störungsflächenverhaltens ähnlich sind. Die Bestimmung der Schliffflächenrichtung in Störungsflächen ermöglicht eine direkte Messung der Gleitbewegung entlang der Störung. Wird angenommen, daß diese Bewegung in der gleichen Richtung verläuft, wie die der durch die Störung erzeugte Scherkraft, und daß die Reibungskennzahlen entlang der Störung nicht abhängig sind von der Lager der Störung oder der Gleitrichtung, dann lassen sich die Euler'schen Winkel, die Hauptrichtung des Spannungstensors und der Formfaktor R bestimmen :

wenn $\underline{f}_i = \underline{\tau}_i \; / \; |\underline{\tau}_i|$ (10a)

worin $\tau = \underset{\sim}{\sigma}\underline{n}_i - (\underset{\sim}{\sigma}\underline{n}_i \cdot \underline{n}_i)\underline{n}_i$ (10b)

dann $\underline{f}_i = \underline{s}_i \; ; \; i = 1, N$ (10c)

f_i ist ein Einheitsvektor

$\underset{\sim}{\sigma}$ ist der "durchschnittliche" Spannungstensor

$\underline{n}_i$ ist die Einheitsnormale zur Störung

$\underline{s}_i$ ist ein Einheitsvektor parallel zur i-ten Schliff-Fläche

N ist die Gesamtzahl

Gleichung (10b) wird zu

$$|\tau_i| = (\sigma_2 - \sigma_1) A_\ell \qquad (11)$$

mit $\ell = 1.2.3$; $A_1 = -kn_1$, $A_2 = (1-k)n_2$, $A_3 = (R-k)n_3$

$$k = n_2^2 + Rn_1^2 \quad .$$

So ist Gleichung (10 c) die Haupthypothese in dieser Spannungsermittlung. Weiter wird angenommen, daß alle Rißflächenbewegungen unabhängig von einander sind.
Ist N groß genug (zwischen 100 und 200 Schliffflächenmessungen), dann ist Gleichung (10 c) das System eines nicht linearen Systems, das durch Linearisierung der Gleichung (10 c) (s. auch Armigio u.A., 1982) verkehrt werden kann.

Bei diesem Verfahren wird der Spannungstensor $\underset{\sim}{\sigma}$ als innerhalb eines Volumens der Größe von einigen Kubikhektometern bezw. selbst einiger Kubikkilometer (Schliffflächenmessungen werden gewöhnlich entlang von Störungen in Steinbrüchen oder grossen Löchern vorgenommen) gleichmäßig angenommen. Schwierigkeiten entstehen, wenn Störungen durch mehr als ein tektonisches Spannungsfeld aktiviert sind; es sind jedoch Umkehrungsverfahren entwickelt worden, mit denen diese verschiedenen Spannungstensoren festgestellt werden können (s. auch Etchecopar u.A., 1981; Armijo u.A., 1982). Da aus Störungsflächenlösungen Hinweise auf die Gleitrichtung entlang Störungen hervorgehen, können diese mit Schliffflächenmessungen verglichen werden. Sind viele Schliffflächen erforderlich, damit die Gleitrichtung mit einiger Sicherheit festzustellen ist, kann eine Störungsflächenlösung sofort über die Gesamtbewegung Aufschluß geben. Sie werden nicht durch Artungleichheit kleinerer Ordnung beeinträchtigt, und sie erteilen in dem gleichen Maßstab Aufschluß, wie der, mit dem der Spannungstensor ermittelt wird. Dementsprechend sollte die Analyse von Störungsflächenlösungen nach einer ähnlichen Methode, wie sie für die Mikrotektonik vorgeschlagen wird, Aufschluß über die Hauptspannungsrichtung in der Tiefe, wie auch über das "Gestaltsverhältnis" des Spannungstensors erteilen. Diese Ermittlungen sollten genauer sein, als die, die zur Zeit nach dem Trescakriterium von Scherbrüchen erzielt werden. Es können jedoch gewisse Schwierigkeiten in Fällen von kleinen Beben auftreten, da hier die Bestimmung der Knotenflächen, die die tatsächlichen Störungsflächen sind, nicht eindeutig ist.

2.2 Einige Ergebnisse :

Vergleich zwischen den mit verschiedenen Verfahren erzielten Ergebnisse.

In einigen Fällen wurden Spannungen am gleichen Ort mit den gleichen Mitteln bestimmt. Haimson (1982) berichtet über Ergebnisse, die mit der hydraulischen Rißbildung und der Überbohrung von Bohrlochverformungskalibern erzielt wurden. Im Nevada-Prüffeld wurden zB. Messungen in zwei Tunnells in 450 m dicker Schicht-Tuff und tuffhaltiger Sandstein-Überlagerung über massivem paleozoischem Gebirge durchgeführt. Die Ergebnisse lassen sich gut mit einander vergleichen (s. Tabelle 1) und auch mit den örtlichen Störungsflächen.

Tabelle 1. Ergebnisse des Nevada-Prüffelds (nach Haimson, 1982).

	σ_H max MPa Richtung	σ_h min MPa Richtung	σ_v MPa Richtung Einfall	Tiefe in m
Hydr. Rißbildung	9,0/N35°O	3,5/N55°W	7.	380
Überbohren	8.0/N45°O	2,5/N45°W	N 42°O 20°	380
	σ_H max MPa Richtung	σ_h min MPa Richtung	σ_v MPa Richtung Einfall	Tiefe in m
Hydr. Rißbildung	7,5/N40°O	3,0/N50°W	7.3	426
Überbohren	8,5/N22°O	N68°W Einfall 83°	N83°W 6,8 Einfall 7	426
Brennpunktmechanik	N45°O	N45°W		7

Die Ergebnisse sowohl der hydraulischen Rißbildung als auch der Überbohrungstechnik von fünf anderen Prüffeldern sind recht reproduzierbar. Die Streuung für die kleinsten waagerechten Hauptspannungsgrößen liegt bei 30%, während die für die maximale Horizontalspannung in einer Größenordnung von 50% liegen kann. Diese große Abweichung bei der Bestimmung von σ_H kann der Schwierigkeit zugeschrieben werden, der man mit der klassischen hydraulischen Rißbildung bei der Bestimmung der Genauigkeit von sowohl der Festigkeit des Gesteins, wie auch des Wiederöffnungsdrucks begegnet. Die mittlere Richtung der Hauptnormalspannungen, die mit beiden Verfahren festgestellt wird, liegt typisch innerhalb von 10°, die Streuung kann jedoch bei der Überbohrungsmethode 120° und bei der hydraulischen Rißbildung 50° erreichen. Ein Vergleich des Stripa-Bergwerks in Schweden für die hydraulische Rißbildung und Überbohrung der dreiachsigen Zelle von Leeman, der weichen C.S.I.R.O.-Einschlußzelle und der dreiachsigen Zelle der Universität von Lulea wird von Doe (1982) vorgestellt. Ein ausführlicher Bericht über die Streuung der Ergebnisse ist darin enthalten. Während das Zementieren einige Probleme mit der C.S.I.R.O.-Zelle aufwarf, ergaben andere Verfahren bei Vergleich der mittleren Werte reproduzierbare Ergebnisse, obschon Abstriche für gewisse bedeutende Streuungen bei einigen bestimmten Ergebnissen gemacht werden mußten.
Li Fang Quan u.A. (1982) berichten auch über einen Vergleich von Spannungsmessungen in der Nähe von Xi Xio in der Provinz Hebe (China), die durch Überbohren hydraulischer Rißbildung erfolgten : die Ergebnisse waren gut reproduzierbar :

σ_{Hmax}=3.4 + 0,03 z MPa, σ_{hmin}=2,5 + 0,031 MPa,

Richtung von σ_{Hmax}=N64°W ± 20° für Überbohren;

σ_{Hmax} = 4,9 + 0,028 z MPa; σ_{hmin} = 2,6 + 0,021

und Richtung von σ_{Hmax} = N80°W ± 10° für hydraulische Rißbildung; Die Tiefe z reicht von 0 bis 90 m unter der Oberfläche. Von Kutznetsow u.A. (1983) wurden die Ergebnisse eines Vergleichs zwischen der durch ein Spannungsausgleichsverfahren und mikrotektonischer Analyse vorgelegt. Beide Verfahren erzielten reproduzierbare Ergebinisse.
Es erscheint so, als würden die von verschiedenen Forschern in unterschiedlichen geologischen Bereichen erworbenen Erfahrungen andeuten, daß verhältnismäßig gleichmäßige Spannungsfelder in Gebirge vorhanden sind, und daß die verfügbaren Verfahren verhältnismäßig befriedigende Wege zur Feststellung bieten, obschon noch für die Verbesserung der Genauigkeit der Ergebnisse einiges zu tun bleibt.

Spannungsveränderung bei zunehmender Tiefe.

Die Kenntnisse über die Größe der Hauptspannungen in der Tiefe haben in den letzten Jahren dank der Tiefenspannungsmessungen, die in Bergwerken in großer Tiefe oder in Bohrungen durchgeführt wurden (s. auch McGarr und Gay, 1978; Brown und Hoek, 1978; Jamison und Cook, 1980).

McGarr (1980) stützte sich auf weltweit vorhandene Unterlagen, als er schloß, daß die maximale Scherspannung bei weichem Gestein, wie Schieferton oder Sandstein im Mittel in Tiefen bis 5,1 km und bei hartem Gestein, wie Granit oder Quarzfels in Tiefen bis 3,7 km linear anwächst (die genannten Werte beziehen sich jeweils auf Messungen in der größten genannten Tiefe). Er bemerkte, daß eine Regressionsanalyse dieser Daten auf einen vertikalen Gradienten der maximalen Scherkraft bis 3,8 MPa/km in weichem Gestein und bis 6,6 MPa/km in hartem Gestein schließen läßt. In der Annahme eines elastischen Verhaltens der oberen Kruste, bewies er, daß Gleichgewichtsbedingungen verbunden mit der Bertrami-Mitchell-Veträglichkeitsgleichungen für die Spannungen in einem homogenen isotrop elastischen Körper zur Erzielung funktionaler Zwänge auf Spannungsänderungen in der Tiefe angewendet werden können. So können für Gebiete, in denen keine seitlichen Spannungsänderungen beobachtet werden (s. den folgenden Abschnitt) diese Bedingungen bei zunehmender Tiefe ein lineares Anwachsen der Hauptspannung, jedoch ein einförmiges Drehen der Hauptrichtungen bedeuten. Dies kann bedeuten, daß selbst für derart einfache Bedingungen die Bestimmung der Oberflächenspannung keinen Aufschluß über Hauptspannungsrichtungen in der Tiefe erteilt. In Bereichen, wo das Spannungsfeld als unabhän-

gig von nur einer der horizontalen Koordinaten (zB. x_2, wobei dann die andere Koordinate x_1 wäre) ist, deuten diese Bedingungen an, daß die diagonalen Komponenten des Spannungstensors linear mit der Tiefe wachsen, daß jedoch die Scherkomponenten σ_{12} und σ_{23} nicht linear sind. Dies läßt zweifellos vermuten, daß selbst bei dem in dieser Analyse berücksichtigten Maßstab, der Rückschluß auf Spannungen in der Tiefe nicht vernachlässigt werden darf, und daß es sich ferner bei der vertikalen Spannungskomponente in großer Tiefe nicht unbedingt um eine Hauptspannung handeln muß.
Für das Erzielen aussagefähiger Daten dürfen offensichtlich bei heterogenen Stoffen zur Zeit nur direkte Messungen vorgenommen werden. So hat zB. Cornet (1982) Unterschiede von bis 2,5 MPa innerhalb weniger als 10 m vorgefunden (der Mittelwert war 13 MPa und die Genauigkeit jeder Einzelmessung war besser als 2 MPa in einer Tiefe von rd. 700 m), als er die verlangsamte Wiedereröffnungstechnik bei hydraulischer Rißbildung zur Bestimmung der kleinsten Hauptspannung in Salzlagerstätten einsetzte.
In diesen weichen Lagerstätten (abwechselnd Salzschichten von 1 bis 10 m Dicke und Schiefertonschichten von 0,01 bis 1 m Dicke) bestehen millimeter- bis zentimeter-starke Anhydritschichten, deren Zahl je Längeneinheit außerordentlich schwankt. Diese starren Anhydritlagen verändern ganz eindeutig das mechanische Verhalten der Lagerstätten und eine gemessene maximale Scherung einer Größe von 5 MPa läßt sich schwer mit dem als plastisch anzusehenden Verhältnis vereinbaren. Die Wirkung dieser örtlichen Aussteifungen wird als Ursache für die abrupten Größenänderungen in der horizontalen Hauptspannung angesehen; zumal die größeren Werte jeweils nahe der Zonen der hohen Dichte der Anhydritlager beobachtet wurden.
Martna u.A. (1983) berichten über weitere drastische beim Überbohren in homogenem Granit-Gneiss gemessene Spannungsänderungen. Ein gleichförmiges Spannungsfeld, dessen größte horizontale Hauptspannung eine Größe von 15 bis 20 MPa aufzeigte, wurde in einem tiefen Bohrloch bei 320 m festgestellt. Dann stieg nach einem Störungsbereich der Wert auf 65 MPa und blieb bis zu einer Tiefe von 500 m gleich. Über eine Drehung der Hauptrichtung wird nicht berichtet. Diese Beispiele beleuchten die Tatsache, daß unabhängig vom Maßstab mit dem diese Spannungen ermittelt werden, Tiefenspannungen sich nicht einfach durch Ableitung von der Oberflächenmessung ermitteln lassen. Jedoch ist grundsätzlich eine monotone Veränderung der maximalen Scherkraft bis in Tiefen von 5 km zu erwarten.

Seitliche Spannungsänderungen.

Abgesehen von den wohlbekannten seitlichen Spannungsänderungen an der Oberfläche, die sich aus der Topographie herleiten lassen (s. auch Scheidegger, 1977; Haimson, 1978; Buyle-Bodin 1980), können in einer gegebenen Tiefe horizontale Veränderungen des Spannungstensors aus einfachen mechanischen Anwendungen vorausgesehen werden. So kann zB. die Artungleichheit in Gebirgseigenschaften bei sonst gleichförmiges Spannungsfeld zu örtlichen Abweichungen führen. Trennflächen, wie zB. Störungsflächen können sowohl räumlich wie zeitlich zu Spannungsänderungen führen. In dieser Hinsicht deuten die Ergebnisse von Li Fang-Quan u.A. (1982) darauf hin, daß die größte Hauptspannung, die mit einer Größe von 7.9 nahe einem Beben gemessen wurde, nach einem Erdbeben der Größe von 2,5 MPa und 6,0 MPa in einiger Entfernung lag.

Jedoch, trotz aller für die fehlende Übereinstimmung vieler bis heute durcgeführten Spannungsmessungen angeführten Gründe, läßt sich doch sagen, daß umfangreiche Gebiete bestehen, in denen die Spannungen verhältnismäßig gleichmäßig sind.

Zoback und Zoback (1980) haben eine Vielzahl in an einander angrenzenden Gebieten der Vereinigten Staaten durchgeführte Spannungsmessungen zusammengetragen, und sie haben hierbei lineare Spannungsbereiche von 100 bis 2000 km, in denen Richtungen wie auch relative Größe der Hauptspannungen verhältnismäßig gleichartig waren vorgefunden (s. Abb. 6). Froideveaux u.A. (1980) und Rummel u.A. (1982) kommen für das westliche Europa zu gleichen Ergebnissen, wohingegen jedoch die Mehrzahl ihrer Ergebnisse in geringen Tiefen erzielt wurden (weniger als 300 m). Diese Ergebnisse sind ein eindeutiger Hinweis dafür, daß noch andere, die breiten durch Spannungswirkung auf seitliche Plattengrenzen beeinflussende Spannungsfelder (s. auch Solomon u.A. (1980), überlagernde Spannungsquellen bestehen. McGarr (1982) schlägt vor, die Zoback und Zoback'schen Spannungsbereiche mit den Feldern von Scherbeanspruchung, die auf den Boden der elastisch-spröden Kruste wirkt, in Beziehung zu setzen, während Fleytout und Froideveaux (1982) ihrerseits vorgeschlagen haben, den Einfluß räumlicher Dichteveränderungen in der Lithosphäre wie auch im äußeren Mantel als Erklärung für diese plattenübergreifenden Spannungsbereiche in Erwägung zu ziehen.
Ganz eindeutig bleiben Spannungsmessungen zZt. erforderliche Daten für geodynamische Modelle, und die Hoffnung geht dahin, daßß weitere Messungen zur Verbesserung dieser Modelle durchgeführt werden. Noch ist es für eine Umkehrung des Verfahrens und zur Verwendung geodynamischer Modelle zur Ermittlung örtlicher Spannungsfelder zu früh. Werden jedoch ausschließlich Größenordnung und die allgemeinen Hauptspannungsrichtungen gesucht, darf für die Ermittlung der Spannungen an gegebenen Orten, auf die benachbarte Spannungsmessung zurückgegriffen werden, sobald Kontinuität und Homogenität zu erwarten stehen. Tatsächlich könnten vielfache Spannungsmessungen, einen Weg für die Zeichnung von Spannungskarten verschiedener Maßstäbe anbieten, wobei diese dann wieder für grobe Schätzungen der an beliebigen Punkten einer Landkarte vorhandenen Spannungen eingesetzt werden könnten. Diese würde sicher die Kosten örtlicher Spannungsbestimmung herabsetzen. Man darf hoffen, daß derartige Karten für breite Gebiete des Kontinents entwickelt werden.

Spannungsveränderung im Laufe der Zeit.

Wenn sich Spannungen nicht im Laufe der Zeit änderten, gäbe es keine Erdbeben zweimal hinter einander am gleichen Ort. Die hierbei eingetretenen zeitliche Variation bedarf einer Erklärung. Während jedoch einige wenige Instrumente entwickelt worden sind (starrer Einschluß, zB.), um zeitliche bedingte Spannungsänderungen zu messen, sind hierüber noch sehr wenig Veröffentlichungen erfolgt (s. auch Clark, 1982).

Ein aufschlußreiches Phänomen hinsichtlich von Erdbeben könnte bald Licht in dieses Dunkel bringen : Gelegentlich sind bedeutende Magnetfeldveränderungen in Gebieten beobachtet worden,

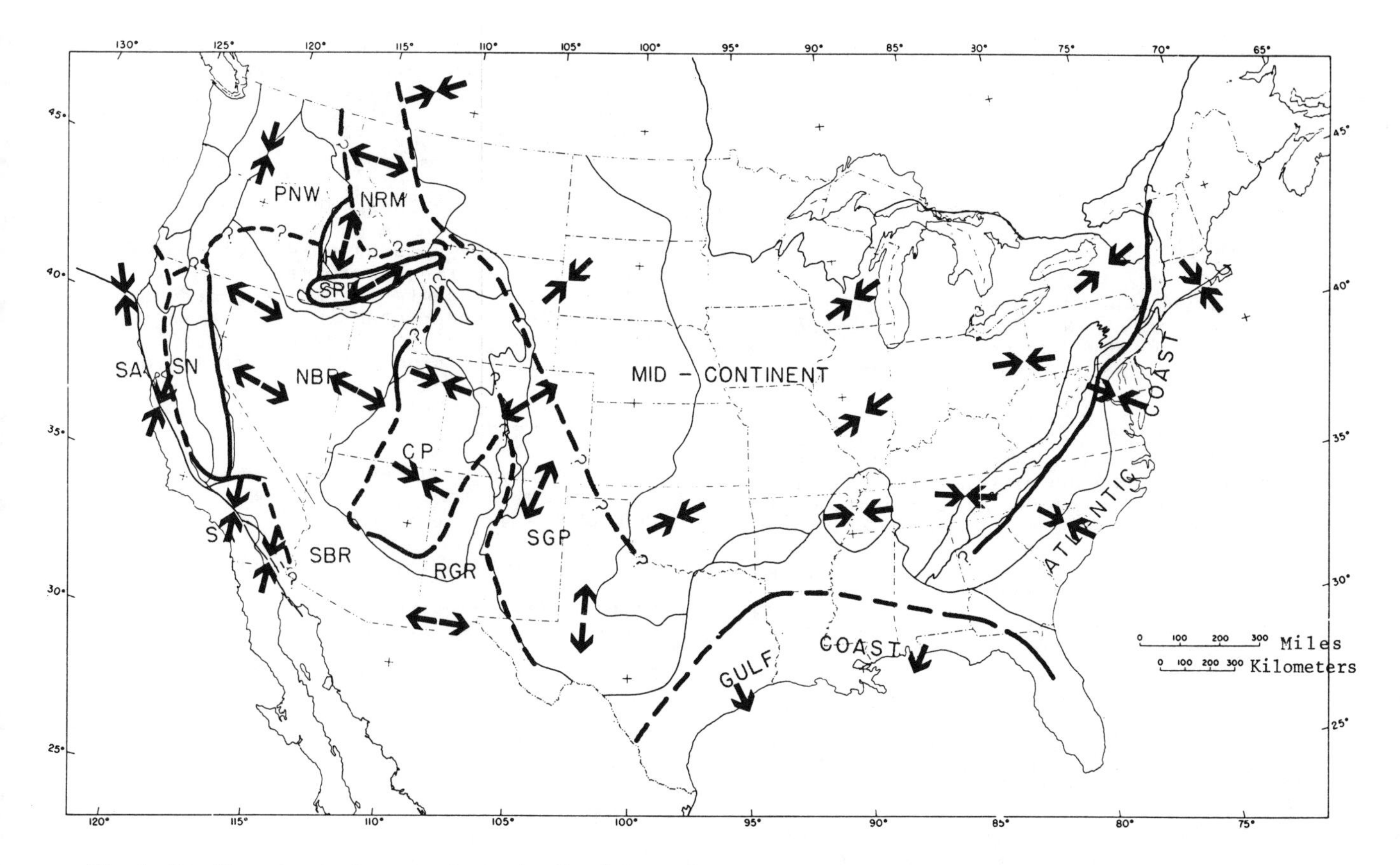

Abb. 6. Verallgemeinerte Spannungskarte der Vereinigten Staaten. Die Pfeile geben die Richtungen sowohl der kleinsten (auswärts gerichtet) als auch der größten (einwärts gerichtet) horizontalen Hauptkompression an. Nach Zoback und Zoback (1980)

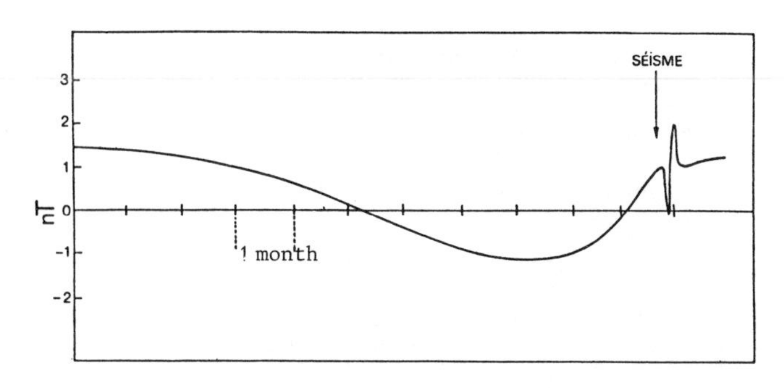

Abb. 7 Magnetische Anomalität, die in der Gegegend von Dushambe (Tadjikistand - UdSSR) vor einem Erdbeben der Größe 5 beobachtet wurde (nach einer persönlichen Mitteilung von Skovorotkin).

in denen schwere Erdbeben vorkommen (s. auch Moore, 1964; Smith und Johnston, 1976; Rikitake, 1979). In Abb. 7 ist ein Beispiel für eine derartige Veränderung gezeigt. Es handelt sich hierbei um ein Beben der Größe 5, das im Bereich von Dushambe in Tadjikistan (UdSSR) auftrat. Eine mögliche Erklärung für eine Magnetfeldveränderung dieser Art wäre im Piezomagnetismus, der in magnetischem Gebirge vorgefunden wird, zu suchen. Die magnetische Empfänglichkeit des Gebirges und seines bleibenden Magnetismus werden sobald sie einer Spannungsveränderung unterworfen werden, verändert (s. auch Martin u.A., 1978; Zlotnitzki u.A., 1978). So stellten zB. Zlotnitzki u.A. fest, daß die Veränderung induzierter Magnetisierung durch Spannungsänderung mit dem folgenden Tensor-Gesetz dargestellt werden kann :

$$\Delta J_i = P_{ijkn} \Delta \sigma_{kN} J_{oj} \qquad (13)$$

J_{oj} ist die j-te komponente der induzierten Magnetisierung durch ein magnetisches Feld $\underline{H}$

($\underline{J}_{or} = \underset{\sim}{S}_o \underline{H}$, $\underset{\sim}{S}_o$ ist die magnetische Empfänglichkeit als Tensor

$\Delta \sigma_{kn}$ ist die Änderung der Spannungskomponente σ_{kn}

P_{ijkn} sind die komponenten eines Tensors der vierten Ordnung mit 36 konstanten im allgemeinsten Fall

Bei Isotropie des Felsens wird P_{ijkn} auf zwei Konstanten reduziert, sodaß (13) wird

$$\Delta J_i = (P_1 \Delta \sigma_{kk}\delta_{ij} + 2P_2 \Delta \sigma_{ij}) J_{oj} \quad (14)$$

Die Bestimmung von P1 und P2 im Labor ergibt die Möglichkeit einigen Aufschluß über Spannungsänderungen zu erzielen : -
Obschon ein einfaches Verhältnis dieser Art noch nicht aus bleibendem Magnetismus abgeleitet worden ist, könnte die Darstellung von Labor-Kurven als Parameter einige quantitative Ergebnisse von zeitbezogenen Spannungsänderungen in weiten Gebieten ergeben. So hat die von Jin-Qi Hao u.A. (1982) durcgeführte digitale Hochrechnung ergeben, daß nach Erdbeben gleicher Größenordnung, wie die von entsprechenden Modellen geschätzten, die gleichen Magnetfeldveränderungen zeigten.

So könnten also eingehende Untersuchungen des Magnetfelds erdbebengefährdeter Störungsbereiche zugleich mit eingehenden Labor-Untersuchungen der piezomagnetischen Eigenschaften örtlicher Gebirge, einen Aufschluß über Spannungsveränderungen im Laufe der Zeit ergeben. Weitere Verfahren könnten für Spannungsänderungen durch menschliche Eingriffe angewendet werden.

3. DAS PROBLEM DER BESTIMMUNG DER GEOMETRISCHEN EIGENSCHAFTEN VON TRENNFLÄCHEN

Trennflächen beeinflussen auf verschiedene Art das mechanische Verhalten von Gebirgen:

- dadurch, daß sie bevorzugte Fließwege darstellen, beeinflussen sie die Porendruckverteilung und dadurch die effektiven Gebirgsspannungen;

- weil sie eine viel geringere Festigkeit aufweisen als das Gesteinsgefüge, sind sie von größter Wichtigkeit bei Stabilitätsuntersuchungen;

- ihr Dilatanzverhalten, kombiniert mit ihren sehr weichen elastischen Eigenschaften, beeinflußt drastisch die Deformation des gesamten Gebirges.

Über einen nunmehr langen Zeitraum wurde große Aufmerksamkeit darauf verwandt, das Verständnis ihrer mechanischen Eigenschaften zu verbessern, und obwohl in dieser Hinsicht noch viel Arbeit erforderlich ist, wurden viele Fortschritte gemacht. Es hat sich jedoch herausgestellt, daß die Charakterisierung ihrer Geometrie ein schwer zu handhabendes Problem darstellt. Erst wenn dieses Problem gelöst ist, werden genauere Modellvorstellungen über das mechanische Gebirgsverhalten (sowie über das Verhalten anderer eingeschlossener Stoffe mit anderen physikalischen Eigenschaften) entwickelt werden können.

Die Bestimmung geometrischer Eigenschaften von Gesteinstrennflächen kann in zwei verschiedenen Zusammenhängen durchgeführt werden:

- Bei dem ersten Zusammenhang ist die Gebirgsoberfläche der Beobachtung zugänglich. Dabei wird versucht, die gemessenen geometrischen Eigenschaften auch auf den dahinterliegenden Bereich zu übertragen. Dieses soll als die lokale Charakterisierung bezeichnet werden.

- Bei dem zweiten Zusammenhang können nur kleine Oberflächen in Bezug auf das in Betracht gezogene Volumen untersucht werden, und diese spärlichen Informationen werden dann auf das gesamte Trennflächengefüge übertragen. Meistens können, zusätzlich zu den Beobachtungen, die von sichtbaren Oberflächen her zu erlangen sind, einige Informationen von Bohrlöchern her erlangt werden. Dieses soll als die globale Charakterisierung bezeichnet werden.

Einige Versuche wurden unternommen, das erste Problem zu klären; das zweite Problem scheint nahezu unangetastet zu sein.

3.1. Die lokale Charakterisierung

Um eine gewisse Einschätzung der Dichte von Trennflächen bei einem gegebenen Volumen zu erhalten, definierte Deere (1964) die Rock Quality Designation (R.Q.D.) (Felsqualitätsbezeichnung) als das Verhältnis von intakten Bohrlochkernen mit einer Länge von größer oder gleich 0,1 m, bezogen auf die Gesamtkernlänge:

$$RQD = 100 \sum_{i=1}^{n} x_i / L \qquad (15)$$

wobei x_i die Länge des i. Stückes des Kerns mit einer Länge größer als 0,1 m ist,
n die Gesamtzahl der Stücke ist, deren Länge größer als 0,1 m ist,
L die gesamte Kernlänge ist.

Ferner schlug er vor, daß eine Bezugs- oder Aufnahmelinie (Meßband), die im Aufschluß benutzt wird, als direkt analog zu einem Bohrlochkern betrachtet werden kann, da der RQD-Wert ebenfalls auf diese Weise bestimmt werden kann. Ein ziemlich ausführlicher Bericht wurde durch die International Society of Rock Mechanics über vorgeschlagene Methoden für die quantitative Beschreibung von Trennflächen bei Gesteinen herausgegeben (Barton 1978). Er bezieht sich hauptsächlich auf direkte Meßtechniken bezüglich Abstand, Durchtrennung, Rauhigkeit, Wandfestigkeit, Öffnungen, Füllungen, Versickerung, Anzahl der Scharen und Blockgröße.

Hagan (1980) schlug einige photogrammetrische Methoden vor, um Trennflächenorientierungen zu kartieren. Er wies darauf hin, daß diese Kartiertechnik die untertägige oder Geländekartierzeit um 90 % reduzieren kann und daß die Flächenabdeckung, die mit dieser technischen Methode erreicht werden kann, mehr oder weniger kostant bleibt, so daß eine nicht repräsentative Trennflächenaufnahme verhindert wird. Bei dieser Methode werden von der Oberfläche her mit zwei normalen Kameras stereographische Bilder aufgenommen. Dann wird ein Computer verwendet für die Berechnung von Koordinaten irgendeiner Anzahl von Punkten, die auf der sich überlappenden Fläche der Photographie liegen, so daß der durchschnittliche lineare Trend berechnet werden kann, unter Verwendung der Methode der kleinsten Quadrate. Der Betrag und die Richtung des Einfallens dieser durchschnittlichen Fläche wird dann ermittelt. Diese Datensammlungstechnik ist direkt anwendbar auf das flächenhafte oder mit Hilfe von Abtastlinien durchgeführte Kartieren von Trennflächen.

Thomas (1981) definiert die richtungsabhängige Trennflächendichte durch:

$$\delta_v = \frac{\Sigma S_{\theta_s}}{V} \quad ; \quad \delta_s = \frac{\Sigma S_{\theta_\ell}}{A}$$

S_{θ_s} = Summe der Trennflächengrößen für Oberflächen im Raumwinkel $_s$

S_{θ_A} = Summe der Trennflächenlängen in Richtung

v = Volumen des untersuchten Bereichs,

A = Fläche des untersuchten Bereichs,

wobei und entsprechend Raumwinkel und Flächenwinkel der Messung sind. Er schlug eine neue Methode für die Ermittlung der richtungsabhängigen Dichte eines Trennflächennetzes, wie an der Oberfläche eines Gebirgskörpers beobachtet, vor.

Der Gedanke besteht darin, eine statistische Abschätzung der richtungsabhängigen Dichte aus einer Berechnung der Anzahl der Verschneidungen zwischen dem Trennflächennetz, wie auf einer Photographie beobachtet, und einem ausgewählten Gitter zu finden. Für ein Netz, das nur eine Schar paralleler Trennflächen aufweist, kann gezeigt werden, daß die Dichte der Verschneidungspunkte, d_i (θ), zwischen den Trennflächen und einer Linie, die unter θ in Bezug auf die Trennflächenrichtung geneigt ist, eine Abschätzung von $\delta_s \sin\theta$ ist; je länger die Linie, um so präziser die Bestimmung. Anstatt eine Abtastlinie zu verwenden, schlug er somit vor, eine Reihe paralleler Linien zu verwenden. Allgemeiner ausgedrückt, für ein Netz von q Scharen (jede Schar ist als F_j bezeichnet) von Trennflächen mit der Dichte J_{α_j}, wodurch sich ein Winkel $_j$ mit einer Bezugsachse ergibt (siehe Abb. 8). wobei die Dichte der Verschneidungspunkte, die man bei einem Gitter aus parallelen Linien beobachtet, einen Winkel θ_i in Bezug auf die Bezugslinie ausmacht, wird angegeben durch:

$$d(\theta_i) = \sum_{j=1}^{q} \delta_j \left|\sin(\theta_i - \alpha_j)\right| \qquad (16)$$

Die Krümmung der Kurve D(θ) gegen θ ist eine stetige Funktion von θ , außer für die Werte, welche mit dem Winkel α_j übereinstimmen (siehe Abb. 9).

Thomas hat zur Ermittlung der Werte α_j ein Verfahren vorgeschlagen, das auf der diskreten Fourierschen Reihe basiert. Diese Methode, die auf einen automatischen Gefügeanalysierer direkt anwendbar ist, wurde in einem Kohlebergwerk angewendet. Bilder wurden aufgenommen auf vertikalen Ebenen, deren Streichrichtung so gewählt wurde, daß ungefähr 20^0 Azimutalrotationszunahmen vorgesehen werden konnten. Eine bestimmte Abschätzung der richtungsabhängigen volumetrischen Dichte konnte in dieser Weise erzielt werden.

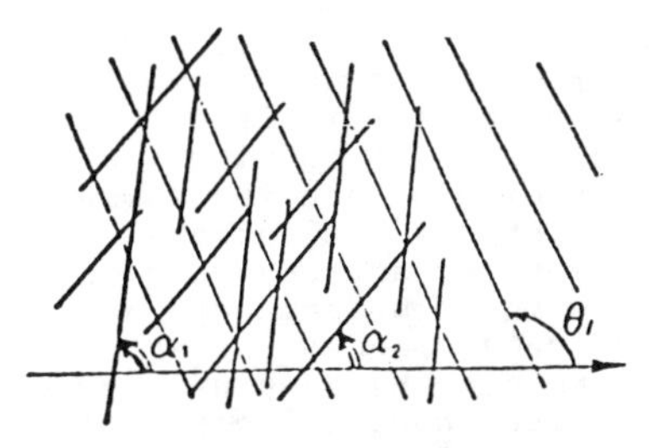

Abb. 8 Trennflächendichtenuntersuchung mit einem Gitter paralleler Linien. α_1 und α_2 sind relative Orientierungen von zwei Kluftscharen bezüglich einer Hilfsachse, θ ist die Neigung des Gitters relativ zur Bezugsachse (nach Thomas 1981).

Warburton (1980 a, b) schlug ein statistisches Modell vor, um eine analytische Voraussage des Musters von Kluftausbissen, die in einer geologischen Kartierung festgestellt wurden, abzuleiten. Die Trennflächen sollen ziemlich flach, sehr dünn, von kreisförmiger oder rechteckiger Gestalt und parallel zu einer gegebenen Richtung sein. Ihre räumliche Verteilung, durch diejenige ihres Flächenmittelpunktes gekennzeichnet, soll zufallsbedingt sein und durch eine Poisson-Verteilung für die Volumendichte der Flächenmittelpunkte beschrieben sein. Die Poisson-Verteilung wird durch experimentelle Beobachtung gestützt, daß Ausbißabstände längs einer Referenzlinie oft exponentiell verteilt sind, und dies ist genau das, was aus den Ergebnissen für Poisson-verteilte Flächen in der geometrischen Wahrscheinlichkeitstheorie erwartet würde. Die Größen von Klüften haben eine statistische Verteilung g(x), die zu der Verteilung der Ausbißlängen in Relation steht. Dieses Verhältnis kann dazu benutzt werden, um die Kluftgrößenverteilung aus der gemessenen Verteilung der Ausbißlängen, wie an einer Aufschlußwand beobachtet, zu ermitteln.

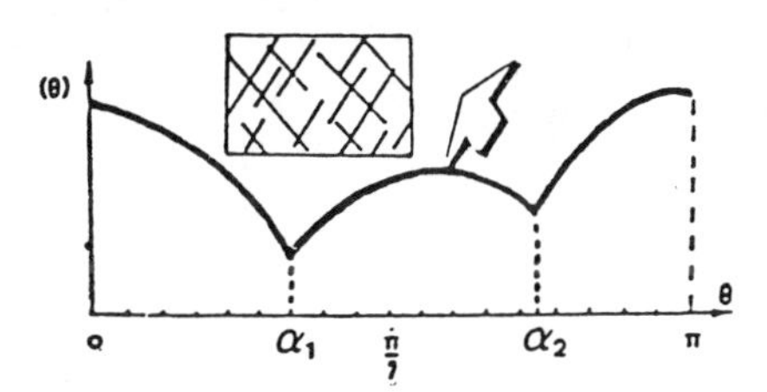

Abb. 9 Theoretische Form einer Dichtekurve D(θ) für zwei Kluftscharen, ausgerichtet nach α_1 und α_2 in Bezug auf eine arbiträre Bezugsachse (nach Thomas 1981).

In der Praxis wird die Stichproben-Häufigkeitsverteilung der Ausbißlänge gewöhnlich an beiden Seiten gekappt. Dies rührt daher, daß einerseits nur Ausbisse ab einer bestimmten Länge kartiert werden, und andererseits daher, daß Ausbisse sich oft bis außerhalb der Untersuchungszone erstrecken. Lösungen sowohl für eine flächenhafte Probenahme als auch einer linearen Probenahme werden vorgeschlagen. Dabei kann auch eine Kappung der Verteilung zugelassen werden, obwohl diese wegen ihres Einflusses auf die Bestimmung von g(x) so niedrig wie möglich gehalten werden sollte.

Für eine flächenhafte Kartierung wird zum Beispiel, wenn keine Kappung der Verteilung auftritt, die Lösung für kreisförmige Klüfte angegeben durch:

$$h(y) = \frac{y}{m} \int_y^\infty \frac{g(x)\,dx}{\sqrt{x^2 - y^2}} \qquad (17)$$

wobei h(y) die Wahrscheinlichkeitsdichte für die Ausbißlänge y ist,

m die mittlere Ausbißlänge ist

$(m = \int_0^\infty x g(x)\,dx$

x der Durchmesser einer kreisförmigen Kluftfläche ist,

g(x) der Wahrscheinlichkeitsdichte entspricht, mit der eine Kluft mit einem Durchmesser x auftritt.

Die Methoden für die Lösung der Gleichung (17) für g(x), wenn h(y) bekannt ist, werden von Warburton erörtert. Aus Gründen der Zweckmäßigkeit schlägt er vor, einen analytischen Ausdruck für g(x) anzunehmen und die logarithmische Normalfunktion zu verwenden.

Priest und Hudson (1976) und Hudson und Priest (1979) untersuchten die Möglichkeit, Meßlinien-Kartierungen zu verwenden, um eine Wahrscheinlichkeitsdichtenverteilung von Blocklänge, Blockfläche und Blockvolumen abzuleiten. Sie zeigten auf, daß der entlang der Meßlinie beobachtete Trennflächenabstand in den meisten Fällen durch eine negative Exponentialverteilung dargestellt werden kann. Die Gültigkeit dieses Vorschlags wurde für acht Aufschlüsse in Sedimentgesteinen und zwei in metamorphen Gesteinen verifiziert. Wallis und King (1980) bestätigten dieses Ergebnis für Granit. Falls f(x) die Wahrscheinlichkeitsdichte ist, λ die mittlere Trennflächenhäufigkeit und x einen Wert für den Kluftabstand darstellt, dann ist der kumulierte Anteil der Trennflächenabstände, die größer als ein bestimmter Abstand ℓ sind, gegeben durch

$$L(\ell) > \int_0^\ell \lambda x f(x)\,dx \qquad (18)$$

Da der Felsqualitätswert RQD als der Prozentanteil der Meßlinie mit Kluftabständen über 0,1 m definiert wird, gilt dementsprechend:

$$RQD = 100^{-0{,}1\lambda}(0{,}1\lambda + 1) \qquad (19)$$

Hierbei wird vorausgesetzt, daß die Wahrscheinlichkeitsdichte durch eine negative Exponentialfunktion ($f(x) = \lambda e^{-\lambda x}$) dargestellt werden kann. Wenn man zwei Satz paralleler, orthogonaler Scharen völlig durchtrennter Klüfte betrachtet, zeigten Hudson und Priest auf, daß der RQD-Wert von der Orientierung der Meßlinie zur Raumlage der Scharen abhängt. Diese Ergebnisse wurden später erweitert (Priest und Hudson, 1981), um die Möglichkeit von Beeinflussung durch Kappung der Verteilungskurve zu berücksichtigen; außerdem zeigten diese Autoren Möglichkeiten auf, sowohl die Genauigkeit und die Präzision der RQD-Bestimmung als auch die Ermittlung der mittleren Kluftausbißlänge zu erhöhen.

Liegt eine statistische Kluftauswertung vor, kann versucht werden, den Zusammenhang zwischen der Klüftigkeit und der Gebirgsdurchlässigkeit zu klären. Hudson und Lapointe (1980) haben vorgeschlagen, elektrische gedruckte Schaltungen für eine derartige Untersuchung zu verwenden, wenn das Problem zweidimensional ist; Priest und Sameniego (1983) schlagen auf diesem Kongreß eine numerische Technik vor. Jedoch wird die Anwendbarkeit dieser zweidimensionalen Modelle auf die dreidimensionalen Bedingungen des wirklichen Lebens nicht erörtert. Man darf hoffen, daß dieser Punkt anläßlich des Kongresses angesprochen wird.

3.2. Die globale Charakterisierung

Bertrand et al. (1982) haben versucht, die Durchlässigkeit und deren Variation innerhalb eines Granitmassivs durch eine Untersuchung der Klüftigkeit, kombiniert mit in-situ-Spannungs- und -Durchlässigkeitsmessungen, zu ermitteln. Obwohl dieses Problem bislang noch nicht gelöst worden ist, werden einige interessante Ergebnisse im Hinblick auf Trennflächenanalysen berichtet.

Die Klüftigkeit wird beschrieben entweder durch eine skalare Größe (Flächen- oder Volumendichte) oder durch einen Vektor (Trennflächendichte in einer gegebenen Richtung). Der Zweck dieser Studie besteht darin, diese Charakteristiken an einigen Lokationen festzustellen und dann Geostatistiken (Statistiken regionalisierter Variablen) zu verwenden, um die Kennwerte auch auf andere Lokationen zu übertragen.

Zwei Arten von Daten werden in Betracht gezogen: Erstens, die ebene Kluftflächenhäufigkeit sowie die Häufigkeit der Raumlagen von Trennflächen, die aus Luftbildaufnahmen ermittelt werden. Hierbei umfaßt jede Luftbildaufnahme einen Quadratkilometer. Zweitens: die räumliche Trennflächendichte, die in Granitaufschlüssen und untertägigen Aufschlüssen (im Größenbereich von 100 bis 1000 m^2) ermittelt wird. Die Ergebnisse ließen erkennen, daß die Klüftigkeit vom Betrachtungsmaßstab abhängig ist, so daß eine Übertragung von einem Maßstab zum anderen nicht ohne weiteres möglich ist.

4. BESCHREIBUNG DES MECHANISCHEN VERHALTENS VON GEBIRGEN

Die Notwendigkeit, das mechanische Verhalten von Gebirgen zu beschreiben, entsteht bei sehr unterschiedlichen Problemen. Manchmal sollen mit Hilfe eines Modells quantitativ Verschiebungen sowie Spannungen in dem gesamten, zu untersuchenden Bereich ermittelt werden.

Diese Daten können dann für eine Stabilitätsuntersuchung, für die Planung großer Bauwerke oder künstlicher Bruchbildungsschemata, oder für das Verständnis natürlicher Prozesse, wie Erdbebenentstehungsmechanismus, oder der Deformation tektonischer Platten, herangezogen werden. Manchmal ist das Modell qualitativ und hilft lediglich dabei, die relevanten Parameter, welche den Deformationsprozeß beeinflussen, festzustellen. Dieser Bericht beabsichtigt nicht, alle Facetten dieses sehr weiten Themas abzudecken; er richtet vielmehr die Aufmerksamkeit auf einige besondere Diskussionspunkte, die quasistatische Deformationen betreffen.

Ein wichtiger Schritt bei der Entwicklung von Modellvorstellungen ist die Auswahl geeigneter konstitutiver Gleichungen für die Darstellung sowohl des entsprechenden Kontinuums als auch der Großtrennflächen. Im Idealfall sollte diese Wahl nur durch Laboratoriums- und in-situ-Prüfungen sowie durch ein gesundes Verständnis der Gebirgsklüftigkeit bestimmt werden. In der Tat sollten nur diejenigen Trennflächen betrachtet werden, die groß sind im Verhältnis zur Größe des Untersuchungsbereiches; die anderen werden ignoriert, aber die konstitutiven Gleichungen für das entsprechende Kontinuum werden so definiert, daß die globalen Wirkungen von Kleintrennflächen, beispielsweise Dilatanz, schwache Scherfestigkeit in einer gegebenen Richtung etc., berücksichtigt werden. Bis vor kurzem wurde praktisch diese Wahl in den meisten Fällen diktiert durch die zur Verfügung stehende mathematische Technik zur Lösung der partiellen Differentialgleichungen: die Notwendigkeit, auf geschlossene analytische Lösungen oder auf linear-elastische, zweidimensionale Modelle zu bauen, brachte unrealistische, allzu große Vereinfachungen mit sich. Heutzutage hat die Entwicklung von Hochgeschwindigkeitscomputern den Weg für präzisere numerische Lösungen eröffnet, so daß nichtlineare, konstitutive Gleichungen und genauere geometrische Formen gehandhabt werden können. Jedoch bleiben viele dieser komplizierten Modelle noch immer nur auf zweidimensionale Situationen anwendbar. Selbst wenn die zweidimensionalen Modelle berechtigt sind, sollte man sich doch immer vor Augen halten, daß die Genauigkeit der Lösung stark von derjenigen der Eingabedaten abhängt. Daher mag es nicht realistisch sein, ein zu kompliziertes Modell anzuwenden, falls die Parameter, auf welchen dieses basiert, direkten Beobachtungen oder einer genauen Auswertung nicht zugänglich sind. Wünschenswerterweise sollten für Probleme des Ingenieurwesens einfache in-situ-Großversuche durchgeführt werden, um ein numerisches Modell zu bestätigen, bevor es angewendet wird. Leider sind derartige Prüfungen oft zu kostspielig, um durchgeführt zu werden; in-situ-Messungen, die während der frühen Entwicklungsstufen eines Projektes durchgeführt werden, werden benutzt, um dann das Modell an die tatsächlichen Verhältnisse anzupassen oder zumindest Mittel vorzusehen, um den "Grad" der Genauigkeit des Modells abzuschätzen.

Ferner ist eine Langzeitbeobachtung immer wünschenswert, um eine Möglichkeit für einen Vergleich zwischen beobachteten und berechneten Ergebnissen vorzusehen. Damit kann das Modell bestätigt oder für eine spätere Benutzung modifiziert werden.

Für geologische Probleme sind Mittel, um eine Lösung zu bestätigen, eher subjektiv. Gewöhnlich wird das Modell als realistisch angesehen, wenn sich seine Ergebnisse mit Beobachtungen an der Oberfläche günstig vergleichen lassen. Jedoch stehen mehr und mehr Angaben über Plattenbewegung, Deformationsgeschwindigkeit oder Spannungsgröße zur Verfügung, so daß eine bestimmte quantitative Bestätigung in einigen Fällen möglich wird.

4.1. Konstitutive Gleichungen

Bei den meisten Situationen im Ingenieurwesen bleibt die Temperatur unter 200 ^{o}C, und die Tiefe ist wesentlich geringer als 5000 m, obwohl höhere Temperaturen in einigen geothermischen Feldern auftreten können (z. B. 350 ^{o} im heißtrockenen Gebirgsbereich bei Los Alamos in den USA) und eine größere Tiefe kann in der Ölindustrie erreicht werden.

Dieser Bereich erscheint noch sehr begrenzt, wenn man sich mit Deformation in der Erdkruste befaßt, wo die Tiefe 50 bis 60 km erreichen kann und die Temperatur bis an 1000 ^{o}C herankommen kann. Ferner kann die Zeitskala quasistatischer Deformation, die im Ingenieurwesen auftritt, von einem halben Tag bis hundert Jahre variieren (z. B. der Simplon-Tunnel in der Schweiz ist mehr als 100 Jahre alt), während bei geologischen Situationen Millionen Jahre berücksichtigt werden müssen. Interessanterweise sind mir relativ wenige Modelle vorgestellt worden, die in der Lage sind, das mechanische Verhalten der vielen Gesteinsarten der Erdkruste zu untersuchen, unter Einfluß des sehr breiten Spektrums von Temperatur-, Spannungs- und Verformungsbedingungen, die vorstehend umrissen wurden. Einige davon liegen nunmehr vor.

Elastische Modelle

Das linear-elastische Modell ist natürlich das älteste, aber immer noch sehr günstige Modell. In der Tat liegen viele Untersuchungsabschnitte im niedrigen Temperatur-Spannungs-Bereich, so daß hier die lineare elastische Hypothese angewendet werden kann. Ferner können mit diesem einfachen Modell zu erwartende Hauptschwierigkeiten, beispielsweise Zonen hoher Spannungskonzentration oder Regionen, in denen Zugspannungen auftreten können, sehr schnell festgestellt werden. Diese Schwierigkeiten werden dann in einer zweiten Stufe mit einem ausgeklügelteren (angepaßteren) Modell untersucht.

Die Elastizität berücksichtigt nicht den Einfluß von Trennflächen großen Ausmaßes; wenn diese dazu neigen, einen signifikanten Einfluß auf den Deformationsprozeß auszuüben, wird das elastische Modell mit einem spezifischen Modell für diese Trennflächen kombiniert, wie nachstehend noch erläutert wird.

Weil Gebirgskörper sehr oft auch durch zahlreiche Trennflächen kleinen Ausmaßes mit eigentlich keiner Zugfestigkeit beeinträchtigt werden, haben Zinckewicz et al. (1968) vorgeschlagen, in Erwägung zu ziehen, daß das entsprechende Kontinuum, (welches in diesem Fall die Trennflächen einschließt), keinen Zugspannungen standhält; es wird dann durch ein bilineares elastisches Modell simuliert (Null Steifigkeit, wenn Zugspannungen unterworfen, normales Elastizitätsverhalten, wenn Druckbeanspruchungen unterworfen). Weil Klüfte eine viel geringere Steifigkeit als das Gestein erkennen lassen, müssen Moduli für das Kontinuum, das dem Gebirge entspricht, die Orientierung von Kluftscharen, den Kluftabstand innerhalb jeder Schar und die Kluftsteifigkeit berücksichtigen. Zum Beispiel schlagen Heuzé et al. (1982) vor, einen Granit mit drei orthogonalen Kluftscharen mit ungefähr gleichem Abstand durch ein entsprechendes isotropes Medium mit folgendem Elastizitätsmodul darzustellen:

$$E_g^{-1} = E_1^{-1} + (S k_n)^{-1} \qquad (21)$$

wobei E_g der Young's Modulus (Elastizitätsmodul) des Gebirges ist,

E_1 der Young's Modulus (Elastizitätsmodul) des intakten Granits ist,

s der Kluftabstand ist,

k_n die Normalsteifigkeit der Klüfte ist.

Gerrard (1982) führte eine detaillierte Untersuchung unter Berücksichtigung des Einflusses von ein, zwei oder drei orthogonalen Kluftscharen durch. Die Klüfte wurden als kontinuierlich mit orthorhombischen elastischen Eigenschaften (9 elastische Konstanten) und mit ungefähr gleichen Abständen angenommen. Ferner wird festgelegt, daß die Senkrechte zu den drei Symmetrieebenen des Materials, in jeder Schicht, entsprechend parallel zu den drei Achsen x_1, x_2, x_3 des Bezugssystems angeordnet sind, (die Klüfte sind lotrecht zu x_3). Es stellt sich dann heraus, daß das äquivalente Material orthorhombisch ist.

Brace et al. (1968) waren die ersten, die den Dilatanzeffekt (d. h., die Zunahme des Hauptmassevolumens) beschrieben haben, der nach Überschreitung der linear-elastischen Grenze des Gesteins, aber vor Erreichen der Spitzenfestigkeit auftritt. Brace und Martin (1968) erörterten den Einfluß dieser Dilatanz auf das mechanische Verhalten flüssigkeitsgesättigter Gesteine und entdeckten einen stabilisierenden Effekt. Stuart und Dietrich (1972) schlugen vor, ein nichtlineares, isotropes, elastisches, äquivalentes Material in Betracht zu ziehen, um diese Volumenzunahme zu berücksichtigen:

$$\underset{\sim}{\varepsilon} = \phi_0 \underset{\sim}{1} + \phi_1 \underset{\sim}{\sigma} + \phi_2 \underset{\sim}{\sigma}^2 \qquad (22)$$

wobei ϕ_0, ϕ_1 und ϕ_2 Polynomfunktionen der

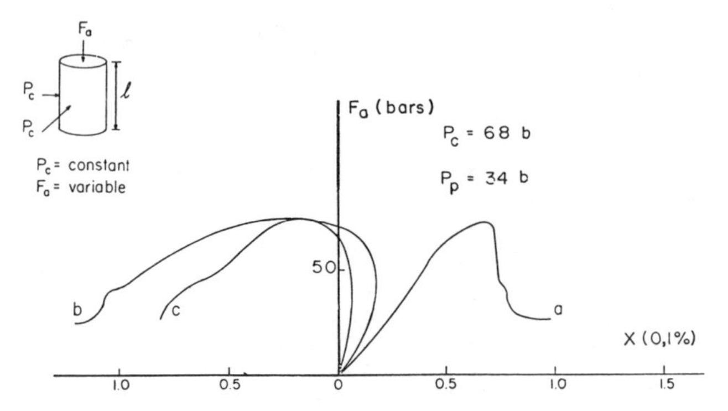

Abb. 10 Ergebnisse von Triaxial-Versuchen an flüssigkeitsgesättigtem Berea-Sandstein;
Kurve a = X Abszisse ist axiale Deformation, während es sich bei Kurve b um eine volumetrische Deformation handelt und um das Porenvolumen bei Kurve c; der effektive Umschließungsdruck ist 3,4 MPa.

ersten Invarianten des Spannungstensors $\underset{\sim}{\sigma}$ sind ($\underset{\sim}{\varepsilon}$ ist der linearisierte Deformationstensor). Nach meiner Kenntnis wurde jedoch eine praktische Anwendung dieses Modells bislang nicht vorgeschlagen.

Elasto-plastische Modelle

Die Entwicklung steifer - oder servogesteuerter - Laboratoriumsversuchssysteme (z. B. Cook und Hojem 1966, Wawersik 1968, Rummel und Fairhurst 1971, Hudson et al. 1972) hat Wege eröffnet zu detaillierten Versuchen über das mechanische Verhalten von Gesteinsproben nach Überwindung der Spitzenfestigkeit. Wawersik und Fairhurst (1970), Crouch (1970), Cornet und Fairhurst (1972), Rummel (1974) und viele andere haben somit die Schwankungsbreite der Dilatanz untersucht und - allgemeiner - die Spannungs-Verformungs-Beziehung, im Bereich nach Uberwindung der Spitzenfestigkeit, für verschiedene Gesteinsarten unter verschiedenen Umschließungsdruckbedingungen. Typische Ergebnisse sind in Abb. 10 bis 13 dargestellt.

Aus Abb. 10 und 12 ist ersichtlich, daß Gesteine nach Überschreiten der Spitzenfestigkeit die wichtige Eigenschaft besitzen, signifikante Belastungen auszuhalten und dabei für einige Spannungsbedingungen ein elasto-plastisches Verhalten bei herabgesetzter Verformung, verbunden mit Dilatanz, erkennen zu lassen.

Für größere Druckspannungen verschwindet die Dilatanz (siehe Abb. 13).

Jedoch entspricht das in Abb. 11 dargestellte Verhalten nicht dieser Regel: Wenn einmal ein bestimmter Deformationsbetrag nach der Spitzenfestigkeit erreicht ist, versagt die Gesteinsprobe in instabiler Weise, durch Wawersik (1968) auch Klasse-II-Verhalten genannt: das Probestück mußte fortschreitend entlastet werden, um so das Versagen quasistatisch zu halten. Dieser Bruchvorgang wurde am Berea-Sandstein bei Umschließungsdrücken von etwa

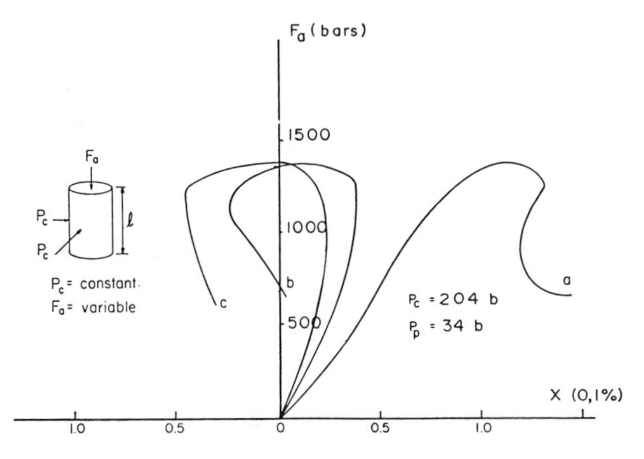

Abb. 11 Wie Abb. 10, der effektive Umschließungsdruck ist jedoch 17,0 MPa.

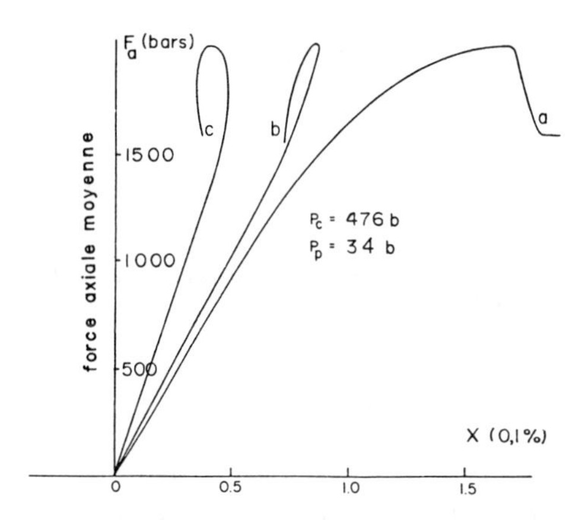

Abb. 12 Wie Abb. 10, der effektive Umschließungsdruck ist jedoch 44,2 MPa.

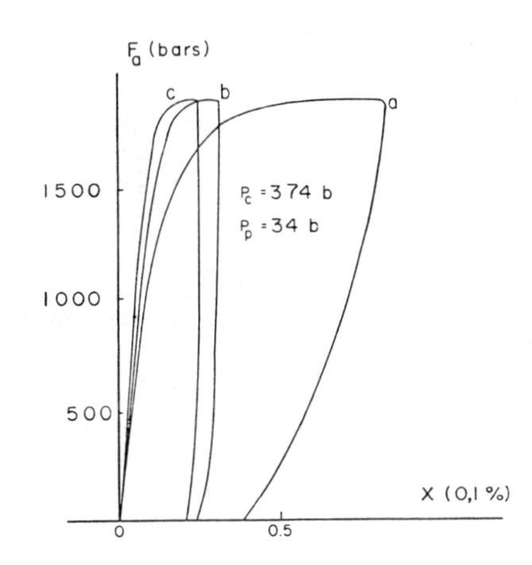

Fig. 13 Wie Abb. 10, das Gestein ist jedoch Indiana-Kalkstein und der effektive Umschließungsdruck beträgt 34 MPa.

7 MPa bis 25 MPa beobachtet, d. h. für Werte, die für viele Situationen im Ingenieurwesen typisch sind, (sie entsprechen einer Tiefe im Bereich von 500 m bis 2000 m). Ähnliche Resultate sind beobachtet worden bei Kalkstein, Kohle oder Granit für verschiedene Bereiche des Umschließungsdruckes. In keinem der durchgesehenen Beiträge, welche Modellvorstellungen für das Verhalten von Gestein nach Überwindung der Spitzenfestigkeit behandeln, wurde dieser Punkt vorgebracht, und man darf hoffen, daß er anläßlich des Kongresses erörtert wird.

Die Folgen dieser Dilatanz für das Verhalten flüssigkeitsgesättigter Gesteine nach Überwindung der Spitzenfestigkeit habe ich bereits früher erörtert (Cornet 1977).

Ich zeigte beispielsweise auf, daß für solche Materialien der Bereich des Umschließungsdrucks, bei dem ein Klasse-II-Bruchvorgang beobachtet wird, größer ist als derjenige für trockene Gesteine, wenn die Entwässerung nur partiell ist: Eine instabile Bruchausbreitung bringt eine Erniedrigung des Porenvolumens und daher eine Erhöhung des Porendrucks mit sich, so daß der effektive Umschließungsdruck reduziert wird.

Jia-Shou Zhuc und Yin-Tang Wang (1983, dieser Kongreß) schlagen drei Arten von Bruchkriterien vor, um die Spitzenfestigkeit zu kennzeichnen:

Reiner Zugbruch: $\sigma_1 - \sigma_t = 0$ wenn $3\sigma_1 + \sigma_3 \leq 0$

Zugtorsionsbruch: $(\sigma_1 - \sigma_3)^2 + 8\sigma_t\,(\sigma_1 + \sigma_3) = 0$

(23) wenn $3\sigma_1 + \sigma_3 > 0$; $\sigma_1 + (3 - 2\sqrt{2})\,\sigma_3 \leq 0$

Drucktorsionsbruch:

$$(\sqrt{1+f^2} + f)\,\sigma_1 - (\sqrt{1+f^2} - f)\,\sigma_3 - 4\sigma_t = 0$$

wenn $\sigma_1 + (\sqrt{1+f^2} - f)^2\,\sigma_3 \geq 0$

wobei Druckspannungen positiv sind, f der Reibungskoeffizient des Gesteins und σ_t seine Zugfestigkeit ist.

Unter Anwendung der Plastizitätszuwachstheorie erhalten sie:

$$\{d\sigma\} = ([D] - (1-r)[D_p])\{d\varepsilon\} \tag{24}$$

wobei r = 1 im elastischen Bereich oder wenn Entlastung auftritt,

r = 0 im plastischen Bereich,

0 < r < 1 im Übergangsbereich,

[D] die elastische Matrix ist;

$$[D_p] = [D]\left\{\frac{\partial F}{\partial\{\sigma\}}\right\}\left(A + \left\{\frac{\partial F}{\partial\{\sigma\}}\right\}^T [D] \left\{\frac{\partial F}{\partial\{\sigma\}}\right\}\right)^{-1}\left\{\frac{\partial F}{\partial\{\sigma\}}\right\}^T [D] \tag{25}$$

F ist die Fließgrenzenfunktion,

A entspricht dem Verformungsfestigungseffekt.

Pierce und Ryder (1983, dieser Kongreß) bevorzugen die Verwendung einer Deformationsformulierung, die durch eine Parameterisierung von Laboratoriumsresultaten erlangt wird. Es wird angenommen, daß diese idealisierten Spannungs-

Verformungs-Kurven von der Belastungsgeschichte und dem Belastungsweg unabhängig sind. Sie nehmen an, daß reines Elastizitätsverhalten bis zur Spitzenfestigkeit fortdauert, welche durch das lineare Gesetz $\sigma_1 = \sigma_c + k\sigma_3$ beschrieben werden kann (σ_c ist der maximale Spannungsabfall, der von σ_3 unabhängig sein soll).

Das Verhalten nach dem Bruch folgt einem Gesetz mit konstantem, negativen Modul; wenn beim Schervorgang die Restfestigkeit erreicht wird, tritt Verformung ohne Dilation ein.

Während bei beiden Modellen angenommen wird, daß das Material in Bezug auf seine Festigkeit isotrop ist, schlagen Guenot und Panet (1983, dieser Kongreß) vor, zwei verschiedene Mohr-Coulomb-Bruchkriterien für Gestein in Betracht zu ziehen, welche eine hohe ebene Festigkeitsanisotropie aufweisen. Das erste Kriterium ($|\tau| = c + \sigma_n \, tg \, \varphi$) wird für alle Bruchoberflächen, außer für diejenigen, welche parallel zu den Schwächeflächen angeordnet sind, definiert; für diese wird ein zweites Kriterium benutzt $|\tau| = \bar{c} + \sigma_n \, tg \, \varphi$

wobei $\bar{c} << c$

Der Hauptnachteil all dieser Modelle besteht darin, daß sie den sehr signifikanten Einfluß der Verformungsrate auf die Spitzenfestigkeit (+) nicht berücksichtigen können. Ferner ist es nicht klar, ob das Fließgrenzkriterium, das durch diese verschiedenen Autoren gewählt wurde, wirklich durch Feld- und Laboratoriumsbeweise erhärtet wurde. (+) sowie das Verhalten nach Erreichen der Spitzenfestigkeit

Zeitabhängige Modelle für niedrige Temperaturen und niedrige minimale Hauptspannungsbedingungen

Zeitabhängige Effekte, die bei niedriger Temperatur (beispielsweise niedriger als 300 °C) und niedrigen minimalen Hauptspannungen (niedriger als 500 MPa) beobachtet wurden, können zwei verschiedene physikalische Ursprünge aufweisen: Verschiebungsbewegung in den Kristallen, die das Gestein bilden, und stabile Mikrorißbildung. Der erste Mechanismus wird zum Beispiel bei Steinsalz angetroffen, selbst unter einaxialen Druckbedingungen und Raumtemperatur, oder bei Calcit und Serpentinit, wenn die minimale Hauptspannung größer ist als einige 10er Megapascal.

Dieser Deformationsprozeß wird oft im Labor bei Kriechversuchen untersucht, bei denen die Last konstant gehalten wird, oder bei Versuchen mit konstanten Belastungsraten (siehe z. B. Langer 1979, Wawersik 1980, Voille et al. 1981). Für gegebene Belastungsbedingungen kann die Kriechgeschwindigkeit abnehmen wegen der Verformungsverfestigung, kann konstant bleiben, oder kann zunehmen und zu einer bestimmten Art von Instabilität führen. Diese verschiedenen Stufen werden primäres, sekundäres und tertiäres Kriechen genannt.

Andrade (1910) schlug eine allgemeine Gleichung für das Kriechverhalten vor:

$$\varepsilon = A + B.E(t) + Ct \qquad (26)$$

wobei A, B und C Konstanten sind, während E(t) eine empirische Funktion ist. Morlier (1966) und Parson und Hedley (1966) fanden ebenso wie Langer (1979) heraus, daß E(t) logarithmisch sein sollte für primäres Kriechen. Für sekundäres Kriechen schlägt Langer vor, daß

$$\dot{\varepsilon} = A \exp(-u/kT) . f(\sigma) \qquad (27)$$

wobei T die Temperatur ist, σ der aufgebrachte einaxiale Druck ist, u die Aktivierungsenergie, k die Boltzmannkonstante und A eine von den Materialeigenschaften abhängige Konstante ist.

Vouille et al. (1981) fanden für den Fall des Kriechens heraus, daß die nach Erreichen der Elastizitätsgrenze beobachtete viskoplastische Verformung ausgedrückt werden kann als

$$\varepsilon_{vp} = \frac{(\sigma_1 - \sigma_3 - 2c)^{\beta}}{k} \, t^{\alpha} \qquad (28)$$

wobei ε_{vp} die viskoplastische Verformung ist,

σ_1 die maximale Hauptspannung ist,

σ_3 die minimale Hauptspannung ist,

c die Kohäsion ist,

k, β, α Konstanten sind, die aus diesen Versuchen ermittelt werden,

t die Zeit ist.

Sie schlagen als rheologisches Gesetz für dieses Steinsalz vor:

$$\langle \sigma_1 - \sigma_3 - 2c \rangle = \frac{K}{(\alpha)^{\alpha/\beta}} (\varepsilon_{vp})^{(1-\alpha)/\beta} (\dot{\varepsilon}_{vp})^{\alpha/\beta} \qquad (29)$$

wobei $\langle x \rangle = \frac{1}{2}(x + |x|)$

Ferner beobachten sie, daß der Winkel der inneren Reibung Null für dieses Gestein ist, während die scheinbare Kohäsion kleiner als 1,5 MPa ist; dieser letzte Wert nimmt ab, wenn Temperatur und Umschließungsdruck zunehmen.

Während des tertiären Kriechens tritt eine geringe Mikrorißbildung auf, die letzlich zum vollständigen Bruch führt (vgl. z. B. Scholtz 1968, Kranz und Scholtz 1977).
Der Einfluß der Verformungsrate auf diesen stabilen Prozeß der Mikrorißbildung ist von primärer Bedeutung für alle Gesteine, sobald die Spitzenfestigkeit überschritten wird. Bieniavski (1970), Peng und Podmieks (1972), Hudson und Brown (1973), Houpert (1974) untersuchten experimentell im Laboratorium diese Abhängigkeit von der Verformungsrate. Die Ergebnisse (siehe Abb. 14) lassen erkennen, daß ein Vermindern der Verformungsrate eine Verminderung der Spitzenfestigkeit hervorruft und ein "Erweichen" des Verhaltens nach Erreichen der Spitzenfestigkeit; je heterogener das Gestein ist, um so größer ist der Erweichungseffekt. Kaiser und Morgenstern (1981) schlugen ein einfaches rheologisches Modell vor, um diesen Effekt phänomenologisch zu erklären.

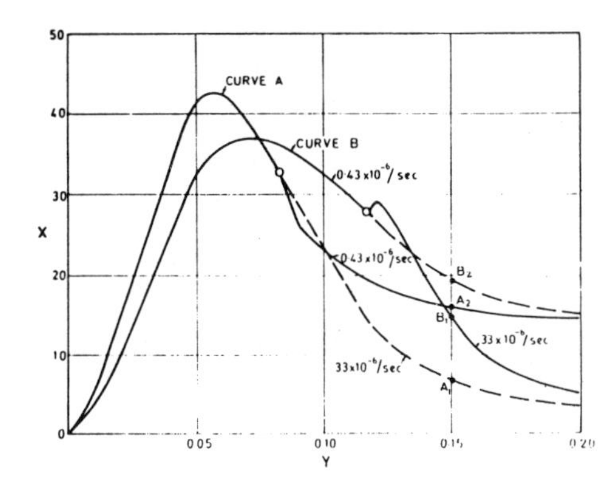

Abb. 14 Effekt der sich ändernden Verformungsrate auf das Verhalten von zerbrochenem Gestein (nach Bieniawski 1970).

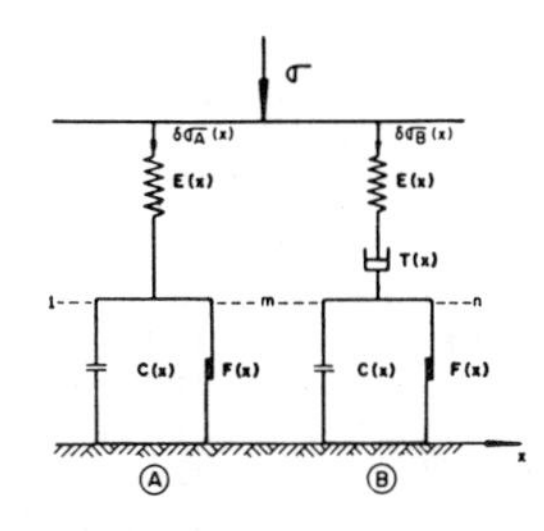

Abb. 15 Elementares Modell für das zeitabhängige Versagensverhalten nach Erreichen der Spitzenfestigkeit, nach Kaiser und Morgenstern (1981).

Sie betrachten ein elementares Modell (siehe Fig. 15), das aus einer sprödelastischen Einheit A, die eine momentane Festigkeit der Größe C_A bei einer Verformung gleich E_A und eine konstante Bruchfestigkeit F_A aufweist, und einer dazu parallel angebrachten zeitabhängigen Einheit B besteht.

Diese Einheit B zeigt einen momentanen Festigkeitsverlust C_B und eine konstante Bruchfestigkeit F_B. Die Einheit B kann vor oder nach der Einheit A versagen. Sie schlagen vor, die zeitabhängige Bruchentwicklung bei Gesteinen dadurch nachzuvollziehen, daß einige dieser elementaren Modelle parallel aufgestellt werden.

Tan und Kang (1983, dieser Kongreß) ziehen in Betracht, daß die Gesamtverformung, die ein dilatantes Material zeigt, als die Überlagerung von drei Komponenten angesehen werden kann:

$$\varepsilon(t)_{ij} = \varepsilon^{E}_{ij} + \varepsilon^{D}_{ij} + \varepsilon^{D}(t)_{ij} \qquad (30)$$

$i, j = 1, 2, 3$

wobei ε^{E}_{ij} die Verformung unterhalb der Elastizitätsgrenze ist,

ε^{D}_{ij} die momentane Dilatationsverformung ist,

$\varepsilon^{D}_{(t)ij}$ die zeitabhängige dilatante Verformung ist.

Es wird angenommen, daß das Einsetzen von Dilatanz eintritt, wenn $\sigma_{oct} > f_3$ wobei f_3 eine Materialkonstante ist.

Sie wenden diese Formulierung auf die Untersuchung einer regionalen Bodenhebung an, die vor einigen großen Erdbeben in China beobachtet wurde.

Dragon und Mroz (1979) schlagen vor, das Kriechverhalten, das in einaxialen Druckversuchen beobachtet wurde, zu einer zeitabhängigen Rißdichte (β) in Beziehung zu setzen:

$$\beta = k_i \varepsilon^{(i)}_{kk} + k_2(\beta)\, g(\sigma_{ke}, \beta) \qquad (31)$$

wobei $\varepsilon^{(i)}_{kk}$ die inelastische, volumetrische Verformungsrate bedeutet;

k_1 eine Materialkonstante ist,

$k_2^{(\beta)}$ und $g(\sigma_{ke}, \beta)$ Funktionen sind, welche die zeitprogressive Bruchbildung $(k, \ell = 1,2,3)$ beschreiben.

Es wird angenommen, daß die Gesamtverformungsrate die Summe einer elastischen und einer inelastischen Verformungsrate ist.

Es wird angenommen, daß die inelastische Verformungsrate eintritt, wenn die folgende Fließgrenzbedingung erfüllt wird:

$$f(\sigma_{ij}, k\beta) = S_{ij} S_{ij} - 2p(k,\beta)(I_1^0 - \sigma_{kk}) = 0 \qquad (32)$$

wobei $p(k,\beta) = p_0 + k - \beta^2$

$k = \sigma_{mn}\varepsilon_{mn}$ = Rate der Verformungsverfestigung,

S_{ij} die Spannungsdeviatorkomponente ist,

I_1^0 eine positive Konstante ist,

P_0 die Konfiguration einer anfänglichen Fließgrenzoberfläche für $k = \beta = 0$ bestimmt.

Martin (1972), Henry et al. (1977), Atkinson (1979) und andere wiesen auf den Einfluß einer spannungsbedingten Korrosion beim tertiären Kriechen hin, das heißt, den Einfluß chemisch aktiver Stoffe beim Mikrorißwachstum, (wobei die Hauptkorrosionssubstanz Wasser ist). Diese Konzeption der Spannungskorrosion wurde durch Das und Scholtz (1981) für die Erklärung einiger mit Erdbeben zusammenhängender Phänomene, wie Mehrfachereignisse, verzögerte Mehrfachereignisse, Vorbeben, Nachbeben, nachseismische Ausbruchvergrößerung oder "langsame" Erdbeben, angewendet. Ihr Gedankengang geht davon aus, daß Erdbeben einer kombinierten Bruchbildung aus Modus II und Modus III zuzuordnen sind, d. h. reine Scherung, und daß der Bruch, Quelle des Erdbebens, sich auf seiner eigenen Ebene fortpflanzt. Diese Annahme wird durch Geländebeobachtungen gestützt, die anzeigen, daß gewöhnlich Erdbeben entlang vorher bestehender Verwerfungen auftreten.

Gemäß klassischer Bruchmechanik (siehe Irwin 1957) breitet sich ein Bruch aus, wenn der Spannungsintensitätsfaktor (k), der das Spannungsfeld am Rand des Risses kennzeichnet, einen kritischen Wert K_c^* erreicht, bei welchem es sich um eine Materialeigenschaft handelt. Bei Modus-I-Bruchbildungen (Zug) beobachtet man oft, daß die Bruchfortpflanzung in stabiler Weise bei solchen Spannungsintensitätsfaktoren auftritt, die viel niedriger als K_c^* liegen. Dies wird der Entwicklung eines Korrosionseffektes zugeschrieben, der den Wert des kritischen Spannungsintensitätsfaktors auf $K_o \leq K_c^*$ (Riß pflanzt sich fort, wenn $k > k_o$) mindert. Diese Bruchfortpflanzung ist quasistatisch, mit einer Ausbreitungsgeschwindigkeit der Form (Atkinson 1979):

$$\dot{X} \propto k^n \qquad (33)$$

wobei n die Konstante ist, die Spannungskorrosionsindex genannt wird.

Das und Scholtz vermuten, daß ähnliche Gesetze für Modus II und Modus III bestehen. Da k proportional zu $\sqrt{X}$ ist, ist ohne weiteres zu erkennen, daß dieser Spannungskorrosionseffekt schließlich zu unstabiler Bruchfortpflanzung führt. Sie argumentieren, daß unstabiles Rißwachstum auftritt, wenn k einen kritischen K_c erreicht, der dicht an dem theoretischen K_c^*-Wert liegt, da bei dynamischer Bedingung für die Korrosion keine Zeit vorhanden ist, sich zu entwickeln. Der Spannungsintensitätsfaktor k für einen kreisförmigen, ebenen Scherriß mit einem Radius X bei einem unendlichen, homogenen, linearelastischen Medium wird aufgestellt, für beide Modi II und III, durch:

$$k = c\,\Delta\tau\,\sqrt{X} \qquad (34)$$

wobei c eine geometrische Konstante und der statische Spannungsabfall ist.

Da für einen kreisförmigen Riß beide Modi II und III gleichzeitig existieren an Punkten entlang des Rißrandes, wird k als die mittlere Quadratwurzel dieser Spannungsintensitätsfaktoren angenommen, so daß

$$C = \frac{20\sqrt{2}}{7\pi} \qquad \text{(Sih, 1973)}$$

Aus experimentellen Ergebnissen (Atkinson 1979) tritt der Beginn der stabilen Fortpflanzung ein, wenn

$$k = k_0 \left(\frac{\dot{X}}{V_0}\right)^{1/n} \qquad (35)$$

wobei K_o und v_o die Werte von k und $\dot{X}$ bei der Initiierung von quasistatischem Wachstum sind. Falls $\Delta\tau$ zeitunabhängig ist, dann gilt, da $k_0 = C\,\Delta\tau\,\sqrt{X_0}$

$$\dot{X} = V_0 \left[\left(\frac{X}{X_0}\right)^{1/2}\right]^n \qquad (36)$$

so daß

$$X = \left[X_0^{\frac{2-n}{2}} - \frac{n-2}{2}\,\frac{V_0 t}{X_0^{n/2}}\right]^{2/(2-n)} \qquad (37)$$

somit wird die Dauer des stabilen Rißwachstums angegeben durch

$$t_f = \frac{X_0}{V_0}\,\frac{2}{n-2} \qquad (38)$$

Die Gleichung (36) erhält man dadurch, daß man den Ausdruck in Klammern in Gleichung (35) gleich Null setzt; tatsächlich entspricht diese Bedingung dem unendlichen Wert X, da K > 2. Die Ableitung des stabilen Rißwachstums, t_f, hängt von den anfänglichen Bedingungen sowie von n, dem Spannungskorrosionsindex, ab. Das und Scholtz haben dieses Ergebnis auf einige typische Bedingungen angewendet und fanden, daß stabiles Bruchwachstum, welches den Erdbeben vorausgeht, variieren kann von 48 Stunden, für eine anfängliche Bruchlänge von 1 km und einen Spannungskorrosionsindex von 12,5, bis zu 278 Tage für eine anfängliche Rißlänge von 0,5 cm und denselben Korrosionsindex; $\Delta\tau$ war gleich 10 MPa in beiden Fällen. Die Rißlänge vor der unstabilen Fortpflanzung war 1,4 km in dem ersten Fall und 14,2 cm in dem zweiten Fall. Ihre Ergebnisse scheinen in Übereinstimmung zu sein mit Dilatometer-Aufzeichnungen, die man vor einem Erdbeben der Größenordnung 3,8 erhielt.

Modelle für das mechanische Verhalten großer, natürlicher Trennflächen

Bei vielen Gesteinsmechanikproblemen wird das Verhalten von Gebirgen durch dasjenige bereits vorhandener großer Trennflächen, wie Klüfte, stark beeinflußt. Klüfte können keine Zugspannung aufnehmen; ihre Schubfestigkeit ist stark variabel, und ihre Scherdeformation wird gewöhnlich von Volumenveränderungen begleitet. Wie bereits vorher in diesem Bericht erwähnt, werden nur diejenigen Trennflächen, die im Vergleich zur Größe des in Untersuchung befindlichen Volumens groß sind, als solche behandelt; die kleineren Trennflächen werden in das entsprechende Kontinuum integriert.

Große Klüfte beeinflussen sowohl die elastische Reaktion eines Gebirges als auch seine Widerstandseigenschaften. Für ihren Einfluß auf die elastische Reaktion wird die Kluft gewöhnlich durch eine spezielle Steifigkeitsmatrix dargestellt, (siehe z. B. Goodman et al. 1968, Heuze et al. 1971, Gerrard 1982, Walsh und Grosenbaugh 1979), welche normale und Scherkräfte, die auf die Kluftoberfläche ausgeübt werden, zu normalen (senkrechten) und Scherverschiebungen an dieser Oberfläche in Beziehung setzt. Desai et al. (1983, dieser

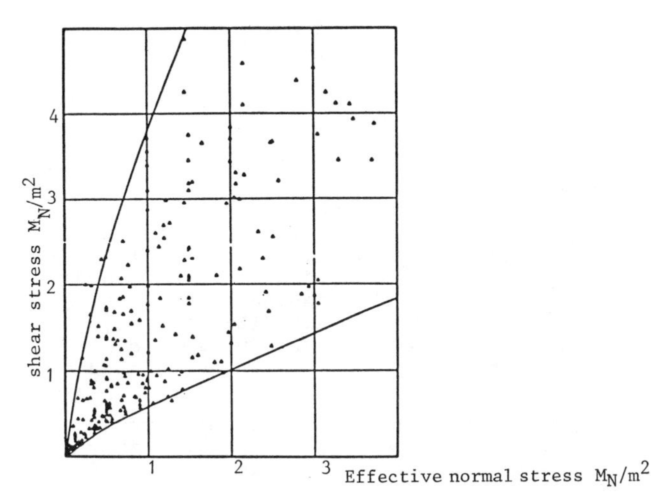

Abb. 16 Spitzenscherfestigkeit nichtgefüllter Felsklüfte aus den veröffentlichten Resultaten direkter Schubversuche im Laboratorium und in situ (nach Barton 1976).

Kongreß) schlagen vor, eine Kluft durch dünne Schichtelemente mit entsprechender Steifigkeitsmatrix abzubilden. Sie schlagen ferner vor, nur die diagonalen Ausdrücke dieser Matrix beizubehalten wegen der Schwierigkeit, die für die praktische Bestimmung nichtdiagonaler Ausdrücke entsteht.

Barton (1976) wies darauf hin, daß das mechanische Verhalten von Klüften stark abhängig ist von dem Wert der auf diese ausgeübten normalen Spannung. Für niedrige, effektive, normale Spannungsgrößen weisen Gesteinsklüfte wegen des großen Einflusses von Oberflächenrauhigkeit und variabler Gesteinsfestigkeit ein breites Spektrum von Scherfestigkeit auf (siehe Abb. 16).

Trotz der großen Variationsbreite der Scherfestigkeit intakter Gesteine ist das Spektrum der Scherfestigkeit von Klüften oder künstlichen Störungen bei hohen effektiven Normalspannungen relativ schmal (Abb. 17). Es wird durch das klassische Coulombsche Reibungskriterium gut dargestellt (Byerlee 1975):

$$\tau = C + \mu\,(\sigma_n - P) \qquad (39)$$

wobei τ die Scherfestigkeit ist, σ_n die Normalspannung, P der Porendruck, μ der Reibungskoeffizient, C die Kohäsion ist, die für große Werte von σ_n üblicherweise vernachlässigbar ist.

Selbst für höhere, effektive Normalspannungen beeinflussen natürliche Klüfte das Scherverhalten von Gebirgen nicht mehr.

Man hat sich bemüht, diese durch die effektive Normalspannung bedingte (+) Variation der Scherfestigkeit sowie die während der Scherdeformation beobachtete Volumenänderung quantitativ zu bestimmen. Barton (1973, 1976) führte direkte Scherversuche durch bei einer Vielzahl künstlicher Spannungsbrüche, die in realistischen, spröden Modellmaterialien erzeugt wurden. Er beobachtete, daß, falls μ der gesamte Reibungswinkel des Materials ($\mu = \arctg \tau | \sigma_n$ wobei τ die Spitzenscherfestigkeit und die effektive normale Spannung ist), und d_n der Spitzendilatationswinkel ist, der als die momentane Neigung des Scherweges definiert ist bei Spitzenfestigkeit, relativ zur mittleren Kluftebene, dann gilt

$$\mu = 2\,d_n + 30 \qquad (40\ a)$$

$$d_n = 10 \log_{10} \left(\frac{\sigma_c}{\sigma_n}\right) \qquad (40\ b)$$

wobei σ_c die einaxiale Druckfestigkeit des intakten Materials ist; der Wert 30 in der Gleichung (40 a) entspricht dem Basisreibungswinkel des Materials.

Für den 0,1 bis 10 MPa Bereich der Normalspannung, der bei diesen Experimenten untersucht wurde, hat sich herausgestellt, daß die Umhüllende der Spitzenzugfestigkeit angegeben werden kann mit

$$\tau = \sigma_n \, tg \left[20 \log_{10} \left(\frac{\sigma_c}{\sigma_n}\right) + 30 \right] \qquad (41)$$

Barton schlug eine generalisierte Form der Gleichung (41) für natürliche Klüfte vor, die niedrigen normalen Spannungen unterworfen sind:

$$\tau = \sigma_n \, tg \left[JRC \log_{10} \left(\frac{JCS}{\sigma_n}\right) + \phi_b \right] \qquad (42)$$

wobei JRC der Kluftrauhigkeitskoeffizient ist, auf einer empirischen Skala definiert, die von 20, für sehr rauhe Klüfte, bis 0 für glatte Oberflächen reicht (siehe Barton und Choubey 1977, Barton und Bandis 1980, in Bezug auf eine ausführliche Erörterung dieser empirischen Skala);

JCS ist die einaxiale Druckfestigkeit der Kluftwand,

ϕ_b ist der Basisreibungswinkel, als der Wert von arctg ($\tau | \sigma_n$) definiert, erhalten aus Restscherversuchen an flachem, unverwitterten Gestein.

Der Wert von JCS kann variieren von σ_c, der einaxialen Druckfestigkeit des Materials, bis $\sigma_c/4$ für verwitterte Oberflächen.

Für größere Normalspannungen schlägt Barton vor, die einaxiale Druckfestigkeit durch die triaxiale Druckfestigkeit zu ersetzen, so daß die Gleichung (42) wird:

$$\tau = \sigma_n \, tg \left[JRC \log_{10} \left(\frac{\sigma_{TC} - \sigma_3}{\sigma_n}\right) + \phi_b \right] \qquad (43)$$

wobei σ_{TC} die Spitzenfestigkeit bei triaxialen Druckversuchen mit Umschließungsdruck gleich σ_3 ist.

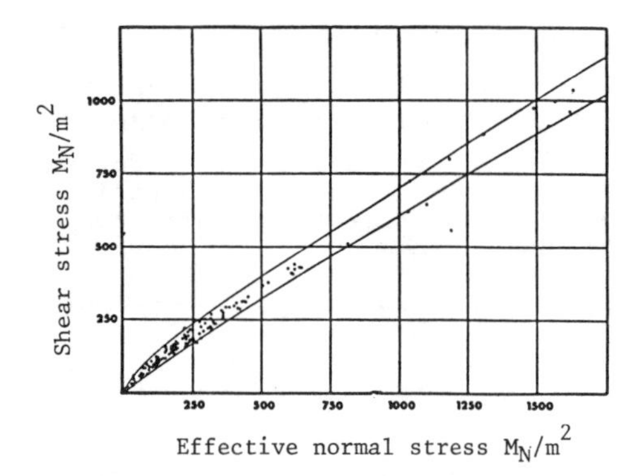

Abb. 17 Die Scherfestigkeit von Verwerfungen und Spannungsrissen im Gestein unter hoher, effektiver Normalspannung (nach Byerlee 1975, 1967, 1968); Angaben vom Westerly-Granit, Solnhofener Plattenkalk, Oak Hale Kalkstein, Mahant Gabbro, Spruce Pine Dunit, Cabramurra Serpentinit, Weber-Sandstein. (Diese Abbildung stammt von Barton 1976).

Die mit der Gleichung (43) erzielten Ergebnisse für Scherung entlang Trennflächen, in Laboratoriums-Triaxialversuchen mit Sandstein-, Kalkstein-, Gabbro-Dunit-, Serpentinit- und Granit-Proben entwickelt, waren in sehr zufriedenstellender Übereinstimmung mit experimentellen Ergebnissen für Umschließungsdrücke bis zu 500 MPa und für JRC-Werte, mit 20 gleichgesetzt.

Während die Gleichung (43) einer Generalisierung der Gleichung (41) entspricht, ist diejenige für die Gleichung (37 b) (die Bestimmung des Dilatationswinkels) für $\sigma_n > \sigma_c$

$$d_n = 10 \log_{10} \frac{\sigma_{TC} - \sigma_3}{\sigma_n} \qquad (44)$$

Ergebnisse, von den oben erwähnten Laboratoriumsversuchen erhalten, Fließgrenzendilatationswinkel der Größenordnung von 2° für Verwerfungen, beim spröd-duktilen Übergang geschert, (duktil kennzeichnet hier Scherdeformation ohne Spannungsabfall).

Pande und Gerrard (1983, dieser Kongreß) schlagen vor, die Kluftscherfestigkeit durch das Coulombsche Kriterium, aber mit einer variablen Kohäsion, C_e, welche von der Kluftöffnung abhängt, darzustellen:

$$C_e = C_o \frac{\langle \varepsilon_c - \varepsilon_t \rangle}{c} \qquad (45)$$

wobei ε_t die viskoplastische Zugbeanspruchung senkrecht zur Kluft ist,

C_o die Kohäsion der Kluft ist, wenn sie geschlossen ist,

ε_c eine kritische Zugbeanspruchung ist, die von der Kluftrauhigkeit abhängt;

$$\langle \varepsilon_c - \varepsilon_t \rangle = \begin{cases} \varepsilon_c - \varepsilon_t & \text{wenn } \varepsilon_c - \varepsilon_t > 0 \\ 0 & \text{wenn } \varepsilon_c - \varepsilon_t \leq 0 \end{cases}$$

Obwohl sich dieser Bericht nicht mit Problemen befaßt, die sich auf Fluidströmung in Gebirgen beziehen, sollte ein interessantes Ergebnis, das durch Walsh (1981) erzielt wurde bezüglich der Effektivspannungskonzeption, auf die Durchlässigkeit von Klüften oder Trennflächen angewendet, erwähnt werden. In seiner Untersuchung zeigt Walsh auf, daß die Änderung im Porenvolumen dv_p innerhalb einer Kluft, hervorgerufen durch Veränderungen sowohl des hydrostatischen Umschließungsdrucks als auch des Porendrucks, ist:

$$dv_p = \frac{\partial v_p}{\partial P}(dP_c - dP_p) + v_p \beta_s dP_p \qquad (46)$$

wobei $\frac{\partial v_p}{\partial P}$ die Änderungsrate des Porenvolumens, unter hydrostatischem Druck für eine Kluft ohne Porenwasser ist,

dP_c die Änderung des hydrostatischen Umschließungsdrucks, auf das Gastgestein ausgeübt, ist,

dP_p die Änderung des Porendrucks ist,

V_P das ursprüngliche Porenvolumen in den Klüften ist,

β_s die Massenzusammendrückbarkeit der Gesteinsmatrix ist.

Aus (46) kann gezeigt werden, daß die effektive Spannung, Pe, von welcher die Kluftdurchlässigkeit abhängt, ist:

$$P_e = P_c - sP_p \qquad (47)$$

wobei $s = 1 - V_p \beta_s / (\partial V_p / \partial P)$

s kann zwischen 0,5 und 5 liegen.

Da die Durchlässigkeit u. a. von der Kluftöffnung abhängig ist, kann vorausgeahnt werden, daß diese Gefügeöffnung von effektiven Spannungen abhängt, durch eine Gleichung ähnlich (47) aufgestellt, eher als durch den klassischen $(\sigma - p_1)$ Ausdruck, der üblicherweise angenommen wird.

4.2. Einige Lösungsbeispiele

Weil die Geometrie gewöhnlich ziemlich komplex ist und weil konstitutive Gleichungen in vielen Fällen nichtlinear sind, kann die Lösung der meisten Gesteinsmechanikprobleme durch geschlossene analytische Verfahren nicht erreicht werden. Wenn das Material ein lineares, isotropes, elastisches Verhalten an den Tag legt, (d. h., wenn die Komplexität des Problems nur aus seiner Geometrie herrührt), kann die photoelastische Methode angewendet werden, für zwei- und dreidimensionale Probleme, um in ziemlich unaufwendiger Weise

Zonen hoher Spannungskonzentration festzustellen. Aber das Erscheinen von Hochgeschwindigkeitscomputern sieht nunmehr Mittel vor, um in einer nahezu genauso unaufwendigen Art genauere numerische Lösungen zu erreichen, so daß numerische Modelle fast immer gegenüber photoelastischen Modellen bevorzugt werden. Jedoch haben sich für die Untersuchung großer Deformationen physikalische Modelle als sehr nützlich erwiesen, um die Hauptaspekte von Deformationsprozessen näher zu untersuchen. Zum Beispiel haben Tapponier et al. (1982) ein plastisches Modell entwickelt, um das Drücken des asiatischen Kontinents durch den indischen Subkontinent zu studieren. Mit diesem Modell waren sie in der Lage, ein globales Verständnis der Kinematik dieses Plattendeformationsprozesses, wie die Öffnung der Südchinesischen See vor der späten Miozänperiode, oder die Rotation und Extrusion von Indochina nach Südosten entlang der damals lateralen Verwerfung des Roten Flusses während der ersten 20 bis 30 Millionen Jahre der Kollision, sowie der Öffnung des Mergui-Beckens und der Andaman-See bis zur Jetztzeit, zu erreichen. Ihr Experiment vermutet, daß ein größeres, übriggebliebenes, laterales Störungssystem sich über Tien-Shan, Mongolei, und Baikal bis zum Okhotk-See fortpflanzen wird. In einem kleineren Maßstab haben Deramont et al. (1983, dieser Kongreß) mit einem Zweimaterialmodell die Wirkung einer tiefen Scherung an einer sedimentären Schicht untersucht. Die untere Schicht besteht aus Plastine, einem kohäsiven, plastischen Material, während die obere Schicht aus einem kohäsionslosen, körnigen Material besteht. Ihr Modell scheint mit einiger Genauigkeit die Merkmale vorauszusagen, die während Gleitungsbewegungen von Überschiebungsdecken beobachtet wurden. Jedoch ist die Hauptschwierigkeit, die bei diesem physikalischen Modell auftrat, die einwandfreie Abwägung der verschiedenen Parameter, die damit verknüpft sind: ein gutes Modell muß Ähnlichkeitsregeln erfüllen.

Dieses Ähnlichkeitsprinzip wird wahrscheinlich besser erfüllt durch sogenannte Basisreibungsmodelle (Bray und Goodman, 1981), bei welchen der Einfluß der Schwerkraft, die auf der Basisfläche eines zweidimensionalen physikalischen Modells wirksam ist, durch eine Reibungskraft, die entlang der Basisfläche wirksam ist, modellartig gebildet wird. Verschiedene Materialien können verwendet werden, um die interessierenden rheologischen Eigenschaften angemessen einzustufen. Ferner gestattet ein einheitlicher Gasdruck, über die gesamte Modelloberfläche aufgebracht, die Simulation von Umschließung (Egger und Gindroz, 1979). Dieses Modell ist gut geeignet für Probleme, bei welchen es um langsame Bewegung geht, für die Massenkräfte im Vergleich zur Schwerkraft gering sind; es ist nicht anwendbar auf allgemeine dynamische Bedingungen. Lessi und Sarda (1980) haben das Problem der Einstufung der Trennflächenzähigkeit von Gesteinen (K_{IC}) für die Modellsimulation hydraulischer Bruchfortpflanzung erörtert; sie zeigen auf, daß, falls λ der geometrische Reduktionsfaktor ist, dann die Trennflächenzähigkeit für Modus I eingestuft werden muß durch:

$$K_{IC\ model} = \sqrt{\lambda}\ K_{IC\ real}$$

und sie entwickelten Modellmaterialien, auf einem Gemisch von Zement und Bentonit basierend, welche diese Bedingung für einen Reduktionsfaktor λ entsprechend ungefähr $2 \cdot 10^{-3}$ erfüllen.

Diese Schwierigkeit, Einstufungsregeln einwandfrei zu erfüllen, ist die Haupteinschränkung physikalischer Modelle, ein Mißstand, der numerische Modelle nicht beeinträchtigt. Ferner sind numerische Modelle direkt geeignet, um den Einfluß dieser oder jener Variablen dadurch zu untersuchen, daß man einfach ihren Wert ändert, um so parametrische Studien durchzuführen. Dies ist insbesondere nützlich, wenn ein Eingabeparameter nicht mit großer Genauigkeit umrissen ist. Darüber hinaus liefert das Erscheinen effizienter graphischer Bildschirm-Terminals nunmehr sehr wirksame Einrichtungen, um die Lösungen direkt sichtbar zu machen. Wahrscheinlich werden deshalb die meisten Probleme jetzt unter Verwendung numerischer Modelle angegangen.

Finite-Elemente-Lösungen

"Finite Elemente" sollen hier entsprechend ihrer üblichen Bedeutung im Ingenieurwesen verstanden werden, d. h., das Äquivalenzkontinuum und die größeren Trennflächen werden in eine endliche Zahl von Teilen diskretisiert, deren Verhalten durch eine endliche Zahl von Parametern spezifiziert wird. Diese Definition steht im Einklang mit derjenigen, die in der angewandten Mathematik dadurch gegeben ist, daß angegeben wird, daß die Lösung des kompletten Systems, als ein Zusammenspiel seiner Elemente, genau denselben Regeln folgt wie diejenigen, die auf ein normales, mathematisches, diskretes Problem anwendbar sind. Eine ausführliche Darstellung der Anwendung dieser Methode auf mechanische Probleme wurde durch Zienckiewiks (1977) angegeben, und hier soll nicht der Versuch unternommen werden, dieses sehr gewaltige numerische Verfahren zu erörtern. Es soll nur daran erinnert werden, daß, wie durch Jaeger und Cook (1979) hervorgehoben, die Vorteile finiter Elemente in folgendem bestehen:

i) Die Elemente können in jeder Weise und von jeder Art gewählt werden, so daß unregelmäßige Abgrenzungen ohne Schwierigkeiten eingepaßt werden können, und eine größere Dichte von Knotenpunkten kann in Bereichen größerer Spannungskonzentration gewählt werden.

ii) Variable Oberflächen- und Körperkräfte können leicht gehandhabt werden.

iii) Jedes Element hat seine eigene Steifigkeitsmatrix, so daß das Material nicht homogen oder isotrop zu sein braucht. Auch nichtlineare und bilineare Stoffe können ohne weiteres studiert werden.

iv) Die Methode ist nicht beschränkt auf elastisches Verhalten, und andere rheologische Moduln können eingeführt werden.

v) Reibung und Dilatanz an Kluftoberflächen können berücksichtigt werden.

vi) Die Ausgabe kann in dem Computer in jeder gewünschten Art und Weise bearbeitet werden.

Die anläßlich dieses Kongresses vorgelegten Papiere sind gute Beispiele für die Möglichkeiten der Methode. Zum Beispiel präsentieren Pande und Gerrard ein elastoviskoplastisches, zweidimensionales Modell für Gesteine, welches auch Scharen paralleler, durchgehender, in gleichem Abstand angeordneter Klüfte berücksichtigen kann. Man vermutet, daß diese eine gegebene Steifheit im elastischen Bereich zeigen, keine Zugfestigkeit besitzen und bei Druck Gleitung gemäß einem Coulombschen Reibungskriterium mit möglicher Dilatanz stattfindet. Armierungselemente können eingeführt werden; sie sollen eine gegebene elastische Reaktion anzeigen und sich dann gemäß einem Verformungsverfestigungsgesetz verhalten. Carol und Alonso schlagen ein neues Kluftelement für eine zweidimensionale Untersuchung von zerklüftetem Gestein vor, welches nicht weniger als elf Parameter benötigt, um sein komplettes mechanisches Verhalten zu beschreiben.

Praktische Anwendungen von Finite-Elemente-Untersuchungen werden ebenfalls präsentiert. Zum Beispiel schlagen Casteleiro et al. eine dreidimensionale Untersuchung von Nichtspannungsmaterialien vor, wie beim Entwurf eines Dammes angewendet. Sousa erläutert die Vorteile, dreidimensionale elastische Lösungen mit zweidimensionalen nichtlinearen Lösungen zu kombinieren, um die Bauphasen eines großen, unterirdischen Kraftwerks sowie eine Untersuchung seiner Langzeitstabilität zu optimieren.

Quiblier und Ngokwey verwenden ein dreidimensionales, viskoelastisches, mehrfachgeschichtetes, orthorhombisches Modell, um, bei großer Deformation, Zonen hoher Wahrscheinlichkeit von starker Klüftigkeit bei deformierten, sedimentären Schichten für die Vorhersage potentieller Ölreserven zu untersuchen.

Grenzelementmethoden

Grenzelementmethoden beziehen sich auf numerische Verfahren, bei denen nur die Abgrenzung des Kontinuums diskretisiert wird; vom mathematischen Standpunkt aus entsprechen sie einer bestimmten Klasse der Finite-Elemente-Methoden. Sie verwenden die Überlagerung bekannter Einflußfunktionen entlang Umrißlinien, um ein System simultaner, integraler Gleichungen, verknüpft mit Grenzbeanspruchungen und Verschiebungen, hervorzubringen. Diese Gleichungen werden numerisch oder analytisch dadurch gelöst, daß die Beanspruchungen, oder die Verschiebungen, oder beide, über kleine Bereiche der Abgrenzung konstant sind. Verschiebungen und Spannungen an Punkten in dem Kontinuum werden dann von den Grenzdaten her berechnet.

Zusätzlich zu der Einfachheit des Diskretisierungsverfahrens besteht der Hauptvorteil der Grenzelementmethoden gegenüber den endlichen Elementen im wesentlichen darin, daß Unendlichkeitsbedingungen durch Grenzelementlösungen implizite berücksichtigt werden können, während für Finite-Element-Lösungen, erstens, das Netz weit weg von dem Bezirk der Spannungsperturbation ausgeweitet werden muß, und, zweitens, eine bestimmte Hypothese in Bezug auf die Verschiebungsbedingungen erstellt werden muß. Ferner sind Grenzelementtechniken besonders gut geeignet, um Probleme mit sich entwickelnden Grenzen, wie Trennflächenfortpflanzungsuntersuchungen, oder Probleme, die spezifisches Kluftverhalten in sich schließen, zu lösen.

Eine grundlegende Behandlung von Integralgleichungsmethoden, auf die Elastostatik angewendet, wird zum Beispiel durch Jawson und Symm (1977) vorgelegt; ausführliche Darstellungen von Grenzelementmethoden sind kürzlich durch Banerjee und Butterfield (1981) und Crouch (1982) vorgeschlagen worden. Nur eine kurze Übersicht über einige Anwendungen dieser Methoden, die in der Felsmechanik-Literatur vorgeschlagen wurden, wird nun gegeben.

Die Verschiebungs-Diskontinuitäts-Methode (Starfield und Crouch 1973, Crouch 1976, 1978) besteht darin, N Verschiebungsdiskontinuitäten unbekannter Größe längs einer Grenze zu placieren, dann ein System linearer, algebraischer Gleichungen aufzustellen und zu lösen, um die Diskontinuitätswerte zu finden, welche die vorgeschriebenen Grenzbedingungen hervorrufen. Wenn die Geometrie geschlossene Konturen in sich schließt, werden das interne Problem und das externe Problem gleichzeitig gelöst. Die Geschlossenformlösung für das Problem einer uniformen normalen oder tangentialen Verschiebungsdiskontinuität längs eines endlichen Grenzflächenelementes werden verwendet, um Einflußkoeffizienten zu erhalten. Zum Beispiel erhält man für zweidimensionale Probleme, bei denen die Grenzbedingungen durch Beanspruchungen angegeben sind, das folgende lineare System:

$$\sum_{j=1}^{M} A_{NN}(i,j)\,DN_j + \sum_{j=1}^{M} A_{NT}(i,j)\,DT_j = TN_i$$

$$\sum_{j=1}^{M} A_{TN}(i,j)\,DN_j + \sum_{j=1}^{M} A_{TT}(i,j)\,DT_j = TT_i \qquad (48)$$

wobei $A_{NN}(i,j)$ die Normalkomponente der Beanspruchung ist, erzeugt in der Mitte des j. Segmentes der Begrenzung durch eine Einheitsnormalverschiebungsdiskontinuität bei dem i. Segment, während sich A_{NT} auf die Tangentialkomponente derselben Beanspruchung bezieht. Die Koeffizienten A_{TN} und A_{TT} beziehen sich auf Normal- und Tangentialkomponenten von Beanspruchungen, durch eine Einheitstangentialverschiebungsdiskontinuität erzeugt. DN_i und DN_j sind die unbekannten Normal- und Tangentialverschiebungsdiskontinuitäten beim j. Segment, während TN_i und TT_i die Komponenten der unbekannten Grenzbeanspruchung in der Mitte des i. Segmentes sind. Das System (48) von 2N Lineargleichungen kann invertiert werden, so daß, wenn DN_j und DT_j bekannt sind, diese dazu verwendet werden können, um das Spannungsfeld, und damit das Verschiebungsfeld, durch den gesamten Bezirk hindurch durch direkte Lösung eines Systems linearer Gleichungen, die ähnlich (48) sind, zu ermitteln.

Man erhält ähnliche Gleichungen, wenn die Grenzbedingungen als Verschiebungen ausgedrückt werden, so daß gemischte Grenzwertprobleme gelöst werden können.

Diese Methode wurde durch Sinha (1979) auf dreidimensionale Fälle angewandt und auf Probleme, die zwei unterschiedliche elastische Stoffe beinhalten, durch Crouch (1982) und Bouchez (1981). Crawford und Curran (1982) haben Variationsverschiebungsdiskontinuitätselemente (Funktion höherer Ordnung) in Betracht gezogen, so daß Beanspruchungen an zwei oder drei Punkten pro Segment angegeben werden können. Diese Elemente höherer Ordnung liefern bessere Genauigkeit für die gleiche Anzahl von Elementen, verglichen mit der Methode der konstanten Verschiebungsdiskontinuität.

Die Verschiebungsdiskontinuitätstechnik ist besonders gut geeignet für die Aufstellung von Abbauschemata bei planaren oder nichtplanaren Flözen (Crouch 1978), oder für die Untersuchung quasistatischer Trennflächenfortpflanzungsprobleme (Cornet 1979). Bei diesen Problemen kann die Reibung entlang den Trennflächen berücksichtigt werden, wenn Gleitung auftritt; eine Anpassung an dilatierende Klüfte ist auch ziemlich unkompliziert, vorausgesetzt, das Dilatationsgesetz ist bekannt.

Ein unterschiedliches Herangehen wird durch Brady und Bray (1978 a und b) und Brady (1979) befolgt. Diese Autoren beobachten, daß häufig die Hauptachsrichtung unterirdischer Hohlräume nicht mit der Hauptspannungsrichtung übereinstimmt, und sie schlagen eine numerische Lösung vor, im Hinblick auf das, was sie vollständige, ebene Verformungsbedingungen nennen, d. h. Situationen, bei welchen die Verschiebungen nur von Koordinaten in der Ebene lotrecht zu dieser Hauptachsenrichtung abhängen. Falls S* die Oberfläche der Öffnungen ist, können diese Probleme als die Überlagerung von zwei separaten Subproblemen angesehen werden:

i) Ein unendliches Medium, einem homogenen Spannungsfeld unterworfen entsprechend demjenigen, das vor der Aushöhlung bestand, (d. h., das ungestörte, vorher existierende Spannungsfeld);

ii) Dasselbe unendliche Medium mit einer gegebenen Verteilung von Lasten $\underline{t}$ und Verschiebung $\underline{u}$, entlang einer Fläche S, identisch mit S*, angewendet.
$\underline{t}$ und $\underline{u}$ stellen Beanspruchungen und Verschiebungen dar, welche an den Oberflächen in dem sonst unbelasteten Kontinuum induziert werden müssen, um die Aushöhlung des Hohlraums zu simulieren.

Die Lösung des zweiten Problems beinhaltet die Ermittlung von

1) Unbekannte Grenzwerte, d. h., aushöhlungsinduzierte Verschiebung, bei der die endgültigen Grenzbeanspruchungen auferlegt sind;

2) Gesamtspannungen um die Aushöhlungsgrenzen herum;

3) Gesamtspannungen und Verschiebungen an ausgewählten internen Punkten des Kontinuums.

Es wird angenommen, daß Y die Hauptthohlraumachse und X und Z die zwei anderen Bezugsrahmenachsen sind. Man erhält die Gesamtspannungen an internen Punkten und um die Aushöhlungsgrenzen herum dadurch, daß man die ungestörten Feldspannungen denjenigen, die aus der Belastung der Fläche S entstehen, superponiert. Wie durch Brady und Bray (1978) aufgezeigt, besteht das Problem ii) aus zwei ungekoppelten Subsidiärproblemen:

a) Ein Ebenenproblem, beinhaltend t_x, t_z, u_x, u_z;

b) ein Anti-Ebenenproblem, beinhaltend t_y, u_y.

Lösungen dieser Probleme werden durch eine Diskretisierung der Oberfläche S gesucht. Sie können auf zwei verschiedene Arten formuliert werden. Bei der ersten Art, als die indirekte Formulierung bezeichnet, wird die Verteilung von Belastungsbesonderheiten über die Oberflächen angewendet, und die Intensitäten dieser Verteilungen werden aufgestellt, um die Grenzbedingungen an allen Elementen der Oberfläche gleichzeitig zu erfüllen. Die zweckmäßigste Form von Singularität wird durch regula falsi ausgewählt (Brady und Bray 1978 a, b). Bei der direkten Formulierung erhält man Lösungen aus Gleichungen, die Beanspruchungen und Verschiebungen an den Elementen der Oberfläche S in Beziehung setzen (Brady 1979), wie nachstehend noch erläutert wird.

Das Ziel besteht darin, Grenzbeschränkungsgleichungen für die Subsidiärprobleme a) und b) aufzustellen und zu lösen. Wir wollen zuerst das Problem a) innerhalb des Bezugsrahmens X, Z in Betracht ziehen; die Beanspruchungen an allen Punkten der Abgrenzung besitzen Komponenten t_x und t_z und lassen Verschiebungen, an allen Punkten, mit den Komponenten u_x, u_z entstehen. Wir wollen nun die lokalen Achsen X', Z' am Punkt i der Grenze S betrachten (X' ist die Tangente zu S, während Z' die Senkrechte dazu ist); Beanspruchungen und Verschiebungen an allen Punkten j von S werden zu t'_{xj}, t'_{zj}, u'_{xj}, u_{zj}, was als Belastungsfall 1 bezeichnet wird. Als nächstes berücksichtigen wir eine Linienbelastung von Einheitsintensität pro Einheitslänge in der Y-Richtung, bei i in der X'-Richtung ausgeübt. Diese Belastung induziert bei j, auf S, Beanspruchungen und Verschiebungen

$T'^{x_i}_{xj}$, $T'^{x_i}_{zj}$, $U'^{x_i}_{xj}$, $U'^{x_i}_{zj}$ (ausgedrückt im Bezugsrahmen X', Z'). Die Oberfläche kann angesehen werden als belastet durch die

Beanspruchungen $T'^{x_i}_{xj}$, $T'^{x_i}_{zj}$, welche Verschiebungen $U'^{x_i}_{xj}$, $U'^{x_i}_{zj}$ entstehen lassen, (dies wird als Fall 2 bezeichnet). Durch Anwendung des reziproken Arbeitstheorems kann man nun schreiben:

$$\int (T'^{x_i}_{xj} u'_{xj} + T'^{x_i}_{zj} u'_{zj}) \, ds = \int (t'_{xj} U'^{x_i}_{xj} + t'_{zj} U'^{x_i}_{zj}) \, ds \qquad (49)$$

oder wenn die Beanspruchung und Verschiebungen von Belastungsfall 1 als konstant angesehen werden bei den kleinen Grenzelementen

$$\sum_{j=1}^{n} \int_{S_j} (T'^{x_i}_{xj} u'_{xj} + T'^{x_i}_{zj} u_{zj}) \, ds_j = \sum_{j=1}^{n} \int_{S_j} (t'_{xj} U'^{x_i}_{xj} + t'_{zj} U'^{x_i}_{zj}) \, ds_j \qquad (50)$$

wobei S_j die Länge des j. Grenzelementes ist.

Wenn wir nun schreiben:

$$F'^{x_i}_{xj} = \int_{S_j} T'^{x_i}_{xj} \, ds_j \, ; \, V'^{x_i}_{xj} = \int_{S_j} U'^{x_i}_{xj} \, ds_j \, ; \, etc. \ldots$$

dann wird die Gleichung (50) zu:

$$\sum_{j=1}^{n} \begin{bmatrix} F'^{x_i}_{xj} & F'^{x_i}_{zj} \end{bmatrix} \begin{bmatrix} u'_{xj} \\ u'_{zj} \end{bmatrix} = \sum_{j=1}^{n} \begin{bmatrix} V'^{x_i}_{xj} & V^{x_i}_{zj} \end{bmatrix} \begin{bmatrix} t'_{xj} \\ t'_{zj} \end{bmatrix} \qquad (51)$$

Das gleiche Verfahren kann für eine Linienbelastung, die bei i, aber in der Z'-Richtung ausgeübt wird, wiederholt werden, und wenn man dieses Verfahren für alle Mittelpunkte der n Elemente von S wiederholt und die Ergebnisse auf der ursprünglichen X-Z-Ebene ausdrückt, erhalten wird:

$$[F] \{u\} = [V] \{t\} \qquad (52)$$

Die Gleichung (52) drückt die erforderliche Grenzbeschränkungsgleichung auf der X-Z-Ebene für das Problem a) aus. Ein ähnliches Verfahren kann für das Anti-Ebenenproblem b) befolgt werden:

$$[F_y] \{u_y\} = [V_y] \{t_y\} \qquad (53)$$

Die Lösung für unbekannte Komponenten für $\underline{u}$ oder $\underline{t}$ bei allen Punkten der Begrenzung wird durch das Lösen des linearen Systems von Gleichungen (52) + (53) erreicht.

Die Quantitäten $\int_{S_j} T'^{x_i}_{xj} ds_j$, $\int_{S_j} U'^{x_i}_{xj} ds_j$, etc.. können gefunden werden entweder durch direkte analytische Integration oder numerisch. Brady schlägt eine numerische Lösung vor, unter Verwendung der Gaußschen Quadratur. Sobald alle Beanspruchungen und Verschiebungen an der Oberfläche S bekannt sind, erhält man ziemlich einfach die Komponenten der Grenzspannungen sowie die Verschiebungen und Spannungen an allen internen Punkten.

Situationen, die nichtlineare, konstitutive Gleichungen in einigen Teilen des untersuchten Bereiches in sich schließen, können durch das Koppeln von Grenzelementen und endlichen Elementen gelöst werden: endliche Elemente werden benutzt, um die nichtlinearen, konstitutiven Gleichungen zu lösen, während Grenzelemente eingeführt werden, um die unendlichen Grenzbedingungen zu behandeln (siehe z. B. Brady und Wassyng 1981, Lovan et al. 1983, dieser Kongreß). Lin DezHang und Liu Baoshen (1983, dieser Kongreß) schlagen vor, die Grenzelementformulierung auf lineare, viskoelastische Stoffe dadurch zu erweitern, daß die Laplace-Transformation angewendet wird, um die Zeitabhängigkeit loszuwerden. Pierce und Ryder (1983, dieser Kongreß) erweitern die zweidimensionale Grenzelementformulierung auf abschnittsweise, lineare, konstitutive Gleichungen durch Einführung von "Initialverformungs"- oder "Initialspannungs"-Konditionen bei einigen Oberflächenelementen. Es wird angenommen, daß diese speziellen Oberflächenelemente Verformungsentfestigungs-Deformationen plastischer Art unterbringen.

Kluftkörpermodelle

Bei Anwendung der Finite- oder Grenzelement-Modelle sind kleine Trennflächen innerhalb des äquivalenten Kontinuums eingeschlossen und nur die großen Trennflächen werden als solche behandelt. Dieses Verfahren ist sehr geeignet für Probleme, bei denen eine Verformung sowohl das Kluftsystem als auch die Gesteinsmatrix beeinflußt, das heißt, wenn das Spannungsfeld groß genug ist. Für geringe Spannungsfelder treten gewöhnlich zwei Phänomene gleichzeitig auf:

- Erstens: die Dichte von Klüften, in verschiedenen Größenordnungen, nimmt drastisch zu, so daß das Gebirge ein Aggregat von Blöcken mit verschiedenen Größen und Formen wird;

- Zweitens: der Deformationsprozeß beeinflußt meistens die Klüfte, so daß die elastische Deformation der Blöcke, verglichen mit derjenigen der Trennflächen, vernachlässigbar wird. Es ist klar, daß unter solchen Bedingungen ein Modell für das Gebirge durch ein äquivalentes Kontinuum nicht zweckmäßig ist; aus diesem Grund sind Kluftkörpermodelle entwickelt worden.

Diese Modelle, ursprünglich entwickelt durch Cundall (1971, 1974) gehen davon aus, daß alle Verschiebungen entlang Klüften oder

Trennflächen auftreten, während die Blöcke massiv bzw. fest bleiben, (obzwar eine elastische Deformation bei den neuesten, durch Cundall, 1980, vorgeschlagenen Modellen eingeführt worden ist). Sie basieren auf zwei Gesetzen:

- dem Kraftverschiebungsgesetz;
- dem Bewegungsgesetz.

Das Kraftverschiebungsgesetz bringt die Kräfte zwischen den Kontaktflächen in Beziehung zu den inkrementellen Verschiebungen quer über die Kontaktflächen. Diese Verschiebungen werden von den inkrementellen Verschiebungen der Blöcke abgeleitet. Das Bewegungsgesetz ist einfach das zweite Newtonsche Gesetz: Kräfte und Momente, die auf jeden Block wirken, rufen Beschleunigungen hervor, die zweimal über eine Zeitstufe t integriert werden, um neue inkrementelle Verschiebungen zu ergeben. Die äquivalenten Kräfte und Momente, die auf jeden Block wirken, sind die Summierung der Kontaktkräfte, die um die Peripherie der Blöcke herum wirksam sind. Das System der Gleichungen wird durch endliche Differenzen in Bezug auf die Zeit gelöst. Diese Modelle handhaben unbegrenzte Rotationen und Translationen für willkürlich geformte zweidimensionale Blöcke. Es wurde darüber hinaus angewendet auf Böschungsstabilitätsprobleme, auf Fluidströmung in einem stark zerklüfteten Medium (Cundall 1982). Warburton (1983, dieser Kongreß) schlägt ein Programm für die Erstellung dreidimensionaler Blockgeometrien und für die Untersuchung der Stabilitäten der einzelnen polyedrischen Blöcke vor.

Die Zeit mag nicht fern sein, wenn, durch Kombination der statistischen Charakterisierung von Kluftgeometrie, entlang Linien ähnlich denjenigen, die in Kapitel 3 dieses Berichtes aufgezeigt sind, und dreidimensionalen Blockmodellen das mechanische Verhalten von stark zerklüfteten Gesteinsmassen auf geeignete Weise untersucht werden kann.

5. SCHLUSSFOLGERUNG

Das natürliche Spannungsfeld, das bei ungestörten Gebirgen existiert, ist ein wichtiger Eingangswert für technische Probleme sowie für ein besseres Verständnis der derzeitigen Tektonik. Es kann durch direkte in-situ-Messungen eingeschätzt werden oder aus seismotektonischen Beobachtungen abgeleitet werden. Die weitgehend angewandten in-situ-Ermittlungsmethoden sind Overcoring- und hydraulische Bruchbildungsmethoden; die erstgenannten Methoden werden im wesentlichen von unterirdischen Hohlräumen her angewendet und sind auf ziemlich flache Bohrlöcher (weniger als 100 m in den meisten Fällen) beschränkt, während die zweite Methode bei beliebiger Tiefe durchgeführt werden kann. Wegen ihrer Kostengünstigkeit und ihrer Tiefenunabhängigkeit kann die hydraulische Bruchbildung in den meisten Fällen die attraktivste Technik sein. Ihre Genauigkeit liegt in der Größenordnung von 15 %, aber eine gewisse Verbesserung kann noch erwartet werden, insbesondere in Bezug auf die Genauigkeit der Einschluß-Druckermittlung.

Alle bisher erzielten Ergebnisse lassen eine Zunahme der deviatorischen Spannungskomponente mit der Tiefe, zumindest bis 5 km hinunter, der Tiefenbegrenzung derzeitiger direkter Messungen, erkennen. Ein bedeutendes Problem muß jedoch noch gelöst werden, nämlich die Schätzung von Spannungsänderungen mit der Zeit, insbesondere in erdbebenanfälligen Gebieten.

Eine zweite wichtige Angabe für das Verständnis des mechanischen Verhaltens von Gebirgen ist die geometrische Charakterisierung von natürlichen Trennflächen. Obwohl einige Fortschritte erzielt worden sind in Bezug auf die örtliche Bestimmung dieser Geometrie, (d. h., die lokale, räumliche Extrapolation von Aufschlußbeobachtungen), dank entsprechender statistischer Untersuchungen, muß eine genaue Methode für die Taxierung der Geometrie von großen Klüften noch entwickelt werden. Die dritte wichtige Frage, die durch Felsmechanikprobleme aufgeworfen wird, ist die Definition geeigneter konstitutiver Gleichungen für die äquivalenten Kontinua sowie für die Trennflächen großen Ausmaßes.

Das am meisten favorisierte Modell ist das der linearen Elastizität, obwohl einige Modelle nun entwickelt werden für Elastoviskoplastizität mit Dilatanz. Die Darstellung von Klüften nimmt Dilatanz mit Scherdeformation an. Eine geeignete Darstellung der Zeitwirkungen für die Gesteinsmatrix und das Verhalten der Klüfte, wenn Mikrorißbildung in Frage kommt, bleibt eine Schwierigkeit, insbesondere, wenn Spannungskorrosion vorhanden ist. Einige Daten hat man im Laboratorium erhalten, aber ihre Anwendung auf in-situ-Bedingungen bleibt problematisch, und einige in-situ-Versuche sind erwünscht. In dieser Hinsicht kann eine Anwendung der neuesten Entwicklung in der Seismologie, soweit es Modelle über den Herdmechanismus betrifft, sehr fruchtbar sein für das Interpretieren der seismoakustischen Aktivität, die während einer Gebirgsdeformation festgestellt wird. Dies könnte zumindest ein besseres Verständnis quasistabiler Bruchbildungsprozesse, die Instabilitäten großen Ausmaßes vorausgehen, mit sich bringen.

Diese Beobachtungen können vorteilhaft gekoppelt werden mit in-situ-Deformationsmessungen und eventuell temporären Spannungsänderungen.

Sobald diese drei Fragen beantwortet worden sind, (d. h., Ermittlung des lokalen Spannungsfeldes, geometrische Charakterisierung von Trennflächen, Definition geeigneter konstitutiver Gleichungen), werden die Felsmechanikprobleme üblicherweise mit numerischen Methoden gelöst, obwohl einige physikalische Modelle manchmal effiziente Problemlöser darstellen, insbesondere, wenn einwandfreie Einstufungskoeffizienten eingeführt worden sind. Wenn lineare Elastizität angenommen wird, hat sich eine Grenzelementformulierung als sehr durchschlagend erwiesen, insbesondere, wenn man sich mit unendlichen oder semiunendlichen Gesteinsmassen befaßt, oder für sich progressiv entwickelnde Grenzgeometrien. Wenn nichtlineare, konstitutive Gleichungen in Betracht gezogen werden müssen, umfaßt die am meisten favorisierte Methode Modelle mit finiten Elementen. Jedoch wird nun eine neue Annäherung für stark zerklüftete Massen, die geringen Spannungen unterworfen sind, entwickelt, nämlich das Kluftkörpermodell, bei welchem eine Deformation nur entlang der Trennflächen auftritt. Obwohl diese Modelle derzeitig zweidimensional bleiben, kann das Erscheinen von dreidimensionalen geometrischen Modellen für Gebirge, mit einer statistischen Definition von Klüftigkeit gekoppelt, bald eine wirksame Handhabe liefern, um das mechanische Verhalten solcher Gebirge zu untersuchen.

Die Rolle von Fluiddrücken wurde in diesem Bericht nicht erörtert, jedoch legen die Ergebnisse, die in Bezug auf Druckabhängigkeit der Durchlässigkeit entlang Trennflächen erreicht wurden, nahe, daß - für dieses Problem - effektive Spannungen nicht der von Terzaghi vorgeschlagenen klassischen Formulierung gehorchen (Terzaghi 1945). Die Konsequenzen dieser Resultate auf Stabilitätsuntersuchungen und - allgemeiner - Deformationsprobleme müssen noch untersucht werden.

DISCUSSIONS / DISCUSSIONS / DISKUSSIONEN

FRACTURE AND FLOW OF THE EARTH'S CRUST, INCLUDING TECTONIC STRESSES

Fracture et écoulement de l'écorce terrestre, y compris les contraintes techniques

Bruch und Kriechen der Erdkruste, einschliesslich tektonischer Spannungen

P. Habib
Chairman

GENERAL REPORT: Presented by F. Cornet, France.

DISCUSSION

Question: Prof. B. Amadei

I would like to make a remark about the influence of rock anisotropy on in-situ stress measurements. The following question often arises when measuring in-situ stresses: how large an error is involved by neglecting rock anisotropy? I was able to answer this question for overcoring techniques that involve instrumented devices such as the US Bureau of Mines gauge, hollow inclusions gauges such as the CSIR or CSIRO gauges or solid inclusion gauges (Amadei and Goodman, 1982).

Consider a hollow inclusion perfectly bonded to the walls of a circular hole drilled in an anisotropic medium. A 3D stress field is applied at infinity. No restrictions are made on the type or orientation of the anisotropy or on the orientation of the principal stress field with respect to the hole. General closed-form solutions were obtained between the six components of strain and the three components of displacement at any point within the inclusion and the stress held components. These closed form solutions can be used to solve the inverse problem, i.e., to calculate in-situ stresses from measured strains and displacements during overcoring and to account for rock anisotropy. It was found that neglecting rock anisotropy by assuming that the rock is isotropic could create large errors when calculating the magnitude and the orientation of the in-situ stress field. Numerical examples have shown errors as large as 100 percent for the magnitude. The orientation could be up to 100 degrees off from the isotropic solution.

Amadei, B and Goodman R.E. (1982). The influence of rock anisotropy on stress measurements by overcoring techniques. Rock Mechanics 15, pp. 167-180 (1983)

Answer: Dr. Cornet

I would like to thank Dr. B. Amadei for his very interesting comment on the role of anisotropy for the interpretation of stress measurements by over coring techniques. In my report I outlined the fact that, most of the time, the hypothesis of isotropy is made and that in fact it is necessary to take into account the rock anisotropy for a more accurate interpretation.

However I am afraid that the main drawback of stress measurements by overcoring methods comes from the fact that stress determinations which are made in volumes of the order of tens of cubic centimeters are extrapolated to volumes of the order of a few cubic meters so that the main source of error comes from the heterogeneity of the rock rather than from its anisotropy. Another question which ought to be raised concerns the accuracy with which the anisotropic behaviour of the rock mass can be appraised.

Detailed results from field measurements are necessary to evaluate the real influence of anisotropy on the accuracy of in-situ stress determination by overcoring techniques.

Question: Prof. R. Goodman, to Prof. Cornet, General Report.

The source mechanism analysis has a problem of uniqueness as the fault plane has one of two possible orientations - it therefore cannot yield a clear indication of $\sigma 1$ and $\sigma 3$ direction (sigma 2 is indicated uniquely, I believe). How do you resolve this problem in determining stresses from such data? Also how would you account for uncertainty in the mode of rock bursting - tensile versus shear) in analyzing data from quarries or mines?

Answer: Dr. Cornet

Thank you Professor Goodman for your very interesting questions. Yes the source mechanism analysis has a uniqueness problem as I pointed out in my report and this is its main drawback. In order to solve this non uniqueness, additional information is required. This additional information may be obtained from synthetic seismograms which give an indication on the directivity of the slip motion along the actual fault plane. By comparison between observed and computed motion it is sometimes possible to identify the fault plane from the nodal planes. Another way to solve for this uniqueness is to compare the direction of the nodal planes with that of the natural fractures wich affect the rock mass. One may assume that the nodal plane which is the closest to a main natural fracture direction is likely to be the actual fault which gave rise to the seismic event.

The P, T and neutral axis which are defined with a fault plane solution must not be confused with the local principal directions of stress because fracture does not occur in the direction where shear is maximum. Very often fracture occurs along preexisting weakness planes when the resolved shear stress reaches a critical value. Accordingly the principal directions of stress can be inclined at 10 to 45^{0} with respect to the P and T axis. A more precise determinaton of these directions can be obtained by assuming that the slip vector which is defined from a focal plane solution is colinear with the resolved shear stress in the fracture plane. When a sufficient number of such slip vectors has been defined, an inversion procedure may be used to identify the local

principal stress direction. However to my knowledge, this has not yet been done on a regular basis and more work is necessary to validate this new stress determination method.
Obviously slip vectors can be defined only when the source mechanism of the seismic event is an unstable shear motion. This is not always the case for events observed in mines and quarries. We conducted a seismic survey for a coal mine in southern France and the analysis of focal mechanisms revealed that most of these events were not shear motion but looked more like "implosion". These cannot help determine the local stress field; however they provide useful informations on the fracture pattern observed with the mining process and this, in itself, is a useful information.

Question: J. C. Roegiers, to Prof. Cornet, General Report

Could the increase in shut-in pressure not also be explained by a change in the pore pressure as the pressurization time increases?

Answer: Dr. Cornet

I presume that Dr. Roegiers' question refers to stress determination by hydraulic fracturing and, more precisely, to my comment that in some instances, we have observed an increase in shut-in pressure as the hydraulic fracture grew.

Dr. Roegiers proposes to explain this increase of shut-in pressure by the change of pore pressure in the rock mass as the pressurization time increases whilst I have proposed to consider that this change of shut-in pressure might be associated with a change of normal stress supported by the fracture plane as it extends away from the well.

If long term shut-in pressure are considered I may agree with Dr. Roegiers' proposition and this is one of the reasons why we do not use long term shut-in pressures. However I am not sure I understand how this mechanism could affect instantaneous shut-in pressure since this value does not depend on the local pore pressure.

PAPER: The determination of the complete state of stress in rock with the flat jack method

AUTHORS: K. Balthasar, E. Wenz

PRESENTED BY: Prof. Natau

DISCUSSION

Question: Prof. W. Wittke, Germany

In your presentation you have shown an example of a successful determination of in situ stresses by means of flat jacks in a rock, which seemed to me to be only slightly jointed. Also the tunnel was excavated by a tunnel boring machine and thus the walls were rather even and smooth. Further no support seemed to be required in the case you referred to. Could you comment on the reliability of the applied method in tunnels excavated by the drilling and blasting technique. Further I would like to hear your opinion on eventual limitations of the applied method, caused by non-homogeneous stress distribution as a consequence of jointing. Could you think of cases in which the jack might be too small in size to reveal reliable results? Finally I would be pleased if you could comment on the interpretation of "flat jack measurements" performed in anisotropic rock as e.g. transversely anisotropic.

Answer: Prof. Natau

The first question: this method I have shown you here is only suitable and practicable in drilled drifts. We can say there is no method in the world which is suitable for all kinds of rock and this system is very suitable for drilled drifts, drilled galleries and shafts with no plastification, so only the surface of the gallery and when you have such gallery it is very simple to make stress measurements. It is not necessary to drill a hole. You can put your equipment in the back box of a BMW for example. It is very important, you don't need over-coring technique, when you have a tunnel with full face machine it is very easy to estimate the state of the stress.

When dealing with jointed rocks, joint spacings, for example of one metre or less this method must be handled very carefully, you have to repeat in this case in short distances your measurements, always seven components, and then you have to compare the results.

And the 3rd questions: in anisotropic rock material this theory is not suitable. Its quite right, therefore you have to make investigations for the anisotropic or isotropic behaviour of rock masses, but in the crystalline material we have in the Alps and sometimes in the German mountains this system will run, in other material not. But in other material doorstopper and triaxial cell is running still much better in our experience. We are working with all kinds of systems.

Question: Mr. Heslop, Australia to Prof. Natau

We have noted that rocks shrink with loss of pore water. One would expect that the rock surrounding the tunnel has been subject to some drying out and therefore would have shrunk and become de-stressed - have you considered how to cater for this?

Answer: Not supplied to Organising Committee.

PAPER: Differential strain curve analysis - a new method for determining the pre-existing in-situ stress state from rock core measurements

AUTHORS: N.K. Ren and J.C. Roegiers

PRESENTED BY: J. Roegiers

DISCUSSION

Question: Dr M. Dunbavan, Australia

I believe this technique may hold great prospect for the future but I was looking at the sedimentary rocks that often associate oil and coal deposits and I was wondering is the assumption of a nonaligned system of microcracks realistic for such a heavily structured fabric, particularly when you consider that swelling clay minerals are often associated with these sediments as well? Would the authors comment on the application of their analysis to sedimentary materials, especially where materials are soft rocks prone to swelling. These materials also may have a relatively high water content which may affect test results

Answer: Dr. Roegiers

We have done so far about 2 dozen cases successfully, with only 2 formations so far where we have real problems, We haven't seen too many problems with swelling clays or even with a shale which is very similar. As long as you try to preserve the rock under good condition, (don't let it dry) - things are okay, if you can just encapsulate it as soon as possible and do your test. This raises an amazing question: it seems like pore pressure is not a very important parameter, in other words we have done some in sandstones for instance where we try to do it under almost the same saturation condition as down hole and then let it dry and do it again, and the same thing results. Now you might ask me, how do you know it is a success, may I just take one minute here to show you how I judge a success. It was a case recently for a very major oil company where we did some work and we predicted a

certain orientation. I know not to divulge any big secret, lets assume that I told them that the fracture was going to go east-west. In that particular case borehole ellipticity (by the way is a very strange technique) predicted north-south. We were literally thrown out of the office of the customer because he felt that it was pore pressure. So he went ahead, - this is 2-3 years ago - , and started his drilling and now as I said he drills closer in that direction and then drills further in this direction. After 2 years of production the holes unfortunately in this direction have declined their pressure while those have not changed.
The customer now is convinced that our technique was okay.

Question: Dr Maury, France

L'hypothese fondamentale de la methode proposee est que le carottage relacae les contraintes et provoque une micro fissuration dont la densite est proportionnelle au niveau initial.

1. Question:

Comment etre sur qu'il n'y avait pas une microfissuration pre-existante, due par exemple a une migration primaire de l'huile par micro oleofracturation? ou autre?

2. Question:

En prelevant la carotte , on relache les constraintes effectives D'abord, puis les pression interstitielies en remontant les carottes. En Industrie petroliere, les niveaux de relachement de constraintes sont tres importants (300-500 bars), et provoquent couramment discage et meme pulverisation des carottes. Comment faire la part de ce qui a ete pris par la fracturation, de ce qui reste pour la microfissuration, de la part prise par le relachement de constrainte effectives, et pression interstitielle?

Que penser de L'application dun modele viscoelastique a ces phenomen?

3. Question:

Il est fait allusion a des methodes de "Surface Mapping" et "Vertical Fracture" sur carottes - pour moi, elles sont a relier a un etat de paleo constraintes qui peut navoir rien avoir avec l'etat actuel - etes-vous d'accord?

Answer: Dr. Roegiers

To the first question which is about pre-existing microcracking. It is certain that if you have a microcrack which is larger than a cracking which is due to the expansion of stresses, well then it won't work. We've only had one case of this at the time being. It was the case of a granite in Canada, which had been subjected in geological times to enormous stresses during the glacial period and we had water tests and had found different orientations than the present ones. Almost certainly this was microcracking problem and they may have an oil-induced (?) microcracking which maybe more important than the expansion microcracking and that won't work. But we've only had one case of this so far.

Your second question, which is about a rock specimen which would be pulverised as a result of effective stresses. It is of course understood that if I don't have an intact material or core, I will not be able to take my measurements.

The third question: I am in complete agreement with you that it is a problem. You saw that there was a problem, which was a time-dependent relation right at the end. We did have a test which was of a visco-elastic model also subjected to plastic, in which case we did not obtain very good measurements. The figures went awry, in all directions. So from this question, of viscoelasticity we will have to try to do better to get a better solution.

Lastly, so far as surface mapping is concerned. Here again I am in agreement that it is very rare when examining cracking or surface discontinuities, we can conclude both the arrangement and direction of discontinuities underground. There are very few cases where there is agreement that can be obtained on this. In the case which I have just mentioned and in my written paper as well, it was particularly the case of course, but otherwise I am in complete agreement with you. It is not a question of the actual stresses today.

NUMERICAL MODELLING OF ROCK BEHAVIOUR

Modelage numérique du comportement des roches

Erstellung numerischer Modelle des Felsverhaltens

H. Einstein
Chairman

PAPER: Extended Boundary Element Methods In The Modelling Of Brittle Rock Behaviour

AUTHORS: A.P. Peirce, J.A. Ryder

PRESENTED BY: J. A. Ryder

DISCUSSION

Question: Dr. F. Cornet, France

In your presentation you mentioned that the post peak behaviour of rock can be modelled by an equivalent continuum the constitutive equation of which is represented by a negative slope straight line.

My first question is whether you considered the possible influence of deformation rate on these constant coefficients. My second question relates to the very nature of your modelling; you assume that the material can be represented by an equivalent continuum whilst in reality the post peak behaviour is characterized by the formation of shear surfaces. Do you consider that an equivalent continuum is still adequate for representing these discontinuities?

Answer: Dr. Ryder

I think I feel confident to answer the first question. Yes, we think that pre-failure, our particular rocks are not time dependent, a lot of laboratory work has confirmed that, but post-failure I agree with you that it seems likely that time dependence is an important fact. That is, the fracturing mechanism within rock doesn't take place immediately, it does take time. The time constants seem to be in the order of a few days or possibly hours. So that if you mine some kind of excavation and wait a couple of days you are going to see an increase in the fracturing that takes place. There is no question about that as well. What we really assumed here is that our constitutive law that we plugged into our model has been established at a suitable loading rate of the order of a day, let us say. If in fact we have that kind of law built in we think we will be probably modelling something reasonably realistic. We don't think the time dependence is particularly strong either, it may be of the order of 50%, lets say: its not 500%, so that if we use appropriate loading rates in our model we are not too far off the mark.

The other question is, does laboratory testing indicate shear type fracturing? That is I think a very contentious and tricky point and I haven't time really to go into my own opinions on this matter, but all I can say is that in the field, if you actually look at the fracturing in the side walls of tunnels and around stopes the bulk of the fracturing is extension type; it's not in fact shear type. There are shear type fractures but it doesn't seem that they are as important in our particular context as the straight extension type fractures. That is the bulk of the rock does not seem to shear.

In other words there is something wrong with our testing; it doesn't really duplicate what we actually see underground, but I did say it's contentious and I don't feel particularly qualified to go into that one any deeper.

Question: Prof. E. T. Brown, London

I would like to comment on Dr. Ryder's plea for those of us who are experimentalists to do some plane strain work. I'd actually attempted to do that many years ago at Townsville. We built a plane strain apparatus which was based on Ian Donald's plane strain apparatus for soils. We found the problem to be fiendishly difficult experimentally and as a result of that didn't ever publish any of the results. However, I suspect that in terms of intact rock and in terms of the brittle quartzites that you are working on the influence of plane strain conditions isn't perhaps quite as important as you suggest. Such evidence as exists on the influence of the intermediate principal stress is that the intermediate principal stress isn't all that important. It has some effect but not a big effect. If you consider a prismatic specimen and you test it in plane strain it will be the intermediate principal stress that you are increasing generally to maintain the plane strain condition. It is $\sigma 3$, the minor principal stress I believe that will be the important one and that will be involved in the confinement that you refer to as a result of the dilation. So therefore I would suggest that it is an interesting problem, its a fascinating problem, but if I were you I wouldn't worry too much about the fact that you don't have plane strain data for your rocks.

Answer: Dr. Ryder

I would agree that the effect on the strength of rock, which is what many investigators have concentrated on in the past, may be relatively small and I mention a figure of about 40%, in terms of Prof. Mogi's work for example, one can infer that the strength will increase, in that sort of order if its a plane strain test. But I'm really talking about the full constitutive behaviour, that is post failure behaviour, and I wouldn't necessarily agree that the effects on the dilation for example are negligible. If you measured dilation in the triaxial tests it is by no means certain that the dilation in a proper plane strain test is going to be within 40%, I guess that it may be more like 100%, but again I would make a plea that someone actually measure this. I agree it isn't necessarily easy to measure, its probably a very difficult technical problem but I think the engineering requirements are so important that someone should make this kind of effort.

PAPER: The Behaviour Of Reinforced Jointed Rock Masses Under Various Simple Loading States

AUTHORS: G.N. Pande and C.M. Gerrard

PRESENTED BY: G.N. Pande

DISCUSSION

Question: Dr. Chappell, Australia

We have experienced some very difficult slopes in various areas and we have used rock bolting anchoring in retaining these slopes. It's been a remedial measure. It hasn't been a fundamental design process, but in the use of this remedial measure we had to apply a rational design technique in evaluating the reinforcement requirements. We measured the modulus of the rock mass by jacking tests and then we rock bolted an area and then remeasured that deformational response and found that for various rock classes we could get an improvement order higher in stiffness than we could without the rock bolts and this, related to heights and various things, stabilized the slope. But what came out in this is that the parameters in the rock mass were very much affected by the rock bolts. The joints themselves were stiffened. I believe that rock bolts have a very interlocking characteristic. We have a lot of arguments about active and passive and all this type of deal, but the moduli was not constant. I see in your models you have used constant moduli. The rheological models are not constant of course. We used a suitable elastic approach. It was difficult to measure the moduli but we did this by varying our field conditions. The question I feel which could help us is:- Have you considered this type of technique in posing a force, creating an active condition, a passive condition, and finding out whether your deformational response is the effect that you are enhancing or of the strength of the joint system itself? I realize its a long question but it's worth answering.

Answer: Dr. Pande

I think that the elastic moduli you have referred to are pseudoelastic values which implies that in elastoplasticity there is a split between elastic and plastic strains. You have a certain stage, certain plastic strains and then you can work out pseudoelastic modulus. I think an overall measurement approach has to be seen and we have to perhaps make back calculations. About your second point, active reinforcements can be easily considered by taking initial stresses into account.

When you put stress on the bolts there will be rock deformation, a stressed state will be created and that has to be taken into account. Using the finite element techniques there is no problem about that. I may also like to say that the stress path approach which I have suggested as an alternative is a well known technique in soil mechanics and perhaps we should do some experiements with that as well.

Question: Dr. Chappell, Australia

In assessing the material properties to use in the analysis, many mechanisms are initiated. Moduli are very much stress dependent and change as the load interaction of the bolt develops, i.e. the moduli change. This I feel sure can be handled by your model but the measurement required to obtain this is difficult. Load paths are in effect brought, these, however, are dependent on mechanisms of slip and geometrical change giving load redistributions. Do you think your model could be expanded to handle this? Active and passive conditions are again mechanistically dependant in that moduli and strengths are changing, field work shows this, here again; do you think your model can handle this?

Answer: Dr. Pande

The theory of plasticity or visco-plasticity used in the model takes into account the stress path dependence. The stress path would dictate the mechanism of slip and the magnitude of slip. The input parameters required for the model are elastic stiffnesses and strength of joint sets. These need to be determined from carefully controlled laboratory experiments.

Question: Mr. M. B. Wold, CSIRO, Australia

There is a contrast between the two papers in the way in which they treat the rock bolt reinforcement. In the paper of Pande and Gerrard, the bolts are "smeared" across the volume of rock considered in a statistical fashion and presumably this is applicable to large volumes of rock. On the other hand, the model of Warburton considers the interaction of individual bolts and blocks, which may act in the fashion of a "keystone", which is discreet within the model and will be of prime importance in maintaining the stability of the blocky system. Would Dr. Pande and Dr. Warburton, please comment on these differences in approach and their relevance to the application of their models?

Answer: Dr. Pande

This is a very relevant question, depending on the type of rock mass you are dealing with. If the rock mass has a blocky nature, has random joints then the approach of kinematic equilibrium is more relevant; but if you have a multi laminate or layered medium of the type we saw in some coal mines (slates, shale), then this type of approach is more relevant. Have I answered your question?

Answer: Dr. Warburton

That seems to be an answer to the question. I think it depends on the type of rock you have as all your models have to be tailored to the types of rock and I think thats very satisfactory.

Question: Mr. Salembier, France

I didn't understand very well your assumptions about the behaviour of the bolt. Did you take into account only the tensile behaviour of the bolt or also the shear behaviour of the bolt?, because laboratory experiments show that especially when the bolt is perpendicular to the fracture the shear behaviour of the bolt is very important!

Answer: Dr. Pande

No, at this stage we have taken the tensile behaviour of rock bolts only into account. However, it is relatively simple to also include shear behaviour and this will be done in the near future.

PAPER: Non-linear Analysis Of Rock Foundations With Soft Interfaces

AUTHORS: J.S. Zhuo and Y.T. Wang

PRESENTED BY: J.S. Zhuo

DISCUSSION

Question: Dr. K. Saari, Finland

You have used an associated plastic model with a Drucker-Prager function, both as yield criterion and plastic potential. That model has been found in soil mechanics to produce excessive dilation. A non-associated model with a hyperbolic plastic potential function of plastic strain, could give more reliable results. Have you looked into this?

Answer: Dr. Zhuo

It is true that an associated plastic model of rock or soil material has been found to produce excessive dilatancy. But it is easy to construct the governing equation of FEM. I have not used an unassociated model, I am planning on going to study its next step. Thank you for your question.

PAPER: Applications Of A New Computer Model For Reconstructing Blocky Rock Geometry-Analysing Single Block Stability And Identifying Keystones

AUTHOR: P.M. Warburton

PRESENTED BY: P.M. Warburton

DISCUSSION

Question: Prof. Einstein

Am I correct to assume you use a vector analysis approach? Could you comment upon the relative advantages and disadvantages of yours and Prof. Goodman's approach?

Answer: Dr. Warburton

Firstly, I think I ought to say what I mean by what I call keystones and what Prof. Goodman calls keyblocks. Now when I talk about a keystone, what I mean in effect is that you have a block for instance, you want to find out whether it is a keystone. What you do is, you go up to the rock and you pull out the block and you see what happens to the rest of the rock. If the rest of the rock falls down, then you can say it's a keystone, but if the rest of the rock stays up you can say that it was not. Of course, I don't suggest that you do this in practice; that's why I work on the computer, where you can do this without any risk. The real point here is that the keystone is defined essentially by the interactions with the surroundings, that is, the other blocks and rock bolts. A block that is a keystone in one set of surroundings may not be a keystone if you change the blocks and the rock bolts that are around that block. To go on to the keyblock theory of Prof. Goodman. As I understand this from the reading that I have done of the papers from Berkeley, it seems that keyblock theory essentially excludes the directions in which blocks cannot move. Now, how I interpret this in a physical interpretation is that you fix all the other blocks apart from the one that you think may be a keyblock and you then go up and you tug or pull the candidate keyblock in every possible direction. If it moves at all then it is a removable block. This is a copy of the figure that you saw the other day: Prof. Goodman's figure (1). I think that if it moves when you pull it (up here) it's a removable block. If you add the forces (down here) I must admit that I can't really see how this part down here differs from an ordinary single-block analysis. The important point is that the interactions are all-important for what I call keystones, and it doesn't seem to me that there is anything down here that really brings in the rest of the rock, the other blocks and so on.

So the interactions are not really here, at this stage anyway. I mean, there could be ways that the Berkeley School could bring these in. The good news about this sort of method from Berkeley is that because the removable blocks at this stage are defined by tugging the block in every direction in every possible way, this means that these sorts of blocks probably include genuine keystones as a subset and so this means that if you support all of the removable blocks, which I understand to be at least part of the Berkeley scheme, I think that you will end up holding up the rock but it may be a case of overkill. You may be putting in far more bolts than you need.

Question: Mr B. P. Knoop, Australia

Is the program available Australia-wide on the CSIRONET system?

Could the author give some indication of the cost of running a model analysis, say of the types shown in the paper?

Has the program been used for the back analysis of stabilized rock slopes?

Answer: Dr. Warburton

It has only been used for the sort of analysis that you saw on the screen. The reason is that development has taken up my time fully, but I understand that at least parts of it, that is the analysis of the blocks, I understand that that part which can also be done on a calculator has been used to analyse blocks in hydro projects in various parts of the world. I've had letters saying it's been used.

Secondly, what I intend doing next. I'm a bit slow in writing up the user's guides for these things, but this is underway. As soon as a user's guide is available anyone who wishes to use the program will be able to get it from CSIRO in exactly the same way as we distribute other programs. Usually, it's a cost of handling and tape, it works out at about $1,000.00. I also hope to publish the listing in full so that if you want to simply use that you can.

In future, the program will be run mainly on a minicomputer, where running costs are negligible. But even on a mainframe, analyses such as those in the paper would cost no more than a dollar or two each. The displays would be more expensive and might cost three dollars or more each.

LECTURES / CONFÉRENCES / KURZREFERATE

REFLECTIONS ON FUTURE DEVELOPMENT OF ROCK MECHANICS

Réflections sur le développement de la mécanique des roches

Gedanken zur Entwicklung der Felsmechanik

Leopold Müller-Salzburg
University of Salzburg, Austria and University of Karlsruhe, FR Germany

If we compare the state of knowledge as it existed in 1964, when our Association was founded, with our present state of knowledge, we may well be proud of what has been achieved. Much progress has been made and many new findings were arrived at.

On the other hand, if we compare the grandiose achievements of rock engineering at the time of the major railway constructions in the Alps and the Rocky Mountains with present practices in rock engineering, we have no reason to be satisfied to the same extent. Without soil mechanics or rock mechanics, indeed even without engineering geology, the most difficult situations were mastered in those days with extraordinary skill. And we have to admit that nowadays construction work is not always better and safer in spite of many scientific aids. In fact, some sensational disasters occurred recently which can be explained only to a small extent by the increased size of our structures and the attendant difficulties.

I have been genuinely disappointed by some developments in the science of rock mechanics since they proceeded along lines similar to those of certain developments in soil mechanics which TERZAGHI so bitterly regretted towards the end of his life. The same applies to publications and is the reason why, after 28 years, I retired as editor of the journal Rock Mechanics.

I might almost refer to a mis-development. What would be the reasons for such a mis-development?

I see the main reason for this in the fact that in most countries theory and practice, research and application, proceed quite separately, and that there are only few interconnections. It is absolutely amazing what sort of matters are quite unknown by some practical people at the site, or worse still, what the designers at the drawing board with their calculators do not know, or perhaps do know, but will not take into account. Examples could be cited in which dozens of millions of dollars were wasted as the result of insufficient transfer of research results to practical applications.

Most researchers have never practiced rock mechanics. If they claim to have been engaged in practical work, they are referring to collaboration in making calculations or being involved in consulting work relating to rock structures. But they have never really come face to face with the rock at the site in the capacity of a partner. They look down on the dirty work at the site with the same disdain with which many practical people think they can look down on the theorists.

For that reason the research topics are seldom initiated by practical requirements. Rather, the choice is for things which can be calculated with the greatest accuracy, or what can be formulated with great elegance, and above all, what can be investigated mainly in the laboratory and with a computer.

Research results, as a consequence, often end up by being very theoretical and solely of academic interest. They have very frequently been gained from excessively idealised and extremely formalised mathematical models, quite removed from nature, and they are dazzling with the elegance of their sophisticated equations, when there is hardly a mention of the enormous problems of the parameters and their topicality, and when it is just that topicality which removes much value from these glowing endeavours.

Many researchers seem to forget that rock mechanics is not an absolute science with its inherent value like physics or mathematics or logic. Rather, it is meant to be only a systematisation of all theoretical bases which may be tools or aids to rock engineering, a systematisation which teaches us to treat rock in a way which causes it to react to our actions without incidents, if possible, and teaches us to work in harmony with nature, not against it.

Conversely, practitioners show far too little interest in research and do not even adopt those results which could be of genuine use in their field. One reason for this is that the results of research are not made known to the practitioners, which includes the designers. And in part it is due to the fact that many research results are not developed to the point of practical use.

In my opinion one of the major reasons for the wide-open gap between theory and practice seems to be that the most elemental insights into rock mechanics, those which explain the fundamental behaviour of rock in its various manifestations, are no longer being noticed. But it is just these elementary laws which are the most important feature in both the theory and the practice of rock engineering. This becomes ever so much clearer the longer one is engaged simultaneously in theoretical geomechanics as well as in rock engineering practice, and if one has to deal with designers and site engineers. How a rock reacts to a technical intrusion, and why it behaves in that way and not in a different manner, that is the most important concept which must be fully understood. That, too, is the prerequisite for any computation which is meant to make sense. Just as an architect must understand his building materials, so a rock engineer must understand his own material: rock.

For the first generation of researchers into rock mechanics the foundations for this behaviour, which were discovered step by step, were surprising, incredibly novel and thus fascinating. They were also in the foreground of their consciousness when rock engineering projects were planned and carried out. But for the next two, younger, generations which have taken their place, this knowledge is no longer novel nor fascinating and is not related to any insight based on experience. It is possible to read about it in textbooks, not in all of them, but in the good ones, and to hear about it in a good lecture.

That does not sink in fully and hence it is no longer in the foreground of consciousness during computing, design work and construction. By contrast, what is fascinating and represents the centre of interest is the further development of geomechanics: computational methods, classification systems, devising indices, etc. These new opportunities provide justifiable grounds for enthusiasm insofar as they concern computational processes, because they represent genuine progress. The indices, too, have their own value, but interest in them is caused less by an inherent mathematical-physical or a logically founded utility. Rather, it originates in the reassuring assumption that there is at last a simple system of quantifiable statements which appears to obviate the practitioner's need to tangle with nature, a nature which is so very complex.

Unfortunately most conferences and congresses are not designed to guide the thoughts of engineers back to the foundations of rock engineering which have escaped their memories. Such meetings deal with the latest developments, with matters which can now be calculated, with critical theoretical concepts of simplified interrelaionships, in one word: with the methodological or scientific superstructure of geomechanics. There is scarcely an opportunity of recalling its foundations since on such occasions it is much more satisfying for speakers to report about new research results than to warm up once more old basic knowledge which has been lost sight of. And case histories are avoided at any cost, the excuse being that they belong to the level of professional journals.

To this must be added that it is impossible at congresses to convey the drama of decision-making which is caused by the problems of the starting data and of geological situations which can frequently be translated only with great difficulty into modelling concepts. Congress participants generally hear only how well it was possible to plan everything and how successfully these designs could be carried out, i.e. how well it all worked out. By being manipulated in this way, listeners have no opportunity of participating in the actual learning process which, as experience has shown, is always based on failures. Because it is only failures from which we have learnt, and shall continue to learn. Small wonder then that the attention of the new generation in our profession is not focussed on matters representing the genuinely creative work of the rock engineer and designer. All those matters which cannot be quantified but which are so very important, matters which cannot be determined cognitively with the brain and which yet have to be recognised.

What should be done? Let's make our congresses and conferences more lively, not a collection of reports of successes, not a forum solely for researchers, but rather an exchange of insights and experiences for everyone who has to deal with rock. But in order to do that we would certainly have to impart an entirely new style to our conferences.

I was speaking of the great importance of the elemental foundations. They are not simple, since rock is the most complex of all construction materials. Many computations (e.g. of slopes) prove to be totally inadequate because the drastically simplified computational models do not do justice to this complexity. It is worst of all when computational models, as for example the concept of the slip circle, are borrowed from soil mechanics, a science which deals with materials whose nature is quite different.

One of the unusual features of rock masses is, e.g. their peculiar behaviour in relation to lateral confinement. The fact that the abutments of the Vajont dam withstood an eightfold overload during the catastrophe is due to lateral confinement by means of elastically flexible prestressed anchors. On the other hand, the fact that the initial salvage operations of an erosion-prone component of a major power station proved unsuccessful during a large-scale rock slide was due to the underestimation of this lateral confinement effect of the tendons, as well as the underestimation of the stress states in three planes, stress states which no longer applied on account of erosion by water.

But the transverse expansion behaviour depends to an extraordinary degree on the anisotropy of a rock mass and on the texture of its plane of weakness. But those very things are two elementary facts to which scant consideration is given by most people active in rock mechanics because understanding these facts involves efforts which are both complicated and uncomfortable. But the much too drastic intellectual simplification of our computer models does not render nature any simpler. It just increases the discrepancy between nature and computation.

Another example of how our vision is blurred by unjustifiable simplifications with regard to nature is the neglect of the effect of time on the behaviour of rock masses. In tunnelling, however, we have learnt to take the effect of time into account, not in calculations as yet, but in our methods as the result of detours, in the form of the so-called New Austrian Tunnelling Method. We have been very successful in this, but have until now unfortunately failed to direct theoretical research towards it. (A few exceptions, e.g. VARDAR's research, confirm the rule).

Increased research in the field of the time-dependence of rock behaviour should in my opinion bring about some major progress. This would benefit in particular the most severely neglected problem in rock engineering, i.e. the stability of slopes. Slopes are basically not problems of statics but of dynamics. Only very few natural slopes are in a state of absolute rest. Just as in tunnelling, we are frequently dealing with unstable intermediate stages between a (fairly) stable initial state and (we hope!) an equally stable final state. A slope can, somewhat sophisticatedly, be defined as a slide which has not yet begun. Just as a landslide is a phenomenon of creep, so slopes should be regarded as phenomena of creep, as was done e.g. by HOFMANN, SHARMA and HAEFELI.

In these matters we are actually dealing with rheological problems, e.g. with creep pressures acting on retaining walls. But the time is not yet ripe for the mathematical solution of such problems. Above all we would not know where to obtain the parameters required as bases for such computations. But by model simulations, by computer simulations and by measurements of the living object in nature, we would be able to achieve in the field of slope engineering results similar to those obtained in tunnelling, irrespective of whether such measurements are entered into the calculations in the form of replacements or as complements.

In spite of an immense volume of research on problems relating to materials and statics, research relating to the effects of time has so far been criminally neglected. Much the same applies also to the effects of primary stresses. I am convinced that primary stresses in rock affect the stability of slopes and the transition from creep to slide of a rock slope as much as is the case with the stability and the driving methods in tunnelling.

I should like to just touch on one further neglected chapter of research: the problem of the representativity of collected rock samples, of large-scale experiments in situ and of parameters in general. How is it possible to find criteria in order to assess the representativity of such things in some reasonable manner?

And an additional chapter of research as well as practice, especially of design practice, which has yet to be dealt with is the rock-mechanical interpretation of geological data, but in particular also the interpretation of total geological situations. This is the source of the mistakes which cause most failures and catastrophes. These are almost never the consequence of incorrect calculations but are usually caused by an incorrect interpretation of the geological situation or by faulty reasoning. (This, by the way, is a general phenomenon of our times, not restricted to rock mechanics.)

This leads us to problems which cannot be solved by geomechanics on its own, but only in close collaboration between geomechanicists and engineering geologists, a collaboration which was far more common in the past than it is now. Let us not forget that, what we call geomechanics or rock mechanics nowadays (or better: rock engineering), has developed from geological engineering. Just as it had been TERZAGHI's lifelong wish that all practitioners of soil mechanics should simultaneously be geologists, we, too, must wish that real teamwork between rock mechanicists and engineering geologists, as it occurs in only a few countries (such as e.g. China and France), will become a reality. This kind of collaboration has generally become much rarer in recent decades. In part the cause for this is that geological descriptions are too academic with little relationship to practice, and in part it is due to the fact that the great distinction between applied geology and engineering geology has to a large extent not yet been realised at all. Another reason is that engineers exhibit little interest and understanding in discussions with engineering geologists because in their studies they received too little training in engineering geology. They are not sufficiently aware how much the fate of their structures and the economic results of their efforts depend on the geological conditions of the subsoil.

I regard rock mechanics without engineering geology as a wrong track. An immense number of examples could be cited to confirm the correctness of this assertion. After all, whether we achieve success in our design and our work relating to rock depends almost exclusively on how well we have recognised the geological situation and how well we have adapted our design and our work to that situation. And the geological situation decides which tools we can employ with advantage at the construction site and which mathematical tools ought to be employed during design.

In which direction should rock mechanics research, teaching and publications be steered in future? Should they be steered at all? I think they should. Because research, as it has developed recently, moves away from practical matters. In my opinion it is vital that the gap between theory and practice should be closed both for the sake of practical rock engineering as well as for rock mechanics research. Those countries, those firms and those engineers are most successful where this gap is smallest. And at present that gap is largest amongst the designers.

In the field of theory let us distance ourselves further from abstractions which are far removed from nature, and from indexing schemes which superimpose factors on one another which cannot validly be superimposed, which are not independent of one another and which are not reproducible.

Let us rather concentrate more strongly than hitherto on the interpretation of geological statements and their preparation for use in geomechanical processes.

Let us give the engineering geologist and the rock mechanicist a greater say than before in the difficult task of transferring the results of geological and geomechanical investigations to the design criteria on the one hand and to the criteria for construction on the other.

Let us train engineers better than hitherto for discussions with geotechnicians, with geomechanicists and with engineering geologists. It is too late to arouse interest in such matters when people are engaged in practical work unless such interest had already been stimulated in the universities.

Let us require of every engineer involved in design a few years of practical construction experience. Let us give him an opportunity in situ to get to know the craftsman's angle of rock engineering and thus become acquainted with the peculiarities of the structural material called rock. Let us also give him an opportunity of learning from failures which are nothing but consequences of the incorrect treatment of rock. This will improve the "quality index" of his work which will have more far-reaching effects than the quality index of rock. It is only during close work with rock that one learns to appreciate those matters which cannot, or cannot yet, be quantified: the stand-up time of the rock, the stage to which a rock slope or a creeping slope has developed. Ignorance of such matters converts all calculations into meaningless theory.

So far we have trained theorists. Let us in future train designers and rock engineers.

Failures are the best teachers. They ought to be publicised more frequently. Those few engineers who possess enough strength of character not to be ashamed of their failures, who recognise that work without mistakes is almost impossible, ought to be our models, models which should be emulated.

Let us move away from our haughty attitude towards the publication of case histories. We can learn a lot more from them than from the most learned theories. Let us put an end to the arrogance of those "academics-only" who are too one-sided.

Let us move back a little from the standards, recommendations and norms, of which there are far too many. And let us not forget that all standards of this world can only be guidelines and not prescriptions, and should be nothing more. They do not relieve the engineer of the obligation of thinking. In the field of geology, where every tunnel differs from every other tunnel, where every foundation and every slope is different from every other foundation or slope, where even one limestone or one granite is not necessarily identical with another limestone or granite, attempts at standardisation and unification which go too far are actually harmful. Rock demands individualized treatment. Damage may well be caused by over-eager homogenisation and standardisation.

Our world is not perfect, and ideals are realisable in it only on a modest scale, whilst progress can be achieved only one step at a time. Everything is a learning process. In order to move forward we ought to assess quite soberly our present state and our past mistakes, and we should not, right from the start, limit our programs for the future to those things which appear to be attainable and executable on superficial considerations. Rather, we ought to aim courageously for what is desirable, even if we are aware that we shall never quite attain it.

GEDANKEN ZUR ENTWICKLUNG DER FELSMECHANIK

Reflections on future development of rock mechanics

Réflections sur le développement de la mécanique des roches

Leopold Müller-Salzburg
Universität Salzburg, Österreich und Universität Karlsruhe, BR Deutschland

Wenn wir den Wissensstand, der zur Zeit der Gründung unserer Gesellschaft 1964 geherrscht hat, mit unserem heutigen Wissensstand vergleichen, dann dürfen wir auf das Erreichte recht stolz sein. Große Fortschritte sind gemacht, viele neue Erkenntnisse gewonnen worden.

Wenn wir aber anderseits die grandiosen Leistungen des Felsbaues zur Zeit der großen Bahnbauten in den Alpen und in den Rocky Mountains dem heutigen Stand der Felsbaupraxis gegenüberstellen, dann können wir nicht im gleichen Maß zufrieden sein. Denn ohne Bodenmechanik, ohne Felsmechanik, ja sogar ohne Ingenieurgeologie sind damals mit außerordentlichem Geschick höchst schwierige Situationen gemeistert worden, indes wir zugeben müssen, daß heute trotz vieler wissenschaftlicher Hilfen nicht immer besser und sicherer gebaut wird, ja daß es in letzter Zeit sogar etliche aufsehenerregende Mißerfolge gegeben hat, welche nur zu einem geringen Teil dadurch erklärt werden können, daß unsere Bauwerke immer größer und dadurch schwieriger werden.

Ich für meinen Teil muß zugeben, daß ich von manchen Entwicklungen der Felsmechanik-Wissenschaft ehrlich enttäuscht bin, weil diese ähnlich verlaufen sind wie gewisse Entwicklungen der Bodenmechanik, welche TERZAGHI gegen Ende seines Lebens so schmerzlich bedauert hat. Dasselbe gilt auf dem Felde der Publikationen und ist der Grund, warum ich nach 28 Jahren meine Tätigkeit als Schriftleiter der Zeitschrift Rock Mechanics zurückgelegt habe.

Ich möchte fast von einer Fehltentwicklung sprechen. Was sind die Gründe dieser Fehlentwicklung?

Den Hauptgrund sehe ich darin, daß in den meisten Ländern Theorie und Praxis, Forschung und Anwendung, völlig getrennt laufen und daß nur wenig Verbindungen zwischen beiden bestehen. Es ist geradezu verblüffend, was alles von den Praktikern auf der Baustelle, aber noch mehr von den Projektanten am Zeichentisch und an der Rechenmaschine nicht gewußt oder aber gewußt und nicht berücksichtigt wird. Beispiele ließen sich anführen, in denen Dutzende von Millionen Dollar verschwendet wurden, weil ein nur ungenügender Transfer von Forschungsergebnissen in die Praxis stattfindet.

Die meisten Forscher waren nie in der Praxis des Rock Engineering tätig. Wenn sie Praxis zu haben behaupten, meinen sie Mitwirkung an Berechnungen oder an der Beratung von Felsbauwerken, aber sie haben nie dem Fels an der Baustelle als Partner wirklich gegenübergestanden und sehen mit dem gleichen Hochmut auf das dirty work der Baustelle herab wie viele Praktiker anderseits auf die Theoretiker herabsehen zu können meinen.

Schon die Forschungsthemen werden daher selten von Bedürfnissen der Praxis angeregt, sondern danach ausgesucht, was sich am exaktesten berechnen, am saubersten formulieren und darstellen läßt, was sich vor allem hauptsächlich im Labor und am Computer untersuchen läßt.

Die Folge ist, daß die Ergebnisse der Forschung oft reichlich hypothetisch ausfallen, nur von akademischem Interesse sind. Sie sind sehr häufig aus überidealisierten und überschematisierten, naturfernen mathematischen Modellen gewonnen und blenden durch die Eleganz sophistizierter Rechenansätze, während von der ungeheuren Problematik der Parameter und ihrer Repräsentativität kaum geredet wird, obgleich doch gerade diese den ganzen schönen Schein entwertet.

Viele Forscher scheinen zu vergessen, daß Felsmechanik keine absolute Wissenschaft ist, die ihren Wert in sich selbst trägt wie Physik oder Mathematik oder Logik, sondern daß sie nichts anderes sein soll als eine Systematisierung all der theoretischen Grundlagen, welche dem Felsbau Werkzeug und Hilfe sein können; welche uns lehren, den Fels so zu behandeln, daß er auf unsere Eingriffe möglichst ohne Zwischenfälle reagiert, und lehren, nicht gegen die Natur, sondern in freundlicher Beziehung zu ihr zu arbeiten.

Umgekehrt ist ihrerseits die Praxis viel zu wenig interessiert an der Forschung und übernimmt selbst diejenigen Resultate nicht, die in ihrem Bereich durchaus von Nutzen sein könnten. Zum Teil aus dem Grunde, weil die Forschungsergebnisse oft gar nicht an die Praktiker herangebracht werden (wobei ich unter Praktikern auch die Projektanten meine); zum Teil deshalb, weil viele Forschungsergebnisse gar nicht bis zur Anwendungsreife ausgearbeitet wurden.

Einer der Hauptgründe aber für die offene Großkluft zwischen Theorie und Praxis scheint mir der zu sein, daß die allerelementarsten Erkenntnisse der Felsmechanik, welche das grundsätzliche Verhalten von Fels in seinen verschiedensten Erscheinungsformen erklären, nicht mehr gesehen werden. Diese elementaren Gesetzlichkeiten sind aber, sowohl in der theoretischen wie in der praktischen Felsbaumechanik das allerwichtigste. Je länger man zugleich in der theoretischen Geomechanik wie im praktischen Felsbau tätig ist und mit Projektanten und Baustell-Ingenieuren zu tun hat, desto deutlicher erkennt man das. Wie sich der Fels einem technischen Eingriff gegenüber verhält und warum er sich so und nicht anders verhält, das ist das wichtigste, das man zutiefst verstanden haben muß; das ist die Voraussetzung auch für eine jede Berechnung, wenn diese sinnvoll sein soll. Wie ein Architekt sein Baumaterial kennen muß, so auch der Felsbauingenieur, dessen Baumaterial der Fels ist.

Für die erste Generation der Felsmechanik-Forscher waren die Grundlagen dieses Verhaltens, welche da Schritt für Schritt entdeckt wurden, überraschend, unerhört neu und daher faszinierend. Sie standen auch vordergründig im Bewußtsein, wenn Felsbauwerke projektiert und ausgeführt wurden. Für die beiden jüngeren Generationen jedoch, welche inzwischen nachgerückt sind, sind diese Kenntnisse nicht mehr neu, nicht mehr faszinierend und mit keinem Erkenntnis=Erlebnis verbunden; denn man kann in Lehrbüchern - zwar nicht in allen, aber in den guten - darüber lesen, in einer guten Vorlesung davon hören.

Das geht nicht unter die Haut und steht daher nicht mehr vordergründig im Bewußtsein, wenn man rechnet, zeichnet oder baut. Faszinierend und im Mittelpunkt des Interesses stehen dagegen die Weiterentwicklungen der Geomechanik: Berechnungsmethoden, Klassifikationssysteme, Indexbildungen usw.. Diese neuen Möglichkeiten begeistern, soweit es die Berechnungsverfahren betrifft, mit Recht, denn sie sind echter Fortschritt. Die Index-Moden haben auch ihren Wert, aber der Gefallen an ihnen ist weniger durch eine innere mathematisch-physikalische oder logisch begründete Brauchbarkeit bedingt, sondern entspringt der beruhigenden Vorstellung, nun endlich ein einfaches System quantifizierbarer Aussagen zu haben, die den Anwender der Mühe, sich mit der so komplizierten Natur auseinanderzusetzen, scheinbar entheben.

Die meisten Tagungen und Kongresse sind nun leider nicht dazu angetan, das Interesse der Ingenieure wieder mehr auf die aus dem Gedächtnis verlorenen Grundlagen der Felsbaumechanik zu lenken: auf ihnen hört man das Neueste, das Berechenbar-gemachte, anspruchsvolle theoretische Darstellungen vereinfachter Zusammenhänge, mit einem Wort, den methodischen bzw. wissenschaftlichen Überbau der Geomechanik. An ihre Grundlagen zu erinnern, ist kaum Gelegenheit; denn es ist natürlich für die Vortragenden sehr viel befriedigender, über neue Forschungsergebnisse zu berichten als alte, in Vergessenheit geratene Grundkenntnisse wieder aufzuwärmen. Und case-histories werden sowieso gemieden - des Tagungs- oder Zeitschriften-Niveaus wegen, wie man meint.

Hinzu kommt, daß in Kongressen jene Dramatik der Entscheidungen nicht vermittelt werden kann, welche die Problematik der Eingangswerte und der oft nur schwer in die Modellvorstellungen zu übersetzenden geologischen Situation mit sich bringt. Der Kongreßteilnehmer hört im allgemeinen nur, wie gut alles geplant werden konnte und wie erfolgreich die Pläne verwirklicht werden konnten, wie alles also bestens stimmt. Solchermaßen manipuliert, hat er keine Gelegenheit, am eigentlichen Lernprozeß teilzunehmen, der sich erfahrungsgemäß immer nur auf Mißerfolge stützt, denn nur von diesen haben wir gelernt und lernen wir. Kein Wunder, wenn der Nachwuchs gerade auf das nicht aufmerksam wird, worin die eigentlich kreative Leistung des Felsbauingenieurs und -planers liegt, nämlich auf alles das, was nicht quantifizierbar und doch so wichtig ist; was nicht kognitiv, mit dem Kopf, ermittelt werden kann und doch erkannt werden muß.

Fazit: Gestalten wir doch unsere Kongresse und Tagungen lebendiger, nicht als eine Sammlung von Erfolgsberichten; nicht als Forum allein für Wissenschafter, sondern als eine Börse von Kenntnissen und Erfahrungen für alle, die mit Fels zu tun haben. Dazu freilich müßten wir unseren Tagungen einen völlig neuen Stil geben.

Ich sprach von der großen Wichtigkeit der elementaren Grundlagen. Sie sind nicht einfach. Denn Fels ist der komplizierteste aller Baustoffe. Viele Berechnungen (z.B. von Böschungen) erweisen sich deshalb oft als völlig unzureichend, weil die drastisch vereinfachten Rechenmodelle dieser Komplexität nicht gerecht werden. Am schlimmsten ist es, wenn Rechenmodelle, z.B. die Gleitkreisvorstellung, von der Bodenmechanik entliehen werden, die es mit einem ganz anders gearteten Material zu tun hat.

Eine der Besonderheiten der Felsmassen ist z.B. ihr eigentümliches Verhalten gegenüber Querstützung (confining). Daß die Widerlager der Staumauer Vajont bei der Katastrophe einer achtfachen Überlastung standgehalten haben, ist der Querstützung durch elastisch bewegliche vorgespannte Anker zu danken. Daß anderseits bei einer großen Felsrutschung und einem erosionsgefährdeten Bauteil eines sehr bedeutenden Großkraftwerkes die ersten Sanierungsarbeiten ohne Erfolg blieben, hatte in der Unterschätzung dieser Querstützungswirkung der Anker und in der Unterschätzung der durch Wasser-Erosion verloren gegangenen dreiachsigen Spannungszustände ihren Grund.

Das Querdehnungsverhalten hängt aber nun in außerordentlich hohem Maße von der Anisotropie einer Felsmasse und von ihrem Trennflächengefüge ab. Aber gerade das sind zwei elementare Tatsachen, welche von den meisten in der Felsbaumechanik Tätigen wenig berücksichtigt werden, weil ihre Erfassung kompliziert und unbequem ist. Durch allzu drastische gedankliche Vereinfachungen unserer Rechenmodelle wird aber die Natur nicht einfacher, nur die Diskrepanz zwischen Natur und Rechnung wird größer.

Ein anderes Beispiel dafür, wie unser Blick durch unerlaubte Vereinfachungen gegenüber der Natur getrübt wird, ist die Vernachlässigung des Zeiteinflusses auf das Verhalten der Felsmassen. Im Tunnelbau allerdings haben wir gelernt, diesen Zeiteinfluß zu berücksichtigen - zwar noch nicht rechnerisch, aber auf Umwegen, methodisch, in der sogenannten Neuen Österreichischen Tunnelbauweise.
Wir haben damit viel Erfolg gehabt, haben aber leider bis jetzt versäumt, die theoretische Forschung darauf abzustellen. (Wenige Ausnahmen, z.B. die Untersuchungen VARDARs bestätigen die Regel).

Vermehrte Forschung auf dem Gebiet der Zeitabhängigkeit des Felsverhaltens sollte, wie ich meine, einen großen Fortschritt bringen. Dies würde insbesondere dem bisher am kraßesten vernachlässigten Problem des Felsbaus zugute kommen, nämlich dem der Standsicherheit von Böschungen. Böschungsaufgaben sind im Grunde genommen kein statisches, sondern ein dynamisches Problem. Denn nur wenige Naturböschungen befinden sich in absoluter Ruhe. Wie beim Tunnelbau, haben wir es häufig mit labilen Zwischenbaustadien zwischen einem (einigermaßen) stabilen Anfangszustand und einem (hoffentlich) ebenso stabilen Endzustand zu tun. Eine Böschung kann, etwas überspitzt, definiert werden als eine Rutschung, die noch nicht abgegangen ist. Wie der Talzuschub ein Kriechphänomen ist, so sollte man auch Böschungen als Kriechphänomene betrachten, wie dies z.B. HOFMANN, SHARMA und HAEFELI getan haben.

Eigentlich handelt es sich dabei um rheologische Probleme, z.B. um Kriechdruck auf Stützmauern, doch für deren mathematische Lösungen ist die Zeit noch nicht reif; vor allem wüßten wir nicht, wo wir die Parameter hernehmen sollten, welche in die Rechnung einzusetzen wären. Wir könnten aber durch Modellsimulationen, Computersimulationen und Messungen am lebenden Objekt in der Natur ebensolche Erfolge im Böschungsbau erzielen, wie wir durch Messungen im Tunnelbau erzielt haben, ob solche Messungen nun stellvertretend oder ergänzend zu Berechnungen eingesetzt werden.

Trotz eines ungeheuren Forschungs-Volumens an Material- und statischen Problemen ist die Forschung über den Einfluß der Zeit bisher sträflich vernachlässigt. Ähnlich auch der Einfluß von Primärspannungen. Ich bin überzeugt, daß Primärspannungen im Fels auf die Standsicherheit von Böschungen und auf den Übergang vom Kriechen zum Abgleiten einer Felsböschung einen ebenso großen Einfluß haben wie auf die Standsicherheit und die Vortriebsmethoden im Tunnelbau.

Ein weiteres vernachlässigtes Kapitel der Forschung möchte ich nur kurz streifen: die Probleme der Repräsentativität von entnommenen Gesteinsproben, von Großversuchen in situ, von Parametern überhaupt. Wie kann man Kriterien finden, um die Repräsentativität dieser Dinge wenigstens einigermaßen einzuschätzen?

Ein weiteres nachzuholendes Kapitel der Forschung wie der Praxis, insbesondere der Entwurfspraxis, ist die felsmechanische Interpretation geologischer Daten, besonders aber auch geologischer Gesamtsituationen. Hier liegt die Fehlerquelle der meisten Mißerfolge und Katastrophen. Diese werden so gut wie nie durch falsche Berechnung, sondern immer durch falsche Interpretation der geologischen Situation, oder durch Denkfehler verursacht.(Eine allgemeine Zeiterscheinung übrigens, nicht nur in der Felsmechanik.)

Damit kommen wir zu Problemen, die von der Geomechanik allein nicht gelöst werden können, sondern nur in inniger Zusammenarbeit zwischen Geomechanikern und Ingenieurgeologen, wie sie früher weit besser geübt wurde als heute. Vergessen wir nicht, daß das, was wir heute Geomechanik oder Felsmechanik (besser: Felsbaumechanik) nennen, aus der Ingenieurgeologie hervorgegangen ist. Wie TERZAGHI sich lebenslänglich gewünscht hatte, daß alle Bodenmechaniker gleichzeitig Geologen sein sollten, so müssen auch wir wünschen, daß eine wirkliche Teamarbeit zwischen Felsbaumechanik und Ingenieurgeologen, wie sie nur in wenigen Ländern (z.B. in China und Frankreich) vorbildlich gepflegt wird, verwirklicht werde. Im allgemeinen ist diese Zusammenarbeit in den letzten Jahrzehnten immer seltener geworden. Dies kommt zum Teil daher, daß die geologischen Darstellungen zu akademisch, zu wenig anwendungsbezogen sind, zum Teil daher, daß man den großen Unterschied zwischen Angewandter Geologie und Ingenieurgeologie vielfach noch gar nicht begriffen hat; zum Teil aber daher, daß die Ingenieure dem Gespräch mit dem Ingenieurgeologen wenig Interesse und Verständnis entgegenbringen, weil sie auf der Schule zu wenig ingenieurgeologisch vorgebildet wurden und sich nicht genügend darüber im klaren sind, wie sehr das Schicksal ihrer Bauwerke und das wirtschaftliche Ergebnis ihrer Arbeit von den geologischen Bedingungen des Untergrundes abhängen.

Felsmechanik ohne Ingenieurgeologie betrachte ich als einen Irrweg. Eine Unmenge von Beispielen könnte angeführt werden, welche die Richtigkeit dieser Behauptung bestätigen. Hängt doch die Entscheidung darüber, ob wir bei unseren Planungen und Arbeiten im Fels erfolgreich sind oder nicht, fast ausschließlich davon ab, wie gut wir die geologische Situation erkannt und wie gut wir unsere Planung und unsere Arbeit derselben angepaßt haben. Welche Werkzeuge wir auf der Baustelle mit Vorteil einsetzen können und welche Berechnungswerkzeuge wir in der Planung verwenden sollten, hängt von einer richtigen Erkenntnis der geologischen Situation ab.

Wohin soll nun die Felsmechanik-Forschung, die Lehre und auch das Publikationswesen in Hinkunft gesteuert werden? Soll es überhaupt gesteuert werden? Ich glaube, ja. Denn so, wie sich die Forschung in der letzten Zeit entwickelt hat, läuft sie der Praxis davon. Die Kluft zwischen Theorie und Praxis zu überwinden, scheint mir lebensnotwendig, sowohl für den praktischen Felsbau als für die Felsmechanik-Forschung. Jene Länder, jene Unternehmungen und Ingenieure sind am erfolgreichsten, bei denen diese Kluft am geringsten ist. Am größten ist sie zur Zeit im Kreise der Projektanten.

Lassen Sie uns in der Theorie von allzu naturfernen Abstraktionen und von der Indexmode, welche Faktoren superponiert, die weder superpositionsfähig, noch voneinander unabhängig, noch konstant sind, weiter abrücken.

Konzentrieren wir uns mehr als bisher auf die Interpretation der geologischen Aussagen und ihre Aufbereitung zur geomechanischen Verarbeitung.

Lassen wir mehr als bisher den Ingenieurgeologen wie den Felsmechaniker mitwirken bei der schwierigen Aufgabe des Transfers geologischer und geomechanischer Untersuchungen in die Entwurfskriterien einerseits und in die Ausführungskriterien anderseits.

Bereiten wir die Ingenieure auf den Schulen besser als bisher auf das Gespräch mit dem Geotechniker, mit dem Geomechaniker, mit dem Ingenieurgeologen vor. Wenn das Interesse für diese Dinge nicht auf den Schulen geweckt wird - in der Praxis ist es zu spät dazu.

Fordern wir ferner von jedem in der Planung tätigen Ingenieur einige Jahre Praxis auf dem Bau. Geben wir ihm dort Gelegenheit, die handwerkliche Seite des Felsbaus und damit die Besonderheiten des Baumaterials Fels kennenzulernen. Geben wir ihm damit auch Gelegenheit, von Mißerfolgen zu lernen, welche nichts anderes sind als die Konsequenz einer falschen Behandlung des Gebirges. Dann wird der "Quality-Index" seiner Arbeit steigen, welcher entscheidender ist als der Quality-Index des

Gebirges. Nur in der unmittelbaren Arbeit im Fels lernt man alle jene Dinge einzuschätzen, welche nicht oder noch nicht quantifiziert werden können: die Standzeit des Gebirges; das Stadium, in welchem sich ein Felshang oder ein Kriechhang befindet und ohne dessen Kenntnis eine jede Berechnnung graue Theorie bleibt.

Wir haben bisher Theoretiker ausgebildet; lassen Sie uns in Hinkunft Projektanten und Felsbauer ausbilden.

Die besten Lehrer sind die Mißerfolge. Diese sollten häufiger publiziert werden. Die wenigen Ingenieure, die Charakter genug haben, sich ihrer Fehler nicht zu schämen, in der Erkenntnis, daß es Arbeit ohne Fehler kaum gibt, sollten uns als rühmliche Vorbilder gelten.

Rücken wir ab von unserer hochmütigen Einstellung gegenüber der Publikation von case-histories. Von diesen kann man mehr lernen als von den gelehrtesten Ableitungen. Brechen wir den Hochmut der allzu einseitigen Nur-Akademiker.

Rücken wir etwas ab von den allzu vielen Standards, Empfehlungen und Normen; und vergessen wir nicht, daß alle Normen dieser Welt nur Richtlinien, nicht Vorschriften sein können und sein sollen, und daß sie den Ingenieur nicht entbinden von der Verpflichtung zu denken. Im Reiche der Geologie, wo jeder Tunnel anders ist als jeder andere, wo jede Fundierung und jede Böschung sich von allen anderen Böschungen unterscheidet, ja wo selbst Kalk nicht gleich Kalk und Granit nicht gleich Granit ist, sind allzu weitgehende Standardisierungsversuche und Vereinheitlichungen von Übel. Fels will individuell behandelt werden. Große Gesellschaften können im Übereifer der Homogenisierung und Standardisierung mitunter auch Schaden anrichten.

Unsere Welt ist nicht perfekt und Ideale können in ihr nur in einem bescheidenen Maß und Fort=Schritt eben nur schrittweise erreicht werden. Alles ist ein Lernprozeß. Um voranzukommen, sollten wir unseren gegenwärtigen Standort und begangene Fehler nüchtern sehen und unsere Programme für die Zukunft nicht von vornherein auf das beschränken, was bei oberflächlicher Betrachtung allein erreichbar und ausführbar erscheint, sondern sollten das, was wünschenswert ist, mutig anstreben, auch wenn wir wissen, daß wir es nie ganz erreichen werden.

FUTURE DEVELOPMENTS AND DIRECTIONS IN ROCK MECHANICS

Développement et voies nouvelles pour la mécanique des roches

Zukünftige Entwicklungen und Entwicklungsrichtungen auf dem Gebiet der Felsmechanik

Tan Tjong-Kie
China

FUTURE TRENDS IN WEAK-ROCK MECHANICS AND ENGINEERING

Rock Mechanics is a branch of science and technology which is indespensible for the design and construction of Rock Engineering works. The requirement of high efficiency in the growing exploration and exploitation of the energy and mineral resources all over the world is putting much more increasing demands on Rock Mechanics and Rock Engineering and the related Technology. The large dams and hydro-electric power stations, the large underground structures and open pit minings and drifting under difficult geological conditions are monuments for the successes of Rock Mechanics. During the last 3 decades the field of Rock Mechanics is rapidly expanding; in the fifties it emerged from civil and mining engineering and geology by the necessity that more efficient and economical solutions are required, which take proper account of the properties of the rock structure. In the beginning the rock is considered as a continuum and the theories of elasticity and plasticity formed its theoretical basis. Gradually a rock mass is considered as a discrete medium, as a rock structure traversed by planes of decreased rigidity: the discontinuities. The rock further is jointed, fissured and cracked; anisotropy and heterogeneity is recognized. In the analysis of rock problems the finite element method is a very suitable tool as it based on the discretation of the medium, whereby every element may have different mechanical properties.
In the course of years a rapidly in creasing number of constructions must be built in less competent, fractured, weak rocks; In general we have to deal with the following types of weak rock formations:
Alternating layers of sandstone and mudstone separated by clayey bedding planes; besides the presence of faults, there are many intercalations, joints, fissures and cracks with argillaceous and mica filling; within the mudstone layers severe folding may occur;
formations of altered andesite, tuff, tuff breccias with thin sheets of clays;
schistosic rock formations as shales, thinly laminated chlorite schists, weathered gneisses, with argillaceous fillings, turbulent foliations are frequently encountered;
potentially swelling rocks as anhydritic rocks, fissured mudstones, decomposed granites, containing a large amount of swelling clay minerals as montmorillonites, illites, chlorites etc.
All these types of rocks have the following properties.

1. The mechanical behaviour of the rock mass is mainly governed by the weak planes of discontinuity and further by the argillised joints, fissures, and cracks;
2. Time effects as experienced from the convergence of tunnels and rock wall deformations, upheavals of the floor in excavations are important;
3. The mechanical properties are more or less strongly influenced by water and electrolytes;
4. Deformations may amount to a few per cent before failure occur;
5. Dilatancy effects i.e. the volume-increase due to deviatoric straining are considerable ;
6. Due to the presence of potentially swelling materials as anhydrites, montmorillonites illites, chlorites etc, these rocks shows swelling phenomena.
7. At larger depths and larger tectonic stresses rock burst may occur for instance in sandstone, gneiss formations etc, rock bursts and gas bursts in coalmines are difficult problems.

The weak rock problem covers a wide variety of rock-types, each with different genetic and tectonic history. Due to the many heterogeneities and anisotropies the geologic picture due to the acting stress field in the past and the weathering may change in a tunnel rapidly from place to place. So for a rational design a minutious geological and structural study of the formation is a conditio sine qua non.
The necessity for more efficient strata control and for a decrease in the yearly costs and the frequency of the repair of tunnels is urgently felt by the mining industry. Tunnel reapair means a decrease in the output of the mine. A further serious problem is the frequent occurence of rock bursts and gas bursts incoal mining. The shotcrete-rock-bolt-iron web-ribbed steel support has proved its ability for better maintenance and strata control; this valuable experience must now be moulded into an unifying

theory and in order to find a solution of this problem I think the following topics are essential:

A_1. A more exact and fundamental investigation of the rheological properties of rock formations including:

Laboratory and field-samples testing:
a. creep test under constant loadings;
b. relaxation tests under constant deformations;
c. constant strain rate tests;

These tests must be carried out over such a long period that extrapolations on logarithmic scale to engineering practice time-scales are justified.
Further the samples must be loaded up to failure and the post-failure characteristics and the long term strength must be carefully studied. The transitional stress from log-t creep phase to constant rate creep phase is important as it denotes the transition from stable to unstable creep;

d. swelling tests, especially for potentially swelling rocks;

A_2. Numerical modeling of rock structure and material behaviour choice of proper constitutive equations, finite element calculations;

A_3. Field convergence, displacement determinations, stress measurements;

A_4. Estimation of reduction factor for mechanical parameters due to size effects of the samples from the results of 1,2,3.

B. Methods of detection, prevention and mitigation of rock burst and gas burst hazards. New techniques for initial stress relief as large borehole method, injection of water, covering with steel webs etc.

C. Back-Analysis of Case records.

D. High efficiency tunneling methods

When the size and shape of the orebody are known and the production and its growth year by year are planned then the tunneling equipment to handle the production must be choosen, which in turn determine the size and shape of the tunnels. Although larger equipment will give a quicker and larger output, the size of the tunnels will depend on their stability, which in jointed, fissured heterogeneous weak rocks will decrease with the size.
So a more efficient output can only be obtained when repair or a minimum of repair can be warranted. Hence a more thorough research in the design of tunnel support is indispensable. In soft rocks for instance the closure of the ring is necessary in order to create an arching effect around tunnels as it is stipulated in the N.A.T. M.. Theoretically the formation of this arch is possible, when the voids, fissures and cracks of the surrounding rock can be closed by inwards motion resulting into increased strength. As the inwards motion is a function of the time, the formation of the arch and its resisting capacity is also a function of the time; dependent on the rheological behaviour of the rock mass.
The forming of this arch can be ensured by some rational method of strengthening. As far bolts are usually placed perpendicular to the rock surface; a crossing of the bolts will enable them to resist shearing stresses and in combination with grouting it will be a stabilising factor in the arch action. Furthermore oblique bolts may be more resistant against loss of capacity as a result of stress relaxation in mud rocks and sandstones.

FUTURE TREND:

MULTIDISCIPLINE COOPERATION WITH GEOPHYSICS AND GEODYNAMICS

A. Initial Stresses (including tectonic stresses)

After more than two hundreds years of development the earth sciences are now obviously approaching a consistent and unifying perception of how the earth works. The earth is now regarded as a dynamic body undergoing convection that is driven by its internal heat. It is expressed at the crustal surface as deformations of large and small scales as shown by deep sedimentary basins, high plateas uplift, continental rifting, orogenetic belts, zones of large faulting, seismic belts, volcanic acticity, diagenesis and lithology, metamorphosis, initial rock stresses, folding, faulting, fissuring and cracking of rock formations. The present creep of the crust at average rates of ca 10^{-16}sec^{-1} and higher, the uplift of the Thibetan plateau and the present upwards motion of the Hymalayan mountain range at a rate of 3 cm/year, volcanic eruptions, and the presence of horizontal stresses exceeding the overburden pressure are obvious indications, that the forces which contributed in the past, are still active at present despite their decrease in intensity.

Initial rock stresses are distributed in the rock mass dependent on many factors reoponsible for their generation.

1) Rock-genesis:

In the genesis of rocks and their metamorphosis crystals are formed from their melt under high temperature and stresses. Due to their different anisotropic mechanical and thermal properties stress inclusions are formed during their cooling.
Due to the heterogeneity and anisotropy of sedimentary materials stress kernels also can be formed during diagenesis and lithology of sedimentary rocks.
Such stress kernels can be released when their constraints are removed; then the energy is suddenly released as in rock bursts.

2) Tectonic history

Rock formations have been subjected to tectonic straining under the action of crustal forces mentionned above. These forces may be due to plate collissions and mantle convection, which generates shearing stresses at the lower **boun**-dary of the plate.

3) Stress accumulation

As the crust is a rheological body with many regions of different relaxation times, the stresses will be accumulated into the rigid regions and simultaneously weaker regions will be released.

4) Topography

Topography has a strong influence on the regional stress distribution; it is known that large stresses exist in the bottom rock of deep valleys.

5) Thermal state

Thermal stresses may be working in heat active regions with different Temperature fields.

6) Body forces due to gravity, vapours, gases and liquids.

In porous rocks and fissured rocks gases, vapours and liquids exerts body forces on the mother rocks. Gas outbursts frequently occur in coal seams. When the gas pressure exceeds the limiting strength of coal during excavations.

Initial stresses distributions certainly must form the basis of our analysis and design. Their release has caused many troubles in the stability of underground works, slopes and in rock bursts and gas bursts. In view of our tendency to go deeper with tunneling ($h > 500$ m) and mining, a more profound knowledge of the following factors is necessary:

1. generation of initial stresses
2. measuremeno of initial stresses and their distribution (hydro fracturing)
3. influence of these stresses on engineering works and prevention of eventual damage. In cases of low competency ratio (< 1) the time dependent dilatancy and the stability of floors and roofs are mainly governed by the horizontal initial stress
4. detection of possible rock and gas bursts sites, prevention and mitigation of their hazards.

B. Geodynamic Research

As far Rock Mechanics is mainly restricted to the research of a limited volume of rock at the skin of the crust covering an extension dictated by the man-induced stress field. However such important problems as initial stress distribution, the possibility of rock and gas bursts in coal mining, possibility of reservoir earthquake after impounding, the estimation of the earthquake risk, heat flow and the analysis by structural discontinuities can be faciliated if we associate them with the geodynamic analysis of the enlarged region. Thereby remote sensing data on the faulting and folding state, structural analysis, stress determinations, and geophysical explorations to a depth of a few km to determine the fine structure of the skin of the crust, geochronology, gravity measurements etc. may supply a wealth of data. These studies are not only very useful in the design of mining and oil industry, but also very helpful for the design of important engineering works. With these large constructions usually an investment of hundreds of millions to billions of U.S. dollars is involved. So it may be economically justified to study rock mechanical problems more thoroughly on a broader scale.

C. Research on Earthquakes

According to an old concept of seismology elastic strain energy is accumulated in some small kernels in the crust and mantle. Due to unknown reasons the kernels can be released. The outgoing diverging seismic energy is regarded to be radiated by a dipole of shearing forces. In the past seismology was involved in the propagation of elastic waves after the earthquake; What is crucial for earthquake prediction however is the sequence of processes active in a limited volume of the lithoshhere in the period prior to the main earthquake. All these processes are active in rocks, so the world's attention is now directed to rock mechanics. As time effects play a dominant role the rheology of igneous and especially plutonic rocks at various stress-levels and temperatures comes into the focus. Prior to a large earthquake there are many precursors in the form of anomalies in gravity, magnetic field, radon content of groundwater, electric resistivity, changes in the ratio between primary and secondary waves velo-cities (V_p/V_s), animal behaviour etc. I will restrict my discussion now to Rock Mechanics only.
The many aspects of precursors in the large Hsingtay, Haycheng and Tangshan earthquakes as observed from Rock-Mechanical point of view can be summarised in the following points:

1. land deformations in the form of upheavals, and sinking, and dislocations along faults have been measured many decades in advance wihin a radius of more than 100 km around the coming epicentre.
2. many years prior to the earthquake the deformations increase in magnitude; in this period groundwater levels in deep wells start to sink, simultaneously an increasing bulging of the crust has been observed.
3. some months prior to the earthquake the above processes proceed with increased intensity. Foreshocks increase in intensity and frequency prior to many earthquakes. In She-Chuan province the temperature of hot waterwells increase with 17°C in epicentral regions.
4. In Northern China the larger earthquakes occur in seismic belts which form an orthogonal belt.

The following conclusions can be made:

1. stresses must have been accumulated to the epicentral region
2. the bulging and simultaneous sinking of the groundwater can be attributed to void formation (thus volume increase) due to dilatancy under deriatoric stresses. Groundwater motion leads to decrease of normal effective stress and lubrication of cracks i.e. an overall time dependent decrease in strength.
3. coalescence and generation of small cracks leads to the formation of the main crack, which in the beginning grows slowly and then suddenly factured with the speed of transversal waves under the liberation of large amounts of accumulated energy.

4. In the stress regions at a depth of h > 10 km where shallow earthquakes occur the failure envelope of rocks is nearly horizontal ($\varphi \approx 0$) and the fault planes must make an angle of $\sim 90^o$. Hence earthquakes in Northern China, where the conditions are of plain strain and the dip of the seismic faults are larger than 81° occur in orthogonal seismic belts; this is a good example of application of Rock Mechanics concepts to Earthquake studies.
5. In all the above processes the time factor is predominant.

From the above analysis I try to show the close correlation between rock mechanics and the mechanics of the earthquakes.

D. Rheological Testing under Crustal and Mantle Conditions

In connection with problems as the fracturing of the earthcrust the creeping land deformations the occurence (or not) of foreshocks, accumulation mechanism of stresses, attenuation of seismic energy, much attention has in the recent years been paid to creep tests on cylindrical specimans at different stress levels and temperatures. These tests are also useful for mining engineering and oil-engineering at great depths. Inspite of considerable experimental work, however, we are only beginning to grasp even the fundamentals of some deformation and fracture mechanism in crystalline rock. These include creep due to gliding of dislocationss within crystals and crystal boundaries, processes of dislocations pile up and climbing resulting in cracking, frictional gliding of grains, recrystallisation, dislocation motion by diffusion at large temperatures; the creep mechanisms are different depending on the applied stresses and temperatures. As a special and simplest case in the an-elasticity of rocks, here creep of cylindrical specimens under axially symmetric loadings at various constant temperatures are studied. The hydrostatic stresses are exerted by gasmedium (Argon up to 1000 MPa and 1400^oC), solid medium (salt or Alumuniumoxide up to 5000 MPa and 1400^oC), or liquid medium (Sillicone oil up to 1000 MPa and 600 C).
I suggest to perform the experiments systematically such that they enable us to set up a set of constitutive equations. The following tests may be suitable:

1. Creep under constant loadings
2. relaxation
3. straining under constant ratcs 10^{-3} to 10^{-9} sec^{-1}
4. Instantaneoues and time increasing dilatancy
5. accoustic emissions
6. wave propagations; velocities of primary V_p and shear waves V_s

The above tests can be carried out at different stresslevels and temperatures.
In order to be able making measurements inside the pressure vessel our Institute of Grophysics has designed and constructed a servo-controlled 800T machine for temperatures up to 600° and servo-controlled hydrostatic pressure up to 800 MPa.

In experiments devised for the study of the study of the lower crust and mantle under hydrostatic stresses > 1500 MPa and temperatures > 800° we are mainly interested in the relationship between strain rate and temperatures; further and the phase transformation and mechanisms of creep.
The frontiers in this branch are rapidly widening and I believe that the results may give rise to new concepts of crustal processes.

DESIRABLE FUTURE DIRECTIONS IN ROCK MECHANICS FOR HARD ROCK MINING

Développement désirable dans la mécanique des roches pour l'exploitation minière en roches dures

Erwünschte neue Entwicklungen auf dem Gebiet der Felsmechanik für den Hartfelsbau

Sten Bjurstrom
Swedish Rock Engineering Research Foundation, Stockholm, Sweden

The desirable future directions and developments of rock mechanics in the area of hard rock mining differ little from the research needs of rock mechanics in general. Mining operations are, however, to a larger extent carried out in the failure stages of the rock-masses. This clearly demonstrates the present shortcomings of Rock Mechanics.

The following short remarks try to stress some of the daily life problems in relation to rock mechanics. To balance the somewhat negative picture in the following, it is of importance to mention that there are numerous exceptions showing excellent work of great scientific and practical value being applied in the mines.

For most mines of today the economic situation is pretty tough, and for many it is absolutely necessary to develop more economical and rational mining methods to survive.

Rock Mechanics is playing an important role in this development.

The key-word for today's mining is optimization, which means planning and control of the whole system and all operations, including rock mechanics and related matters.

Hard rock mining concerns a variety of caving methods, room and pillar mining, open pits, vein mining and not least long hole open stoping and cut-and-fill.

One of the problems concerning today's mining methods is too much waste rock dilution or ore losses. For the future we are looking for more flexible methods allowing selective mining. This we want to achieve through inexpensive rock breakage, good fragmentation and by solving the stability problems disturbing production or causing safety hazards. The need for large open rooms deep down does not make the situation easier.

The orebodies show large variations with regard to geometry, rock conditions and mineral constituents. Different types of orebodies require different methods of mining. Much of the interest in future mining concerns long hole stoping including vein mining where open rooms and delayed filling is used.

The Rock Mechanical aspects for the planning of this kind of operation include the design of stopes and drifts, the determination of certain blasting or excavation procedures, the evaluation of the stability of openings and surroundings and the design of support systems.

Fig. 1 Long hole open stopes at the experimental mine in Kiruna, Sweden

Fig..2 Subsidence due to mining operations

Fig. 3 Cable bolting of open stopes

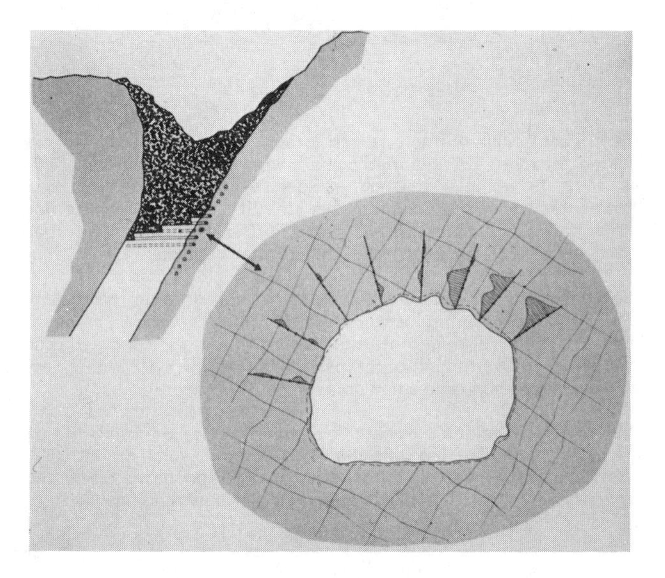

Fig. 4 Experimental drift in LKAB-mine in Kiruna, Sweden where instrumented rock bolts are used

How is this done today and where do we have our shortcomings?

Today the approach to this problem in the mines varies quite a lot, the whole range being used: from rules of thumb, empirical guidelines, to more advanced mathematical modelling. From the rock mechanical point of view we must admit that in practice things are often rather primitive, quite a lot of trial and see, not seldom trial and fail.

To cope with the mentioned demands for a controlled mining situation we need relevant input data, understanding of hard rock mass behaviour and realistic design models.

In the past much of the interest regarding input data concerned instruments and measuring methods. In relation to the often massive efforts put into this, surprisingly little work has been done to decide the validity of measured values and the application of the findings and data. This is particularly emphasized regarding rock stress measurements with often a high precision in the measuring point but also often poor possibilities of extrapolating measured stresses to a general situation and relating the stresses to the strength of the rock mass. For the crude mining situation there is, as a rule, no use for a better precision in measuring rock stresses than the best possible estimation of the strength of the rock masses.

Fig. 5 Luossavaara Research Mine in northern Sweden

As to other data from investigations, much of the interest of today is focused on small scale observations, disregarding that often such values are not representative for the rock mass. This is well known to everyone, but despite this too little is done, in fact, to relate measured data to large scale strength and the behaviour of rock masses. For the future massive efforts are needed regarding scale factors, heterogenity, statistical evaluation methods etc. to improve the situation.

As has been mentioned hard rock mining is to a larger degree taking place in a failure stage situation compared to a lot of other civil engineering work. Mining people also have to cope with changing stress fields and excavation towards a factor of safety around unity.

In such a mining situation a detailed understanding is needed of the failure mechanism and post failure strength of rock masses. This we are usually not able to model or estimate today, certainly not with the commonly used elastic approaches.

To obtain controlled excavation we need detailed information about the influences on the failure behaviour of the various geological structures, blasting vibrations, redistribution of stresses etc.

We should also try to find other ways of designing. Today a kind of mining-NATM is often used in the sense of designing by observation. This can certainly be more systematically developed.

As regards the inverse question, fragmentation - of great importance for rational mining - a deeper understanding of the interaction between the dynamic forces of the explosives and the jointed rock mass is very important for optimal blasting with improved breakage and the use of less explosives.

Stabilization methods. The introduction of cable bolting in hard rock mining has opened new possibilities of safe and flexible mining operations, and it has a wide use today.

Thousands of cables and millions of traditional rock bolts are being successfully used today and we have found a practical way to deal with them.

Do we understand the function of such bolting or traditional bolting? The answer is yes and no.

To put it extremely in slightly more complex cases in hard jointed rock, we know how to bolt but we do not know how much to bolt. Thus, the theory of bolting and the interaction of bolts with surrounding rock masses does not balance with practice. The same is true for cable bolts. It is probably true to say that there does not exist a reliable method for the design of cable bolts of e.g. a hanging wall.

What about our knowledge of backfill as a measure of support?

Fill is regarded as one of the most important measures of support for cut and fill mining. But here again practice and analysis differ a lot and in many cases we are not able to explain its support functions and desirable properties. Cement stabilized fill is very costly and much more research work ought to be devoted to the appropriate design of various types of fill.

Before I finish, some words on the appreciation of rock mechanics among miners!

With increasing demands and problems there has been a change of attitude. The rock mechanical people sometimes sense a greater appreciation of their work and a better understanding of the possibilities of rock mechanics.

But still there is also much frustration in the discussion. And what else could be expected? To solve the very complicated problems in the mines there is often, on the one side, a very practical engineer put there to keep up production and keep down the costs, and, on the other side, a quite specialized engineer or scientist often using quite sophisticated and academic means for the solutions!

Those two worlds seldom match, although the situation has improved in the last few years. If the differences in outlook were more frequently considered in proposals and presentations from the rock mechanical side, there might be a better chance of mutual understanding.

CONCLUSIONS

For the realization of controlled and optimal hard rock mining, there is strong need of:

- accurate rock mechanical design predictions for the evaluation of stability of stopes and surrounding rocks, support requirements and fragmentation;
- methods for a realistic description, the evaluation and use of investigation and test data;
- improved knowledge of the large scale failure behaviour and strength of hard jointed rock masses including post failure strength;
- improved design methods and understanding of support, in particular as regards cable bolts and fill.

DEVELOPMENTS IN DRILLING AND PETROLEUM ENGINEERING

Développements dans le forage et l'exploitation pétrolière

Entwicklungen in der Bohr- und Erdöltechnik

K. Hadley
Exxon Company, Houston, Texas, USA

SYNOPSIS:

Techniques for making hole better faster such as PCD bits, downhole motors, measurement while-drilling systems and custom muds will find increasing application. Drilling programs, completion designs and production performance prediction schemes are beginning to recognize the importance of effective stress. Improvements are needed in determination of subsurface heterogeneity and its characterization for design purposes.

Leistungsfähigere Tiefbohrverfahren mit grösserer Präzision werden zunehmend angewendet. Als Beispiele gelten Verwendung von PCD Bohrmeissel, Bohrlochmotoren, Systeme ausgerüstet mit MWD (Messung-während-Bohren) und speziale Spülungen. Bohrprogramme, Fertigstellungspläne und Methoden zur Vorhersage des Produktionsverhaltens fangen an, die Wichtigkeit der effektiven Spannung zu berücksichtigen. Bessere Methoden zur Kennzeichnung des Untergrundes und Erfassung seiner Heterogenität sind erforderlich.

Des techniques nouvelles permettant d'améliorer la qualité du trou et la vitesse de forage sont de plus en plus largement utilisées; on peut citer notamment trépans a diamants polycristallins, moteurs et telemetrie de fond, boues speciales de forage. L'importance de la contrainte effective pour les programmes de forage, les projets de completion et les programmes de production commence à être reconnue. Des progrès dans la détermination de l'hétérogénéité de subsurface et sa caractérisation sont la condition d'un meilleur planning.

I. Drilling

Economics are becoming more constraining as the cost of exploration and exploitation catches up with the price of oil. Therefore, it is becoming critical to make hole better faster. One promising means of accomplishing this in soft to medium formations is the use of polycrystalline diamond compact bits. The concept behind these is that rock fails more easily in shear than in compression. A 31.1 cm bit recently drilled a total of 4690 m in 3 separate wells at an average rate of penetration just under 24.4 m/hr. This is more than 4 times the footage usually realized from a single conventional bit.

These bits are usually run at very high speed and low weight, often with a downhole turbine or positive-displacement motor, rather than rotary table power. Over 80% of the footage drilled in the USSR reportedly is drilled with downhole motors, but for the US, the figure is less than 1%. The usual application of these devices is in high-speed, low weight-on-bit drilling. However, the lower-speed, high-torque characteristics of multilobe, Moineau-type motors may permit sustained high bit weights and the use of mill-tooth bits. Optimization of motor-bit combinations offers the possibility of further increases in drilling rate and footage drilled in the future.

Drilling rapidly ahead safely and in the right direction will be facilitated by increasing use and sophistication of measurement-while-drilling. Current tools in use in 44.5 cm and 31.1 cm holes in the Gulf of Mexico and the North Sea cut directional surveying time by an order of magnitude. Extension of usage onshore requires service simplification and cost reduction. The usual means of transmitting downhole information to the surface is via mud pressure pulse telemetry. The spreading of the pressure pulse limits the amount of information that can be transmitted in a given time. Recently, Tele-Dril successfully tested an electromagnetic telemetry system that offers the possibility of more rapid and non-mechanical information transfer. Whilst service companies appear eager if not ready to enter the directional surveying and logging-while-drilling markets, systems designed to improve drilling economy and safety have received relatively little attention.

Most drilling in the oil industry is done with a circulating mud system. The mud cools and lubricates the bit, carries rock cuttings out of the hole and acts to hold back the formation and the pore fluid pressure. Specialized mud systems can inhibit or reduce shale swelling or other causes of borehole instability or formation damage. Usually, these are oil base, but among others, potassium-polymer systems constitute attractive alternatives. While the exact mechanisms by which they work are debated, potassium-polymer drilling fluids can inhibit shale swelling, while providing excellent rheological properties at moderate cost. They circumvent the environmental problems associated with oil-base muds.

Industry is moving into deeper water. Deepwater drilling with a riser involves walking a fracture-gradient/pore-pressure-gradient tightrope. We are going to have to unravel effective stress laws to do this with confidence.

II. Formation Evaluation

In formation evaluation, the properties of the rock mass and its contained fluids are interpreted from well logs and well tests. Well logging is one of the few areas in which the state of practice is very nearly state of the art if the operator is willing to pay for it.

New tools can be divided into two classes, those which measure heretofore unmeasurable properties, and those which measure more information about old properties. In the former category are the magnetic resonance log, and electromagnetic propagation logs. These respectively offer the possibility of measuring permeability, and hydrocarbon saturation in zones in which formation water salinity is not known. In the latter category are radioactive logs that measure energy spectra of emitted particles, and full waveform sonic logs that allow the extraction of shear arrivals and attenuation information. While measurement of inelastic and capture gamma ray spectra is a promising tool for applications ranging from clay typing to cased hole determination of hydrocarbon saturation, it is less certain what practical advantages the sonic tool offers over conventional transit time devices.

Ongoing problems for research include fracture detection, logging in irregular boreholes, evaluation of low resistivity sands and low porosity rocks, and determination of formation properties remote from the wellbore.

Well testing will benefit from incremental progress, but no breakthroughs are foreseen owing to the fundamental non-uniqueness of any model producing pressure-time behavior. The state of the art is to match the observed pressure-time response with the pressure-time response predicted by some model, and thereby determine anything and everything from the reservoir limits to the magnitude of formation damage at the wellbore. Unfortunately, inverse problems have non-unique solutions.

Apart from the geometry of and permeability distribution within the reservoir, one of the major needs in estimating oil recovery is a reliable value for rock effective pore space compressibility. Measuring changes in head or shut-in bottomhole pressure produced by earth tides may enable determination of aquifer or reservoir elastic constants from well tests. In hydraulically-fractured wellbores, such measurements can be used to constrain fracture strike and dip.

In general, evaluation of naturally and artificially fractured reservoirs continues to be plagued by non-uniqueness of pressure-time behavior to choice of model. Isolated gas caps or other inhomogeneous gas-liquid distributions may affect system response in unforeseen ways. Accuracy and reliability of flow rate measurement widens the error envelope within which admissible models must fall. Especially in low permeability reservoirs, reservoir size often cannot be estimated from tests of economically reasonable duration.

III. Completions

In gravel packing, understanding of the process, from perforation washing to the action of the sand carrier fluid may improve to the point that a mathematical description becomes possible. Certain myths, such as the certainity of shale mobility, will be exploded.

With regard to mitigating formation damage in better consolidated formations porosity and permeability decrease associated with increasing effective stress will come to be more appreciated, and back-pressures designed accordingly. Elucidation of the causal mechanisms of near wellbore permeability degradation in clay-bearing formations will allow countermeasures to be more effectively applied. Tying laboratory studies of rock-fluid interaction to field behavior requires development of chemical models of formation damage incorporating kinetics and allowing for differentiation of stable and metastable phases. Only then can the relative importance of chemical and mechanical effects confidently be addressed through model studies.

No discussion of rock mechanics in the petroleum industry would be complete without mention of massive hydraulic fracturings. The explosion in MHF technology is a direct result of increased gas prices and a tight supply. Gas prices have moderated, and currently even in the US there is a gas glut. Worldwide, capacity far exceeds demand. However, when its time comes around again techniques and knowledge not generally applied that will influence design and operations in the future include:

(i) Use of pressure-time curves to interpret whether fracture propagation is radial or vertical.

(ii) Minifracs to ascertain the stress contrast between the pay zone and the bounding layer and thereby the likelihood of fracture containment.

(iii) Prevalence of elliptical or even more sophisticated models to relate fracture width, length and treatment parameters.

Despite these advances reservoir heterogeneity will continue to complicate treatment design. Estimates of fracture length will be low if the layered nature of a bedded formation is not considered. If natural fractures are intersected, a design based on matrix permeability only also will result in shorter than intended fracture length.

Operationally, frictional losses, and the strength of the barrier controlling fracture height will continue to impose limits on fracture length. Fluid optimization under field conditions requires the development of proppant transport media that are not shear history dependent.

IV. Production

The rapid increase in secondary and tertiary recovery projects has produced a greater need for detailed geologic modeling. We have some cause for optimism. Knowledge of sedimentology and diagenesis has reached a stage at which realistic prototype models of sand bodies can be designed based on depositional environment and compositional analyses extracted from relatively limited log and core data. Laboratory measurements of oil water relative permeability at temperature are improving our understanding of the mechanisms whereby residual oil saturation decreases, and consequently, our predictive capability for thermal enhanced oil recovery (EOR) schemes.

At the same time, geophysical techniques of increasing variety and resolution allow us to image the subsurface as never before. Seismic techniques for the investigation of reservoir geometry and continuity and contained fluid boundaries range from trace inversion to 3-D surveying. Confidence in models developed from these data is improved by ties to downhole measurements. Surface-based electromagnetic techniques such as CSAMT have been applied to shallow EOR projects. The technique at present is qualitative, but has a potential advantage over seismic methods in that the shape, asymmetry and progress of the process are readily apparent in the raw resistivity data. Tomographic techniques using electromagnetic waves as the imaging energy require too close a test well spacing to be practical in most instances although the validity of the method has been proven in pilot studies.

Geophysical techniques best suited to mapping EOR progress depend upon which physical parameters change during the process. For firefloods and CO_2 floods in which gas displaces the liquids, a significant change in compressional velocities should result and consequently, seismic techniques apply. When injection of displacing fluids under pressure should cause deformation of the earth's surface, tiltmeter techniques may apply. When significant changes in electrical resistivity of the pay zone are expected because of changes in temperature, water saturation, and/or resistivity of the process fluids, electrical and/or electromagnetic techniques may be appropriate.

Continued incremental improvement in reservoir simulation will occur as experience discloses more accurate classes of simplifying hypotheses and adjustment parameters. Examples of recent significant advances include:

(i) Nine-point finite-difference formulations that reduce grid orientation effects.

(ii) Use of equations of state that promise improvement in compositional simulation.

(iii) Gains in machine speed and processing efficiencies that allow larger, more detailed field studies to be economical to undertake.

(iv) Improvements in fracture-matrix or two-tier permeability algorithms. Several models are available that include capillarity, gravity, pressure drop between matrix blocks and fractures, and pressure gradients across the matrix blocks themselves.

The major problem in reservoir simulation will continue to be adequate characterization and, therefore, model treatment, of reservoir geometry and heterogeneity.

ENTWICKLUNGEN IN DER BOHR- UND ERDÖLTECHNIK

Developments in drilling and petroleum engineering

Développements dans le forage et l'exploitation pétrolière

K. Hadley
Exxon Company, USA

ZUSAMMENFASSUNG

Es werden zunehmend Techniken zur Optimierung von Tiefbohrverfahren angewandt. Als Beispiele gelten: Die Verwendung von PCD-Bohrmeißeln, Bohrlochmotoren, von MWD-Systemen, bei denen während des Bohrvorganges gemessen wird und spezielle Spülungen. Bohrprogramme, Vervollständigungspläne und Produktionsleistungsprognosen beginnen, die Bedeutung der effektiven Spannung zu berücksichtigen. Verbesserungen sind erforderlich bei der Bestimmung der Heterogenität des Untergrundes und dessen Charakterisierung für Planungszwecke.

I. Bohrtechnik

(1) Wirtschaftlichkeit wird immer zwingender, da die Erkundungs- und Gewinnungskosten den Ölpreis einholen. Daher ist entscheidend, die Bohrfortschrittsgeschwindigkeit zu erhöhen.

(a) PCD-Meißel: Das dafür grundsätzliche Prinzip ist, daß Gestein leichter durch Scherbeanspruchung als durch Druck zerfällt. Ein Meißel von 31,1 cm Ø hat kürzlich 4.690 m an drei getrennten Bohrlöchern gebohrt; dies bei einem durchschnittlichen Bohrfortschritt, der knapp unter 24,4 m/h lag.

(b) Bohrlochmotoren: Bei mehr als 80 % der Gesamtbohrlänge in der UDSSR wird, Berichten zu Folge, mit Bohrlochmotoren gearbeitet; in den USA liegt der Prozentsatz jedoch unter 1 %. Die langsamen, mit hoher Drehmomentcharakteristik ausgerüsteten Moineau-Motoren erlauben den Dauereinsatz hoher Meißelgewichte.

(c) Messungen beim Bohren: Heutige Bohrwerkzeuge, die beim Bohren von 44,5 cm und 31,1 cm großen Löchern im Golf von Mexiko und in der Nordsee Verwendung finden, verkürzen die Bohrerkundungszeit um ein Vielfaches. Eine Ausbreitung des Einsatzes an Land erfordert eine Vereinfachung der Bedienung und eine Kostenminderung. Während Serviceunternehmen bereits eifrig bemüht sind - wenn nicht gar schon geschehen - in die Märkte der Bohrlocherkundung und des Messens während des Bohrens einzudringen, fanden Systeme zur Verbesserung zum wirtschaftlichen und sicheren Bohren relativ wenig Beachtung.

(d) Spezielle Spülungen: Zwar werden die genauen Vorgänge, nach denen diese Spülungen funktionieren, noch erörtert. Man weiß aber, daß Kalium-Polymer-Flüssigkeiten Quellerscheinungen von Schiefer verhindern können. Dabei werden ausgezeichnete Fließeigenschaften zu bescheidenen Kosten erzielt. Sie unterbinden Umweltprobleme, die mit Spülungen auf Öl-Basis zusammenhängen.

(2) Die Industrie begibt sich zunehmend in tiefere Gewässer. Tiefwasserbohrungen mit einer Steigrohrleitung erlauben den Einsatz einer Leitung, um den Bruchgradienten-/Porendruckgradienten zu ermitteln. Wir werden Gesetze über die tatsächlichen Spannungen zu entwirren haben, um die nötige Sicherheit dafür zu erzielen.

II. Formationsauswertung

(1) Messungen: Das ist einer der wenigen Bereiche, in denen der Stand der Praxis etwa dem Stand der Technik entspricht, vorausgesetzt, der Unternehmer bezahlt dafür. Ohne ein brauchbares Bohrloch kann man jedoch keine Messungen durchführen.

(a) Neue Geräte dazu können in zwei Kategorien eingeteilt werden; in solche, die früher nicht meßbare Eigenschaften messen, und in jene, die weitere Informationen über bereits bekannte Eigenschaften erfassen.

(i) In der zuerst genannten Kategorie gibt es das neutronenmagnetische Zerfallszeitlog und das elektromagnetische Wellenausbreitungslog. Diese beiden Gruppen eröffnen die Möglichkeit, sowohl die Durchlässigkeit als auch die Kohlenwasserstoffsättigung in Bereichen zu messen, in denen der Salzgehalt des Schichtwassers unbekannt ist.

(ii) Zur zweiten Kategorie gehören Radioaktivitätsmessungen, die die Energiespektren von ausgestoßenen Teilchen messen, und Schallwellenmessungen, womit ankommende Scherwellen entzerrt und Informationen über Dämpfung zu erhalten sind. Das Messen des „inelastic" und „capture" - Gammastrahlenspektrum scheint eine vielversprechende Methode zur Typisierung von Tonmineralien und der Ermittlung der Kohlenwasserstoffsättigung bei verrohrten Bohrungen zu sein. Dagegen ist es nicht so sicher, welche praktischen Vorteile die Durchschallungsmethoden gegenüber den konventionellen Laufzeitmeßgeräten aufweisen.

b) Bleibende Probleme für die Forschung umfassen das Auffinden von Brüchen, Messungen in unregelmäßigen Bohrlöchern und Ermittlung der Fließgeschwindigkeit, speziell in abgewichenen Bohrlöchern oder unter vielphasigen Fließbedingungen.

(2) Bohrlochteste profitieren vom technischen Fortschritt. Es ist jedoch kein Durchbruch bei der Ermittlung des Druck-Zeitverhaltens zu erwarten, weil jedes Modell grundsätzlich von der Natur abweicht.

a) Neue „Werkzeuge"
Eine der Haupterfordernisse ist heute ein verläßlicher Wert für die Zusammendrückbarkeit des wirksamen Felsporenraumes. Messungen von Änderungen des Drucks im Bohrlochkopf oder als geschlossenes System auf der Bohrlochsohle, die durch Erdbewegungen entstehen, ermöglichen die Bestimmung der Elastizitätskonstanten eines Aquifers oder einer Lagerstätte. In Bohrungen, in denen Hydraulic-Fracturing durchgeführt wurde, kann man durch diese Methode Streichen und Einfallen von Spalten erfassen.

b) Bleibende Probleme

(i) Isolierte Gaskappen oder andere inhomogene Verteilung von Gas-Flüssigkeit können die Zusammendrückbarkeit eines Systems auf unvorhergesehene Weise beeinträchtigen.

(ii) Sorgfalt und Zuverlässigkeit bei den Messungen der Fließgeschwindigkeit erhöhen die Fehlerschwelle, unterhalb derer die zulässigen Modelle sich befinden müssen.

III. Verschlüsse

(1) Kiespackungen

(a) Das Verständnis des Vorganges vom Freispülen der Perforation bis zur Wirkung der Sandträgerflüssigkeit kann bis zu einer mathematischen Beschreibung verbessert werden.

Bestimmte Vorstellungen, wie z.B. über eine Beweglichkeit des Schiefers sind überholt.

(2) Abschwächung des Skin-Effektes

(a) Die Abnahme der Porosität und der Durchlässigkeit zusammen mit einer Zunahme der wirksamen Spannung werden besser eingeschätzt und Gegendruck entsprechend aufgezeichnet.

(b) Die Aufklärung der ursächlichen Mechanismen von Durchlässigkeitsabminderungen in der Nähe des Bohrloches ermöglicht es, Gegenmaßnahmen wirksamer anzuwenden. Die Verbindung von Laborstudien über die Gesteins-Flüssigkeits-Wechselwirkung mit dem Verhalten des Feldes erfordert die Entwicklung von chemischen Modellen des Schichtaufreissens einschließlich der Bewegungsabläufe, wobei zwischen stabilen und metastabilen Phasen zu unterscheiden ist. Erst dann kann die Bedeutung chemischer und mechanischer Auswirkungen durch Modellstudien zuverlässig angegangen werden.

(3) Massives hydraulisches Zerbrechen

Die Explosion in der MHF-Technologie ist ein direktes Ergebnis der angestiegenen Gaspreise und einer knappen Versorgung. Die Gaspreise sind gesunken und gleichzeitig gibt es sogar in den USA einen Gasüberschuss. Weltweit übertrifft das Angebot die Nachfrage.

(a) Techniken und Kenntnisse, die nicht überall angewendet werden und die die Planung und Ausführung in der Zukunft beeinflussen, umfassen folgende:

(i) Verwendung von Druck-Zeit-Diagrammen zur Prüfung radialer oder vertikaler Rissausbreitung

(ii) Kleinstrisse zur Ermittlung der Spannungskontraste zwischen der Abbauzone und der begrenzenden Schicht. Damit kann die Wahrscheinlichkeit von Brüchen festgestellt werden.

(iii) Das Überhandnehmen von elliptischen oder sogar höher entwickelten Modellen, um Rissbreite, Länge und Verfahrenskennwerte zueinander in Beziehung zu bringen.

(b) Bleibende Probleme

(i) Die Ausbreitung der Flüssigkeit über den gesamten Ausschnitt hinweg hängt von der Wirksamkeit des angewandten Perforationsgerätes ab.

(ii) Reibungsverluste und die Mächtigkeit der die Bruchgröße beeinflussende Felsbarriere grenzen die Risslängen verfahrensmäßig ein.

(iii) Speicherheterogenität wird weiterhin die Verfahrensplanung erschweren. Schätzungen über die Risslänge fallen niedrig aus, wenn die Schichtung einer Sedimentformation nicht berücksichtigt wird.

Werden natürliche Klüfte angeschnitten, würde eine Planung, die nur auf der Gesteinsdurchlässigkeit beruht, auch bei kürzeren Risslängen als beabsichtigt, gelingen.

(iv) Fließverbesserungen unter Feldbedingungen erfordern die Entwicklung von Transportmedien, die nicht von der Scherbeeinflussung abhängen.

IV. Produktion:

Der schnelle Anstieg bei sekundären und tertiären Gewinnungsprojekten hat die Notwendigkeit detaillierter geologischer Modelle bewirkt.

(1) Anwendung von bisher zuwenig benutzter Hilfsmittel zur Speicherbeschreibung

(a) Der heutige Kenntnisstand läßt es zu, daß wirklichkeitsnahe Prototypmodelle aus Sandkörpern auf der Grundlage des Ablagerungsumfeldes und zusammengefaßter Daten aus begrenzten Messungen und Bohrkerndaten entworfen werden können.

(b) Labormessungen über temperaturabhängige relative Durchlässigkeit von Ölwasser steigern das Verständnis für die Mechanismen, wodurch die Restölsättigung abnimmt. Folglich wird die Voraussagefähigkeit für termisch betriebene Ölgewinnung (EOR) planmäßiger.

(c) Geophysikalische Methoden, die sich am besten zur Aufzeichnung des EOR-Fortschrittes eignen, hängen davon ab, welche physikalischen Parameter sich während des Vorganges ändern. Bei Öl-in-situ - und Kohlendioxid-Verbrennungen, in denen Gas Flüssigkeiten ersetzt, sollte sich eine signifikante Änderung der Kompressionswellengeschwingkeit ergeben und somit seismische Techniken zulassen. Wenn Verdrängungsflüssigkeiten injiziert werden, und sich dabei Deformationen an der Erdoberfläche einstellen, sollten Neigungsmeßgeräte eingesetzt werden. Wenn in der Abbauzone Änderungen des elektrischen Widerstandes infolge Temperaturänderungen, der Wassersättigung und/oder der Prozessflüssigkeit auftreten, können elektrische und/oder elektromagnetische Verfahren sinnvoll sein.

(i) Seismische Verfahren zur Untersuchung der Speichergeometrie und -stetigkeit sowie der Erkundung der Flüssigkeitsgrenzen reichen von der „Trace Inversion" bis hin zur dreidimensionalen Erkundung. Die Zuverlässigkeit solcher Modelle wird durch Bohrlochmessungen erhöht.

(ii) Oberflächenhafte elektromagnetische Techniken, wie z.B. CSAMT, wurden für flache EOR-Projekte angewandt. Diese Methode erbringt zur Zeit nur qualitative Ergebnisse, hat jedoch einen deutlichen Vorteil gegenüber den seismischen Methoden dadurch, daß die Form, die Asymmetrie und der Gang des Verfahrens sofort aus den Rohdaten der Widerstandsmessungen hervor gehen. Tomographische Techniken, die elektromagnetische Wellen als Abbildungsenergie verwenden, erfordern einen sehr geringen Abstand der Testbohrungen, der in den meisten Fällen nicht mehr praktikabel sein wird, obgleich die Gültigkeit der Techniken in Pilotstudien bewiesen wurde.

(2) Speichersimulation:

Fortgesetzte, schrittweise Verbesserung ist zu erwarten, da durch Versuche mehr und mehr vereinfachende Hypothesen sowie Parameter zur besseren Anpassung entdeckt bzw. aufgestellt werden können.

(a) Beispiele von bereits erzielten Fortschritten umfassen:

(i) Berechnungen nach der finiten Differenzmethode mit 9 Punkten, um die Gitterorientierungseffekte zu vermindern

(ii) Benützung von Zustandsgleichungen, die eine Verbesserung bei der komplexen Simulation versprechen.

(iii) Fortschritte in der Geschwindigkeit der Maschinen und zunehmende Leistungsfähigkeit der Verfahren, wodurch umfangreichere, detaillierte Geländestudien wirtschaftlicher durchgeführt werden können.

(iv) Verbesserungen bei der Rissbildungsmatrix oder bei zweireihigen Durchlässigkeitsalgorithmen. Es sind verschiedene Modelle verfügbar, die Kapillarität, Schwerkraft, Druckverlust zwischen Matrixblöcken und Rissen und Druckgefällen über die Matrixblöcke selbst umfassen.

(b) Das Hauptproblem bei der Speichersimulierung wird weiterhin in der zutreffenden Charakterisierung und in der modellhaften Behandlung der Speichergeometrie und der Heterogenität liegen.